AF804904

MAIN LIBRARY WHITEKNIGHTS

90 0402008 3

EN

The Neurobiology of Dopamine

The Neurobiology of Dopamine

Editors

A. S. Horn J. Korf
B. H. C. Westerink

University of Groningen,
Groningen, The Netherlands

1979

ACADEMIC PRESS
London · New York · San Francisco
A Subsidiary of Harcourt Brace Jovanovich, Publishers

ACADEMIC PRESS INC. (LONDON) LTD.
24/28 Oval Road,
London NW1

United States Edition published by
ACADEMIC PRESS INC.
111 Fifth Avenue
New York, New York 10003

Copyright © 1979 by
ACADEMIC PRESS INC. (LONDON) LTD.

All Rights Reserved
No part of this book may be reproduced in any form by photostat, microfilm, or any other means, without written permission from the publishers

British Library Cataloguing in Publication Data
The neurobiology of dopamine
1. Dopamine
I. Horn, A S II. Korf, J
III. Westerink, B H C
547'.75 QP563.D66 78–72547

ISBN 0-12-354150-6

402,008

Printed in Great Britain by
J. W. Arrowsmith, Bristol

Contents

Dopamine Receptors

Release, Uptake and Metabolism of Dopamine

Neurophysiology of Dopamine Systems

List of Contributors

R. N. Adams, University of Kansas, Department of Chemistry, Lawrence, Kansas, U.S.A.

N. E. B. Andén, University of Göteborg, Fack, S-400 33 Göteborg 33, Sweden.

Brigitte Berger, Laboratoire de Neuropathologie Charles Foix, Hôpital de la Salpêtrière, 47 Boulevard de l'Hôpital, 75013 Paris, France.

B. S. Bunney, Department of Psychiatry, Yale University of Medicine, 34 Park Street, New Haven, Connecticut 06508, U.S.A.

A. Cheramy, Unité de Neuropharmacologie Biochimique, Laboratoire de Neurophysiologie Générale, Collège de France, 11, Place Mercelin—Berthelot, 75005 Paris, France.

B. Costall, University of Bradford, School of Studies in Pharmacology, Bradford, West Yorkshire BD7 1DP, U.K.

A. C. Cuello, University Department of Pharmacology, Oxford, U.K.

H. C. Fibiger, Department of Psychiatry, Kinsmen Laboratory of Neurological Research, University of British Colombia, Vancouver 8, B.C., Canada.

M. F. Giorguieff, Unité de Neuropharmacologie Biochimique, Laboratoire de Neurophysiologie Générale, Collège de France, 11, Place Mercelin-Berthelot, 75005 Paris, France.

J. Glowinski, Unité de Neuropharmacologie Biochimique, Laboratoire de Neurophysiologie Générale, Collège de France 11, Place Mercelin-Berthelot, 75005 Paris, France.

L. I. Goldberg, Committee on Clinical Pharmacology, University of Chicago, Chicago, Illinois 60637, U.S.A.

C. J. Grol, Department of Pharmacy, University of Groningen, Groningen, The Netherlands.

H. C. Guldberg, Department of Pharmacology and Clinical Pharmacology with Toxicology, University of Trondheim, The Regional Hospital, N-7000 Trondheim, Norway.

M. Holzbauer, A.R.C. Institute of Animal Physiology, Babraham, Cambridge, U.K.

A. S. Horn, Department of Pharmacy, University of Groningen, Groningen, The Netherlands.

O. Hornykiewicz, Institute of Biochemical Pharmaclogy, University of Vienna, Vienna, Waehringer Str. 17, A-1090 Vienna, Austria.

F. Karoum, Laboratory of Clinical Psychopharmacology, St. Elizabeth's Hospital, William A. White Building, Washington D.C., 20032, U.S.A.

J. Korf, University Psychiatric Clinic, University of Groningen, Groningen, The Netherlands.

B. Libet, Department of Physiology, School of Medicine, University of California, San Francisco, U.S.A.

W. Lichtensteiger, Pharmacology Institute, University of Zurich, Gloriastrasse 32, 8006, Zurich, Switzerland.

O. Lindvall, Department of Histology, University of Lund, Biskopsgatan 5, S-223 62 Lund, Sweden.

J. McDermed, Wellcome Research Laboratories, 3030 Cornwallis Road, Research Triangle Park, N.C. 27709, U.S.A.

E. McGeer, The University of British Columbia, Faculty of Medicine, Kinsmen Laboratory of Neurological Research, Vancouver, B.C., Canada.

P. McGeer, The University of British Columbia, Faculty of Medicine, Kinsmen Laboratory of Neurological Research, Vancouver B.C., Canada.

R. J. Miller, Department of Pharmacological and Physiological Sciences, University of Chicago, Chicago, U.S.A.

R. Naylor, School of Studies in Pharmacology, University of Bradford, Bradford, West Yorkshire, BD7 1DP, U.K.

N. H. Neff, Laboratory of Preclinical Pharmacology, St. Elizabeth's Hospital, William A. White Building, Room 129, Washington D.C. 20032, U.S.A.

J. Nguyen-Legros, Laboratoire de Neuropathologie Charles Foix, Hôpital de la Sopêtrière, 47 Boulevard de l'Hôpital, 75013 Paris, France.

M. Palkovits, Department of Anatomy, Semmelweis Medical University, 1450 Budapest, Tüzoltó-u 58, Hungary.

A. G. Phillips, Department of Psychiatry, Kinsmen Laboratory of Neurological Research, University of British Colombia, Vancouver 8, B.C., Canada.

H. M. van Praag, University Psychiatric Clinic, University of Utrecht, Utrecht, The Netherlands.

R. Roth, Department of Pharmacology (Psychiatry), Yale University School of Medicine, 333 Cedar Street, New Haven, CT 06510, U.S.A.

P. Seeman, Department of Pharmacology, University of Toronto, Toronto, Ontario, Canada.

D. Sharman, A.R.C. Institute of Animal Physiology, Babraham, Cambridge, U.K.

T. Sourkes, Department of Psychiatry, McGill University, 1033 Pine Avenue West, Montreal 112, Quebec, Canada.

K. Taylor, Roche Research Institute of Marine Pharmacology, 4–10 Inman Road, Dee Why, N.S.W. Box 255, Australia.

K. F. Tipton, Biochemistry Department, Trinity College, University of Dublin, Dublin, Ireland.

M. Titeler, Department of Pharmacology, University of Toronto, Toronto, Canada.

U. Ungerstedt, Department of Histology, Karolinska Institutet, Stockholm, Sweden.

J. S. Wassenaar, Department of Neurophysiology, Bloemsingel 10, State University of Groningen, Groningen, The Netherlands.

B. H. C. Westerink, Department of Pharmacy, University of Groningen, Groningen, The Netherlands.

G. N. Woodruff, University of Southampton, Department of Physiology and Biochemistry, Bassett Crescent East, Southampton, S09 3TU, U.K.

D. H. York, Department of Physiology, University of Missouri, Columbia, Missouri 65201, U.S.A.

Preface

This book is an attempt to bring together in one volume as many important facets of dopamine research as possible. By the very nature of the subject and its explosive growth in recent years, however, it is inevitable that certain areas have not been covered.

The order of the chapters was very loosely based on the concept of increasing complexity of the system dealt with in the chapter. Thus in simple terms the two ends of the spectrum of complexity are the chemistry of the dopamine molecule itself and the functional role of dopamine-containing neuronal systems in man.

We would like to take this opportunity to thank sincerely all of our authors (both the fast and the slow writers!) for making this book possible.

Alan Horn
Jakob Korf
Ben Westerink

CHAPTER 1

Historical Introduction

Nils-Erik Andén

University of Göteborg, Sweden.

The interest in dopamine (DA) has been great in the past decade, partly due the spectacular results of the L-DOPA treatment of Parkinson's disease and also to the demonstration of the probable connection between the antischizophrenic effect and the DA receptor blockade of neuroleptic drugs. Dopamine was first synthesized by Mannich and Jacobsohn (1910). It was, however, long neglected in comparison to the closely related biogenic catecholamines adrenaline and noradrenaline. One reason for this obscurity was in all likelihood that DA has but small sympathomimetic properties (Barger and Dale, 1910).

The history of DA is linked to that of its immediate precursor L-DOPA. DOPA was first synthesized in its racemic form by Funk (1911), who also coined the term "vitamin" because he considered all compounds of that kind as amines. He had the correct idea that adrenaline is formed *in vivo* from this amino acid. Soon afterwards, the L-form of DOPA was isolated from the bean *Vicia faba* and chemically characterized by Guggenheim (1913). He experienced nausea and vomiting when he administered 2.5 g of this compound orally to himself. Unfortunately, he gave rather small oral doses to animals and, therefore, he did not succeed in observing any remarkable effects.

After these early discoveries, it took more than 20 years before the next important step was taken. It was the demonstration of L-DOPA decarboxylase in animal tissues (Holtz *et al*., 1938). The same group also found that DA is a normal constituent of the human urine and that L-DOPA is decarboxylated *in vivo* with subsequent changes in the blood pressure (Holtz *et al.*, 1942). Soon after the discovery of L-DOPA decarboxylase, L-DOPA and DA were proposed as intermediates in the biosynthesis of noradrenaline and adrenaline from L-tyrosine (Blaschko, 1939).

Most of our present knowledge of DA has been obtained in the last 25 years with the rapid development of the field of neuropsychopharmacology. Soon after the introduction of the antipsychotic drugs at the beginning of the fifties, it was realized that they produce two peculiar side-effects in humans. Firstly, the patients often developed parkinsonism, i.e., motor disturbances indistinguishable from those seen in Parkinson's disease. Secondly, it was not unusual that the psychosis was converted to an indifferent and sedated condition but with intact intellectual abilities, a so-called neuroleptic syndrome (Delay *et al.*, 1952). These changes could be observed also in animals, e.g., as catalepsy and decreased motor activity in rodents (Courvoisier *et al.*, 1953). A further development that was of great importance was that one of the clinically active neuroleptic drugs, reserpine, was found to interfere with the storage of two monoamines, 5-hydroxytryptamine and noradrenaline. These compounds had previously been demonstrated to occur in the brain (Twarog and Page, 1953; Amin *et al.*, 1954; Vogt, 1954). Initially, the depletion of the body stores of 5-hydroxytryptamine by reserpine was reported (Plestcher *et al.*, 1955). Somewhat later, it was detected that reserpine similarly influences the peripheral and central stores of noradrenaline (Bertler *et al.*, 1956; Holzbauer and Vogt, 1956).

It was tempting to connect the observed changes in the concentrations of 5-hydroxytryptamine and noradrenaline following treatment with reserpine to the pharmacological effects. It was not known, however, if reserpine caused an excess or lack of monoamines at the receptors of the effector cells and if 5-hydroxytryptamine or noradrenaline or both were involved. In order to investigate these problems, Carlsson *et al.* (1957) gave the 5-hydroxytryptamine precursor 5-hydroxytryptophan and the noradrenaline precursor DOPA to reserpine-treated mice and rabbits. They observed that the reserpine-induced sedation was readily counteracted by DOPA, but not by 5-hydroxytryptophan. These data indicated that the reserpine-induced sedation is due to a lack of noradrenaline rather than 5-hydroxytryptamine but, surprisingly, the DOPA treatment did not induce any noticeable increase in the noradrenaline concentration of the reserpine-treated animals. Therefore, interest was directed to DA, the intermediate in the formation of noradrenaline from DOPA. At that time, there was no good method to determine DA so Carlsson and Waldeck (1958) had to develop a sensitive and specific fluorimetric method. By means of this method it was discovered that not only do large amounts of DA accumulate in the brain following treatment with DOPA, but also that DA is normally present in the brain at a concentration of about the same magnitude as that of noradrenaline (Carlsson *et al.*, 1958). The normal occurrence of DA in the brain was independently discussed by Montagu (1957) and Weil-Malherbe and Bone (1957).

The fact that the concentration of DA in the normal brain is at least as high as that of noradrenaline suggested that DA does not serve only as a precursor of noradrenaline, but that it might have an action of its own. Further evidence for this view was obtained when it was demonstrated that DA has a remarkable distribution in the brain. The caudate nucleus and putamen, i.e. the neostriatum, contains 70–80% of the brain's DA (Bertler and Rosengren, 1959, Sano *et al.*, 1959). Significant amounts of DA were also detected in the substantia nigra (Bertler, 1961).

These findings indicated that DA is involved in the extrapyramidal motor functions and in disorders such as Parkinson's disease and Huntington's chorea. The arguments supporting these hypotheses were summarized as follows (Carlsson, 1959):

1. The presence of large amounts of dopamine in the corpus striatum, which forms an important part of the extrapyramidal system.
2. The extrapyramidal actions of reserpine, which depletes dopamine from the corpus striatum.
3. The ability of DOPA to counteract the hypokinetic action of reserpine. Whether this action of DOPA is entirely due to formation of dopamine, or whether formation of noradrenaline contributes to the effect, remains an open question.

Carlsson's theories soon received substantial support by important clinical observations in Vienna, Austria. First, it was found that DA is more or less completely depleted from the corpus striatum in patients with Parkinson's disease (Ehringer and Hornykiewicz, 1960). The noradrenaline and the 5-hydroxytryptamine in the brain were also lowered, but to a much less marked degree (Bernheimer *et al.*, 1963). The disappearance of DA from the brains of parkinsonian patients prompted trials with intravenous administration of small doses of L-DOPA (Birkmayer and Hornykiewicz, 1961). At about the same time, small oral doses of L-DOPA were given in Canada (Barbeau *et al.*, 1962). Some beneficial effects on the parkinsonian signs and symptoms, particularly the hypokinesia and the rigidity, were obtained but side-effects such as nausea and vomiting were troublesome. Therefore, the clinically useful treatment of Parkinson's disease with DOPA was accomplished only by using large oral doses following slow increases in the dosage (Cotzias *et al.*, 1967. Cotzias *et al.*, 1969).

The understanding of the function of DA in the brain has been facilitated by the visualization of monoamines in tissue sections under a fluorescence microscope (Falck *et al.*, 1962). By this method, the occurrence of DA in nerve cell bodies and nerve terminals was demonstrated (Carlsson *et al.*, 1962; Dahlström and Fuxe, 1964; Fuxe, 1965). Several ways to map out neuronal systems were also described. The largest DA-containing neuronal pathway originates from cell bodies in the zona compacta of the substantia nigra and has numerous nerve terminals with a high concentration of DA in the caudate nucleus and putamen, i.e. the neostriatum (Andén *et al.*, 1964). A degeneration of this nigroneostriatal DA pathway can explain the most typical morphological and biochemical findings in Parkinson's disease, i.e., the cell loss in the substantia nigra and the DA disappearance from the corpus striatum. Another DA pathway originates from the medial area of the mesencephalon and terminates in different parts of the limbic system such as the nucleus accumbens, the olfactory tubercle and certain regions of the limbic cortex (Andén *et al.*, 1966). Stimulation of DA mechanisms in the nucleus accumbens increases motor activity (Pijnenburg and van Rossum, 1973) but other mesolimbic DA neurons may participate in mental processes (Andén, 1972). A tuberoinfundibular DA neurone system is involved in the hypothalamic control of the secretion of hormones from the anterior pituitary gland (Fuxe, 1964).

The formation, storage, release, metabolism and uptake of DA have been extensively studied and clarified in the last twenty years, as seen in the different chapters of this book. In recent years, the greatest interest has perhaps been focused on DA receptors. The DA receptors of the effector cells appear to be distinct from other monoamine receptors to judge from pharmacological studies on behaviours such as turning following unilateral manipulations, stereotypies and motor activity. Thus, the emetic agent apomorphine selectively and directly stimulates DA receptors (Ernst, 1967), whereas the common denominator of all neuroleptic drugs appears to be blockade of DA receptors (Andén *et al.*, 1970). These results have been amply supported in more sophisticated models such as the DA-induced stimulation of the adenylate cyclase in brain homogenates (Kebabian *et al.*, 1972) and the binding of DA and neuroleptic drugs to synaptic membranes (Creese *et al.*, 1975; Seeman *et al.*, 1975).

Apart from regulating the activity of the effector cells, DA receptors also mediate changes in the turnover of DA. The synthesis and utilization of DA are stimulated by DA receptor antagonists whereas they are inhibited by agonists (Carlsson and Lindqvist, 1963; Andén *et al.*, 1967). These effects were originally supposed to be controlled by the postsynaptic receptors. Later results indicate, however, that DA receptors located on the cell bodies (Aghajanian and Bunney, 1973) and on the nerve terminals (Kehr *et al.*, 1972) of the DA neurones are probably also involved. The latter receptors have been called autoreceptors, since they are sensitive to the neurotransmitter of the neurone on which they are localized.

These efforts expended on DA research have produced a large amount of information within a relatively short period of time. As can be seen from the chapters in this book, it ranges from molecular biology to the clinic. It is not unlikely that we now know more about DA than about any other neurotransmitter. Hopefully, the future of neurobiology will benefit from this.

REFERENCES

Aghajanian, G. K. and Bunney, B. S. (1973). *In* "Frontiers in Catecholamine Research" (Eds. E. Usdin and S. H. Snyder), 643–648. Pergamon, New York.

Amin, A. H., Crawford, T. B. B. and Gaddum, J. H. (1954). *J. Physiol. (London)* **126,** 596–618.

Andén, N.-E. (1972). *J. Pharm. Pharmacol.* **24,** 905–906.

Andén, N.-E., Butcher, S. G., Corrodi, H., Fuxe, K. and Ungerstedt, U. (1970). *Eur. J. Pharmacol.* **11,** 303–314.

Andén, N.-E., Carlsson, A., Dahlström, A., Fuxe, K., Hillarp, N.-Å. and Larsson, K. (1964). *Life Sci.* **3,** 523–530.

Andén, N.-E., Dahlström, A., Fuxe, K., Larsson, K., Olson, L. and Ungerstedt, U. (1966). *Acta Physiol. Scand.* **67,** 313–326.

Andén, N.-E., Rubenson, A., Fuxe, K. and Hökfelt, T. (1967). *J. Pharm. Pharmacol.* **19,** 627–629.

Barbeau, A., Sourkes, T. L. and Murphy, G. F. (1962). *In* "Monoamines et Système Nerveux Centrale" (Ed. J. de Ajuriaguerra), 247–262. George, Genève and Masson, Paris.

Barger, G. and Dale, H. H. (1910). *J. Physiol. (London)* **41,** 19–59.

Bernheimer, H., Birkmayer, W. and Hornykiewicz, O. (1963). *Klin. Wschr.* **41,** 465–469.

Bertler, Å. (1961). *Acta Physiol. Scand.* **51,** 97–107.

Bertler, Å., Carlsson, A. and Rosengren, E. (1956). *Naturwissenschaften* **22,** 521.

Bertler, Å. and Rosengren, E. (1959). *Experientia* **15,** 10–11.

Birkmayer, W. and Hornykiewicz, O. (1961). *Wien. Klin. Wschr.* **73,** 787–788.

Blaschko, H. (1939). *J. Physiol. (London)* **96,** 50P–51P.

Carlsson, A. (1959). *Pharmacol. Rev.* **11,** 490–493.

Carlsson, A., Falck, B. and Hillarp, N.-Å. (1962) *Acta Physiol. Scand.* **56,** *Suppl.* 196, 1–28.

Carlsson, A. and Lindqvist, M. (1963). *Acta Pharmacol. Tox.* **20,** 140–144.

Carlsson, A., Lindqvist, M. and Magnusson, T. (1957). *Nature (London)* **180,** 1200.

Carlsson, A., Lindqvist, M., Magnusson, T. and Waldeck, B. (1958). *Science* **127,** 471.

Carlsson, A. and Waldeck, B. (1958). *Acta Physiol. Scand.* **44,** 293–298.

Cotzias, G. C., van Woert, M. H. and Schiffer, L. M. (1967). *New Engl. J. Med.* **276,** 374–379.

Cotzias, G. C., Papavasiliou, P. S. and Gellene, R. (1969). *New Engl. J. Med.* **280,** 337–345.

Courvoisier, S., Fournel, J., Ducrot, R., Kolsky, M. and Koetschet, P. (1953). *Archs Int. Pharmacodyn. Thér.* **92,** 305–361.
Creese, I., Burt, D. R. and Snyder, S. H. (1975). *Life Sci.* **17,** 993–1002.
Dahlström, A. and Fuxe, K. (1964). *Acta Physiol. Scand.* **62,** Suppl. 232, 1–55.
Delay, J., Deniker, P. and Harl, J.-M. (1952). *Ann. Méd.-psychol.* **110,** 112–117.
Ehringer, H. and Hornykeiwicz, O. (1960) *Klin. Wschr.* **38,** 1236–1239.
Ernst, A. M. (1967). *Psychopharmacol.* (*Berlin*) **10,** 316–323.
Falck, B., Hillarp, N.-Å., Thieme, G. and Torp, A. (1962). *J. Histochem. Cytochem.* **10,** 348–365.
Funk, C. (1911). *J. Chem. Soc.* **99,** 554–557.
Fuxe, K. (1964). *Z. Zellforsch.* **61,** 710–724.
Fuxe, K. (1965). *Acta Physiol. Scand.* **64,** *Suppl.* 247, 37–85.
Guggenheim, M. (1913). *Hoppe-Seyler's Z. Physiol. Chem.* **88,** 276–284.
Holtz, P., Credner, K. and Koepp, W. (1942).*Naunyn-Schmiedeberg's Arch. Exp. Path. Pharmak.* **200,** 356–388.
Holtz, P., Heise, R. and Lüdtke, K. (1938). *Naunyn-Schmiedeberg's Arch. Exp. Path. Pharmak.* **191,** 87–118.
Holzbauer, M. and Vogt, M. (1956). *J. Neurochem.* **1,** 8–11.
Kebabian, J. W., Petzold, G. L. and Greengard, P. (1972). *Proc. Natl Acad. Sci., U.S.A.* **69,** 2145–2149.
Kehr, W., Carlsson, A., Lindqvist, M., Magnusson, T. and Atack, C. (1972). *J. Pharm. Pharmacol.* **24,** 744–747.
Mannich, C. and Jacobsohn, W. (1910). *Ber. Dt. Chem. Ges.* **43,** 189–197.
Montague, K. A. (1957). *Nature* (*London*) **180,** 244–245.
Pijnenburg, A. J. J. and Rossum, J. M. van (1973). *J. Pharm. Pharmacol.* **25,** 1003–1005.
Pletscher, A., Shore, P. and Brodie, B. B. (1955). *Science* **122,** 374–375.
Sano, I., Gamo, T., Kakimoto, Y., Taniguchi, K., Takesada, M. and Nichinuma, K. (1959). *Biochem. Biophys. Acta* **32,** 586–587.
Seeman, P., Chau-Wong, M., Tedesco, J. and Wong, K. (1975). *Proc. Natl Acad. Sci., U.S.A.* **72,** 4376–4380.
Twarog, B. M. and Page, I. H. (1953). *Am. J. Physiol.* **175,** 157–161.
Vogt, M. (1954). *J. Physiol.* (*London*) **123,** 451–481.
Weil-Malherbe, H. and Bone, A. D. (1957). *Nature* (*London*) **180,** 1050–1051.

CHAPTER 2

The Chemistry of Dopamine

Cor J. Grol

State University, Groningen, The Netherlands

SYNTHESIS

Synthesis of Dopamine

Dopamine (1,2-benzenediol-4(2-aminoethyl); 2 or β-(3,4-dihydroxyphenyl)-ethylamine; 3,4-dihydroxyphenethylamine) hydrochloride (8) was synthesized by Barger and Ewins in 1910 who demethylated the 3,4-dimethoxyphenethylamine (5) with concentrated HCl. This dimethoxy derivative was made following a reaction sequence as developed by Pictet and Finkelstein (1909). After a Perkin condensation of veratrylaldehyde (1) with acetic anhydride the

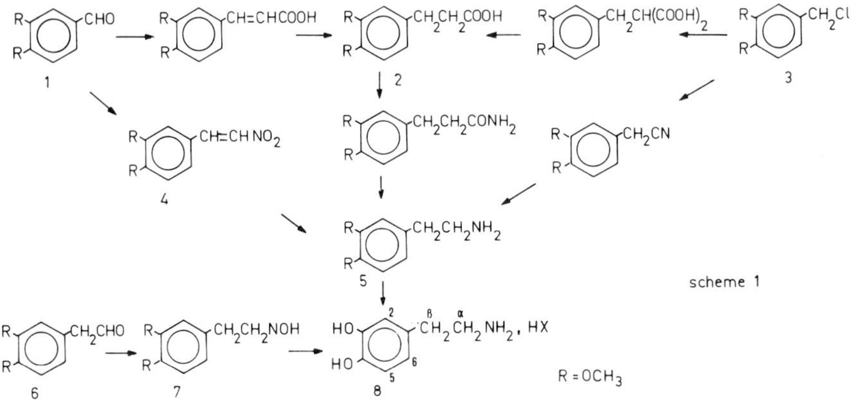

scheme 1

product is reduced to the propionic acid (2). A Hoffman reaction on the amide of this acid yields the phenethylamine (5). The intermediate propionic acid (2) can also be synthesized from the benzylchloride (3) and a sodiomalonic ester followed by hydrolysis and decarboxylation (Binns *et al.*, 1970a). Starting with the veratrylaldehyde (1) Rosenmund (1910) synthesized the nitrostyrene (4) with nitromethane and potassium hydroxide (Gairand and Lapin, 1953). From this nitrostyrene the phenethylamine is easily obtained by reduction with lithium aluminium hydride (Swan and Wright, 1954). Mannich (Mannich and Jacobson, 1910) performed his synthesis for dopamine HCl via the oxime (7) which was made from the homoveratrylaldehyde (6).

In 1924 dopamine HBr was synthesized by the introduction of a hydroxylgroup into the tyramine nucleus by nitration, reduction, diazotation and hydrolysis (Waser and Sommer, 1923). A more direct synthesis was developed in 1933 by Buck who catalytically reduced the mandelonitrile (9), obtained by a reaction of 3,4-dihydroxybenzaldehyde with hydrogen cyanide (scheme 2).

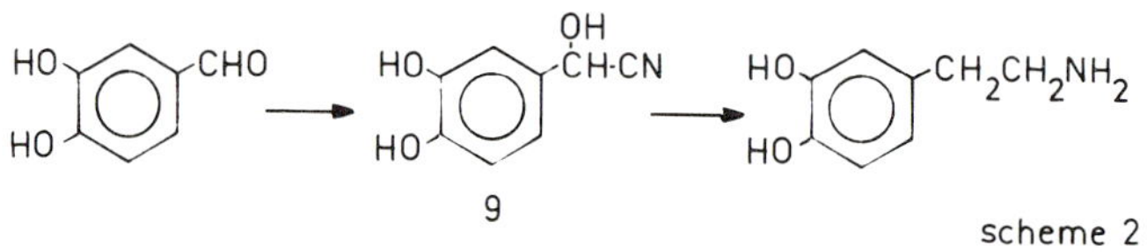

scheme 2

It is obvious that the ethylamino side chain could also be obtained from a phenylacetonitrile by reduction which may be effected catalytically or with reducing agents like lithium aluminium hydride or diborane. The phenylacetonitriles can be synthesized from benzylchlorides with potassium cyanide in dimethyl sulphoxide. The benzylchlorides are prepared from benzylalcohols by treatment with thionylchloride. In these reactions, however, the phenolic groups must be protected as ethers. Another practical procedure for the synthesis of dopamine is outlined in scheme 3. In this reaction benzylamine is used in the place of benzylchloride to form the nitrile (Short *et al.*, 1973).

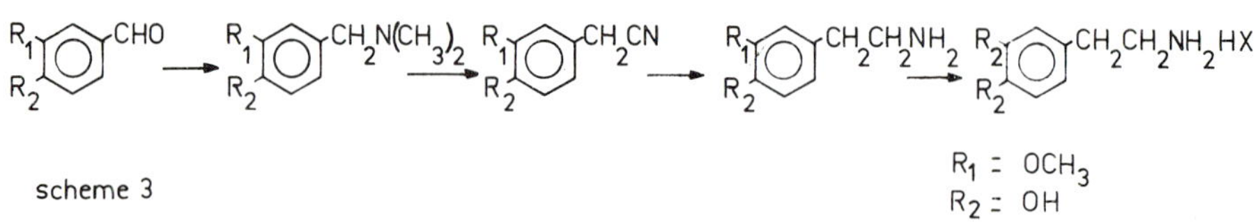

scheme 3

Finally dopamine can be synthesized by thermal decarboxylation of DOPA and *O*-protected DOPA derivatives. Dopamine HBr is thus obtained in 58% yield from veratrylglycine by treatment with 0.05% tetralinperoxide and demethylation with HBr (Kanao *et al.*, 1975).

Synthesis of Labelled Dopamine

Carbon isotopes

^{11}C-α-Dopamine is synthesized by a reduction of the cyanohydrin obtained by a reaction of the bisulphite addition product of 3,4-dihydroxybenzaldehyde with ^{11}C-labelled sodium cyanide prepared by a recoil synthesis (Christman *et al.*, 1970). A more efficient synthesis of carrier-free ^{11}C-dopamine with a specific activity of 40 mCi mg^{-1} can be effected with the use of carrier-free H^{11}CN of high activity in an on-line process (Fowler *et al.*, 1973). A recent patent describes the synthesis of ^{11}C-dopamine from di-*O*-benzyl benzylchloride and Na^{11}CN via catalytic reduction of the intermediate cyanide (Fowler *et al.*, 1977).

^{14}C-Labelled dopamine can be synthesized by most of the routes described for dopamine, provided a ^{14}C atom has been introduced in some preliminary reaction step. α-^{14}C-Dopamine, for instance, is synthesized by a reduction of 3-methoxy-4-benzyloxyphenyl (α-^{14}C) acetonitrile with boron hydride (Battersby *et el.*, 1976).

α- and β-^{14} C-Dopamine were made by the route via the appropriate nitrostyrenes, starting with the benzaldehyde and ^{14}C-nitromethane or with labelled benzaldehyde and nitromethane (Swan and Wright, 1954). Dopamine labelled in the ring positions 3,4 or 5 with ^{14}C was synthesized by Swan and Wright (1956) in an eleven-stage synthesis starting with 5-nitrosalicyclic acid, the condensation product of nitromalondialdehyde and labelled ethylacetoacetate in the α, β or γ position.

Dopamine labelled with deuterium or tritium

It is obvious that practically all procedures in which hydrogen atoms are introduced during the synthesis of dopamine could in principle give labelled compounds when a label is present in the reagent. Besides this introduction of hydrogen isotopes, exchange procedures are available when activated hydrogen atoms are present. The following reactions are examples of both approaches.

α, α′ D-Labelled dopamine (87.5% incorporation of D) is synthesized via reduction of 3,4-dimethoxyphenylacetonitrile with lithium aluminium deuteride.

The β, β′-dideuterodopamine (95% incorporation) is prepared from 3,4-dimethoxyphenylacetic acid by incorporating deuterium in the side chain in an exchange process with NaOD and D_2O, followed by the formation and reduction of the amide (Perel *et al.*, 1972). β, β' di-D-Dopamine (96% incorporation) can also be synthesized starting with a condensation of deuterated benzylchloride with a malonic ester (scheme 1, Binns *et al.*, 1970a). The classical route via the nitrostyrene synthesis was used for the preparation of α- or β-mono deuterated dopamine with only tracer concentrations of deuterium (Binns *et al.*, 1970a). Following the methods of Kirby and Ogunkoya (1965) for the specific deuteration of 3,4-dihydroxyphenylalanine in the unsubstituted ring positions by exchange reactions in alkaline solutions, Binns (1970a) adapted these methods for the synthesis of ring-labelled dopamine from labelled aldehydes. 6-D-, 2-D- and 5-D-Dopamine were obtained in 80, 90 and 91% isotopic purity, respectively.

Most of the routes described for deuterium-labelled dopamine should also be applicable for the tritium-labelled molecule. Additional reactions were developed for β, β'-di-T-dopamine which was prepared by base-catalysed exchange of the methylene protons of 4-benzyloxy-3-methoxyphenyl acetonitrile with tritiated methanol followed by reduction and cleavage of the protecting groups (Battersby *et al.*, 1976). The same authors developed an elegant synthesis for stereospecific tritium-labelled (2R)—and (2S)—dopamine starting with labelled or unlabelled *O*-methylsulphoxyl vanillin in which a hydrogen or tritium atom, respectively, was introduced by stereospecific transfer from an alcohol in a reaction mediated by a liver alcohol dehydrogenase and its coenzyme NAD^+.

PHYSICOCHEMICAL PROPERTIES

Dopamine is mostly used as its HCl (m.p. 241–243) or HBr salt (m.p. 218–220°C).

It is even in its salt form sensitive to air and light, a reason to test its purity prior to use. The purity of a sample of a dopamine salt can be established with thin layer chromatography on cellulose plates in the solvent system,MeOH-*n*-BuOH-C_6H_6-H_2O (4:3:2:1) with 0.01% Na_2 EDTA added. After visualization with $K_3Fe(CN)6$-ethylene-diamine and heating, a green fluorescent spot can be observed (Perel *et al.*, 1972). The identity can be verified with the aid of infrared spectroscopy. The IR spectrum of dopamine-HBr in a KBr-disc is shown in Fig. 1.

The relative intensities of the absorptions at 3380 and 3260 cm^{-1} can vary depending upon the concentration and the mode of recording the spectrum. These absorptions are due to intramolecular and intermolecular hydrogen-bonded O—H stretching vibrations. The NMR spectrum of dopamine HCl in D_2O consists of a multiplet for the aromatic protons at δ 6.8 and a multiplet for the side chain protons at δ 3.25 ppm. The chemical

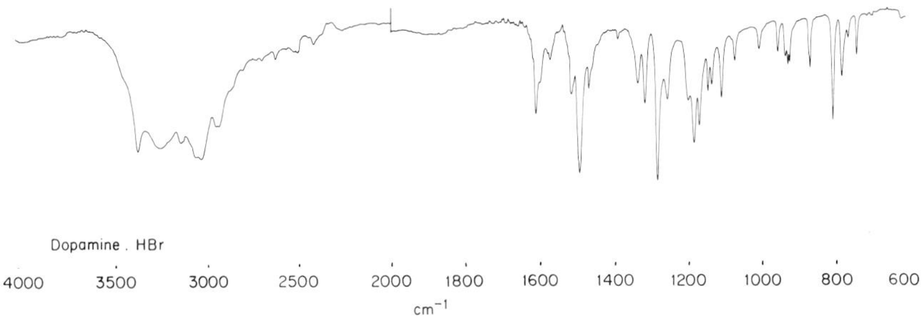

Fig. 1. Infrared spectrum of dopamine-H Br in a K Br disc.

shifts and the coupling constants of the aromatic protons were analysed by computer simulation (Lambert *et al.*. 1975). An important shift towards high field was observed for the aromatic protons as compared with those of benzene (δ 7.27), due to the influence of the hydroxyl groups. The chemical shifts were found to be for H_2—δ 6.83, for H_5—6.88, and for H_6—6.73 ppm. The two pairs of methylene protons of the ethane fragment, which give rise to an AA′ BB′ multiplet, were analysed by Bustard and Egan (1971). The down-field half of the multiplet is assigned to the —CH_2—N protons and the up-field half to the Ar—CH_2 protons.

The EI mass spectrum of dopamine at 70 eV shows the following fragments 153 (M^+,4), 124(55), 123(25), 105(1), 77(15), 30(100) (Reisch *et al.*, 1968). The fragments are due to a β-fission of the side chain and to the loss of COH from the phenolic portion.

Fragmentation occurs so readily that the molecular ion has a low intensity. This is dramatically changed in the chemical ionization (CI) spectrum where the base peak is the protonated molecular ion at *m/e* 154; the only fragment present is found at *m/e* 137 and may be due to loss of ammonia (Milne *et al.*, 1973).

The UV spectrum of dopamine in 0.05 M H_2SO_4 consists of an absorption at 279.5 nm ($E^{1\%}_{1\,cm}$ 133) (Clarke, 1974). As the molecule contains phenolic OH groups the UV spectrum is dependent upon the pH of the solution due to the formation of phenolate ions. Lewis (1954) used the pH dependency of the UV spectrum to determine the dissociation constant of one phenolic OH group (pK_1 8.92). The dissociation constant of the second OH group is difficult to determine due to oxidation reactions at high pH, but is in the order of pK_3 $\approx$ 13. The higher second ionization constant compared to the first one is attributed mainly to the inhibiting effect of the nearby negative charge rather than to intramolecular H-bonding (Boggess and Martin, 1975).

The acidity of the first phenolic hydrogen and the protonated amine group are comparable and their deprotonations often occur in the same pH range. The determination of the proton acidities of catecholamines by potentiometric titrations affords therefore two stoichiometric pK values in the same range. Some authors attributed the lower pK_a to the phenolic OH-group. Such assignments were shown to be in error for the measured pK_a values are composite values due to the fact that the catecholamines exist as equilibrium mixtures of different ionic spècies. The deprotonations from the phenolic and ammonium groups of phenolic amines may be described by the equilibria shown in scheme 4.

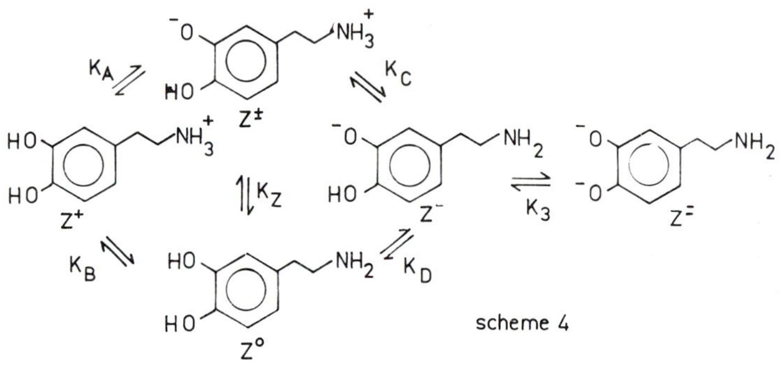

scheme 4

Adams (1916) showed that the microscopic constants (K_A–K_D) are related to the measured macroscopic constants K_1 and K_2 by the following equations:

$$K_1 = K_A + K_B$$
$$K_2^{-1} = K_C^{-1} + K_D^{-1}$$
$$K_1 K_2 = K_A K_C = K_B K_D$$

Edsall (1958) developed a method for calculating the microscopic constants with spectroscopic data from which the fractional ionization of the phenol group as a function of the pH is measured. The results formerly obtained from spectrophotometric studies in the literature were reinterpreted and the microscopic constants for dopamine were calculated (Martin, 1971). From these microscopic constants the macroscopic constants can be calculated independently of the potentiometric determined values; both values were found to be in good agreement. The microscopic constants were determined by Granot (1976) in a PMR study, in which the chemical shifts of the ring protons and the side chain protons were measured as a function of the pD. From this study it could be clearly demonstrated that the phenolic group was more acidic. The same conclusion was reached by Antikainen and Witikainen (1973) on the basis of the relationship between the complex formation of dopamine with boric acid and the potentiometric determined pK_a value.

From the measured microconstants K_A and K_B, the ratio (K_z) between the concentration of the zwitterionic form to the neutral form can be calculated and this was found to be 15 (Martin, 1971). Armstrong and Barlow (1976) calculated in various ways the K_z values with the aid of model compounds in which one of the two deprotonations is blocked by substitution, so that one microconstant can be determined independently. It should be noted that K_z is independent of the pH and describes a tautomeric equilibrium between the zwitterion and neutral form. The real fraction of the diverse species present as a function of the pH can be obtained with equations like the one shown for the zwitterion.

$$f_z = \frac{K_z}{1+K_z} \times \frac{K_1[H^+]}{[H^+]^2 + K_1[H^+] + K_1K_2}$$

The relative populations of the species for dopamine can be calculated at pH 7.4 with the values for K_1, K_2 and K_z from Martin and are shown to be 95% cation, 3% zwitterion, 0.2% neutral form.

The experimental values of the microscopic and the potentiometric determined macroscopic dissociation constants are summarized in Table I.

TABLE I Microconstants and macroscopic dissociation constants of dopamine.

pK_A	pK_B	pK_C	pK_D	K_z	pK_1	pK_2	pK_3	T	μ
8.90	10.06	10.60	9.44	15	8.87	10.63		20	0.10[a]
9.6	10.2	11.0	10.4		9.5	11.1		27[b]	
					8.555	10.312		25	0.1[c]
				5.8–7.7–10	8.81	10.52		25[d]	
					9.06	10.60	12.05	20	0.5[e]
					8.96	10.50		25	1.0[f]

[a] Martin (1971) [b] Granot (1976) [c] Antikainen and Witikainen (1973) [d] Armstrong and Barlow (1976) [e] Grgas-Kužnar *et al.* (1974) [f] Rajan *et al.* (1971).

In the same way that the values of the pKs' determine the deprotonations of dopamine a pK value can be formulated describing the dissociation of the ion pair dopamine.HCl into the cationic protonated dopamine and its anion. This ion pair formation constant was determined in *n*-octanol and *n*-octanol–water mixtures (Lippold and Biedermann, 1974) and found to be 4.41 indicating a low dissociation tendency. Related to this dissociation constant of ion pairs are the partition coefficient and the extraction constant which were determined for the dopamine–perchlorate ion pair between pentanol-1 and perchloric acid (Persson, 1971).

REACTIONS OF DOPAMINE

Cyclization Reactions Leading to Isoquinoline Derivatives

Dopamine forms tetrahydroisoquinolines in the Pictet–Spengler reaction—a condensation between a carbonyl compound and a phenethylamine (Whaley and Govindachary, 1951). The first step in this reaction is the formation of a Schiff base followed by a cyclization reaction, the ease of which is dependent upon the substituents in the aromatic ring. For a facile cyclization to isoquinolines nuclear activation in the *para* position to the carbon atom where the ring closure will occur is necessary. The activation by a hydroxyl group is greater than for a methoxyl group and ring closure is then possible even under physiological conditions (Kovács and Fodor, 1951). Although the main product is formed from the attack on the *para* position it was found by King *et al.* (1974) that in the reaction of dopamine with acetaldehyde the product of an *ortho* attack was also formed in a yield of 10% at pH 4.5.

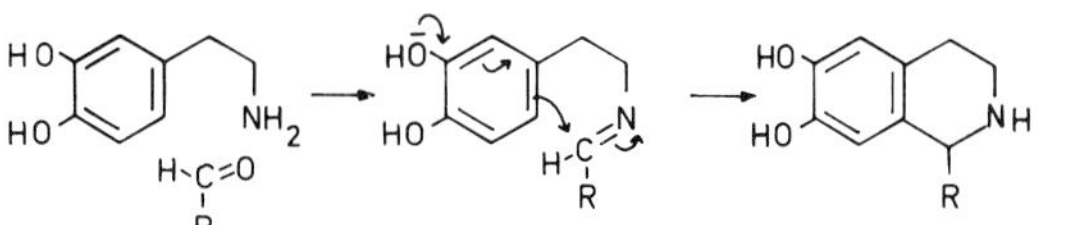

10

scheme 5

The reaction between dopamine and acetaldehyde has received considerable interest because acetaldehyde is the primary metabolite of ethanol, and salsolinol (10, $R = CH_3$) can be formed in mammals after the ingestion of alcohol. It was shown by Greenberg and Cohen (1973) that the compound is taken up by catecholamine binding sites where it may possibly act as a false neurochemical transmitter. The formation under physiological conditions from ethanol and acetaldehyde was shown to occur in homogenates of rat liver and brain-stem (Yamanaka *et al.*, 1970). Salsolinol was also detected in rat brain after administration of ethanol to pyrogallol (a COMT inhibitor) pretreated rats (Collins and Bigdeli, 1975). Its formation in man was demonstrated by its detection in urine from Parkinsonian patients on L-DOPA treatment after ethanol ingestion and even before the alcohol administration (Sandler *et al.*, 1973).

Another well-known tetrahydroisoquinoline derivative is norsalsolinol (10, R = H), the product of the reaction between dopamine and formaldehyde. This product is the precursor for the formation of highly fluorescent end-products of use in the histochemical detection of dopamine at the cellular level (see Berger and Ngyen Legros, Chapter 19). Although the tetrahydroisoquinoline is by itself non-fluorescent or only weakly fluorescent it is oxidized in a second step to a dihydroisoquinoline which has an intense fluoresence in a solid matrix. This dehydrogenation reaction is catalysed by protein but can also take place on silica gel thin-layer plates (Björklund *et al.*, 1970).

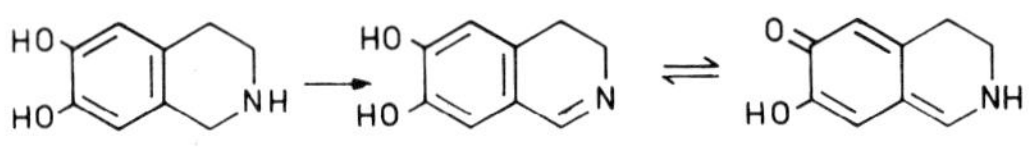

10 11

scheme 6

In a solid matrix the dihydroisoquinoline (11) exists predominantly in its quinoidal form which has a strong absorption at 390 nm and seems responsible for the fluorescence of 480 nm (Corrodi and Hillarp, 1964). A strong fluorescent product is also formed in the reaction of dopamine with glyoxylic acid. The reaction path is visualized in scheme 7. After a first condensation to a tetrahydroisoquinoline-1-carboxylic acid there is a further reaction with glyoxylic acid followed by a decarboxylation to the 2-carboxymethyl-3,4-dihydro-7-hydroxy-isoquinolin-6-one (Svensson *et al.*, 1975).

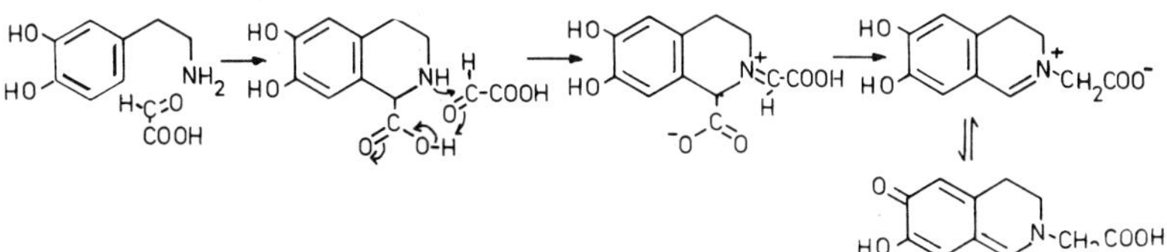

scheme 7

Besides the above-described most familiar isoquinoline derivatives a number of other compounds were prepared from dopamine and various carbonyl compounds. A description of all these derivatives lies beyond the scope of this chapter (see Morita *et al.*, 1975).

Functional Group Derivatives of Dopamine

From the phenol groups and the primary alkylamine function the usual derivatives may be prepared. The amine group can be alkylated to form secondary or tertiary amines. For epinine (12) and the *N*,*N*-dimethyl amine a synthesis is outlined in scheme 8 (Borgman *et al.*, 1973). Because the

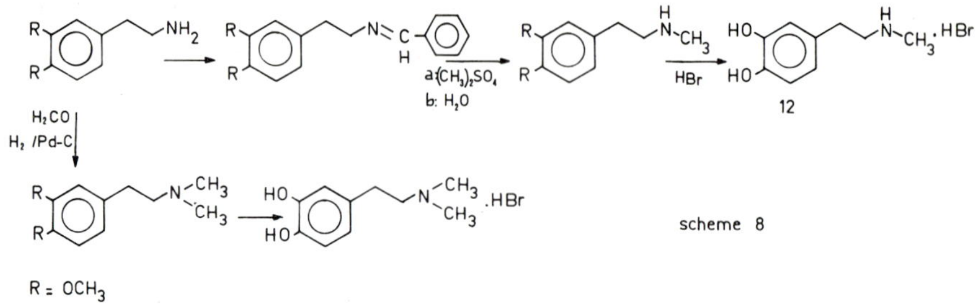

scheme 8

catechol function and the NH_2 group have a similar reactivity in some reactions, it is sometimes necessary to protect one of these groups. As the acetylation of nitrogen occurs more readily than that of oxygen, amides can be obtained by treatment of dopamine with acid chlorides (Symchowicz and Faro, 1976). In the selective acetylation of the phenolic groups it is on the other hand necessary to protect the nitrogen. This can be done through preparation of the *N*-carbobenzyloxy derivative followed by acetylation and hydrogenolysis (Casagrande and Ferrari, 1973).

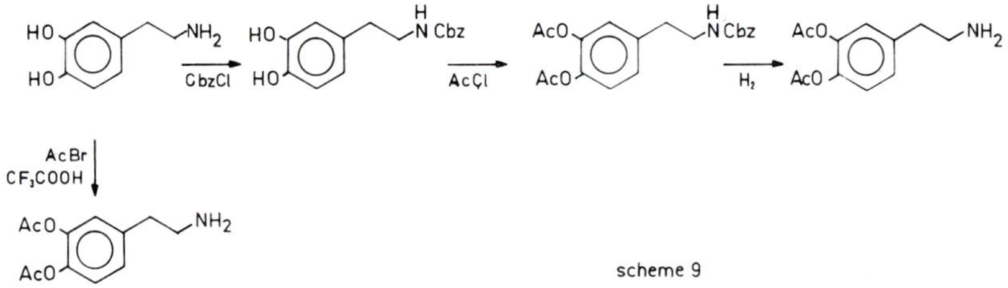

scheme 9

A more direct method for the preparation of dopamine esters is the acetylation in trifluoracetic acid with acetylbromide. Acetylation of the nitrogen is prevented as it is completely protonated by the strong acid (Borgman *et al.*, 1973).

Complex Formation of Dopamine

Catecholamines form rather stable chelate complexes with metal ions wherein the two oxygen atoms of the phenolic groups are the donating atoms. From potentiometric titrations with and without metal ions the stability constants of these complexes can be calculated on the basis of the shift occurring in the titration curves due to a competition between the metal ion and the added base for the hydrogen atoms of the ligand.

Grgas-Kužnar *et al.* (1974) determined the stability constants of metal complexes of dopamine with Ni^{2+}, Cu^{2+}, Zn^{2+}, Cd^{2+} and Pb^{2+}. For the more stable Ni and Cu complexes it was found that various species of protonated and unprotonated mono- and bischelates were formed.

Because of the possible involvement of metal chelation in the process of the storage, release and reuptake *in vivo* of catecholamines the nature of interactions of dopamine with metal ions and ATP is a field of great interest (Maas and Colburn, 1965). In a number of studies Rajan investigated potentiometrically the complex formation between dopamine and various bivalent ions. For the DA-Mg^{2+} and DA-Cu^{2+} (1:1) complexes values for the log K of 3.59 and 10.55 were found. Due to precipitation and polymerization no values could be calculated for Fe^{2+}, Zn^{2+} and Ca^{2+}. In the same way the equilibrium constants for the formation of complexes of these ions with ATP were determined. In the pH region 3–6.5 it was found that they were strongly bonded to ATP. Finally a quantitative study was made on the interaction of ATP, dopamine and the metal ions. In the pH region 6.5–11 ternary chelates were formed between dopamine and the ion-ATP complex. The order of the binding of dopamine to the various metal-ATP (1:1) complexes was found to be $Zn > Cu > Fe > Mg > Ca$ (Rajan *et al.*, 1971, 1972, 1974, 1976a,b). The absolute values of the determined stability constants should be viewed with caution because of the uncertainty in some of the curves obtained due to polymerization and because the acidity dissociation constants used for dopamine were the macroconstants and as stated by Boggess and Martin (1975) the use of the first two acidity constants to calculate the free ligand concentration is patently wrong. The free ligand concentration can only be correctly calculated with a knowledge of the acidity microconstants.

The role of complex formation between catecholamines and nucleotides as being responsible for storage mechanisms was questioned by Granot and Fiat (1977). In an NMR study where the ring protons and side chain protons can be separately studied the interactions between catecholamines and adenine nucleotides were investigated in the absence and presence of divalent metal ions. The induced chemical shifts revealed the formation of binary catecholamine–nucleotide complexes and in the presence of metal ions the formation of ternary metal–nucleotide–dopamine complexes. Their results indicated that at $pH < 7$ the association of dopamine with the nucleotide–metal complex is without direct association to the metal ion which is strongly bound to the nucleotide. Only in the ionized state does dopamine bind strongly to the complex via the metal ion. However this is of minor importance in biological systems, because of the small fraction of ionized dopamine at $pH \sim 7$. They concluded that due to the overall weak interaction of dopamine with nucleotides an appreciable fraction would not be bound when the catecholamines are in excess in solution and that the influence of divalent metal ions on this association is insignificant.

Oxidation of Dopamine

The oxidation reaction of dopamine is a much-studied process because of its possible involvement in biological electron transfer processes. The oxidation reactions can be divided into direct and indirect oxidations. In the direct reactions dopamine is oxidized with chemical oxidizers or electrochemically.

On the basis of several investigations the following scheme can be developed for the oxidation of dopamine.

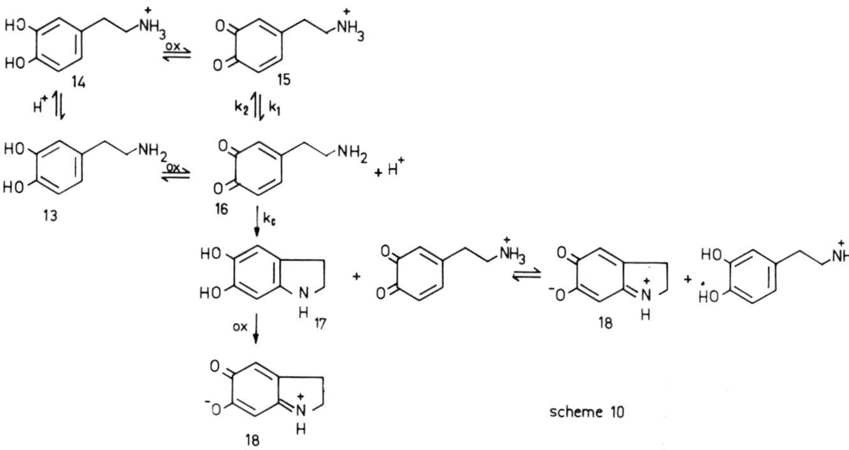

scheme 10

The first step consists in an oxidation of dopamine to the open chain quinone (15), this reaction is strongly dependent upon the pH of the medium until dopamine is completely ionized. At low pH it was shown by cyclic voltammetry (Hawley *et al.*, 1967) that the oxidation to the quinone is reversible. The existence of the *O*-quinone as an intermediate was shown by trapping the quinone with cyclopentadiene to form an adduct (Omote and Komatsu, 1974). At higher pH, when a small quantity of the unprotonated quinone is formed, an intramolecular nucleophilic attack can occur resulting in the product of a 1,4-Michael addition, the leucochrome (17). In the last step the leucochrome is oxidized to the aminochrome (18). In the electrochemical investigation of Hawley this step was assumed to proceed, with the quinone, in a second redox couple, the overall conversion of (14–18) comprises then an ECE (electrochemical–chemical–electrochemical) mechanism.

In the oxidation with ferricyanide (Harrison *et al.*, 1968) the aminochrome was formed with a total release of $5H^+$ and the reduction of four molecules of ferricyanide indicating a final oxidation step by the reagent. Besides the intramolecular cyclization reaction of the *O*-quinone, yielding the aminochrome, two other pathways might be important for the oxidation *in vivo* namely the addition of strong nucleophiles (like sulphydryl groups) concurrent with the intramolecular cyclization and a re-reduction of dopamine by endogenous reductants like ascorbic acid (Tse *et al.*, 1976). For the overall reaction (14–18) the following expression for the observed rate constant was derived:

$$k_{obs} = \frac{k_1 k_c}{k_2[H^+] + k_c}$$

At low pH this equation simplifies to

$$k_0 = \frac{k_1 k_c}{k_2[H^+]} = K_1 \frac{k_c}{[H^+]}$$

while at high pH $k_0 = k_1$ corresponding to a rate-limiting deprotonation reaction and a pH-independent rate constant for the cyclization (Hawley *et al.*, 1967). The rate of the cyclization reaction was determined chronoamperometrically at pH 7.0 to be $1.6 \times 10^{-2} s^{-1}$. Pelizzetti *et al.* (1976) followed the oxidation reaction of dopamine with periodic acid at a pH range of 5.5–8.5 spectrophotometrically. In this region the observed rate was still pH-dependent and was expressed as

$$k_{obs} = k_c \frac{K_1}{K_1 + [H^+]}.$$

In contrast to the work of Hawley at high pH $k_{obs} = k_c$ and cyclization is rate determining at higher pH (k_c at $25° \approx 17 s^{-1}$). The rate constants obtained in both experiments should be viewed with caution due to the contradiction concerning the rate-limiting step and because both authors assume the dissociation constant of the ammonium group in the side chain of the *O*-quinone is the *macro*constant pK_1 for dopamine.

The indriect oxidations, in which molecular oxygen is the electron acceptor, are more complex involving the formation of several intermediates and products. Ponomareff *et al.* (1972) have shown that the oxidation is influenced by temperature, pH, oxygen concentration and light. Besides these influences the reaction is catalysed by various transition metal ions, the order being $Mn^{2+} > Co^{2+} > Cu^{2+}$ at pH 8.1 as measured from the oxygen uptake in the reaction (Gillette *et al.*, 1954).

The catalytic activity of these ions may take place through their metal chelates as being precursors in the formation of semiquinone radicals

(Mentasti *et al.*, 1975). The reaction is also catalysed by a polyphenol oxidase isolated from mushrooms (Binns *et al.*, 1970b).

The ultimate products in these indirect reactions are polymers of the melanin-type. According to the Raper–Mason theory (Mason, 1959) the melanins from tyrosine, DOPA, dopamine and 5,6-dihydroxyindole should all be identical because they were all presumed to be formed via 5,6-dihydroxyindole (scheme 11). Much work has been done by the group of

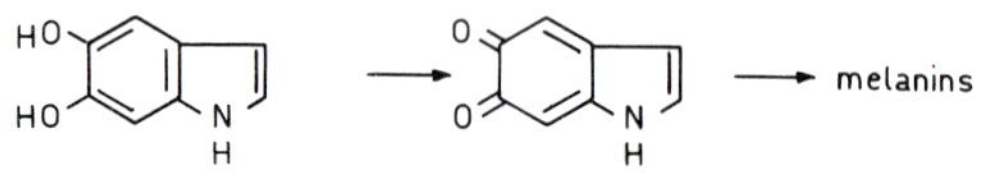

scheme 11

Swan (Swan and Wright, 1954, 1956) to elucidate the nature of the melanin formed from dopamine. It soon became clear that there were, however, considerable differences between the melanins formed from DOPA, dopamine and 5,6-dihydroxyindole.

With the aid of methylation, benzoylation and other chemical experiments such as alkaline peroxide oxidation and hydrolysis as well as studies with melanin formed from deuterium-labelled and ^{11}C-labelled dopamine a preliminary structure was put forward representing autoxidative dopamine-melanin (Binns *et al.*, 1970b). This consists of a combination of units as shown in scheme 12.

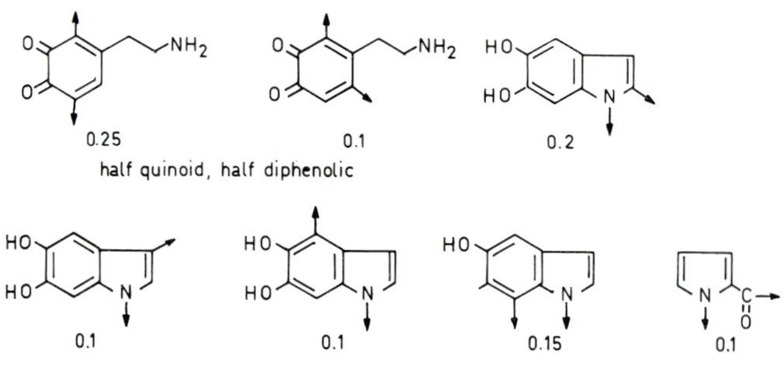

scheme 12

Because dopamine is found in the substantia nigra in high concentrations it is tempting to speculate that dopamine may be the precursor for the melanin of the brain. Indications for such an assumption were found in the fact that dopamine is readily oxidized by an enzyme of nervous tissue to a black pigment which exhibited the same visual appearance, spectral and solubility properties as dopamine-melanin (VanderWende, 1964, 1963).

CONFORMATION OF DOPAMINE

The conformation of biological molecules *in vivo* is a determinant of their biological activity. Dopamine is a flexible molecule and may exist in solution in a number of conformations.

Hypotheses about the conformation of dopamine at the receptor assume that the initial contact between the drug molecule and the receptor is dependent upon an optimum spatial orientation of essential atoms in the drug molecule. Although the molecule may alter its conformation on binding the formation of the drug–receptor complex must involve a process of conformational selection which will influence the energetics and kinetics of the complex formation. Thus a knowledge of the energy of the preferred and non-preferred conformations is important and it is not surprising that the study of these conformations has received considerable attention.

The conformation of dopamine has been studied with X-ray analysis, nuclear magnetic resonance and theoretical calculations.

The structure of dopamine hydrochloride was determined three dimensionally and was shown to be an extended conformation whereby the structure in the crystal is stabilized by a network of hydrogen bonds with the chloride ions as acceptor (Bergin and Carlström, 1968). These

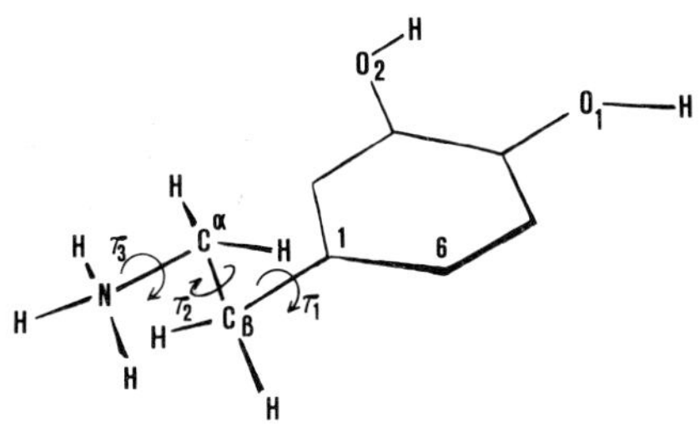

Fig. 2. Dopamine in a *trans* conformation with the aromatic ring perpendicular to the plane through the side chain. The three torsion angles are τ_1 (C_α–C_β–C_1–C_6), τ_2(N–C_α–C_β–C_1) and τ_3(H–N–C_α–C_β).

intermolecular forces will not act in aqueous solution. So there is no guarantee that the crystal conformation corresponds to that occurring under physiological conditions.

Nuclear magnetic resonance seems the method of choice for the study of conformations in solution. The use of the method is based on the fact that the spin–spin coupling constants between protons attached to two adjacent carbon atoms depends upon the dihedral angle about the connecting carbon–carbon bond.

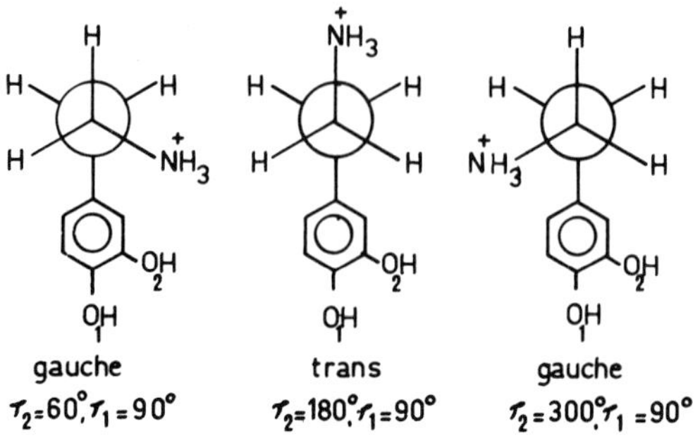

Fig. 3. Newman projections of the two *gauche* and the *trans* conformations of dopamine.

Bustard and Egan (1971) determined from the analysis of the AA′ BB′ spectrum of the side chain protons that the *trans* conformer of dopamine hydrochloride has the lowest energy and that the population in this conformation was 0.43 ± 0.05 over a temperature range of 60°. Their results show that there is a small increase in the *trans* form population with an increase in temperature. This finding, not expected from the Boltzmann law, was explained as a solvent-mediated intra- or intermolecular interaction between the N and the OH group stabilizing the *gauche* form. An increase in temperature will break this weak bonding resulting in an increase in *trans* conformers.

A number of theoretical investigations have been conducted to understand the conformational behaviour of various drug molecules. In the molecular orbital (MO) theory the electrons, being associated with the entire molecule, are described in terms of molecular orbitals and are subject to the potentials of all nuclei and the potentials of the charge distributions of all other electrons.

The energy of this system of interacting particles is the sum of a kinetic energy contribution and a potential energy function. Schrödinger described the wave character of the particles with the equation $H\psi = E\psi$, the solution to this equation, the wave function ψ, describing the spatial motion of the particles of the system. An exact solution to the Schrödinger equation is impossible for organic molecules with the currently known mathematical tools. The methods used are gross approximations. The quantity, however, which we are looking for, the energy difference between certain conformations, is a small difference between two large energies. In this light, discussions about a few kcal difference between conformers seem exaggerated, and the results therefore should always be correlated with experimental observations. The theoretical methods which are applied differ in their degree of sophistication, the simplest being the Extended Hückel

Theory (EHT) followed by the method of Complete Neglect of Differential Overlap (CNDO), the PCILO (Perturbative Configuration Interactions using Localized Orbitals) and the INDO Method (Intermediate Neglect of Differential Overlap). In *ab initio* methods all electrons including the core electrons are considered, but due to the enormous computer time needed the conformations of only small molecules or parts of molecules are calculated. In most of these studies nowadays the energies of the differential conformations are calculated as a function of two torsion angles and are depicted as energy maps with isoenergy curves describing conformers with the same energy.

Extended Hückel Theory calculations on dopamine were done by Bustard and Egan (1971) and Rekker *et al.* (1972) with only the variable methylene–methylene bond angle. They found the *trans* conformation to be preferred. Kier (1973) calculated with the EHT method the energies with two variable angles and found as an energy minimum the *trans* conformation of the side chain with the chain perpendicular to the aromatic ring (see Fig. 2).

Katz *et al.* (1973) performed CNDO calculations on protonated dopamine and found three minimum energy conformations but now the *gauche* forms were about 2 $kcal\,mol^{-1}$ more stable than the *trans* but with low energy barriers between the conformers. Pullman *et al.* (1972) using the PCILO method also found from the energy map small differences between the *gauche* and *trans* forms. A small preference (1–2 $kcal\,mol^{-1}$) was obtained for the *gauche* form using an *ab initio* STO 36 computation for the essential parts of the map.

Finally, calculations were made with the INDO method (Grol and Rollema, 1977). Besides the two *gauche* and *trans* forms as minima two extra energy minima were found, probably because the energy conformation map was carefully constructed with small increments in the torsion angles.

As a result of these various calculations it can be stated that the flexibility of the dopamine molecule was confirmed and that a range of possible conformations are possible.

An important aspect of the MO calculations is the fact that the calculations are performed on gas phase molecules and that the effect of solvation is neglected. A treatment for this problem was developed by Pullman and Pullman (1975) and consists in fixing water molecules on the most favourable hydration sites and calculating the conformational map of this so-called "supermolecule". For dopamine this solvation effect was found to result in an increase in the probability of the occurrence of the *trans* form but the energy differences remained small (Pullman *et al.*, 1974). The effect of the medium was also considered in empirical calculations,

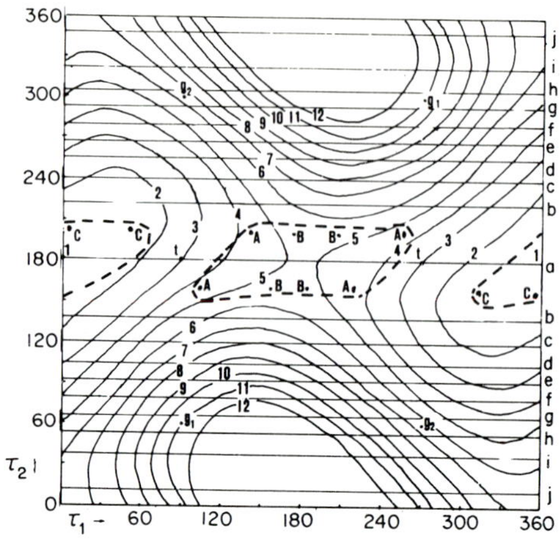

Fig. 4 Isodistance map for dopamine where each line gives a fixed value for the N–O_2H distance (1–12) in dopamine. The N–O_1H distances (a–j) are 25 pm (0.25 Å) beyond the maximal distance of 785 pm (7.85 Å) (a) and the N–O_2H distances are 25 pm (0.25 Å) beyond the maximal distance of 7.35 Å (1). A, B and C indicate the conformations of dopamine in which the key distances N–O_1H and N–O_2H are the same as the comparable distances in resp. apomorphine, 5,6-dihydroxy-2-aminotetralin and isoapomorphine as well as 6,7-dihydroxy-2-aminotetralin.

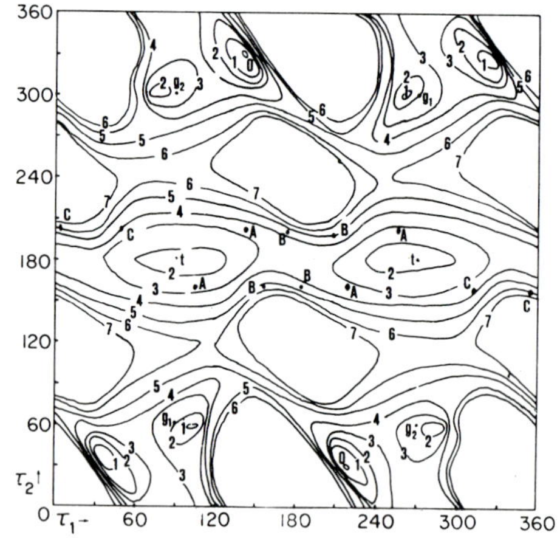

Fig. 5. Conformational energy map of dopamine constructed with isoenergy curves 1 kcal mol^{-1} above the global minimum at $\tau_1 = 220°$, $\tau_2 = 30°$. Points A, B and C are the same as in Fig. 4.

TABLE II Calculated distances in dopamine and rigid analogues

	$N\text{-}O_1H$	$N\text{-}O_2H$
Dopamine ($\tau_1 = 90°$, $\tau_2 = 60°$) g_1	6.19	4.89
Dopamine ($\tau_1 = 90°$, $\tau_2 = 180°$) t	7.85	6.87
Dopamine ($\tau_1 = 90°$, $\tau_2 = 300°$) g_2	6.21	5.99
Apomorphine calc. A	7.76	6.44
X-ray	7.78	6.48
5,6-Dihydroxy-2-aminotetralin B	7.74	6.20
Isoapomorphine C	7.77	7.31
6,7-Dihydroxy-2-aminotetralin C	7.77	7.32

where the conformational energy was calculated as the sum of the non-bonded interaction energy, the torsial potential energy, the electrostatic energy and the hydrogen bond energy (Kumbar, 1976). The electrostatic term contained the effective dielectric constant. Three minima were found with the two *gauche* forms preferred above the *trans*. The relative populations shifted when ε increased from 1.0 to 5.0. At $\varepsilon = 80$ the population percentage was increased to 32% *trans*. Ison *et al.* (1973) calculated from the basis of the energy maps from Pullman with the Boltzmann equation a *trans* population of 35%.

One approach to settle the question of the dopamine conformation at the receptor might be the study of rigid dopamine analogues having a biological action similar to that of dopamine.

This approach was followed by Grol and Rollema (1977) who determined from a distance map (Fig. 4) the conformations of dopamine in which dopamine had the same $N\text{-}O_1H$ and $N\text{-}O_2H$ distances as the rigid dopamine analogues. On the energy map (Fig. 5) it can then be seen whether these conformations are energetically favourable. Using two kinds of rigid compounds, namely apomorphine and 2-amino-5,6-dihydroxy tetralin on the one side and isoapomorphine and 2-amino-6,7-dihydroxy tetralin (ADTN) on the other side, it can be seen that dopamine can achieve the $N\text{-}O_1H$ and $N\text{-}O_2H$ distances of the analogues in each group with four different conformations. In both clusters the side chain is more or less *trans* ($\tau_2 = 180°$) but in the first cluster τ_1 (determining the position of the aromatic ring) is centred around $\tau = 0°$ and in the other cluster τ_1 is around 180°, indicating that dopamine might be active in two conformations. This picture agrees well with the suggested α and β dopamine rotamers as developed by Cannon (1975).

REFERENCES

Adams, E. Q. (1916) *J. Am. Chem. Soc.* **38,** 1503–1510.

Antikainen, P. J. and Witikainen, U. (1973). *Act Chem. Scand.* **27,** 2075–2082.

Armstrong, J. and Barlow, R. B. (1976). *Br. J. Pharmacol.* **57,** 501–516.

Barger, G. and Ewins, A. J. (1910). *J. Chem. Soc.* **97,** 2253–2261.

Battersby, A. R., Sheldrake, P. W., Stannton, J., Williams, D. C. (1976). *J. C. S. Perkin I*, 1056–1062.

Bergin, R. and Carlström, D. (1968). *Acta Cryst. B* **24,** 1506–1510.

Binns, F., King, J. A. G., Percival, A., Robson, N. C., Swan, G. (1970a). *J. Chem. Soc.* (*C*) 1134–1138.

Binns, F., King, J. A. G., Mishra, S. N., Percival, A., Robson, N. C., Swan, G. A., Waggot, A., (1970b). *J. Chem. Soc.* (*C*) 2036–2070.

Björklund, A., Falck, B., Håkanson, R. (1970). *Anal. Biochem.* **35,** 264–276.

Boggess, R. K. and Martin, R. B. (1975). *J. Am. Chem. Soc.* **97,** 3076–3081.

Borgman, R., Mc Phillips, J. J., Stitzel, R. E., Goodman, J. J. (1973). *J. Med. Chem.* **16,** 630–633.

Buck, J. S. (1933). *J. Am. Chem. Soc.* **55,** 3388–3390.

Bustard, T. M. and Egan, R. S. (1971). *Tetrahedron* **27,** 4457–4469.

Casagrande, C. and Ferrari, G. (1973). *Il Farmaco Ed. Sci.* **28,** 143–148.

Cannon, J.G. (1975). In "Advances in Neurology", Vol. 9. (Eds D. B. Calne, T. N. Chase and A. Barbeau), Raven Press, New York.

Christman, O. R., Hoyle, R. M., Wolf, A. P. (1970). *J. Nucl. Med.* **11,** 474–478.

Clarke, E. G. C. (1974). In "Isolation and Identification of Drugs, Vol. 3, 321. Pharmaceutical Press, London.

Collins, E. G. C. and Bigdeli, M. G. (1975). *Life Sci.* **16,** 585–602.

Corrodi, H. and Hillarp, N. A. (1964). *Helv. Chim. Acta* **47,** 911–917.

Edsall, J. T., Martin, R. B., Hollingworth, B. R. (1958). *Proc. Natal. Acad. Sci., U.S.A.* **44,** 505–518.

Fowler, J. S., Ansari, A. N., Atkins, H. L., Bradley-Mosse, P. R., McGregor, R. R., Wolf, A. P. (1973). *J. Nucl. Med.* **14,** 867–869.

Fowler, J. S., Wolf, A. P., McGregor, R. R. (1977). *Chem. Abstr.* **86,** 5851 p.

Gairand, C. B. and Lapin, G. R. (1953). *J. Org. Chem.* **18,** 1–3.

Gillette, J. R., Watland, D., Kalnitsky, G. (1954). *Biochem. Biophys. Acta* **15,** 526–532.

Granot, J. (1976). *FEBS Lett.* **67,** 271–276.

Granot, J. and Fiat, D. (1977). *J. Am. Chem. Soc.* **99,** 4963–4968.

Greenberg, R. S. and Cohen, G. (1973). *J. Pharmacol. Exp. Ther.* **184,** 119–128.

Grgas-Kužnar, B., Simeon, Vl. Weber, O. A. (1974). *J. Inorg. Nucl. Ch.* **36,** 2151–2154.

Grol, C. J. and Rollema, H. (1977). *J. Pharm. Pharmacol.* **29,** 153–156.

Harrison, W. H., Whisler, W. W., Hill, B. J. (1968). *Biochemistry* **7,** 3089–3094.

Hawley, M. D., Tatawadi, S. V., Piekarski, S., Adams, R. N. (1967). *J. Am. Chem. Soc.* **89,** 447–450.

Ison, R. R., Partington, P., Roberto, G. C. K. (1973). *Molec. Pharm.* **9,** 756–765.

Kanao, S., Toyota, T., Matsumi, M., Suyama, T. (1975). *Chem. Abstr.* **82,** 97817 a.

Katz, R., Heller, S. R., Jacobson, A. E. (1973). *Molec. Pharm.* **9,** 486–494.

Kier, L. B. (1973). *J Theor. Biol.* **40,** 211–217.

King, G. S., Goodwin, B. L., Sandler, M. (1974). *J. Pharm. Pharmacol.* **26,** 476–478.

Kirby, G. W. and Ogunkoya, L. (1965). *J. Chem. Soc.* 6914–6922.

Kovács, D. and Fodor, G. (1951). *Chem. Ber.* **84,** 795–801.
Kumbar, M. (1976). *J. Med. Chem.* **19,** 1232–1238.
Lambert, F., Ellenberger, M., Martin, L., Cohen, Y. (1975). *Org. Magnet. Res.* **7,** 266–293.
Lewis, G. P. (1954). *Br. J. Pharmacol.* **9,** 488–493.
Lippold, B. C. and Biedermann, J. (1974). *Pharmazie* **29,** 393–395.
Maas, J. W. and Colburn, R. W. (1965). *Nature (London)* **208,** 41–46.
Mannich, C. and Jacobson, W. (1910). *Chem. Ber.* **43,** 189–197.
Martin, R. B. (1971). *J. Phys. Chem.* **75,** 2657–2661.
Mason, H. S. (1959). *In* "Pigment Cell Biology". (Ed. M. Gordon), 563. Academic Press New York and London.
Mentasti, E., Pelizetti, E., Pramauro, E. (1975). *Gazetta* **105,** 551–558.
Milne, G. W. A., Fales, H. M., Colburn, R. W. (1973). *Anal. Chem.* **45,** 1952–1954.
Morita, S., Ito, T., Tono, T. (1975). *Agr. Biol. Chem.* **39,** 547–549.
Omote, Y. and Komatsu, T. (1974). *Bull. Chem. Soc. Jpn* **47,** 3128–3130.
Pelizzetti, E., Mentasti, E., Pramauro, E. (1976). *J. C. S. Perkin II,* 1651–1655.
Perel, J. M., Dawson, D. K., Dayton, P. G., Goldberg, L. I. (1972). *J. Med. Chem.* **15,** 714–716.
Persson, B. (1971). *Acta Pharm. Suec.* **8,** 193–204.
Pictet, A. and Finkelstein, M. (1909). *Chem. Ber.* **42,** 1979–1996.
Ponomareff, M., Soliva, M., Speiser, P. (1972). *Helv. Chim. Acta* **47,** 467–479.
Pullman, A. and Pullman, B. (1975). *Quart. Rev. Biophys.* 510–565.
Pullman, B., Coubeils, J. L., Courrière, Ph., Gervois, J. P. (1972). *J. Med. Chem.* **15,** 17–23.
Pullman, B., Berthod, H., Courrière, Ph. (1974). *Int. J. Quant. Chem.* **1,** 93–108.
Rajan, K. S., Davis, J. M., Colburn, R. W. (1971). *J. Neurochem.* **18,** 354–364.
Rajan, K. S., Davis, J. M., Colburn, R. W., Jarke, F. H. (1972). *J. Neurochem.* **19,** 1099–1116.
Rajan, K. S., Davis, J. M., Colburn, R. W. (1974). *J. Neurochem.* **22,** 137–147.
Rajan, K. S. and Davis, J. M. (1976a). *J. Inorg. Nucl. Chem.* **38,** 897–905.
Rajan, K. S., Skripkus, A., Marks, G. E., Davis, J. M. (1976b). *Bioinorg. Chem.* **6,** 93–117.
Rekker, R. F., Engel, D. J. C., Nijs, G. S. (1972) *J. Pharm. Pharmacol.* **24,** 589–591.
Reisch, J., Pagnucco, R., Alfes, H., Jantos, N., Möllmann, H. (1968). *J. Pharm. Pharmacol.* **20,** 81–86.
Robin, Ganellin C. (1977). *J. Med. Chem.* **20,** 579–581.
Rosenmund, K. W. (1910). *Chem. Ber.* **43,** 4312–3417.
Sandler, M., Bonham-Carter, S., Hunter, K. R., Stern, G. M. (1973). *Nature (London)* **241,** 439–443.
Short, J. H., Dunnigan, D. A., Ours, C. W. (1973) *Tetrahedron* **29,** 1931–1939.
Svensson, L., Björklund, A., Lindvall, O. (1975). *Acta Chem. Scand.* **B29,** 341–348.
Swan, G. A. and Wright, D. (1954). *J. Chem. Soc.* 381–384.
Swan, G. A. and Wright, D. (1956). *J. Chem. Soc.* 1549–1557.
Symchowicz, S. and Faro, H. P. (1976). *Chem. Abstr.* **84,** P179904 g.
Tse, D. C. S., McCreery, R. L. Adams, R. N. (1976). *J. Med. Chem.* **19,** 37–40.
Yamanaka, Y., Walsh, M. J., Davis, V. E. (1970). *Nature (London)* **227,** 1143–1144.
VanderWende, C. and Spoerlein, M. T. (1963). *Life Sci.* **6,** 386–392.
VanderWende, C. (1964). *Arch. Int. Pharmacodyn.* **152,** 433–444.
Waser, E. and Sommer, H. (1923). *Helv. Chim. Acta* **6,** 54–60.
Whaley, W. M. and Govindachari, T. R. (1951). *In* "Organic Reactions", Vol. VI, (Ed. R. Adams), 151–190. Wiley, New York.

Analysis of Dopamine and its Metabolites

CHAPTER 3

Classical Fluorimetry

K. M. Taylor

Roche Research Institute of Marine Pharmacology,
Dee Why, N.S.W., Australia.

Unlike adrenaline and noradrenaline, dopamine is not easily measured by bioassay techniques and does not possess favourable native fluorescent characteristics, as does serotonin, that allow simple quantitation. Therefore the development and application of fluorimetric techniques was directly responsible for much of our knowledge of the role of dopamine, especially in brain tissue. The classical techniques for quantitating dopamine and its metabolites in biological samples involve a separation procedure followed by a simple chemical reaction to form a highly fluorescent product. Since dopamine was first measured quantitatively in the brain by a method based on condensation with 1,2-diaminomethane (Montague, 1957) and its identity confirmed by a sensitive and specific assay based on the production of a hydroxyindole derivative (Carlsson and Waldeck, 1958), there have been numerous improvements and modifications of the fluorescence techniques (Bertler *et al.*, 1958; Euler, 1959; Drujan *et al.*, 1959; Anton and Sayre, 1964; Chang, 1964; Laverty and Sharman, 1965; Laverty and Taylor, 1968; Atack and Magnusson, 1970).

In the initial separation process dopamine can be effectively separated from biological samples after protein precipitation (Euler, 1959) by solvent extraction (Shore and Olin, 1958), absorption onto alumina (Anton and Sayre, 1964) or by ion exchange chromatography using a strong cation exchange resin such as Dowex 50-W (Carlsson and Waldeck, 1958). The last separation technique is the most efficient and reliable and provides the best recovery of dopamine. However, solvent separation is a useful technique if simultaneous determination of metabolites and other biogenic amines in a single tissue sample is desired (Chang, 1964; Maickell *et al.*, 1968; Murphy *et al.*, 1969; Haubrich and Denzer, 1973). All catechols are absorbed by alumina at pH 8.4 and thus it provides a useful technique to concentrate dopamine or separate the acid metabolite 3,4-dihydroxyphenylacetic acid or 3,4-dihydroxyphenethanol in the presence of the corresponding 3-methoxylated metabolites.

The three separation techniques all have application to the separation of metabolites of dopamine. The acid metabolites 3,4-dihydroxyphenylacetic acid and homovanillic acid can be separated by anion exchange chromatography (Andén *et al.*, 1963) or by solvent extraction (Murphy *et al.*, 1969). 3-Methoxytyramine can be separated by cation exchange chromatography (Carlsson and Waldeck, 1958). The neutral alcohol metabolites 3,4-dihydroxyphenethanol and 3-methoxy-4-hydroxyphenethanol can be obtained in the effluent from anion and cation exchange columns in series (Taylor and Laverty, 1969).

Once separated, the fluorimetric methods available to determine dopamine and metabolites are

(i) the hydroxyindole reaction (Carlsson and Waldeck, 1958) which is applicable to phenylethylamines.
(ii) the 1,2-diaminoethane condensation for catechols (Laverty and Sharman, 1965; Sharman, 1971).
(iii) the ferricyanide reaction (Andén *et al.*, 1963) which is applicable to 3-methoxy-4-hydroxyphenyl compounds.

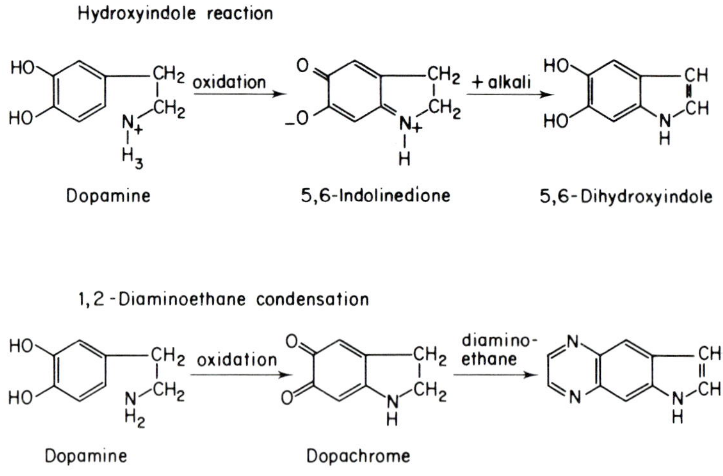

Fig. 1. Fluorescent products of dopamine and metabolites.

The chemistry of these reactions and the nature of the fluorophore is described in Fig. 1. The ferricyanide reaction product is omitted as it has not yet been confirmed or isolated (Corrodi and Werdinius, 1965).

It is impossible to detail all of the many fluorimetric methods and variations available for dopamine and its metabolites. The method described represents a collection into a comprehensive yet broadly applicable technique of many of the improvements applied by other workers to the fluorimetric determination of dopamine and other catecholamines. Basically the method involves separation of dopamine from other catecholamines and metabolites using Dowex 50-W cation exchange resin followed by fluorimetric analysis using the hydroxyindole technique (Laverty and Taylor, 1968; Taylor and Laverty, 1969).

MATERIALS

All glassware must be acid washed and all solutions prepared in freshly distilled deionized water. Analytical grade reagents were used and the purity of standard samples of catecholamines and metabolites were checked by

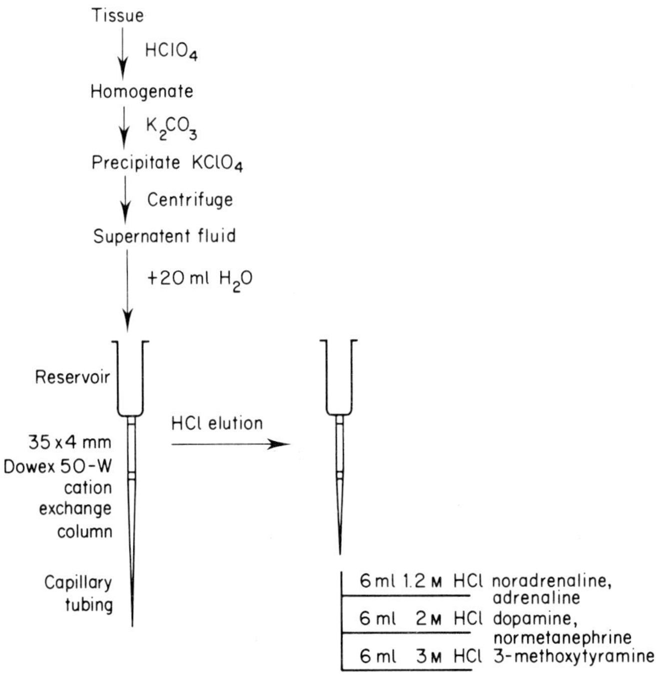

Fig. 2. Separation of dopamine by ion-exchange chromatography.

thin layer or column chromatography. Dowex 50W-X8 (200-400 mesh—chloride form) cation exchange resin (Bio-Rad) was washed successively with 20 volumes of 2 M HCl, twice with water, 2 M NaOH, twice with water, and finally stored in 2 M HCl. Dowex 50-W columns (3.5 × 0.4 cm) were prepared by the method of Juorio *et al.* (1966, Fig. 2). The columns were prepared by successively applying 20 ml volumes of 2 M HCl, water, 1 M sodium acetate and finally water. The column chromatography system relies on capillary attraction to maintain an even flow rate. Flow rate can be increased by positive pressure (Bertler *et al.*, 1958) using a modified syringe system. Columns can be prepared in suitable pipettes (Atack, 1971) if glassblowing facilities are unavailable. The post-mortem decomposition of dopamine is rapid so it is necessary to quickly freeze tissue samples on dry ice immediately and store at −14°C until estimation.

PROCEDURE

Tissue homogenates were prepared in 10 volumes of cold 0.4 M perchloric acid containing 0.1% ascorbic acid and 0.01% EDTA (sodium salt) using a Potter–Elvejham ground-glass mortar and pestle or a Polytron homogenizer. After centrifugation at 7000 × *g* for five minutes, the supernatant fluid was removed and adjusted to pH 4.0 with 2.5 M K_2CO_3. Samples were then centrifuged at 7000 × *g* for five minutes to remove the precipitated potassium perchlorate, and the supernatant fluid was carefully applied to the ion exchange columns and washed through with 20 ml water. Increasing concentrations of hydrochloric acid were applied to the column to elute catecholamines and metabolites in the following sequence:

6 ml of 1.2 M HCl—noradrenaline, adrenaline
6 ml of 2.0 M HCl—dopamine, normetanephrine
6 ml of 3.0 M HCl—3-methoxytyramine

The 2.0 M HCl fraction containing dopamine was adjusted to pH 7.0 with 2.5 M K_2CO_3. For fluorimetric analysis 0.5 ml samples in duplicate were removed for sample, blank and internal standard determination as follows:

(a) 0.5 ml of 0.1 M sodium phosphate buffer pH 7.0 was added. Internal standards of dopamine (20 ng) were added at this stage.
(b) 0.05 ml of 0.02 M iodine solution in 0.3 M sodium iodide was added and the oxidation reaction allowed to proceed for three minutes.
(c) Oxidation was stopped and the solution made alkaline to allow tautomerization by adding 0.25 ml of a solution containing 0.1 M sodium sulphite, 2.5 M sodium hydroxide and 0.02 M disodium EDTA.

TABLE 1 Structure–reactivity relationships for the hydroxyindole reaction

Amine	Structure					Oxidation			Tautomerization		Acidification		Development		Fluorescence		Sensitivity
	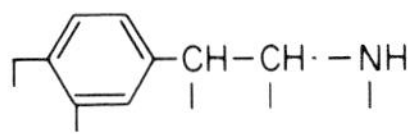					pH	I_2 soln (ml)	time (min)	alk. sulphite (ml)	time (min)	acetic (ml)	final pH	time (min)	temp. (°C)	excit. (mμ)	emiss. (mμ)	twice blank (ng)
DA	OH	OH	H	H	H	7.0	0.05	3	0.25	5	0.1	4.4	40	100	320	375	2
NA	OH	OH	OH	H	H	6.5	0.05	3	0.5	5	0.15	4.8	25	20	380	480	0.5
A	OH	OH	OH	H	CH_3	4.0	0.05	3	0.5	1	0.1	5.0	0	20	410	500	2
NM	OH	OCH_3	OH	H	H	8.6	0.05	3	0.5	5	0.3	4.2	25	20	380	475	4
M	OH	OCH_3	OH	H	CH_3	9.1	0.05	3	0.5	1	0.1	5.6	0	20	410	490	3
3MT	OH	OCH_3	H	H	H	8.8	0.05	3	0.25	5	0.25	4.5	40	100	320	375	8
OA	OH	H	OH	H	H	9.0	0.05	3	0.4	5	0.2	4.7	30	100	375	475	200
TA	OH	H	H	H	H	9.9	0.05	3	0.4	5	0.15	4.5	40	100	320	375	200
TS	OH	H	H	COOH	H	10.6	0.05	3	0.4	5	0.1	5.0	40	100	330	380	50
DOPA	OH	OH	H	COOH	H	5.4	0.05	3	0.25	5	0.05	5.7	40	100	330	380	4

TABLE II The recovery of added amines from pure solutions and tissue samples estimated by the ion exchange hydroxyindole fluorimetric procedure

	Mean percentage recovery ± S.E.M. ($n = 12$)	
	pure solution	tissue sample
Dopamine	80 ± 2	69 ± 3
3-Methoxytyramine	85 ± 6	78 ± 3
Noradrenaline	95 ± 2	88 ± 2
Normetanephrine	80 ± 3	74 ± 3

(d) Five minutes later the solution was acidified with 0.1 ml of glacial acetic acid to achieve a final pH of 4.4

(e) The fluorophore was then developed by placing the solutions contained in stoppered test tubes in a boiling water bath for 40 minutes.

(f) The solutions were cooled and the concentration of the fluorescent product determined in an Aminco–Bowman spectrophotofluorimeter at excitation wavelength 320 nm and emission wavelength 375 nm. The fluorescent product was stable at room temperature for 60 min.

The most reliable blank determination is that obtained by adding the reactant solutions in reverse. The fluorescence of the blank and internal standard solutions allows the calculation of the concentration of dopamine in the sample and thus, by allowing for the volume of the 2.0 M HCl fraction sampled and the tissue weight, the concentration of dopamine in the original tissue sample can be calculated. The technique can be used to estimate 3-methoxytyramine, but under different reaction conditions applied to the 3 M HCl fraction. Normetanephrine and metanephine occur in the same fraction as dopamine but there is no cross-interference in the hydroxyindole reaction (Table 1). The recoveries of dopamine, noradrenaline, normetanephrine and 3-methoxytyramine added to pure solutions and tissue samples and estimated by the above technique are listed in Table 2. Overlap into adjacent fractions is less than 5% of the applied catecholamine.

The procedure can measure dopamine in the picomolar range and is sensitive enough to determine the concentration of dopamine in brain regions (Table 3). Usually 24–48 tissue samples can be estimated in one day. The cation exchange columns can be reused. Variations on the basic technique include the use of ferricyanide as oxidant, ascorbic acid as antioxidant (Smith and Weil-Malherbe, 1962), the use of ultraviolet light to

TABLE III Endogenous catecholamine levels in various regions of the brain of the rat[a] as determined by the hydroxyindole technique after separation on a Dowex-50W cation exchange column.

		Amine concentration ($\mu g\, g^{-1}$)			
Brain region	Weight (g)	NA	NM	DA	3MT
Whole brain	1.75 ± 0.06	0.38 ± 0.02	0.05 ± 0.01	0.46 ± 0.05	0.04 ± 0.01
Brainstem	0.25 ± 0.01	0.44 ± 0.01	0.05[b]	0.08 ± 0.01	ND[c]
Cerebellum	0.23 ± 0.04	0.26 ± 0.02	0.03[b]	0.03[b]	ND
Cortex	0.53 ± 0.02	0.22 ± 0.01	0.04 ± 0.01	0.04[b]	ND
Corpus striatum	0.24 ± 0.03	0.28 ± 0.01	0.03[b]	3.08 ± 0.07	0.39 ± 0.13
Thalamus–hypothalamus–midbrain	0.50 ± 0.03	0.58 ± 0.02	0.07 ± 0.01	0.13 ± 0.01	ND

[a] Sprague–Dawley male rats (180–200 g) were used. Brain regions were dissected by the method of Taylor and Laverty (1969). Each value is the mean ± S.E.M. of eight determinations.

[b] Amine detectable but not consistently measurable.

[c] ND = Metabolite not detectable.

develop the fluorophore (Carlsson and Waldeck, 1958; Chang, 1964) and ethanolic hydrochloric acid as the eluting agent (Atack, 1971).

The method can be applied to the study of dopamine metabolism. The effluent from Dowex 50-W cation exchange column can be directly applied to a Dowex AGI-X2 anion exchange column in series (Taylor and Laverty, 1969). Homovanillic and 3,4-dihydroxyphenylacetic acids are retained on the anion exchange column and can be differentially eluted with hydrochloric acid. The alcohol metabolites of dopamine are obtained in the effluent from the cation and anion exchange column in series. For the application of this technique tissues are homogenized in acetic not perchloric acid which interferes with anion exchange chromatography. The two-column technique is particularly useful for the study of the metabolism of radioactively labelled dopamine. However the endogenous concentration of the acidic metabolites of dopamine is usually very low and solvent extraction techniques provide greater sensitivity than ion exchange chromatography.

HOMOVANILLIC ACID

The most applicable fluorimetric technique to measure homovanillic acid is the ferricyanide reaction (Anden *et al.*, 1963) despite the lack of confirmation of the chemical nature of the fluorophore produced. The method involves the reaction of homovanillic acid with alkaline ferricyanide which is terminated by the addition of cysteine. The fluorescent product which may be a dimer (Corrodi and Werdinius, 1965) has an excitation wavelength of 315 nm and an emission wavelength of 430 nm and interference filters can be used to minimize blank readings. As the reaction is adaptable to other catecholamine metabolites with the 3-methoxy-4-hydroxyphenyl group, prior separation of homovanillic acid is necessary. This can be achieved by anion exchange chromatography (Anden *et al.*, 1963) or solvent extraction using *n*-butylacetate (Murphy *et al.*, 1969). The latter method is more sensitive and provides a higher recovery.

3,4-DIHYDROXYPHENYLACETIC ACID

After solvent separation, 3,4-dihydroxyphenylacetic acid can be fluorimetrically estimated after condensation with 1,2-diaminoethane (Rosengren, 1960; Sharman *et al.*, 1967; Murphy *et al.*, 1969). The fluorescent product has an excitation wavelength of 385 nm and emission wavelength of 450 nm. The sensitivity and recovery of 3,4-dihydroxyphenylacetic acid is adequate for determination of this metabolite in individual brain tissue samples.

3-METHOXYTYRAMINE

This metabolite of dopamine can be fluorimetrically estimated after ion exchange chromatography by either the hydroxyindole technique (Carlsson and Waldeck, 1964) or by reaction with alkaline ferricyanide (Kehr, 1974). Despite high sensitivity, neither technique can reliably determine the very low endogenous levels of 3-methoxytyramine in brain tissue.

3-METHOXY-4-HYDROXYPHENYLETHANOL, 3,4-DIHYDROXYPHENYLETHANOL

The alcohol metabolites of dopamine have not been as extensively studied as the corresponding glycol metabolites of noradrenaline (Meek and Neff, 1972). No adequate separation technique is available to allow the possible application of the ferricyanide reaction or the ethylenediamine condensation reaction to the fluorimetric determination of 3-methoxy-4-hydroxyphenylethanol and 3,4-dihydroxyphenylethanol respectively.

SIMULTANEOUS ESTIMATION OF DOPAMINE, OTHER BIOGENIC AMINES AND METABOLITES

Methods involving fluorimetric analysis have been adapted to the separation and determination of dopamine, noradrenaline, serotonin and metabolites in individual tissue samples. Generally there is a loss in recovery and sensitivity for each substance determined in a combined technique as compared to the techniques that have optimal conditions for the extraction of each substance alone. Solvent (Maickel *et al.*, 1968; Haubrich and Denzer, 1973; Metcalf, 1974; Smith *et al.*, 1975) or ion exchange chromatography (Atack and Magnusson, 1970) techniques followed by fluorimetric analysis of individual amines or metabolites are the basis of these intricate methods.

REFERENCES

Andén, N.-E., Roos, B.-E. and Werdinius, B. (1963). *Life Sci.* **2,** 448–458.

Anton, A. H. and Sayre, D. F. (1964). *J. Pharmacol. Exp. Ther.* **145,** 326–336.

Atack, C. V. (1971). *Br. J. Pharmacol.* **48,** 699–714.

Atack, C. V. and Magnusson, T. (1970). *J. Pharm. Pharmacol.* **22,** 625–627.

Bertler, A., Carlsson, A. and Rosengren, E. (1958). *Acta Physiol. Scand.* **44,** 273–292.

Carlsson, A. and Waldeck, B. (1958). *Acta Physiol. Scand.* **44,** 293–298.

Carlsson, A. and Waldeck, B. (1964). *Scand. J. Clin. Lab. Invest.* **16,** 133–138.

Chang, C. C. (1964). *Int. J. Neuropharmacol.* **3,** 643–649.

Corrodi, H. and Werdinius, B. (1965). *Acta. Chem. Scand.* **19,** 1854–1858.

Drujan, B. D., Sourkes, T. L., Lane, D. S. and Murphy, F. G. (1959). *Can. J. Biochem. Physiol.* **37,** 1153–1159.

Euler, U. S. von (1959). *Pharmacol. Rev.* **11,** 262–268.

Haubrich, D. R. and Denzer, J. S. (1973). *Anal. Biochem.* **55,** 306–312.

Juorio, A. V., Sharman, D. F. and Trajkov, T. (1966). *Br. J. Pharmacol.* **26,** 385–392.

Kehr, W. (1974) *Naunyn-Schmiedeberg's Arch. Pharmak.* **284,** 149–158.

Laverty, R. and Sharman, D. F. (1965). *Br. J. Pharmacol.* **24,** 538–548.

Laverty, R. and Taylor, K. M. (1968). *Anal. Biochem.* **22,** 269–279.

Maickel, R. P., Cox, R. H., Saillant, J. and Miller, F. P. (1968). *Int. J. Neuropharmacol.* **7,** 275–285.

Meek, J. L. and Neff, N. H. (1972). *Br. J. Pharmacol.* **45,** 435–441.

Metcalf, G. (1974). *Anal. Biochem.* **57,** 316–320.

Montague, K. A. (1957). *Nature (London)* **180,** 244–245.

Murphy, G. F., Robinson, D. and Sharman, D. F. (1969). *Br. J. Pharmacol.* **36,** 107–115.

Rosengren, E. (1960). *Acta Physiol. Scand.* **49,** 370–375.

Sharman, D. F. (1971). In "Methods of Neurochemistry" Vol. 1, (Ed. Fried, R.), 88–127. Marcel Dekker, New York.

Sharman, D. F., Poirier, L. J., Murphy, G. F. and Sourkes, T. L. (1967). *Can. J. Physiol. Pharmacol.* **45,** 57–62.

Shore, P. A. and Olin, J. S. (1958). *J. Pharmacol. Exp. Ther.* **122,** 295–306.

Smith, J. E., Lane, J. D., Shea, P. A., McBride, W. J. and Aprison, M. H. (1975). *Anal. Biochem.* **64,** 149–169.
Smith, E. R. and Weil-Malherbe, H. (1962). *J. Lab. Clin. Med.* **60,** 212–220.
Taylor, K. M. and Laverty, R. (1969). *J. Neurochem.* **16,** 1361–1366.

CHAPTER 4

Semiautomated Fluorimetry

B. H. C. Westerink

Department of Pharmacy, University of Groningen, Groningen, The Netherlands

INTRODUCTION

During the last 20 years fluorimetric detection techniques have been used for the analysis of monoamines and related compounds in nervous tissue. The limited sensitivity of the manual fluorimetric procedures, however, often prevents a detailed analysis of regional aspects of monoamine metabolism. Methods for a concurrent determination of various compounds have been described (e.g. Metcalf, 1974; Karasawa *et al.*, 1975; Smith *et al.*, 1975; Holman *et al.*, 1976), but their use is limited because various purification procedures have to be used before a fluorimetric detection method can be applied.

To overcome the limited sensitivity of the manual fluorimetric detection techniques, semiautomated methods based on the continuous flow principle (Autoanalyzer®) have been developed. Adaptation of a manual assay to an automated method, based on the continuous flow technique, usually results in an increased accuracy. As smaller volumes can be used, combination with a miniaturized isolation procedure will result in analysis of relatively concentrated solutions. In addition continuous flow analysis gives a very reproducible reagent blank on top of which the fluorescence caused by low levels of the compounds can be reproducibly measured.

Waldmeier *et al.* (1974) developed a simultaneous semiautomated assay for noradrenaline (NA) and dopamine (DA). For this method in which the amines were purified on alumina, a detection limit of 2 ng ml^{-1} of extract was claimed by the authors.

Westerink and Korf (1975a; 1976b; 1977) have developed a concurrent semiautomated assay for NA, DA, 3-methoxytyramine (3-MT), 3,4-dihydroxyphenylacetic acid (DOPAC) and homovanillic acid (HVA), based on a simple and rapid isolation procedure of these compounds on small Sephadex G10 columns, followed by automated fluorimetric detection. With this method DA and its metabolites 3-MT, DOPAC and HVA can be analysed in the ng range. Using these sensitive and rapid assays we were able to study DA metabolism in various brain areas including the striatum, nucleus accumbens, tuberculum olfactorium, frontal cortex, substantia nigra and retina of the rat (Westerink and Korf, 1975b; 1976a; 1976b; 1976c; 1976d; 1976e; Korf *et al.*, 1976; Rollema *et al.*, 1976).

PRINCIPAL STEPS

The principal steps of the concurrent assays are depicted in Fig. 1. After homogenization of the brain samples and precipitation of the perchlorate, a small amount of perchloric acid (PCA) is added to the samples in order to bind the amines reproducibly to the Sephadex G10 columns. To remove interfering fluorescence produced by the Sephadex resin the columns are washed with diluted ammonia and formic acid. After the samples have been applied to the columns the interfering substances are washed away with the diluted formic acid. Hereafter the monoamines are eluted with dilute formic acid. The acidic metabolites leave the column when the pH of the eluent is neutralized with an Na_2HPO_4 solution. The two fractions are collected in test tubes or sample cups.

The eluates are then distributed among the various flow systems for automated fluorimetric detection, based on modifications of classical manual procedures. The eluents are applied to the columns with automated pipettes. Samples are run in parallel with standard solutions. Usually at the beginning and at the end of a series of samples, standard solutions are placed in the line of samples. A standard curve is constructed by measuring the peak height of the standard solutions. For the determination of NA, 3-MT, DA (DHI-method) and HVA the samples pass the flow system twice in order to record the background fluorescence (blank). Typical recordings of the HVA assay including pure standards ($ng\,ml^{-1}$), brain samples and blank recordings are depicted in Fig. 2. The background fluorescence is subtracted from the total fluorescence of the brain samples and the concentration of the compound is graphically read from the standard curve. The background fluorescence for the DOPAC and DA (EDA-method) assay is obtained by analysing 20–50 mg cerebellar tissue. Occasionally internal standards added to cerebellar tissue samples are run in parallel with standards in order to check the recovery of the various assays.

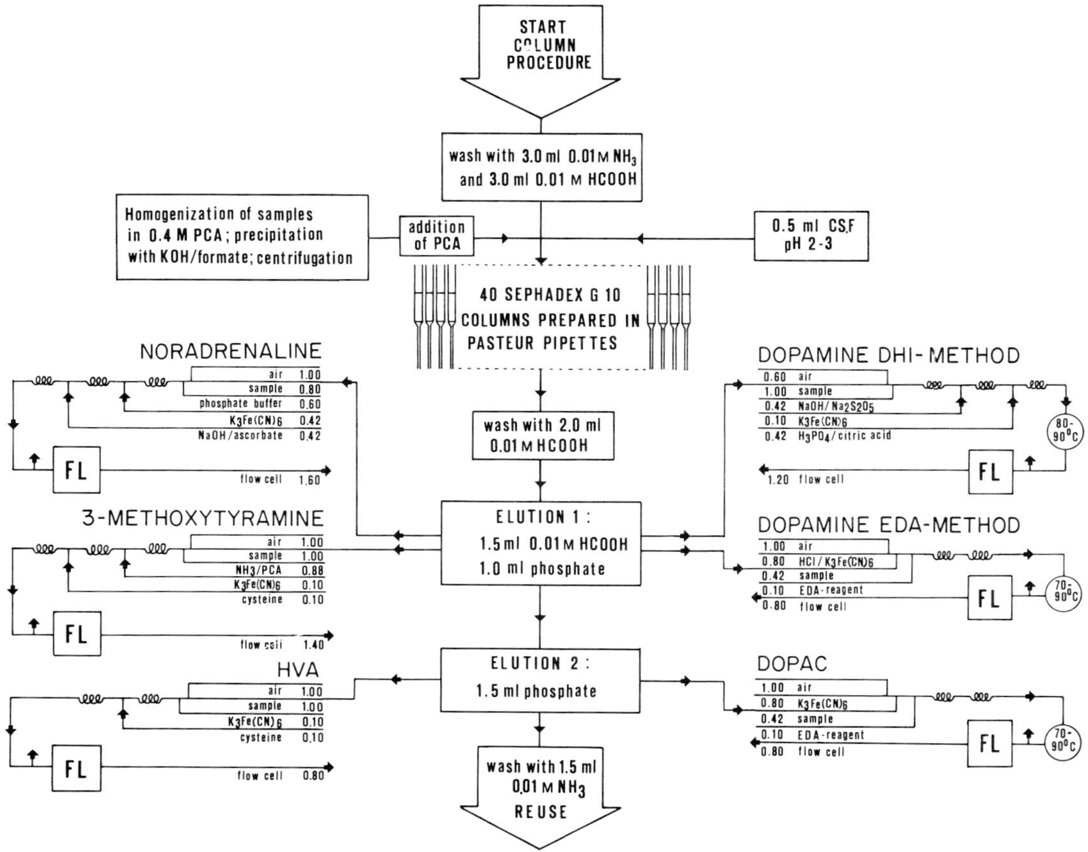

Fig. 1. Flow chart for the concurrent assay. Fl refers to fluorimeter.

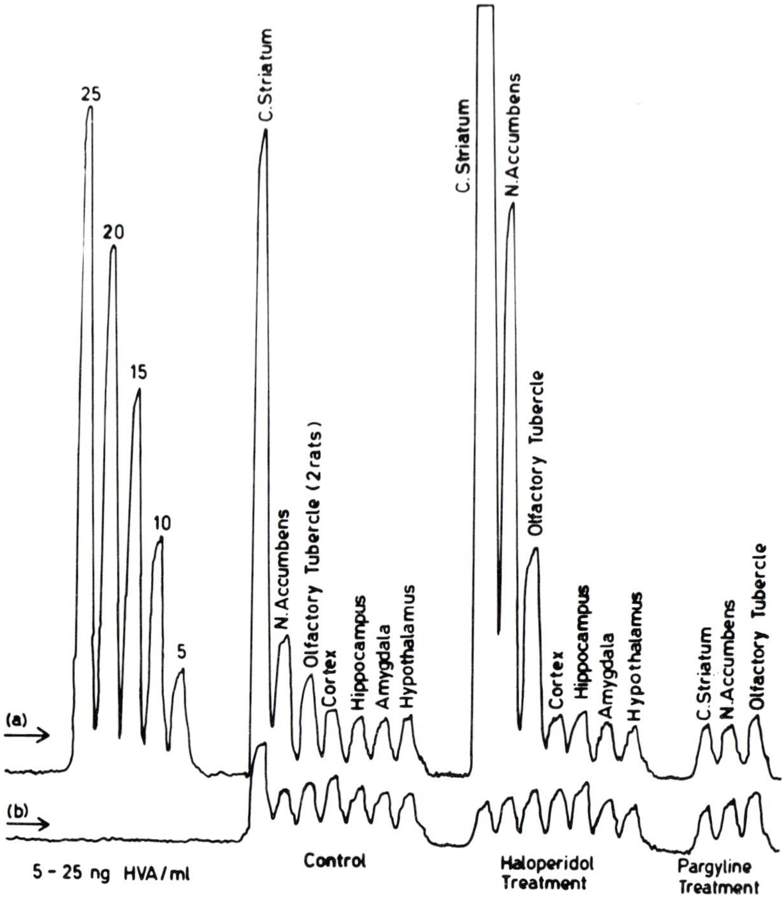

Fig. 2. HVA recordings (a) of different brain samples compared with blanks. (b) Unilateral striatal tissue was applied. After haloperidol treatment HVA content in the striatum exceeded the recording capability.

MATERIALS AND METHODS

Instrumentation

A continuous-flow system equipped with a filter fluorimeter (Autoanalyzer®, Technicon Inc.) is used. The light source is a Mercury lamp (85W). Good results were also obtained with the Perkin Elmer 1000 fluorimeter equipped with a Mark I flow cell (600 μl) and a xenon light source, when the monochromator was replaced by a secondary filter.

Standard Solutions

Stock solutions of NA, DA, 3-MT, DOPAC and HVA contain 10 mg per 100 ml in 0.01 M formic acid, and are stored in the freezer (−20°C). Standard solutions (5–500 ng ml^{-1}) are prepared from the stock solutions with 0.001 M formic acid (NA, DA, 3-MT) or 100 mg per 100 ml Na_2HPO_4 (DOPAC, HVA), and can be used for several days when kept at 4°C. All solutions are made with glass double distilled water.

Isolation Procedures

A phosphate solution (about 0.005 M) is freshly prepared by dissolving 1 g $Na_2HPO_4 . 2H_2O$ in 1 litre of water. The KOH-formate solution used for the precipitation of perchlorate, is prepared by adding carefully 60 ml of concentrated formic acid (98%) to 200 ml 10 M KOH. This solution can be used for several weeks when kept at 4°C.

Tissue samples are homogenized in 1 ml 0.4 M perchloric acid (PCA). The excess of perchlorate is precipitated by addition of 100 µl of the KOH-formulate solution (final pH: 2–3). After centrifugation (20 min, 4000 g, 4°C) the supernatants are transferred to glass tubes and 60 µl of 6% PCA is added. The tissue extracts are then put on Sephadex G10 columns. The Sephadex G10 columns (5 × 70 mm) are prepared in long-size Pasteur pipettes. The pipettes are extended with a glass tube (length 50 mm) of the same diameter. A piece of glass wool is placed at the bottom of the pipette and spontaneous flow through the column is prevented with Parafilm. The pipette is filled with an aqueous suspension of Sephadex G10, prepared the day before use. The flow rate of the columns is adjusted to 0.15–0.20 ml min^{-1} by varying the density of the glass wool pellet.

Before use the columns are washed with 3 ml 0.01 M ammonia and 3 ml 0.01 M formic acid. Forty columns are handled in one run. After the tissue extracts have passed through the columns, 2 ml 0.01 M formic acid is added. NA, DA and 3-MT are eluted with 1.5 ml 0.01 M formic acid and 1.0 ml phosphate solution and are collected in glass tubes.

Homovanillic acid and DOPAC are subsequently eluted with 1.5 ml of the phosphate solution and collected in plastic Autoanalyzer® cups. The columns are washed with 1.5 ml 0.01 M ammonia and, when not reused, stored in 0.01 M ammonia. During the wash and isolation procedure the columns are allowed to run dry.

When only DA has to be determined in the amine fraction, the samples are washed with 3.0 ml 0.01 M formic acid and the DA is eluted with 1.0 ml formic acid and 1.0 ml phosphate, and collected in Autoanalyzer® cups.

Detection Procedures

NA detection

The NA detection is based on the trihydroxyindole reaction as described by Smith *et al.* (1975). A phosphate buffer (pH 7.0) is prepared by dissolving 12g$Na_2HPO_4.2H_2O$ and 6gKH_2PO_4 in 1 litre water; it can be used for several days. The $K_3Fe(CN)_6$ solution contains 30 mg per 100 ml and is stable at 4°C for several weeks. The NaOH reagent is freshly prepared daily by dissolving 120 g NaOH in 300 ml water. After cooling 600 mg ascorbic acid dissolved in 2 ml water is added to the NaOH solution. Sample blanks are prepared by replacement of the $K_3Fe(CN)_6$ solution by water. The filters for the NA assay are: primary filter Turner 47B and secondary filter Corning C.S. 3-69. Sampling and washing times are both 40 s.

DA detection based on the ethylenediamine (EDA) method

In this method a fluorophore is formed by a condensation reaction with EDA. The EDA-reagent is prepared by carefully mixing 2 parts EDA (1 × distilled) with 1 part concentrated HCl (36%) and 9 parts water.

The EDA reagent could be used for several weeks when kept refrigerated. The $K_3Fe(CN)_6$ reagent contains 0.5 mg per 100 ml $K_3Fe(CN)_6$ prepared in 0.35 M HCl. In order to obtain maximal fluorophore formation the samples are heated at 70–80°C for 4–6 min in a heating bath. Sample blanks are prepared by analysing 20–50 mg cerebellar tissue. The linearity of the calibration curve depends on the composition of the $K_3Fe(CN)_6$ reagent. If the calibration curve is not linear the amount of HCl and $K_3Fe(CN)_6$ should be adjusted. As the fluorophore also occurs with other catecholamines and related compounds, the method is only applicable to those areas in the brain known to have a very high DA/NA ratio (striatum, nucleus accumbens and tuberculum olfactorium). The filters for the assay are: primary filter Turner 47B and secondary filter Corning C.S. 3–69. Sampling and washing times are both 40 s.

DA detection based on the dihydroxyindole method

The method is based on the dihydroxyindole procedure described by Atack (1973). The NaOH reagent is freshly prepared daily and contains 1.5% NaOH and 1.4% $Na_2S_2O_5$. The $K_3Fe(CN)_6$ solution contains 100 mg per 100 ml which is stable at 4°C for several weeks. The citric acid–phosphoric acid mixture is prepared by dissolving 7.5 g citric acid in 100 ml water and 0.5 ml concentrated phosphoric acid. The reaction mixture is treated at 75–85°C for 4–6 min in a heating bath.

Sample blanks are prepared by replacement of the $K_3Fe(CN)_6$ solution by a 5 mg per 100 ml cysteine solution. The filters for the DA assay are: primary filter C.S.0-69 and 313 nm interference filter and secondary filter Corning C.S. 3–73. Sampling and washing time are both 40 s.

3MT detection

The 3MT detection is based on a modification of the procedure described by Kehr (1974), in which a fluorophore was developed with $K_3Fe(CN)_6$ under alkaline conditions.

The NH_3 reagent is prepared daily by mixing 6 ml per 100 ml PCA with 100 ml concentrated NH_3 (25%). The $K_3Fe(CN)_6$ reagent contains 40 mg per 100 ml and is stable at 4°C for several weeks. The 20 mg per 100 ml cysteine solution is used on the same day only. Sample blanks are prepared by replacement of the $K_3Fe(CN)_6$ solution by cysteine. The filters for the 3MT assay are: primary filter 313 nm interference filter, and secondary filter Corning C.S. 3–73. Sampling and washing times are both 40 s.

DOPAC detection

The method is based on a modification of the procedure described by Spano and Neff (1971). In this method a fluorophore of DOPAC is formed by a condensation reaction with EDA. The EDA reagent is similar to the reagent used for the DA (EDA-method) detection. The $K_3Fe(CN)_6$ reagent contains 0.2 mg per 100 ml and is prepared daily from a stock solution (25 mg%), stored at 4°C. The fluorophore is formed in a heating bath (70–80°C; 4–6 min). The fluorescence of 20–50 mg cerebellar tissue is used as the blank for the DOPAC assay. The calibration curve of the DOPAC assay should be linear. The filters for the DOPAC assay are: primary filter Kodak Wratten 47B and secondary filter Corning C.S. 3-69. Sampling and washing times are both 40 s. More details about the DOPAC assay are given elsewhere (Westerink and Korf, 1976b).

The DOPAC assay gains in specificity and sensitivity when modified according to Murphy *et al.* (1969) (D. F. Sharman, personal communication).

HVA detection

The method is based on the manual fluorimetric procedure of Andén *et al.* (1963). In this method a fluorophore of HVA is formed by oxidative dimerization by $K_3Fe(CN)_6$ under alkaline conditions. The $K_3Fe(CN)_6$ reagent is prepared by dissolving 10 mg of $K_3Fe(CN)_6$ in a mixture of 85 ml water and 15 ml concentrated ammonia (25%). Concentrated ammonia (25%) (15 ml) diluted to 100 ml with water is used as blank reagent. The filters for the assay are: primary filter 313 nm, interference filter and secondary filter Corning C.S. 3-73. Sampling time is 30 s and washing time 40 s. As the determination of small amounts of HVA requires a very accurate blank procedure, addition of cysteine to the flow system is recommended when 2–5 ng HVA has to be measured. In that case the $K_3Fe(CN)_6$-NH_3 reagent and the cysteine reagent are exchanged for the blank procedure. Standard solutions are prepared in the phosphate solution used in the column procedure. To improve the flow of samples through the Autoanalyzer® manifold the wash compartment of the sampler is filled with the phosphate solution.

Preservation of Samples

Homogenized samples (pH 2–3), before addition of PCA can be preserved for at least 6 weeks when kept at −80°C. When stored for more than 2 weeks at −20°C a serious loss of NA, DA and DOPAC, but not of HVA, was observed. Column eluates collected in Autonalyzer® cups or glass tubes can be stored at −20°C for at least one week, except for the NA and DA samples which are susceptible to decomposition. If the latter samples have to be preserved till the next day, they are stored at 4°C.

RESULTS AND DISCUSSION

The small Sephadex G10 column appears to be a very useful tool for a miniaturized isolation procedure. The Sephadex G10 material is easy to handle and can be used for many months. Sephadex G10 introduces a negligible column blank fluorescence and it can be used for the isolation and separation of various monoamines and related metabolites. The wash and elution procedures can be achieved with very diluted solutions (not exceeding 0.01 M), which is an additional advantage when used in combination with a continuous flow analysis.

The monoamines NA, DA and 3-MT bind optimally to Sephadex G10 when a small amount of PCA is added to the PCA still present after perchlorate precipitation of the tissue extracts. Perchloric acid is known to form ion pair complexes with monoamines (Persson and Karger, 1974), and it is possible that such a complex is weakly bound to the Sephadex resin. The binding of the various amines to the Sephadex was sufficient for the separation from interfering substances mainly present in the void volume. This separation was illustrated by the finding of very low sample blank values. The elution of the monoamines with diluted formic acid is an additional advantage because of the relative stability of these compounds at lower pH.

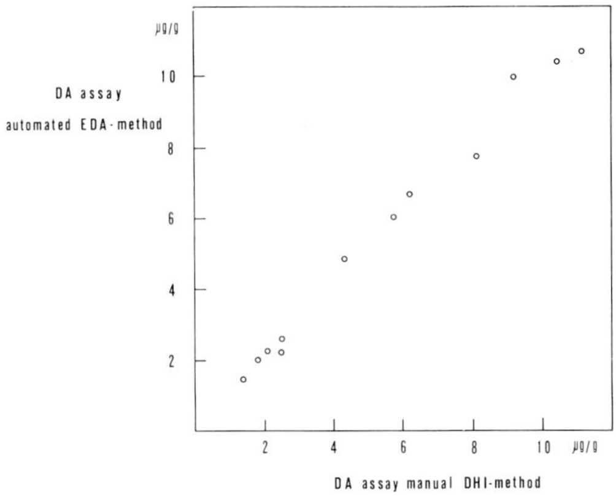

Fig. 3. Correlation between the manual DA assay (Roth and Surh, 1970) after purification on alumina and the automated assay (EDA-method) after purification of Sephadex G10. Identical samples of striatal tissue of saline or reserpine treated rats were used for both assays. Values were corrected for the recoveries (manual assay: $\bar{x} = 5.46\,\mu g\,g^{-1}$ and automated assay: $\bar{x} = 5.57\,\mu g\,g^{-1}$; $r = 0.99$).

The described detection procedures are very specific when applied to nervous tissue (see for detailed discussion: Westerink and Korf, 1977). The good correlation found between the currently used manual method for DA (after purification on alumina) is shown in Fig. 3. The DA assay based on the EDA reaction lacks specificity but can be used in combination with the Sephadex G10 separation procedure to analyse brain areas with a high DA/NA ratio such as the striatum, nucleus accumbens and tuberculum olfactorium of rat brain (Table I).

The main advantage of the EDA-method is the use of the same flow scheme, except for an exchange of the $K_3Fe(CN)_6$ reagent, for the detection of both DA and DOPAC which allows a very rapid screening of 40 samples containing the amine and its main metabolite within 2 hours.

Recoveries of the total procedures are usually between 75% and 90%. Hundreds of samples of human CSF (0.25–0.50 ml) were analysed for HVA with the semi-automated method. The weekly controlled recovery of 50 ng HVA added to CSF samples varied between 90% and 100%. The results of analysis of various brain areas are shown in Table I. Samples containing no more than 2–5 ng (NA, DA, DOPAC and HVA) and 10 ng (3-MT) can be reproducibly assayed. When necessary, the detection limit of the amines' determination can be decreased four- to five-fold when a Sephadex G10 column is directly connected with an Autoanalyzer[R] system (flow rate $0.42\,ml\,min^{-1}$), and elution profiles are recorded. In this case the

TABLE I The content ($\mu g\, g^{-1}$ wet weight; mean $\pm$ S.E.M.; N) of NA, DA, 3MT, DOPAC and HVA from different areas of rat brain.

	NA	DA(DHI-method)	DA(EDA-method)	3MT[a]	DOPAC	HVA
Nucleus accumbens	0.75 ± 0.09 (3)	5.53 ± 0.31 (8)	5.82 ± 0.37 (9)	0.61 ± 0.21 (3)	0.90 ± 0.9(9)	0.60 ± 0.06 (3)
Cerebellum	0.20 ± 0.01 (5)	0.10 ± 0.03 (3)	N.A.[b]	<0.10	<0.05	<0.02
Frontal cortex	0.27 ± 0.01 (6)	0.12 ± 0.03 (3)	N.A.	N.D.[c]	0.05 ± 0.02 (4)	0.04 ± 0.02 (4)
Hypothalamus	1.17 ± 0.07 (6)	N.A.	N.A.	<0.10	<0.05	N.D.
Striatum	0.38 ± 0.10 (3)	8.30 ± 0.42 (9)	8.38 ± 0.39 (9)	0.66 ± 0.09 (5)	0.85 ± 0.07 (9)	0.63 ± 0.05 (8)
Substantia nigra	0.39 ± 0.05 (3)	0.38 ± 0.10 (3)	N.A.	0.12 ± 0.03 (3)	0.08 ± 0.02 (5)	0.05 ± 0.01 (8)
Tuberculum olfactorium	0.29 ± 0.02 (3)	4.46 ± 0.29 (6)	4.33 ± 0.26 (9)	0.43 ± 0.09 (3)	0.70 ± 0.04 (9)	0.33 ± 0.04 (4)

[a] After pargyline treatment (75 mg kg^{-1}, i.p.; 120 min).

[b] N.D. = not determined.

[c] N.A. = method not applicable.

semiautomated method is transformed into a liquid chromatograph with fluorimetric detection (Westerink, 1979).

Some aspects of the analysis warrant special attention:

(a) Only filter fluorimeters in combination with appropriate filters will result in the described sensitivity. Monochromators (e.g. Perkin Elmer 1000) should be replaced by filters. An optimum primary filter is of great importance for the HVA assay. As the excitation and emission wavelengths of the DA-dihydroxyindole fluorophore partly overlap, the detection requires an optimum filter combination.

(b) The sensitivity of the assays usually increases with the volume of the flow cell. This volume however, should not exceed 600 μl.

(c) The separation between the amines and the acidic metabolites is a function of the amount of PCA added to the sample.* Although this function has a broad optimum, the separation between DA and DOPAC (EDA-method) should be controlled and when necessary adjusted with PCA when new columns are prepared, especially when a new batch of Sephadex G10 is used. As DA is present in 10 times higher concentrations than DOPAC in the dopaminergic structures, a complete separation of the two compounds is necessary. Columns are controlled with cerebellar samples to which a certain amount of DA is added. For optimal recovery in the DA assay, columns were not used during the first 10 days after preparation.

(d) Care should be taken that reagents used for sample preparation and column procedure do not contribute to the fluorescence of the eluates. Column blanks should be negligible. If sample blanks are unacceptably high, samples may be homogenized in 0.05 M H_2SO_4. The use of H_2SO_4 usually results in lower sample blanks, although the homogenization is less easy to perform.

In conclusion, the semiautomated method provides a rapid and sensitive assay of nanogram amounts of NA, DA, 3-MT, DOPAC or HVA in one sample of nervous tissue, with recoveries higher than 80%. At least 100 assays of each compound can be performed daily. With respect to sensitivity and reproducibility, the semiautomated method can compete with gas chromatographic techniques. In a recent study on DA metabolism in rat brain using gas chromatography–mass spectrometry (Karoum *et al.*, 1977), results were obtained which were very similar to an earlier study based on semiautomated fluorimetry (Westerink and Korf, 1976d). The simplicity of the equipment used, with minimal maintenance required, means that this technique is easy to teach to students who have had little or no analytical

* For a good separation between DA and DOPAC it is necessary that the columns should be packed homogeneously.

experience. The method has been rapidly replicated in other laboratories. The small Sephadex G10 column has proven to be an excellent alternative to the currently used columns of alumina or ion exchange resins. The purification step on the Sephadex G10 column can also be combined with manual fluorimetric methods.

REFERENCES

Andén, N.-E., Roos, B. E. and Werdinius, B. (1963). *Life Sci.* **2,** 448–458.

Atack, C. V. (1973). *Br. J. Pharmacol.* **48,** 699–714.

Holman, R. B., Angwin, P. and Barchas, J. D. (1976). *Neuroscience* **1,** 147–150.

Karasawa, T., Furukawa, K., Yoshida, K. and Shimizu, M. (1975). *Japan J. Pharmacol.* **25,** 727–736.

Karoum, F., Neff, N. H. and Wyatt, R. J. (1977). *Eur. J. Pharmacol.* 311–318.

Kehr, W. (1974). *Naun. Schmied. Arch. Pharmacol.* **284,** 149–158.

Korf, J., Zieleman, M. and Westerink, B. H. C. (1976). *Nature (London)* **260,** 257–258.

Metcalf, G. (1974). *Anal. Biochem.* **57,** 316–320.

Murphy, G. F., Robinson, D. and Sharman, D. F. (1969). *Br. J. Pharmacol.* **36,** 107–115.

Persson, B.-A. and Karger, B. L. (1974). *J. Chromatogr. Sci.* **12,** 521–528.

Rollema, H., Westerink, B. H. C., and Grol, C. J. (1976). *J. Pharm. Pharmacol.* 321–323.

Roth, R. H. and Suhr, Y. (1970). *Biochem. Pharmacol.* **10,** 3001–3012.

Smith, J. E., Lane, J. D., Shea, P. A., McBride, W. J. and Aprison, M. H. (1975). *Anal. Biochem.* **64,** 149–169.

Spano, P. F. and Neff, N. H. (1971). *Anal. Biochem.* **42,** 113–118.

Von Euler, U. S. and Lishajko, F. (1961). *Acta Physiol. Scand.* **51,** 348–355.

Waldmeier, P., De Herdt, P. and Maître, L. (1974). *Clin. Chem.* **20,** 81–83.

Westerink, B. H. C. and Korf, J. (1975a). *Biochem. Med.* **12,** 106–115.

Westerink, B. H. C. and Korf, J. (1975b). *Eur. J. Pharmacol.* **33,** 31–40.

Westerink, B. H. C. and Korf, J. (1976a). *Eur. J. Pharmacol.* **37,** 249–255.

Westerink, B. H. C. and Korf, J. (1976b). *Eur. J. Pharmacol.* **38,** 281–291.

Westerink, B. H. C. and Korf, J. (1976c). *Brain Res.* **113,** 429–434.

Westerink, B. H. C. and Korf, J. (1976d). *Eur. J. Pharmacol.* **40,** 131–137.

Westerink, B. H. C. and Korf, J. (1976e). *Eur. J. Pharmacol.* **40,** 175–178.

Westerink, B. H. C. and Korf, J. (1977). *J. Neurochem.* **29,** 697–706.

Westerink, B. H. C. (1978), (submitted).

CHAPTER 5

Gas-liquid Chromatography in the Analysis of Dopamine and its Metabolites

D. F. Sharman

Agricultural Research Council Institute of Animal Physiology, Babraham, Cambridge, U.K.

INTRODUCTION

At this time, reviewing the application of gas-chromatographic techniques to the analysis of dopamine and its metabolites is mainly counting the foundation stones of gas chromatography–mass spectrometry. However, I would like to discuss those aspects of gas-chromatographic procedures that can still be put to good use in the neurochemical laboratory.

The main advantages of the gas-chromatographic methods that have been applied to the biochemistry of dopamine lie in the great sensitivity that can be achieved and the rapidity with which separations can be made. Unfortunately, suitable conditions for an adequate separation of the substance under consideration are sometimes difficult to attain. As with most analytical procedures the specificities of the final detecting devices that have been employed are not absolute but if used with care, gas chromatography can still be used to advantage in the study of dopamine. In my own experience, the technique has proved most valuable as one method of analysis in experiments where two methods based on different chemical or physical procedures have been employed to obtain greater specificity.

This approach is analogous to the use of two biological assay systems used in parallel to determine a biologically active substance.

The details of column support materials, stationary phases and operating conditions will not be given in full but can be obtained from the original reports. Where methods have been successfully applied to tissues or body fluids, the proportion of stationary phase used, the column temperature and the detection system will be given in parentheses in some cases to illustrate the different conditions that have been employed.

THE AMINES: DOPAMINE AND 3-METHOXYTYRAMINE

3-Methoxytyramine (4-hydroxy-3-methoxyphenylethylamine) is the major amine metabolite of dopamine in those situations where dopamine is formed to be used as a biologically active substance. Adrenaline and noradrenaline and their metabolic products might also be regarded as metabolites of dopamine but will not be considered in detail here.

The separation and estimation of catecholamines by gas-liquid chromatography (GLC) was investigated by Sen and McGeer (1963) who converted the catecholamines and their methoxy metabolites to trimethylsilyl derivatives. They showed, using flame ionization detection (FID) that some separations could be achieved on an S.E. 30 column. In 1964, Lindstedt also showed that the trimethylsilyl derivatives of catecholamines could be seperated by GLC and in the same year Brooks and Horning (1964) and Horning *et al.* (1964) described the GLC separation of different derivatives of biologically occurring amines. These included condensation products with acetone, fully acetylated derivatives, trimethylsilyl (TMS) ethers, eneamines or Schiff bases, propionyl derivatives and pentafluoropropionyl derivatives. Several different column stationary phases were investigated and it was clearly shown that the derivatives of the different biologically occurring amines could be separated from each other. Clear separations of adrenaline and noradrenaline and their methoxy derivatives were obtained when the pentafluoropropionyl derivatives were prepared. Although Horning *et al.* (1964) used FID to detect these derivatives they drew attention to the suitability of halogenated derivatives for electron capture detection.

During the following years many different derivatives of biologically occurring amines were prepared and suggested as bases for methods that could be applied to biological material. Capella and Horning (1966) extended the observations on TMS derivatives, Schiff bases and acetone condensation products; Horning *et al.* (1967) and Horning *et al.* (1968) investigated the formation of suitable derivatives using trimethylsilylimidazole and *N*-heptafluorobutyrylimidazole. Studies of the electron capturing properties of halogenated derivatives of amines (e.g. Clarke *et al.*, 1966; Kawai and Tamura, 1968a; Moffat and Horning, 1970 and Walle and Ehrsson, 1970) showed that fluoroacyl and fluoroaryl derivatives could allow the estimation of very small quantities of the parent amine. Of the derivatives examined, trifluoroacetyl derivatives have found most use in applications to biological material. However, pentafluoropropionyl derivatives have proved extremely valuable in mass-fragmentographic analyses. Gas-liquid chromatographic methods have been successfully applied to the estimation of dopamine in plasma, urine and in tissues. The routine application to the estimation of 3-methoxytyramine appears to be more elusive. Cashaw *et al.* (1971) and Schwedt and Bussemas (1975) have described GLC methods for methoxytyramine, the former using a TMS derivative and the latter a TMS-trifluoroacetyl derivative. However, the sensitivity of these methods as described seems to be insufficient for the estimation of 3-methoxytyramine in small pieces of tissue. The gas-chromatographic properties of the trifluoroacetyl derivative of 3-methoxytyramine has been described by Sugiura (1973) who developed an extraction procedure that could be applied to urine, plasma and brain tissues (1% XF-1105; 160°; ECD: 3%OV17; 150°; ECD). Narasimhachari and Vouros (1972a, b) have studied the GLC properties of dopamine and 3-methoxytyramine as isothiocyanate derivatives.

The excretion of dopamine in human urine has been measured using the *O,O,N*-tri-trifluoroacetyl derivative. Clarke *et al.* (1967) (6% QF 1; 160°; ECD) estimated the excretion of free dopamine to be $225\,\mu g\,24\,h^{-1}$ ($0.15\,\mu g\,mg^{-1}$ creatinine), Wong, *et al.* (1973) (10% SE-54; 200°; ECD) measured both free and conjugated dopamine in human urine; the mean free dopamine excretion was estimated to be $402.4\,\mu g\,24\,h^{-1}$ and conjugated dopamine comprised 50–75% of the total excretion. In both reports, adsorption onto and elution from alumina was used, thus excluding methoxytyramine from the extraction procedure.

Adsorption and elution from alumina appears to be important in most of the extraction methods that have been developed for the GLC estimation of dopamine in blood plasma and tissues. This method of isolating catecholamines was used by Kawai and Tamura (1968b) who then converted the catecholamines to their trifluoroacetyl derivatives for GLC (2% XF-1105; 175°; ECD). These authors estimated the concentrations of noradrenaline and dopamine in human urine and in phaechromocytoma tumour tissue. A similar procedure was employed by Imai *et al.* (1970) to show that conjugated dopamine was present in human plasma and by Imai *et al.* (1971) to measure L-3,4-dihydroxyphenylalanine (L-DOPA) and dopamine in the blood plasma of patients treated with L-DOPA. In

1973, Martin and Ansell (1973) described a method based on GLC of the trifluoroacetyl derivatives of noradrenaline and dopamine that could be applied to brain tissue (5% SE-52; 115° then increasing by 1.5° min^{-1}; ECD). In addition to an alumina extraction step the method included extraction into di-(2-ethylhexyl) phosphate. Recently Wang *et al.* (1975a) have extended their methods to enable noradrenaline, adrenaline and dopamine to be measured in blood plasma. These authors and also Mizuno (1977) employed a double column system (Imai and Tamura, 1969) in which the outflow from the first column was allowed to vent to the air for a short period following the injection of the sample. This prevented the solvent vapour and other substances from contaminating the detector. The chromatographic separation of the substances in the sample was then allowed to continue on the second column by closing the outflow vent (e.g. 10% QFl precolumn, 2% XF-1105 main column; 190° ECD). Stanley and Wilk (1977) used the tri-pentafluoropropionyl derivative to measure tissue concentrations of dopamine (3% OV 17, 132°; ECD; 3% SE30; 140°; ECD). The use of trimethylsilyl derivatives for GLC of catecholamines has been used by Maruyama and Takemori (1972) to measure the concentration of dopamine in a single mouse brain (3% OV-17; 170°; FID). It has also been possible to measure the concentration of both free and conjugated 3-methoxytyramine in human plasma in addition to normetanephrine and metanephrine following their isolation on ion exchange resin columns and conversion to their trifluoroacetyl derivatives (Wang *et al.*, 1975b; (5% QF-1 precolumn, 2% XF-1105 main column; 163°; ECD)).

ACIDIC METABOLITES

Although routine GLC analysis of dopamine and 3-methoxytyramine has not been extensively employed, the estimation of the two major acidic metabolites of dopamine, 3,4-dihydroxyphenylacetic acid (DOPAC) and 4-hydroxy-3-methoxyphenylacetic acid (homovanillic acid; HVA) by GLC has found wider acceptance.

The estimation of HVA and DOPAC is necessary in the analysis of the phenolic and other metabolic products found in urine. Thus most GLC methods which have been developed for this purpose have been concerned with measuring other substances in urine in addition to these two metabolites of dopamine. Sweeley and Williams (1961) converted urinary aromatic acids to their methyl esters with diazomethane for GLC analysis (3% SE 30; 191°; argon ionization detector, (AID)). These observations were extended (Williams, 1962) and it was found that dihydroxy aromatic acids could be separated and estimated as their *O*-acetyl-methyl esters (15%

SE 30; 190°; AID) (Williams and Leonard, 1963). Horning *et al.* (1967b) studied the GLC characteristics of TMS-methyl ester derivatives of urinary acids and applied their method to rat and human urine (5% OV-17; 70°—increasing by 2° min^{-1}; FID). Detailed investigations of the GLC analysis of urinary acids using TMS-methyl esters and TMS ether/ester derivatives are reported in the papers of Karoum *et al.* (1968), Karoum *et al.* (1969) and Karoum and Sandler (1971). Some difficulties in separating the derivatives of certain of the acids found in urine might be overcome by using sulphonyl-TMS ester derivatives (Scott, 1972). The publications of Begue *et al.* (1971) and more recently, Wadman *et al.* (1976) and Muskiet *et al.* (1977), are examples of research reports in which the estimation of HVA, among other urinary acids, using TMS-derivatives and GLC is described. Two rapid methods for studying the excretion of phenolic acids in the urine of patients treated with L-DOPA have been described by Fukaya *et al.* (1971) and Narasimhachari *et al.* (1974), the former being based on the GLC separation of TMS-derivatives, the latter on the formation of ethyl esters or TMS-ethyl esters. An interesting approach to the formation of suitable derivatives is that reported by Roginsky *et al.* (1974) who methylated HVA in the flash heater zone of the chromatography column by injecting the acid together with trimethylanilinium hydroxide (3% QF-1; 170°; FID).

Although the amounts of HVA and DOPAC present in urine are relatively large and usually do not require the high sensitivity of the electron capture detector for their measurement, Dziedzic *et al.* (1972) and Dziedzic *et al.* (1973) have used the hexafluoroisopropyl esters of trifluoroacetyl HVA and isoHVA to measure the concentrations of HVA and isoHVA in urine (3% XE-60; 120°; ECD; 3% OV-17; 115°; ECD), a result that illustrates one advantage of GLC methods. This sensitive method is a development of the observation of Smith and Tsai (1971) that acids could be esterified with a halogenated alcohol in the presence of trifluoroacetic anhydride. Dziedzic *et al.* (1972) showed that HVA could be esterified with hexafluoroisopropanol and that a trifluoroacetyl group could be introduced at the phenolic hydroxyl group. This derivative has very good electron capture properties and has enabled GLC methods to be applied to tissues and to cerebrospinal and ocular fluids (Dziedzic and Gitlow, 1974; Pearson and Sharman, 1974, 1975; Hutchins *et al.*, 1975). The last three papers illustrate the improved specificity of the analysis that can be achieved by carrying out two different analytical procedures for the same substance. In examining aqueous humour, a material was found to be present that behaved similarly to DOPAC. However no DOPAC could be detected when a fluorimetric method was used. In the study of Pearson and Sharman (1974) on the concentrations of acidic metabolites of dopamine in the superior cervical ganglion, the concentration of HVA determined by GLC could be confirmed by fluorimetry, and a small change in the concentration of DOPAC in mouse striatum, observed using a fluorimetric method, was confirmed by GLC (Hutchins *et al.*, 1975). A similar GLC method has also been used by Hamilton and McKay (1976) to measure HVA and DOPAC in different areas of the human brain.

The use of fluorinated derivatives for the GLC estimation of acidic metabolites was furthered by the work of Watson *et al.* (1974) who found that a mixture of pentafluoropropanol and pentafluoropropionic anhydride would yield derivatives of phenolic acids that showed excellent GLC and electron capture properties. Further studies (Watson *et al.*, 1975; Wilk *et al.*, 1975c) have shown that the preparation of derivatives of organic acids using pentafluoropropionic anhydride and a perhalogenated alcohol forms the basis of an extremely useful method for the estimation of organic acids particularly when pentafluoropropanol or 1-chloro-1,1,3,3,3-pentafluoropropan-2-ol is used. The method has been applied to cerebrospinal fluid (Watson *et al.*, 1974) and to studies on the metabolism of dopamine in specific regions of the brain (Wilk *et al.*, 1975a; Wilk *et al.*, 1975b; Wilk and Glick, 1976; Wilk and Stanley, 1977; Stanley and Wilk, 1977). The method has also been successfully employed by Braestrup *et al.* (1975). Homovanillic acid in human cerebrospinal fluid has also been estimated by GLC using the heptafluorobutyryl-methyl ester derivative (1% OV-1; 140°; ECD) (Sjöquist and Änggård, 1972).

ALCOHOLIC METABOLITES OF DOPAMINE

Gas-liquid chromatographic analysis can also be applied to the measurement of the main alcoholic metabolites of dopamine, 3,4-dihydroxyphenylethanol (DHPE) and 4-hydroxy-3-, methoxyphenylethanol (MHPE). Sharman (1969) showed that the acetyl-heptafluorobutyryl derivative of MHPE was a derivative suitable for GLC with ECD. The corresponding derivative of DHPE was used to obtain evidence for the presence of this metabolite in sheep blood plasma following the i.v. injection of dopamine (Kelly *et al.*, 1970). Karoum *et al.* (1971) tested many different derivatives of phenolic alcohols and showed that the methyl-pentafluoropropionyl derivatives of phenolic acids and the pentafluoropropionyl derivatives of alcohols could be used for GLC analysis. They also demonstrated that their methods were applicable to urine, cerebrospinal fluid and brain tissue (6% QF-1; 180°–195°; ECD). The pentafluoropropionyl derivative of MPHE was also used by Braestrup (1973) to determine this metabolite in rat brain and Wilk

(1971) has developed a method based on the formation of a trifluoroacetyl derivative (2% XF-1105+3% OV-17 (3:1 w/v); 125°; ECD). There is however some disagreement on the concentration of MHPE in the rat brain which, if present, appears to be mainly in a conjugated form (Karoum *et al.*, 1971; Braestrup, 1973; Wilk and Zimmerberg, 1973). Goodwin *et al.* (1972) have indicated some possible sources of error in GLC methods for phenolic alcohols but this particular problem will probably be best resolved by GLC–MS studies.

CONCLUSIONS

The reports cited above show that many approaches have been made to the problems of estimating dopamine and its metabolites using GLC methods. Some of these have been developed to the stage where they can be used as reliable, routine analytical procedures. The greatest advantage of such methods almost certainly lies in the great sensitivity of the electron capture detector, but the full potential of this device has not yet been attained since none of the methods are able to use the total extract for the analysis. Usually less than 5% is taken for this purpose. In spite of this there are few other methods that can compete in respect of sensitivity. In addition it is not usually possible to process as many samples at one time as can be analysed by other methods such as automated fluorimetric analysis. However, GLC methods form an important part of the neurochemist's armoury and can be employed when required for their sensitivity or when no other method is easily available. The problem of absolute specificity is insoluble without using the mass spectrometer but comparison of results obtained by GLC with those obtained by other methods increases the confidence that can be placed in them. More specific separations and reactions to form volatile derivatives will, no doubt, be developed in the future.

REFERENCES

Begue, R. J., Bazerolle, O., Desgres, J., Lallement, C. and Padieu, P. (1971). *Path. Biol.* **19,** 719–725.

Braestrup, C. (1973). *J. Neurochem.* **20,** 519–527.

Braestrup, C., Andersen, H. and Randrup, A. (1975). *Eur. J. Pharmacol.* **34,** 181–187.

Brooks, C. J. W. and Horning, E. C. (1964). *Anal. Chem.* **36,** 1540–1545.

Capella, P. and Horning, E. C. (1966). *Anal. Chem.* **38,** 316–321.

Cashaw, J. L., Walsh, M. J., Yamanaka, Y. and Davis, V. E. (1971). *J. Chromatogr. Sci.* **9,** 98–104.

Clarke, D. D., Wilk, S. and Gitlow, S. E. (1966). *J. Gas Chromatog.* **4,** 310–313.

Clarke, D. D., Wilk, S., Gitlow, S. E. and Franklin, M. J. (1967). *J. Gas Chromatog.* **5,** 307–310.

Dziedzic, S. W., Bertani-Dziedzic, L. and Gitlow, S. E. (1973). *J. Lab. Clin. Med.* **82,** 829–835.

Dziedzic, S. W., Bertani, L. M., Clarke, D. D. and Gitlow, S. E. (1972). *Anal. Biochem.* **47,** 592–600.

Dziedzic, S. W. and Gitlow, S. E. (1974). *J. Neurochem.* **22,** 333–335.

Fukaya, J., Imai, K. and Tamura, Z. (1971). *Clin. Chim. Acta* **33,** 470–472.

Goodwin, B. L., Karoum, F., Ruthven, C. R. J. and Sandler, M. (1972). *Clin. Chim. Acta* **40,** 269–272.

Hamilton, P. N. and Mackay, A. V. P. (1976). *Brain Res.* **118,** 161–166.

Horning, M. G., Boucher, E. A. and Moss, A. M. (1967a). *J. Gas Chromatogr.* **5,** 297–302.

Horning, E. C., Horning, M. G., Vanden Heuvel, W. J. A., Knox, K. L., Holmstedt, B. and Brooks, C. J. W. (1964). *Anal. Chem.* **36,** 1546–1549.

Horning, M. G., Moss, A. M., Boucher, E. A. and Horning, E. C. (1968). *Anal. Lett.* **1,** 311–321.

Horning, M. G., Moss, A. M. and Horning, E. C. (1967b). *Biochim. Biophys. Acta* **148,** 597–600.

Hutchins, D. A., Pearson, J. D. M. and Sharman, D. F. (1975). *J. Neurochem.* **24,** 1151–1154.

Imai, K., K., Sugiura, M. and Tamura, Z. (1970). *Chem. Pharm. Bull.* **18,** 2134.

Imai, K., Sugiura, M., Tamura, Z., Hirayama, K. and Narabayashi, H. (1971). *Chem. Pharm. Bull.* **19,** 439–440.

Imai, K. and Tamura, Z. (1969). *Chem. Pharm. Bull.* **17,** 1076–1077.

Karoum, F., Anah, C. O., Ruthven, C. R. J. and Sandler, M. (1969). *Clin. Chim. Acta* **24,** 341–348.

Karoum, F., Ruthven, C. R. J. and Sandler, M. (1968). *Clin. Chim. Acta* **20,** 427–437.

Karoum, F., Ruthven, C. R. J. and Sandler, M. (1971). *Biochem. Med.* **5,** 505–514.

Karoum, F. and Sandler, M. (1971). *Clin. Chim. Acta* **32,** 391–397.

Kawai, S. and Tamura, Z. (1968a) *Chem. Pharm. Bull.* **16,** 699–701.

Kawai, S. and Tamura, Z. (1968b). *Chem. Pharm. Bull.* **16,** 1091–1094.

Kelly, M., Sharman, D. F. and Tegerdine, P. (1970). *J. Physiol.* (*London*) **210,** 130P.

Lindstedt, S. (1964). *Clin. Chim. Acta* **9,** 309–310.

Martin, I. L. and Ansell, G. B. (1973). *Biochem. Pharmacol.* **22,** 521–533.

Maruyama, Y. and Takemori, A. E. (1972). *Anal. Biochem.* **49,** 240–247.

Mizuno, Y. (1977). *Clin. Chim. Acta* **74,** 11–19.

Moffat, A. C. and Horning, E. C. (1970). *Anal. Lett.* **3,** 205–216.

Muskiet, F. A. J., Fremouw-Ottevangers, D. C., Wolthers, B. G. and De Vries, J. A. (1977). *Clin. Chem.* **23,** 863–867.

Narasimhachari, N., Leiner, K., Plaut, J. M. and Lin, R.-L. (1974). *Clin. Chim Acta* **50,** 337–344.

Narasimhachari, N. and Vouros, P. (1972a). *Anal. Biochem.* **45,** 154–163.

Narasimhachari, N. and Vouros, P. (1972b). *J. Chromatogr.* **70,** 135–140.

Pearson, J. D. M. and Sharman, D. F. (1974). *J. Neurochem.* **22,** 547–550.

Pearson, J. D. M. and Sharman, D. F. (1975). *Br. J. Pharmacol.* **53,** 143–148.

Roginsky, M. S., Gordon, R. D. and Bennett, M. J. (1974). *Clin. Chim. Acta* **56,** 261–264.

Schwedt, G., and Bussemas, H. H. (1975). *J. Chromatogr.* **106,** 440–445.

Scott, P. H. (1972). *J. Chromatogr.* **70,** 67–72.
Sen, N. P. and McGeer, P. L. (1963). *Biochem. Biophys. Res. Comm.* **13,** 390–393.
Sharman, D. F. (1969). *Br. J. Pharmacol.* 523–534.
Sjöquist, B. and Änggård, E. (1972). *Anal. Chem.* **44,** 2297–2301.
Smith, R. V. and Tsai, S. L. (1971). *J. Chromatogr.* **61,** 29–34.
Stanley, M. and Wilk, S. (1977). *Eur. J. Pharmacol.* **44,** 292–302.
Sugiura, M. (1973). *Jikei Med. J.* **20,** 125–142.
Sweeley, C. C. and Williams, C. M. (1961). *Anal. Biochem.* **2,** 83–88.
Wadman, S. K., Ketting, D. and Voûte, P. A. (1976). *Clin. Chim. Acta* **72,** 49–68.
Walle, T. and Ehrsson, H. (1970). *Acta Pharm. Suecica,* **7,** 389–406.
Wang, M.-T., Imai, K., Yoshioka, M. and Tamura, Z. (1975a). *Clin. Chim. Acta* **63,** 13–19.
Wang, M.-T., Yoshioka, M., Imai, K. and Tamura, Z. (1975b). *Clin. Chim. Acta* **63,** 21–27.
Watson, E., Travis, B. and Wilk, S. (1975). *Life Sci.* **15,** 2167–2178.
Watson, E., Wilk, S. and Roboz, J. (1974). *Anal. Biochem.* **59,** 441–451.
Wilk, S. (1971). *Biochem. Pharmacol.* **20,** 2095–2096.
Wilk, S. and Glick, S. D. (1976). *Eur. J. Pharmacol.* **37,** 203–206.
Wilk, S. and Stanley, M. (1977). *Eur. J. Pharmacol.* **41,** 65–72.
Wilk, S., Watson, E. and Glick, S. D. (1975a). *Eur. J. Pharmacol.* **30,** 117–120.
Wilk, S., Watson, E. and Stanley, M. E. (1975b). *J. Pharmacol. Exp. Ther.* **195,** 265–270.
Wilk, S., Watson, E. and Travis, B. (1975c). *Eur. J. Pharmacol.* **30,** 238–243.
Wilk, S. and Zimmerberg, B. (1973). *Biochem. Pharmacol.* **22,** 623–628.
Williams, C. M. (1962). *Anal. Biochem.* **4,** 423–432.
Williams, C. M. and Leonard, R. H. (1963). *Anal. Biochem.* **5,** 362–366.
Wong, K. P., Ruthven, C. R. J. and Sandler, M. (1973). *Clin. Chim. Acta* **47,** 215–222.

CHAPTER 6

Analysis of Dopamine and its Metabolites in Biological Materials by Mass Fragmentography

F. Karoum and N. H. Neff

National Institute of Mental Health,
Saint Elizabeth's Hospital, Washington, U.S.A.

INTRODUCTION

The possible involvement of dopaminergic neurones in the aetiology of some psychiatric illnesses (Randrup and Munkvad, 1967; Snyder, 1972; Angrist and Gershon, 1974) and the established connection between reduced dopamine (DA) content of the basal ganglia and parkinsonism (Hornykiewicz, 1966) have created a great interest in the physiological role of DA in the CNS. Consequently, there is a demand for specific methods sensitive enough to measure DA and its major metabolites, homovanillic acid (HVA) and 3,4-dihydroxyphenylacetic acid (DOPAC) (Rosengren, 1969; Andén *et al.*, 1963; Sharman, 1963; Wilk *et al.*, 1975a; Wilk *et al.*, 1975b) in ever decreasing concentrations in biological samples. A number of methods have recently been described which appear ideal for the assay of sub-nanomole concentrations of DA and its metabolites in a variety of biological materials (Sjöquist and Anggård, 1972; Koslow *et al.*, 1972; Sjöquist *et al.*, 1973; Wiesel *et al.*, 1973; Koslow *et al.*, 1974; Karoum *et al.*, 1975; Wiesel, 1975; Watson and Wilk, 1975; Wilk *et al.*, 1975a; Wilk *et al.*, 1975b; Westerink and Korf, 1976; Karoum *et al.*, 1976). Of all of the methods available, mass fragmentography (MF) is the most specific and sensitive (Costa *et al.*, 1975). The technique incorporates the high separation properties of gas chromatography (GC) together with high sensitivity and specific structural information about compounds derived from mass spectrometry. This unique combination distinguishes MF from the other techniques.

Mass fragmentography is gas chromatography–mass spectrometry (GC–MS) where the mass spectrometer is used to detect and identify molecules in a GC column effluent. The following is a brief description of the molecular identification process as it applies in our GC–MS instruments. Within the mass spectrometer system called the analyser, sample molecules are ionized by collision with a beam of electrons. The beam of electrons may knock electrons off the sample molecule giving rise to molecular ions ($M^{\dot{+}}$) or shatter the molecule producing positively charged fragments. The mixture of positively charged ions is then separated according to the ions' mass-to-charge ratio (*m/e*). The relative abundance of each group of ions of a particular *m/e* is the output (mass spectrum) of the mass spectrometer. Each molecule has a characteristic mass spectrum. There are several methods for ionizing sample molecules. The method described is known as electron impact ionization. The energy of the electron beam is such that only a single electron is lost from a sample that has been fragmented by the beam and therefore the *m/e* of the fragment is its atomic mass unit (a.m.u.). In the MF mode, the mass spectrometer detector is operated to record preselected a.m.u. for a sample molecule.

During the analysis of mixtures of structurally related compounds, such as the biogenic amines and their metabolites, it is possible that they might elute from the GC together and produce similar ion fragments in the mass spectrometer. Interference of structurally related compounds can be evaluated by monitoring several fragments of the same molecule as it emerges from the GC. For any particular molecule the ratio of one fragment to another is constant. If the ratio of fragments in a standard sample and a biological sample are different, it indicates interference of similar fragments derived from different parent molecules in the biological sample. This type of analysis, termed multiple ion detection, is an essential requirement for the absolute identification of an endogenous compound. Moreover, it emphasizes the importance of understanding the principles of GC so that suitable conditions can be devised to separate interfering materials.

For the purposes of using MF for quantitating an endogenous material, known quantities of an internal standard containing a stable isotope, for example DA containing deuterium, are added to a biological sample before extraction and chemical processing. Quantitation of samples is achieved by plotting a linear curve measuring the ratio of the peak heights of a fragment of the authentic compound in different concentrations, and the equivalent fragment of its deuterated internal standard in constant concentration, versus the quantity of the known authentic substance. The ratio of fragments for each biological sample is similarly calculated. The quantity of substance which this latter ratio represents is read from the standard curve.

In this chapter, MF methods as they are presently used in our laboratories will be described for the assay of DA and its metabolites.

INSTRUMENTATION

Our studies were performed using a Finnigan model 3000D or 4000 gas chromatograph–quadrupole mass spectrometer with programmed multiple ion monitoring. Gas-chromatographic separation was achieved on an 8 ft, $\frac{1}{8}$ in o.d. steel column packed with 3% SE 54 (methylphenylvinyl silicone gum) coated on 80–100 mesh chromosorb G. The column parameters were: helium flow rate 30 ml min^{-1}; injector 250°C; oven temperature 190°C for HVA and DOPAC or 200°C for DA.

REAGENTS

Reagents were purchased from the following: pentafluoropropionic anhydride (PFPA) and pentafluoropropionic imidazole (PFPI), Pierce Chemical Company, Rockford, Illinois, USA; SE 54 on chromosorb G, Analabs, North Haven, Connecticut, USA; "Lipopure" methanol, Applied Sciences Laboratories, State College, Pennsylvania, USA; deuterated isomers of HVA ($^2H_2$-HVA), DOPAC ($^2H_5$-DOPAC) and DA ($^2H_4$-DA), Merck, Sharpe and Dohme, Canada, Ltd., Montreal, Quebec, Canada. All other reagents and materials used were of the highest purity available.

Derivatizations were in 1 ml Microflex tubes purchased from Kontes Glass Co., Pierce, Illinois, USA.

DERIVATIZATION

It is necessary to convert DA and its metabolites to volatile derivatives for separation in the GC. Homovanillic acid and DOPAC are converted to their methyl esters by the addition of 20% HCl in methanol, 100 µl (prepared by the dropwise addition of acetylchloride, 1 ml to "lipopure" methanol, 4 ml). The samples remain at room temperature for 5–10 min and are then evaporated to dryness under a gentle stream of N_2. The methylesters are then acylated to the pentafluoropropionyl derivatives by heating at 70° for 15 min with 10% PFPI in ethyl acetate, 20–50 μl. One or two microlitres of this solution are injected into the GC–MS for analysis.

Dopamine is converted to a pentafluoropropionyl derivative by adding ethyl acetate, 10 μl (dried and stored over CaH_2), PFPA, 100 μl, and heating for 15 min at 70°. The mixture is cooled, evaporated to dryness under N_2 and reconstituted in ethyl acetate, 20 μl. One or two μl of this solution are injected into the GC–MS.

The derivatization of the methylesters of HVA and DOPAC with PFPI gives higher product yields than PFPA. In contrast, PFPA gives a higher yield for DA.

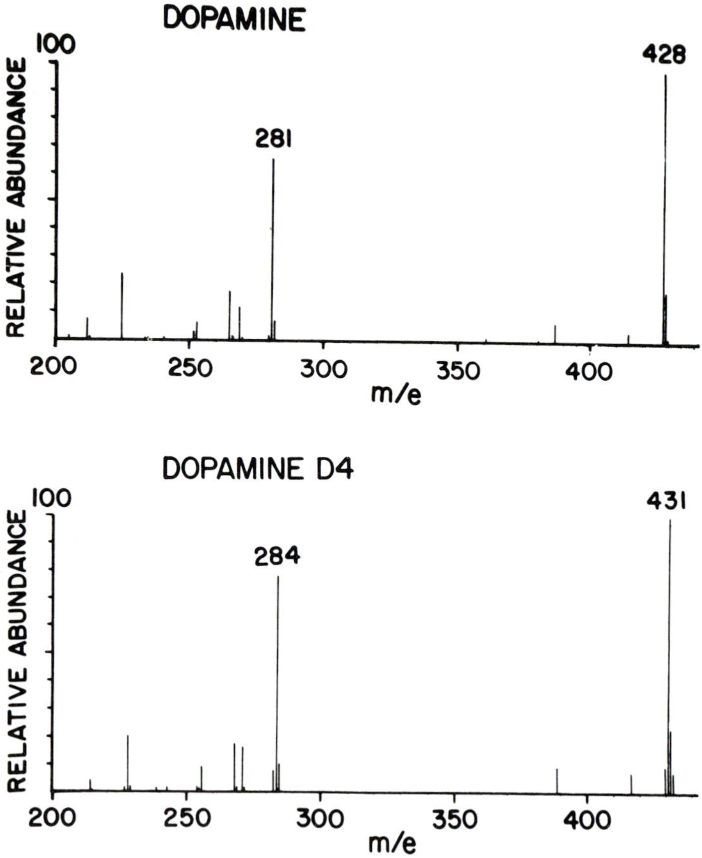

Fig. 1. Mass spectra of the pentafluoropropionyl derivatives of dopamine and dopamine containing four atoms of deuterium. Fragments that could be used for multiple ion detection for the absolute identification of endogenous dopamine are marked.

MASS FRAGMENTOGRAPHY

The spectra for DA, HVA and DOPAC together with the spectra for their corresponding deuterated isomers are shown in Figs. 1–3. The fragments selected for MF are summarized in Table 1. The choice of a suitable fragment for MF is governed by two factors: its relative intensity in comparison with other fragments and the frequency with which it appears in the spectra of other compounds. These two factors together with the choice of the GC column determine the overall specificity of an MF method. The fragments shown in this table are ideally suited for the MF of DA and its metabolites. The mass-fragmentographic peaks corresponding to these fragments during the analysis of biological samples were found to be pure as judged by multiple ion detection (Karoum *et al.*, 1977b).

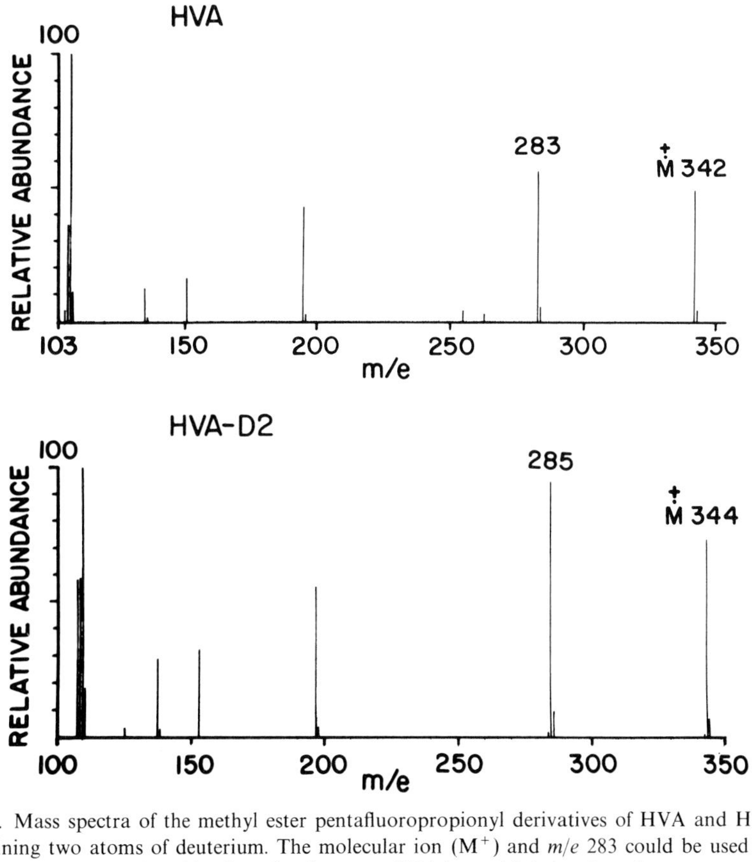

Fig. 2. Mass spectra of the methyl ester pentafluoropropionyl derivatives of HVA and HVA containing two atoms of deuterium. The molecular ion (M^+) and *m/e* 283 could be used for absolute identification of endogenous HVA by multiple ion detection.

PREPARATION OF BIOLOGICAL SAMPLES FOR MF

We have followed two procedures for the analysis of DA and its metabolites in biological samples. The first, used for neuronal tissue, requires no isolation of DA and its metabolites prior to MF. The second, used for biological fluids, involves solvent extraction of HVA and DOPAC prior to analysis.

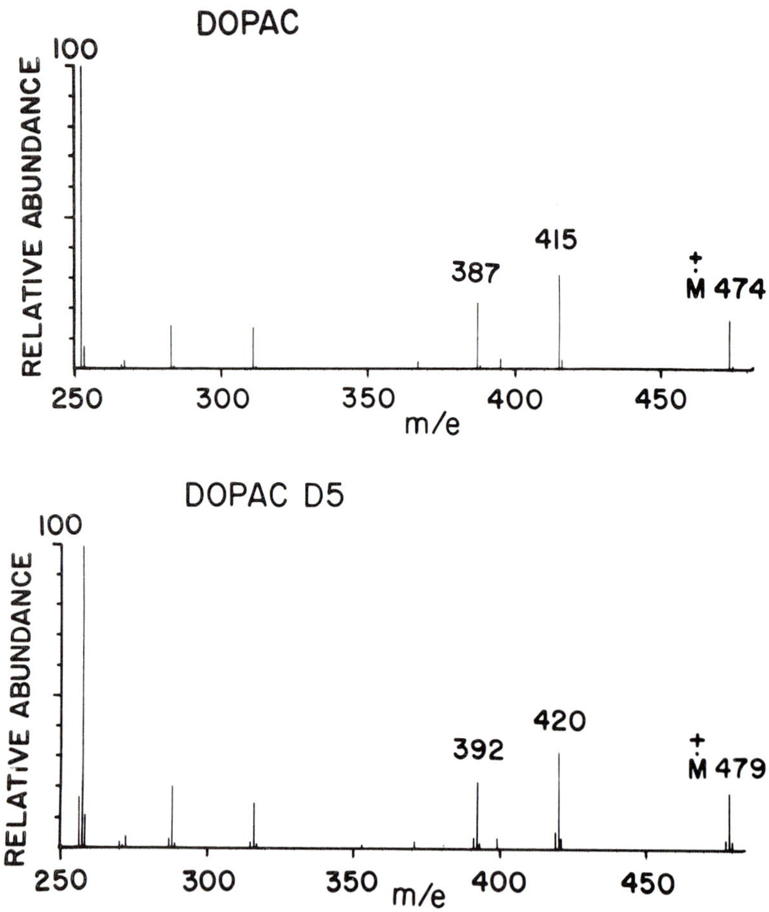

Fig. 3. Mass spectra of the methyl ester pentafluoropropionyl derivatives of DOPAC and DOPAC containing five atoms of deuterium. The molecular ion (M^+) and the fragments marked could be used for absolute identification of endogenous DOPAC by multiple ion detection.

TABLE I Atomic mass units of fragments employed for the mass fragmentography of the pentafluoropropionyl derivative of dopamine and the methylester pentafluoropropionyl derivatives of HVA and DOPAC

		Atomic mass units[a]
Compound	Molecular ion $M^{+\cdot}$	Fragments[b] (origin of fragment)
Dopamine (DA)	591	428 ($M^{+\cdot}-NH_2COC_2F_5$)
Homovanillic acid (HVA)	342	342 283($M^{+\cdot}-COOCH_3$)
Dihydroxyphenylacetic acid (DOPAC)	474	387 ($M^{+\cdot}-COOCH_3-CO$) 415($M^{+\cdot}-COOH_3$)
$^2H_4$-DA	595	431 ($M^{+\cdot}-NH_2COC_2F_5$)
$^2H_2$-HVA	344	344 283($M^{+\cdot}-COOCH_3$)
$^2H_5$-DOPAC	479	392 ($M^{+\cdot}-COOCH_3-CO$) 420 ($M^{+\cdot}-COOH_3$)

[a] The notations in parenthesis indicate the fragments lost from the molecular ion to produce the atomic mass unit shown.
[b] Fragments we have used for the analysis of dopamine and its metabolites.

Assay of DA and its Metabolites in Neuronal Tissues

Rat brain regions (Karoum *et al.*, 1977) or a superior cervical ganglion (Karoum *et al.*, 1977b) are homogenized with 1 M HCl, 0.3–1 ml, centrifuged for 10 min at 15 000 × *g* and the clear supernatant, 0.1–0.2 ml, transferred to a 1 ml Microflex tube. Tissues are weighed before homogenization or the pellet is used for protein assay. The contents of the Microflex tube are evaporated to dryness under a stream of N_2 and the dried residue derivatized as already described for DA or for HVA and DOPAC.

Internal standards of $^2H_4$-DA, $^2H_2$-HVA and $^2H_5$-DOPAC, 50 pmol each, are added to each Microflex tube prior to evaporation to correct for differences in derivation among the samples. Alternatively, the internal deuterated standard can be added directly in the homogenizing solution of 1 M HCl. In addition, four tubes, each containing 0.2 ml of pooled supernatant, deuterated internal standards and increasing quantities (25, 50 and 75 nmol) of authentic non-deuterated material are also included with the samples and used for the construction of a standard curve.

Assay of HVA and DOPAC in Biological Fluids

Cerebrospinal fluid, 0.3 ml, plasma, 0.2 ml, or urine, 0.1 ml (Karoum *et al.*, 1977a) is mixed with 1 M HCl, 1 ml, in a 20 ml round-bottom centrifuge tube. After adding $^2H_2$-HVA and $^2H_5$-DOPAC, 50 nmol each, to all tubes, the metabolites are extracted twice with ethyl acetate, 5 ml, by vortex mixing for one minute. Following centrifugation at 1000 × *g* for 10 min, the ethyl acetate extracts are combined (4 ml from the first and 5 ml from the second extraction). The combined extracts are evaporated *in vacuo* and the dried residue dissolved in ethyl acetate, 0.3 ml. This solution, 0.2 ml, is transferred to a microflex tube, evaporated under a stream of N_2 and the acidic metabolites derivatized as already described.

Four samples, each with an increasing quantity of authentic HVA and DOPAC and deuterated standards, are also included and used for the construction of a standard curve as described previously for the direct assay.

Quantification of DA and its metabolites is based on the recovery of the deuterated standards and on a standard curve.

DISCUSSION

The methods described can be adapted to measure other biogenic amines and their metabolites (Karoum *et al.*, 1977c) as well. Specificity and sensitivity of MF methods are based on the fragments selected and the efficiency and selectivity of the GC column. The GC column should not be overlooked and reliance placed solely on the mass spectrometer for specificity.

Stainless steel columns meet the GC requirements for the MF of biogenic amines and their metabolites. In contrast to glass columns which are commonly used in some laboratories, they are easily customized to meet almost any desired specification. Steel columns are cheap, indestructable and once made, last almost indefinitely. In spite of these advantages, steel columns are not commonly used due to the fear of possible decomposition of the pentafluoropropionyl derivatives by metal columns. No on-column decomposition has been observed for any of the various biogenic amines or their metabolites studied in our laboratories. On those occasions when decomposition was suspected, i.e. loss of sensitivity during an analysis, the problem was traced to other causes. We believe that the apprehension concerning the use of steel columns for biogenic amines and their metabolites is not well founded.

With the analytical methods described, quantities as little as 5 pmol of DA or its metabolites can be assayed in most biological materials. Generally, tissue sample sizes should be selected to contain at least twice the above minimally detectable quantities. Moreover, sample preparation is simple and therefore 50 or more samples can be processed in a day.

The methods described have been used to evaluate the dynamics of DA metabolism in the rat brain (Karoum *et al.*, 1977c), superior cervical ganglion (Karoum *et al.*, 1977b) and in human biological fluids (Karoum *et al.*, 1977a).

Table II summarizes the regional concentrations of HVA and DOPAC in the rat brain. Using the methods described, we have studied the characteristics of the transport system that rids the rat brain of HVA and

TABLE II Concentration of HVA and DOPAC in various regions of rat brain

Brain region	HVA (nmol gm^{-1} ± S.E.M. (N))	DOPAC (nmol gm^{-1} ± S.E.M. (N))
Whole Brain	0.40 ±0.02 (13)	0.37±0.04 (13)
Medulla	0.12 ±0.02 (8)	0.33±0.03 (8)
Hypothalamus	0.46 ±0.06 (8)	1.4 ±0.6 (8)
Midbrain	0.21 ±0.03 (5)	0.60±0.09 (5)
Cerebellum	0.088±0.009 (5)	0.10±0.02 (5)
Cortex	0.67 ±0.02 (8)	0.76±0.08 (3)
Striatum	5.5 ±0.5 (8)	6.6 ±0.8 (3)
Hippocampus	0.16 ±0.01 (4)	0.18±0.01 (3)

DOPAC, and evaluated the rates of formation of these materials in selected regions of brain (Karoum *et al.*, 1977c). Our studies revealed that DOPAC is the major metabolite of DA, that the acid transport system for HVA and DOPAC is not equally affected by the acid transport blocking drug probenecid and that the mode of metabolism of DA in neuronal cell bodies is different to that in nerve endings. Moreover, we found that the steady-state concentrations of the metabolites of DA are a relative indication of their rates of formation and elimination from the brain.

Dopamine apparently plays a modulating role in peripheral sympathetic ganglionic transmission (Libet, 1970). Because of the size of the ganglion and the low concentrations of DA and metabolites, a high degree of sensitivity and specificity are required to evaluate biochemical events that occur in the ganglion.

Tables III and IV summarize our results on the dynamics of DA metabolism in surgically decentralized and sham-operated rat superior cervical ganglia in response to administering probenecid or cholinergic drugs (Karoum *et al.*, 1977b). The response in decentralized ganglia

TABLE III Blockade of acid transport in rat superior cervical ganglion by treatment with probenecid

Treatment	HVA (pmol ganglion^{-1} ± S.E.M. (N))	DOPAC (pmol ganglion^{-1} ± S.E.M. (N))
Solvent	2.7 ± 0.5 (20)	41 ± 3 (25)
Probenecid[a]	4.4 ± 0.6 (10)[b]	64 ± 6 (10)[b]

[a] Probenecid (200 mg kg^{-1}, i.p.) was administered 1 h before the rats were killed.
[b] $P < 0.01$ when compared with solvent-treated animals using a two-tailed "t" test.

differentiates a direct response of the ganglia to drugs from those that originate in supraganglionic sites. 3,4-Dihydroxyphenylacetic acid (DOPAC) and not HVA is the major metabolite of DA in ganglia, and a transport system that is blocked by probenecid facilitates the removal of both acids from the ganglion (Table III). Furthermore, treatment with carbachol alone or in combination with atropine or hexamethonium revealed that DA metabolism is modulated by nicotinic and muscarinic mechanisms. Muscarinic stimulation increases while nicotinic stimulation decreases DA metabolism. These receptors appear to be located within the ganglion as decentralization of the ganglion does not abolish these responses. In other studies not presented here, we found the rate of formation of DA to be about 15 times faster than that of DA in the striatum.

TABLE IV DOPAC and dopamine concentrations in decentralized and sham surgery rat superior cervical ganglion after drugs that act at cholinoceptive sites

Drug treatment	DOPAC (pmol ganglion^{-1} ± S.E.M.)		Dopamine (pmol ganglion^{-1} ± S.E.M.)	
	Decentralized	Sham	Decentralized	Sham
Control	34 ± 2	63 ± 5	19 ± 3	17 ± 2
Carbachol (1 mg kg^{-1}, i.p.)	59 ± 5^a	59 ± 9	26 ± 5	23 ± 6
Atropine (20 mg kg^{-1}, i.p.)	29 ± 4	37 ± 5^a	22 ± 3	23 ± 3
Hexamethonium (7 mg kg^{-1}, i.p.)	32 ± 2	56 ± 3	18 ± 2	18 ± 3
Atropine + carbachol	20 ± 4^a	29 ± 3^a	22 ± 2	21 ± 2
Hexamethonium + carbachol	86 ± 14^a	99 ± 14^a	32 ± 8^a	26 ± 3^a

Data presented as the mean ± S.E.M. for 5–8 experiments.

[a] $P < 0.05$ when compared with control samples.

Carbachol was injected 15 min before the animals were killed while atropine or hexamethonium was injected 20 or 25 min, respectively, before killing. The same intervals were maintained when the drugs were given in combination.

TABLE V Concentration of HVA and DOPAC in human CSF; plasma and urine.

Sample	HVA Mean ± S.E.M.	Range	DOPAC Mean ± S.E.M.	Range
Ventricular CSF[a] pmol ml^{-1} ($n = 8$)	840 ± 80	370–1200	57 ± 9	9 – 59
Lumbar CSF[b] pmol ml^{-1} ($n = 16$)	200 ± 80	130– 260	7.1 ± 0.6	2.9– 11
Plasma[c] pmol ml^{-1} ($n = 10$)	92 ± 10	29– 160	61 ± 9	15 –120
Urine[c] umol 24 h^{-1} ($n = 10$)	51 ± 4	34– 66	47 ± 7	22 – 92

[a] Samples were collected from patients with ventricular shunts for hydrocephalus or presenile dementia.
[b] Samples were collected from schizophrenic subjects (Karoum *et al.*, 1977a).
[c] Samples were collected from normal volunteers.

Table V summarizes the concentrations of HVA and DOPAC in the CSF, plasma and urine of man (Karoum *et al.*, 1976; Karoum *et al.*, 1977a). Homovanillic acid is the predominant DA metabolite in the CSF. In contrast to CSF, HVA and DOPAC are present in about equal concentrations in plasma and urine. There are, however, wide variations in the urinary excretion of HVA and DOPAC between subjects. Whether these variations are related to differences between subjects or to daily differences of the same subject is currently being evaluated. There also seems to be a decreasing concentration gradient for HVA and DOPAC along the ventriculo-lumbar space. A concentration gradient has been reported by other workers for HVA and 5-hydroxyindoleacetic acid for man and other species (Garelis and Sourkes, 1973; Maira *et al.*, 1975; Sjöstrom *et al.*, 1975; Weir *et al.*, 1973).

In conclusion, this chapter is not meant to be a detailed description of GC–MS. It is meant to demonstrate the sensitivity and specificity of MF and how it can be applied to biological problems. For example, MF can be used to analyse DA and its metabolites where they are found in extremely low concentrations, such as the rat superior cervical ganglion, and in biological fluids which contain interfering materials such as urine. Advances in analytical methodology generally precede advances of our understanding of biological mechanisms.

REFERENCES

Anden, N. E., Roos, B. E. and Werdinius, B. (1963). *Life Sci.* **7,** 448–458.

Angrist, B. and Gershon, S. (1974). *In* "Neuropsychopharmacology of Monoamines and Their Regulatory Enzymes" (Ed. E. Usdin), 211–219. Raven Press, New York.

Costa, E., Koslow, S. H. and LeFevre, H. (1975). *In* "Handbook of Psychopharmacology" (Eds. L. Iversen, S. Iversen and S. Snyder), 1–24. Plenum, New York.

Garelis, E. and Sourkes, T. L. (1973). *J. Neurol. Neurosurg. Psychiat.* **36,** 625–629.

Hornykiewicz, O. (1966). *Pharmacol. Rev.* **18,** 925–964.

Karoum, F., Gillin, J. C. and Wyatt, R. J. (1975). *J. Neurochem.* **25,** 653–658.

Karoum, F., Gillin, J. C., Wyatt, R. J. and Costa, E. (1976). *Biomed. Mass Spectrom.* **2,** 183–189.

Karoum, F., Bunney, Jr., W., Gillin, J. C., Jimerson, D., Van Kammen, D. and Wyatt, R. J. (1977a). *Biochem. Pharmacol.* **26,** 629–632.

Karoum, F., Garrison, C. K., Neff, N. H. and Wyatt, R. J. (1977b). *J. Pharmacol. Expl. Ther.* **201** (3), 654–661.

Karoum, F., Neff, N. H. and Wyatt, R. J. (1977c). *Eur. J. Pharmacol.* **44,** 311–318.

Koslow, S. H., Cattabeni, F. and Costa, E. (1972). *Science* **176,** 177–180.

Koslow, S. H., Racagni, G. and Costa, E. (1974). *Neuropharmacology* **13,** 1123–1130.

Libet, B. (1970). *Fed. Proc.* **29,** 1945–1956.

Maira, G., Bareggi, S. R., DiRocco, C., Galderini, G. and Morselli, P. L. (1975). *J. Neurol. Neurosurg. Psychiat.* **38** (2), 123–128.

Randrup, A. and Munkvad, I. (1967). *Psychopharmacologia* **11,** 300–310.
Rosengren, E. (1960). *Acta Physiol. Scand.* **49,** 364–369.
Sharman, D. F. (1963). *Br. J. Pharmacol.* **20,** 204–213.
Sjöquist, B. and Anggård, E. (1972). *Anal. Chem.* **44,** 2297–2301.
Sjöquist, B., Lindstrom, B. and Anggård, E. (1973). *Life Sci.* **13,** 1655–1664.
Sjöstrom, R. Ekstedt, J. and Anggård, E. (1975). *J. Neurol. Neurosurg. Psychiat.* **38,** 666–668.
Snyder, S. H. (1972). *Arch. Gen. Psychiat.* **27,** 169–179.
Watson, E. and Wilk, S. (1975). *Psychopharmacologia* **42,** 57–62.
Westerink, B. H. C. and Korf, J. (1976). *Eur. J. Pharmacol.* **37,** 249–255.
Weir, R. L., Chase, T. N., Ng, L. K. Y. and Kopin, I. J. (1973). *Br. Res.* **52,** 409–412.
Wiesel, F. (1975). *Neurosci. Lett.* **1,** 219–224.
Wiesel, F. A., Fri, C. G. and Sedvall, G. (1973). *Eur. J. Pharmacol.* **23,** 104–106.
Wilk, S., Watson, E. and Glick, S. D. (1975a). *Eur. J. Pharmacol.* **30,** 117–120.
Wilk, S., Watson, E. and Travis, B. (1975b). *Eur. J. Pharmacol.* **30,** 238–243.

CHAPTER 7

The Radioenzymatic Assay of Catecholamines

A. C. Cuello

University Department of Pharmacology,
South Parks Road, Oxford, UK

INTRODUCTION

Several radioenzymatic methods for the assay of catecholamines at picogram levels have been described in recent years. These assays employ methyltransferase enzymes in conjunction with labelled (methyl-^{14}C or methyl-^{3}H,S-adenosyl methionine, SAMe). These techniques are becoming increasingly popular since they allow the measurement of the catecholamine content from a small amount of tissue and body fluids with high accuracy and sensitivity as compared with the more conventional fluorimetric methods.

Techniques employing the enzyme phenyl-ethanolamine-*N*-methyl-transferase (PNMT) have been reported to measure noradrenaline (Saelens *et al.*, 1967; Iversen and Jarrot, 1970; Henry *et al.*, 1975; Weise and Kopin, 1976), octopamine (Molinoff *et al.*, 1969; Saavedra, 1974; Harmar and Horn, 1976) and β-phenylethylamine (Saavedra, 1974b). However, PNMT, cannot be applied to the assay of dopamine or adrenaline. In contrast dopamine (DA), noradrenaline (NA) and adrenaline (A) are equally good substrates for the methylating enzyme catechol-*O*-methyltransferase (COMT) (Axelrod and Tomchick, 1958; Guldberg and Marsden, 1975). Engelman *et al.* (1968) and Nikodijevik *et al.* (1969) reported the first attempts to use COMT for the determination of catecholamines from plasma and tissue (heart) respectively. Developments of this approach

TABLE I Main steps, applications and sensitivity of current

Extraction of catecholamines	Incubation pH	Organic extraction of *O*-methylated derivatives	Acid extraction	Products dried down and resuspended	Conversion of M and NM to methyl vanillin
5% TCA, washed in ether	9.1	T/I (7:3)	—	–	–
Adsorption by column chromatography	8.4	—	—	+	+ (after chromatography)
0.1 M PCA	9.6	EA/M (10:1) twice	0.1 M HCl plus wash in EA/M	–	+
0.1 M PCA	9.1	T/I (3:2)	0.1 M HCl	–	–
Whole plasma	8.2	T/I (3:2)	—	–	+ (after chromatography)
0.1 M PCA	9.1	—	1 M PCA followed by 3 washes in EA	–	–
Whole plasma	8.9	T/I (3:2)	0.1 M HCl washes in EA	–	–
0.1 M PCA	9.1	T/I (3:2)	0.1 M HCl	–	–
0.1 M PCA plus 10 mM thioglycollic acid, alumina adsorption eluates dried down	9.1	T/I (1:1)	0.1 M HCl	+	+
0.3 M PCA, plus 0.1% ascorbic acid	8.25	TPB,DE	0.1 M HCl	+	+ (after chromatography)
Deproteinized plasma, alumina adsorption,	9.6	EA/M (10:1)	0.1 M HCl	–	+
Whole plasma, tissue in 0.1 M PCA	8.6	T/I (3:2) plus washes in EA	0.1 M HCl	–	–
0.1 M PCA plus 10 mg% EGTA	9.1	T/I (3:2)	0.1 M HCl	–	–
Whole plasma	8.2	T/I (3:2) twice	0.1 M Acetic acid, plus wash in T/I		+ (after chromatography)

TCA, trichloroacetic acid; PCA, perchloric acid; T/I, toluene/isoamylalcohol; TPB, tetraphenylborate; DE, diethyl ether; EA/M, water-saturated ethylacetate-methanol; EA, water-saturated ethylacetate; (3:2, 10:1, etc, refer to v/v relations; TLC, thin-layer chromatography; PC, paper chromatography; CSF, cerebrospinal fluid.

[a] Catecholamine-*O*-methylated derivatives cannot be discriminated.

[b] Double isotopic technique. Tracer ^{3}H-catecholamines are incubated in presence of COMT and ^{14}C-methy-*S*-adenosyl methiomine.

techniques using COMT for catecholamine determination

Acetylation	Separation of O-methylated metabolites	Application	Sensitivity	Reference
–	—	tissue (heart)	Catecholamines = 300 pg[a]	Nikodijevic *et al.* (1969)
–	Column chromatography followed by TCL of lyophylized eluates	plasma, urine	[b]	Engelman and Portnoy (1970)
–	Aqueous/organic solvent separation	tissue	DA = 200 pg NA = 100 pg[c]	Coyle and Henry (1973)
–	PC	tissue	DA = 10–40 pg[d] NA = 70–100 pg[d] A = 10–40 pg[d]	Cuello *et al.* (1973)
–	TCL	plasma	NA,A = 170 pg	Passon and Peuler (1973)
+	PC	tissue		Fry *et al.* (1974)
+	PC	plasma	DA = 25–50 pg NA = 100–200 pg A = 25–50 pg	Callingham and Barrand (1976)
–	Ion-exchange chromatography	tissue	DA = 100 pg NA = 100 pg	Zschaeck and Ramirez (1976)
–	Aqueous/organic solvent separation	CSF, plasma, tissue	DA = 24 pg NA = 17 pg[c]	Gauchy *et al.* (1976)
–	TLC	plasma tissue	DA = 5 pg NA = 1 pg A = 1 pg	De Prada and Zurcher (1976)
–	Aqueous/organic solvent separation plus washes	plasma	DA = 84 pg NA = 14 pg[e] A = 11 pg[e]	Weise and Kopin (1976)
–	bidimensional TLC	tissue plasma	DA,NA,A = 10–30 pg	Ben-Jonathan and Porter (1976)
+	TLC	tissue	DA = 20–40 pg NA = 40–100 pg A = 10–30 pg	Martin *et al.* (1978)
–	TLC	plasma	DA = 6 pg NA = 1 pg A = 1 pg	Peuler and Johnson (1977)

[c] Noradrenaline-*O*-methylated derivatives cannot be discriminated from adrenaline derivatives.
[d] As described here, originally reported 100 pg for DA and NA (Cuello *et al.*, 1973).
[e] Noradrenaline, adrenaline values obtained by comparing COMT results in combination with *N*-methylation of catecholamines by PNMT assay (Weise and Kopin, 1976).

allowed the measurement of catecholamines from tissue (Coyle and Henry, 1973; Cuello *et al.*, 1973) and plasma (Passon and Peuler, 1973) in the picogram range. Several versions of this procedure have since been published (see Table 1).

CHARACTERISTICS OF THE ASSAY

The derivatives which result from the *O*-methylation of the catecholamine by COMT are: 3-methoxytyramine (MT) for dopamine, normetanephrine (NM) for noradrenaline and metanephrine (M) for adrenaline. The enzymatic reaction common to these techniques proceeds as shown in Fig. 1.

The *O*-methylated derivatives from standards and biological samples obtained from the enzymatic incubation are extracted, separated and the radioactivity counted by liquid scintillation spectrometry. The initial amount of the catecholamines present as substrates in the unknown samples is inferred from the radioactivity obtained for the corresponding *O*-methylated product.

The combination of enzymatic incubation and separation of the enzymatic products results in highly specific methods for the assay of the catecholamines. When the labelled derivatives are separated by oxidation and organic solvent extractions, only dihydroxyphenyl glycol among endogenous metabolites interferes with the noradrenaline–adrenaline assay (Coyle and Henry, 1973). This is largely corrected when a chromatographic step is applied (Da Prada and Zurcher, 1976; Peuler and Johnson, 1977). Nevertheless even after chromatography isoproterenol and α-methyl noradrenaline interfere up to 100% with the assay of noradrenaline and adrenaline respectively (Da Prada and Zurcher, 1976; Peuler and Johnson, 1977). Therefore when drugs structurally related to the catecholamines are

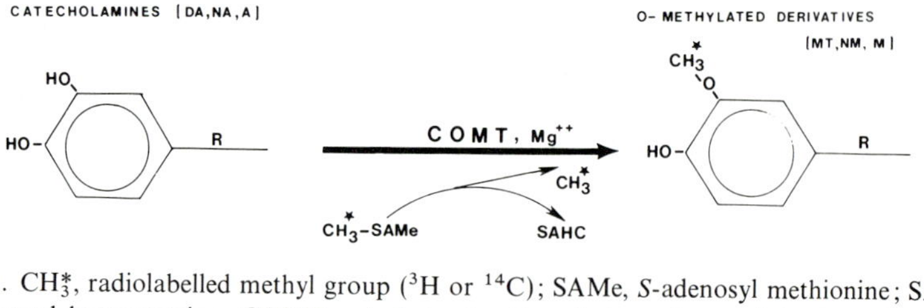

Fig. 1. CH_3^*, radiolabelled methyl group (^3H or ^{14}C); SAMe, *S*-adenosyl methionine; SAHC, *S*-adenosyl homocysteine; COMT, catecholamine-*O*-methyltransferase; DA, dopamine; NA, noradrenaline; A, adrenaline; MT, 3-methoxytyramine; NM, normetanephrine. R is $-CH_2-CH_2-NH_2$ for DA and MT, $-CH.OH-CH_2-NH_2$ for NA and NM, $-CH.OH-CH_2-NH-CH_3$ for A and M.

administered it is advisable to test for possible cross-reactivity. The cross-reactivity is total for noradrenaline and adrenaline in assays based on the formation of methyl vanillin from metanephrine and normetanephrine followed by organic extraction without a previous chromatographic step (see Table I). Otherwise the overlapping of the three catecholamines is very small. In the procedure described here nanogram amounts of dopamine interfere by 1–2% in the adrenaline and noradrenaline assay. The overlapping of noradrenaline on the other amines is negligible. Adrenaline produces slightly greater cross-reactivity with the other catecholamines but the amounts of brain adrenaline are insufficient to interfere. Ascorbic acid is also a substrate for the enzyme and in high amounts it may interfere with the catecholamine assay (Blaschke and Hertting, 1971). Dihydroxyphenylalanine (DOPA) can contribute to the dopamine values if DOPA decarboxylase activity is present in the COMT preparations (Ben-Jonathan and Porter, 1975). In this case a DOPA-decarboxylase inhibitor should be added to the incubation mix.

In our procedure (see below) we extract the *O*-methylated derivatives in toluene–isoamyl alcohol (3:2 v/v) (Weinshilboum and Axelrod, 1971). More than 80% of the formed 3-methoxytyramine is recovered in this solvent (Gauchy *et al.*, 1976). When noradrenaline is the main catecholamine to be assayed, extraction with water-saturated ethyl acetate–methanol (10:1 v/v) is recommended (Gauchy *et al.*, 1976). Paper chromatography, although simple to apply, results in low recoveries (20–30%) as compared with methods using thin-layer chromatography (50–80%).

The intra-assay variation is in the range of 2–5% for the present method. Somewhat larger interassay variations may be expected when different enzyme or methyl donor batches are used. When concentrated tissue extracts are assayed (1:1 to 1:10, v/v) tissue components partially inactivate the enzyme. In these cases internal catecholamine standards (500–1000 pg) should be added to duplicate or triplicate samples and values corrected accordingly. No enzyme inactivation is found when larger dilutions are prepared. To avoid the Ca^{2+} inhibition of the enzyme (Axelrod and Tomchick, 1958) ethylene glycol-bis-(β-aminoethyl ether) *N,N'*-tetraacetic acid (EGTA) is added during homogenate extraction as recommended by De Champlain *et al.* (1976).

ASSAY PROTOCOL (after Cuello *et al.*, 1973)

Specimen Preparation

Tissue samples are quickly frozen in dry ice after removal, weighed, homogenized in 0.1 M perchloric acid (PCA) containing 10 mg per 100 ml of

EGTA and centrifuged. When tissue samples are too small for accurate weighing, or other simultaneous assays are to be carried out, the tissue may alternatively be homogenized in water. Aliquots are taken for protein determination or other assays and the rest of the homogenate is brought to 0.1 M PCA and centrifuged. When tissue samples in the microgram range of weight are studied in order to extract the catecholamines it is sufficient to freeze and thaw the tissue several times with known volumes of PCA while disrupting it with the tip of a needle.

Enzyme Preparation

Catechol-*D*-methyltransferase (COMT) preparation obtained from bovine (Cuello *et al.*, 1973), rat (Nikodijevik *et al.*, 1969, 1970; Coyle and Henry, 1973) and pig liver (Fry *et al.*, 1974) have been applied successfully to the radioenzymatic assay of catecholamines.

We have found that relatively crude COMT preparations are suitable for amine determination. We have adopted a simplified purification procedure following that described by Axelrod and Tomchick (1958). Fifty grams of fresh liver is homogenized in 200 ml of 1.1% (w/v) ice-cold potassium chloride. Fats are removed from the homogenate by filtering through glass wool and centrifuging at 100 000 *g* for 60 min at 4°C. The whole procedure should be carried out on ice or in a cold room. The supernatant from the first centrifugation is filtered again through glass wool and adjusted to pH 5.3 by adding drop-wise 1 M acetic acid. After 10–20 min the precipitate is removed by centrifugation at 14 000 × *g* for 15 min at 4°C. The supernatant is adjusted to pH 6.8 with freshly prepared 0.5 M sodium phosphate buffer pH 7.0. To this solution dry ammonium sulphate (enzyme grade) is added *slowly* while gently stirring to obtain 30% saturation (17.7 g per 100 ml of supernatant). The solution is allowed to stand for 10–20 min and centrifuged for 15 min at 14 000 *g*. The resulting precipitate is discarded and the supernatant adjusted (as above) to 45% ammonium sulphate saturation (28 g per 100 ml). The precipitate is removed by centrifugation for 15 min at 14 000 *g*. The pellets obtained are gently dissolved with the help of a pasteur pipette in 5–6 ml of 0.1 M sodium phosphate buffer pH 7.0 containing 1 mM dithiothreitol. Avoid bubbling during this step. This is dialized overnight against two or three changes of 500 volumes of the same buffer. Aliquots (0.5–1 ml) of this enzyme preparation are kept at −20°C. This partially purified preparation contains approximately 50 mg of protein ml^{-1}. Monoamine oxidase activity is undetectable or slightly above blank when measured with the method of McCaman *et al.* (1965). To inhibit any possible interference from this enzyme pargyline is added to the incubation medium. Alternatively more purified preparations can be obtained following the method of Nikodijevik *et al.* (1970). The COMT activities of these preparations are assayed by using 3,4-dihydroxybenzoic acid as substrate according to McCaman (1965). The specific activity of the preparations varies from batch to batch. Preparations rendering 300–500nmol of product h^{-1} mg^{-1} of protein give optimum results for catecholamine assays. For practical purposes the enzyme can be tested simply for its efficiency in measuring the catecholamine conversion to labelled *O*-methylated derivatives by counting aliquots of the acid phase (Step 3). It is convenient to test the enzymes with the previous batch in order to determine the dilution which yield the optimal standard/blank ratio.

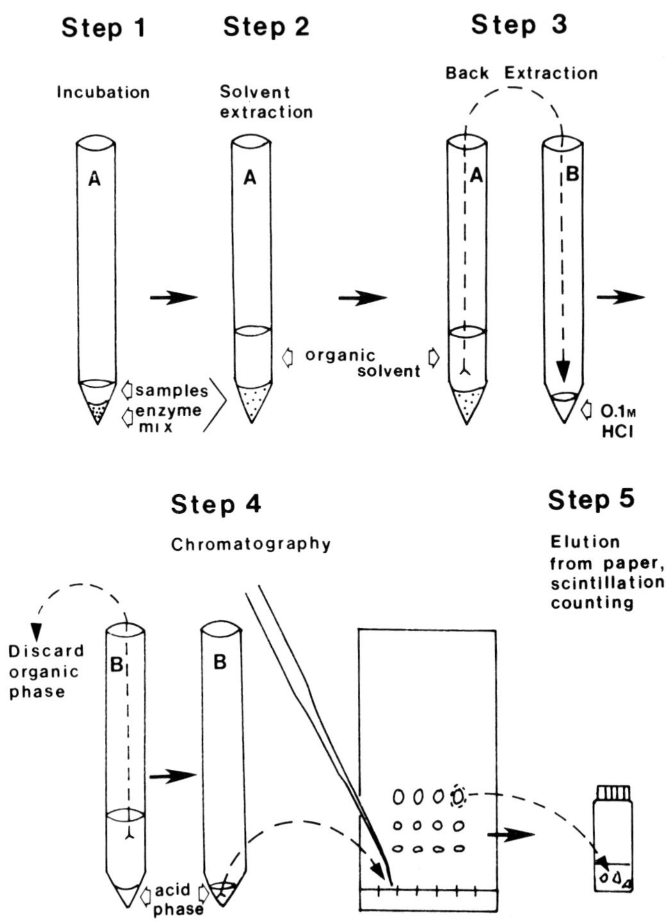

Fig. 2. Schematic representation of the steps for the radioenzymatic assay of catecholamines.

Step 1: Incubation

Incubation mix for 30 assays (to be prepared in cold tubes):

200 mM Ethylene glycol-bis-(β-aminoethyl ether) *N,N'*-tetraacetic acid (EGTA) 10 µl

^{3}H-Methyl SAMe (Sp. ac. 11.4 Ci mmol^{-1}, The Radiochemical Centre, Amersham) (kept in aliquots at −20 to −80°C) 80 µl

COMT preparation 250 µl

Just before use add:

3 M Tris base 220 µl
300 mM MgCl 10 µl

Vortex this mix gently for a few seconds.

Have ready set A of tubes (see Fig. 2) in an ice-cold bath. Prepare blanks (50 µl containing 0.1 M PCA, standards (50 µl containing 250–1000 pg of dopamine, noradrenaline and adrenaline in 0.1 M PCA) and the tissue samples (50 µl of 0.1 M PCA supernatant). When internal standards are introduced pipette 2 µl of 0.1 M PCA containing 500 or 1000 pg of the catecholamines.

Add 20 µl of the enzyme–methyl donor mix to each tube and shake. Final volume of incubation: 70 µl. The enzymatic incubation is carried out in a shaking bath during 40 min at 37°C. The reaction is stopped by immersing the tubes in an ice–salted water bath and to each is added 0.45 M borate buffer/carrier solution (20:1, v/v) (carrier solution: 1 mg of each methoxyamine, M, NM and M in 0.1% sodium metabisulphite).

Step 2: Solvent extraction

Add 500–1000 µl (larger volumes for plasma) of the organic solvent toluene–isoamyl alcohol (3:2, v/v). When noradrenaline is the main catecholamine to be assayed better recoveries are obtained with water-saturated ethyl acetate–methanol (10:1 v/v). Vortex vigorously for 30 s and centrifuge to separate the phases.

Step 3: Back extraction

Have ready a new set of conical pointed tubes (tubes B of Fig. 2) containing 25 µl of 0.1 M HCl. Remove from tubes A 400 µl (800 µl if 1 ml was added in Step 2) of the organic phase and transfer to the set of B tubes. Vortex for 30 s and centrifuge.

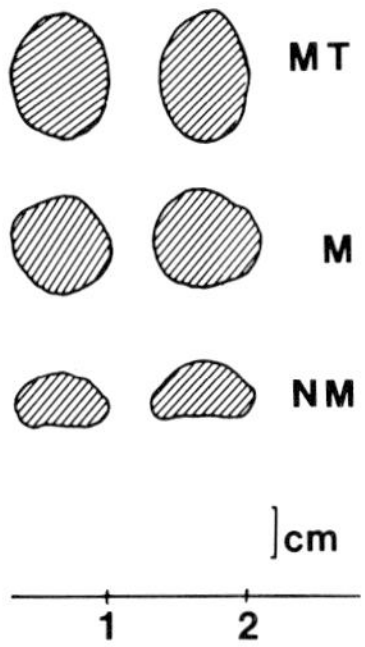

Fig. 3. Separation of *O*-methylated derivatives by paper chromatography 1,2, origin; MT, 3-methoxytyramine-NM, normetanephrine; M, metanephrine scale/centrimeter.

Step 4: Chromatography

Discard the organic phase by careful suction with Pasteur pipettes. A black background helps to distinguish the two phases. Three to four 5 μl aliquots of the aqueous phase are transferred to the origin of Whatman 1 chromatographic paper previously spotted with 5 μl of the "carrier" solution (MT, NM, M, $10\,mg\,ml^{-1}$ each dissolved in 0.1% sodium metabisulphite). If glass micropipettes are used they should be rinsed in distilled water or 0.1 M PCA after each delivery. Alternatively 5 μl Drummond microcapillaries may be used. It is important to obtain regular and small-sized origin spots. The papers are then developed by descending chromatography in 33–40% methylamine/ter-amyl alcohol (1:4 v/v) overnight. The front should reach three-quarters of the paper length. The tanks should be clean, saturated with the solvent mix and properly sealed. We also cover them with black plastic bags to prevent direct light during chromatography. The developed chromatograms are allowed to dry in a fume cabinet until the spots can be visualized. The spots can also be visualized more quickly by exposing the paper chromatograms under a UV lamp (254 nm). Under these conditions MT and M produce a weak violet fluorescence and NM an intense yellow fluorescence. The *O*-methylated derivatives are separated with the following R_f: MT, 0.46; M, 0.33; NM, 0.22. The size and shape of the spots obtained are illustrated in Fig. 3 as drawn from an actual chromatogram.

Step 5: Scintillation counting

The spots are cut out from the paper, shredded and placed in counting vials. The radioactivity is eluted from the paper in 3 ml of ethanol–concentrated ammonia (10:22, v/v) for several hours (minimum 3 h). Add 2 ml of ethoxy ethanol and 10 ml of 0.4% butyl PBD in toluene and count by liquid scintillation spectrometry.

Comments and applicability of the assay

The enzyme can be obtained in a single day and up to 40 samples can be assayed in a single run for one operator. Steps 1–3 can be carried out in a couple of hours while the chromatography requires approximately one hour of preparation and 24 hours for the developing and drying of the chromatograms. The assay can be safely interrupted at Step 3 and the aqueous acid phase kept frozen for further analysis.

Typical results are illustrated in Fig. 4. For this particular assay the sensitivity (estimated as twice blank) for the different catecholamines was: DA, 23 pg; NA, 63 pg; and A, 30 pg, though a sensitivity of 10 pg can be obtained for DA and A. Results are calculated by reference to the amount of radioactivity obtained from 1000 pg of each catecholamine simply by applying the following formula:

pg sample (DA, NA, A)

$$= \frac{\text{cpm sample} - \text{cpm blank (mean)}}{\text{cpm 1000 pg standard (mean)} - \text{cpm blank (mean)}} \times 1000$$

When the standard curves are not linear (old COMT batches, very extended standard curves) values should be calculated by reference to the plotted standard curve. When appropriate, sample values should be corrected for the tissue inactivation of the enzyme (internal standards).

The method described above has permitted the study of catecholamines

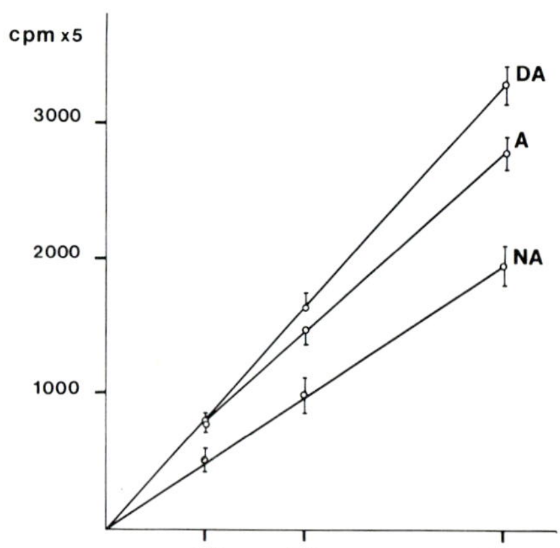

Fig. 4. Standard curve for the assay of dopamine (DA), noradrenaline (NA) and adrenaline (A). Each value represents the mean of four determinations ±S.E.M.

TABLE II Endogenous content of dopamine (DA), noradrenaline (NA) and adrenaline (A) in microdissected layers of the rat olfactory bulb

Microdissected layers	DA ($pg\,mg^{-1}$ prot.)	NA ($pg\,mg^{-1}$ prot.)	A ($pg\,mg^{-1}$ prot.)
Olfactory nerve layer	195 ± 9.4	116 ± 7	—
Glomerular layer	$783 + 6$	261 ± 9	85 ± 8
Plexiform layer	155 ± 7	567 ± 11	—
Granule cells layer	266 ± 12	214 ± 9	45 ± 3
Medullaris	75 ± 8	339 ± 12	—

Each value corresponds to the mean ± S.E.M. of 3–6 determinations. From E. Jaffe and A. C. Cuello, unpublished.

in large sample or microdissected areas of the brain (Cuello *et al.*, 1973; Horn *et al.*, 1974; Ben-Ari *et al.*, 1975; Kelly *et al.*, 1975 and see Table II). The application of this present method provided further biochemical evidence for the existence of adrenaline containing neurones in the rat nervous system (Van der Gugten *et al.*, 1976; Moore and Phillipson, 1975; and see Table II). The spontaneous and potassium-stimulated release of endogenous dopamine from single microdissected substantia nigra tissue slices was also demonstrated with this methodology (Cuello and Iversen, 1978), thus giving further evidence for the concept of a dendritic release of monoamines. More recently we have been able to measure dopamine and noradrenaline from cannulated median eminence portal vessels in the rat (Fink and Cuello, preliminary observations). With this technique catecholamine turnover studies have been performed in very discrete areas of the hypothalamus (Versteeg *et al.*, 1975).

The present and other radiochemical enzymatic methods (see Table I) based on the use of COMT are relatively easy to develop in conventional laboratories and they provide a sensitive and reliable tool for the study of catecholamines and some of their metabolites.

ACKNOWLEDGEMENTS

I would like to thank Drs I. Martin and M. Barrand for revising the manuscript.

REFERENCES

Argiolas, A., Fadda, F., Stefanini, E. and Gessa, G. C. (1977). *J. Neurochem.* **29,** 599–601.
Axelrod, J. and Tomchick, R. (1958). *J. Biol. Chem.* **233,** 702–705.
Ben-Ari, Y., Zigmond, R. E. and Moore, K. E. (1975). *Brain Res.* **87,** 96–101.
Ben-Jonathan, N. and Porter, J. C. (1976). *Endocrinology* **98,** 1497–1507.

Blaschke, E. and Hertting, G. (1971). *Biochem. Pharmacol.* **20,** 1363–1370.
Coyle, J. T. and Henry, D. (1973). *J. Neurochem.* **21,** 61–67.
Callingham, B. A. and Barrand, M. A. (1976). *J. Pharm. Pharmacol.* **28,** 356–360.
Cuello, A. C. and Iversen, L. L. (1978). *In* "International Symposium on Interactions among Putative Neurotransmitters in the Brain" (Eds. S. Garatini, J. F. Pujol and R. Samanin). 127–149, Raven Press, New York.
Cuello, A. C., Hiley, R. and Iversen, L. L. (1973). *J. Neurochem.* **21,** 1337–1340.
Cuello, A. C., Horn, A., Mackay, A. V. P. and Iversen, L. L. (1973). *Nature* (*London*) **243,** 465–467.
Da Prada, M. and Zurcher, G. (1976). *Life Sci.* **19,** 375–384.
De Champlain, J., Farley, L., Cousineau, D. and van Ameringen, M.-R. (1976). *Circ. Res.* **38,** 109–114.
Engelman, K. and Portnoy, B. (1970). *Circ. Res.* **26,** 53–57.
Engelman, K., Portnoy, B. and Lovenberg, W. (1968). *Am. J. Med. Sci.* **225,** 259–268.
Fry, J. P., House, C. R. and Sharman, D. F. (1974). *Br. J. Pharmacol.* **51,** 116–117.
Gauchy, C., Tassin, J. P., Glowinski, J. and Cheramy, A. (1976). *J. Neurochem.* **26,** 471–480.
Guldberg, H. C. and Marsden, Ch. A. (1975). *Pharmacol. Rev.* **27,** 135–205.
Harmar, A. J. and Horn, A. S. (1976). *J. Neurochem.* **26,** 987–993.
Henry, D. P., Starman, B. J., Johnson, D. G. and Williams, R. H. (1975). *Life Sci.* **16,** 375–384.
Horn, A. S., Cuello, A. C. and Miller, R. (1974). *J. Neurochem.* **22,** 265–270.
Iversen, L. L. and Jarrot, B. (1970). *Biochem Pharmacol.* **19,** 1841–1843.
Kelly, P. H., Seviour, P. W. and Iversen, S. D. (1975). *Brain Res.* **94,** 507–522.
Martin, I. L., Baker, G. B. and Fleetwood-Walter, S. M. (1978). *Biochem. Pharmacol.* (in press).
McCaman, R. E. (1965) *Life Sci.* **4,** 2553–2559.
McCaman, R. E., McCaman, M., Hunt, J. M. and Smith, S. M. (1965). *J. Neurochem.* **12,** 15–23.
Molinoff, P. B., Landsberg, L. and Axelrod, J. (1969). *J. Pharmacol. Exp. Ther.* **170,** 253–261.
Moore, K. E. and Phillipson, O. T. (1975). *J. Neurochem.* **25,** 289–294.
Nikodijevik, B., Daly, J. and Creveling, C. R. (1969). *Biochem. Pharmacol.* **18,** 1577–1584.
Nikodijevik, B., Senoh, S., Daly, J. W. and Creveling, C. R. (1970). *J. Pharmacol. Exp. Ther.* **174,** 83–93.
Passon, P. G. and Peuler, J. D. (1973). *Anal. Biochem.* **51,** 618, 631.
Peuler, J. D. and Johnson, G. A. (1977). *Life Sci.* **21,** 625–636.
Saavedra, J. M. (1974). *Anal. Biochem.* **59,** 628–633.
Saavedra, J. M. (1974b). *J. Neurochem.* **22,** 987–993.
Saelens, J. K., Schoen, M. S. and Kovacsics, G. B. (1967). *Biochem. Pharmacol.* **16,** 1043–1049.
Van der Gugten, J., Palkovitz, M., Wijnen, H. Z. Y. and Versteeg, D. H. G. (1976). *Brain Res.* **107,** 171–175.
Versteeg, D. H. G., Van der Gugten, J. and van Ree, J. M. (1975). *Nature* (*London*) **256,** 502–503.
Weinschilboum, R. and Axelrod, J. (1971). *Circ. Res.* **28,** 307.
Weise, V. K. and Kopin, I. J. (1976). *Life Sci.* **19,** 1673–1686.
Zschaeck, L. L. and Ramirez, V. D. (1975). *J. Neural Trans.* **39,** 291–300.

CHAPTER 8

Electrochemical Methodology

R. N. Adams

University of Kansas, Lawrence, Kansas, USA.

Electrochemical techniques for the analysis of dopamine (DA), while very recent in application, are derived, in principle, from Faraday's time. They also bear an unappreciated relationship to the familiar hydroxyindole fluorimetry method in that they employ the same initial step—oxidation of the catechol moiety of DA. Rather than use a chemical oxidant such as ferricyanide or iodine, DA is oxidized at an inert electrode surface with a suitable applied potential. In contrast to the fluorescence procedures, which oxidize DA and then promote intracyclization to the hydroxyindole and its subsequent further conversion to the fluorophore, the electrochemical methods ordinarily stop at the initial oxidation to the *o*-quinone, as:

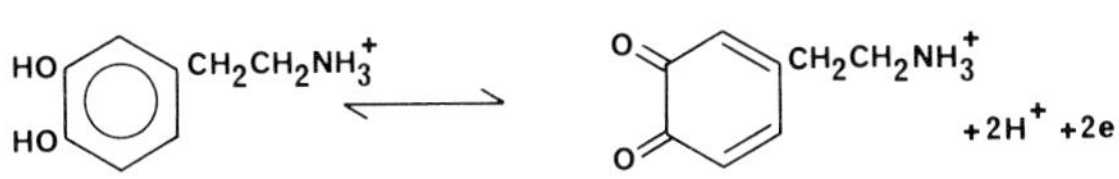

Quantitation is achieved by counting the electrons released in the oxidation, i.e., by measuring the current. (Dopamine is shown in the above equation in the protonated form since all the in-vitro electrochemical assays are carried out at pH's where the amino group is fully protonated.)

In theory the electrochemical oxidation of DA can be carried out at any inert electrode. Platinum, gold and carbon are typical inert materials often used by electroanalytical chemists and DA does indeed oxidize at these electrodes. However, platinum and gold undergo fairly easy surface oxidation and this involves the passage of finite electrochemical currents, even in the absence of analysable species (DA) in solution. This is called residual current. While small, it can amount to a noise level far in excess of the current representing DA oxidation in neurochemical applications. Fortunately, high-purity graphite, mixed into a relatively dry paste with an inert liquid, serves as an electrode surface with extremely low residual current. All of the high-sensitivity electrochemical methods employ graphite electrodes of various compositions. For a fuller discussion of electrodes and electrochemical procedures, the reader is referred to a review particularly pertinent to this section (Adams, 1969) and references therein.

Unfortunately, the ease with which electrons can be removed from DA (i.e. the electrochemical potential at which the reaction written above occurs) is primarily a function of the catechol structure and is only slightly influenced by substituents on the side chain. Hence, norepinephrine (NE), with a β-hydroxyl on the side chain, is oxidized at almost the same potential as DA. While it is possible with sophisticated electroanalytical techniques and special conditions to differentiate between DA and NE, it is not feasible for any practical, high-sensitivity analysis. Some sort of separation procedure must precede the electrochemical determination. In developing the assays, we chose two separation techniques which were well established in brain catecholamine analyses. The first, an alumina treatment, separates catechol-like compounds, by adsorption onto alumina, from any methoxylated metabolites. It also, in our case, becomes an efficient preconcentration step when the catechol compounds are next desorbed into a micro-volume of acid. The compounds are then quickly and efficiently separated using high performance liquid chromatography (HPLC). High performance liquid chromatography is ordinary column chromatography which uses very small particle size stationary phases (resins) so that excellent separations are achieved with relatively short columns and rapid elution. Such columns will not operate with gravity flow of eluent, but always require pressure pumping. The designation HPLC is often taken to mean high pressure liquid chromatography. Much HPLC is actually done at high pressures (2–4000 psi), but very efficient separations can also be achieved at lower pressures. The catecholamine LC is easily done with simple apparatus at pressures usually 600 psi or less.

As the NE, DA and other components, now well separated, emerge from the LC column, they immediately pass a sensitive electrochemical detector. The detector is a very simple device—a block of lucite with a very thin channel in the centre. The eluent flows through this channel at a high rate past a smooth graphite paste electrode. The electrode potential is maintained at a value such that each catecholamine is oxidized as it passes. The recording of current vs. time (or eluate volume) is completely analogous to any other chromatographic recording system and the peak

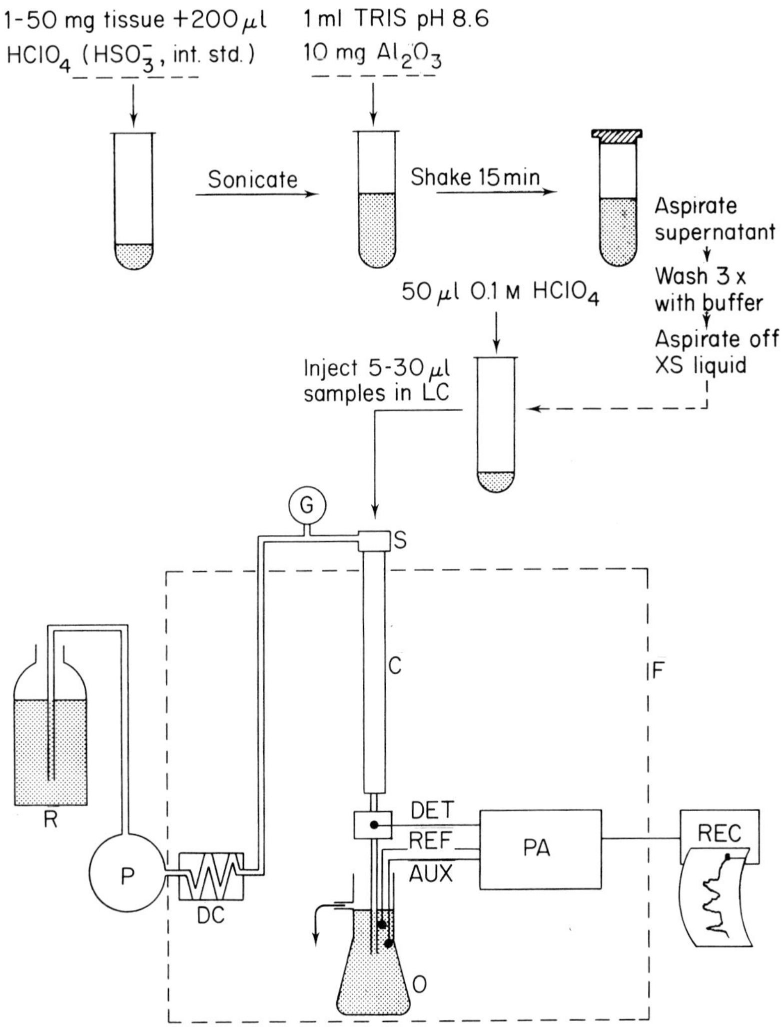

Fig. 1. R, reservoir for eluent; P, pump; DC, damping coil to limit pumping pulsations; G, pressure gauge; S, on-line septum injection port; C, column packed with resin; DET, detector electrode; REF, reference electrode; AUX, auxiliary electrode; O, overflow flask; PA, potentiostat-amplifier; REC, X-time recorder, slow chart speed; F, copper Faraday cage.

heights are used for quantitative measurements. It is possible to adjust separation conditions to observe only DA, but this is so wasteful of additional information that it is difficult to justify doing so. In the usual catecholamine assay, NE and DA are determined. By a slight change in eluent composition, NE, DA and epinephrine (E) are obtained. Other choices of column packing and eluent provide analyses for various metabolite mixtures.

Neuroscientists unfamiliar with electrochemical techniques may find it difficult to understand how such a simple detector can have a sensitivity comparable to that of some of the more sophisticated systems used in catecholamine assays. Electrodes set in a high velocity, relatively laminar flow stream (or with other efficient means of hydrodynamic mass transport) can measure concentrations of electroactive species in the range 10^{-8}–10^{-9} M. The combination of alumina preconcentration, efficient chromatographic separation and tiny detector cell volume (*c.* 1 µl) promotes the high sensitivity.

Figure 1 is a flow sheet illustrating the overall sample preparation and chromatography. Details of the apparatus and methodology have been described thoroughly in the literature (Refshauge *et al.*, 1974; Keller *et al.*, 1976). The following remarks concern points of emphasis, data concerning solution stabilities, or problems which have become apparent since the previous publications.

REAGENTS AND SOLUTIONS

One of the essential features of the assay is that all manipulations are designed to minimize oxidation of catecholamines prior to the detector reaction. Solutions are made from distilled water which has been redistilled from alkaline permanganate. The common laboratory deionized water is not satisfactory for high sensitivity electrochemical usage.

The perchloric acid for sample dissolution is prepared by adding 400 µl of 1 M sodium bisulphite ($NaHSO_3$) to 1 litre of 0.1 M $HClO_4$. Kept stoppered, this solution may be used for about one month without difficulty.

After preparation of the 0.5 M stock solution of Tris buffer (pH 8.6), it is deoxygenated for 15 min with nitrogen gas. Then 3 ml of freshly prepared 1.0 M $NaHSO_3$ and 2 ml of 0.1 M disodium ethylenediamine-tetraacetate (EDTA) are added. This stock solution is stored in a refrigerator. An aliquot sufficient for one day's work is removed and allowed to warm to room temperature before use. Unused portions are discarded at the end of the day.

The "dilute buffer" for the alumina wash is made by mixing 100 ml of the double distilled water, 100 µl of 0.1 M $NaHSO_3$ and 1.0 ml of the stock Tris buffer. This solution is freshly made each day.

The acetate–citrate eluent (pH 5.2) is prepared as described (Keller *et al.*, 1976). Initially this solution is vacuum degassed or deoxygenated thoroughly with nitrogen. It has been found unnecessary to maintain the eluent reservoir on a hot plate as described earlier.

The acid-washed alumina is prepared essentially according to Weil-Malherbe (1971) and Anton and Sayre (1962). The washing to remove fine particles is particularly important. Ultrafine particles not only lower recovery, but clog syringe needles. A convenient washing procedure involves placing the HCl-washed Al_2O_3 in a large beaker. Distilled water is run into the bottom of the beaker and allowed to flush over the sides. The beaker is stirred so that the Al_2O_3 "fines" wash over the sides and heavier particles accumulate in the bottom. Washing is continued until the pH of the distilled water reaches 3.4. Following filtration and drying, the Al_2O_3 is stored permanently in a desiccator and only small portions are removed for a day's work as required. Continual exposure to air will degrade the adsorption properties of the alumina.

APPARATUS

No changes in the equipment described previously have been made other than elimination of heating for the eluent reservoir. As seen in the bottom of Fig. 1, the chromatography apparatus is easily assembled from commercial components. The potentiostat-amplifier and detector block can easily be laboratory built but, especially for those inexperienced with electrochemical equipment, the units available from Bioanalytical Systems, Inc., West Lafayette, Indiana, are highly recommended. Our laboratory no longer finds it a saving of time or money to construct the potentiostat units. The commercial detector block contains integral auxiliary and reference electrodes, unlike the overflow arrangement of Fig. 1. The Bioanalytical Systems equipment is highly reliable and comes complete with operating instructions, application sheets, etc. (Complete LC systems with electrochemical detection are also available.)

Whether the commercial or home-built detector block is used, the most critical requirement for the highest sensitivity work is the smoothness of the graphite paste electrode surface. This point cannot be over-emphasized. The electrode must be polished very smooth and have no visible surface striations or indentations. The paste also must not extend beyond the edges of the electrode well. Preparing smooth electrodes requires no magic—just practice and attention to detail as with any high-sensivity technique.

Freshly prepared detector electrodes require a drift stabilization period of several hours with the applied potential set and eluent pumping. Drift will also occur when the unit is first started each day; our practice is to start the equipment each morning before the coffee—the drift will normally stabilize within half an hour.

The small pumps used (Milton-Roy No. 396–57) are very reliable, but if the system is unused for some time the buffer components will crystallize and may cause leaks in the pump. To prevent this, pumps should be operated at least 1–2 h every week and never allowed to "dry out".

CHROMATOGRAPHIC CONDITIONS FOR VARIOUS SEPARATIONS

The following section lists the detailed conditions for several separations in addition to the regular assay for NE and DA. Slight variations in column length, pump-rate, or eluent composition will not alter results appreciably. The recommended resins appear to work best for the particular separations.

1. Regular DA and NE determination
   Column: 1 m, Corasil CX resin
   Eluent: acetate–citrate (pH 5.2) buffer as described herein
   Flow rate: 20–30 ml h^{-1}
   The preferred internal standard is dihydoroxybenzylamine (DHBA, Aldrich); elution order: NE, internal standard, DA.
2. Simultaneous NE, E, DA determination
   Column: 1 m, Corasil CX resin
   Eluent: same as above
   Flow rate: 30–40 ml h^{-1}
   The preferred internal standard is α-methyl-dopamine (Merck); elution order: NE, E, DA, internal standard; see Mefford *et al.* (1977).
3. Simplified DA-5-HT determination (direct injection)
   Column: 30 cm, DuPont Zipax SCX cation resin
   Eluent: acetate–citrate buffer "double ionic strength"—prepare as before but with twice the amount of each reagent.
   Flow rate: 50–60 ml h^{-1}
   No internal standard, direct calibration. Following the usual tissue homogenization, samples are spun at 17 000 rpm for 20 min, followed by direct injection of 20–30 μl samples. Elution order: DA, 5-HT; ideal for synaptosome preparations.
4. CSF assay for DA acidic metabolites
   Column: 40 cm, DuPont Zipax SAX anion resin

Eluent: pH 4.0 acetate buffer (0.3 M acetic acid and 0.054 M sodium acetate).

Flow rate: 36 ml h^{-1}

Used for direct injection of CSF samples, no internal standard used; elution order: HVA, DOPAC, 5-HIAA; see Wightman *et al.* (1977). This separation will not work on tissue because of overlap of the solvent front with the HVA peak.

RESULTS

The electrochemical technique has been applied by now to a wide range of small animal and human tissue samples. In 3–4 years of operation in our laboratory, we have found no examples where it gives unexpected or inconsistent results. It has been specifically compared with fluorescence and GCMS assays and found to give equivalent results (Weintraub *et al.*, 1975). In a very thorough examination, Sasa and Blank (1977) compared whole mouse brain DA and 5-HT determined by the electrochemical LC technique with five earlier fluorescence methods and GCMS; again the various methods gave consistent results. We have no experience with samples other than CNS tissue or CSF, but Kissinger's group has examined DA and other catecholamines in serum and urine using the electrochemical method and apparently found reliable results (Riggin *et al.*, 1976; Kissinger *et al.*, 1975).

To test the reproducibility of the whole procedure used in the "regular" NE, DA assay, a whole rat brain was homogenized and nine portions taken as repetitive samples. Internal standard was then added and the entire Al_2O_3 adsorption, etc. carried out to give nine separate samples injected on the column. Each of these samples contained approximately 3 ng of NE and 5 ng of DA actually injected—hence, these are quite large samples run at only moderate sensitivity settings. The measurements test the *reproducibility* of the entire assay (excluding brain sampling inhomogeneities, etc.) and are summarized below in terms of ng g^{-1} wet tissue.

These results may be considered typical of what can be expected with moderate amounts to be determined—many of our determinations, even for

	NE (ng g^{-1})	DA (ng g^{-1})
Mean	274	610
S.D.	±2.0	±5.1
S.E.M.	±0.68	±1.7

$N = 9$

brain parts, fall in this category. Recoveries, although not particularly pertinent with the use of the internal standard (Refshauge *et al.*, 1974) are about 70%. Using a single LC unit, approximately 20 samples per day can be processed.

An interesting check on the high sensitivity operation was made during the preparation of this manuscript. Our four chromatographs had all been operating on fairly large quantities (i.e. 1–3 ng input as in the example above) for some 6 months. I requested, "Let's see what sort of reproducibility we get for multiple injections of about 35 pg each of DA and NE," pushing the high sensitivity operation close to limits. The first several runs produced highly disappointing results—the electrode was very insensitive and barely produced measurable peaks. A fresh electrode surface was prepared, allowed to stabilize overnight, and the following morning six repetitive injections were made. The reproducibility of these high-sensitivity results (obtained with a stock Bioanalytical Systems detector block and potentiostat-amplifier) were:

NE (pg) 35.0 ± 1.3 (S.D.) or ± 0.48 (S.E.M.)
DA (pg) 35.0 ± 0.8 (S.D.) or ± 0.30 (S.E.M.)
$N = 6$

These data, of course, verify only the reproducibility of the chromatography and detection system—not the manipulations of the entire assay. One could not expect such precision on real tissue samples containing only 35 pg of DA or NE. However, it is safe to say there are no difficulties in analysing samples whose injectable content of DA or NE exceeds 35–50 pg. Below 20 pg (*c.* 0.1 pmol) there are problems such as possible syringe carryover, etc. One is obviously best advised to adjust, whenever possible, sampling conditions so that *c.* 100–500 pg or more of catecholamine are injected on the column.

One cannot determine the *accuracy* of the method for small tissue samples, because it is obvious from the data above that the precision of the assay ordinarily exceeds that of neuroanatomical considerations (punching errors, sub-nuclei delineation, etc.). Hence, the actual CNS tissue sampling determines the accuracy. For whole brains, or parts where sampling is not a limiting factor, one can state that the electrochemical technique agrees with all previous assay procedures.

ACKNOWLEDGEMENTS

I am especially indebted to R. Keller, I. Mefford and A. Oke for valuable discussions and new data during the course of preparing this review.

REFERENCES

Adams, R. N. (1969). *J. Pharm. Sci.* **58,** 1171–1184.
Anton, A. H. and Sayre, D. F. (1962). *J. Pharmacol. Exp. Ther.* **138,** 360–374.
Keller, R., Oke, A. Mefford, I. and Adams, R. N. (1976). *Life Sci.* **19,** 995–1004.
Kissinger, P. T., Riggin, R. M., Alcorn, R. L. and Rau, L-D. (1975). *Biochem. Med.* **13,** 299–306.
Mefford, I., Oke, A., Adams, R. N. and Jonsson, G. (1977). *Neurosci. Lett.* **5,** 141–145.
Refshauge, C., Kissinger, P. T., Dreiling, R., Blank, L., Freeman, R. and Adams, R. N. (1974). *Life Sci.* **14,** 311–322.
Riggin, R. M., Alcorn, R. L. and Kissinger, P. T. (1976). *Clin. Chem.* **22,** 782–784.
Sasa, S. and Blank, C. L. (1977). *Anal. Chem.* **49,** 354–359.
Weil-Malherbe H. (1971). *In* "Analysis of Biogenic Amines and their Related Enzymes" (Ed. D. Glick) 119–152. Wiley-Interscience, New York.
Weintraub, S. T., Stavinoha, W. B., Pike, R. L., Morgan, W. W., Modak, A. T., Koslow, S. T. and Blank, L. (1975). *Life Sci.* **17,** 1423–1428.
Wightman, R. M., Plotsky, P. M., Strope, E., Delcore, R. and Adams, R. N. (1977) *Brain Res.* **131,** 345–349.

Note Added in Proof

During the writing and publication of this chapter several papers of importance to dopamine in brain function have appeared and are listed below.

Gibbs, D. M. and Neill, J. D. (1978). *Endocrinology* **102,** 1895–1900.
Hallman, H., Farnebo, L. O., Hamberger, B. and Jonsson, G. (1978). *Life Sci.* **23,** 1049–1052.
Oke, A., Keller, R. and Adams, R. N. (1978). *Brain Res.* **148,** 245–250.
Plotsky, P. M., Gibbs, D. M. and Neill, J. D. (1978). *Endocrinology* **102,** 1887–1894.

Enzymes Involved in the Biosynthesis and Metabolism of Dopamine

CHAPTER 9

Tyrosine Hydroxylase

Robert H. Roth

Yale University School of Medicine, New Haven, U.S.A.

DISCOVERY

In 1939, after demonstration of the decarboxylation of dihydroxyphenylalanine to dopamine (Holtz, 1939), the pathway for biosynthesis of norepinephrine from tyrosine was proposed. Although by the mid-1950s evidence for this pathway was quite well established by radioisotopic techniques (Udenfriend *et al.*, 1953, 1965) it was not until 1964 that Udenfriend's group (Nagatsu *et al.*, 1964) succeeded in the isolation and characterization of the first enzyme in this biosynthetic pathway, tyrosine hydroxylase (L-tyrosine, tetrahydropteridine, oxygen oxidoreductase (3-hydroxylating), EC 1.14.16.2).

Tyrosine hydroxylase is an enzyme which requires atmospheric oxygen and a reduced pterin cofactor for activity (Nagatsu *et al.*, 1964; Brenneman and Kaufman, 1964). It is classified as a mixed function oxygenase because during the oxidation process catalysed by this enzyme, the two atoms of oxygen in the oxygen molecule utilized in this reaction have different fates: one atom is incorporated in the substrate tyrosine while the other is reduced to water.

Tetrahydrobiopterin (6,7-dihydroxypropyltetrahydropterin) is considered to be the natural cofactor for tyrosine hydroxylation in both bovine adrenal and rat brain (Lloyd and Weiner, 1971; Bullard *et al.*, 1978; Gal and Sherman, 1977). However, in many instances, because of the limited availability of this pterin most of the assays for tyrosine hydroxylase are carried out with 6,7-dimethyltetrahydropterin or 6-methyltetrahydropterin which are commercially available.

ASSAY PROCEDURES

There are three major radiometric methods for determination of tyrosine hydroxylase activity *in vitro*. One method involves the measurement of tritiated water, one the measurement of labelled DOPA and the other the measurement of carboxyl-labelled CO_2. The first useful method was described by Nagatsu *et al.* (1964) and is based on the formation of labelled DOPA from tyrosine. The labelled DOPA formed is usually isolated by alumina chromatography and the amount of radioactive DOPA formed determined by liquid scintillation spectrometry. A second method involves the use of $3',5'$-^{3}H-tyrosine as the substrate and measurement of the formation of ^3H_2O as the product. The formation of H_2O is proportional to the rate of hydroxylation of tyrosine to DOPA. The third assay is a coupled enzymatic assay that takes advantage of the high degree of specificity that enzymes exhibit for their substrates. In this assay the DOPA formed from L-(1-^{14}C) tyrosine is decarboxylated by the addition of hog kidney decarboxylase, and the radioactive CO_2 formed is trapped and measured by liquid scintillation spectrometry (Waymire *et al.*, 1971). This enzyme assay retains the necessary specificity when employed with low concentrations of tyrosine since DOPA decarboxylase has a hundred-fold higher affinity for the product of the hydroxylase reaction, DOPA, than for tyrosine. Although most tyrosine hydroxylase assays employ one of the above methods almost every investigator has made some modifications in these methods to suit his or her specific needs. Since this complex enzyme system is influenced by numerous factors (see p. 107), it is not surprising that absolute values for tyrosine hydroxylase activity measures made on crude tissue preparations vary considerably from laboratory to laboratory.

Several assays are also available for the in-vivo estimation of tyrosine hydroxylase. One method involves following the short-term accumulation of DOPA after inhibition of DOPA decarboxylase (Kehr *et al.*, 1972a; Walters and Roth, 1974). Another method involves measurement of tritiated water after direct injection or superfusion of discrete brain regions with 3,5-^{3}H-L-tyrosine (Agid *et al.*, 1974).

TISSUE DISTRIBUTION

Tyrosine hydroxylase is found in all peripheral sympathetically innervated tissue, brain and adrenal glands. The brain enzyme is concentrated in areas

TABLE I Regional distribution of tyrosine hydroxylase in the rat brain[a]

	Tyrosine Hydroxylase Activity (nmol DOPA mg protein^{-1} h^{-1})
Substantia nigra	17.5
Caudate nucleus	12.0
Olfactory tubercle	11.1
Nucleus accumbens	7.7
Locus coeruleus	4.6
Amygdala	4.3
Hypothalamus	3.1
Pons medulla	0.7
Frontal cortex	0.3
Cerebellum	0.2
Hippocampus	0.2

[a] Data taken from Bacopoulos and Bhatnagar, 1977.

which are known to contain high levels of the catecholamines. In general, the highest tyrosine hydroxylase activity is found in areas rich in dopaminergic neurones (Bacopoulos and Bhatnagar, 1977). In brain regions containing predominantly dopamine terminals, tyrosine hydroxylase activity is positively correlated with dopamine concentration and turnover rates. Likewise there also appears to be a high degree of correlation between tyrosine hydroxylase activity, norepinephrine* concentration and norepinephrine turnover in regions of brain containing predominantly noradrenergic terminals. Regions containing predominantly dopamine or norepinephrine cell bodies are characterized by higher activities of tyrosine hydroxylase than regions containing mostly nerve terminals rich in the corresponding catecholamines. Table I illustrates the regional distribution of tyrosine hydroxylase in rat brain.

SUBCELLULAR DISTRIBUTION

Tyrosine hydroxylase in catecholaminergic nerve terminals does not appear to be associated with any particular type of subcellular organelle and there is now considerable evidence to indicate that the enzyme in most tissues is soluble (Musacchio, 1968; Stjärne and Lishajko, 1967; Laduron and Belpaire, 1968). The original controversy concerning whether tyrosine hydroxylase is soluble or particulate is probably best explained by the observation that the enzyme readily aggregates and adsorbs to subcellular particles and becomes sedimentable. In brain preparations the majority of

*In this Volume, Chapters by American authors refer to epinephrine and norepinephrine whereas Chapters by non-American authors refer to adrenaline and noradrenaline.—Editor's note.

the enzyme (about 60%) is found in the synaptosomal fraction, presumably present in the cytoplasm entrapped within the synaptosomes. When the synaptosomal fraction is lysed, the bulk of the tyrosine hydroxylase is found in the soluble fraction (Coyle, 1972; Kuczenski and Mandell, 1972). The study by Kuczenski and Mandell (1972) pointed to a difference in the subcellular distribution of tyrosine hydroxylase found in the striatum and the midbrain; the midbrain enzyme was largely soluble whereas in the striatum the enzyme was localized to the synaptosomal fraction. This biochemical observation is probably related to the fact that the striatum contains primarily catecholaminergic nerve terminals which result in a preparation with a larger proportion of synaptosomes, while the midbrain is enriched in cell bodies and upon fractionation leads to a smaller proportion of synaptosomes. Kuczenski (1973) has further demonstrated that the two forms of striatal tyrosine hydroxylase, soluble and particulate, have different kinetic properties. The particulate enzyme has a higher affinity for substrate tyrosine ($K_m = 6\text{–}10\,\mu\text{M}$) and for 6,7-dimethyltetrahydropterin (0.15 mM) than the soluble enzyme (K_m tyrosine = 56 μM; K_m 6,7-dimethyltetrahydropterin = 0.75 mM).

PROPERTIES

The catalytic behaviour of tyrosine hydroxylase isolated from tissue appears to differ considerably depending upon the conditions of the assay, the source and manner of preparation of the enzyme and the concentration and type of the pterin cofactor employed in the assay. Completely valid observations on the kinetic properties of an enzyme can only be obtained from homogenous enzyme preparations. However, isolation of purified tyrosine hydroxylase has been difficult due to its moderate instability and its tendency to aggregate with other proteins. Because of the difficulties in obtaining highly purified tyrosine hydroxylase most kinetic studies have been carried out on crude preparations of this enzyme. Despite this limitation many of these studies have provided useful and seemingly relevant information.

It was suggested many years ago that this enzyme might be an iron-containing protein because in such crude enzyme preparation ferrous ion was found to increase tyrosine hydroxylase activity and a wide variety of iron-chelating agents inhibited activity (Nagatsu *et al.*, 1964; Petrack *et al.*, 1968; Taylor *et al.*, 1969). Recent studies have shown that tyrosine hydroxylase purified from bovine adrenal medulla is an iron-containing enzyme and that iron is necessary for catalysis (Hoeldtke and Kaufman, 1977). Although little is known about the role of the protein-bound iron in the tyrosine hydroxylase reaction it is likely that it will prove to play a role similar to that played by iron in the phenylalanine reaction.

The stoichiometry of the tyrosine hydroxylase catalysed hydroxylase reaction is summarized in the following reaction sequence:

L-tyrosine + tetrahydrobiopterin + O_2 = L-DOPA + quinoid form of dihydrobiopterin H_2O.

Substrate Specificity

Although tyrosine hydroxylase shows a fairly high degree of substrate specificity and absolute stereospecificity for the L isomer of tyrosine, the enzyme prepared from bovine adrenal will also catalyse the hydroxylation of L-phenylalanine to tyrosine. In fact, Shiman and coworkers (1971) demonstrated that adrenal tyrosine hydroxylase will hydroxylate phenylalanine and tyrosine at comparable rates when tetrahydrobiopterin rather than 6,7-dimethyltetrahydropterin is employed as the cofactor. Also, striatal synaptosomes will catalyse the synthesis of catecholamines from both phenylalanine and tyrosine (Karobath and Baldessarini, 1972). Thus, it appears that, *in vivo*, phenylalanine can serve as an initial precursor for catecholamine formation in brain.

Cofactors

The hydroxylation of tyrosine to DOPA is considered to be the rate-limiting step in catecholamine biosynthesis. Since this reaction is quite complex, it is still unclear which of its components is limiting to the tyrosine hydroxylase reaction *in vivo*. As indicated above, this enzyme requires molecular oxygen and a reduced pteridine cofactor for activity. Dihydropteridine reductase (EC 1.6.99.7) is the enzyme that catalyses the reduction of the quinoid dihydropterin which is oxidized during the hydroxylation of tyrosine to DOPA. Since reduced pteridines are essential for tyrosine hydroxylation, dihydropteridine reductase should also be considered part of the hydroxylation system since this enzyme is responsible for maintenance of the cofactor in the reduced state and allowing it to function catalytically. It seems likely that any alteration in the activity of dihydropteridine reductase could effectively influence the activity of tyrosine hydroxylase. Thus, in addition to providing a potential regulatory site in the tyrosine hydroxylation step, this enzymatic step might be a potential site for drug intervention in catecholamine biosynthesis.

The apparent affinity of tyrosine hydroxylase for its substrate tyrosine is influenced by the type of pterin cofactor used in the assay system. The affinity of the enzyme for tyrosine in the presence of tetrahydrobiopterin is greater than when one of the synthetic cofactors, 6-methyltetrahydropterin or 6,7-dimethyltetrahydropterin, are used (Kaufman, 1975). Also, high concentrations of tyrosine inhibit the enzyme when tetrahydrobiopterin is used as the cofactor. Although not a great deal is known about the interaction of oxygen with tyrosine hydroxylase, it has been shown that the apparent K_m of the enzyme for oxygen is also dependent upon the type of pterin cofactor used in the assay. The K_m of the brain enzyme for oxygen is about 0.01 mM in the presence of tetrahydrobiopterin (Fisher and Kaufman, 1972). When present at very high concentrations, oxygen inhibits tyrosine hydroxylase.

Different Molecular Forms

Evidence from sucrose density gradient centrifugation studies suggests that striatal tyrosine hydroxylase can exist in four catalytically active forms of different molecular weights ranging from approximately 49 000 to 186 000. The molecular weights of the four forms are approximately multiples of the smallest form. The relative amounts of each of these forms depends upon the pH and ionic strength of the medium. Experiments on bovine and guinea-pig adrenal and rat striatal tyrosine hydroxylase have demonstrated that treatment with trypsin produces a dramatic decrease in the sedimentation coefficient of the native form of the enzyme. The molecular weight of the trypsin-treated enzyme was estimated to be about 34 000 while the native adrenal and striatal enzymes appear to have molecular weights in excess of 100 000 (Musacchio *et al.*, 1973). Most recently, Joh and Reis (1975) have suggested that different forms of tyrosine hydroxylase exist in different brain areas. These authors reported that the enzyme in noradrenergic tissue has an approximate molecular weight of 130 000 while that in dopaminergic tissue has a molecular weight of 65 000.

Effects of Small Molecules

There is a growing body of evidence which indicates that tyrosine hydroxylase can be allosterically regulated. Membrane binding, polyanions, heparin, lysolecithin, phosphatidylserine, cAMP, calcium or limited proteolytic treatment increases the activity of the enzyme. The increase in enzyme activity is associated in many instances with a decreased K_m for cofactor and an increased K_i for catecholamines and under some experimental conditions a decrease in the K_m for tyrosine (Table II). Despite the fact that numerous factors have been shown to alter the kinetic properties of tyrosine hydroxylase *in vitro* it is still unclear whether any of these factors are

TABLE II In-vitro treatments or conditions modifying the kinetic properties of tyrosine hydroxylase

	K_m for tyrosine	K_m for cofactor	K_i for catecholamines	Reference sources
Tryptic proteolysis	↓	↓		Kuczenski (1975)
Binding to membranes	↓	↓	↓	Kuczenski (1973)
Heparin	No change	↓	↓	Kuczenski (1975); Kuczenski and Mandel (1972)
Lysolecithin	No change	↓	↑	Raese *et al.* (1976)
Phosphatidylserine	No change	↓	↑	Lloyd and Kaufman (1974), Raese *et al.* (1976)
Polyanions	No change	↓		Katz *et al.* (1976)
cAMP (+ATP and Mg^{2+})	↓ or no change	↓	↑	Ames *et al.* (1978), Rubio (1977) Goldstein *et al.* (1976), Harris *et al.* (1974, 1975), Lloyd and Kaufman (1975), Lovenberg *et al.* (1975), Morgenroth *et al.* (1975b)
Calcium	↓	↓	↑	Morgenroth *et al.* (1975a), Osborne and Neuhoff (1976)
EGTA (dopaminergic tissue, reversed by Ca^{2+})	↓ or no change	↓ or no change	↑ or no change	Morgenroth *et al.* (1976a), Roth *et al.* (1974), Lerner *et al.* (1977)
Magnesium (dopaminergic tissue, reversed by Ca^{2+})		↓		Bustos and Roth (1979)[a]
Gel filtration	No change	↓	No change	Ames *et al.* (1978)

[a] Slices were incubated with 12 mM Mg^{2+} prior to homogenization and kinetic analysis.

involved in the in-vivo regulation of enzyme activity. The observations made with cAMP and calcium seem most relevant to the in-vivo activation of tyrosine hydroxylase observed during periods of enhanced nerve activity. It is known that neuronal depolarization results in an increase in the levels of endogenous cAMP (Kakiuchi *et al.*, 1969) and a mobilization of calcium (Baker, 1972) but it is still uncertain whether either or both events play a role in the stimulation-induced increase in tyrosine hydroxylase activity. However, it should be pointed out that although under some conditions calcium addition to brain homogenates can be shown to increase tyrosine hydroxylase activity (Gutman and Segal, 1973; Morgenroth *et al.*, 1975; Osborne and Neuhoff, 1976; Roth *et al.*, 1975), calcium has little or no effect on rat brain striatal homogenates (Roth *et al.*, 1974; Lerner *et al.*, 1977). Nevertheless, if striatal slices are incubated in a sodium-free medium, a condition thought to increase intracellular calcium levels (Baker, 1975), a marked increase in tyrosine hydroxylase activity and a kinetic activation of the enzyme is observed (Simon and Roth, unpublished data).

In conclusion, although the mechanisms of activation produced by these various agents and conditions listed in Table II have been extensively studied in recent years it is still unclear whether any of these activation processes are identical to the in-vivo activation of tyrosine hydroxylase which occurs in catecholaminergic neurones during periods of enhanced neuronal activity.

Inhibitors

Prompted by the therapeutic potential of inhibiting catecholamine biosynthesis and the fact that tyrosine hydroxylase is the rate-limiting step, there has been an intensive search for effective tyrosine hydroxylase inhibitors. At present tyrosine hydroxylase inhibitors can be broken down into four major classes: catechols, amino acid analogues, divalent metal chelating agents and the naphthoquinones. A partial list of tyrosine hydroxylase inhibitors in these classes is presented in Table III and the approximate molar concentration of these agents necessary to produce 50% inhibition is provided.

Catechols

Catecholamines and other catechol compounds act as inhibitors of tyrosine hydroxylase. Despite their obvious similarity to tyrosine, the observed inhibition is competitive with reduced pteridine cofactors and not with substrate. The most effective inhibitor of tyrosine hydroxylase in this series is 3,4-dihydroxyphenyl-propylacetamide. This compound is effective

TABLE III Tyrosine hydroxylase inhibitors

	IC 50%[a]	Reference
Catechols		
Norepinephrine	10^{-4}	Petrack *et al.* (1968), Musacchio *et al.* (1973)
Epinephrine	10^{-4}	Petrack *et al.* (1968), Musacchio *et al.* (1973)
3,4-Dihydroxyphenyl-*n*-propylacetamide (Hassle 22/54)	2×10^{-5}	Udenfriend *et al.* (1965), Petrack *et al.* (1968)
n-Propylgallate	2×10^{-6}	Levitt *et al.* (1967)
Amino acid analogues		
L-α-Methyl-*p*-tyrosine	2×10^{-5}	Nagatsu *et al.* (1964), Udenfriend *et al.* (1965)
L-3-Iodotyrosine	5×10^{-7}	Udenfriend *et al.* (1965)
D,L-3-Iodo-α-methyl-*p*-tyrosine	3×10^{-7}	Udenfriend *et al.* (1965)
D,L-α-Methylphenylalanine	6×10^{-5}	Weissman *et al.* (1966)
L-α-Methyl-5-hydroxytryptophan	7×10^{-5}	Zhelyaskov *et al.* (1968), McGeer *et al.* (1967)
D,L-5-Iodotryptophan	9×10^{-7}	Zhelyaskov *et al.* (1968), McGeer *et al.* (1967)
D,L-5-Bromotryptophan	3×10^{-6}	McGeer *et al.* (1967)
Divalent metal chelating agents		
O-Phenanthroline	10^{-5}	Taylor *et al.* (1968, 1969)
α-α-Dipyridyl	10^{-5}	Nagatsu *et al.* (1964), Taylor *et al.* (1968, 1969)
Diethyldithiocarbamate	10^{-4}	Taylor *et al.* (1968, 1969)
Naphthaquinones		
Aquayamycin	4×10^{-7}	Ayukawa *et al.* (1968)
Chrothiomycin	10^{-8}	Ayukawa *et al.* (1969)
Deoxyfrenolicin	8×10^{-5}	Taylor *et al.* (1970)
Spinochrome A	4×10^{-6}	Mizutani *et al.* (1972)
Equinochrome A	2×10^{-4}	Mizutani *et al.* (1972)

[a] The approximate molar concentrations that are required to produce 50% inhibition of tyrosine hydroxylase. Data taken in part from Mussachio, 1975.

both *in vivo* and *in vitro*. However, because this compound also acts as a competitive inhibitor with pteridine cofactor it does not possess a high degree of specificity and inhibits tryptophan and phenylalanine hydroxylase as well (Carlsson *et al.*, 1963; Jequier *et al.*, 1969).

In the original studies catecholamines were found to be relatively weak inhibitors of tyrosine hydroxylase when tested in the presence of high concentrations of artificial cofactor (Nagatsu *et al.*, 1964). However, when experiments were run in the presence of low cofactor concentrations or at higher pH, catecholamines were found to be very effective inhibitors of tyrosine hydroxylase (Musacchio *et al.*, 1971; Simon *et al.*, 1978). The former is expected since the inhibitory effect of catechols was shown to be competitive with reduced pteridine cofactor. Since tissue levels of endogenous reduced pteridine cofactor are quite low (10^{-5} M) it is likely that endogenous catecholamine may effectively control tyrosine hydroxylase activity. The observation that the endogenous end-product of a biosynthetic pathway may effectively inhibit the rate limiting enzyme in the pathway provided an attractive mechanism for the physiological regulation of transmitter synthesis by catecholamine-containing neurones. Over the past decade much support has been given to the concept that feedback control is an important mechanism by which catecholamine-containing neurones regulate synthesis of transmitter (Weiner, 1970). However, recent studies have demonstrated that other mechanisms are also operative (Roth *et al.*, 1978; see p. 112). Since, as indicated earlier, tyrosine hydroxylase is largely a cytoplasmic enzyme, in dopaminergic neurones only the free cytoplasmic dopamine should be effective in exerting an inhibitory effect on tyrosine hydroxylase. The metabolites of dopamine that retain the catechol nucleus must also be considered as possible physiological inhibitors of tyrosine hydroxylase. Metabolites such as 3,4-dihydroxyphenylacetic acid (DOPAC) which are not taken up and retained by vesicles are therefore present primarily in the cytoplasm and thus should have easy access to tyrosine hydroxylase. In the rat, where DOPAC is the major metabolite of dopamine (Wilk *et al.*, 1975), this metabolite might play a significant role in the regulation of tyrosine hydroxylase. However, in the primate, HVA rather than the DOPAC is the major metabolite of dopamine (Bacopoulos and Roth, unpublished) and it is unlikely that the concentrations of DOPAC present in brain would be high enough to influence tyrosine hydroxylase activity.

Amino acid analogues

Among this group of inhibitors, α-methyltyrosine is the drug which has been studied most intensively *in vivo*. In both man and experimental animals this agent is effective in reducing catecholamine formation. However, the usefulness of this drug is limited to some extent by its low potency and insolubility. The problem of insolubility has been in part overcome by synthesis of the methyl and ethyl esters which are highly water soluble. However, use of the esters does not alleviate the problem of renal toxicity caused by high doses of the drug. The renal toxicity is most likely related to the low solubility of the parent compound, α-methyltyrosine, since the esters are rapidly hydrolysed in the body.

Most amino acid analogues such as the L-isomers of a number of tyrosine and phenylalanine analogues are competitive inhibitors of tyrosine hydroxylase. 5-Halotryptophans are also competitive inhibitors with tyrosine but 5-hydroxytryptophan and its α-methyl analogue produce an inhibition of tyrosine hydroxylase that is non-competitive with both tyrosine and pteridine cofactor.

Divalent metal chelating agents

The divalent metal chelating agents inhibit tyrosine hydroxylase activity *in vitro*. The trivalent metal chelating agents are inactive. The in-vitro inhibition observed with the divalent chelators is non-competitive with tyrosine and reduced pteridine cofactors and is reversed by Fe^{2+}. Compounds like α,α-bipyridyl when administered to rats causes a reduction in the catecholamine content of heart, adrenal and brain (Nagatsu *et al.*, 1964). However, the observation that the content of norepinephrine was reduced at a similar rate in all tissues, despite the fact that normally the turnover of norepinephrine in brain is much more rapid, suggests that the depletion of norepinephrine produced by this agent is not exclusively the result of tyrosine hydroxylase inhibition (Bapna *et al.*, 1970).

Napthaquinones

Several of the compounds in this class are quite potent inhibitors of tyrosine hydroxylase. Aquayamycin and chrothiomycin cause about 50% inhibition of tyrosine hydroxylase at concentrations of 4×10^{-7} and 1×10^{-8} M, respectively. Some of these compounds like the antibiotic, deoxyprenolicin, are active *in vivo* (Taylor *et al.*, 1970).

REGULATION

A large number of observations have indicated that the activity of tyrosine hydroxylase is responsive to alterations in the level of physiological activity within catecholamine-containing neurones.

There are essentially two types of mechanisms whereby tyrosine hydroxylase activity can be regulated *in vivo*: (1) alterations in the activity of pre-existing enzyme molecules and (2) alterations in the amount of enzyme protein by synthesis of new enzyme molecules or by slowing of the degradation of enzyme protein. These two mechanisms have different time courses and will be dealt with separately below. The activity of existing tyrosine hydroxylase molecules can be altered quickly via removal of end-product inhibition or by adjustment of the kinetic parameters of the enzyme. A net increase in the amount of enzyme via new protein synthesis or by slowed degradation of enzyme protein takes hours or days to manifest itself.

Short-term Regulation

Effects of increased impulse flow

The majority of studies on central dopaminergic neurones have focused on the nigro-neostriatal dopaminergic neurones. Recent studies have demonstrated that these dopamine neurones behave in a fashion very similar to both peripheral and central noradrenergic neurones in their response to an increase in impulse flow. Stimulation of the nigro-neostriatal dopamine neurones produces a significant increase in the formation of labelled dopamine from exogenous tyrosine which appears to be dependent upon impulse flow (Murrin and Roth, 1976). Studies in which the short-term accumulation of endogenous DOPA following administration of a DOPA decarboxylase inhibitor was used as an index of tyrosine hydroxylase activity demonstrated that stimulation for varying periods of time at frequencies of 5–30 Hz resulted in a stimulus-dependent increase in the accumulation of DOPA in the striatum on the stimulated side. The increase in apparent tyrosine hydroxylase activity induced by electrical stimulation persisted following the termination of the stimulation period. Since there was no observed depletion of striatal dopamine, these in-vivo experiments suggested that some process other than dopamine depletion and therefore removal of feedback inhibition must be responsible for activating tyrosine hydroxylase. Later studies demonstrated that if tyrosine hydroxylase is isolated from the neostriatum on the stimulated side and its in-vitro activity compared to the activity of tyrosine hydroxylase isolated from the unstimulated contralateral side, nearly a 50% increase in activity is observed (Murrin *et al*., 1976). Ten minutes of continuous stimulation at 15 Hz is required to obtain a maximal activation of tyrosine hydroxylase. This activation persists for about 15 min following the termination of the stimulation period. The mechanism responsible for this activation of tyrosine hydroxylase, as in noradrenergic neurones, appears to be related to alterations in the kinetic properties of the enzyme. The tyrosine hydroxylase isolated from the neostriatum on the stimulated side appears to have an increased affinity for the substrate tyrosine and for pterin cofactor and a decreased affinity for the natural end-product inhibitor, dopamine (Murrin *et al.*, 1976).

Several laboratories have shown that cyclic AMP also causes an increase in tyrosine hydroxylase activity when added to high-speed supernates or homogenates prepared from several brain regions known to be rich in dopaminergic neurones (Goldstein *et al.*, 1975, 1976; Harris *et al.*, 1974, 1975; Lovenberg *et al.*, 1975). This cAMP-induced activation has an absolute requirement for ATP and Mg^{2+} and appears to involve a cAMP-dependent protein kinase (Goldstein *et al.*, 1976; Lovenberg *et al.*, 1975; Morgenroth *et al.*, 1975b). The observed activation appears to be mediated by a change in the affinity of the enzyme for pterin cofactor and end-product inhibitor dopamine (Harris *et al.*, 1974, 1975; Goldstein *et al.*, 1976). Conflicting reports have appeared regarding whether the K_m for tyrosine is also reduced by the cAMP activation process (Goldstein *et al.*, 1976; Harris *et al.*, 1974; Lovenberg *et al.*, 1975; Roth *et al.*, 1975a, c).

The activation of striatal tyrosine hydroxylase produced by electrical stimulation of the nigro-neostriatal pathway is very similar kinetically to the activation produced by addition of cAMP to striatal homogenates. This observation is consistent with the possibility that both treatments may be activating tyrosine hydroxylase by a similar or identical mechanism (Murrin *et al.*, 1976).

In-vitro studies on tissue slices also provided similar results. Depolarization of striatal slices by incubation in the presence of high concentrations of potassium ions results in an increase in the tyrosine hydroxylase activity measured in the homogenate prepared from these slices (Bustos *et al.*, 1976). This increase in activity appears to be a result of changes in the kinetic properties of the isolated enzyme. These effects on tyrosine hydroxylase produced by depolarization are similar to the effects produced by preincubation of the slices with dibutyryl cAMP prior to homogenization or by addition of cAMP (together with ATP, Mg^{2+} and protein kinase) to the isolated enzyme preparation. However, in studies carried out in synaptosomes or striatal slices the effects produced by cAMP and potassium depolarization are additive (Goldstein *et al.*, 1976; Simon and Roth, unpublished). These results could be interpreted as indicating that different rather than similar mechanisms of activation are involved. However, they are far from conclusive and further experimentation is necessary before any definitive conclusions can be made.

Effects of inhibition of impulse flow

Numerous studies have now demonstrated that conditions which cause a cessation of impulse flow in the nigro-neostriatal pathway result in a rapid accumulation of dopamine in the neostriatum. This increase can be observed in the rat neostriatum after lesion of the substantia nigra or median forebrain bundle (Faull and Laverty, 1969; Walters *et al.*, 1973), cerebral hemisection (Andén *et al.*, 1971; Kehr *et al.*, 1972), direct injection of a local anesthetic into the nigro-striatal pathway (Andén *et al.*, 1973), and after administration of gamma-butyrolactone (GBL) (Agid *et al.*, 1974; Stock *et al.*, 1973; Walters *et al.*, 1973). Gamma-butyrolactone is a precursor of gamma-hydroxybutyrate (GHB) which has been shown to inhibit impulse flow in the nigro-neostriatal neurones (Roth *et al.*, 1973). Similar changes are also observed in the nucleus accumbens and olfactory tubercles after impulse flow is blocked in the dopamine neurones originating in the ventral tegmental area. Thus, the dopamine system when compared to other monoamine systems appears to respond in a rather unique and paradoxical fashion to an inhibition of impulse flow by rapidly increasing the transmitter content of its terminals. An analogous increase in the steady state levels of norepinephrine or serotonin or an activation of the initial biosynthetic enzyme is not observed when impulse flow in these respective neuronal systems is interrupted either pharmacologically or by placement of a lesion in the locus coeruleus (Korf *et al.*, 1973) or median raphe (Herr and Roth, 1976), respectively. Interestingly, a rapid and marked increase in the steady state levels of acetylcholine in the hippocampus is observed when impulse flow is interrupted in the septohippocampal cholinergic pathway (Sethy *et al.*, 1973).

The rapid increase in the accumulation of dopamine seen after pharmacological or mechanical interruption of impulse flow in the nigro-neostriatal pathway appears to be explainable in part by a decrease in release coupled with a marked increase in dopamine biosynthesis (Agid *et al.*, 1974; Roth *et al.*, 1973; Walters *et al.*, 1973). The increased conversion of labelled tyrosine to dopamine and the increased accumulation of DOPA after inhibition of DOPA decarboxylase are both suggestive that treatments which abolish impulse flow in dopamine neurones result in an increase in tyrosine hydroxylase activity. The observation that tyrosine hydroxylase activity is apparently increased during a time in which the endogenous levels of dopamine in the nerve terminals are rapidly increasing seems, however, quite paradoxical. Tyrosine hydroxylase is usually responsive to end-product inhibition and thus small increases in endogenous catecholamine within the nerve terminal usually result in a reduction in tyrosine hydroxylase activity. These unexpected results suggested that inhibition of impulse flow in dopaminergic neurones might somehow decrease the ability of endogenous dopamine to exert an inhibitory effect on the activity of tyrosine hydroxylase.

When impulse flow in dopamine neurones was blocked by treatment with GBL or by placement of an electrothermic lesion in the nigro-neostriatal pathway the activity of the tyrosine hydroxylase measured in the presence of subsaturating concentrations of tyrosine and pterin cofactor was markedly increased (Roth *et al.*, 1974, 1975a; Morgenroth *et al.*, 1976). In addition, the kinetic properties of the isolated enzyme were altered. The striatal tyrosine hydroxylase obtained from GBL treated rats or rats with unilateral electrothermic lesions in the nigro-neostriatal pathway had a significant increase in its affinity for tyrosine and pterin cofactor and a dramatic decrease in its affinity for the end-product inhibitor, dopamine. No change was observed in the V_{max} for tyrosine or pterin cofactor. These results have not been replicated by experiments performed in Dr Costa's laboratory (Zivkovic *et al.*, 1975) perhaps due to somewhat different conditions used in the assay of tyrosine hydroxylase activity.

Thus, it appears as if a blockade of impulse flow in dopaminergic neurones leads to a change in the physical properties of pre-existing enzyme molecules. The fact that the tyrosine hydroxylase obtained from the neostriatum of rats in which impulse flow is blocked has a marked decrease in its sensitivity to inhibition by dopamine may well explain why tyrosine hydroxylase activity appears to be enhanced even under conditions in which the endogenous levels of dopamine are rapidly increasing within the nerve terminals.

Role of presynaptic receptors

Recent studies conducted on both peripheral and central noradrenergic systems have given rise to the concept that there may exist on the presynaptic side of the synapse a receptor sensitive to the amount of transmitter in the synaptic cleft and capable of somehow regulating the release of transmitter from the presynaptic terminals (Langer, 1974, 1977).

Experiments by many laboratories have suggested that such a presynaptic receptor might exist on central dopamine nerve terminals and, in addition to regulating transmitter release, might also be involved in the modulation of dopamine synthesis (for review see Nowycky and Roth, 1978). Receptor activation appears to exert a damping effect on synthesis and release, while receptor blockade appears to facilitate both processes. If such presynaptic receptors exist and do indeed exert some modulatory influence on tyrosine hydroxylase it might be expected that administration of a dopamine agonist would appreciably attentuate the post-stimulation increase in tyrosine hydroxylase activity observed following periods of increased impulse flow. In contrast, administration of dopamine receptor blockers would be expected to enhance the stimulus-induced activation of tyrosine hydroxylase by preventing the braking effect exerted by interaction of endogenous synaptic dopamine with the presynaptic receptor.

Experiments with dopamine agonists and antagonists support this theory. Administration of dopamine receptor agonists such as trivastal 1-(2″-pyrimidyl)-4-piperonyl-piperazine (ET495) caused a significant reduction in the stimulus-induced activation of tyrosine hydroxylase. Trivastal antagonized both the increase in tyrosine hydroxylase activity and the kinetic alterations in the enzyme induced by stimulation of the nigro-neostriatal pathway (Roth *et al.*, 1975a, b).

Administration of antipsychotic drugs such as haloperidol, which appear to have potent presynaptic dopamine receptor-blocking capabilities (Walters and Roth, 1975), causes a further increase in the stimulation-induced activation of tyrosine hydroxylase (Roth *et al.*, 1975a, b). This is reflected both by an increase in tyrosine hydroxylase activity and by an increase in the magnitude of the alteration in the kinetic properties of tyrosine hydroxylase when compared to stimulation in drug-free rats. In these experiments stimulation of the nigro-neostriatal pathway was carried out for a time period and at a frequency which produced a maximal activation of tyrosine hydroxylase. Thus, no further activation of tyrosine hydroxylase should be produced by the increase in dopaminergic impulse flow normally caused as a result of the feedback loop mediated postsynaptic blocking effects of haloperidol (see Bunney, this Volume, p. 417). However, haloperidol, by blocking the interaction of synaptic dopamine with this presynaptic receptor, may effectively remove any modulatory influence on tyrosine hydroxylase exerted by presynaptic receptor activation, thus leading to the observed enhanced activation of tyrosine hydroxylase.

Presynaptic receptor modulation of the activation of tyrosine hydroxylase produced by inhibition of impulse flow

Studies from several laboratories have now shown that stimulation of presynaptic dopamine receptors also effectively counteracts the changes in dopamine synthesis induced by a blockade of impulse flow in the nigro-neostriatal pathway (Kehr *et al.*, 1972; Walters and Roth, 1974). Thus, administration of amphetamine, apomorphine or trivastal prevents both the increase in endogenous levels of neostriatal dopamine and the increase in DOPA accumulation following decarboxylase inhibition, which normally occurs when impulse flow is blocked pharmacologically or by placement of a lesion in the nigro-neostriatal pathway. Furthermore, the in-vivo kinetic activation of tyrosine hydroxylase produced by a block of impulse flow can be completely reversed by concurrent administration of a dopamine agonist (Morgenroth *et al.*, 1976). The ability of dopamine agonists to reverse the activation of tyrosine hydroxylase produced by a blockade of impulse flow can also be demonstrated in in-vivo studies in which an estimation of tyrosine hydroxylase activity is obtained by following the short-term accumulation of DOPA after administration of a decarboxylase inhibitor. For example, administration of apomorphine or trivastal alone normally causes only a slight inhibition of in-vivo tyrosine hydroxylase activity (Walters and Roth, 1976). However, if endogenous levels of dopamine are increased to a new steady state by causing a blockade of impulse flow by administration of GBL (a condition which changes tyrosine hydroxylase to a form which is insensitive to inhibition by endogenous dopamine) then subsequent administration of either apomorphine or trivastal now causes a much more potent inhibition of DOPA formation (Walters and Roth, 1976; Roth and Nowycky, 1977). This enhanced inhibitory effect of the dopamine agonists is presumably due to a conversion of the GBL-activated striatal tyrosine hydroxylase back to a form which is now sensitive to inhibition by the elevated levels of endogenous dopamine. It is generally believed that these effects of apomorphine and trivastal are mediated by the interaction of the dopamine receptor agonist with receptors on the presynaptic side of the synapse. However, it is premature to rule out the possibility that a trans-synaptic event mediated by an interaction of these dopamine receptor stimulants with postsynaptic dopamine receptors might also be involved.

If an interaction with dopamine receptors is indeed responsible for the ability of dopamine agonists to block or reverse the increase in tyrosine hydroxylase activity observed during a cessation of impulse flow in dopamine neurones, then dopamine receptor blocking agents should be able to nullify the inhibitory effects of dopamine receptor stimulants. Administration of antipsychotic drugs such as haloperidol effectively reverses the inhibitory effects exerted by dopamine agonists on the increase in tyrosine hydroxylase activity observed after inhibition of impulse flow by GHB administration or electrothermic lesions. Phenothiazines devoid of antipsychotic effects and lacking dopamine receptor-blocking capabilities, such as promethazine and diethazine, are ineffective in reversing the inhibitory effects of the dopamine receptor stimulants (Roth *et al.*, 1975b; Walters and Roth, 1974; Walters and Roth, 1975).

These experiments provide further evidence for the contention that activation of a dopamine receptor is responsible for the inhibitory effects exerted by the dopamine agonists on the increase in synthesis produced by a cessation of impulse flow. These observations are also suggestive that a presynaptic receptor mechanism may be involved since the effects of neuronal feedback and subsequent alterations in the firing rate of dopaminergic neurones normally produced by the antipsychotic drugs (Bunney *et al.*, 1973) are prevented by the blockade in the impulse flow produced by treatment of the rats with GBL.

Long-term Regulation

In addition to the mechanisms for immediate adaptation to increased transmitter utilization discussed above, a second mechanism comes into play after prolonged increases in the activity of catecholaminergic neurones. This latter mechanism, in contrast to short-term regulation, is reflected by an increase in the activity of tyrosine hydroxylase measured *in vitro* under *saturating* conditions of substrate and cofactor. For a while it was unclear whether this increase in activity observed *in vitro* was due to the disappearance of an inhibitor, the appearance of an activator, or an actual increase in the number of active enzyme sites. However, over recent years much evidence has accumulated which indicates that in the adrenal medulla and in some noradrenergic nerves there is a measurable increase in the formation of new enzyme molecules. Thus, prolonged neuronal activity actually results in an induction of tyrosine hydroxylase protein. This process has been termed trans-synaptic induction and refers to the increase in the synthesis of specific proteins by persistent changes in the rate of synaptic transmission. This process has been most successfully studied in the adrenal medullary model (for review see Guidotti and Costa, 1977). In this model system it has been demonstrated that persistent release of acetylcholine from the splanchnic nerves can modify the nuclear synthesis of specific messenger RNA. It has been postulated that a sustained increase in the release of acetylcholine activates postsynaptic nicotinic receptors located in the chromaffin cell membrane and this receptor activation initiates a chain of metabolic events that alters the rate of protein synthesis and results in an increase in the formation of tyrosine hydroxylase. A cAMP-dependent protein kinase has been implicated in this process. This process of trans-synaptic induction does not appear to be specific for tyrosine hydroxylase since an increase in the formation of dopamine-β-hydroxylase is also observed. No significant increase in the formation of DOPA-decarboxylase has been reported. Despite the elegant elucidation of the molecular mechanisms mediating the trans-synaptic regulation of gene expression in the adrenal medulla it is premature to extend this model to catecholaminergic neurones although attempts to verify this model in noradrenergic neurones have met with partial success. As yet there is no direct evidence to indicate that a similar process occurs in dopaminergic neurones.

ROLE IN NEURONAL FUNCTION

Tyrosine hydroxylase plays a key role in controlling the synthesis of the transmitter associated with catecholamine neurones. Since this enzyme is rate limiting in catecholamine biosynthesis, it is most susceptible to physiological control and pharmacological manipulation. This enzyme is responsive to alterations in neuronal function providing the neurone with the needed flexibility to meet altered periods of demands for transmitter. Short periods of increased impulse flow in central dopaminergic neurones results in an increase in tyrosine hydroxylase activity primarily by alterations in the kinetic properties of pre-existing enzyme molecules. From a regulatory standpoint, the most important changes are the increase in affinity of the enzyme for pterin cofactor and the decrease in affinity for end-product inhibitor. Both alterations result in an increase in the functional activity of tyrosine hydroxylase which is independent of any changes in the concentration of reduced pterin cofactor or endogenous dopamine. Any change in the concentration of reduced pterin cofactor or dopamine in the nerve terminal may act to further modulate this regulatory process. During prolonged periods of increased functional activity it is likely that the neurone develops an enhanced capacity for synthesis of neurotransmitter primarily by increasing the amount of tyrosine hydroxylase via enzyme induction. However, it has not yet been demonstrated that such a mechanism is functional in central dopaminergic neurones.

Drugs or other conditions which influence neuronal activity have a dramatic effect on the activity of tyrosine hydroxylase and transmitter formation. Drugs which increase impulse flow cause an increase in tyrosine hydroxylase activity and drugs which cause a cessation of impulse flow also result in an increase in tyrosine hydroxylase activity and transmitter accumulation. The physiological significance of the paradoxical response of dopaminergic neurones to a cessation of impulse flow is unclear. However, it is conceivable that these neurones may achieve some operational advantage by being able to activate tyrosine hydroxylase and rapidly increase their supply of transmitter during periods of quiescence.

ACKNOWLEDGEMENTS

This research was supported in part by a grant from the US Public Health Service (MH-14092).

REFERENCES

Agid, Y., Javoy, F. and Glowinski, J. (1974). *Brain Res.* **74,** 41–49.

Ames, M. M., Lerner, P. and Lovenberg, W. (1978). *J. Biol. Chem.* **253,** 27–31.

Anden, N.-E., Magnusson, T., and Stock, G. (1973). *Naunyn-Schmiedeberg's Arch. Pharmacol.* **278,** 363–372.

Ayukawa, S., Takeuchi, T., Sezaki, M., Hara, T., Umezawa, H., and Nagatsu, T. (1968). *J. Antibiot.* **21,** 350–355.

Ayukawa, S., Hamada, M., Kojiri, K., Takeuchi, T., Hara, T., Nagatsu, T., and Umezawa, H. (1969). *J. Antibiot.* **22,** 303–308.

Bacopoulos, N. G. and Bhatnagar, R. K. (1977). *J. Neurochem.* **29,** 639–643.

Baker, P. F. (1972). *Progr. Biophys. Molec. Biol.* **24,** 177–223.

Baker, P. F. (1975). *In* "Calcium Movements in Excitable Cells" (Eds. P. F. Baker and H. Reuter). Pergamon, New York.

Bapna, J., Neff, N. H., and Costa, E. (1970). *Neuropharmacology* **9,** 333–340.

Brenneman, A. R. and Kaufman, S. (1964). *Biochem. Biophys. Res. Commun.* **17,** 177.

Bullard, W. P., Guthrie, P. B., Russo, P. V. and Mandel, A. J. (1978). *J. Pharmacol. Exp.* Ther. **206,** 4–20.

Bunney, B. S., Walters, J. R., Roth, R. H. and Aghajanian, G. K. (1973). *J. Pharmacol. Exp. Ther.* **185,** 560–571.

Bustos, G. and Roth, R. H. (1979). *Biochem. Pharmacol.* **27,** in press.

Bustos, G., Roth, R. H., and Morgenroth, V. H., III (1976). *Biochem. Pharmacol.* **25,** 2493–2497.

Carlsson, A., Corrodi, H., and Waldeck, B. (1963). *Helv. Chim. Acta* **46,** 2271–2285.

Coyle, J. T. (1972). *Biochem. Pharmacol.* **21,** 1935–1944.

Faull, R. L. M. and Laverty, R. (1969). *Expl. Neurol.* **23,** 332–340.

Fisher, D. B. and Kaufman, S. (1972). *J. Neurochem.* **19,** 1359–1365.

Gal, E. M. and Sherman, A. D. (1977). *In* "Structure and Function of Monoamine Enzymes" (Eds E. Usdin, N. Weiner, and M. Youdim), 23–42. Marcel Dekker, New York.

Goldstein, M., Ebstein, B., Bronaugh, R. L. and Roberge, C. (1975). *In* "Chemical Tools in Catecholamine Research II" (Eds. Almgren, Carlsson and Engel), 257–265. Elsevier, New York.

Goldstein, M., Bronaugh, R. L., Ebstein, B., and Roberge, C. (1976). *Brain Res.* **109,** 563–574.

Guidotti, A. and Costa, E. (1977). *Biochem. Pharmacol.* **26,** 817–823.

Gutman, Y. and Segal, J. (1973). *Biochem. Pharmacol.* **22,** 865–868.

Harris, J. E., Morgenroth, V. H., III, Roth, R. H., Baldessarini, R. J. (1974). *Nature (London)* **252,** 156–158.

Harris, J. E., Baldessarini, R. J., Morgenroth, V. H., III, Roth, R. H. (1975). *Proc. Natl. Acad. Sci., U.S.A.* **72,** 789–793.

Herr, B. E. and Roth, R. H. (1976). *Brain Res.* **110,** 189–193.

Hoeldtke, R. and Kaufman, S. (1977) *J. Biol. Chem.* **252,** 3160–3169.

Holtz, P. (1939). *Naturwissenschaften* **27,** 724–725.

Jequier, E., Robinson, D. S., Lovenberg, W., and Sjoerdsma, A. (1969). *Biochem. Pharmacol.* **18,** 1071–1081.

Joh, T. H. and Reis, D. J. (1975). *Brain Res.* **85,** 146–151.

Kakiuchi, S., Rall, T. W., McIlwain, H. (1969). *J. Neurochem.* **16,** 485–491.

Karobath, M. and Baldessarini, R. (1972). *Nature (London)* **236,** 206–208.

Katz, I. R., Yamauchi, T., and Kaufman, S. (1976). *Biochim. Biophys. Acta* **429,** 84–95.

Kaufman, S. (1975). *In* "Neurobiological Mechanisms of Adaptation and Behavior" (Ed. A. J. Mandel), 127–136. Raven Press, New York.
Kehr, W., Carlsson, A., Lindqvist, M., Magnusson, T., and Atack, C. V. (1972a). *J. Pharm. Pharmacol.* **24,** 744–747.
Kehr, W., Carlsson, A., and Lindqvist, M. (1972b). *Naunyn-Schmiedeberg's Arch. Pharmacol.* **274,** 273–279.
Korf, J., Roth, R. H., and Aghajanian, G. K. (1973). *Eur. J. Pharmacol.* **23,** 276–282.
Kuczenski, R. (1973). *Life Sci.* **13,** 247–255.
Kuczenski, R. (1973). *J. Biol. Chem.* **248,** 5074–5080.
Kuczenski, R. (1975). *In* "Neurobiological Mechanisms of Adaptation and Behavior" (Ed. A. J. Mandell), 103–125. Raven Press, New York.
Kuczenski, R. T. and Mandell, A. J. (1972). *J. Biol. Chem.* **247,** 3114–3122.
Kurosawa, A., Guidotti, A., and Costa, E. (1976). *Science* **193,** 691–693.
Laduron, P. and Belpaire, F. (1968). *Biochem. Pharmacol.* **17,** 1127–1140.
Langer, S. Z. (1974). *Biochem. Pharmacol.* **23,** 1793–1796.
Langer, S. Z. (1977). *Br. J. Pharmacol.* **60,** 481–497.
Lerner, P., Ames, M. M., and Lovenberg, W. (1977). *Molec. Pharmacol.* **13,** 44–49.
Levitt, M., Gibb, J. W., Daly, J. W., Lipton, M., and Udenfriend, S. (1967). *Biochem. Pharmacol.* **16,** 1313–1321.
Lloyd, T. and Kaufman, S. (1974). *Biochem. Biophys. Res. Commun.* **59,** 1262–1269.
Lloyd, T. and Kaufman, S. (1975). *Biochem. Biophys. Res. Commun.* **66,** 907–913.
Lloyd, T. and Weiner, N. (1971). *Molec. Pharmacol.* **7,** 569–580.
Lovenberg, W., Bruckwick, E. A., and Hanbauer, I. (1975). *Proc. Natl. Acad. Sci.* **72,** 2955–2958.
McGeer, E. G., McGeer, P. L., and Peters, D. A. (1967). *Life Sci.* **6,** 2221–2232.
Mizutani, K., Nagatsu, T., Asashima, M., and Kinoshita, S. (1972). *Biochem. Pharmacol.* **21,** 2463–2468.
Morgenroth, V. H., III, Boadle-Biber, M. C. and Roth, R. H. (1975a). *Molec. Pharmacol.* **11,** 427–435.
Morgenroth, V. H., III, Hegstrand, L. R., Roth, R. H. and Greengard, P. (1975b). *J. Biol. Chem.* **250,** 1946–1948.
Morgenroth, V. H., III, Boadle-Biber, M. C., and Roth, R. H. (1976a). *Molec. Pharmacol.* **12,** 41–48.
Morgenroth, V. H., III, Walters, J. R. and Roth, R. H. (1976b). *Biochem. Pharmacol.* **25,** 655–661.
Murrin, L. C., Morgenroth, V. H., III and Roth, R. H. (1976). *Molec. Pharmacol.* **12,** 1070–1081.
Murrin, L. C. and Roth, R. H. (1976). *Molec. Pharmacol.* **12,** 463–475.
Musacchio, J. M. (1968). *Biochem. Pharmacol.* **17,** 1470–1473.
Musacchio, J. M. (1975). *In* Handbook of Pharmacology, Vol. 3. "Biochemistry of Biogenic Amines", (Eds L. L. Iversen, S. D. Iversen and S. H. Snyder), 1–35. Plenum, New York.
Musacchio, J. M., McQueen, C. A., and Crabsio, G. L. (1973). *In* "New Concepts in Neurotransmitter Regulation" (Ed. A. J. Mandell), 69–88. Plenum, New York.
Nagatsu, T., Levitt, M., Udenfriend, S. (1964). *J. Biol. Chem.* **239,** 2910–2917.
Nowycky, M. C. and Roth, R. H. (1978). *Prog. Neuro-Psychopharmacol.* **2,** 139–158.
Osborne, N. N. and Neuhoff, V. (1976). *Hoppe Syler's Z. Physiol. Chem.* **357,** 1271–1276.
Petrack, B., Sheppy, F., and Fetzer, V. (1968). *J. Biol. Chem.* **243,** 743–748.
Raese, J., Patrick, R. L., and Barchas, J. D. (1976). *Biochem. Pharmacol.* **25,** 2245–2250.
Roth, R. H. and Nowycky, M. C. (1977). *Adv. Biochem. Pharmacol.* **16,** 465–470.

Roth, R. H., Walters, J. R., and Aghajanian, G. K. (1973). *In* "Frontiers in Catecholamine Research" (Eds. S. H. Snyder and E. Costa), 567–574. Pergamon, New York.
Roth, R. H., Walters, J. R., and Morgenroth, V. H., III. (1974). *In* "Neuropsychopharmacology of Monoamines and their Regulatory Enzymes" (Ed. E. Usdin), 369–384. Raven Press, New York.
Roth, R. H., Walters, J. R., Murrin, L. C., and Morgenroth, V. H., III (1975a). *In* "Pre- and Postsynaptic Receptors" (Eds. E. Usdin and W. E. Bunney, Jr.), 5–48. Marcel Dekker, New York.
Roth, R. H., Morgenroth, V. H., III and Murrin, L. C. (1975b). *In* "Antipsychotic Drugs, Pharmacodynamics and Pharmacokinetics". Wenner-Gren Center International Symposium Series, 133–145. Pergamon, New York.
Roth, R. H., Morgenroth, V. H., III and Salzman, P. M. (1975c). *Naunyn-Schmiedeberg's Arch. pf Pharmacol.* **289,** 327–343.
Roth, R. H., Salzman, P. M., and Nowycky, M. C. (1978). *In* "Psychopharmacology: A Generation of Progress" (Eds. M. A. Lipton, A. DiMascio and K. F. Killam), 185–198. Raven Press, New York.
Rubio, M. C. (1977). *Naunyn-Schmiedeberg's Arch. Pharmacol.* **299,** 69–75.
Sethy, V. H., Kuhar, M. J., Roth, R. H., Van Woert, M. H. and Aghajanian, G. K. (1973). *Brain Res.* **55,** 481–484.
Shiman, R., Akino, M., and Kaufman, S. (1971). *J. Biol. Chem.* **246,** 1330–1340.
Simon, J. R., Hegstrand, L. R., and Roth, R. H. (1978). *Life Sci.* **22,** 421–428.
Stjärne, L. and Lishajko, F. (1967). *Biochem. Pharmacol.* **16,** 1719–1728.
Stock, G., Magnusson, T. and Anden, N.-E. (1973). *Naunyn-Schmiedeberg's Arch. Pharmacol.* **278,** 347–361.
Taylor, R. J., Stubbs, C. S., and Ellenbogen, L. (1968). *Biochem. Pharmacol.* **17,** 1779–1788.
Taylor, R. J., Stubbs, C. S., and Ellenbogen, L. (1969). *Biochem. Pharmacol.* **18,** 587–594.
Taylor, R. J., Stubbs, C. S., and Ellenbogen, L. (1970). *Biochem. Pharmacol.* **19,** 1737–1741.
Udenfriend, S. and Wyngaarden, J. B. (1956). *Biochim. Biophys. Acta* **20,** 48–52.
Udenfriend, S., Cooper, J. R., Clark, C. T., and Baer, J. E. (1953). *Science* **117,** 663–670.
Udenfriend, S., Zaltzman-Nirenberg, P. and Nagatsu, T. (1965). *Biochem. Pharmacol.* **14,** 837–845.
Walters, J. R. and Roth, R. H. (1974). *J. Pharmacol. Exptl Ther.* **191,** 82–91.
Walters, J. R. and Roth, R. H. (1975). *In* "Antipsychotic Drugs, Pharmacodynamics and Pharmacokinetics", (Eds G. Sedvall, B. Uvnas and Y. Zotterman), 147–160. Wenner-Gren Center International Symposium Series. Pergamon, New York.
Walters, J. R. and Roth, R. H. (1976). *Biochem. Pharmacol.* **25,** 649–654.
Walters, J. R., Roth, R. H., and Aghajanian, G. K. (1973). *J. Pharmacol. Expl Ther.* **186,** 630–639.
Waymire, J. C., Bjur, R., and Weiner, N. (1971). *Anal. Biochem.* **43,** 588–600.
Weiner, N. (1970). *A. Rev. Pharmacol.* **10,** 273–293.
Weissman, A., Koe, B. and Tenen, S. (1966). *J. Pharmacol. Expl Ther.* **151,** 339–448.
Wilk, S., Watson, E. and Travis, B. (1975). *Eur. J. Pharmacol.* **30,** 238–243.
Zhelyaskov, D. K., Levitt, M., and Udenfriend, S. (1968). *Molec. Pharmacol.* **4,** 445–451.
Zivkovic, B., Guidotti, A. and Costa, E. (1975). *Naunyn-Schmiedeberg's Arch. Pharmacol.* **291,** 193–200.

CHAPTER 10

DOPA Decarboxylase (aromatic amino acid decarboxylase)

Theodore L. Sourkes

Faculty of Medicine, McGill University,
Montreal, Canada.

INTRODUCTION

For many years tyrosine was regarded as the probable precursor of the catecholamines but a suitable enzymic mechanism for the transformation was not known. With the discovery of DOPA (3,4-dihydroxyphenylalanine) decarboxylase in 1938 (Holtz *et al.*, 1938) in extracts of mammalian organs, an enzyme was now available that could accomplish the conversion of the amino acid to an amine in a single step. It took another 15–20 years until DDC (DOPA decarboxylase) and the decarboxylation product dopamine (DA) were demonstrated to play a major role in the biosynthesis of noradrenaline (NA) and adrenaline (AD), based on the proposals of Blaschko (1939, 1942) and Holtz (1939). The study of DDC was taken up by Blaschko, who distinguished this enzyme from the bacterial tyrosine decarboxylase (Blaschko, 1950), and by Sourkes and Clark who studied its action on isomers of DOPA and 3,4-dihydroxyphenylserine, and its inhibition by a series of analogues (Clark, 1959; Hartman *et al.*, 1955a; Sourkes *et al.*, 1952; Sourkes, 1954b, 1955). This work led to the description of the first pharmacologically useful inhibitors of the enzyme, namely, α-methyldopa and α-methyl-*m*-tyrosine (Sourkes, 1954a). The former was soon afterwards shown to be effective in vivo against the action of DDC (Dengler and Reichel, 1958), and within a relatively short time came into use as an antihypertensive drug in man (Oates *et al.*, 1960).

NOMENCLATURE

DOPA decarboxylase was early found to catalyse the decarboxylation of *m*-tyrosine and *o*-tyrosine (Blaschko, 1950) and all the positional isomers of dihydroxyphenylalalanine (Sourkes *et al.*, 1952; Sourkes, 1955). Soon after the discovery of 5-hydroxytryptamine and the synthesis of 5-hydroxytryptophan (5HTP), the latter compound proved to be a substrate for DDC (Westermann *et al.*, 1958). Inasmuch as many aromatic amino acids are its substrates the enzyme has come to be known as aromatic-L-amino-acid decarboxylase (AADC). Its systematic name is aromatic-L-amino acid carboxylase, with the number EC 4.1.1.28.

DISTRIBUTION

Aromatic-L-amino-acid carboxylase is widely distributed in mammalian tissues, especially in the kidney cortex, liver, adrenals and intestinal mucosa (Sourkes, 1966; Boeker and Snell, 1972). Concentrations of the enzyme are low in skeletal muscle and brain. Within the brain there is considerable regional variation, as well as differences between species. In the rat brain (Saavedra, 1975) 5HTP decarboxylase activity is highest in the dorsal raphe nucleus; there is about half as much activity in the central superior nucleus, the area postrema, substantia nigra and the suprachiasmatic nucleus. Caudate nucleus and locus coeruleus have very low amounts of the enzyme. In rabbit brain (McCaman *et al.*, 1965) the area postrema, caudate nucleus and inferior colliculi contain much enzyme activity; there is less in the medulla oblongata, thalamus and cerebral cortex. In human brain (Langemann and Ackermann, 1961; Robins *et al.*, 1967; Lloyd and Hornykiewicz, 1972) DDC activity is generally low. It is highest in the striatum, reticular tegmentum, pineal gland, hypothalamus and dorsal raphe region. The activity is lower in the pigmented nuclei, periaquaductal grey, cortical tissue, thalamus and globus pallidus. No activity is detectable in the hippocampus, internal capsule or white matter.

DOPA decarboxylase occurs in the endothelial cells of the capillaries of the brain and seems to represent a barrier to unrestricted passage of DOPA (or other aromatic amino acid) from blood to brain. This blood–brain barrier is of varying efficacy, depending upon the species: it is very important in any consideration of the rat, but less so for the cat and monkey (Langelier *et al.*, 1972; Wade and Katzman, 1975). Nevertheless, it is possible to demonstrate the transport of DOPA across the blood–brain barrier (Yoshida *et al.*, 1963; Oldenorf, 1971; Pardridge, 1977). In addition to this saturable system, a non-saturable, diffusional component operates in the uptake of L-DOPA (Wade and Katzman, 1975). Neuronal perikarya contain AADC in the cytoplasm, as demonstrated by the immunofluorescence technique (Hökfelt *et al.*, 1973). In regard to the terminations of the fibres, immunofluorescence is readily detected for dopaminergic terminals, but the reaction is weak for serotonergic endings, and not at all apparent in the case of presumed noradrenergic terminations (Goldstein *et al.*, 1973; Hökfelt *et al.*, 1975).

Aromatic-L-amino-acid carboxylase occurs in sympathetic postganglionic fibres and, in fact, its rate of movement along the fibre by axoplasmic transport has been measured (Dahlström and Jonason, 1968; Dairman *et al.*, 1973).

PURIFICATION AND PROPERTIES

Aromatic-L-amino-acid carboxylase has been purified from guinea-pig kidney (Werle and Aures, 1959), hog kidney (Christenson *et al.*, 1970; Lancaster and Sourkes, 1972), beef adrenals (Fellman, 1959; Goldstein, 1972), and small intenstine of monkey (Murali and Radhakrishnan, 1970) as well as from the larvae of two *Calliphora* species (Sekeris, 1963; Fragoulis and Sekeris, 1975). The purified kidney enzyme has a molecular weight of 85 000–90 000 by gel filtration and consists of two unequal fragments (Lancaster and Sourkes, 1972). The preparation of Christenson *et al.* (1970), made by a slightly different procedure, includes a third fragment that augments the molecular weight to 112 000.

Several arguments have been adduced, besides kinetic evidence (Yuwiler *al.*, 1960), for the view that a single enzyme catalyses the decarboxylation of 5HTP and DOPA.

1. Antibodies raised against the purified pig kidney enzyme and beef adrenal enzyme, respectively (Christenson *et al.*, 1972; Goldstein *et al.*, 1973), titrate DDC and 5HTP decarboxylase activities from various sources to the same extent.
2. Intracisternal injection of the neurotoxins 5,6-dihydroxytryptamine and 6-hydroxydopamine leads to relatively specific destruction of serotonin and catecholamine-containing neurones, respectively. Despite this selective action the decarboxylase activities with the two substrates 5HTP and DOPA are reduced in parallel in spinal cord, brain (Dairman *et al.*, 1975) and striatum (Goldstein *et al.*, 1973).

3. During purification of guinea-pig kidney there is simultaneous enrichment of activity toward DOPA, 5HTP and 3,4-dihydroxyphenylserine (Werle and Aures, 1959). The same is true for adrenal AADC (Fellman, 1959) with respect to DOPA, 5HTP and σ-tyrosine and for rat-brain AADC (Corgier *et al.*, 1973) with respect to DDC and 5HTP decarboxylase activities.

Despite this evidence for the identity of the two enzymic activities Goldstein and his colleagues (Goldstein *et al.*, 1973) consider the possibility that certain regions of the brain contain enzymes specific for these substrates, and Sims is unconvinced about the identity of the two activities (Sims *et al.*, 1973; Sims, 1974). Bender and Coulson (1972) have observed certain kinetic differences between the decarboxylase activities but they regard these differences as arising from separate locations for fixing the two substrates on the apoenzyme, which would have a single catalytic site. One could just as well postulate the existence of two homologous enzymes possessing equivalent structure and conformation at the immunocompetent regions of the peptide chain.

An electrophoretic study of AADC from mouse and beef brain and from human and beef liver has shown that there is only a single species acting on 5HTP; for beef liver and brain this was the same one. Rat brain contained more than one species of the enzyme (Cavalli-Sforza *et al.*, 1974) as did rat liver (Coulson *et al.*, 1969).

In a remarkable and concisely written footnote to a paper on transaminases published in 1945, Green, Leloir and Nocito described the purification and resolution of pig-kidney DDC and its reactivation in the presence of yeast codecarboxylase (i.e. pyridoxal phosphate) and "reducing substances such as cysteine or glutathione" (Green *et al.*, 1945). Pyridoxal phosphate has long been accepted as the coenzyme of DDC; the presence of an essential sulphydryl group on the apoenzyme has been confirmed recently (Christenson *et al.*, 1970).

The nature of the association of coenzyme with AADC protein is not yet understood [for discussion see Awapara *et al.*, (1962) and Christenson *et al.*, (1970)]. The problem is complicated by the non-enzymic reaction of DOPA substrate with pyridoxal and pyridoxal phosphate; this reaction can result in inhibition of the enzymic activity as coenzyme is progressively bound in a Schiff's base reaction, with subsequent rearrangement to a stable polycyclic compound (Schott and Clark, 1952). The reaction product of pyridoxal phosphate with D,L-DOPA is essentially without effect on DDC (Sourkes, 1954a), but repeated administration of pridoxine to cats may yield an inhibitory compound, formed with DOPA, in the tissues (Roberge, 1977).

INHIBITORS

The antihypertensive action of α-methyldopa prompted the synthesis of large numbers of compounds for testing as decarboxylase inhibitors. Many proved useful in that regard (Sourkes, 1966; Sourkes and Rodriquez, 1967), but none was effective in the treatment of essential hypertension. With the laevo-DOPA treatment of Parkinson's disease there has been a resurgence of interest in clinically effective decarboxylase inhibitors as adjunctive drugs. Two are currently in use: carbidopa (Porter *et al.*, 1962) and benserazide (Burkard *et al.*, 1964). These may be regarded as derivatives of hydrazine. They selectively inhibit extracerebral AADC, although if high doses of benserazide are administered, some of the drug penetrates into the brain substance and inhibits AADC there. Other experimentally useful inhibitors are derived from hydroxylamine (Sourkes and Rodriguez, 1967; Ellenbogen *et al.*, 1973; Huszti *et al.*, 1973; Huszti and Sourkes, 1975). These pass the blood–brain barrier readily.

REGULATION

Aromatic-L-amino-acid carboxylase of liver, kidney and heart increases after birth to a maximal level (Huang *et al.*, 1960; Inagaki and Tanaka, 1974; Awapara and Saine, 1975) and then declines slowly in the case of heart (Inagaki and Tanaka, 1974) and kidney (Awapara and Saine, 1975). Cerebral cortex shows a gradual increase over the first two weeks of life but the enzyme appears in the cerebellum only during the second week (Eberle and Eiduson, 1968). There is disagreement as to whether the brain AADC activity increases after birth or not. During development, AADC increases in the sympathetic ganglia of the avian embryo, but after hatching occurs the enzyme decreases somewhat in activity (Filogamo *et al.*, 1971).

Although stress causes a trans-synaptic increase in the activity of tyrosine hydroxylase and dopamine β-hydroxylase in sympathetically innervated tissues, the activity of DDC, the other member of the trio of catecholamine-synthesizing enzymes, does not alter (Thoenen, 1972). In rats subjected to stress by restraint, DDC of the glandular portion of the stomach increases (Schauer *et al.*, 1974). However, Davis found that cortisone caused an increase of AADC activity of rat organs and had some conserving effect in regard to the coenzyme in tissues of the pyridoxine-deficient rat (Davis, 1963). Thyroxine administration decreases AADC activity, perhaps by reduction of the rate of formation of the apoenzyme (Davis, 1963). In insects the enzyme is induced by administration of ecdysone (Fragoulis and Sekeris, 1975; Schlaeger *et al.*, 1975).

Chronic administration of L-DOPA to cats causes an increase in DDC activity in the striatum, thalamus, hypothalamus and mesencephalon (Roberge and Poirier, 1973), which may be modified by selective lesions (Roberge and Poirier, 1974). Similar treatment of rats leads to a decrease of AADC activity of the liver (Dairman *et al.*, 1971).

DETECTION AND DETERMINATION

The investigator has a wide selection of methods available to him for the detection and measurement of AADC. Many of these are described in detail in reviews by the author (Sourkes and Lancaster, 1972; Duncan and Sourkes, 1978). A new isotopic method allows determination of the enzyme activity in very small tissue samples (Hefti and Lichtensteiger, 1976). Another has been described for the detection of the enzyme in electrophoretic media (Landon, 1977).

PHYSIOLOGICAL ROLE

The early proposals about the obligatory role of DDC (AADC) in the biosynthesis of catecholamines are now canonical wisdom (Blaschko, 1959; Holtz, 1959) and by extension apply to the biosynthesis of many other amines, e.g. 5-hydroxytryptamine. Lesser amounts of phenethylamine, tyramine and tryptamine are found in organ extracts and these probably are formed essentially in one step from the corresponding amino acid. The claim that this reaction is quantitatively significant for phenylalanine, tyrosine and tryptophan (David *et al.*, 1974a, b; David, 1975) has been successfully rebutted in the case of the second and third compounds (Fellman *et al.*, 1976; Sourkes, 1977; Young *et al.*, 1978) and probably does not hold for phenylalanine either.

O-Methyldopa, a physiological metabolite of DOPA, is a substrate for AADC, albeit a poor one (Sourkes *et al.*, 1952; Ferrini and Glässer, 1964). Decarboxylation proceeds much more rapidly if compounds of this type bear a free hydroxyl group *meta* to the side chain.

α-Methyldopa, a synthetic derivative, is also a substrate for AADC (Weissbach *et al.*, 1960). Indeed, its decarboxylation in the brain is a necessary condition for its antihypertensive effect (Rubenson, 1971; Waldmeier *et al.*, 1975).

3,4-Dihydroxyphenylserine (DOPS), another synthetic compound, has been proposed in the past as an intermediate in the biosynthesis of noradrenaline, its immediate decarboxylation product, but experimental evidence runs contrary to this view. The compound has two asymmetric centres and, therefore, four diastereoisomers. *erythro*-DOPS gives rise to the unnatural form of noradrenaline (Porter *et al.*, 1972; Hirai *et al.*, 1975; Constantinidis *et al.*, 1975), whereas *threo*-DOPS, a poorer substrate for AADC (Lancaster and Sourkes, 1972; Goodwin *et al.*, 1972), yields natural laevo-rotatory noradrenaline (Fujiwara *et al.*, 1976).

Normally AADC is not rate-limiting in the biosynthesis of catecholamines and 5-hydroxytryptamine. In the former case, tyrosine hydroxylase and dopamine β-oxidase are present in much lower concentrations; and in the case of the indoleamine, tryptophan hydroxylase is generally regarded as the limiting factor. However, when the enzymes earlier in the sequence are by-passed, as during the administration of L-DOPA or L-5HTP, the activity of AADC becomes important. This fact must be taken into consideration in the use of these amino acids in therapeutics, as in the treatment of Parkinson's disease with DOPA, or the attempts to influence myoclonic phenomena with 5HTP. In "typical" cases of Parkinson's disease the AADC activity of the striatum is reduced, but never wholly absent (Hornykiewicz, 1973). However, it is possible to produce a complete loss of enzyme activity in the cat striatum by introducing a lesion of the nigrostriatal tract (Lancaster *et al.*, 1970).

ACKNOWLEDGEMENT

Research in the author's laboratory was supported by the Medical Research Council (Canada).

DEDICATION

The author dedicates this chapter to the memory of William Gilbert Clark (April 17, 1909–February 13, 1979). He made important contributions to our knowledge of amino acid decarboxylases, and their inhibition for therapeutic ends.

REFERENCES

Awapara, J., and Saine, S. (1975). *J. Neurochem.* **24,** 817–818.

Awapara, J., Sandman, R. P. and Hanly, C. (1962). *Arch. Biochem. Biophys.* **98,** 520–525.

Bender, D. A. and Coulson, W. F. (1972). *J. Neurochem.* **19,** 2801–2810

Blaschko, H. (1939). *J. Physiol (London)* **96,** 50P–51P.
Blaschko, H. (1942). *J. Physiol. (London),* **101,** 337–349.
Blaschko, H. (1950). *Biochim Biophys. Acta* **4,** 130–137.
Blaschko, H. (1959). *Pharmacol. Rev.* **11,** 307–316.
Boeker, E. A. and Snell, E. E. (1972). *In* "The Enzymes" (Ed. P. D. Boyer), 3rd edn, **6,** 217–253.
Burkard, W. P., Gey, K. F. and Pletscher, A. (1964). *Arch. Biochem. Biophys.* **107,** 187–196.
Cavalli-Sforza, L. L., Santachiara, S. A., Wang, L., Erdelyi, E. and Barchas, J. (1974). *J. Neurochem.* **23,** 629–634.
Christenson, J. G., Dairman, W. and Udenfriend, S. (1970). *Arch. Biochem. Biophys.* **141,** 356–367.
Christenson, J. G., Dairman, W. and Udenfriend, S. (1972). *Proc. Natl. Acad. Sci., U.S.A.* **69,** 343–347.
Clark, W. G. (1959). *Pharmacol. Rev.* **11,** 330–349.
Constantinidis, J., Geissbühler, F., Gaillard, J. M., Aubert, C., Hovaguimian, T. and Tissot, R. (1975). *Psychopharmacologia (Berlin),* **41,** 201–209.
Corgier, M., Trappaz, M. and Pacheco, H. (1973). *Compt. Rend. Acad. Sci. (Paris) Série D,* **276,** 1065–1068.
Coulson, W. F., Bender, D. A. and Jepson, J. B. (1969). *Biochem. J.* **115,** 63P.
Dahlström, A. and Jonason, J. (1968). *Eur. J. Pharmacol.* **4,** 377–383.
Dairman, W., Christenson, J. G. and Udenfriend, S. (1971). *Proc. Natl Acad. Sci., U.S.A.* **68,** 2117–2120.
Dairman, W., Geffen, L. and Marchelle, M. (1973). *J. Neurochem.* **20,** 1617–1623.
Dairman, W., Horst, W. D., Marchelle, M. and Bautz, G. (1975). *J. Neurochem.* **27,** 619–623.
David, J. C. (1975). *FEBS Lett.* **55,** 81–83.
David, J. C., Dairman, W. and Udenfriend, S. (1974a). *Arch. Biochem. Biophys.* **160,** 561–568.
David, J. C., Dairman, W. and Udenfriend, S. (1974b). *Proc. Natl Acad. Sci., U.S.A.* **71,** 1771–1775.
Davis, V. E. (1963). *Endocrinology* **72,** 33–38.
Dengler, H. and Reichel, G. (1958). *Naunyn-Schmiedeberg's Arch. Exp. Path. Pharmakol.* **234,** 275–281.
Duncan, R. J. S. and Sourkes, T. L. (1978). Res. Methods Neurochem **4,** in press.
Eberle, E. D. and Eiduson, S. (1968). *J. Neurochem.* **15,** 1071–1083.
Ellenbogen, L., Kelly, R. G., Taylor, R. J. Jr. and Stubbs, C. S. Jr. (1973). *Biochem. Pharmacol.* **22,** 939–947.
Fellman, J. H. (1959). *Enzymologia,* **20,** 366–376.
Fellman, J. H., Roth, E. S. and Fujita, T. S. (1976). *Arch. Biochem. Biophys.* **174,** 562–567.
Ferrini, R. and Glässer, A. (1964). *Biochem. Pharmacol.* **13,** 798–801.
Filogamo, G., Giacobini, E., Giavovini, G. and Nore, B. (1971). *J. Neurochem.* **18,** 1589–1591.
Fragoulis, E. G. and Sekeris, C. E. (1975). *Biochem. J.* **146,** 121–126.
Fragoulis, E. G. and Sekeris, C. E. (1975). *Arch. Biochem. Biophys.* **168,** 15–25.
Fujiwara, H. Inagaki, C., Ikeda, Y. and Tanaka, C. (1976). *Folia Pharmacol. Japonica,* **72,** 891–898.

Goldstein, M. (1972). *Res. Methods Neurochem.* **1,** 317–340.

Goldstein, M., Anagnoste, B., Freedman, L. S., Roffman, M., Ebstein, R. P., Park, D. H., Fuxe, K. and Hökfelt, T. (1973). *In* "Frontiers in Catecholamine Research" (Eds. E. Usdin and S. Snyder), Pergamon, New York.

Goodwin, B. L., Johnson, R. D., Leask, B. G. S., Ruthven, C. R. J. and Sandler, M. (1972). *Experientia*, **28,** 1298–1299.

Green, D. E., Leloir, L. F. and Nocito, V. (1945). *J. Biol. Chem.* **161,** 559–582.

Hartman, W. J., Pogrund, R. S., Drell, W. and Clark, W. G. (1955a). *J. Am. Chem. Soc.* **77,** 816–817.

Hartman, W. J., Akawie, R. I. and Clark, W. G. (1955b). *J. Biol. Chem.* **216,** 507–529.

Hefti, F. and Lichtensteiger, W. (1976). *J. Neurochem.* **27,** 647–649.

Hirai, M., Matsuoka, Y., Nakajima, T. and Sano, I. (1975). *Med. J. Osaka Univ.* **26,** 51–59.

Hökfelt, T., Fuxe, K. and Goldstein, M. (1973). *Brain Res.* **53,** 175–180.

Hökfelt, T., Fuxe, K., Goldstein, M., Johansson, O., Park, D., Fraser, H. and Jeffcoate, S. L. (1975). *In* "Anatomical Neuroendocrinology" (Eds. W. E. Stumpf and L. D. Grant), 381–392 Karger, Basle.

Holtz, P. (1939). *Naturwissenschaften*, **27,** 724–725.

Holtz, P. (1959). *Pharmacol. Rev.* **11,** 317–329.

Holtz, P. Heise, R. and Lüdtke, K. (1938) *Naunyn-Schmiedeberg's Arch. Exp. Path. Pharmakol.* **191,** 87–118.

Hornykiewicz, O. (1973). *Adv. Neurol.* **2,** 1–11.

Huang, I., Tannenbaum, S. and Hsia, D. Y. Y. (1960). *Nature (London)*, **186,** 717–718.

Huszti, Z. and Sourkes, T. L. (1975). *J. Pharmacol. Exp. Ther.* **192,** 432–440.

Huszti, Z., Kasztreiner, E., Szilagyi, G., Kosary, J. and Borsy, J. (1973). *Biochem. Pharmacol.* **22,** 2267–2275.

Inagaki, C. and Tanaka, C. (1974). *Jap. J. Pharmacol.* **24,** 439–446.

Lancaster, G. A. and Sourkes, T. L. (1972). *Can. J. Biochem.* **50,** 791–797.

Lancaster, G., Larochelle, L., Bédard, P., Missala, K., Sourkes, T. L. and Poirier, L. J. (1970). *J. Neurol. Sci.* **11,** 265–274.

Landon, M. (1977). *Anal. Biochem.* **77,** 293–297.

Langemann, H. and Ackermann, H. (1961). *Helvet. Physiol. Pharmacol. Acta* **19,** 399–406.

Langelier, P., Parent, A. and Poirier, L. J. (1972). *Brain Res.* **45,** 622–629.

Lloyd, K. G. and Hornykiewicz, O. (1972). *J. Neurochem.* **19,** 1549–1559.

McCaman, R. E., McCaman, M. W., Hunt, J. M. and Smith, M. S. (1965). *J. Neurochem.* **12,** 15–23.

Murali, D. K. and Radhakrishnan, A. N. (1970). *Indian J. Biochem.* **7,** 13–18.

Oates, J. A., Gillespie, L., Udenfriend, S. and Sjoerdsma, A. (1960). *Science*, **131,** 1890–1891.

Oldendorf, W. H. (1971). *Am. J. Physiol.* **221,** 1629–1639.

Pardridge, W. M. (1977). *J. Neurochem.* **28,** 103–108.

Porter, C. C., Watson, L. S., Titus, D. C., Totaro, J. A. and Byer, S. S. (1962). *Biochem. Pharmacol.* **11,** 1067–1077.

Porter, C. C., Torchiana, M. L. and Stone, C. A. (1972). *Life Sci.* **11,** 787–795.

Roberge, A. G. (1977). *J. Neurochem.* **28,** 479–485.

Roberge, A. G. and Poirier, L. J. (1973). *J. Neur. Trans.* **34,** 171–185.

Roberge, A. G. and Poirier, L. J. (1974). *Brain Res.* **76,** 401–412.

Robins, E., Robins, J. M., Croninger, A. B., Moses, S. G., Spencer, S. J. and Hudgens, R. W. (1967). *Biochem. Med.* **1,** 240–251.
Rubenson, A. (1971). *J. Pharm. Pharmacol.* **23,** 228–230.
Saavedra, J. M. (1975). *J. Neurochem.* **26,** 585–589.
Schauer, A., Kunze, E., Feifel, G., Permanetter, W. and Fraps, P. (1974). *Digestion,* **11,** 12–24.
Schlaeger, D. A., Fuchs, M. S. and Kang, S. H. (1975). *J. Cell Biol.* **61,** 454–465.
Schott, H. and Clark, W. G. (1952). *J. Biol. Chem.* **196,** 449–462.
Sekeris, C. E. (1963). *Hoppe-Seyler's Z. Physiol. Chem.* **332,** 70–78.
Sims, K. L. (1974). *Adv. Biochem. Psychopharmacol.* **11,** 43–50.
Sims, K. L., Davis, G. A. and Bloom, F. E. (1973). *J. Neurochem.* **20,** 449–464.
Sourkes, T. L. (1954a). *Arch. Biochem. Biophys.* **51,** 444–456.
Sourkes, T. L. (1954b). *Can. J. Biochem. Physiol.* **32,** 515–518.
Sourkes, T. L. (1955). *Rev. Can. Biol.,* **14,** 49–63.
Sourkes, T. L. (1966). *Pharmacol. Rev.* **18,** 53–60.
Sourkes, T. L. (1978). *In* "Structure and Function of Monoamine Enzymes" (Eds. E. Usdin, N. Weiner and M. B. H. Youdin), Dekker, New York, 477–496.
Sourkes, T. L. and Lancaster, G. A. (1972). *In* "Methods of Investigative and Diagnostic Endocrinology" (Ed. I. J. Kopin) Vol. 1, part 2, 519–530, North Holland Publishing Co., Amsterdam.
Sourkes, T. L. and Rodriguez, H. R. (1967). *In* "Antihypertensive Agents" (Ed. E. Schlittler), 151–189. Academic Press, New York, and London.
Sourkes, T. L., Heneage, P. and Trano, Y. (1952). *Arch. Biochem. Biophys.* **40,** 185–193.
Thoenen, H. (1972). *Pharmacol. Rev.* **74,** 255–267.
Wade, A. and Katzman, R. (1975). *Am. J. Physiol.* **228,** 352–359.
Waldmeier, P., Hedwall, P. R. and Maitre, L. (1975). *Naunyn-Schmiedeberg's Arch. Pharmacol.* **289,** 303–314.
Weissbach, H., Lovenberg, W. and Udenfriend, S. (1960). *Biochem. Biophys. Res. Commun.* **3,** 225–227.
Werle, E. and Aures, D. (1959). *Hoppe-Seyler's Z. Physiol. Chem.* **316,** 45–60.
Westermann, E., Balzer, H. and Knell, J. (1958). *Naunyn-Schmiedeberg's Arch. Exp. Path. Pharmakol.* **234,** 194–205.
Yoshida, H., Kaniki, K. and Namba, J. (1963). *Nature* (*London*), **198,** 191–192.
Young, S. N., St.-Arnaud-McKenzie, D. and Sourkes, T. L. (1978). *Biochem. Pharmacol.,* in press.
Yuwiler, A., Geller, E. and Eiduson, S. (1960). *Arch. Biochem. Biophys.* **89,** 143–147.

CHAPTER 11

Catechol-*O*-Methyltransferase

Hans C. Guldberg

University of Trondheim, Regional Hospital,
Trondheim, Norway.

INTRODUCTION

Catechol-*O*-methyl transferase (COMT, EC 2.1.1.6) catalyses the transfer of methyl groups to substrates containing the catechol grouping in the presence of *S*-adenosylmethionine (SAM) as methyl donor and magnesium ions as activator. In 1958 the enzyme was first described and partially purified by Axelrod and Tomchick. This work demonstrated that COMT was present in a number of animal species examined, that the enzyme had a wide tissue distribution and that it was localized predominantly to the soluble fraction of the cell (Axelrod and Tomchick, 1958).

Parallel to the work which led to the discovery of COMT, Axelrod and his co-workers showed that the enzyme *in vivo* catalysed the *O*-methylation of catecholamines, such as dopamine and (nor-) adrenaline (Axelrod *et al.*, 1958). The consensus of opinion on the biological role of COMT supported the view that the enzyme contributed to the removal of circulating catecholamines, thereby taking part in the process of detoxication not only of endogenous but also of xenobiotic substances. On the other hand, no definite evidence pointed to the importance of COMT in the inactivation of catecholamines at receptor sites of neuronal systems (Crout *et al.*, 1961).

A possible approach to studies of the role of COMT in humans is based on the observation that COMT was found in the erythrocyte (Assicot and Bohuon, 1971; Axelrod and Cohn, 1971) and hence the properties of erythrocytic COMT may be used as a basis for an experimental model of the enzyme in nervous tissue in order to enlarge our understanding of COMT in human physiology and disease.

DISTRIBUTION

Species and Tissues

Catechol-*O*-methyltransferase has a wide distribution both in animals (Axelrod and Tomchick, 1958) and plants (Finkle and Nelson, 1963). The greatest activity is found in the liver where it is particularly high in the rat (Axelrod and Tomchick, 1958). The enzyme is found in most mammalian tissues including spleen, kidney, intestine, skin, peripheral nerves, heart and skeletal muscles, blood vessels, adrenal, thyroid, pituitary and salivary glands, uterus, ovaries, testis, adipocytes, fibroblasts, red and white blood cells (for review see detailed references: Guldberg and Marsden, 1975) and more recently breast tissue (Assicot *et al.*, 1977).

In the central nervous system (CNS) of the monkey there is some evidence of an unequal distribution of COMT in that the activity is highest in the area postrema and lowest in the cerebellum (Axelrod *et al.*, 1959) while in the human and rat brains a specific regional distribution of the enzyme is not detected (Vogel *et al.*, 1968; Broch and Fonnum, 1972). There is no correlation between the distribution of catecholamines and COMT as there is with tyrosine hydroxylase and DOPA decarboxylase in the rat brain (Broch and Fonnum, 1972).

Cellular and Subcellular Distributions

Catechol-*O*-methyltransferase is an intracellular enzyme, for instance, demonstrated in the peripheral blood by the fact that there is no activity in blood plasma but there is COMT activity in erythrocytes (Axelrod and Cohn, 1971). In relation to neuronal tissue of the peripheral system some of the COMT activity is located to the sympathetic nerve endings, as evidenced by a fall in enzyme activity after surgical sympathectomy of certain tissues such as cat nictitating membrane (Jarrot and Langer, 1971) and rat or rabbit vas deferens (Jarrot and Iversen, 1971) and submaxillary gland (Jarrot, 1971; Marsden *et al.*, 1971). In other tissues such as the kidney, liver and arteries the fall in COMT activity after denervation is not significant (Nagatsu *et al.*, 1969; Jarrot, 1971; Head *et al.*, 1977) and this may be related to the lower density of sympathetic innervation in these tissues with COMT mostly located to the extraneuronal parenchymal cells. In the brain direct evidence for the occurrence of COMT in glial cells was furnished by the demonstration of enzyme activity in cultured rodent astrocytes (Silberstein *et al.*, 1972). Synaptosmal COMT in rat brain contains an occluded form which is disintegrated by Triton (Broch and Fonnum, 1972) but if this is associated with membranes of nerve endings proper or glial cells is not known. Activities of COMT in the rat striatum are not significantly changed by chemical or electrolytic lesions of the dopaminergic pathways to the striatum (Uretsky and Iversen, 1970; Marsden *et al.*, 1972), which is indicative of a non-specific localization of the brain enzyme.

Catechol-*O*-methyltransferase is a cytoplasmic enzyme that is free or loosely bound to membranes in the cytoplasm of the cell, the highest activity residing in the "soluble" form of the liver parenchymal cells (Axelrod and Tomchick, 1958). The microsomal COMT, which is firmly bound to membranes, represents only about 0.1% of the total COMT activity in the liver (Borchardt *et al.*, 1974). In the erythrocytes, Assicot and Bohuon (1971) found two distinct *O*-methylating activities, one associated with the supernatant fraction and the other with erythrocyte "ghosts" and so membrane bound. Catechol-*O*-methyltransferase is also assumed to be a cytoplasmic enzyme in the brain but some of the enzyme is recovered in a particle-bound form (Alberici *et al.*, 1965) which may be accounted for by the cytoplasm being trapped within the nerve terminals and thus appearing in the synaptosomal fraction after differential centrifugation (Broch and Fonnum, 1972).

Regulation of Tissues

There is no convincing evidence in the literature that COMT belongs to the type of enzymes showing the characteristics of induction and/or suppression, though some reports suggest some small changes in liver COMT activities following pyrogallol (Nukada *et al.*, 1962) and glucocortiocoid (Parvez and Parvez, 1973) administrations or hypophysectomy (Landsberg *et al.*, 1969) and perinatal adrenalectomy (Parvez *et al.*, 1977) in rats.

ENZYME PROPERTIES

Assays and Purification

The most sensitive assays for COMT are based on a radiochemical mode of end-product detection, using most frequently SAM-^{14}C-methyl or ^{3}H-methyl as the methyl donor to various types of catechol substrates (for review see Guldberg and Marsden, 1975).

Catechol-*O*-methyltransferase has proved instable during purification procedures. The enzyme was first partially purified (about 30-fold) from rat liver (Axelrod and Tomchick, 1958) and a 450-fold purification was achieved by applying ammonium sulphate and gel filtrations, dithiothreitol treatment and diethylaminoethyl (DEAE)-Sephadex chromatography (Assicot and Bohuon, 1970). More recently the use of affinity chromatography after the filtration steps resulted in a 1400-fold increase in COMT activity compared to that of the rat liver homogenate (Gulliver and Wharton, 1976). The purified enzyme was shown to be homogeneous by disc gel electrophoresis.

Molecular Weight and Multiple Forms of COMT

The molecular weight of rat liver COMT calculated from gel filtration experiments is 24 000 (Assicot and Bohuon, 1970) which is in good agreement with results obtained from human liver (Ball *et al.*, 1971). In addition to the main fraction of the enzyme protein ($23\,000 \pm 700$) at least two other forms of COMT with molecular weights of 11 250 and 37 000 are present in the soluble fraction from rat liver (Creveling *et al.*, 1974).

Different forms of COMT have been observed during the purification of the enzyme with electrophoretic or chromatographic techniques (Anderson and D'Iorio, 1968; Flohé and Schwabe, 1970). Axelrod and Vesell (1970) identified biochemically distinguishable forms of COMT using starch gel electrophoresis at pH 8.6. The rat liver exhibited two isoenzymes of COMT with different heat stability and K_m values. Later it is reported that when stabilizing the enzyme using dithiothreitol in gels and buffers and performing the experiments at pH 7.9 or below, only a single form of COMT is detected (White and Wu, 1975) and it was considered likely that the second peak of activity was that of an oxidized form of the enzyme. Moreover, a membrane-bound form of rat liver microsomal COMT is reported to have similar physiochemical properties to the soluble COMT from rat liver (Borchardt *et al.*, 1974). Thus, the occurrence of isoenzymes of COMT in rat liver preparations is by no means certain and the studies have not provided kinetic data to support the notion that COMT exists in different functional forms. However, there is some evidence for the existence of multiple forms of COMT in the soluble and "ghost" fractions of rat erythrocytes (Assicot and Bohuon, 1971). Although there are both similarities and differences in physiochemical properties of rat erythrocyte COMT in particulate and soluble enzyme preparations (Roffman *et al.*, 1976) it is, perhaps, likely that two distinct forms exist in that membrane-bound and soluble COMT from red blood cells showed no immunochemical cross-reactivity (Assicot and Bohuon, 1971). Additional studies will be necessary to resolve the question of isoenzymes of COMT and their possible functional significance.

Substrate Specificity and Kinetics

In spite of the broad substrate specificity of COMT, substrates for the enzyme are characterized by a catechol configuration. Physiological substrates include the catecholamines and a number of simple phenolic acids with a catechol grouping (Axelrod and Tomchick, 1958) as well as foreign catechols (Daly *et al.*, 1960). Catechol-*O*-methyltransferase shows no stereospecificity with respect to D- and L-isomers of catecholamines (Axelrod and Tomchick, 1958) but for various isomeric catechols the distribution of *m*- versus *p*-methylated catechols was shown to be quite different (Creveling *et al.*, 1972). *O*-Methylation of catecholamines *in vivo* results almost exclusively in the formation of *m*-*O*-methylated derivatives while the administration of xenobiotic catechols results in the formation of *p*-*O*-methylated metabolites (Daly *et al.*, 1960). Physiological substrates tested *in vitro* produce significant amounts of *p*-*O*-methylated derivatives and the ratio of *m*- and *p*-*O*-methylated products depends strongly on the relative polarities of the substituent groups, with non-polar substituents the ratio was found to be close to unity (Creveling *et al.*, 1972). The magnitude of this ratio depends on the pH of the medium, the concentration of the divalent cation and the nature and position of the substituent on the catechol ring.

The kinetic data on commonly employed catechol substrates do not seem to differ markedly in that K_m values for amines such as noradrenaline, adrenaline and dopamine (2.6, 3.8 and 1.6×10^{-4} M respectively) do not deviate much (Creveling *et al.*, 1970). On the other hand, the kinetic data of COMT reported from different laboratories is variable, and an extreme example illustrating this is the K_m values for noradrenaline reported to be 6×10^{-6} M (Assicot and Bohuon, 1971) and 1.6×10^{-3} M (Creveling *et al.*, 1972). Some of the factors which can affect the K_m value for a substrate have been investigated in recent years and may explain the variations. The tissue source of the enzyme does not appear to affect the K_m for noradrenaline (Crout, 1961) while heterogeneity of COMT from erythrocyte soluble and "ghost" cell fractions may affect significantly the rate constants for noradrenaline as substrate (Assicot and Bohuon, 1971). It is evident that discrepancies in K_m values can arise as a result of the instability of COMT (Flohé and Schwabe, 1970) and the addition to the experiments of a protective reagent for COMT (Creveling *et al.*, 1972). An important finding, which can affect kinetic analyses of COMT catalysed reactions, is that *S*-adenosylhomocysteine, a product of SAM, is a potent inhibitor of COMT (Coward *et al.*, 1972). Errors in kinetic analyses may result when working under non-saturation conditions for SAM such that conversion to product is greater than 5%. Finally, it has been shown that the nature of the ionic medium, pH and requirements for Mg^{2+} of the COMT catalysed reactions have effects on the reaction rate. In 1972, it was shown that COMT was inhibited at high concentrations of Mg^{2+} (Flohé and Schwabe, 1972) and the importance of Ca^{2+} as an inhibitor of COMT has been established (Weinschilboum and Raymond, 1976).

INHIBITORS

Pyrogallol and Derivatives

Pyrogallol was reported several years ago to inhibit COMT (Axelrod and LaRoche, 1959). Pyrogallol acts mainly as a competitive substrate to form a methylated derivative (Daly *et al.*, 1960) but a non-competitive component of its action has been demonstrated *in vitro* (Crout, 1961). Pyrogallol is a potent inhibitor of COMT *in vitro* (Crout, 1961) but it shares with COMT inhibitors as a whole a number of limiting factors for its pharmacological usefulness. The most important of these are its low potency *in vivo* which is linked to its pharmacokinetic properties and its relatively high toxicity. The inhibitory effects *in vivo* are of short duration and effective inhibition of the enzyme can be achieved for only about 0.5 h with a single, high dose (Crout *et al.*, 1961). The relative high toxicity, adrenergic or otherwise in nature, of pyrogallol (Sollmann, 1944; Wylie *et al.*, 1960) makes it quite unsuitable for use in man. Another trihydroxybenzene derivative, *gallic acid*, is an inhibitor of COMT (Booth *et al.*, 1959) with more favourable physiochemical properties as regards cellular toxicity when compared with pyrogallol. Esters of gallic acid have been used as antioxidants in foods (Lehman *et al.*, 1951) and *N*-butylgallate has been administered to man as a COMT inhibitor (Ericsson, 1971) but its duration of action *in vivo* is very short (Dorris and Dill, 1977) and this limits its usefulness. N^1-(D,L-seryl)-N^2-(2,3,4-Trihydroxybenzyl) hydrazine (R04-4602), a decarboxylase inhibitor, is also an active inhibitor of COMT (Baldessarini, 1972) and its in-vivo inhibition may be partly responsible for some of its beneficial effects in patients treated with L-DOPA (Baldessarini and Greiner, 1973).

Catechol and Derivatives

Many catechols and derivatives have been reported to be inhibitors of COMT. The parent compound *catechol* is a competitive substrate for COMT and an in-vivo COMT inhibitor (Ross and Haljasmaa, 1964). A disadvantage of catechol is its high toxicity, the most noteworthy of which is its convulsive property which is unrelated to COMT inhibition (Rogers *et al.*, 1968). The *dopacetamides* are catechol derivatives which inhibit COMT *in vitro* and are relatively potent and non-toxic *in vivo* (Carlsson *et al.*,

1962). A wide range of *flavonoids* of different basic structure can inhibit COMT. The most powerful inhibitors are catechol-configurated flavenoids such as quercetin and rutin which inhibit the enzyme competitively by serving as substrates (Gugler and Dengler, 1973) while flavonols such as caempferol, which lack neighbouring OH-groups, are not metabolized by COMT, but effectively inhibit the enzyme in a mixed type non-competitive manner (Schwabe and Flohé, 1972). Catechol ketones inhibit COMT and U-0521 (3,4-dihydroxy-2-methyl-propiophenone) is a potent inhibitor of the enzyme (Moffet *et al.*, 1964). A number of commonly used drugs, which are catechol derivatives, are inhibitors of COMT and these include: L-DOPA or α-methyldopa (Axelrod and Tomchick, 1958), L-α-methyldopa-α-hydrazine (Baldessarini, 1972), isoprenaline (Giles and Miller, 1967) and apomorphine (Missala *et al.*, 1973). All these compounds would be expected to inhibit COMT by competing with the substrate for the active sites of the enzyme.

Tropolones

The tropolones are isosteric with catechols and are inhibitors of COMT and a competitive mechanism for tropolone inhibition was proposed (Belleau and Burba, 1963). Careful kinetic studies with COMT indicate that tropolones are linear non-competitive inhibitors with respect to various commonly used substrates as shown in the extensive results of Borchardt (1973). Studies of tropolones as inhibitors of COMT *in vivo* indicate that they are short acting (Broch, 1972). However, the measurement of COMT *in vitro* may not be an adequate measure of enzyme inhibition *in vivo*. Tropolone in high doses given to rats caused a modest inhibition when measured by essentially in-vitro assay techniques (Broch, 1973) while a small dose of tropolone in mice pretreated with L-DOPA lowered the *O*-methylated metabolites of dopamine in the striatum (Murphy *et al.*, 1969).

Miscellaneous Group

These compounds include: 8-hydroxyquinolines, iodophenol derivatives, 3-mercaptotyramine, *S*-adenosylhomocysteine, pyridoxal-5^1-phosphate, ascorbic acid and others (for reviews see Guldberg and Marsden, 1975).

ROLE IN NEURONAL FUNCTIONS

Peripheral System

The importance of neuronal uptake in the regulation of the noradrenaline concentration in the region of the sympathetic receptor is established and has been carefully reviewed (Trendelenburg, 1972). In addition to re-uptake into the sympathetic terminals, uptake into non-neuronal sites (extraneuronal uptake) is also known to take place (Iversen, 1965). Catechol-*O*-methyltransferase is an intracellular enzyme mostly found in extraneuronal cells, though a proportion of the activity is neuronal in tissues with a dense adrenergic innervation. Thus *O*-methylation is preceded by uptake mostly into extraneuronal sites. It is possible that the extraneuronal uptake -*O*-methylation series is the primary means of regulating levels of circulating adrenaline (Kopin, 1972) while in the case of noradrenaline neuronal uptake also influences its concentration (Iversen *et al.*, 1966). Neuronal uptake is the primary method for removing noradrenaline from the region of adrenergic receptors and this is followed by retention of the amine, rather than intraneuronal metabolism, while under certain experimental conditions extraneuronal uptake and subsequent metabolism by COMT is at least an alternative to neuronal uptake (Hughes, 1972). In some tissues like the nictitating membrane, vas deferens and salivary gland, intraneuronal *O*-methylation occurs, as well as amine retention, subsequent to neuronal uptake (Langer *et al.*, 1972; Burnstock *et al.*, 1972; Marsden *et al.*, 1971).

Central Nervous System

That *O*-methylation of cerebral catecholamines plays a major role in their metabolic inactivation, is demonstrated by the presence in brain of 3-methoxytyramine (Carlsson and Waldeck, 1964) and homovanillic acid (Andén *et al.*, 1963) from dopamine and normetanephrine (Häggendal, 1963) and 3-methoxy-4-hydroxyphenylethylene glycol (Schanberg *et al.*, 1968) from noradrenaline. Stimulation of the substantia nigra after the injection of ^{3}H-dopamine into the ventricle increased the efflux of ^{3}H-3-methoxytyramine (Voigtlander and Moore, 1971) while tropolone infused together with ^{14}C-tyrosine into the caudate nucleus, transitorily decreased the resting release of ^{14}C-homovanillic acid (Riddel and Szerb, 1971). Furthermore, amphetamine, which releases catecholamines, markedly increased 3-methoxytyramine concentrations in nialamide pretreated animals (Scheel-Krüger, 1972a) while neuroleptic drugs, which block dopamine receptors, increased 3-methoxytyramine levels in brain (Scheel-Krüger, 1972b). These results indicate the role of *O*-methylation in the enzymatic inactivation of dopamine released from nerve terminals. One may speculate that it is extraneuronally located COMT that functions in the inactivation of catecholamines released from nerve endings while one does not know if intraneuronal COMT in brain has any specialized function.

Human Pathophysiology

It is probable that erythrocyte COMT activity reflects the activity of this enzyme in several other tissues (Quiram and Weinshilboum, 1976). Since blood is an easily obtained human tissue, the measurement of erythrocyte COMT activity offers an opportunity to study the role of individual differences in COMT activity in normal human physiology and disease states. Erythrocyte COMT activity has been reported to be decreased in blood of females with mental depression (Cohn *et al.*, 1970) and to be increased in children with Down's syndrome (Gustavson *et al.*, 1973). Familial factors have been shown to be important in determining the individual COMT activity levels in blood (Weinshilboum and Raymond, 1977) and this raises the possibility of significant genetically determined differences in enzymatic *O*-methylation of catecholamines in man.

REFERENCES

Alberici, M., Roderiguez De Lores Arnaiz, G. and De Robertis, E. (1965). *Life Sci.* **4,** 1951–1960.
Andén, N.-E., Roos, B.-E. and Werdinius, B. (1963). *Life Sci.* **2,** 448–458.
Anderson, P. J. and D'Iorio, A. (1968). *Biochem. Pharmacol.* **17,** 1943–1949.
Assicot, M. and Bohuon, C. (1970). *Eur. J. Biochem.* **12,** 490–501.
Assicot, M. and Bohuon, C. (1971). *Biochimie* **53,** 871–874.
Assicot, M., Contesso, G. and Bohuon, C. (1977). *Eur. J. Cancer* **13,** 961–966.
Axelrod, J., Albers, W. and Clemente, C. D. (1959). *J. Neurochem.* **5,** 68–72.
Axelrod, J. and Cohn, C. K. (1971). *J. Pharmacol. Expl Ther.* **176,** 650–654.
Axelrod, J. and LaRoche, M.-J. (1959). *Science* **130,** 800.
Axelrod, J., Senoh, S. and Witkop, B. (1958). *J. Biol. Chem.* **233,** 697–701.
Axelrod, J. and Tomchick, R. (1958). *J. Biol. Chem.* **233,** 702–705.
Axelrod, J. and Vesell, E. S. (1970). *Molec. Pharmacol.* **6,** 78–84.
Baldessarini, R. J. (1972). *J. Pharm. Pharmacol.* **24,** 78–80.
Baldessarini, R. J. and Greiner, E. (1973). *Biochem. Pharmacol.* **22,** 247–256.
Ball, P., Knuppen, R. and Breuer, H. (1971). *Eur. J. Biochem.* **21,** 517–525.
Belleau, B. and Burba, J. (1963). *J. Med. Chem.* **6,** 755–759.
Booth, A. N., Masri, M., Robbins, D. J., Emerson, O. H., Jones, F. T., and DeEds, F. (1959). *J. Biol. Chem.* **234,** 3014–3016.
Borchardt, R. T. (1973). *J. Med. Chem.* **16,** 377–381.
Borchardt, R. T., Cheng, C.-F., Cooke, P. H. and Creveling, C. R. (1974). *Life Sci.* **14,** 1089–1100.
Broch, O. J. (1972). *Acta Pharmacol. Toxicol.* **31,** 217–225.
Broch, O. J. (1973). *Acta Pharmacol. Toxicol.* **33,** 417–428.
Broch, O. J. and Fonnum, F. (1972). *J. Neurochem.* **19,** 2049–2055.
Burnstock, G., McCulloch, M. W., Story, D. F. and Wright, M. E. (1972). *Br. J. Pharmacol.* **46,** 243–253.
Carlsson, A., Lindqvist, M., Fila-Hromadko, S. and Corrodi, H. (1962). *Helv. Chim. Acta* **45,** 270–276.

Carlsson, A. and Waldeck, B. (1964). *Scand. J. Clin. Lab. Invest.* **16,** 133–138.
Cohn, C. K., Dunner, D. L. and Axelrod, J. (1970). *Science* **170,** 1323–1324.
Coward, J. K., D'Urso-Scott, M. and Sweet, W. (1972). *Biochem. Pharmacol.* **21,** 1200–1203.
Creveling, C. R., Borchardt, R. T. and Isersky, C. (1974), *Biochem. Pharmacol. Suppl.* (part 1), 72–74.
Creveling, C. R., Dalgard, N., Shimizu, H. and Daly, J. W. (1970). *Molec. Pharmacol.* **6,** 691–696.
Creveling, C. R., Morris, N. Shimizu, H., Ong, H. H. and Daly, J. (1972). *Molec. Pharmacol.* **8,** 398–409.
Crout, J. R. (1961). *Biochem. Pharmacol.* **6,** 47–54.
Crout, J. R., Creveling, C. R. and Udenfriend, S. (1961). *J. Pharmacol. Expl Ther.* **132,** 269–277.
Daly, J. W., Axelrod, J. and Witkop, B. (1960). *J. Biol. Chem.* **235,** 1155–1159.
Dorris, R. L. and Dill, R. E. (1977). *Neuropharmacology* **16,** 631–634.
Ericsson, A. D. (1971). *J. Neurol. Sci.* **14,** 193–197.
Finkle, B. J. and Nelson, R. F. (1963). *Biochim. Biophys. Acta* **78,** 747–749.
Flohé, L. and Schwabe, K. P. (1970). *Biochim. Biophys. Acta* **220,** 469–476.
Flohé, L. and Schwabe, K. P. (1972). *Hoppe-Seyler's Z. Physiol. Chem.* **353,** 463–475.
Giles, R. E. and Miller, J. W. (1967). *J. Pharmacol. Exp. Ther.* **156,** 201–206.
Gugler, R. and Dengler, H. J. (1973). *Naun. Schmied. Arch. Pharmacol. Exp. Pathol.* **276,** 223–233.
Guldberg, H. C. and Marsden, C. A. (1975). *Pharmacol. Rev.* **27,** 135–206.
Gulliver, P. A. and Wharton, C. W. (1976). *Biochem. Pharmacol.* **25,** 2033–2038.
Gustavson, K.-H., Wetterberg, L., Bäckstrøm, M. and Ross, S. B. (1973). *Clin. Genet.* **4,** 279–280.
Häggendal, J. (1963). *Acta Physiol Scand.* **59,** 261–268.
Head, R. J., Stitzel, R. E., Lande, I. S. de la and Johnson, S. M. (1977). *Blood Vessels* **14,** 229–239.
Hughes, J. (1972). *Br. J. Pharmacol.* **44,** 472–491.
Iversen, L. L. (1965). *Br. J. Pharmacol. Chemother.* **25,** 18–33.
Iversen, L. L., Glowinski, J. and Axelrod, J. (1966). *J. Pharmacol. Exp. Ther.* **151,** 273–284.
Jarrot, B. (1971). *J. Neurochem.* **18,** 17–27.
Jarrot, B. and Iversen, L. L. (1971). *J. Neurochem.* **18,** 1–6.
Jarrot, B. and Langer, S. Z. (1971). *J. Physiol.* (*London*) **212,** 549–559.
Kopin, I. J. (1972). *In* "Handbook of Experimental Pharmacology". (Eds. H. Blaschko and E. Muscholl) Vol. 33, 270–279. Springer, Berlin.
Landsberg, L., de Champlain, J. and Axelrod, J. (1969). *J. Pharmacol. Exp. Ther.* **165,** 102–107.
Langer, S. Z., Stefano, F. J. E. and Enero, M. A. (1972). *J. Pharmacol. Exp. Ther.* **183,** 90–102.
Lehman, A. J., Fitzhugh, O. G., Nelson, A. A. and Woodard, G. (1951). *Adv. Fd Res.* **3,** 197–208.
Marsden, C. A., Broch, O. J. and Guldberg, H. C. (1971). *Eur. J. Pharmacol.* **15,** 335–342.
Marsden, C. A., Broch, O. J. and Guldberg, H. C. (1972). *Eur. J. Pharmacol.* **19,** 35–42.
Missala, K., Lal, S. and Sourkes, L. (1973). *Eur. J. Pharmacol.* **22,** 54–58.
Moffet, R. B., Hanze, A. R. and Seay, P. H. (1964). *J. Med. Chem.* **7,** 178–186.

Murphy, G. F., Robinson, D. and Sharman, D. F. (1969) *Br. J. Pharmacol.* **36,** 107–115.
Nagatsu, T., Rust, L. A. and DeQuattro, V. (1969). *Biochem. Pharmacol.* **18,** 1441–1446.
Nukada, T., Matsuoka, M. and Imaizumi, R. (1962). *Jap. J. Pharmacol.* **12,** 57–61.
Parvez, H. and Parvez, S. (1973). *Experientia* **29,** 1259–1262.
Parvez, S., Ismahan, G. and Parvez, H. (1977). *Hormone Res.* **8,** 159–170.
Quiram, D. R. and Weinschilboum, R. M. (1976). *J. Neurochem.* **27,** 1197–1203.
Riddel, D. and Szerb, J. C. (1971). *J. Neurochem.* **18,** 989–1006.
Roffman, M., Reigle, T. G., Orsulak, P. J. and Schildkraut, J. J. (1976). *Biochem. Pharmacol.* **25,** 208–209.
Rogers, K. J., Angel, A. and Butterfield, L. (1968). *J. Pharm. Pharmacol.* **20,** 727–729.
Ross, S. B. and Haljasmaa, Ø. (1964). *Acta Pharmacol. Toxicol.* **21,** 215–225.
Schanberg, S. M., Breese, G. R., Schildkraut, J. J., Gordon, E. K. and Kopin, I. J. (1968). *Biochem. Pharmacol.* **17,** 2006–2008.
Scheel-Krüger, J. (1972a). *Eur. J. Pharmacol.* **18,** 63–73.
Scheel-Krüger, J. (1972b). *Arch. Int. Pharmacodyn. Ther.* **195,** 372–378.
Schwabe, K. P. and Flohé, L. (1972). *Hoppe-Seyler's Z. Physiol. Chem.* **353,** 476–482.
Silberstein, S. D., Shein, H. M. and Berv, K. R. (1972). *Brain Res.* **41,** 245–248.
Sollmann, T. (1944). "A Manual of Pharmacology" 6th edn, 658. W. B. Saunders, Philadelphia and London.
Trendelenburg, U. (1972). *In* "Handbook of Experimental Pharmacology" Vol. 32, 270–279, Springer, Berlin.
Uretsky, N. J. and Iversen, L. L. (1970). *J. Neurochem.* **17,** 268–278.
Vogel, W. H., Orfei, V. and Century, B. (1968). *J. Pharmacol. Exp. Ther.* **165,** 196–203.
Voigtlander, P. F. von and Moore, K. E. (1971), *Neuropharmacology* **10,** 733–741.
Weinschilboum, R. M. and Raymond, F. A. (1976). *Biochem. Pharmacol.* **25,** 573–579.
Weinschilboum, R. M. and Raymond, F. A. (1977). *Am. J. Hum. Genet.* **29,** 125–135.
White, H. L. and Wu, J. C. (1975). *Biochem. J.* **145,** 135–143.
Wylie, D. W., Archer, S. and Arnold, A. (1960). *J. Pharmacol. Exp. Ther.* **130,** 239–244.

CHAPTER 12

Monoamine Oxidase

K. F. Tipton

University of Dublin, Trinity College, Dublin, Ireland.

DISCOVERY AND CLASSIFICATION

The ability of rat liver homogenates to catalyse the oxidation of tyramine was first reported by Hare (1928) who used the name tyramine oxidase for the enzyme responsible for this activity. Subsequent work showed this enzyme to have a wide specificity in catalysing the oxidative deamination of aliphatic and aromatic amines including the catecholamines (Blaschko *et al.*, 1937; Pugh and Quastel, 1937). The name monoamine oxidase was introduced by Zeller (1938) to distinguish this enzyme from the enzyme that catalysed the oxidative deamination of histamine and short-chain aliphatic diamines which he called diamine oxidase. Since this latter enzyme was inhibited by carbonyl reagents and by cyanide whereas monoamine oxidase was not, Zeller (1938) suggested that this could be used as a criterion for distinguishing between the two types of enzyme. This distinction based on inhibitor sensitivity is widely used, but the nomenclature of the enzymes has resulted in a great deal of confusion since the cyanide-sensitive diamine oxidases include enzymes that act on monoamines (see Buffoni, 1966, for review) and monoamine oxidase has been shown to act on long-chain aliphatic diamines (Blaschko and Hawkins, 1950).

A number of other criteria have been suggested for distinguishing between the two types of enzyme. Monoamine oxidase differs from diamine oxidase in being tightly bound to the mitochondrion (see later) and being active towards *N*-methyl-substituted secondary and tertiary amines (see, for example, Zeller, 1963). The cofactor requirements are also different since monoamine oxidase contains bound FAD whereas the cyanide-sensitive enzymes contain bound copper (see Yasunobu *et al.*, 1976, for review). This criterion has been adopted by the Commission on Enzyme Nomenclature who have introduced the systematic names amine : O_2 oxidoreductase (deaminating) (flavin-containing) (EC 1.4.3.4) and amine:O_2 oxidoreductase (deaminating) copper-containing) (EC 1.4.3.6) for these two enzymes and recommended the shortened names amine oxidase (flavin-containing) and amine oxidase (copper-containing) for general use (see Dixon and Webb, 1979). This chapter will be concerned with the former of these two classes of enzymes.

SPECIFICITY AND REACTION CATALYSED

The wide specificity of monoamine oxidase has already been mentioned and has been reviewed in detail by Blaschko (1952) and by Tipton (1975). The reaction catalysed follows the general equation:

$$RCH_2NH_2 + O_2 + H_2O \rightarrow RCHO + NH_3 + H_2O_2$$

where R can be an aliphatic or an aromatic group.

Steady-state kinetic studies on the enzyme from a number of sources (Tipton, 1968a; Fischer *et al.*, 1968; Oi *et al.*, 1970; Houslay and Tipton, 1973a) have shown the enzyme-catalysed reaction to follow a double-displacement (or ping-pong) kinetic mechanism in which the reaction proceeds in two separable steps which can be represented by the overall equations:

$$E + RCH_2NH_2 + H_2O \rightarrow EH_2 + RCHO + NH_3$$

$$EH_2 + O_2 \rightarrow E + H_2O_2$$

where E represents the free enzyme and EH_2 represents the form of the enzyme in which the bound flavin is in the reduced form. It has been suggested that a significant feature of this kinetic mechanism is that it results in the activity of the enzyme being relatively insensitive to fluctuations in the cellular oxygen concentration at physiological concentrations of its amine substrates (Tipton, 1972). In contrast to its wide specificity for amine substrates the enzyme appears to be highly specific for oxygen and no alternative hydrogen acceptor has been found that will act in its place (see Tipton, 1975, for review).

DISTRIBUTION AND LOCALIZATION

Monoamine oxidase has been shown to be widely distributed in mammalian tissues (see Blaschko, 1952 and 1963 and Tipton, 1975, for detailed reviews). It has been shown to be present in all glandular tissues studied with particularly high activities being associated with liver, parotid and salivary glands. It also occurs in nervous tissue, smooth muscle, heart muscle, kidney, gonads and blood platelets but it is apparently absent from blood plasma, erythrocytes and skeletal muscle. It is fairly uniformly distributed throughout the different regions of the brain where it has been shown to be present in glial cells (Silberstein *et al.*, 1972; Sinha and Rose, 1972) as well as in nerve cells (Arniaz and De Robertis, 1962). The presence of extraneuronal monoamine oxidase has also been demonstrated in other tissues, including kidney (Boadle and Bloom, 1969), rabbit artery (De La Lande *et al.*, 1970) and heart (Horita and Lowe, 1972) by the use of histochemical techniques.

Chemical or surgical sympathetic denervation has been used to study the distribution of the enzyme in a number of organs, and the results have been interpreted as indication a predominantly extraneuronal location of the enzyme in heart (Horita and Lowe, 1972; Jarrott, 1971), brain (Uretsky and Iversen, 1970), salivary gland (Snyder *et al.*, 1965; Goridis and Neff, 1971). Estimation of the proportion of extraneuronal monoamine oxidase using this method relies on the simple assumption that neuronal enzymes will be lost on denervation whereas extraneuronal activities will be unaffected. Marsden *et al.* (1971) have, however, pointed out that such an interpretation may be an oversimplification, since a fall in the activity of an extraneuronal enzyme might occur if it depended upon an intact nerve supply for its activity. Such a mechanism might explain the fall in the monoamine oxidase activity after sympathectomy of the liver, which is only very poorly innervated, that has been reported by Klingman (1966). In addition the observation that monoamine oxidase is associated with non-adrenergic, as well as with adrenergic, neurones (Woods, 1970; Furness and Costa, 1971) suggests that not all of the neuronal enzyme may be affected by sympathetic denervation.

Within the cell, the enzyme appears to be mainly localized in the mitochondria although varying proportions of the enzyme activity have been found associated with the microsomal fraction prepared from different tissues. Sub-fractionation of mitochondria has shown the enzyme to be associated with the outer membrane of that organelle in liver (Schnaitman *et al.*, 1967; Tipton, 1967; Beattie, 1968), brain (Tipton, 1967) and heart (Maisterrena *et al.*, 1974). Histochemical studies have indicated a similar localization of the enzyme in kidney (Boadle and Bloom, 1969; Bloom *et al.*, 1972), and in liver, kidney and heart (Shannon *et al.*, 1974) although the latter study also detected the presence of some monoamine oxidase activity associated with the endoplasmic reticulum and with the nuclear envelope.

The possibility that the enzyme activity associated with the microsomal fraction might result from mitochondrial breakage was first discussed by Hawkins (1952) and received support from density-gradient fractionation studies of vas deferens (Jarrott and Iversen, 1968), but other studies have suggested that the situation is more complicated than this (see, for example, De Champlain *et al.*, 1969; Stjärne *et al.*, 1968). The possibility that some of the microsomal monoamine oxidase represents a precursor of the mitochondrial enzyme (see, for example, Greenawalt, 1972) receives support from the observation that the recovery of monoamine oxidase activity after irreversible inhibition occurs more rapidly in the microsomal fraction than in the mitochondrial fraction (Erwin and Simon, 1969).

SOLUBLE PREPARATIONS

The tight association of monoamine oxidase with the mitochondrial outer membrane necessitates the use of vigorous treatments, such as sonication, detergent treatment, and extraction with organic solvents, to bring the enzyme into solution (see Tipton, 1975; Youdim, 1976, for reviews). Soluble enzyme preparations obtained in such ways have been shown to differ from the native, membrane-bound, enzyme in their sensitivity towards inhibitors (Tipton *et al.*, 1976) and in the kinetic mechanism of substrate oxidation (Houslay and Tipton, 1975). In addition the specificity of soluble preparations of the enzyme has been shown to depend upon the method used to bring it into solution (Kandaswami *et al.*, 1977). It is uncertain whether these changes are a consequence of removal of the enzyme from its native environment or whether they result from the vigor of the solubilization procedures, but these observations suggest that soluble preparations of the enzyme may not provide satisfactory models for the behaviour of the enzyme in its native environment.

The molecular properties of the enzyme have recently been reviewed in detail by Yasunobu *et al.* (1976). The molecular weights of purified preparations vary from about 100 000 to over one million and the enzyme from ox liver has been shown to exist in a polymerizing equilibrium (Gomes *et al.*, 1969). A molecular weight of about 100 000 has been reported for a number of preparations after treatment with detergents (see Yasunobu *et al.*, 1976) and polyacrylamide gel electrophoresis of the purified rat liver enzyme gave a subunit-molecular weight of about half this value (Youdim and Collins, 1971). Although there have been various reports of the presence of non-covalently bound FAD in purified preparations of the enzyme (Tipton, 1968b; Youdim and Sourkes, 1972; Harada *et al.*, 1971) this appears to have been due to contamination and the FAD cofactor has been shown to be bound covalently to the enzyme by way of a thioether linkage from the 8-position of the isoalloxazine ring to a cysteinyl residue in the protein (Kearney *et al.*, 1971). Enzyme preparations from a number of sources have been shown to contain one mole of FAD per 100 000 g of protein, there is no evidence for the involvement of any metal-ion cofactor but sulphydryl groups in the enzyme appear to be essential for activity (see Yasunobu *et al.*, 1976).

INHIBITORS

Monoamine oxidase is irreversibly inhibited by derivatives of hydrazine, cyclopropylamine and propargylamine (see, for example, Pletscher, 1966; Zirkle and Kaiser, 1964 for reviews) and several compounds of pharmacological interest, including the tricyclic antidepressants (see Achee *et al.*, 1977 for review), amphetamine (Mann and Quastel, 1940) and the harmala alkaloids (see Fuller, 1972, for review), have been shown to be reversible inhibitors of the enzyme. As will be discussed later a number of these inhibitors have been shown to inhibit the activity more potently towards some substrates than towards others.

MULTIPLE FORMS

The idea that there might be more than one form of monoamine oxidase arose from studies which showed that the activities of the enzyme towards different substrates were inhibited to different extents by elevated temperatures and by a number of inhibitors (see Houslay *et al.*, 1976, and Achee *et al.*, 1977, for reviews). In 1966 Johnston showed that clorgyline (*N*-methyl-*N*-propargyl-3-(2,4 dichlorophenoxy) propylamine) irreversibly inhibited the activity of rat brain monoamine oxidase towards serotonin at considerably lower concentrations than were required to inhibit the oxidation of benzylamine and that the activity towards tyramine was inhibited in a biphasic manner in which about half of the activity resembled the inhibition of serotonin oxidation in its sensitivity whereas the remainder resembled that of benzylamine. He suggested that there were two forms of the enzyme, one of which (the A-form) was sensitive to inhibition by clorgyline and active towards serotonin and tyramine whereas the other (the B-form) was insensitive to clorgyline inhibition and active towards benzylamine and tyramine. More recently several other irreversible inhibitors, including Lilly 51641 (*N*-[2 (*O*-chlorophenoxyl) ethyl] cyclopropylamine) (Fuller, 1968) and P.C.O. (5-phenyl-3-(*N*-cyclopropyl) ethylamine 1,2,4-oxadiazole) (Mantle *et al.*, 1975) have been shown to resemble clorgyline in their selectivity whereas deprenil (phenylisopropylmethylpropinylamine) (Knoll and Magyar, 1972) and pargyline (*N*-methyl-*N*-2-propynylbenzylamine) (Fuller *et al.*, 1970) have been shown to be more potent inhibitors of the B-form of the enzyme.

Use of these inhibitors together with some reversible inhibitors that showed a selective action allowed the specificities of these two forms to be investigated in more detail. In rat liver it was found that adrenaline, noradrenaline and their methylated derivatives were substrates for the A-form, dopamine resembled tyramine in being a substrate for both forms and phenylethylamine was a substrate only for the B-form, and a generally similar behaviour was observed with the enzyme from several other sources (see Neff and Yang, 1974; Houslay *et al.*, 1976, and Achee *et al.*, 1977, for reviews). More detailed investigations have, however, shown that the specificities of the two forms that were determined using the enzyme from rat liver were not applicable to the situation in all other organs and animal species (see Fowler *et al.*, 1978, for review), for example serotonin and benzylamine have been shown to be substrates for both forms of the enzyme in ox and rat heart respectively (Mantle *et al.*, 1976a; Lyles and Callingham, 1974) and dopamine has been shown to be a substrate for only the B-form in human brain (Glover *et al.*, 1977) whereas it is a substrate for only the A-form of the enzyme in rat striatum (Waldmeier and Maitre, 1975). Such variations make it difficult to draw exact parallels between the situation in experimental animals and that in human patients.

The effects of selective inhibitors can also be demonstrated *in vivo* although the results are often not as clear-cut as those from in-vitro experiments (see Houslay *et al.*, 1976, for review). There have been some reports that the selective action may decrease on prolonged administration of the inhibitors but, at least in the case of deprenil, this effect only occurs at high dose levels (M. Sandler, personal communication). The use of deprenil to decrease the rate of dopamine breakdown to a greater extent than that of other amines that are normally regarded as being neurotransmitters was first suggested as a treatment for Parkinsonism by Neff and Yang (1974) and subsequent clinical studies on the use of this drug in combination with L-DOPA and a peripherally acting decarboxylase inhibitor have given encouraging results (Birkmayer *et al.*, 1975 and 1977).

The nature of the two forms of the enzyme is still uncertain. On the basis of experiments with agents that weaken the interactions between lipids and proteins, it was suggested that they might arise from a single enzyme existing in more than one membrane-bound environment (Houslay and Tipton, 1973b; Tipton *et al.*, 1973) and this view has been supported by experiments demonstrating the immunological identity of the two forms (Dennick and Mayer, 1977; Chau and Hackenbrock, 1975; but see McCauley and Racker, 1973) and by electron spin resonance studies

(Huang and Eiduson, 1977). (See Houslay *et al.*, 1976 and Achee *et al.*, 1977 for more detailed discussions of the nature of the two forms of the enzyme.) The dependence of the properties of the enzyme on its membrane environment could result in its activity being affected by conditions that affect lipid metabolism and membrane structure, and it may provide an explanation for the inhibitory action of local anaesthetics (see Blaschko, 1952, for review).

The proportions of the two forms have been shown to vary widely between different organs and animal species (see Houslay *et al.*, 1976, for review). In brain they appear to be relatively uniformly distributed despite the wide differences in the types of innervation in different areas of brain (Neff and Yang, 1974; Youdim and Holman, 1975). The results from denervation experiments suggested a predominantly neuronal localization for the A-form in the pineal gland (Goridis and Neff, 1971), but such a preferred localization does not appear to apply to other tissues (see Houslay *et al.*, 1976; Achee *et al.*, 1977, for reviews). The B-form of the enzyme appears to be absent from cultured rat glioma and neuroblastoma cells (Murphy *et al.*, 1976; Donnelly *et al.*, 1976) and from early foetal rat brain (Mantle *et al.*, 1976b).

Electrophoretic studies of soluble preparations of the enzyme showed that up to five bands with monoamine oxidase activity could be separated (see Youdim, 1972; Diaz Borges and D'Iorio, 1972, for reviews). One of these bands, which had an opposite charge from the others at pH 9.1, showed a particularly high activity towards dopamine, but subsequent work has shown these electrophoretically separable forms to be artefacts of the methods used (see Houslay *et al.*, 1976; Achee *et al.*, 1977, for reviews) and there is no evidence for a dopamine specific form in membrane-bound preparations of the enzyme.

REGULATION

There is no evidence for allosteric regulation of monoamine oxidase activity, but it is possible that inhibition by the aldehyde product might be important in some circumstances (Turner *et al.*, 1974). Although several workers have observed inhibition of the enzyme by high concentrations of its amine substrate it appears that this may have resulted from contamination of the substrate by the aldehyde product (Houslay and Tipton, 1973a). The activity of the enzyme is affected by a number of hormones. Thyroidectomy causes an increase in the activity of rat liver monoamine oxidase whereas administration of the hormone has the opposite effect. The response of the rat heart enzyme to this hormone, however, appears to depend upon the age of the animal (Ho-Van-Hap *et al.*, 1967). Changes in the activity in rat heart in response to thyroid hormones have been shown to be confined to the A-form of the enzyme and in this organ and in salivary gland the effects appear to be due to alterations in the rate of enzyme synthesis (Lyles and Callingham, 1974; Goridis and Neff, 1973). Rat heart is unusual in that the specific activity of its monoamine oxidase increases with the age and weight of the animal, this increase appears to involve only the A-form of the enzyme (Callingham and Lyles, 1975) and to result from a decreased rate of degradation (Lyles and Callingham, 1974).

Adrenalectomy results in increase in the activity of the enzyme in rat heart (Avakian and Callingham, 1968), vas deferens (Sampath *et al.*, 1976) and small intestine (Csaba *et al.*, 1972) and these effects in heart and vas deferens have been shown to be prevented by treatment with hydrocortisone or dexamethasone (Avakian and Callingham, 1968; Sampath and Clarke, 1972). Adrenalectomy appears to have no effect on the activity of the enzyme in rat liver, spleen and brain (Csaba *et al.*, 1972; Callingham and Laverty, 1973; Caesar *et al.*, 1970; but see Parvez and Parvez, 1973). The effects in rat heart appear to involve both the A- and B-forms of the enzyme equally and to result from a change in the rate of enzyme biosynthesis (Della Corte and Callingham, 1977a and b). In contrast to the situation in the rat, adrenalectomy does not appreciably affect the monoamine oxidase activity of mouse tissues (Laverty *et al.*, 1973; Went and Csaba, 1973). The activity of the enzyme is also affected by variations of the levels of the gonadal sex hormones (see, for example, Wurtman and Axelrod, 1963; Holzbauer and Youdim, 1973). In the uterus the activity of the enzyme has been shown to vary during the oestrus cycle (Holzbauer and Youdim, 1973; Grosso and Gawienowski, 1975) and testicular monoamine oxidase activity has been shown to vary with age correlating with the onset of sexual maturation (Urry *et al.*, 1975).

FUNCTION

In the nerve terminal monoamine oxidase will degrade free amines that arise by leakage from storage vesicles or as a consequence of re-uptake processes. In the nerve terminal there will be a dynamic balance between free and bound transmitter amines that will be affected by the activities of the enzymes catalysing their synthesis and degradation as well as by release and re-uptake mechanisms. The importance of monoamine oxidase in maintaining this balance is shown by the rise in the concentration of free catecholamines after inhibition of the intraneuronal enzyme (Spector *et al.*,

1963) which, in turn results in an impairment of the presynaptic reuptake process (Trendelenburg *et al.*, 1972).

The role of post- and extrasynaptic monoamine oxidase may be largely protective in the degradation of amines taken into the cells by low affinity uptake processes (see, for example, Hendley *et al.*, 1970). Such a function in the destruction of dietary amines would be in accord with the relatively high concentrations of the enzyme in stomach, intestine, liver and blood platelets, although metabolism of circulating adrenaline will also be important. The relationship between monoamine oxidase and catechol-*O*-methyltransferase in terminating the action of the catecholamines is still uncertain (see Sharman, 1973, for review) but it has been shown that methylation by the latter enzyme does not impair the activity of monoamine oxidase towards these substrates.

REFERENCES

Achee, F., Gabay, S. and Tipton, K. F. (1977). *Prog. Neurobiol.* **8,** 325–348.
Arniaz, G. R. del and De Robertis, E. D. P. (1962). *J. Neurochem.* **9,** 503–508.
Avakian, V. M. and Callingham, B. A. (1968). *Br. J. Pharmacol.* **33,** 211P.
Beattie, D. S. (1968). *Biochem. Biophys. Res. Commun.* **31,** 901–908.
Birkmeyer, W., Riederer, P., Youdim, M. B. H. and Linauer, W. (1975). *J. Neural Transmiss.* **36,** 303–326.
Birkmeyer, W., Riederer, P., Ambrozi, L. and Youdim, M. B. H. (1977). *Lancet i,* 439–443.
Blaschko, H. (1952). *Pharmacol. Rev.* **4,** 415–453.
Blaschko, H. (1963). *In* "The Enzymes" (Eds. P. D. Boyer, H. Lardy and K. Myrbäck) Vol. 8, 337–351. Academic Press, New York and London.
Blaschko, H. and Hawkins, J. (1950). *Br. J. Pharmacol.* **5,** 625–632.
Blaschko, H., Richter, D. and Schlossman, H. (1937). *Biochem. J.* **31,** 2187–2196.
Bloom, F. E., Sims, K. L., Weitsen, H. A., Davis, G. A. and Hanker, J. S. (1972). *Adv. Biochem. Psychopharmacol.* **5,** 243–262.
Boadle, M. C. and Bloom, F. E. (1969). *J. Histochem. Cytochem.* **17,** 331–340.
Buffoni, F. (1966). *Pharmacol. Rev.* **18,** 1163–1199.
Caesar, P. M., Collins, G. G. S. and Sandler, M. (1970). *Biochem. Pharmacol.* **19,** 921–926.
Callingham, B. A. and Laverty, R. (1973). *J. Pharm. Pharmacol.* **25,** 940–947.
Callingham, B. A. and Lyles, G. A. (1975). *Brit. J. Pharmacol.* **53,** 458P.
Chau, R. M. W. and Hackenbrock, C. R. (1975). *J. Cell. Biol.* **67,** 63a.
Csaba, B., Went, M., Muszbek, L. and Kovács, K. (1972). *Acta Physiol. Hung.* **42,** 327–331.
De Champlain, J. R., Mueller, A. and Axelrod, J. (1969). *J. Pharmacol. Exp. Ther.* **166,** 339–345.
Della Corte, L. and Callingham, B. A. (1977a). *Biochem. Pharmacol.* **26,** 407–415.

Della Corte, L. and Callingham, B. A. (1977b). *J. Pharm. Pharmacol.* **29,** 657–663.
De La Lande, I. S., Hill, B. D., Jellet, L. B. and McNeil, J. M. (1970). *Brit. J. Pharmacol.* **40,** 249–256.
Dennick, R. G. and Mayer, R. J. (1977). *Biochem. J.* **167,** 167–174.
Diaz Borges, J. M. and D'Iorio, A. (1972). *Adv. Biochem. Psychopharmacol.* **5,** 74–90.
Dixon, M. and Webb, E. C. (1979). "Enzymes." Longman, London. In the press.
Donnelly, C. E., Richelson, E. and Murphy, D. L. (1976). *Biochem. Pharmacol.* **25,** 1639–1643.
Erwin, V. G. and Simon, R. J. (1969). *J. Pharm. Sci.* **58,** 1033–1035.
Fischer, A. G., Schulz, A. R. and Oliner, L. (1968). *Biochim. Biophys. Acta* **159,** 460–471.
Fowler, C. J., Callingham, B. A., Mantle, T. J. and Tipton, K. F. (1978). *Biochem. Pharmacol.* **27,** 97–101.
Fuller, R. W. (1968). *Biochem. Pharmacol.* **17,** 2097–2106.
Fuller, R. W. (1972). *Adv. Biochem. Physchopharmacol.* **5,** 339–354.
Fowler, C. J., Callingham, B. A., Mantle, T. J. and Tipton, K. F. (1978). *Biochem. Pharmacol.* **27,** 97–101.
Furness, J. B. and Costa, M. (1971). *Histochemie* **28,** 324–336.
Glover, V., Sandler, M., Owen, F. and Riley, G. J. (1977). *Nature (London)* **265,** 80–81.
Gomes, B., Igaue, I., Kloepfer, H. and Yasunobu, K. T. (1969). *Arch. Biochem. Biophys.* **132,** 16–27.
Goridis, C. and Neff, N. H. (1971). *Neuropharmacol.* **10,** 557–564.
Goridis, C. and Neff, N. H. (1973). *Biochem. Pharmacol.* **22,** 2501–2510.
Greenawalt, J. W. (1972). *Adv. Biochem. Psychopharmacol.* **5,** 207–226.
Grosso, D. S. and Gawienowski, A. M. (1975). *J. Endocr.* **65,** 207–214.
Harada, M., Mizutani, K. and Nagatsu, T. (1971). *J. Neurochem.* **18,** 549–569.
Hare, M. L. C. (1928). *Biochem. J.* **22,** 968–979.
Hawkins, J. (1952). *Biochem. J.* **50,** 577–581.
Hendley, E. D., Taylor, K. M. and Snyder, S. H. (1970). *Eur. J. Pharmacol.* **12,** 167–179.
Holzbauer, M. and Youdim, M. B. H. (1973). *Br. J. Pharmacol.* **48,** 600–608.
Horita, H. and Lowe, M. C. (1972). *Adv. Biochem. Psychopharmacol.* **5,** 227–242.
Houslay, M. D. and Tipton, K. F. (1973a). *Biochem. J.* **135,** 735–750.
Houslay, M. D. and Tipton, K. F. (1973b). *Biochem. J.* **135,** 173–186.
Houslay, M. D. and Tipton, K. F. (1975). *Biochem. J.* **145,** 311–321.
Houslay, M. D., Tipton, K. F. and Youdim, M. B. H. (1976). *Life Sci.* **19,** 467–473.
Ho-Van-Hap, A., Babineau, L. and Berlinguet, L. (1967). *Can. J. Biochem.* **45,** 355–362.
Huang, R. H. and Eiduson, S. (1977). *J. Biol. Chem.* **252,** 284–290.
Jarrott, B. (1971). *J. Neurochem.* **18,** 7–16.
Jarrott, B. and Iversen, L. L. (1968). *Biochem. Pharmacol.* **17,** 1619–1625.
Johnston, J. P. (1968). *Biochem. Pharmacol.* **17,** 1285–1297.
Jonason, J. (1969). *Acta Physiol. Scand.* Suppl. **320,** 1–50.
Kandaswami, C., Diaz Borges, J. M. and D'Iorio, A. (1977). *Arch. Biochem. Biophys.* **183,** 273–280.
Kearney, E. B., Salach, J. I., Walker, W. H., Seng, R. L., Kenney, W., Zeszotek, E. and Singer, T. P. (1971). *Eur. J. Biochem.* **24,** 321–327.
Klingman, G. I. (1966). *Biochem. Pharmacol.* **15,** 1729–1736.
Knoll, J. and Magyar, K. (1972). *Adv. Biochem. Psychopharmacol.* **5,** 393–408.

Laverty, R., Browne, B. J. and Callingham, B. A. (1973). *J. Pharm. Pharmacol.* **25,** 1001–1002.
Lyles, G. A. and Callingham, B. A. (1974). *J. Pharm. Pharmacol.* **26,** 921–930.
McCauley, R. and Racker, E. (1973). *Molec. Cell. Biochem.* **1,** 73–81.
Maisterrena, B., Comte, J. and Gautheron, D. C. (1974). *Biochim. Biophys. Acta* **367,** 115–126.
Mann, P. J. G. and Quastel, J. H. (1940). *Biochem. J.* **34,** 414–431.
Mantle, T. J., Wilson, K. and Long, R. F. (1975). *Biochem. Pharmacol.* **24,** 2031–2038.
Mantle, T. J., Houslay, M. D., Garrett, N. J. and Tipton, K. F. (1976a). *J. Pharm. Pharmacol.* **28,** 667–671.
Mantle, T. J., Garrett, N. J. and Tipton, K. F. (1976b). *FEBS Lett.* **54,** 227–230.
Marsden, C. A., Broch, A. J. and Guldberg, M. C. (1971). *Eur. J. Pharmacol.* **15,** 335–342.
Murphy, D. L., Donnelly, C. H. and Richelson, E. (1976). *J. Neurochem.* **26,** 1231–1235.
Neff, N. H. and Yang, H-Y. T. (1974). *Life Sci.* **14,** 2061–2074.
Oi, S., Shimada, K., Inamasu, M. and Yasunobu, K. T. (1970). *Arch. Biochem. Biophys.* **139,** 28–37.
Parvez, H. and Parvez, S. (1973). *Acta Endocr., Copen.* **73,** 509–517.
Pletscher, A. (1966). *Pharmacol. Rev.* **18,** 121–129.
Pugh, C. E. M. and Quastel, J. H. (1937). *Biochem. J.* **31,** 288–291.
Sampath, S. S. and Clarke, D. E. (1972). *Life Sci.* **11,** 1037–1048.
Sampath, S. S., Shih, T. M. and Clarke, D. E. (1972). *Fed. Proc.* **31,** 544.
Schnaitman, C., Erwin, V. G. and Greenawalt, J. W. (1967). *J. Cell. Biol.* **32,** 719–735.
Shannon, W. A., Wasserkrug, H. C. and Seligman, A. M. (1974). *J. Histochem. Cytochem.* **22,** 170–182.
Sharman, D. F. (1973). *Br. Med. Bull.* **29,** 110–115.
Silberstein, S. D., Shein, H. M. and Berv, K. R. (1972). *Brain Res.* **41,** 245–248.
Sinha, A. K. and Rose, S. R. (1972). *J. Neurochem.* **19,** 1607–1610.
Snyder, S. H., Fisher, J. and Axelrod, J. (1965). *Biochem. Pharmacol.* **14,** 363–365.
Spector, S., Hirsch, C. W. and Brodie, B. B. (1963). *Int. J. Neuropharmacol.* **2,** 81–93.
Stjärne, L., Roth, R. H. and Giarman, N. J. (1968). *Biochem. Pharmacol.* **17,** 2008–2012.
Tipton, K. F. (1967). *Biochim. Biophys. Acta* **135,** 910–920.
Tipton, K. F. (1968a). *Eur. J. Biochem.* **5,** 316–320.
Tipton, K. F. (1968b). *Biochim. Biophys. Acta* **159,** 451–459.
Tipton, K. F. (1972). *Adv. Biochem. Psychopharmacol.* **5,** 11–24.
Tipton, K. F. (1975). *In* "Handbook of Physiology" (Eds. H. Blaschko and A. D. Smith). Sect. 7., Vol. 6, 677–697. American Physiological Society, Washington.
Tipton, K. F., Houslay, M. D. and Garrett, N. J. (1973). *Nature (London)* **246,** 213–214.
Tipton, K. F., Houslay, M. D. and Mantle, T. J. (1976). *Ciba Found. Symp.* **39,** 5–16.
Trendelenberg, U., Draskoczy, P. R. and Graefe, K. H. (1972). *Adv. Biochem. Psychopharmacol.* **5,** 371–371.
Turner, A. J., Illingworth, J. A. and Tipton, K. F. (1974). *Biochem. J.* **144,** 353–360.
Uretsky, N. J. and Iversen, L. L. (1970). *J. Neurochem.* **17,** 269–278.
Urry, R. L., Dougherty, K. A. and Ellis, L. C. (1975). *Proc. Soc. Exp. Biol. Med.* **148,** 805–807.

Waldmeier, P. C. and Maitre, L. (1975). *Arch. Exp. Path. Pharmac.* **287,** R2-R3.

Went, M. and Csaba, B. (1973). *Acta Physiol. Hung.* **43,** 233–237.

Woods, R. J. (1970). *Proc. Roy. Soc. B* **176,** 43–54.

Wurtman, R. J. and Axelrod, J. (1963). *Biochem. Pharmacol.* **12,** 1417–1419.

Yasunobu, K. T., Ishizaki, H. and Minamiura, N. (1976). *Molec. Cell. Biochem.* **13,** 3–29.

Youdim, M. B. H. (1972). *Adv. Biochem. Psychopharmacol.* **5,** 67–78.

Youdim, M. B. H. (1976). *In* "M.T.P. International Review of Science—Biochemistry" (Ed. H. Blaschko) Series 1, Vol. 12, 189–209. Butterworth, London.

Youdim, M. B. H. and Collins, G. G. S. (1971). *Eur. J. Biochem.* **18,** 73–78.

Youdim, M. B. H. and Holman, B. (1975). *J. Neur. Transmiss.* **37,** 11–24.

Youdim, M. B. H. and Sourkes, T. L. (1972). *Adv. Biochem. Psychopharmacol.* **5,** 45–53.

Zeller, E. A. (1938). *Helv. Chim. Acta* **21,** 880–890.

Zeller, E. A. (1963). *In* "The Enzymes" (Eds. P. D. Boyer, H. Lardy and K. Myrbäck) Vol. 8, 313–335. Academic Press, New York and London.

Zirkle, C. L. and Kaiser, C. (1964). *In* "Psychopharmacological Agents" (Ed. M. Gordon), 445–454. Academic Press, New York and London.

Dopamine Receptors

CHAPTER 13

Dopamine-sensitive Adenylate Cyclase

Richard J. Miller* and John McDermed

Wellcome Research Laboratories,
Research Triangle Park,
North Carolina 27709, U.S.A.

INTRODUCTION

Of all the areas of pharmacology, the neuropharmacology of biogenic amines has probably been the most intensively investigated during the last decade. The reason for this is clearly the involvement of aminergic systems in several pathological states. For example, in the case of dopamine (DA) the degeneration of the nigrostriatal pathway and the resulting deficit in dopaminergic transmission in the basal ganglia are clearly responsible for many of the symptoms of Parkinson's disease (Hornykiewicz, 1973). In addition there is now strong evidence linking the pathogenesis of schizophrenia to some alteration of central dopaminergic function (Matthysse, 1973). Considerable progress has been made in analysing the various aspects of dopaminergic function at the molecular level. Thus the control of the various steps regulating DA synthesis, release and uptake are currently being elucidated with considerable success (this volume). Moreover, our understanding of the mechanism of action of DA at its receptor site(s) has also greatly advanced over the last few years. This is due in large measure to two developments. The first of these was the realization that dopaminergic

* Present address: Department of Pharmacological and Physiological Sciences, University of Chicago, Chicago, Illinois 60637, U.S.A.

transmission may be mediated by cyclic adenosine-3′,5′-monophosphate (cAMP) (Kebabian and Greengard, 1971; Kebabian *et al.*, 1972; Iversen, 1975; Brown and Makman, 1972). The second was the development of receptor-binding assays which enabled direct analysis of the interaction of dopaminergic agonists and antagonists with DA receptors (Burt *et al.*, 1975; Creese *et al.*, 1975a; Seeman *et al.*, 1976; Titeler *et al.*, 1977). This article is concerned with the former area of research.

The discovery that the metabolic effects of noradrenaline, acting via β-adrenergic receptors, were mediated by cAMP (Robinson *et al.*, 1971) and the observation that the specific activity of adenylate cyclase in the CNS was extremely high (DeRobertis *et al.*, 1967) led several groups to speculate that noradrenergic neurotransmission might be mediated by cAMP (Rall and Sattin, 1970; Shimizu *et al.*, 1970). In support of this hypothesis it was found that β-agonists and depolarizing agents were extremely effective in increasing cAMP levels in slices of certain brain areas (Daly, 1977). This hypothesis was also supported by histological and electrophysiological data (Hoffer *et al.*, 1972). Similar evidence for the involvement of cAMP in dopaminergic neurotransmission followed shortly afterwards. In particular it was found that areas containing putative DA receptors contained an adenylate cyclase which was stimulated by DA. This article reviews the physiological and pharmacological evidence linking DA receptors with this enzyme. Discussion of the subsequent role of cAMP in neural function can be found in Nathanson's recent review (1977).

DISTRIBUTION OF THE DOPAMINE-SENSITIVE ADENYLATE CYCLASE

Superior Cervical Ganglion

The first experimental system which was used to link cAMP to dopaminergic transmission was the superior cervical ganglion (SCG). Most of the initial research in this area was carried out by Greengard and his collaborators. It was found that orthodromic stimulation of rabbit SCG increased the cAMP content of this tissue (McAfee *et al.*, 1971; Kalix *et al.*, 1974). It was subsequently found that DA would increase cAMP concentration in bovine SCG and in homogenates of this tissue (Kebabian and Greengard, 1971; Tomasi *et al.*, 1977). Dopamine was effective in stimulating adenylate cyclase in this tissue at fairly low concentration (e.g., >1 μM). Although β-adrenergic agonists such as isoproterenol also increased cAMP levels in these ganglia, the response to DA was clearly not mediated via a β-receptor. The response to isoproterenol but not to DA was inhibited by propranolol, whereas the reverse was found to be true with relatively high concentrations of the α-adrenergic agonist phentolamine (Kebabian and Greengard, 1971; J. Kebabian, personal communication). It was therefore concluded that the effects of DA were mediated by a DA receptor distinct from the β-receptor (Kebabian and Greengard, 1971; Greengard, *et al.*, 1972). It was known from previous histological data that the SCG contained "small intensely fluorescent" (SIF) interneurones. These interneurones had been shown to contain DA (Björklund *et al.*, 1970). Consequently the effects of exogenous DA on cyclic nucleotide metabolism were interpreted as being in-vitro manifestations of dopaminergic transmission normally occurring within the ganglion.

Further support for this concept came from elegant histofluorescence experiments in bovine SCG using fluorescent antibodies to cAMP. Thus, whereas the cAMP accumulation produced by β-adrenergic agonists was shown to occur primarily in glial cells, the response to dopaminergic stimulation was localized solely within postganglionic neurons (Kebabian *et al.*, 1975).

Although some of the electrophysiological data from work in the SCG is now somewhat controversial (Dun and Karczmar, 1977) and large interspecies differences have been found in the response of adenylate cyclase to DA (Cramer *et al.*, 1973), it is reasonable to say that work in this system represents the most complete set of evidence that cAMP mediates dopaminergic neurotransmission.

The Central Nervous System

Although the initial studies on DA-stimulated adenylate cyclase were done on rabbit and bovine SCG, most subsequent studies have been done on the CNS. This is particularly true of the extensive pharmacological studies in which the DA-sensitive adenylate cyclase has been used as a model of the DA receptor (Iversen *et al.*, 1975; Miller and Hiley, 1975). Dopaminergic synapses are localized to certain specific areas of the CNS. From classical histological work performed primarily in Sweden dopaminergic terminals have been described in the caudate nucleus, nucleus accumbens, olfactory tubercule, central amygdaloid nucleus, and median eminence (Miller and Hiley, 1975; Livett, 1973; Ungerstedt, 1971). The cell bodies from which these terminals arise are located in the mesencephalic A9 and A10 areas except in the case of the median eminence, where the cell bodies are found in the arcuate nucleus. More recently Swedish and French workers have demonstrated that dopaminergic systems which innervate the cerebral cortex also arise from cell bodies within the A9 and A10 areas (Lindvall and Björklund, 1974; Thierry *et al.*, 1973). In addition, the retina is known to contain small dopaminergic interneurones (Haggendal and Malmfors, 1963).

Following the work on the SCG two reports appeared which showed that homogenates of bovine retina and rat caudate nucleus contained an adenylate cyclase that was stimulated by DA (Kebabian *et al.*, 1972). Various pharmalogical criteria (see below) indicated that the effects of DA were mediated by a "DA receptor". These reports generated a flurry of activity and the observations were rapidly confirmed and extended in many laboratories (Iversen, 1975). A very large amount of pharmacological data has now been generated, especially using the cyclase in homogenates of caudate (Iversen, 1975). These data, which will be discussed below, have been used both to support the notion that DA-sensitive adenylate cyclase represents the "real" DA receptor(s) (Iversen, 1975) and to contest that notion (Seeman, 1977).

Since the original observations in the caudate and retina, various workers have demonstrated a DA-sensitive adenylate cyclase in all the other areas receiving dopaminergic input, i.e. the olfactory tubercule, nucleus accumbens, median eminence, central amygdaloid nucleus and cortex (Clement-Cormier and Robison, 1977; Clement-Cormier *et al.*, 1974; Horn *et al.*, 1974; Mishra *et al.*, 1975; Trabucchi *et al.*, 1976a; Bockaert *et al.*, 1977). Homogenates of tissue from relevant parts of human brain have also been shown to respond in an analogous fashion (Carenzi *et al.*, 1975).

More recently some groups have performed more detailed studies on the precise distribution of the enzyme within these structures. Bockaert *et al.* (1976) assayed the effects of DA on adenylate cyclase in homogenates of microdiscs punched out of serial frozen sections of rat striatum. There was a progressive decrease in the amount of cAMP produced in the presence of dopamine in the rostrocaudal plane of this structure. The topographical distribution of the DA-sensitive adenylate cyclase was comparable to both the distribution of endogenous DA and the high-affinity synaptosomal uptake system. Interestingly, isoproterenol-stimulated adenylate cyclase in microdisc homogenates was found to be uniformly distributed within the striatum. Similar experiments have been carried out using microdisc punches of the rat frontal cortex (Bockaert *et al.*, 1977). In this case DA had its greatest effect on adenylate cyclase activity in homogenates of the medial portion of the frontal cortex. Another study has examined the distribution of adenylate cyclase activity within the various layers of the olfactory tubercule of the rat (Krieger *et al.*, 1977). Substantial levels of DA-sensitive adenylate cyclase activity were found at all depths of the tubercule. From a consideration of the various cell types and processes found at different laminae in this structure the authors concluded that the enzyme is localized on the apical dendrites of pyramidal cells in the plexiform layer as well as the cell bodies of these neurones.

Several papers have also appeared demonstrating DA-sensitive adenylate cyclase activity in the substantia nigra, particularly in the zona reticulata (Seeber and Kuschinsky, 1976; Spano *et al.*, 1976; Gale, *et al.*, 1977; Phillipson and Horn, 1976; Premont *et al.*, 1976; Kebabian and Saavedra, 1976). Whereas the other areas discussed above all contain classical dopaminergic synapses, the substantia nigra contains only dopaminergic cell bodies and dendrites. One may therefore wonder just what the function of such dopaminergic receptors might be. However, it is known that the localization of the DA-sensitive adenylate cyclase in the nigra is not on the dopaminergic neurones or cell bodies themselves, but probably on neurones passing through or terminating within this area (see below). Now it has been shown that the dendrites of dopaminergic neurones within the nigra may store, take up and release DA in a manner very similar to that of axons (Geffen *et al.*, 1976). Consequently it is felt that such dendrites within the nigra may serve as a store of DA for the stimulation of nigral DA-sensitive adenylate cyclase, although the role of such a receptor remains undefined.

In order to define further the localization of the DA-sensitive adenylate cyclase, lesioning experiments and investigations of the subcellular distribution of the enzyme have been made. Although one study has suggested a glial localization of the enzyme (Henn *et al.*, 1977), initial observations on subcellular distribution reported that the highest specific activity of the enzyme was in the synaptosomal fraction (Clement-Cormier *et al.*, 1975). Subsequent studies have been interpreted to mean that the adenylate cyclase activity is localized in membranes, possibly postsynaptic, apart from presynaptic structures (Laduron *et al.*, 1976; Sieghart *et al.*, 1976). Such an interpretation is in accord with data from lesioning experiments. Recently an electron microscopic examination of structures within the rat caudate has also located the DA-sensitive adenylate cyclase activity attached to the postsynaptic membrane (Rechardt and Härkönen, 1977).

Lesioning experiments utilizing 6-hydroxy-DA (6-OH-DA) or kainic acid have also been carried out, both in the striatum and in the substantia nigra. In neither case did lesions using 6-OH-DA have any effect on enzyme activity, indicating that the enzyme was not associated with dopaminergic neurones themselves, and in particular not with dopaminergic "autoreceptors" which may function in the feedback control of DA synthesis and release (Premont *et al.*, 1976; Von Voigtlander *et al.*, 1973; Krueger *et al.*, 1976; Mishra *et al.*, 1974). However, in the case of kainic acid, which is claimed to selectively destroy neuronal perikarya, a large decrease of DA-sensitive adenylate cyclase activity in the striatum was observed following instrastriatal injection, indicating the presence of adenylate cyclase on neuronal cell bodies. However, these remain to be further identified

(Schwarcz and Coyle, 1977). It should be mentioned that one report suggested that striatal adenylate cyclase activity became more responsive to DA following 6-OH-DA lesions (Mishra *et al.*, 1974). This would fit in nicely with the reported supersensitivity to DA agonists found in animals following nigrostriatal dopaminergic lesions or chronic neuroleptic treatment. However, subsequent studies on DA-sensitive adenylate cyclase from animals rendered behaviorally supersensitive either by lesions or drug treatment have failed to confirm these observations (Krueger *et al.*, 1976; Rotrosen *et al.*, 1975; Von Voigtlander *et al.*, 1975). On the other hand it is interesting to note that binding studies in which DA receptors were labelled with tritiated neuroleptics did seem to reflect the supersensitivity produced by the above-mentioned manipulations (Creese *et al.*, 1977a; Burt *et al.*, 1977).

Other Localizations

In the field of DA pharmacology the renal artery preparation as developed by Goldberg and colleagues has been frequently used to assess dopaminergic activity (Goldberg, 1975). In this preparation DA causes vasodilation via action on a specific dopamine receptor whose topography has been well defined through extensive structure–activity studies (Volkman *et al.*, 1977). While it is not clear that the DA receptor in the renal artery is precisely the same as that found in the CNS (Goldberg *et al.*, 1978), it has been shown that homogenates of this tissue contain a DA-sensitive adenylate cyclase. The effect of DA was specifically blocked by haloperidol but not by propranolol. In contrast the effects of isoproterenol on adenylate cyclase were blocked by propranolol but not by haloperidol (Murthy *et al.*, 1976; Nakajima *et al.*, 1977). However, to date extensive structure–activity studies on this enzyme have not been reported.

Dopamine-sensitive adenylate cyclase has also been reported in ganglia from insects. For example, homogenates of cockroach ganglia contain adenylate cyclase stimulated by isoproterenol, DA and octopamine, all by apparently independent receptor mechanisms. The characteristics of the DA-sensitive adenylate cyclase in such systems appear similar to the mammalian CNS in so far as they have been determined (Nathanson and Greengard, 1973).

PHARMACOLOGY

Apart from its intrinsic interest in the field of neurobiology, the DA-sensitive adenylate cyclase found considerable favour for a time as an in-vitro mechanism for defining the structural requirements of agonists and antagonists at the DA receptor and even as a screening procedure for the pharmaceutical industry. Although such methods may now be generally superseded by the use of DA receptor binding assays, a large body of interesting data has been generated which we shall attempt to summarize below.

Dopamine and Phenylethylamine-related Structures

The fundamental observations made in homogenates of striatum and SCG illustrate one of the major characteristics of the DA-sensitive adenylate cyclase: DA is more potent in this system than noradrenaline and isoproterenol, which have very weak activity (Kebabian *et al.*, 1972). Thus the order of potency is reversed in comparison to the β-adrenergic receptor, for example. Half-maximal stimulation is observed with approximately 5 μm DA, and maximal stimulation at 50 μm or above. These characteristics for DA-sensitive adenylate cyclase in homogenates of tissues ranging from rat caudate to cockroach ganglion seem to be rather consistent. Minor variations do occur from laboratory to laboratory, but these may be rather trivial in nature (Kebabian *et al.*, 1972; Nathanson and Greengard 1973; Miller *et al.*, 1974a; Brown and Makman, 1972; Bucher and Schorderet, 1975). On the other hand certain differences in the effects of DA antagonists in different brain areas may be more significant (see below). Table I shows the results of one study on the structural requirements for DA agonists in the striatal DA-sensitive adenylate cyclase system (Miller *et al.*, 1974a). It can be seen that DA was the most potent agent tested. Only epinine of the simpler structures was comparable with DA in potency. But addition of a second or third methyl group greatly reduced potency. The catechol group also appears to be important, as removal or modification of one or both hydroxyl groups again reduced activity. Modification of the ethylamine side chain as in *l*-noradrenaline also reduced potency. It is interesting to note that *l*-noradrenaline was considerably more potent than *d*-noradrenaline.

The study of semi-rigid analogues of DA has been a useful method for inferring the conformation of DA when binding to its receptor(s). The consequence of constraining the ethylamine side chain of DA into a second ring may be seen in Table II. Compounds 1–3 contain a DA moiety with the ethylamine side chain projected in three different fashions. Compounds 1 and 2 incorporate the folded (or *gauche*) and extended (or *trans*) conformations, respectively. Compounds 2 and 3 both have extended ethylamine groups but represent two possible rotameric extremes, the "β" and "α" rotamers, respectively. The high potency of compound 2 in this

TABLE I Effects of phenylethylamines on rat striatal cAMP production (taken from Miller *et al.*, 1974a).

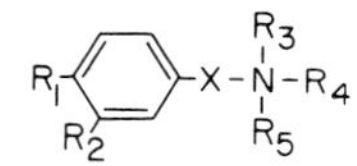

	R_1	R_2	X	R_3	R_4	R_5	Max. Stim. (%)[a]	EC50M[b]
Dopamine	OH	OH	$-(CH_2)_2-$	H	H	—	100	2×10^{-6}
Epinine	OH	OH	$-(CH_2)_2-$	H	CH_3	—	100	1.5×10^{-6}
N,N-Dimethyl dopamine	OH	OH	$-(CH_2)_2-$	CH_3	CH_3	—	48	2×10^{-5}
N,N,N-Trimethyl dopamine	OH	OH	$-(CH_2)_2-$	CH_3	CH_3	CH_3+	30	3×10^{-5}
l-Noradrenaline	OH	OH	$-CH(OH)-CH_2-$	H	H	—	97	4×10^{-5}
D-L-α-Methyl dopamine	OH	OH	$-CH_2-CH(CH_3)-$	H	H	—	58	1×10^{-4}
m-Tyramine	H	OH	$-(CH_2)_2-$	H	H	—	—	—
p-Tyramine	OH	H	$-(CH_2)_2-$	H	H	—	—	—
3,4-Dihydroxybenzylamine	OH	OH	$-CH_2-$	H	H	—	—	—
3,4-Dihydroxypropylamine	OH	OH	$-(CH_2)_3-$	H	H	—	—	—
3,4-Dihydroxybutylamine	OH	OH	$-(CH_2)_4-$	H	H	—	—	—
d-Noradrenaline	OH	OH	$-CH(OH)-CH_2-$	H	H	—	—	—
						—	—	—
m-Methoxytyramine	OH	CH_3O	$-(CH_2)_2-$	H	H	—	—	—
p-Methoxytyramine	CH_3O	OH	$-(CH_2)_2-$	H	H	—	—	—
L-DOPA‡	OH	OH	$-CH_2-CH(COOH)-$	H	H	—	—	—
d,L-Amphetamine	H	H	$-CH_2-CH(CH_3)-$	H	H	—	—	—

[a] Stimulation produced by 100 μM dopamine is taken as 100%.

[b] EC50 refers to concentration required to give half maximal stimulation produced between 10^{-6} and 10^{-3} M. Agents shown as giving no stimulation of cyclic AMP production were not significantly effective ($P > 0.05$) at concentrations between 10^{-6} and 10 M.

preparation compared to both 1 and 3 implies that the preferred conformation of the flexible DA molecule in stimulating its adenylate cyclase is the extended form (Miller *et al.*, 1974a; Sheppart and Burghardt, 1974a,b), and more particularly the "β" rotamer (Woodruff *et al.*, 1977; Kotake *et al.*, 1977).

TABLE II Effects of bicyclic amines on rat striatal cAMP production

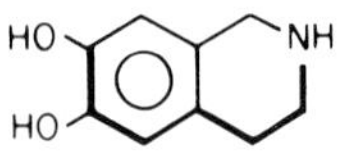

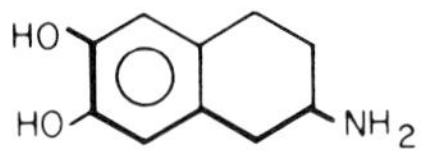

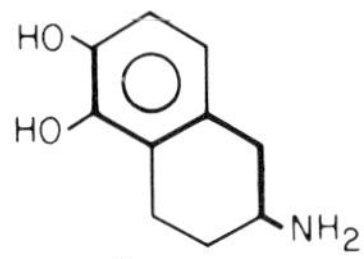

	1	2	3
	(*gauche*)	(*trans*, "β")	(*trans*, "α")
EC_{50}(M)	2×10^{-5} [a]	3.5×10^{-6} [b]	2.0×10^{-4} [b]

[a] Miller, *et al.* (1974a).
[b] Woodruff, *et al.* (1977).

Some discrepancies appear to exist between in-vivo activity of several molecules and their action on the DA-sensitive adenylate cyclase. For example, certain monohydroxyl analogues of 2-amino-1,2,3,4-tetrahydronaphthalene (McDermed *et al.*, 1976) and of aporphine (Neumeyer *et al.*, 1974, 1976) have quite reasonable in-vivo activity in tests of dopaminergic agonist activity but are rather weak in the DA-sensitive adenylate cyclase model (M. Schorderet, personal communication; Miller *et al.*, 1976). In addition, analogues based on the 5,6-dihydroxyl compound 3 (Table II), have a smaller effect on the DA-sensitive adenylate cyclase model than might be expected from their activity in the CNS (Costall *et al.*, 1977; McDermed *et al.*, 1975).

Aporphines

Table III illustrates the fact that the aporphine molecule embodies a phenylethylamine profile in both the folded and extended forms. Thus it is not surprising that properly substituted aporphines such as apomorphine show DA agonist activity. The ability of apomorphine to cross the blood–brain barrier has made it the most widely used DA agonist in investigations of central dopaminergic functions (Anden *et al.*, 1967). The homologous *N*-(*n*-propyl)-norapomorphine is even more potent *in vivo* (Schoenfeld *et al.*, 1975; Kelly *et al.*, 1976).

TABLE III Effects of aporphines on rat striatal cAMP production

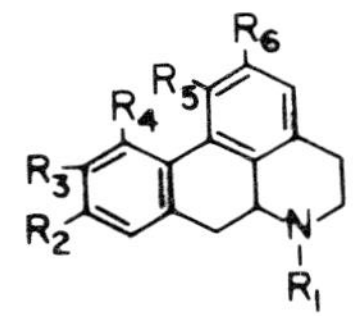

Name	Configuration	R_1	R_2	R_3	R_4	R_5	R_6	Max. stim. (%)	EC_{50} (M)	% Blockade[a]
Aporphine	(±)	CH_3	H	H	H	H	H	—	—	11±4 NS
11-*OH*-*N*-*n*-Propylnoraporphine	(±)	C_3H_7	H	OH	H	H	H	—	—	22±7 ($p<0.02$)
Apomorphine	(−)	CH_3	H	OH	OH	H	H	45	2×10^{-6}	38±5 ($p<0.001$)
N-*n*-Propylnorapomorphine	(±)	C_3H_7	H	OH	OH	H	H	41	4×10^{-6}	30±6 ($p<0.01$)
Isoapomorphine	(±)	CH_3	OH	OH	H	H	H	—	—	21±2 ($p<0.05$)
N-*n*-Propylnorapocodeine	(±)	C_3H_7	H	CH_3O	OH	H	H	—	—	15±3 ($p<0.05$)
Dimethoxyapomorphine	(−)	CH_3	H	CH_3O	CH_3O	H	H	—	—	NT
10,11-Methylenedioxyaporphine	(−)	CH_3	H	—O—CH_2—O—		H	H	—	—	NT
1,2-Dihydroxyaporphine	(−)	CH_3	H	H	H	OH	OH	—	—	26±4 ($p<0.02$)
Nuceferine	(−)	CH_3	H	H	H	CH_3O	CH_3O	—	—	24±3 ($p<0.02$)
Bulbocapnine	(+)	CH_3	H	CH_3O	OH	—O—CH_2—O—		—	—	62±4 ($p<0.001$)

[a] Blockade refers to % inhibition of response to 100 μM dopamine when drugs were tested for antagonist properties at 10 μM. Figures represent means for at least 4 separate incubations. Figures in parentheses show degree of significance of this blockade against dopamine (10^{-4} M) stimulated activity. Mean basal activity was 31.8±2.3 pmoles cyclic AMP per incubation (2 mg wet weight) and in the presence of 10^{-4} M dopamine 66.2±5.5 per incubation. EC_{50} refers to agonist concentration giving half maximal stimulation of adenylate cyclase. NS = not significant; NT = not tested.

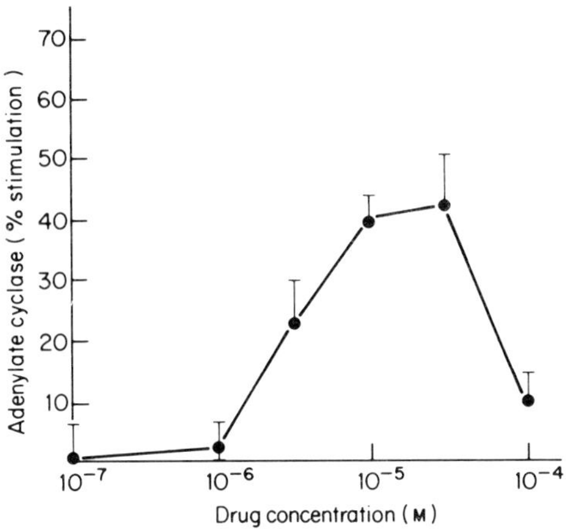

Fig. 1. Effects of *N*-(*n*-propyl)-norapomorphine on rat striatal cAMP production. (Miller *et al.*, 1976.)

Table III illustrates the activity of apomorphine and several related compounds on DA-sensitive adenylate cyclase in homogenates of rat striatum. It can be seen that only apomorphine and its *N*-(*n*-propyl) homologue are effective in stimulating the cyclase. Figure 1 shows that *N*-(*n*-propyl)-norapomorphine is not a full agonist in this preparation, but in fact behaves as a partial agonist (Kebabian *et al.*, 1972; Miller, *et al.*, 1974a). Apomorphine behaves the same way. At low concentrations they stimulate the cyclase, but as higher concentrations are reached inhibitory effects are apparent. It is interesting to note that evidence for partial agonist activity of apomorphine has also been obtained in DA receptor binding studies (Creese *et al.*, 1975b).

It may be seen in Table III that the 1,2-dihydroxyaporphine and the 9,10-dihydroxyaporphine (isoapomorphine) are ineffective, a fact that is consistent with their inactivity *in vivo*. The former compound displays the DA side chain in the unfavourable folded conformation and thus is consistent with the results in Table II for the corresponding bicyclic structure. However, the inactivity of isoapomorphine is remarkable. Taking as models the bicyclic compounds 2 and 3 of Table II, one would predict precisely the opposite order of activity for apomorphine and isoapomorphine. This enigma has been the subject of considerable speculation, but to date no fully satisfying explanation has been put forth. On the whole the results in Table III correlate well with in-vivo activity of these molecules. An exception is 11-hydroxy-*N*-(*n*-propyl)-noraporphine, which has significant activity *in vivo* (Neumeyer, *et al.*, 1976).

Bulbocapnine produces profound cataleptic effects, an activity typical of dopaminergic antagonists (Tseng *et al.*, 1973). Table III shows that this molecule is not an agonist in the cyclase system and in fact shows higher antagonist activity than apomorphine.

Ergot Alkaloids

Several reports have appeared in the literature indicating that LSD and ergot alkaloids such as bromoergocryptine can stimulate central DA receptors as assessed by various behavioural tests and also by criteria such as inhibition of prolactin release (Corrodi *et al.*, 1973; Fuxe *et al.*, 1975; Meltzer *et al.*, 1977; Sachdev *et al.*, 1975; Silbergeld and Pfeiffer, 1977; Pieri and Pieri, 1974). In addition several ergot alkaloids including bromoergocryptine have been shown to be therapeutic in Parkinson's disease, again indicating their DA agonist effects (Calne *et al.*, 1974; Lieberman *et al.*, 1975). From these observations it would be reasonable that such drugs would act as agonists in the DA-sensitive adenylate cyclase preparations. Indeed it has been shown that LSD is effective in this respect, although it is not immediately obvious from inspection of its structure which parts of this molecule are likely to be involved in binding to DA receptors for agonist activity. Stimulation of striatal adenylate cyclase by LSD is effectively blocked by neuroleptic drugs (Von Hungen *et al.*, 1975; Spano *et al.*, 1975). In addition, Bockaert *et al.* (1976) found that the distribution of LSD-sensitive adenylate cyclase in homogenates of striatal microdiscs exactly corresponded to that of DA-sensitive adenylate cyclase. However, LSD clearly does not function as a full agonist in this system, but acts as a partial agonist. Again, as in the case of apomorphine, partial agonist activity would be predicted from dopamine receptor binding studies (Creese *et al.*, 1975b). Several other ergot alkaloids (e.g. lergotrile, bromoergocryptine, and ergotamine) which are also DA agonists by several in-vivo criteria, do not stimulate striatal adenylate cyclase at all. In fact these substances appear to act as DA antagonists in this cyclase system (Govoni *et al.*, 1977; Trabucchi *et al.*, 1976b; Kebabian *et al.*, 1977).

Clearly there is a major discrepancy here between in-vivo data and the DA-sensitive adenylate cyclase system. The usual explanation given for this discrepancy is that there may be more than one type of DA receptor in the CNS and that *in vivo* these compounds are acting on DA receptors which are not linked to adenylate cyclase.* However no direct biochemical evidence supporting this view has yet appeared. It must also be kept in mind that apart from their affects in dopaminergic systems, the ergot alkaloids also interact strongly with serotoninergic systems (Aghajanian, 1972).

* See note added in proof, p. 177.

ET 495

A few years ago it seemed possible that the drug piribedil (ET 495) might be useful in treating Parkinson's disease. This drug shows a typical DA agonist profile "*in vivo*". It was hypothesized that the agonist action of ET 495 *in vivo* might result from metabolism of the molecule to the corresponding catechol S 584 (Miller and Iversen, 1974a). In accord with this proposal it was reported that S 584 but not ET 495 was effective in stimulating striatal DA-sensitive adenylate cyclase (Miller and Iversen, 1974a; Mishra *et al.*, 1974). However subsequent experiments by other workers have failed to confirm the agonist activity of S 584. The reason for this discrepancy is not clear (Schorderet, 1975).

Dopamine Antagonists

The advent of neuroleptic drugs with specific antipsychotic effects has revolutionized the treatment of schizophrenia. The mechanism of action of such drugs has been the subject of very intensive investigation, as it has been hoped that elucidation of their mode of action would illuminate the aetiology of schizophrenia and ultimately lead to design of better drugs. The most generally held hypothesis suggests that since neuroleptic drugs act as DA antagonists, their therapeutic effect must therefore imply hyperactivity of some dopaminergic system(s) in schizophrenia. The evidence for this hypothesis has been evaluated (Hornykiewicz, 1977).

The evidence that neuroleptic drugs are generally DA antagonists is very good. Neuroleptics increase serum prolactin levels (Login and MacLeod, 1977), block the central effects of L-DOPA and apomorphine (Kelly and Miller, 1975), inhibit the effects of DA in the renal artery (Yeh *et al.*, 1969), and cause a syndrome resembling Parkinson's disease as a common side-effect in man (Calne, 1976). It is consistent with these findings that neuroleptics of varying structural types are effective inhibitors of the DA-sensitive adenylate cyclase (Kebabian *et al.*, 1972; Miller *et al.*, 1974b; Iversen *et al.*, 1976; Miller *et al.*, 1975; Miller and Iversen, 1974b). Thus, phenothiazines, butyrophenones, thioxanthenes, and dibenzodiazepines are all effective. In general drugs with high neuroleptic activity are potent inhibitors of DA-sensitive adenylate cyclase. However the overall correlation between inhibition of the adenylate cyclase and neuroleptic activity is highly variable, ranging from good in some structural types to non-existent in others. In some cases the positive correlations expected are clearly seen. Drugs that exist in different isomeric forms exhibit the same stereospecificity on the DA-sensitive adenylate cyclase as they do behaviourally. For example the *cis* geometrical isomer of flupenthixol is potent in both respects while the *trans* isomer is inactive in both (Miller *et al.*, 1974b). Similarly, the *dextro* enantiomer of butaclamol displays both types of activity and the *laevo* enantiomer displays neither (Miller *et al.*, 1975; Lippman *et al.*, 1975). There is a fair correlation between clinical potency of phenothiazines as antischizophrenics and their ability to inhibit DA-sensitive adenylate cyclase (Seeman, 1977).

However when considering other drug classes, difficulties arise. Some of these are as follows: (a) some drugs of the tricyclic antidepressant series which are not believed to have significant dopaminergic blocking activity are as effective as inhibitors of DA-sensitive adenylate cyclase as are the butyrophenones (Karobath, 1975); (b) the drugs metoclopramide and sulpiride appear to have central DA receptor antagonist activity but do not block DA-sensitive adenylate cyclase (Roufogalis, *et al.*, 1976; Peringer, *et al.*, 1976); (c) certain butyrophenones (haloperidol) and diphenylbutylamines (pimozide) are especially potent neuroleptics but are only marginally effective inhibitors of adenylate cyclase. Their potency against the adenylate cyclase seems far too low when compared to the phenothiazines (Kebabian *et al.*, 1972; Miller *et al.*, 1974b; Iversen *et al.*, 1976). However, it must be cautioned that the high clinical potency of the butyrophenones may owe significantly to presynaptic events not associated with adenylate cyclase. These compounds have been reported to be potent inhibitors of the evoked release of DA from terminals, and such an effect should be additive to postsynaptic blockade (Seeman and Lee, 1975). There is an interesting report that haloperidol is much more effective in inhibiting the stimulation of adenylate cyclase by DA in the cortex than in the striatum (Bockaert *et al.*, 1977). Such a finding could be very significant if it can be confirmed, as it is likely that schizophrenia involves DA effects in the cortex rather than in the basal ganglia.

In contrast to these observations the relative effects of neuroleptics on ^{3}H-haloperidol binding to DA receptor preparations show a spectacularly good correlation with clinical efficacy (Creese *et al.*, 1976; Seeman *et al.*, 1975). It should be noted that if the DA receptor is labelled with ^{3}H-DA instead of an antagonist, displacement studies with neuroleptics generate a potency series more like that for the DA-sensitive adenylate cyclase, i.e. butyrophenones are rather weak (Creese *et al.*, 1976). It is possible that when the adenylate cyclase is stimulated by DA, the receptor is in its agonist conformation, and that is why DA-sensitive adenylate cyclase data and ^{3}H-DA binding data are similar. On the other hand a much more precise correlation of antagonist effects is seen when a system that examines effects of neuroleptics on the antagonist conformation of the receptor is used, i.e. DA receptor binding assays using a labelled antagonistic such as ^{3}H-haloperidol. However other evidence suggests that the DA-sensitive adenylate cyclase and the DA receptor as measured in binding assays are not part of the same entity. Thus striatal kainic acid injections which reduce DA-sensitive adenylate cyclase activity reduce ^{3}H-haloperidol binding far less. In addition the two entities appear to have a different subcellular distribution (Leysen and Laduron, 1977).

CONCLUSION

It is clear from the preceding discussions that a DA-sensitive adenylate cyclase exists in many of the places that one would expect to find DA receptors. However, too much contrary pharmacological evidence exists to identify comfortably DA-sensitive adenylate cyclase as "the" DA receptor in these locations. Thus the effects of some ergot alkaloids and several classes of neuroleptics are difficult to explain.

It must be noted here that the role of cAMP in neural function is not understood (Nathanson, 1977). Some histological (Kebabian *et al.*, 1975), electrophysiological (Siggins *et al.*, 1974), and behavioural data (Miller and Kelly, 1975) suggest that cAMP may play a role in neurotransmission. However, other data do not support this explanation. Alternatively, it is tempting to think that cAMP may serve to maintain the ongoing activity of the neurone by regulating glycogen metabolism or something of the sort (Anchors and Garcia-Rill, 1977).

We must therefore conclude that at present the physiological significance of DA-sensitive adenylate cyclase stimulation as a manifestation of DA receptor activation is unclear and will likely remain so until the relationship of cAMP accumulation to neural function is better defined.

REFERENCES

Aghajanian, G. K. (1972). *Fed. Proc.* **31,** 91–96.

Anden, N. E., Rubenson, A., Fuxe, K. and Hokfelt, T. (1967). *J. Pharm. Pharmacol.* **19,** 627–629.

Anchors, J. M. and Garcia-Rill, E. (1977). *Brain Res.* **133,** 183–189.

Björklund, A., Cegrell, L., Falck, B., Ritzen, M. and Rosengen, E. (1970). *Acta Physiol. Scand.* **78,** 334–338.

Bockaert, J., Premont, J., Glowinski, J., Thierry, A. and Tassin, J. (1976). *Brain Res.* **107,** 303–315.

Bockaert, J., Tassin, J., Thierry, A., Glowinski, J. and Premont, J. (1977). *Brain Res.* **122,** 71–86.

Brown, J. and Makman, M. (1972). *Proc. Natl Acad. Sci., U.S.A.* **69,** 539–543.

Bucher, M. B. and Schorderet, M. (1975). *Naunyn-Schmiedeberg's Arch. Pharmacol.* **288,** 103–107.

Burt, D., Enna, S., Creese, I. and Snyder, S. (1975). *Proc. Natl Acad. Sci., U.S.A.* **72,** 4655–4659.
Burt, D., Creese, I. and Snyder, S. (1977). *Science* **196,** 326–328.
Calne, D. B., Teychenne, P. F., Claveria, L. E., Eastman, R., Greenacre, J. and Petrie, A. (1974). *Br. Med. J.* **4,** 442–444.
Calne, D. B. (1976). *In* "Biochemistry and Neurology" (Eds H. F. Bradford and C. B. Marsden), 21–27. Academic Press, New York and London.
Carenzi, A., Gillin, J. C., Guidotti, A., Schwartz, M. A., Trabucchi, M. and Wyatt, R. J. (1975). *Arch. Gen. Psych.* **32,** 1056–1059.
Clement-Cormier, Y. C., Kebabian, J. W., Petzold, G. L. and Greengard, P. (1974). *Proc. Natl Acad. Sci. U.S.A.* **71,** 1113–1117.
Clement-Cormier, Y. C., Parrish, R. G., Petzold, G. L., Kebabian, J. W. and Greengard, P. (1975). *J. Neurochem.* **25,** 143–149.
Clement-Cormier, Y. C. and Robison, G. A. (1977). *Biochem. Pharmacol.* **26,** 1719–1722.
Corrodi, H., Fuxe, K., Hokfelt, T., Lidbrink, P. and Ungerstedt, U. (1973). *J. Pharm. Pharmacol.* **25,** 409–411.
Costall, B. Naylor, R. J., Cannon, J. G. and Lee, T. (1977). *Eur. J. Pharmacol.* **41,** 307–319.
Cramer, H., Johnson, D. G., Hanbauer, I., Silberstein, S. D. and Kopin, I. J. (1973). *Brain Res.* **53,** 97–104.
Creese, I., Burt, D. R. and Snyder, S. H. (1975a). *Life Sci.* **17,** 993–1002.
Creese, I., Burt, D. R. and Snyder, S. H. (1975b). *Life Sci.* **17,** 1715–1720.
Creese, I., Burt, D. R. and Snyder, S. H. (1976). *Science* **192,** 481–483.
Creese, I., Burt, D. R. and Snyder, S. H. (1977a). *Science* **197,** 596–598.
Creese, I., Schwarcz, R., Coyle, J. and Snyder, S. H. (1977b). *Society for Neurosciences, Abst.* **3,** 454.
Daly, J. W. (1977). "Cyclic Nucleotides in the Nervous System." Plenum, New York.
DeRobertis, E., Arnaiz, G., Butcher, R. W. and Sutherland, E. W. (1967). *J. Biol. Chem.* **242,** 3487–3493.
Dun, N. J. and Karczmar, A. G. (1977). *J. Pharmacol. Exp. Ther.* **202,** 89–96.
Elliot, P. N. C., Jenner, P., Hurzing, G., Marsden, C. D. and Miller, R. J. (1977). *Neuropharmacol.* **16,** 333–342.
Fuxe, K., Ågnati, L. F., Corrodi, H., Everitt, B., Hokfelt, T., Logstrom, A. and Ungerstedt, U. (1975). *In* "Advances in Neurology", Vol. 9, 223–242. (Eds. D. B. Calne, T. N. Chase and A. Barbeau) Raven Press, New York.
Gale, K., Guidotti, A. and Costa, E. (1977). *Science* **195,** 503–505.
Geffen, L. B., Jessell, T. M., Cuello, A. C. and Iversen, L. L. (1976). *Nature (London)* **260,** 258–260.
Goldberg, L. I. (1975). *Biochem. Pharmacol.* **24,** 651–653.
Goldberg, L. I., Volkman, P. H. and Kohli, J. D. (1978). *A. Rev. Pharmacol. Toxicol.* **18,** 57–79.
Govoni, S., Iuliano, E., Spano, P. F. and Trabucchi, M. (1977). *J. Pharm. Pharmacol.* **29,** 45–47.
Greengard, P., McAfee, D. A. and Kebabian, J. W. (1972). In "Advances in Cyclic Nucleotide Resezrch", Vol. I (Eds P. Greengard and G. A. Robison), 337–355. Raven Press, New York.
Haggendal, J. and Malmfors, T. (1963). *Acta Physiol. Scand.* **59,** 295–296.
Henn, F. A., Anderson, D. J. and Sellstrom, A. (1977). *Nature (London)* **266,** 637–638.

Hoffer, B. J., Siggins, G. R., Oliver, A. D. and Bloom, F. E. (1972). *In* "Advances in Cyclic Nucleotide Research" Vol. **2,** 411–423 (Eds G. A. Robison, R. Paoletti and P. Greengard). Raven Press, New York.
Horn, A. S., Cuello, A. C., and Miller, R. J. (1974). *J. Neurochem.* **22,** 265–270.
Hornykiewicz, O. (1973). *Br. Med. Bull.* **29,** 172–178.
Hornykiewicz, O. (1977). *A. Rev. Pharmacol. Toxicol.* **17,** 545–559.
Iversen, L. L., (1975). *Science* **188,** 1084–1089.
Iversen, L. L., Horn, A. S., and Miller, R. J. (1975). *In* "Advances in Neurology" Vol. 9, (Eds D. B. Calne, T. N. Chase, and A. Barbeau), 197–212. Raven Press, New York.
Iversen, L. L., Rogawski, N. J. and Miller, R. J. (1976). *Molec. Pharmacol.* **12,** 251–262.
Kalix, P., McAfee, D. A., Schorderet, M. and Greengard, P. (1974). *J. Pharmacol. Exp. Ther.* **188,** 676–687.
Karobath, M. D. (1975). *Eur. J. Pharmacol.* **30,** 159–163.
Kebabian, J. W. and Greengard, P. (1971). *Science* **174,** 1346–1349.
Kebabian, J. W. and Saavedra, J. M. (1976). *Science* **193,** 683–685.
Kebabian, J. W., Petzold, G. L. and G. L. and Greengard, P. (1972). *Proc. Natl Acad. Sci., U.S.A.* **69,** 2145–2149.
Kebabian, J. W., Bloom, F. E., Steiner, A. L. and Greengard, P. (1975). *Science* **190,** 157–159.
Kebabian, J. W., Calne, D. B. and Kebabian, P. R. (1977). *Commun. Psychopharmacol.* **1,** 311–318.
Kelly, P. H. and Miller, R. J. (1975). *Br. J. Pharmacol.* **54,** 115–121.
Kelly, P. H., Miller, R. J. and Neumeyer, J. L. (1976). *Eur. J. Pharmacol.* **35,** 85–92.
Krieger, N. R., Kauer, J. S., Shepherd, G. M. and Greengard, P. (1977) *Brain Res.* **131,** 303–312.
Krueger, B. K., Forn, J., Walters, J. R., Roth, R. H. and Greengard, P. (1976). *Molec. Pharmacol.* **12,** 639–648.
Kotake, C., Goldberg, L., Hoffmann, P. and Cannon, J. (1977). *Pharmacologist* (in press).
Laduron, P., Verwimp, M., Janssen, P. F. M., and Leysen, J. (1976). *Life Sci.* **18,** 433–440.
Leysen, J. and Laduron, P. (1977). *Life Sci.* **20,** 281–288.
Lieberman, A., Miyamoto, T., Battista, A. F., and Goldstein, M. (1975). *Neurology,* **25,** 459–462.
Lindvall, O. and Björklund, A. (1974). *Acta Physiol. Scand. Suppl. 412,* 1–48.
Lindvall, O. and Björklund, A., Moore, R. Y. and Stenevi, U. (1974). *Brain Res.* **81,** 325–331.
Lippmann, W., Pugsley, T., and Merker, J. (1975). *Life Sci.* **16,** 213–224.
Livett, B. G. (1973). *Br. Med. Bull.* **29,** 93–99.
Login, I. S. and McLeod, R. M. (1977). *Brain Res.* **132,** 477–483.
Matthysse, S. (1973). *Fed. Proc.* **32,** 200–205.
McAfee, D. A., Schorderet, N. and Greengard P. (1971). *Science* **171,** 1156–1158.
McDermed, J. D., McKenzie, G. M. and Phillips, A. P. (1975). *J. Med. Chem.* **18,** 362–368.
McDermed, J. D., McKenzie, G. M. and Freeman, H. S. (1976). *J. Med. Chem.* **19,** 547–549.
Meltzer, H. Y., Fessler, R. G., Simanovic, M., Doherty, J. and Fang, V. S. (1977). *Psychopharmacologia* **54,** 39–44.

Miller, R. J. (1976). *Biochem. Pharmacol.* **25,** 537–545.
Miller, R. J. and Hiley, C. R. (1975). *In* "Advances in Neurology" Vol. 9, (Eds D. B. Calne, T. N. Chase, and A. Barbeau), 197–212. Raven Press, New York.
Miller, R. J. and Iversen, L. L. (1974a). *Naunyn-Schmiedeberg's Arch. Pharmacol.* **282,** 213–217.
Miller, R. J. and Iversen, L. L. (1974b). *J. Pharm. Pharmacol.* **26,** 142–145.
Miller, R. J. and Kelly, P. H. (1975). *Nature (London)* **255,** 163–167.
Miller, R. J., Horn, A. S., Iversen, L. L., and Pinder, R. (1974a). *Nature (London)* **250,** 238–241.
Miller, R. J., Horn, A. S. and Iversen, L. L. (1974b). *Molec. Pharmacol.* **10,** 759–766.
Miller, R. J., Horn, A. S., and Iversen, L. L. (1975). *J. Pharm. Pharmacol.* **27,** 212–215.
Miller, R. J., Kelly, P. H. and Neumeyer, J. L. (1976). *Eur. J. Pharmacol.* **35,** 77–83.
Mishra, R. K., Gardner, E. C., Katzman, R. and Makman, M. H. (1974). *Proc. Natl Acad. Sci. U.S.A.* **71,** 3883–3887.
Mishra, R. K., Demirjian, C. and Katzman, R. (1975). *Brain Res.* **96,** 395–399.
Munday, K. A., Poat, J. A. and Woodruff, G. N. (1974). *J. Physiol.* **241,** 119–120.
Murthy, V. V., Gilbert, J. C., Goldberg, L. I. and Kuo, J. F. (1976). *J. Pharm. Pharmacol.* **28,** 567–571.
Nakajima, T., Naitoh, F. and Kuruma, I. (1977). *Eur. J. Pharmacol.* **41,** 163–169.
Nathanson, J. A. (1977). *Physiol. Rev.* **57,** 157–256.
Nathanson, J. A. and Greengard, P. (1973). *Science* **174,** 1346–1349.
Neumeyer, J. L., Granchelli, F. E., Fuxe, K., Ungerstedt, U. and Corrodi, H. (1974). *J. Med. Chem.* **17,** 1090–1094.
Neumeyer, J. L., Reinhard, J., Dafeldecker, W., Kosersky, D., Fuxe, K. and Agnati, L. (1976). *J. Med. Chem.* **19,** 25–29.
Peringer, E., Jenner, P., Donaldson, M., Marsden, C. D. and Miller R. J. (1976). *Neuropharmacology* **15,** 463–469.
Phillipson, O. and Horn, A. S. (1976). *Nature (London)* **261,** 418–420.
Pieri, L. and Pieri, H. (1974). *J. Pharmacol. Suppl. 2,* **5,** 77.
Premont, J., Thierry, A. M., Tassin, J. P., Glowinski, J. Blanc, G. and Bockaert, J. (1976). *FEBS Lett.* **68,** 99–104.
Rall, T. W. and Sattin, A. (1970). *In* "The Role of cAMP in Cell Function", (Eds P. Greengard and E. Costa) 113–134. Raven Press, New York.
Rechardt, L. and Härkönen, M. (1977). *Histochemistry* **51,** 113–119.
Robison, G., Butcher, R. W. and Sutherland, E. W. (1971). "Cyclic AMP." Academic Press, New York and London.
Rotrosen, J., Friedman, E. and Gershon, S. (1975). *Life Sci.* **17,** 563–568.
Roufogalis, B. D., Thornton, M. and Wade, D. N. (1976). *Life Sci.* **19,** 927–934.
Sachdev, Y., Turnbridge, W. R. G., Weightman, D. R., Gomez-Pan, D. A., Duns, A., Hall, R. and Gollamali, S. K. (1975). *Lancet* ii, 1164–1168.
Schoenfeld, R. I., Neumeyer, J. L., Dafeldecker, W. and Roffer-Tarlov, S. (1975). *Eur. J. Pharmacol.* **30,** 63–68.
Schorderet, M. (1975). *Experientia* **31,** 1325–1327.
Schwarcz, R. and Coyle, J. T. (1977). *Life Sci.* **20,** 431–436.
Seeber, U. and Kuschinsky, K. (1976). *Experientia* **32,** 1558–1559.
Seeman, P. (1977). *Biochem. Pharmacol.* **26,** 1741–1748.
Seeman, P. and Lee, T. (1975). *Science* **188,** 1217–1219.
Seeman, P., Lee, T., Chau-Wong, M. Tedesco, J. and Wong, K. (1976). *Proc. Natl Acad. Sci., U.S.A.* **73,** 4354–4358.

Sheppard, H. and Burghardt, C. R. (1974a). *Res. Commun. Chem. Pathol. Pharmacol.* **8,** 527–534.
Sheppard, H. and Burghardt, C. R. (1974b). *Molec. Pharmacol.* **10,** 721–726.
Sheppard, H., Burghardt, C. R. and Teitel, S. (1976). *Molec. Pharmacol.* **12,** 854–861.
Shimzu, H., Creveling, C. R. and Daly, J. W. (1970). *In* "The Role of cAMP in Cell Function", (Eds P. Greengard and E. Costa), 135–155. Raven Press, New York.
Sieghart, W., Leitch, H. and Karobath, M. (1976). *Brain Res.* **109,** 418–422.
Siggins, G. R., Hoffer, B. J. and Ungerstedt, U. (1974). *Life Sci.* **15,** 779–792.
Silbergeld, E. K. and Pfeiffer, R. F. (1977). *J. Neurochem.* **28,** 1323–1326.
Spano, P. F., Kumakura, K., Tonon, G., Govoni, S. and Trabucchi, M. (1975). *Brain Res.* **93,** 164–167.
Spano, P. F., DiChiara, G., Tonon, G. and Trabucchi, M. (1976). *J. Neurochem.* **27,** 1565–1568.
Thierry, A. M., Blanc, G., Sobel, A., Stinus, L. and Glowinski, J. (1973). *Science* **182,** 499–501.
Titeler, M., Weinrich, P. and Seeman, P. (1977). *Proc. Natl Acad. Sci., U.S.A.* **74,** 3750–3753.
Tomasi, V., Biondi, C., Trevisani, A., Martini, M. and Perri, V. (1977). *J. Neurochem.* **28,** 1289–1297.
Trabucchi, M., Govoni, S., Tonon, G. C. and Spano, P. F. (1976a). *J. Pharm. Pharmacol.* **28,** 244–245.
Trabucchi, M., Spano, P. F., Tonon, G. C. and Frattola, L. (1976b). *Life Sci.* **19,** 225–231.
Tseng, L. S., Wei. E. and Loh, H. (1973). *Eur. J. Pharmacol.* **22,** 363–367.
Ungerstedt, U. (1971). *Acta Physiol. Scand. Suppl.* **357,** 1–18.
Volkman, P. H., Kohli, J. D., Goldberg, L. I., Cannon, J. G., and Lee, T. (1977). *Proc. Natl Acad. Sci., U.S.A.* **74,** 3602–3606.
Von Hungen, K., Roberts, S. and Hill, D. F. (1975). *Brain Res.* **94,** 57–66.
Von Voigtlander, P. F., Boukma, S. J. and Johnson, G. A. (1973). *Neuropharmacology* **12,** 1081–1086.
Von Voigtlander, P. F., Losey, E. G. and Triezenberg, H. J. (1975). *J. Pharmacol. Exp. Ther.* **193,** 88–94.
Woodruff, G. N., Elkawad, A. O. and Pinder, R. M. (1974). *Eur. J. Pharmacol.* **25,** 80–86.
Woodruff, G. N., Watling, K. J., Andrews, C. D., Poat, J. A. and McDermed, J. D. (1977). *J. Pharm. Pharmacol.* **29,** 422–427.
Yeh, B. K., McNay, J. and Goldberg, L. I. (1969). *J. Pharmacol. Exp. Ther.* **168,** 303–313.

For a full discussion of DA receptors which are and are not linked to an adenylate cyclase, as well as a classification of multiple DA receptors, see Kebabian, J. W. and Calne, D. B. (1979). *Nature (London)* **277,** 93–96.

CHAPTER 14

In-Vitro Measurement of Brain Receptors for Dopamine and Neuroleptics

Milt Titeler and Philip Seeman

University of Toronto, Toronto, Ontario, Canada.

CRITERIA FOR DOPAMINE RECEPTORS

Abnormal function of neurotransmitter receptors has been implicated in a variety of disease states such as schizophrenia, tardive dyskinesia, Parkinson's disease and Huntington's chorea. The development of the direct binding assays offers a promising new tool for investigating the aetiology of such diseases.

Direct binding assays offer several distinct advantages over other in-vivo or in-vitro approaches to studying neurotransmitter receptors. The binding properties of the receptor can be examined in detail uncomplicated by subsequent biochemical and physiological reactions. The logistics of these in-vitro assays permits the investigation of many drugs over large concentration ranges, thus expanding the scope of pharmacological and physiological experiments.

The location of neurotransmitter receptors in the neuronal membrane presents several technical problems: (a) the specific receptor sites are present in vanishingly small quantities in the membrane (of the order of 1 site μm^{-2}) in comparison to the amount of lipids, proteins, and other macromolecules present; (b) most drugs and neurotransmitters which are used to label such receptors are highly lipid-soluble and thus readily adsorb onto the membranes. This non-specific binding to the membrane must be quantitatively and qualitatively separated from the specific binding to membrane receptors; (c) many of the drugs used to label receptors are capable of specifically binding to more than one type of receptor. Thus, it is necessary to find conditions under which only the appropriate receptor is being labelled.

The nicotinic acetylcholine receptor was the first neurotransmitter receptor investigated and many of the current techniques and theories grew out of that work (Barnard *et al.*, 1973; Chang and Lee, 1963; Changeux *et al.*, 1970; Lee and Chang, 1966). The direct binding assay for this receptor was greatly aided by the use of a specific irreversible ligand, α-bungarotoxin. Unfortunately, no such toxin-ligand has been found for other neurotransmitter receptors.

The work on opiate receptors (Goldstein *et al.*, 1971; Pert *et al.*, 1973; Pert and Snyder, 1973; Snyder, 1975) resulted in several new approaches to overcome these problems. The first of the above problems was partially surmounted by the use of radioactive ligands with high specific activity (1–40 Ci $mmol^{-1}$), permitting study in the nM and pM concentration range, and detection of small quantities of specific binding as low as 10^{-14} mol per mg of tissue.

Drug enantiomers are used to differentiate non-specific binding (due to lipid solubility) from specific binding to receptors. The lipid solubilities of two stereoisomers (enantiomers) are identical, and thus equal amounts of the stereoisomers bind non-specifically to the membranes. On the other hand, neurotransmitter receptors have stereo-selectivity towards drugs and neurotransmitters. Thus, "*stereospecific binding*" is defined as that amount of radioligand bound in the presence of inactive enantiomer (in excess) minus that amount of radioligand bound in the presence of active enantiomer (in excess), as in Fig. 1. Often, however, it is simpler to work with "*specific binding*", which is defined as the total amount of radioligand bound minus that which occurs in the presence of the active enantiomer (in excess). Ideally specific and stereospecific binding should be identical. Generally, however, stereospecific binding is about 10–25% smaller than specific binding; this may occur because the inactive enantiomer may contain 0.5% or more of active enantiomer (see Fig. 1).

In order to identify the receptor being investigated, a proper pharmacological profile must be obtained. The potencies of a series of drugs on the stereospecific radioactive signal should correlate with the potencies of these drugs to a biological or clinical response thought to be mediated via that receptor.

A series of biochemical and pharmacological criteria have been developed to characterize the "displaceable binding" as truly representing binding of a radioactive ligand to its receptor (Caron *et al.*, 1978). These criteria include

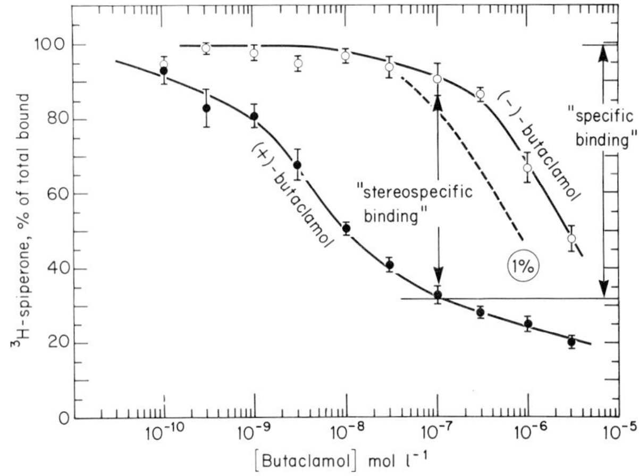

Fig. 1. Effect of (+)-butaclamol (active enantiomer) and (−)-butaclamol (inactive enantiomer) on the binding of ^{3}H-spiperone to calf caudate membranes. One nanomolar of ^{3}H-spiperone was used as this concentration gives an optimal signal/noise ratio, arbitrarily defined as [specific binding/standard deviation of specific binding]. The dotted line represents the theoretical displacement curve for the inactive enantiomer if the latter contains 1% contamination by the active enantiomer.

(1) saturability; (2) an affinity that approximates the in-vivo affinity for the receptor; (3) stereoselectivity; (4) a distribution of binding sites that corresponds to the distribution of nerve terminals in the brain; and (5) a pharmacological profile that is identical to the in-vivo profile. If a radioactive signal fulfills all of these criteria then that signal may be properly assumed to represent a specific physiological receptor.

In this chapter we will discuss the various radioactive ligands that have been used to label the dopamine receptor, with emphasis on and analysis of the more controversial aspects of the binding data.

NEUROLEPTIC AND ERGOT RADIOLIGANDS FOR BRAINDOPAMINE RECEPTORS

At present three radioligands fulfill the criteria expected for a specific dopaminergic tag: ^{3}H-haloperidol (Creese *et al.*, 1975a, b; Seeman, 1977; Seeman *et al.*, 1975), ^{3}H-spiroperidol (Fields *et al.*, 1977; Leysen *et al.*, 1977; Titeler *et al.*, 1978), and ^{3}H-dihydroergocryptine [in the presence of excess phentolamine (Caron *et al.*, 1978; Titeler *et al.*, 1977)]. Figure 2 illustrates the high affinity, saturable binding (at 21°C) of these ligands to calf caudate (homogenized total particulate). The dopaminergic characteristics of these radioligands are also shown in Fig. 2. In all three instances the relative potency of apomorphine is greater than that for dopamine which in turn is greater than that for noradrenaline. Table I is a compilation of IC_{50}

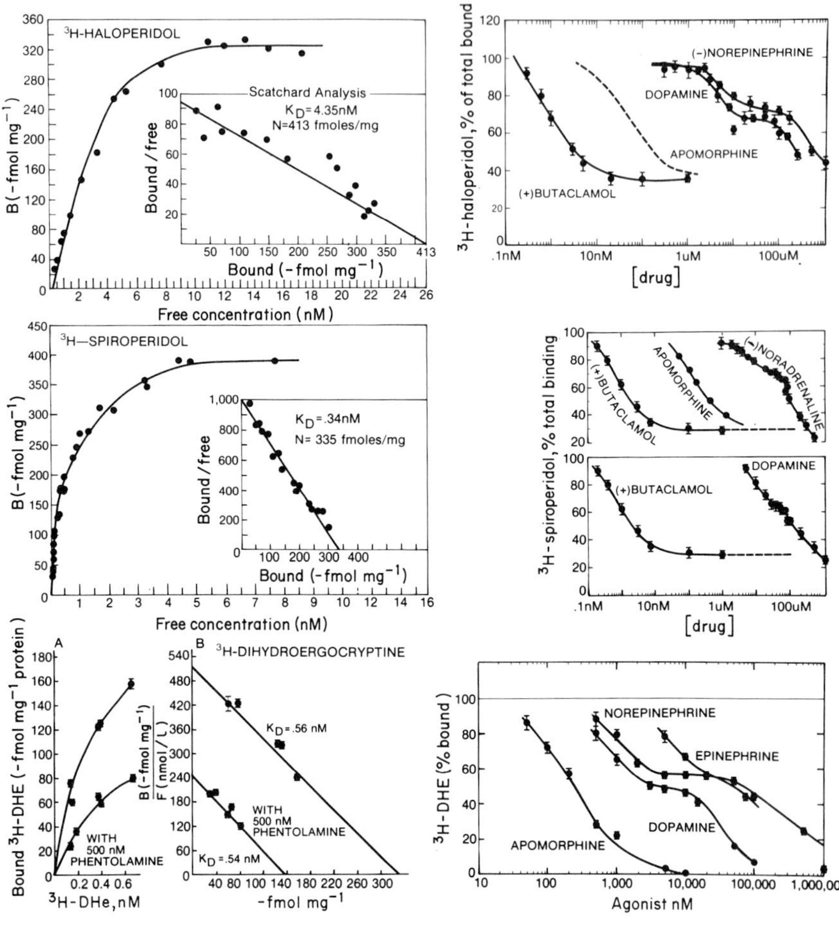

Fig. 2. Characteristics of the saturable binding of ^{3}H-haloperidol, ^{3}H-spiroperidol, and ^{3}H-dihydroergocryptine to calf caudate membranes. The figures on the left demonstrate the specific saturable binding of ^{3}H-haloperidol and ^{3}H-spiroperidol (as defined by 1 μM (+)-butaclamol (see text)) and ^{3}H-DHEC [in the presence and absence of 500 nM phentolamine, and as defined by 500 nM (+)-butaclamol (see text)]. Scatchard analyses of the saturation data are also presented. The figures on the right demonstrate the inhibition of the specific radioactive signals by (+)-butaclamol, apomorphine, dopamine, norepinephrine and epinephrine. The ^{3}H-haloperidol (1 nM) and ^{3}H-spiroperidol (0.5 nM) curves represent total binding while the ^{3}H-DHEC (0.7 nM) curve is specific binding (in the presence of 500 nM phentolamine) as defined by 500 nM (+)-butaclamol. The order of potency (apomorphine > dopamine > norepinephrine) is characteristic of a dopamine receptor.

values (concentrations which inhibit specific binding by 50%) for various dopamine agonists and antagonists on the specific binding of ^{3}H-haloperidol, ^{3}H-spiroperidol and ^{3}H-DHEC.

The specific binding of ^{3}H-haloperidol (Burt *et al.*, 1976; Seeman *et al.*, 1975) and ^{3}H-spiroperidol (Fields *et al.*, 1977) is highest in dopamine-rich regions of the brain. The distribution of specific ^{3}H-DHEC binding (in the presence of excess phentolamine) under conditions in which only dopamine receptors are labelled has not yet been investigated.

Finally, there is a correlation between the clinical antipsychotic potencies of various dopamine antagonists (neuroleptics) and their IC_{50} values versus ^{3}H-haloperidol or ^{3}H-DHEC (Fig. 3).

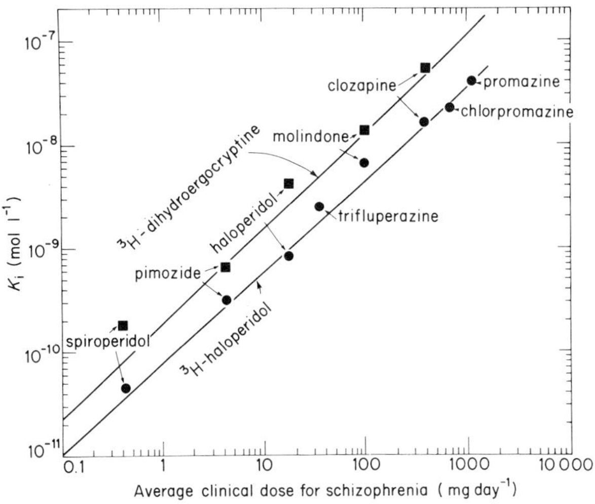

Fig. 3. The K_i values (ordinate) are the concentrations of the antipsychotic drugs which reduce the specific component of ^{3}H-haloperidol (1 nM) or ^{3}H-DHEC (0.7 nM) (in the presence of 500 nM phentolamine) by 50%. The abscissa indicates the average values (and ranges) of doses used for schizophrenia.

Another very important correlation is found between the clinical anti-Parkinsonian doses of various dopamine-mimetic drugs and their IC_{50} values against ^{3}H-haloperidol or ^{3}H-spiroperidol (Fig. 4).

Thus, these three radioactive ligands (^{3}H-haloperidol, ^{3}H-spiroperidol and ^{3}H-DHEC) display the properties expected of specific dopamine receptors: saturability, high affinity, stereospecificity (demonstrated by (+)- and (−)-butaclamol), dopaminergic distribution within the brain, and a direct correlation with clinical data.

TABLE I Equilibrium dissociation constants of dopamine agonists and antagonists as determined by direct-binding procedures. The tritiated ligand, used to determine the constant are listed at the top of the columns. The competing ligands are listed in the column at the left of the table. For publications in which IC_{50} values (concentrations of competing ligand that reduce the specific signal by 50%) were reported, these were converted to K_D by the equation $K_D = IC_{50}/(1+K_A/[A]$ where K_A is the equilibrium dissociated constant of the radioactive ligand for the binding site, and [A] is the concentration of radioactive ligand used. K_D values with no reference number are from unpublished results from our laboratory.

	^{3}H-Haloperidol	^{3}H-Spiroperidol	^{3}H-DHEC	^{3}H-DA	^{3}H-APO
Neuroleptics					
Spiroperidol	0.05 nM (1)[a] 0.25 nM (2) 0.24 nM (3)	0.43 nM (3) 0.32 nM (4)	0.29 nM (5)	720.0 nM (6) 1400.0 nM (2)	
(+)-Butaclamol	0.57 nM 0.54 nM (2) 6.0 nM (3)	3.6 nM (3) 0.64 nM (4)	2.2 nM (5)	80.0 nM (2)	
(−)-Butaclamol	2.7 μM 700.0 nM	6.4 μM (4)	2.1 μM (5)	10.0 μM (2)	
Haloperidol	0.67 nM (1) 1.4 nM (2) 1.9 nM (3)	3.6 nM (3) 3.2 nM (4)	6.7 nM (5)	920.0 nM (2)	350.0 nM (6)

table I—*continued*

Chlorpromazine	20.1 nM (1) 10.2 nM (2) 78.0 nM (3)	25.2 nM (3)	42.0 nM (5)	440.0 nM (6) 900.0 nM (2)	1050.0 nM (7)
Pimozide	0.27 nM (1) 0.81 nM (2) 1.9 nM (3)	2.9 nM (3) 4.8 nM (4)	1.0 nM (5)		
Fluphenazine	0.40 nM (1) 0.88 nM (2) 9.6 nM (3)	7.0 nM (3) 2.2 nM (4)		55.0 nM (6) 230.0 nM (2)	
Trifluperidol	0.20 nM (1) 0.95 nM (2) 0.96 nM (3)	2.5 nM (3)		330.0 nM (6) 740.0 nM (2)	
Clozapine	13.4 nM (1) 120.0 nM (2) 268.0 nM (3)	227.0 nM (3)	84.0 nM (5)	2050.0 nM (2)	
Agonists					
Bromocryptine	2.5 nM (2) 25.0 nM	30.0 nM		106.0 nM (2)	
Apomorphine	34.0 nM (3) 51.0 nM (2)	34.0 nM (3) 128.0 nM (4)	105.0 nM (5)	0.44 nM (6) 8.6 nM (2)	0.91 nM (6)
Dopamine	670.0 nM (2) 600.0 nM (3)	1130.0 nM (3)	273.0 nM (5)	2.0 nM (6) 17.5 nM (2)	2.5 nM (6)
Noradrenaline	5600.0 nM (3) 7000.0 nM (3)	32 000.0 nM (3)	504.0 nM (5)	10.0 nM (6) 200.0 nM (2)	10.5 nM (6)

[a] References: (1) Seeman *et al.*, 1976; (2) Burt *et al.*, 1976; (3) Leysen *et al.*, 1977; (4) Fields *et al.*, 1977; (5) Titeler *et al.*, 1977; (6) Seeman *et al.*, 1976; (7) Seeman *et al.*, 1975.

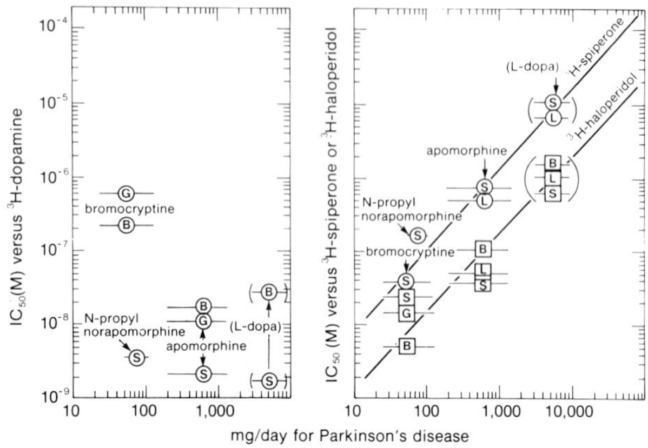

Fig. 4. Correlation between the clinical anti-Parkinsonian doses and the drug potencies for inhibiting the binding of ^{3}H-haloperidol or ^{3}H-spiperone, but not ^{3}H-dopamine. The IC_{50} values are the in-vitro concentrations which inhibit binding by 50% in crude homogenates of rat or calf brain striatum. The IC_{50} values labelled G are from Lew *et al.* (1977) (calf for ^{3}H-dopamine; rat for ^{3}H-haloperidol), B are from Burt *et al.* (1976) L are from Leysen *et al.* (1977) (rat), while those labelled S are from this laboratory (calf: unpublished or from Seeman *et al.*, 1976). The anti-Parkinsonian range of dosages are: apomorphine: 200–1400 mg day^{-1}, with a mean of 600 mg day^{-1} (Cotzias *et al.*, 1976a, b; Duby *et al.*, 1972); *N*-propylnorapomorphine: 60–90 mg day^{-1} (Cotzias *et al.*, 1976b); bromocryptine: 42 mg day^{-1} (Calne *et al.*, 1974), 79 mg day^{-1} (Kartzinel *et al.*, 1976), 70 mg day^{-1} (Lieberman *et al.*, 1976), 120 mg day^{-1} (Gerlach, 1976), 26 mg day^{-1} (Lieberman *et al.*, 1976), 26 mg day^{-1} (Parkes *et al.*, 1976). Lergotrile has not been included because it is not considered clinically effective (Lieberman *et al.*, 1975). The standard clinical doses for L-DOPA are generally 3–9 gm day^{-1} (without a DOPA-decarboxylase inhibitor). The IC_{50} values are for dopamine in this case, not for L-DOPA. Rigorously, the data for L-DOPA should not be mixed with the data for dopamine since they are two different chemicals; however, dopamine is the active metabolic of L-DOPA and the main point is that this drug is weak both *in vivo* and *in vitro*.

^{3}H-DOPAMINE AND ^{3}H-APOMORPHINE AS RADIOLIGANDS FOR DOPAMINE RECEPTORS

While the three radioactive ligands mentioned above fulfill the criteria for specific dopamine receptors. the ^{3}H-dopamine and ^{3}H-apomorphine radioligands themselves surprisingly have not fulfilled all these criteria. ^{3}H-Dopamine (Burt *et al.*, 1975; Seeman *et al.*, 1974; Seeman *et al.*, 1975) and ^{3}H-apomorphine (Seeman *et al.*, 1976) were obvious choices as specific radioactive tags for the dopamine receptor. ^{3}H-Dopamine is, of course, the neurotransmitter itself and apomorphine is a well-characterized specific direct-acting dopaminergic agonist (Anden *et al.*, 1967; Ernst, 1967; Horn *et al.*, 1975). Both of these radioactive agonists reveal a high affinity

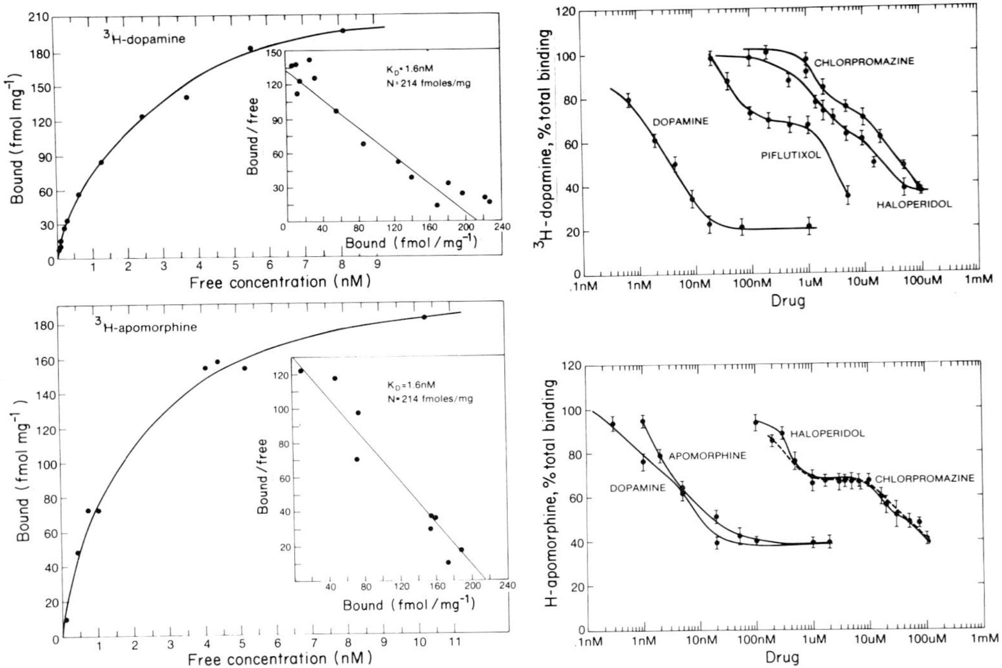

Fig. 5. Characteristics of the saturable binding of ^{3}H-dopamine and ^{3}H-apomorphine are displayed in the figures on the left. Scatchard analyses of the data are inserted. The figures on the right represent the displacement of the ^{3}H-dopamine (3 nM) and ^{3}H-apomorphine (3 nM) with various agonists and antagonists.

saturable set of sites in calf caudate homogenates (Fig. 5). Furthermore, the specific binding of both ^{3}H-ligands occurred principally in dopamine-rich regions of the brain (Burt *et al.*, 1975; Seeman *et al.*, 1976).

Unfortunately, neither ^{3}H-dopamine or ^{3}H-apomorphine fulfills the most crucial criterion, i.e. a pharmacological profile (for inhibition of binding) for agonists and antagonists that correlates to in-vivo potencies of dopamine agonists and antagonists (Fig. 5). To account for this apparent anomaly, Creese *et al.* (1975a, b) proposed that ^{3}H-haloperidol was labelling the antagonist conformation of the dopamine receptor and ^{3}H-dopamine was labelling the agonist conformation, a so-called "two-state" hypothesis.

However, if this two-state hypothesis has merit, the IC_{50} values (against ^{3}H-dopamine or ^{3}H-apomorphine) of the two key dopamine agonists, apomorphine and bromocryptine (Besser and Thorsen, 1976; Colse *et al.*, 1976; Loew *et al.*, 1976), should correlate with their clinical potencies. The clinical potencies of bromocryptine and apomorphine or Parkinson symptoms correlate well with their potencies on ^{3}H-haloperidol binding (Fig. 4), but not at all with their potencies on ^{3}H-dopamine or ^{3}H-apomorphine binding.

Further evidence that the two-state hypothesis is untenable is presented in Fig. 6. Figure 6A is a diagrammatic representation of the two-state conversion hypothesis. Figures 6A and 6B are the competition curves that would be obtained if the agonist or antagonist conformation were "preferred", respectively. The important point to note is that no matter which conformation is theoretically preferred, one diagram shows one low and one high potency competition curve, and one diagram shows two high potency competition curves. In neither case (A or B) can *both* diagrams display a low potency competition curve. Figure 6D shows the actual data for ^{3}H-haloperidol and ^{3}H-dopamine. Neither 6A nor 6B display the types of competition curves that are observed experimentally and thus the two-state hypothesis appears unsupported by the evidence. As shown in Fig. 6C, only a model based on two separate and independent binding sites can account for the observed competition curves.

Another piece of evidence that is contrary to the agonist–antagonist conformation theory is that (+)-butaclamol, a potent dopamine antagonist, and 2-bromocryptine, a potent dopamine agonist (Caron *et al.*, 1978) have identical binding properties when competing for ^{3}H-haloperidol sites and ^{3}H-dopamine sites. This high affinity of an agonist for the ^{3}H-haloperidol and low affinity for the ^{3}H-dopamine site is contrary to the agonist–antagonist conformation theory.

One further piece of evidence incompatible with the two-state hypothesis is the difference in relative potencies obtained with neuroleptics depending upon whether ^{3}H-dopamine or ^{3}H-haloperidol is the radioligand. Halo-

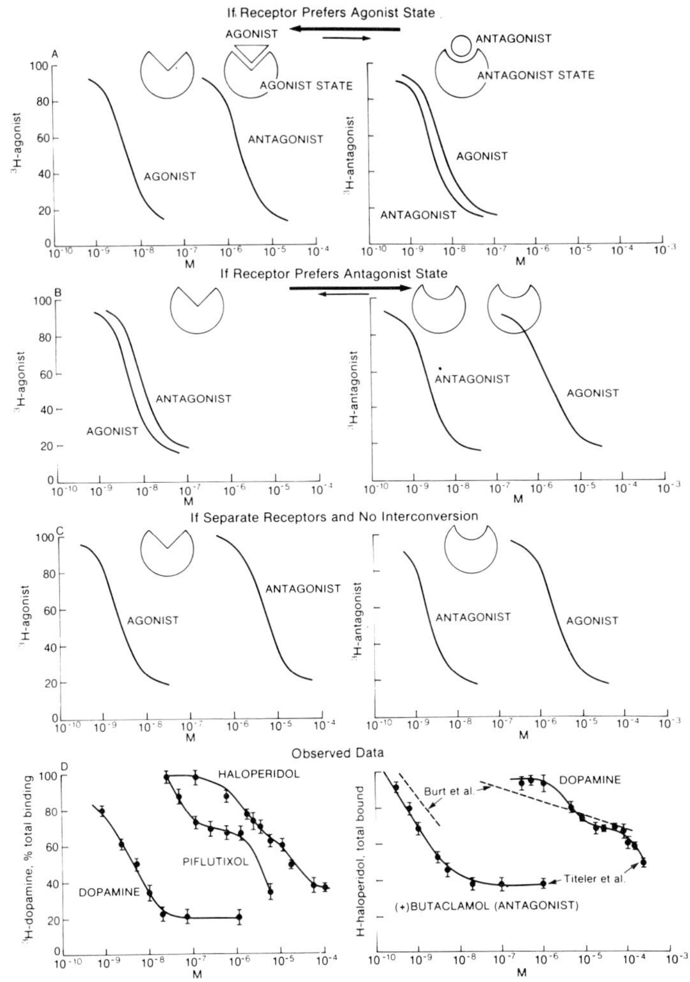

Fig. 6. A (Top row): these competition curves represent the theoretical results obtained if the dopamine receptor exists in two interconvertible conformations, with the agonist state thermodynamically more stable. B: the next competition curves (second from the top) represent the results that would be obtained if the receptor interconverts and prefers the antagonist state. Neither A or B resemble the experimentally observed competition data (section D at bottom). C (Second from bottom): here there is no interconversion of dopamine receptor conformations, but rather two separate, non-interconvertible sites. The expected data here do fit the experimentally observed competition curves (section D, bottom).

peridol is 21 times more potent than chlorpromazine in competing for ^{3}H-haloperidol (Table I), while equipotent in competing for ^{3}H-dopamine. As shown in Fig. 6A these ligands should display identical relative potencies (although different absolute potencies) in competing for either ^{3}H-dopamine or ^{3}H-haloperidol, if the two-state hypothesis holds. Along these same lines, apomorphine is 4.4 times more potent than dopamine in competing for ^{3}H-haloperidol binding, but 12 times more potent than dopamine in competing for ^{3}H-haloperidol.

It seems reasonable, therefore, to conclude that both ^{3}H-dopamine and ^{3}H-apomorphine are labelling a site distinct from the postsynaptic dopamine receptor that is labelled by ^{3}H-haloperidol, ^{3}H-spiroperidol or ^{3}H-dihydroergocryptine (in the presence of excess phentolamine).

This high-affinity site ($K_D = 1$ nM) for ^{3}H-dopamine binding is not in accordance with biochemical, physiological and clinical studies on the effect of neurotransmitters on postsynaptic receptors in general, as follows:

(a) For instance, direct binding studies on the nicotinic acetylcholine receptor have revealed a dissociation constant (for acetylcholine) in the μM concentration range, not the nM range (Dolly and Barnard, 1974). The concentrations of epinephrine and (-)norepinephrine that half maximally contract uterine smooth muscle are 1.6 μM and 4 μM respectively (Anderson *et al.*, 1972). These concentrations (Furchgott and Bursztyn, 1967) are in close agreement with the K_D values for the α-receptor as determined by direct binding studies. Parasympathetic receptors of rabbit smooth muscle have been shown to have 2 μM affinity for acetylcholine (Ariens, 1967). β-Receptors of human coronary vessels have been shown to have an affinity in the μM range for adrenaline and noradrenaline (Williams and Lefkowitz, 1976).

(b) Dopamine receptors in aplysia have been shown to have threshold concentrations in the μM range (Swann *et al.*, 1977) rather than in the nM range.

(c) This apparently low affinity of neurotransmitters for their receptors (i.e. K_D in the μM range) is probably reflected in the huge clinical doses of transmitter precursors needed to replete dopaminergic and serotoninergic function, using L-DOPA (Yahr, 1973) and tryptophan (Hartman, 1977), respectively, even in the presence of metabolic inhibitors. The neuroanatomy of the synapse assures high specificity for the neurotransmitter. Exogenous drugs, however, must possess high affinity (e.g. in the nM range) for their respective receptors if they are to possess specificity and clinical potency.

Thus, the approximate value of 1 μM for the K_D of dopamine (as measured by ^{3}H-haloperidol binding) appears plausible for dopamine acting on the classical type of postsynaptic dopamine receptor. On the other hand, therefore, the very low dissociation constant of 1.6 nM for the binding of ^{3}H-dopamine (see Fig. 5) then probably represents the action of dopamine on some other neuronal structure.

An anomalous and interesting facet in the competition experiments for both ^{3}H-agonists (Fig. 5) and ^{3}H-antagonists (Fig. 2) are the bi-phasic modes of competition produced when agonists compete with ^{3}H-antagonists (Fig. 2) (Titeler, 1978), or when antagonists compete with ^{3}H-agonists (Fig. 5). Two possibilities may account for this phenomenon. (1) There may be two high-affinity sites for ^{3}H-haloperidol, ^{3}H-DHEC, and ^{3}H-spiroperidol as well as for ^{3}H-dopamine and ^{3}H-apomorphine. (2) A second possibility is that high concentrations (1 μM) of dopamine can "expose" new ^{3}H-apomorphine and ^{3}H-haloperidol binding sites, as recently found by McManus *et al.*, 1978. This latter effect could account for the bi-phasic nature of these competition curves. As competition occurs at higher and higher dopamine concentrations, new dopamine sites become progressively available, offsetting further inhibition of binding and resulting in a "plateau" in the competition curve.

MODULATION OF RECEPTOR LEVELS

Dopamine receptor supersensitivity (Creese and Iversen, 1975) has been implicated in schizophrenia (Matthysse, 1973; Meltzer, 1976; Snyder, 1976), untreated Parkinson's disease (Hornykiewicz, 1966), and neuroleptic induced tardive dyskinesia (Jackson *et al.*, 1974).

Long-term treatment of rats with neuroleptic drugs results in more dopamine receptors in the striatum (Burt *et al.*, 1977; Kobayashi *et al.*, 1978; Muller and Seeman, 1977). Since ^{3}H-apomorphine binding generally increases by 80%, while ^{3}H-haloperidol only increases by 19% to 35%, this is further circumstantial evidence that ^{3}H-haloperidol binding is measuring a different neurone site than ^{3}H-apomorphine.

One possible basis of tardive dyskinesia may be an increased number of dopamine receptors resulting from long-term neuroleptic treatment (Muller and Seeman, 1977). Friedhoff *et al.* (1977) postulated that L-DOPA treatment can reduce the elevated number of receptors and possibly eliminate the dyskinetic signs. They found an increase in ^{3}H-dopamine binding sites after long-term neuroleptic treatment of rats; subsequent L-DOPA therapy rapidly lowered these excess ^{3}H-dopamine binding sites. Unfortunately, as mentioned above, there is convincing evidence that the high-affinity site for ^{3}H-dopamine ($K_D \approx 1$ nM) does not measure the classical postsynaptic dopamine receptors. Hence the study by Friedhoff *et al.* should be repeated using ^{3}H-haloperidol, ^{3}H-spiroperidol, or ^{3}H-DHEC (in presence of excess phentolamine).

Additional evidence that ^{3}H-apomorphine and ^{3}H-haloperidol are tagging different neurone sites comes from lesion experiments where 6-*OH*-dopamine has been injected to destroy the nigrostriatal neurones. Snyder *et al.* (1976) found an increase in the number of striatal ^{3}H-haloperidol sites. In a similar study Nagy *et al.* (1977) also found an increase of 22% in ^{3}H-haloperidol binding, but a decrease of 56% in the amount of specific ^{3}H-apomorphine binding. This provides further evidence that ^{3}H-apomorphine and ^{3}H-haloperidol measure different neurone sites.

Recent data from post-mortem human brain analysis further support the possibility that ^{3}H-haloperidol and ^{3}H-apomorphine are preferentially binding to somewhat different sites. Lee *et al.* (1977; see also Reisine *et al.*, 1977) found that ^{3}H-haloperidol binding was elevated in the striatum in Parkinson's disease but that ^{3}H-apomorphine binding was decreased. This is compatible with ^{3}H-apomorphine tagging the presynaptic dopamine receptors, while ^{3}H-haloperidol tags the denervated and supersensitive postsynaptic sites. Finally, Lee and Seeman (1977) and Lee *et al.* (1977) have also found that ^{3}H-haloperidol binding was increased by 50–80% in schizophrenic striatal and limbic tissues, while ^{3}H-apomorphine binding was normal.

The tuberoinfundibular dopamine neurones have an inhibitory effect on prolactin secretion from the pituitary gland (Hökfelt and Fuxe, 1972). The issue of whether or not this is a direct control mediated by dopamine receptors was investigated by examining the pituitary tissues for ^{3}H-dopamine-binding sites. Brown *et al.* (1976) used ^{3}H-haloperidol and ^{3}H-dopamine to assay for dopamine receptors and found substantial binding in the pituitary but none in the basal hypothalamus. Caron *et al.* (1978), using ^{3}H-DHEC, also detected dopamine receptors in pituitary cell membranes. A surprising result of long-term haloperidol treatment was the discovery by Friend *et al.* (1977) that ^{3}H-haloperidol binding decreased in the pituitary cell membrane, rather than increasing as in the caudate (see above). Thus the level of dopamine receptors in these two tissues is under different control mechanisms, although they are otherwise pharmacologically identical (Caron *et al.*, 1978; Titeler *et al.*, 1977).

ACKNOWLEDGEMENTS

We thank Mrs Margaret Chau-Wong for excellent technical assistance. This work was supported by the Ontario Mental Health Foundation, the Medical Research Council of Canada (grant MT-2951) and the Connaught Foundation of the University of Toronto.

REFERENCES

Andén, N. E., Rubenson, A., Fuxe, K., and Hökfelt, T. (1967). *J. Pharm. Pharmacol.* **19,** 627–629.

Anderson, R., Holmberg, S., Svedmyr, N., and Äberg, G. (1972). *Acta Med. Scand.* **191,** 241–244.

Ariens, A. J. (1967). *Ann. N.Y. Acad. Sci.* **139,** 606.

Barnard, E. A., Chiu, T., Jedrzejcyzk, J., Porter, C. W., and Wieckowski, J. (1973). *In* "Drug Receptors" (Ed. H. P. Range), 225–240. University Park Press, Baltimore.

Besser, G. M., and Thorsen, M. D. (1976). *Postgrad. Med. J.* **52,** 64–70.

Brown, G. M., Seeman, P., and Lee, T. (1976). *Endocrinology* **99,** 1407–1410.

Burt, D. R., Enna, S. H., Creese, I., and Snyder, S. H. (1975). *Proc. Natl Acad. Sci.. U.S.A.* **72,** 4655–4659.

Burt, D. R., Creese, I., and Snyder, S. H. (1976). *Molec. Pharmacol.* **12,** 800–812.

Burt, D. R., Creese, I., and Snyder, S. H. (1977), *Science* **196,** 326–328.

Calne, D. B., Teychenne, P. D., Leigh, P. N., Bamji, A. N., and Greenacre, J. K. (1974), *Lancet* **ii,** 1255–1256.

Caron, M. G., Beaulieu, M., Raymond, V., Gogné, B., Drouin, J., Lefkowitz, R. J. and Labrie, F. (1978). *J. Biol. Chem.* **253,** 2244–2253.

Chang, C. C., and Lee, C. Y. (1963). *Arch. Int. Pharmacodyn.* **144,** 241.

Changeux, J. P., Kasai, M., and Lee, C. H. (1970). *Proc. Natl Acad. Aci. U.S.A.* **67,** 2141.

Colse, D. B., Kotziscl, R., and Shoulson, I. (1976). *Postgrad. Med. J.* **52,** 81–82.

Cotzias, G. C., Papavasiliou, P. S., and Ginos, J. Z. (1976a). *In* "The Basal Ganglia" (Ed. M. D. Yahr), 305–315. Raven Press, New York.

Cotzias, G. C., Papavasiliou, P. S., Tolosa, E. S., Mendez, J. S., and Bell-Midura, M. (1976b). *New Engl. J. Med.* **294,** 567–572.

Creese, I., Burt, D. R., and Snyder, S. H. (1975a). *Life Sci.* **17,** 993–1002.

Creese, I., Burt, D. R., and Snyder, S. H. (1975b). *Life Sci.* **17,** 1715–1720.

Creese, I., and Iversen, S. O. (1975). *Brain Res.* **82,** 419–436.

Dolly, J. O., and Barnard, E. A. (1974). *FEBS Lett.* **46,** 145–148.

Duby, S. E., Cotzias, G. C., Papavasiliou, P. S., and Lawrence, W. H. (1972). *Arch. Neurol.* **27,** 474–480.

Ernst, A. M. (1967). *Psychopharmacologia* **10,** 316–323.

Fields, J. Z., Reisine, T. D., and Yamamura, H. I. (1977). *Brain Res.* **136,** 578–584.

Friedhoff, A. J., Bonnet, K., and Rosengarten, H. (1977). *Res. Commun. Chem. Pathol. Pharmacol.* **18,** 411–423.

Friend, W. C., Brown, G. M., Kawahir, G., Lee, T., and Seeman, P. (1978). *Am. J. Psychiat.* **135,** 839–841.

Furchgott, R. F., and Bursztyn, P. (1967). *Ann. N.Y. Acad. Sci.* **144,** 882–899.

Gerlach, J. (1976). *Acta Neurol. Scand.* **53,** 189–200.

Goldstein, A., Lowney, L. I., and Pal, B. K. (1971). *Proc. Natl Acad. Sci., U.S.A.* **68,** 1742–1747.

Hartman, E. (1977). *Am. J. Psychiat.* **134,** 366–370.

Hökfelt, T., and Fuxe, K. (1972). "Brain–Endocrine Interaction, Median Eminence—Structure and Function" (Eds. K. M. Knogyi, D. E. Scott and H. Weidle), 181–223.

Horn, A. S., Post, M. L., and Kennard, O. (1975). *J. Pharm. Pharmacol.* **27,** 553–563.

Hornykiewicz, O. (1966). *Pharmacol. Rev.* **18,** 925–964.

Jackson, G., Baldessarini, R. J., and Monsheck, T. (1974). *Am. J. Psychiat.* **131,** 910–913.

Kartzinel, R., Perlow, M., Teychenne, P., Gielen, A. C., Gillespie, M. M., Sadowsky, D. A., and Calne, D. B. (1976). *Lancet* **ii,** 272–275.

Kobayashi, A. M., Fields, J. Z., Hruska, R. F., Beaumont, K., and Yamamura, H. I. (1978). *In* "Animal Models of Psychiatry" (Ed. E. Usdin), Pergamon, Oxford.

Lee, C. Y., and Chang, C. C. (1966). *Mem. Inst. Butanton. Simp. Int.* **33,** 555.

Lee, T., and Seeman, P. (1977). *Soc. Neurosci.* **7,** 443.

Lee, T., and Seeman, P., Bird, E. D., Spokes, E. M., Tourtellotte, W. W., Farley, I., and Hornykiewicz, O. (1977). Submitted for publication.

Lee, T., Seeman, P., Rajput, A., Farley, I., and Hornykiewicz, O. (1977). Submitted for publication.

Lew, J. Y., Hata, F., Ohashi, T., and Goldstein, M. (1977). *J. Neural Trans.* **41,** 109–221.

Leysen, J. E., Gommeren, W., and Laduron, P. M. (1977). *Biochem. Pharmacol.* In press.

Lieberman, A., Kupersmith, M., Estey, E., and Goldstein, M. (1976). *New Engl. J. Med.* **295,** 1400–1404.

Lieberman, A., Miyamoto, T., Battista, A. F., and Goldstein, M. (1975). *Neurology (Minneap.)* **25,** 459–462.

Lieberman, A., Zolfaghari, M., Boal, D., Hassouri, H., Vogel, B., Battista, A., Fuxe, K., and Goldstein, M. (1976). *Neurology (Minneap.)* **26,** 405–409.

Loew, D. M., Vigouret, J. M., and Jaton, A. L. (1976). *Postgrad. Med. J.* **52,** 40–46.

Matthysse, S. (1973). *Fed. Proc.* **32,** 200–204.

McManus, C., Hartley, E., and Seeman, P. (1978). *Life Sci.* In press.

Meltzer, H. Y. (1976). *Arch. Gen. Psychiat.* **33,** 279–286.

Muller, P., and Seeman, P. (1977). *Life Sci.* **21,** 1751–1758.

Nagy, J. I., Lee, T., Seeman, P., and Fibiger, H. C. (1977). Submitted for publication.

Parkes, J. D., Marsden, C. D., Donaldson, I., Galae-Debono, A., Walters, J., Kennedy, G., and Asselman, P. (1976). *J. Neurol. Neurosurg. Psychiat.* **29,** 184–193.

Pert, C. B., Pasternak, G., and Snyder, S. H. (1973). *Science* **182,** 1359–1361.

Pert, C. B., and Snyder, S. H. (1973). *Science* **179,** 1011–1014.

Reisine, T. D., Fields, J. Z., Yamamura, H. I., Bird, E., Spokes, E., Schreiner, P. S., and Enna, S. J. (1977). *Life Sci.* **21,** 335–344.

Seeman, P., Wong, M. and Lee, T. (1974). *Fed. Proc.* **33,** 246.

Seeman, P. (1977). *Biochem. Pharmacol.* **26,** 1741–1748.

Seeman, P., Chau-Wong, M., Tedesco, J., and Wong, K. (1975). *Proc. Natl Acad. Sci., U.S.A.* **72,** 4376–4380.

Seeman, P. Lee, T., Chau-Wong, M., Tedesco, J., and Wong, K. (1976). *Proc. Natl Acad. Sci., U.S.A.* **72,** 4354–4358.

Seeman, P., Lee, T., Chau-Wong., and Wong, K. (1976). *Nature (London)* **261,** 717–719.

Snyder, S. H. (1975). *In* "Predictability in Psychopharmacology: Preclinical and Clinical Correlations" (Ed. A. Sudilovsky), 269–283. Raven Press, New York.

Snyder, S. H. (1976). *Am. J. Psychiat.* **133,** 197–202.

Snyder, S. H., Burt, D. R., and Creese, I. (1976). *Soc. Neurosci. Symposia*, 1.

Swann, J. W., Siback, C. N., and Carpenter, D. O. (1977). *Soc. Neurosci.* **7,** 189.

Titeler, M., Weinreich, P. and Seeman, P. (1977). *Proc. Natl Acad. Sci., U.S.A.* **74,** 3750–3753.

Titeler, M., Weinreich, P., Sinclair, D., and Seeman, P. (1978). *Proc. Natl Acad. Sci., U.S.A.* **75,** 1153–1156.
Williams, L. T., and Lefkowitz, R. J. (1976). *Science* **192,** 791–793.
Yahr, M. D. (Ed.) (1973). "Advances in Neurology". Vol. 2: "Treatment of Parkinsonism—The Role of Dopa-decarboxylase Inhibitors". Raven Press, New York.

Release, Uptake and Metabolism of Dopamine

CHAPTER 15

In-Vivo and In-Vitro Release of Dopamine

J. Glowinski, A. Cheramy
and M. F. Giorguieff

Collège de France, Paris, France.

INTRODUCTION

Most in-vitro and in-vivo studies of the release of dopamine (DA) from nerve terminals have been done on the nigrostriatal DA system. This is due to the high density of DA terminals in the striatum and to the well-known anatomical organization of this pathway. In-vitro studies have generally been carried out on rat striatal slices or synaptosomes to analyse the basic mechanisms involved in the process of the transmitter release and their control by presynaptic receptors. In-vivo studies allow the investigation of the interneuronal regulations of the activity of the nigrostriatal DA systems in pharmacological and physiological states. In several cases, the rate of DA utilization has been estimated by turnover studies or by measuring the changes in DA metabolites in the rat striatum. Despite the interest of these indirect approaches we will mainly discuss studies made on the cat caudate nucleus in which the transmitter release into the extracellular space was directly measured. It has recently been shown both *in vitro* and *in vivo* that DA is released not only from nerve terminals but also from dendrites. These studies will also be discussed since the dendritic release of DA in the substantia nigra (SN) may be a critical process in the regulation of the activity of the nigrostriatal DA neurones.

IN-VITRO STUDIES OF THE RELEASE OF DOPAMINE FROM NERVE TERMINALS

Methodological Considerations

Striatal slices and synaptosomal preparations maintain respiratory metabolism and resting membrane potential. They can thus be used to study the release process of DA. Slices preserve the histological integrity of the tissue. Synaptosomes allow one to distinguish between direct and indirect effects of drugs on DA release.

In the presence of an MAO inhibitor, it is possible to estimate endogenous DA released in the incubating medium of a crude striatal synaptosomal fraction. Under these conditions the release of DA is enhanced under depolarization by potassium (K^+) (De Belleroche *et al.*, 1976; Lanne and Aprison, 1977) or veratridine (Patrick and Barchas, 1976) through a calcium-dependent process. However, due to the small amount of DA released from this preparation, or from slices, most authors have used isotopic techniques which consist of labelling DA terminals with exogenous ^{3}H-DA or with ^{3}H- or ^{14}C-DA synthesized from labelled tyrosine.

Owing to the high affinity uptake process of DA, it is widely assumed that the exogenous ^{3}H-amine added in a small concentration (about 10^{-7} M) is selectively transported into DA terminals. Tissues are then incubated or, most often, superfused in appropriate chambers and the efflux of total tritium is measured in several superfusate fractions. According to several authors, the ^{3}H-outflow which is enhanced by K^+ (Bustos and Roth, 1972) or veratridine (Mulder *et al.*, 1975) or by electrical field stimulation (Farnebo and Hamberger, 1971) through a calcium-dependent process reflects the physiological release of the transmitter. However, this method may have some limitations. Indeed, even when loading with exogenous ^{3}H-DA was done in the presence of an MAO inhibitor ^{3}H-DA represented only 77% of the K^+-induced ^{3}H overflow (Dismukes and Mulder, 1977). Moreover, the spontaneous efflux of the radioactivity measured is not calcium dependent (Mulder *et al.*, 1975). Finally, ^{3}H-DA taken up in tissues does not always behave as newly synthesized DA. For instance, ACh stimulated the release of newly formed ^{3}H-DA but not that of exogenous ^{3}H-DA (Besson, unpublished observations). Similarly, γ-hydroxybutyrate was shown to reduce the K^+-induced release of endogenously synthesized DA but not that of exogenous ^{3}H-DA (Bustos and Roth, 1972).

Two experimental approaches were used to study the release of labelled DA synthesized from labelled tyrosine (labelled DOPA is a poor precursor since it is decarboxylated not only in DA terminals but also in other aromatic amino acid decarboxylase-containing cells (Ng *et al.*, 1971)). (1) Isolated striata or slices were first incubated in the presence of L-^{3}H- or ^{14}C-tyrosine and then transferred to superfusion chambers to estimate the efflux of labelled DA in superfusate fractions. (2) Since the newly synthesized transmitter is preferentially used in the release process, we developed a method for the measurement of ^{3}H-DA released during the continuous superfusion of striatal slices with L-^{3}H-tyrosine. (A sensitive separation procedure is required to isolate the small quantity of ^{3}H-DA from the large amount of ^{3}H-tyrosine present in superfusates.) The rate of synthesis of ^{3}H-amine can be simultaneously monitored by measuring ^{3}H-H_2O in superfusate fractions, one molecule of ^{3}H-H_2O being formed during the conversion of each molecule of L-3,5-^{3}H-tyrosine into ^{3}H-DOPA. Under these conditions, the spontaneous release of ^{3}H-DA which rapidly reaches a steady state level, is calcium dependent and is reduced by 50% at least in the presence of tetrodotoxin (Giorguieff *et al.*, 1977a). It is markedly enhanced during K^+ depolarization (Westfall *et al.*, 1976). Since tyrosine hydroxylase is selectively located in DA terminals, the spontaneous and evoked releases of newly formed ^{3}H-DA are specific processes of these terminals. In fact, they are reduced by more than 90% after destruction of the DA neurones (Giorguieff *et al.*, unpublished observations).

Effect of Some Psychotropic Drugs on DA Release

The stimulatory effect of *d*-amphetamine on DA release, first observed in isolated striata prelabelled with L-^{3}H-tyrosine (Besson *et al.*, 1969), was then investigated by various authors who used different approaches. *d*-Amphetamine, which is more active than the *l*-form (Heikkila *et al.*, 1975a) enhanced ^{3}H-DA released from striatal slices when added to their superfusion medium or when previously injected into the animals (Besson *et al.*, 1969b). This effect is still observed in tissues of rats pretreated with reserpine and nialamide suggesting that the extravesicular DA may also be sensitive to the releasing action of the drug (Farnebo, 1971). In contrast to that observed with DA uptake inhibitors such as cocaine, benztropine or nomifensine, the *d*-amphetamine-induced efflux of exogenous ^{3}H-DA persisted in superfused synaptosomes (Raiteri *et al.*, 1975). Therefore, the releasing effect of *d*-amphetamine is independent from its action of DA uptake. However, since the *d*-amphetamine-induced release of exogenous ^{3}H-DA was blocked by cocaine and benztropine (Heikkila *et al.*, 1975a), the drug must be transported into DA terminals to be effective.

Several phenylethylamines such as *p*- and *m*-tyramine or norepinephrine released ^{3}H-DA previously taken up in synaptosomes (Raiteri *et al.*, 1977). Amantadine, an antiparkinson drug, also stimulated ^{3}H-DA release

(Scatton *et al.*, 1970; Heikkila *et al.*, 1975b) but its effect was much less pronounced than that of *d*-amphetamine.

Involvement of Presynaptic DA Receptors in the Control of DA Release

Farnebo and Hamberger (1971) first reported that the ^{3}H-overflow induced by electrical field-stimulation in striatal slices preloaded with exogenous ^{3}H-DA was reduced by apomorphine (10^{-7} M), a DA agonist, and enhanced by DA antagonists such as chlorpromazine (10^{-7} M) and pimozide (10^{-6} M). This suggested that presynaptic DA receptors were involved in the control of DA release from DA terminals in a way similar to that observed for the control of noradrenaline release by presynaptic noradrenergic receptors in peripheral or central noradrenergic neurones (Langer, 1977). However, Dismukes and Mulder (1977) could not reproduce these experiments. In fact, they observed that neuroleptics decreased the ^{3}H-overflow induced by electrical field-stimulation, agreeing with the results of Seeman and Lee (1975) obtained on striatal synaptosomes. Some neuroleptics were also shown to reduce the veratridine- (Raiteri *et al.*, 1978) or protoveratrine- (Iversen *et al.*, 1975) induced ^{3}H-overflow from synaptosomes preloaded with ^{3}H-DA in the presence of an MAO inhibitor. These paradoxical effects may be related to the anaesthetic properties of neuroleptics which could mask their action on DA presynaptic receptors. In fact, while measuring the release of newly synthesized ^{3}H-DA we observed a marked potentiation of the K^{+}-induced release of ^{3}H-DA by fluphenazine (10^{-6} M) (Westfall *et al.*, 1976). Complementary experiments are still required to establish more firmly the role of DA presynaptic receptors in the control of DA release. Since the influence of these receptors in the control of DA synthesis has been well established both in striatal slices (Westfall *et al.*, 1976) and synaptosomes (Christiansen and Squires, 1974; Iversen *et al.*, 1975) it will be surprising if they were not also involved in the local regulation of the transmitter release.

Influence of Various Transmitters on the Release of DA from DA Terminals

Several types of presynaptic receptors are involved in the control of noradrenaline release from peripheral noradrenergic neurones (Langer, 1977). This also occurs in the control of DA release from striatal DA terminals.

In earlier studies, we reported that acetylcholine (ACh) (5×10^{-5} M) in presence of eserine stimulated the release of ^{3}H-DA from isolated striata previously incubated with L-^{3}H-tyrosine (Besson *et al.*, 1969a). In more recent experiments made on rat striatal slices continuously superfused with the labelled amino acid, the effect of ACh (10^{-5} M) could be seen even in the absence of eserine (Giorguieff *et al.*, 1977a). It is calcium dependent and can be reproduced by various cholinergic agonists, such as carbachol (10^{-5} M), oxotremorine (10^{-5} M) (Giorguieff *et al.*, 1977a) or nicotine (10^{-5} M). The involvement of muscarinic and nicotinic receptors was confirmed by the reduction of the ACh-induced release of ^{3}H-DA in the presence of atropine (10^{-6} M), a muscarinic blocker, or pempidine (10^{-5} M), a nicotinic antagonist (Giorguieff *et al.*, 1977a). According to Westfall (1974a), ACh (10^{-5} M) reduces the release of exogenous ^{3}H-DA induced by nicotine (10^{-3} M), K^+ or electrical field-stimulation and these effects are abolished by muscarinic antagonists (Westfall, 1974b). This could suggest that the stimulation of muscarinic receptors inhibits the evoked release of ^{3}H-DA in contrast to that observed for the spontaneous release. However, under our experimental conditions the nicotine-induced release of newly synthesized ^{3}H-DA was not reduced by oxotremorine (10^{-5} M).

Binding studies with specifically labelled ligands have also revealed the presence of muscarinic (Yamamura and Snyder, 1974) and nicotinic (Speth *et al.*, 1977) receptors in the striatum. Some of these should be located on DA nerve terminals. Indeed, the stimulatory effects of ACh, oxotremorine or nicotine on the spontaneous release of newly synthesized ^{3}H-DA were still observed when slices were superfused in the presence of tetrodotoxin (Giorguieff *et al.*, 1977a), a neurotoxin which should abolish interneuronal effects by interrupting action potentials in nerve fibres. The cholinergic presynaptic receptors located on the DA terminals may play a physiological role. Indeed, since the nigrostriatal DA neurones exert a tonic inhibitory influence on some striatal cholinergic neurones, these receptors could be involved in a local feedback regulation of DA release by cholinergic neurones innervating the striatum.

L-Glutamic acid, but not the D-form of the amino acid, enhanced the release of ^{3}H-DA continuously synthesized from L-^{3}H-tyrosine in striatal slices. This effect was still observed in the presence of tetrodotoxin and was not abolished by cholinergic antagonists (Giorguieff *et al.*, 1977b). These results suggest that this excitatory amino acid acts on specific glutaminergic receptors located on DA terminals. This is not surprising since an important corticostriatal glutamatergic pathway has been described (Divac *et al.*, 1977), and since the electrical stimulation of the motor cortex in the cat enhanced the release of newly synthesized ^{3}H-DA in the ipsilateral caudate nucleus (Nieoullon *et al.*, 1978a).

Binding studies with ^{3}H-leu-enkephalin and ^{3}H-naloxone have suggested the presence of enkephalin presynaptic receptors on DA terminals. In fact, a permanent reduction of these binding sites was observed after destruction of the DA neurones (Pollard *et al.*, 1977). These enkephalin presynaptic receptors may be involved in the control of DA release since met-enkephalin (Subramanian *et al.*, 1977), morphine and β-endorphin (Loh *et al.*, 1976) were shown to reduce the K^+-evoked release of ^{3}H-DA previously taken up in striatal slices. However, these data could not be confirmed by Arbilla and Langer (1978).

GABAergic neurones could also be involved in the local control of DA release within the striatum. Indeed, pretreatment of rats with amino-oxyacetic acid, an inhibitor of GABA transaminase, enhanced the K^+-evoked release of exogenous ^{3}H-DA from striatal slices (Stoof *et al.*, 1977). Moreover, we demonstrated that GABA (from 10^{-5} to 10^{-3} M) itself, as well as muscimol (10^{-5} M) which is a potent GABAergic agonist, enhanced the release of ^{3}H-DA continuously synthesized from L-^{3}H-tyrosine. The GABA stimulatory effect which is calcium dependent, was partially prevented by picrotoxin (Giorguieff *et al.*, 1978). γ-Aminobutyric acid could act on GABAergic receptors located on interneurones or neuronal afferences projecting on DA terminals. Indeed, the GABA-induced release of ^{3}H-DA was no longer seen in the presence of tetrodotoxin, suggesting an indirect effect on DA terminals.

In conclusion, these in-vitro studies indicate that several interneuronal processes could be involved in the regulation of DA release from DA terminals independently of events occurring at the SN level.

IN-VIVO STUDIES OF THE RELEASE OF DOPAMINE FROM NERVE TERMINALS

Methodological Considerations

Most acute in-vivo experiments on DA release from the caudate nucleus were performed in anaesthetized cats or in "encéphale isolé" or gallamine-immobilized preparations. Few chronic studies were performed in the cat or the monkey. Dopamine release was estimated either in perfusates of the lateral ventricle or in superfusates from a cup placed on the dorsal ventricular surface of the caudate nucleus or from a push-pull cannula inserted in the structure. The origin of released DA is very well defined when the last two procedures are used, but tissue damage is completely avoided with the first one.

McLennan (1964) was the first to detect DA in superfusates from a push-pull cannula. Using a fluorimetric method he estimated the spontaneous release of the transmitter to be about 1000 pg min^{-1}. This value was overestimated since a resting release of DA of about 20 pg min^{-1} was found by McKenzie and Szerb (1968) who used similar experimental conditions, and by Bartholini *et al.* (1975) and our group (Chéramy *et al.*, 1975a), using a sensitive radioenzymatic assay. This low quantity of DA released (two to four times the blank value) is in agreement with estimations made by Portig and Vogt (1969) in ventricular perfusates (35–70 pg min^{-1}). As in in-vitro studies isotopic methods were used. After labelling with exogenous ^{3}H-DA, the ^{3}H-amine efflux or the total tritium was measured in ventricular perfusates (Von Voigtlander and Moore, 1971). More specific methods consisting of the continuous superfusion of the tissues with ^{14}C- or ^{3}H-L-^{3}H-tyrosine allowed the detection of the labelled amines in superfusates from a cup (Besson *et al.*, 1971) or from a push-pull cannula (Riddell and Szerb, 1971; Nieoullon *et al.*, 1977a). In our hands, using L-^{3}H-tyrosine (40 Ci mmol^{-1}) the resting release of newly formed ^{3}H-DA (0.2 nCi min^{-1}) was about 100 times the blank value. It was reduced by the omission of calcium or addition of tetrodotoxin in the superfusion fluid and completely abolished shortly after either transection of the DA pathway or electro-coagulation of the SN (Nieoullon *et al.*, 1977a).

Release of DA Evoked by Depolarization of DA Neurones

Several authors have shown the stimulatory effect of K^+ on DA release. Under electrical stimulation of the SN, McLennan (1965) observed an enhanced release of DA in the putamen but not in the caudate nucleus. An effect was seen in the latter structure under stimulation of the thalamic centro-medianus nucleus (McLennan, 1964). Portig and Vogt (1969) detected an increase in homovanillic acid, in most cases, but inconsistent changes in DA in ventricular perfusates during stimulation of the SN. Better responses were seen under stimulation of the sciatic nerve.

More convincing data were obtained with isotopic methods. While stimulating the medial forebrain bundle, Riddell and Szerb (1971) observed a two- to ten-fold increase in the release of ^{14}C-DA synthesized from ^{14}C-tyrosine in push-pull cannula superfusates, Moore and his colleagues reported that exogenous ^{3}H-DA (Von Voigtlander and Moore, 1971) as well as endogenously formed ^{3}H-catecholamines (Chiueh and Moore, 1973) were released in ventricular perfusates during stimulation of the SN, but better results were obtained by stimulating DA fibres with a frequency of 30 Hz (400 μA, 1 ms). We observed a marked stimulation in the release of newly formed ^{3}H-DA in a cup or push-pull cannula superfusates during stimulation of the SN (Nieoullon *et al.*, 1977a), the red nucleus, the mesencephalic reticular formation (Besson *et al.*, 1974) the motor or visual cortices (Nieoullon *et al.*, 1977a), the ipsilateral fastigial nucleus and the contralateral dentate nucleus of the cerebellum (Nieoullon *et al.*, 1978b). Marked effects were also seen under ipsilateral sensory stimuli (Nieoullon *et al.*, 1977b).

Changes in DA Release Induced by Systemic Injections of Psychotropic Drugs

Injected intravenously, *l*-amphetamine ($2\,mg\,kg^{-1}$) or tranylcypromine ($10\,mg\,kg^{-1}$) stimulated the release of newly formed ^{3}H-DA in cup superfusates (Besson *et al.*, 1974). Harmaline ($20\,mg\,kg^{-1}$) and its metabolite harmine ($20\,mg\,kg^{-1}$), two reversible MAO inhibitors exhibiting tremorigenic effects, reduced both the release of labelled and unlabelled DA (Chéramy *et al.*, 1975a). These latter effects could be mediated by cerebello-nigral pathways involved in the control of DA neurones. Drugs acting on DA receptors such as chlorpromazine ($10\,mg\,kg^{-1}$), haloperidol ($0.5\,mg\,kg^{-1}$) and to a lesser extent clozapine ($20\,mg\,kg^{-1}$) stimulated the release of DA for several hours in push-pull cannula perfusates of the caudate nucleus in gallamine-immobilized or unrestrained cats; clozapine was more effective than other neuroleptics in releasing DA from the nucleus accumbens (Bartholini *et al.*, 1976). Although apomorphine was not tested in these elegant studies of Bartholini *et al.*, LSD ($1\,mg\,kg^{-1}$) a partial dopaminergic agonist reduced the release of the transmitter in the caudate nucleus (Da Prada *et al.*, 1975). Haloperidol ($10\,mg\,kg^{-1}$, i.v.) was without effect on the release of exogenous ^{3}H-DA (Von Voigtlander and Moore, 1973) but, as shown in our cup studies, thioproperazine ($5\,mg\,kg^{-1}$, i.v.) stimulated both ^{3}H-DA release and $^3H\text{-}H_2O$ efflux during the continuous delivery of L-^{3}H-tyrosine. These effects were prevented by prior transection of the DA neurones indicating that nerve activity is required (Chéramy *et al.*, 1975b).

Halothane (2%) anaesthesia which increases the firing rate of nigral DA cells in the rat enhanced the spontaneous release of newly synthesized ^{3}H-DA in push-pull cannula superfusates of the cat caudate nucleus (when compared with "encéphale isolé" animals) (Nieoullon *et al.*, 1977a). Injected in large doses (300–$700\,mg\,kg^{-1}$) γ-hydroxybutyrate, often used as an anaesthetic, reduces the activity of DA neurones in the rat (Roth *et al.*, 1973). However, at doses of 30 to $300\,mg\,kg^{-1}$ (i.p.) it stimulated the release of ^{3}H-DA in our cat experiments (Chéramy *et al.*, 1977a).

The GABAergic antagonist picrotoxin (0.25–$2.5\,mg\,kg^{-1}$, i.v.) enhanced the release of newly formed ^{3}H-DA and this effect was prevented or abolished by diazepam ($10\,mg\,kg^{-1}$, i.p.) which is without effect when given alone (Chéramy *et al.*, 1977b). These biochemical data are in agreement with the electrophysiological observations of Schaffner and Haefely (1975), since diazepam prevented the suppressing effect of bicuculline on the inhibitory potential evoked in the SN under stimulation of the cat caudate nucleus. They support the hypothesis of an inhibitory control of the DA neurones by striatonigral GABAergic neurones.

To our knowledge few data are available concerning the effects of cholinergic drugs. Oxotremorine ($0.15\,mg\,kg^{-1}$, i.v.) was shown to enhance the release of DA in push-pull cannula superfusates (Bartholini *et al.*, 1976). The cholinergic presynaptic receptors located on DA terminals could be involved in this effect.

Changes in DA Release Induced by Local Application of Transmitters or drugs in the Caudate Nucleus

Using different approaches, several authors have shown the stimulatory effect of the local application of *d*-amphetamine on DA release using concentrations ranging from 10^{-6} M to 3×10^{-3} M. The effect of *d*-amphetamine on newly formed ^{3}H-DA was even seen repeatedly for at least 5 days in cup superfusates in the awake monkey (Gauchy *et al.*, 1974). According to Von Voigtlander and Moore (1973) *d*-amphetamine was more active than the *l*-form in releasing exogenous ^{3}H-DA in ventricular perfusates of the cat and its effect was reduced shortly after lesion of the nigrostriatal DA neurones. Although less active, amantadine also required nerve activity to exert its optimal effect. Chiueh and Moore (1974) indicated that α-methylparatyrosine did not alter the *d*-amphetamine- (10^{-5} M) induced release of ^{3}H-DA previously taken up or synthesized from ^{3}H-tyrosine. This is in contrast with our own results since *d*-amphetamine (10^{-6} M) was much more effective (about six-fold) in releasing ^{3}H-DA in cup superfusates during its continuous synthesis from L-^{3}H-tyrosine than shortly (40 min) after α-methylparatyrosine injection ($100\,mg\,kg^{-1}$, i.v.) despite the presence of large amounts of the ^{3}H-amine in tissues. Similar observations were made with pheniprazine (10^{-5} M) (an MAO inhibitor exhibiting *d*-amphetamine properties) but not under strong depolarization with K^{+} (30 mM) which acted as well before as after the application of the DA synthesis inhibitor (Glowinski, 1974). These data, and the rapid and complete disappearance of the spontaneous release of ^{3}H-DA induced by α-methylparatyrosine (10^{-4} M or $100\,mg\,kg^{-1}$, i.v.) support the hypothesis of a preferential release of the newly synthesized transmitter in physiological states (Besson *et al.*, 1973).

Other drugs acting on DA metabolism were tested. Benztropine (10^{-6} M), a DA uptake inhibitor, induced a progressive increase in the release of newly formed ^{3}H-DA in cup superfusates. This effect differed from the immediate action of *d*-amphetamine. Tranylcypromine (10^{-5} M) acted as *d*-

amphetamine. Much less pronounced effects were seen with irreversible or reversible MAO inhibitors such as pargyline (10^{-6} M) or harmaline and harmine (10^{-4} M), the two latter drugs acting in an opposite way to that observed after their intravenous injection (Besson *et al.*, 1974).

In agreement with in-vitro data on the role of cholinergic presynaptic receptors in the control of DA release, we observed that ACh (10^{-5} M), carbachol (10^{-5} M) or oxotremorine (10^{-5} M) stimulated the release of newly formed ^{3}H-DA in cup or push-pull cannula superfusates and that hexamethonium (10^{-5} M) or mecamylamine (10^{-5} M) partially antagonized the ACh effect (Giorguieff *et al.*, 1977a). A stimulation of ^{3}H-DA release was also seen during the 15 min application of GABA (10^{-5} M) and several related compounds including muscimol (10^{-6} M), γ-hydroxybutyrate (10^{-5} M), GABA-choline (10^{-5} M) and aminooxyacetic acid (10^{-5} M), the GABA-transaminase inhibitor. The initial activation of ^{3}H-DA release was followed by an inhibitory phase which lasted until the end of the application, when GABA was added for one or two hours in the superfusion fluid delivered to the push-pull cannula (Chéramy *et al.*, 1978a). Such an inhibition of DA release was also observed by Bartholini and Stadler (1975) in the second 60 min superfusate fraction collected during a 120 min GABA (10^{-3} M) application. Moreover GABA (10^{-3} M) reduced or abolished the picrotoxin- (10^{-3} M) and bicuculline- (10^{-3} M) induced release of DA. This inhibiting effect of GABA, not seen *in vitro*, could be mediated by a complex neuronal loop through pathways projecting to the SN.

CHANGES IN DA RELEASE INDUCED BY NIGRAL APPLICATION OF TRANSMITTERS OR DRUGS

Cats were implanted with a push-pull cannula in the caudate nucleus to study the release of ^{3}H-DA continuously formed from L-^{3}H-tyrosine and with another one in the SN to deliver transmitters or drugs locally for various time intervals.

In agreement with electrophysiological studies revealing an inhibition of firing in DA cells (Aghajanian and Bunney, 1975), the nigral application of DA (10^{-6} M) reduced the release of ^{3}H-DA from nerve terminals (Chéramy *et al.*, 1977c). A similar effect was seen with *d*-amphetamine (10^{-6} M) or benztropine (10^{-6} M), drugs which enhance the nigral release of DA. Haloperidol (10^{-6} M) induced an opposite effect. These data support the concept of the presence of autoreceptors on the DA cell as initially suggested by Aghajanian and Bunney (1975). As it will be discussed later DA receptors located on nigral neuronal afferences may also be involved.

The nigral application of substance P (10^{-8} M) stimulated the release of ^{3}H-DA in the ipsilateral caudate nucleus suggesting that the striatonigral neurones containing substance P exert a facilitatory control on the DA pathway (Chéramy *et al.*, 1977c). In agreement with the effect seen after its systemic administration, picrotoxin (10^{-5} M) added in the SN induced a potent stimulation of ^{3}H-DA release suggesting that the blockade of GABA transmission activates the DA neurones (Chéramy *et al.*, 1977d). However, striatonigral GABAergic neurones may not only act on DA neurones by inhibiting the activity of the DA pathway. Indeed, the initial effect of GABA (10^{-5} M) and other GABA-related compounds such as muscimol (10^{-6} M), GABA-choline (10^{-5} M), γ-hydroxybutyrate (10^{-5} M) and lioresal (10^{-5} M) was to stimulate the release of ^{3}H-DA in the caudate nucleus. With GABA and GABA-choline only, this initial stimulation (15 min) was followed by a transcient inhibition of the ^{3}H-transmitter release (Chéramy *et al.*, 1978a). If this inhibitory action was expected, the initial stimulation seen in all cases suggested an indirect action of GABA and other related substances on DA neurones. This surprising effect involved GABAergic receptors since the release of ^{3}H-DA remained unchanged during the combined nigral application of picrotoxin and GABA. These receptors could be located on glycinergic inhibitory interneurones controlling the activity of the DA neurones. Indeed, glycine (10^{-5} M) slightly inhibited ^{3}H-DA release and the nigral application of strychnine (10^{-5} M), a glycinergic antagonist, induced an opposite effect. Furthermore, the inhibitory effect of glycine was antagonized by strychnine (Chéramy *et al.*, 1978b). Another hypothesis is that the enhanced release of ^{3}H-DA induced by GABA is mediated by neurones of the pars reticulata, for example the nigrothalamic neurones which are sensitive to GABA and GABAergic agonists and controlled by striatonigral GABAergic neurones. These neurones may indirectly exert presynaptic influences on DA terminals through complex polysynaptic pathways (thalamocorticostriatal or thalamostriatal neuronal loops).

Reciprocal Control of the Activity of the Two DA Pathways

A reciprocal control of the activity of the two DA pathways has been shown in in-vivo release studies made in halothane-anaesthetized or "encéphale isolé" cats implanted with several push-pull cannulae using the continuous labelling of DA terminals with L-^{3}H-tyrosine.

The various results obtained will be briefly summarized since they have been extensively described elsewhere.

1. By comparing the variations in the spontaneous release of ^{3}H-DA in the right and left caudate nuclei in successive 15 min collected fractions, we observed that the increased release of ^{3}H-DA in one side was associated

with a decreased release of ^{3}H-transmitter on the other side and vice versa in 70% of 120 paired estimations made in several animals (Nieoullon *et al.*, 1977c).

2. The imbalance between the activity of the two DA pathways was confirmed by measuring the effects of the interruption of nerve activity in one pathway on the release of ^{3}H-DA in both caudate nuclei. The electrocoagulation of one substantia nigra or the nigral application of DA (10^{-6} M), *d*-amphetamine (10^{-6} M) or benztropine (10^{-6} M) which reduced the release of ^{3}H-DA in the ipsilateral caudate nucleus induced a concomitant increase in ^{3}H-DA release in the contralateral side (Nieoullon *et al.*, 1977c). Conversely, the activation of one DA pathway induced by the unilateral nigral application of haloperidol (10^{-5} M) was associated with a reduction of ^{3}H-DA release in the contralateral caudate nucleus.

3. The asymmetric response of the two DA pathways was also observed when the role of cerebellar nuclei and of sensory messages in the regulation of their activity was studied.

(a) As revealed by anatomical studies, a neuronal pathway originating in the cerebellar dentate nucleus projects into the pars compacta of the contralateral SN (Snider *et al.*, 1976). The unilateral electrical stimulation of this nucleus led to a long-lasting activation of the contralateral DA pathway (as shown by the increase in ^{3}H-DA release in the corresponding caudate nucleus) and to an opposite response in the ipsilateral structure (Nieoullon *et al.*, 1978b). Thus, messages delivered to one SN induced opposite effects on the activity of the two DA pathways.

(b) The electrical stimulation of the right forelimb or the delivery of light flashes to the right eye enhanced the activity of the ipsilateral DA pathway but reduced that of the contralateral side as indicated by the opposite changes in ^{3}H-DA release seen in the two caudate nuclei (Nieoullon *et al.*, 1977b).

The mechanisms involved in the reciprocal control of the two DA pathways are not yet well understood. Indeed, to our knowledge, direct anatomical connections between the two SN have not yet been described. However, this phenomenon could be initially triggered by messages originating from DA dendrites. Indeed, both the stimulation of the cerebellar dentate nucleus or sensory stimuli affected the relase of ^{3}H-DA in both SN. In all cases the effects were opposite to those observed in the corresponding caudate nuclei but their amplitude and duration were similar. This imbalance in the dendritic release of ^{3}H-DA between the two SN could be responsible for the opposite changes in the activity of the two DA pathways.

RELEASE OF DOPAMINE FROM DENDRITES OF NIGROSTRIATAL DOPAMINERGIC NEURONES

The first data suggesting a dendritic release of DA was first obtained *in vitro* by Geffen *et al.* (1976). Potassium evoked the release of ^{3}H-DA previously taken up in slices of the rat SN through a calcium-dependent process. This effect was more pronounced in slices prepared from the pars reticulata rich in DA dendrites than in those obtained from the pars compacta which mainly contains the DA cell bodies. Similar results were obtained when endogenous DA was estimated in the superfusing medium (Cuello and Iversen, 1978). Indirect evidence for an in-vivo dendritic release of DA was obtained by Korf *et al.* (1976) who observed an increase in DA metabolites in the rat SN after stimulation of the medial forebrain bundle. This could reflect an enhanced release of DA from dendrites by antidromic activation.

We demonstrated directly the in-vivo release of DA in the SN in the cat using the push-pull cannula method and the continuous labelling of DA dendrites with L-^{3}H-tyrosine (Nieoullon *et al.*, 1977d). The spontaneous release of ^{3}H-DA (only about three to four times lower than observed in the caudate nucleus) was enhanced by the local application of *d*-amphetamine (10^{-6} M) and benztropine (10^{-6} M) and was markedly stimulated by K^+ (30 mM). Messages delivered by neuronal pathways projecting into the SN modulate the release of DA from dendrites (some of these effects have already been described, p. 203). The electrical stimulation of the ipsilateral cortex (motor or visual) (Nieoullon *et al.*, 1977a) and of the ipsilateral cerebellar dentate nucleus (Nieoullon *et al.*, 1978b) enhanced the release of ^{3}H-DA from dendrites, these effects being mediated by cortico- and cerebellonigral projections. In contrast a decrease in ^{3}H-DA dendritic release was seen under stimulation of the ipsilateral fastigial nucleus and contralateral dentate nucleus (Nieoullon *et al.*, 1978b). As already indicated, modulations in the dendritic release of ^{3}H-DA were also induced by sensory messages since unilateral somatic or visual stimuli decreased the release of ^{3}H-DA in the ipsilateral SN and induced an opposite effect in the contralateral side (Nieoullon *et al.*, 1977b).

If DA is released from recurrent collaterals and not from dendrites a close parallel between the effects detected in the caudate nucleus and the SN should always be seen. Tetrodotoxin (5×10^{-7} M) reduced DA release from nerve terminals in the caudate nucleus but not in the SN (Nieoullon *et al.*, 1977d). Furthermore, in most cases when DA release was enhanced in the caudate nucleus, it was reduced in the SN and vice versa. This also indirectly reveals that DA is not released from nerve terminals but from dendrites. In fact, tetrodotoxin stimulated the nigral release of newly synthesized ^{3}H-DA suggesting that sodium channels are not involved in the dendritic release process and that inhibitory inputs normally regulate DA released from dendrites.

Close temporal relationships were seen between the opposite changes in ^{3}H-DA release detected in the caudate nuclei and their respective SN under either stimulation of cerebellar nuclei or sensory stimuli, the effects persisting for 40 min at least after the end of the stimuli. It can thus be postulated that the changes in dendritic release of DA modulate the activity of the DA neurones, this activity being inversely correlated to the extent of the dendritic release of DA. By acting on DA autoreceptors or on DA receptors located on the terminals of neuronal pathways controlling the DA neurones, DA released from dendrites may regulate the activity of DA neurones. Dopamine receptors coupled to an adenylate cyclase are present on nerve terminals of striatonigral neurones (Prémont *et al.*, 1976; Gale *et al.*, 1977). Since DA stimulated the release of ^{3}H-GABA but not that of substance P in slices of the rat SN (Reubi *et al.*, 1977) DA may act on DA receptors located on GABAergic terminals. These various data emphasize the physiological significance of the dendritic release of DA.

CONCLUSION

The resting release of DA from nerve terminals and dendrites of the nigrostriatal DA neurones is very low ($\leqslant 20\,\mathrm{pg\,min^{-1}}$). The amine can be detected in superfusates with a radioenzymatic assay, but isotopic methods seem to be much more suitable for in-vitro or in-vivo release studies. To ensure specificity, and since the newly synthesized transmitter is stored in a pool which plays a critical role in the release process, the estimation of ^{3}H-DA released during the continuous delivery of L-^{3}H-tyrosine appears to be the most appropriate approach.

The spontaneous and evoked releases of DA from nerve terminals are calcium dependent and sensitive to tetrodotoxin. Various depolarizing agents enhance the in-vitro release of DA. Changes in nerve firing in DA neurones directly induced or evoked by stimulation of various neuronal pathways projecting to the SN affect the in-vivo release of DA in the caudate nucleus. Some, but not all, in-vitro data suggest that DA presynaptic receptors are involved in the control of DA release. Strong arguments in favour of the presence of cholinergic, glutaminergic and enkephalinergic presynaptic receptors on the DA terminals have also been obtained. Changes in DA release are seen under stimulation of these receptors by their agonists. γ-Aminobutyric acid and related substances also locally affect the release of DA, but the GABAergic receptors involved do not seem to be located on DA nerve terminals. The physiological significance of these local regulations has still to be demonstrated. However, an activation of DA release has already been shown *in vivo* during the stimulation of the corticostriatal glutamatergic pathway.

Dopamine is not only released from nerve terminals but also from dendrites. This has been shown *in vivo* and *in vitro*. The dendritic release of DA is affected by DA drugs such as *d*-amphetamine and benztropine in a way similar to the amine release from nerve terminals. However, it is not reduced by tetrodotoxin and generally affected in an opposite way to the release from nerve terminals by messages reaching the SN. Through DA autoreceptors or DA receptors located on nigral neuronal afferences such as the GABAergic terminals, the dendritic release of DA may contribute to the regulation of the activity of the DA neurones.

Finally in-vivo studies allowing the simultaneous monitoring of DA release from terminals and dendrites of the two DA pathways have revealed a close interdependence between these pathways. Opposite changes in the dendritic release of DA seen in the two SN under various experimental conditions may be responsible for the reciprocal control of the activity of the two DA pathways.

REFERENCES

Aghajanian, G. K. and Bunney, B. S. (1975). *In* "Neuropsychopharmacology" (Eds. J. R. Boissier, H. Hippius, and P. Pichot), 444–452. Excerpta Medica, Amsterdam, and American Elsevier, New York.

Arbilla, S. and Langer, S. Z. (1978). *Nature London* **271,** 559–561.

Bartholini, G. and Stadler, H. (1975). *In* "Chemical Tools in Catecholamine Research", Vol. II (Eds. O. Almgren, A. Carlsson and J. Engel), 235–241. North Holland Publishing Co., Amsterdam.

Bartholini, G., Stadler, H. and Lloyd, K. G. (1975). *In* "Neuropsychopharmacology" (Eds J. R. Boissier, H. Hippius and P. Pichot), 487–492. Excerpta Medica, Amsterdam, and American Elsevier, New York.

Bartholini, G., Stadler, H., Gadea-Ciria, M. and Lloyd, K. G. (1976). *Neuropharmacology*, **15,** 515–519.

Besson, M. J., Chéramy, A., Feltz, P. and Glowinski, J. (1969a). *Proc. Natl Acad. Sci., U.S.A.* **62,** 741–748.

Besson, M. J., Chéramy, A. and Glowinski, J. (1969b). *Eur. J. Pharmacol.* **7,** 111–114.

Besson, M. J., Chéramy, A., Feltz, P. and Glowinski, J. (1971). *Brain Res.* **32,** 407–424.

Besson, M. J., Chéramy, A., Gauchy, C. and Glowinski, J. (1973). *Naun. Schmied. Arch. Pharmacol.* **278,** 101–105.

Besson, M. J., Chéramy, A., Gauchy, C. and Glowinski, J. (1974). *Adv. Neurol.* **5,** 69–78.

Bustos, G. and Roth, R. H. (1972). *Br. J. Pharmacol.* **46,** 101–115.

Chéramy, A., Bioulac, B., Besson, M. J., Vincent, J. D., Glowinski, J. and Gauchy, C. (1975a). *In* "Neuropsychopharmacology" (Eds. J. R. Boissier, H. Hippius and P. Pichot), 493–498. Excerpta Medica, Amsterdam, and American Elsevier, New York.

Chéramy, A., Besson, M. J., Gauchy, C., Javoy, F., Agid, Y., Guyenet, P. and Glowinski, J. (1975b). *In* "Antipsychotic Drugs, Pharmacodynamics and Pharmacokinetics" (Eds G. Sedvall, B. Uvnas and Y. Zotterman), 161–173, Pergamon, Oxford.

Chéramy, A., Nieoullon, A. and Glowinski, J. (1977a). *J. Pharmacol. Expl. Ther.* **203,** 283–293.

Chéramy, A., Nieoullon, A. and Glowinski, J. (1977b). *Life Sci.* **20,** 811–816.

Chéramy, A., Nieoullon, A., Michelot, R. and Glowinski, J. (1977c). *Neurosci. Lett.* **4,** 105–109.

Chéramy, A., Nieoullon, A. and Glowinski, J. (1977d). *Naun. Schmied. Arch. Pharmacol.* **297,** 31–37.

Chéramy, A., Nieoullon, A. and Glowinski, J. (1978a). *Eur. J. Pharmacol.* **47,** 141–147.

Chéramy, A., Nieoullon, A. and Glowinski, J. (1978b). *Eur. J. Pharmacol.* **48,** 281–295.

Chiueh, C. C. and Moore, K. E. (1973). *Brain Res.* **50,** 221–225.

Chiueh, C. C. and Moore, K. E. (1974). *J. Pharmacol. Expl. Ther.* **190,** 100–108.

Christiansen, J. and Squires, R. F. (1974). *J. Pharm. Pharmacol.* **26,** 367–369.

Cuello, A. C. and Iversen, L. L. (1978). *In* "Interactions Between Putative Neurotransmitters in the Brain" (Eds. S. Garattini, J. F. Pujol and R. Samanin) Raven Press, New York (in press).

DaPrada, M., Saner, A., Burkard, W. P., Bartholini, G. and Pletscher, A. (1975). *Brain Res.* **94,** 67–73.

De Belleroche, J., Bradford, H. F. and Jones, D. G. (1976). *J. Neurochem.* **26,** 561–571.

Dismukes, K. and Mulder, A. H. (1977). *Naun. Schmied. Arch. Pharmacol.* **297,** 23–29.

Divac, I., Fonnum, F. and Storm-Mathisen, J. (1977). *Nature* (*London*) **266,** 377–378.

Farnebo, L. O. (1971). *Acta Physiol. Scand. Suppl. 271*, 45–52.

Farnebo, L. O. and Hamberger, B. (1971). *Acta Physiol. Scand. Suppl. 271*, 35–44.

Gale, K., Guidotti, A. and Costa, E. (1977). *Nature* (*London*) **260,** 258–260.

Gauchy, C., Bioulac, B., Chéramy, A., Besson, M. J., Glowinski, J. and Vincent, J. D. (1974). *Brain Res.* **77,** 257–268.

Geffen, L. B., Jessel, T. M., Cuello, A. C. and Iversen, L. L. (1976). *Nature* (*London*) **260,** 258–260.

Giorguieff, M. F., Le Floc'h, M. L., Westfall, T. C., Glowinski, J. and Besson, M. J. (1976). *Brain Res.* **106,** 117–131.

Giorguieff, M. F., Le Floc'h, M. L., Glowinski, J. and Besson, M. J. (1977a). *J. Pharmacol. Expl. Ther.* **200,** 535–540.

Giorguieff, M. F., Kemel, M. L. and Glowinski, J. (1977b). *Neurosci. Lett.* **6,** 73–77.

Giorguieff, M. F., Kemel, M. L., Glowinski, J. and Besson, M. J. (1978). *Brain Res.* **139,** 115–130.

Glowinski, J. '1974). *In* "Handbook of Psychopharmacology", Vol. 3 (Eds. L. L. Iversen, S. Iversen and S. H. Snyder), 139–167, Plenum, New York.

Heikkila, R. E., Orlansky, H., Mytilineou, C. and Cohen, G. (1975a). *J. Pharmacol. Expl. Ther.* **194,** 47–56.
Heikkila, R. E., Orlansky, H. and Cohen, G. (1975b). *Biochem. Pharmacol.* **24,** 847–852.
Iversen, L. L., Rogawski, M. A. and Miller, R. J. (1975). *Molec. Pharmacol.* **12,** 251–262.
Korf, J., Zieleman, M. and Westerink, B. H. C. (1976). *Nature (London)* **260,** 257–258.
Langer, S. E. (1977). *Br. J. Pharmacol.* **60,** 481–497.
Lanne, J. D. and Aprison, M. N. (1977). *Life Sci.* **20,** 665–672.
Loh, H. H., Brase, D. A., Sampath-Khanna, S., Mar, J. B., Way, E. L. and Li, C. H. (1976). *Nature (London)* **264,** 567–568.
McKenzie, G. M. and Szerb, J. C. (1968). *J. Pharmacol. Expl. Ther.* **162,** 302–308.
McLennan, H. (1964). *J. Physiol.* **174,** 152–161.
McLennan, H. (1965). *Experientia* **21,** 725–726.
Mulder, A. H., Van den Berg, W. B. and Stoof, J. C. (1975). *Brain Res.* **99,** 419–424.
Ng, K. Y., Chase, T. N., Colburn, R. W. and Kopin, I. J. (1971). *Science* **172,** 487–489.
Nieoullon, A., Chéramy, A., and Glowinski, J. (1977a). *J. Neurochem.* **28,** 819–828.
Nieoullon, A., Chéramy, A. and Glowinski, J. (1977b). *Nature (London)* **269,** 340–342.
Nieoullon, A., Chéramy, A. and Glowinski, J. (1977c). *Science,* **198,** 416–418.
Nieoullon, A., Chéramy, A. and Glowinski, J. (1977d). *Nature (London)* **266,** 375–377.
Nieoullon, A., Chéramy, A. and Glowinski, J. (1978a). *Brain Res.* **145,** 69–83.
Nieoullon, A., Chéramy, A. and Glowinski, J. (1978b). *Brain Res.* **148,** 143–152.
Patrick, R. L. and Barchas, J. D. (1976). *J. Pharmacol. Expl. Ther.* **197,** 89–96.
Premont, J., Thierry, A. M., Tassin, J. P., Glowinski, J., Blanc, G. and Bockaert, J. (1976). *FEBS Lett.* **68,** 99–104.
Pollard, H., Llorens-Cortes, C. and Schwartz, J. C. (1977). *Nature (London)* **268,** 745–746.
Portig, P. G. and Vogt, M. (1969). *J. Physiol. (London)* **204,** 687–715.
Raiteri, M., Bertollini, A., Angellini, F. and Lévi, G. (1975). *J. Pharmacol.* **34,** 189–195.
Raiteri, M., Delcarmine, R., Bertollini, A. and Lévi, G. (1977). *Eur. J. Pharmacol.* **41,** 133–143.
Raiteri, M., Cerrito, F., Cervoni, A. M., Del Carmine, R., Ribera, M. T. and Lévi, G. (1978). *In* "Dopamine". Advances in Biochemical Pharmacology, Vol. 19, (Eds P. Roberts, G. Woodruff and L. Iversen), 35–56, Raven Press, New York.
Reubi, J. C., Iversen, L. L. and Jessel, T. M. (1977). *Nature (London)* **268,** 652–654.
Riddell, D. and Szerb, J. C. (1971). *J. Neurochem.* **18,** 989–1006.
Roth, R. H., Walters, J. R. and Aghajanian, G. K. (1973). *In* "Frontiers in Catecholamine Research" (Eds. E. Usdin and S. Snyder), 567–574. Pergamon, New York.
Scatton, B., Chéramy, A., Besson, M. J. and Glowinski, J. (1970). *Eur. J. Pharmacol.* **13,** 131–133.
Schaffner, R. and Haefely, W. (1975). *Experientia* **31,** 732.
Seeman, P. and Lee, T. (1975). *Science* **188,** 1217–1219.
Snider, R. S., Maiti, A. and Snider, S. R. (1976). *Expl. Neurol.* **53,** 714–728.

Speth, R. C., Chen, F. M., Lindstrom, J. M., Kobayashi, R. M. and Yamamura, H. I. (1977). *Brain Res.* **131,** 350–355.

Stoof, J. C. and Mulder, A. H. (1977). *Eur. J. Pharmacol.* **46,** 177–180.

Subramanian, N., Mitzivegg, P., Sprügel, W., Domschke, W., Domschke, S., Wünsch, E. and Demling, L. (1977). *Naun. Schmied. Arch. Pharmacol.* **299,** 163–166.

Von Voigtlander, P. F. and Moore, K. E. (1971). *Brain Res.* **35,** 580–583.

Von Voigtlander, P. F. and Moore, K. E. (1973). *J. Pharmacol. Expl. Ther.* **184,** 542–552.

Westfall, T. C. (1974a). *Life Sci.* **14,** 1641–1652.

Westfall, T. C. (1974b). *Fed. Proc.* **33,** 524.

Westfall, T. C., Besson, M. J., Giorguieff, M. F. and Glowinski, J. (1976). *Naun. Schmied. Arch. Pharmacol.* **292,** 279–287.

Yamamura, H. I. and Snyder, S. H. (1974). *Proc. Natl. Acad. Sci., U.S.A.* **71,** 1725–1729.

CHAPTER 16

Characteristics of Dopamine Uptake

Alan S. Horn

Department of Pharmacy, University of Groningen, Groningen, The Netherlands.

In 1932 it was suggested by Burn that catecholamines might be taken up by storage sites in tissue but it was not until the introduction of tritium-labelled (^3H) catecholamines of high specific activity that the process could be studied with relative ease. Axelrod *et al.* (1959) and Whitby *et al.* (1961) showed that after intravenous injection of small amounts of ^3H-noradrenaline (NA) a large proportion of the catecholamine was taken up in various peripheral tissues. The characteristics of the noradrenergic uptake system in the peripheral and central nervous systems have been extensively reviewed (Iversen, 1967; Iversen, 1975).

Direct evidence for an accumulation of dopamine (DA) in brain tissue was obtained by Glowinski and Iversen in 1966 who showed that one hour after intraventricular injection of ^3H-DA the highest concentration was found in the corpus striatum followed by the hypothalamus, hippocampus, midbrain, medulla oblongata, cortex and cerebellum. The level of ^3H-DA in the hypothalamus was about 62% of that in the striatum. Various groups of workers then showed that both *in vivo* (Glowinski *et al.*, 1966; Carlsson *et al.*, 1966; Fuxe and Ungerstedt, 1968) and *in vitro* (Ross and Renyi, 1967; Hamberger, 1967; Jonason and Rutledge, 1968) the DA uptake system differed from the NA system in that it was not potently inhibited by the tricyclic antidepressants such as desipramine, imipramine and protriptyline.

A significant advance in this area was made when Snyder and Coyle (1969) and Coyle and Snyder (1969a) showed that homogenization of brain tissue in isotonic sucrose solution and centrifugation at $1000 \times g$ resulted in a supernatant that was both enriched in synaptosomes (Whittaker, 1966, DeRobertis, 1967) and sufficiently "pure" to be suitable to study the neurochemical properties of this transport system *in vitro*.

At this point it should be made clear that with regard especially to the in-vitro studies the term *uptake* should be taken to mean the passage or transport of a molecule across a cell membrane rather than a simple binding process of the molecule to the external membrane surface.

FUNCTIONAL IMPORTANCE OF DA UPTAKE

There is now compelling evidence that the neuronal uptake mechanism is probably the major means of terminating the synaptic actions of NA (Iversen, 1975). It has in fact been suggested that in peripheral tissues with a dense sympathetic innervation, neuronal uptake of NA could account for the recapture of as much as 70–80% of the released amine under normal conditions. It is not definitely known whether this is also true of the DA uptake mechanisms in the CNS but considering some of the similarities between the two neuronal systems it is quite likely to be the case. For a discussion of the subsequent events occurring after neuronal uptake of DA see Westerink, Chapter 18.

ANATOMICAL ASPECTS OF NEURONAL DA UPTAKE

The neuroanatomy of the dopaminergic systems has already been dealt with elsewhere in this volume (Lindvall, Chapter 20 and Palkovits, Chapter 21). The dopaminergic areas of the nervous system in which the uptake process has been studied *in vitro* using ^{3}H-DA include:

(a) The corpus striatum (Snyder and Coyle, 1969; Harris and Baldessarini, 1973a; Holz and Coyle, 1974).
(b) The substantia nigra (Geffen *et al.*, 1976).
(c) The median eminence (Cuello *et al.*, 1973). (For in-vivo studies in this brain area see Lichtensteiger and Langeman, 1966; Fuxe *et al.*, 1967; Cuello and Iversen, 1973.)
(d) The olfactory tubercle (Horn *et al.*, 1974).
(e) The nucleus accumbens (Horn *et al.*, 1974).
(f) The septal nuclei (Garey and Heath, 1974; Garey, 1976).
(g) The retina (Ehringer and Falck, 1971; Kramer *et al.*, 1971).

It has also been studied *in vitro* in the following areas:

(h) The limbic cortex (Tassin *et al.*, 1974).

(i) The cerebral cortex, hypothalamus midbrain, medulla oblongata–pons and the cerebellum (Snyder and Coyle, 1969).

BIOCHEMICAL PROPERTIES OF THE NEURONAL DA UPTAKE SYSTEM

Although early work on the properties of the DA uptake system employed tissue slices or synaptosome-enriched homogenates, recently Holz and Coyle (1974) have published a detailed study of the properties of this system in purified striatal synaptosomes. The DA transport system has the following biochemical properties.

Temperature

It is temperature sensitive. Studies with purified striatal synaptosomes have shown that there is a break in the Arrhenius plot at 30°; above this temperature the Q_{10} was 1.7 whereas below it the value was 4.5. The discontinuity in the Arrhenius plot may be due to a change in the state of the hydrocarbon interior of the membrane or it may result from a change at the lower temperature in the rate-determining step to one with a higher activation energy (Holz and Coyle, 1974).

Linearity

In synaptosomes the uptake is linear for two minutes (Holz and Coyle, 1974) and for 5–10 min in synaptosome-enriched homogenates from the corpus striatum (Snyder and Coyle, 1969; Harris and Baldessarini, 1973a). It is also proportional to the amount of synaptosomal protein.

Tissue-to-Medium Ratio

After two minutes of incubation with ^{3}H-DA the tissue-to-medium ratio (T/M) reached a value of 170. However, as with ^{3}H-NA uptake (Iversen, 1975) the actual concentration gradient achieved in the dopaminergic synaptosomes may be in the order of 1000 : 1 (Holz and Coyle, 1974).

Kinetics

The system in purified synaptosomes displays Michaelis–Menten kinetics and has a $K_m = 1.3 \times 10^{-7}$ M for DA and a $V_{max} = 25.3$ pmol per 100 μg protein per 2 min at normal physiological concentrations of Na^+ and K^+ (Holz and Coyle, 1974). The affinity for (−) NA is lower, $K_m = 2.0 \times 10^{-6}$ M (Iversen *et al.*, 1971).

The term K_m in the above context is generally assumed to be a measure of the dissociation constant for the binding of substrate (DA) to the membrane uptake sites. Thus the K_m value may be taken as a reciprocal measure of the affinity of the substrate for the uptake sites. The V_{max} is a measure of the number of amine uptake sites present and it is dependent on the type and amount of tissue present. The K_m, however, is independent of the total number of uptake sites present (Iversen, 1972).

Effect of Na^+ and K^+

The uptake of ^{3}H-DA is dependent on the concentration of Na^+ and to a certain extent on that of K^+ (Holz and Coyle, 1974).

Other Ions and Ca^{2+}

It is possible to replace K^+ by Rb^+ but Cs^+ is less effective; the system does not appear to require Ca^{2+} (Harris and Baldessarini, 1973a).

Na^+ Permeability

Veratridine (50 μM) and batrachotoxin (0.5 μM) both effectively inhibited ^{3}H-DA uptake. This inhibition was partly blocked by tetrodotoxin (1.0 μM). The effects with veratridine and batrachotoxin may be due to an increased Na^+ permeability (Holz and Coyle, 1974).

Uptake of Other Analogues

By using radioactively labelled compounds it has been shown (Ross, 1976) that various other β-phenylethylamine analogues can be taken up with varying degrees of efficiency by dopaminergic neurones. Their structures are shown in Table I.

Development

Developmental studies have demonstrated that homogenates of the corpus striatum from newborn rats show only 10% of the activity of the adult system in transporting ^{3}H-DA. By four weeks after birth, however, it reaches 75% of the adult level of activity. Kinetic studies show that this increase in uptake ability is related to an increase in the V_{max} (Coyle and Campochiaro, 1976).

Table I β-Phenylethylamine analogues that are taken up by dopaminergic neurones.

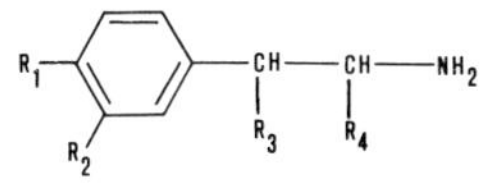

Drug	Structure R_1	R_2	R_3	R_4
Dopamine	OH	OH	H	H
Noradrenaline	OH	OH	OH	H
Tyramine	OH	H	H	H
m-Tyramine	H	OH	H	H
α-Methyl-*m*-tyramine	H	OH	H	CH_3
Octopamine	OH	H	OH	H
m-Octopamine	H	OH	OH	H
Metaraminol	H	OH	OH	CH_3

Data taken from Ross (1976).

Topography of the Uptake System

In an attempt to investigate the possible involvement of certain cell surface glycoproteins and peptides in the DA transport system, synaptosomes have been treated with concanavalin A and trypsin. Neither agent was found to have an effect on ^{3}H-DA uptake (Wang *et al.*, 1975).

INHIBITION OF DOPAMINE UPTAKE

Although the uptake or transport of DA is an accepted and reasonably well characterized phenomenon certain aspects of the *inhibition* of this uptake process by various drugs have been a subject of controversy and debate in the literature during the past few years. The intention of this Section is to attempt to clarify some of these points of contention.

Inhibition of Uptake or Release?

Several authors (Ferris *et al.*, 1972, Hendley *et al.*, 1972) have pointed out the difficulties in clearly distinguishing between a drugs ability to inhibit the uptake and/or cause the release of ^{3}H-biogenic amines. The best example of this is, of course, amphetamine. Heikkila *et al.* (1975) have clearly stated the problem in general terms as follows:

> Consider first the measurement of ^3H-amine uptake. During the time that the ^3H-amine is taken up, a releasing agent could cause egress of a portion of the accumulated ^3H-amine. The resultant diminution in accumulated ^3H-amine could be interpreted as decreased "uptake". Therefore, a releasing agent could be misclassified as an uptake blocker. A second difficulty exists in the experimental verification of a presumed releasing action. If tissue-containing ^3H-amine is incubated in fresh medium at 37°, there is spontaneous efflux of a portion of the ^3H-amine. An agent that blocks the reuptake of ^3H-amines could, in theory, augment the net loss from the tissue. Since increased net loss is generally construed as constituting a "releasing" action, it is apparent that an uptake blocker could be misclassified as a releasing agent. Thus, there is reason to suspect that uptake blockade and release can become confounded.

These authors determined dose–response curves for in-vitro release and inhibition of the uptake of ^3H-DA by various drugs. In the case of (+)-amphetamine they are of the opinion that it is mainly a releaser of ^3H-DA in the striatum. Using somewhat similar techniques other workers have concluded that amphetamine has a stronger effect on ^3H-DA uptake than on release (Ferris *et al.*, 1972; Holmes and Rutledge, 1974). In experiments using synaptosomal "beds" prepared from a purified synaptosomal fraction obtained from sheep corpus striatum De Belleroche *et al.* (1976) also concluded that amphetamine was operating largely as an inhibitor of ^3H-DA uptake.

The above in-vitro studies on release versus inhibition of uptake all suffer, to a greater or lesser extent depending on the experimental design, from the problem of re-uptake of previously released ^3H-DA (Raiteri *et al.*, 1976). In an attempt to eliminate the re-uptake problem Raiteri *et al.* (1975) have developed an elegant superfusion technique for synaptosomes.

This technique has several advantages over the simpler incubation procedures, i.e. synaptosomes can be used immediately after prelabelling with radioactive substrate, one can obtain a complete profile of the release process, and the composition of the medium can be quickly charged. The main advantage, however, is that the released radioactive substrate (e.g. ^3H-DA) is removed by the stream of superfusion fluid as soon as it is released, thus its re-uptake is prevented. Thus one can clearly distinguish between the process of drug-induced release and drug-induced inhibition of uptake.

Using this technique Raiteri *et al.* (1975, 1976) have shown that (+)amphetamine has a mixed action, i.e. it both releases and inhibits the uptake of ^3H-DA in synaptosomes from rat corpus striatum.

Benztropine (Fig. 1) has been shown to be a potent inhibitor of ^3H-DA uptake into synaptosomal enriched homogenates of the corpus striatum (Coyle and Snyder, 1969b; Horn *et al.*, 1971). More recent studies have confirmed that both in-vitro and in-vivo, benztropine is predominantly an inhibitor of ^3H-DA uptake (Orlansky and Heikkila, 1974; Goodale and

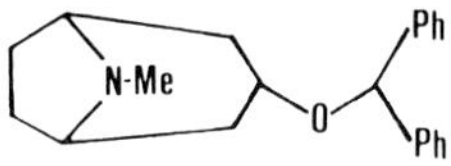

a

Fig. 1 Benztropine.

Moore, 1975; Raiteri *et al.*, 1977). The last group of workers have shown convincingly that benztropine has only a modest releasing action on ^{3}H-DA even at a high concentration (10^{-5} M).

Nomifensine (Fig. 2) is another potent inhibitor of ^{3}H-DA uptake (Hunt *et al.*, 1974; Koe, 1976) and Raiteri *et al.* (1978) have shown that it does not produce a release of ^{3}H-DA even at 10^{-5} M. They therefore classify nomifensine as a pure DA uptake inhibitor.

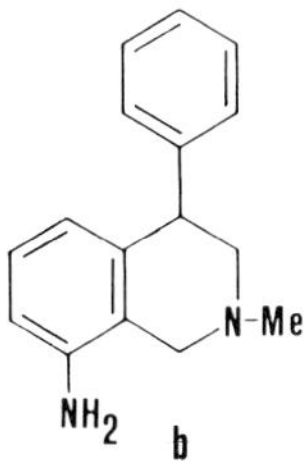

b

Fig. 2 Nomifensine.

Cocaine as well as being a potent inhibitor of NA uptake (Hendley *et al.*, 1972) is also an inhibitor of ^{3}H-DA uptake and it has been shown that even at a concentration of 10^{-4} M it does not produce a significant release of ^{3}H-DA (Heikkila *et al.*, 1975).

Thus it would appear that the drugs nomifensine, benztropine and cocaine have minimal or no activity (nomifensine) in bringing about ^{3}H-DA release and they are therefore predominantly inhibitors of ^{3}H-DA uptake.

The reported IC_{50} values for the inhibition of DA uptake by benztropine in striatal synaptosomes range from $1.1–5.4 \times 10^{-7}$ M (Horn *et al.*, 1971, Hunt *et al.*, 1974; Koe, 1976; Friedman *et al.*, 1977; Hyttel, 1977; Randrup and Braestrup, 1977). In striatal slices the IC_{50} value is higher, i.e. values of 1 μM (Ross and Renyi, 1975) and 4 μM (Orlansky and Heikkila, 1974) have been reported. The IC_{50} values in synaptosomes for nomifensine range from $1.0–4.1 \times 10^{-7}$ M (Hunt *et al.*, 1974; Koe, 1976; Randrup and Braestrup, 1977). Although nomifensine would appear to be the drug of choice for inhibiting DA uptake without causing DA release it should be borne in mind that whilst it is a less potent inhibitor of 5-HT uptake ($IC_{50} = 5.4 \times 10^{-6}$ M) it is a more potent inhibitor of NA than of DA uptake (NA uptake inhibition, $IC_{50} = 2.0 \times 10^{-8}$ M) (Koe, 1976).

Amphetamine Enantiomers and the Inhibition of ^{3}H-DA Uptake

Amphetamine has an asymmetric carbon atom and can therefore exist as two enantiomers, i.e. the (+) and (−) forms which, of course, have identical chemical and physical properties. They only differ in their effect on the sign of the rotation of plane-polarized light which they produce. Initial in-vitro studies with the enantiomers of amphetamine showed that they were equally potent in inhibiting ^{3}H-DA uptake into homogenates of rat corpus striatum (Coyle and Snyder, 1969a). More recent in-vitro studies by several groups of workers have shown that the (+) form is approximately four to six times more potent than the (−) isomer in this brain area (Ferris *et al.*, 1972; Harris and Baldessarini, 1973b; Thornburg and Moore, 1973; Holmes and Rutledge, 1974; Horn *et al.*, 1974 and De Belleroche *et al.*, 1976). It must be borne in mind, however, that, as previously discussed, amphetamine has a dual action, i.e. it both releases DA and inhibits its uptake.

Inhibition of DA Uptake by the Tricyclic Antidepressants

There is a substantial body of in-vivo (Carlsson *et al.*, 1966; Glowinski *et al.*, 1966; Fuxe and Ungerstedt, 1968) and in-vitro evidence (Ross and Renyi, 1967; Hamberger, 1967; Jonason and Rutledge, 1968; Horn *et al.*, 1971; Ferris *et al.*, 1972; Koe, 1976; Horn, 1976a) that the tricyclic antidepressants are much weaker inhibitors of DA than of NA or 5-HT uptake. The comparative effects of the most commonly used tricyclic antidepressants on the DA, NA and 5-HT uptake systems *in vitro* are shown in Table II (Koe, 1976).

In recent years several groups of workers (Halaris *et al.*, 1975; Friedman *et al.*, 1977; Randrup and Braestrup, 1977) have attempted to place a greater emphasis on the possible clinical importance of the ability of these drugs to inhibit, to some extent, the uptake of DA. However, in view of the in-vivo findings of the relative ineffectiveness of these drugs on DA transport the above suggestions remain speculative.

In-Vivo—in-Vitro Inhibition of ^{3}H-DA Uptake

It has been claimed (Bauman and Maitre, 1976) that although one can demonstrate an inhibition of ^{3}H-DA uptake *in vitro* by a drug this either

Table II Inhibition of biogenic amine uptake into rat brain synaptosomes from striatum and hypothalamus by tricyclic antidepressants.

	DA	$IC_{50} \times 10^{-6}$ M NA	5-HT
Imipramine	20.0	0.06	0.81
Desipramine	21.0	0.005	3.4
Chlorimipramine	8.1	0.11	0.09
Amitriptyline	13.0	0.13	1.2
Nortriptyline	11.0	0.25	1.7
Doxepin	24.0	0.12	2.9

Data taken from Koe (1976). Dopamine and 5-HT uptake were studied in synaptosomal preparations of rat corpus striatum and NA uptake in similar preparations from the hypothalamus. The IC_{50} is the concentration of drug required to inhibit the uptake of ^{3}H-biogenic amine by 50%.

Table III In-vivo–in-vitro inhibition of ^{3}H-DA into the corpus striatum.

Drug	ED_{50} (μmol kg^{-1})	Ref.
(+) Amphetamine (sl)	60	(a)
(−) Amphetamine (sl)	240	(a)
(+) Amphetamine (syn)	<17	(b)
Amphetamine (syn)	<14	(c)
Benztropine (sl)	60	(a)
Benztropine (syn)	~78	(c)
Prolintane (sl)	55	(a)

Data are taken from (a) Ross and Renyi (1975), (b) Carruba *et al.*, 1977, (c) Kannengiesser *et al.*, 1976. Rats or mice received i.p. injections of the drugs before isolation of striatal tissue and incubation with ^{3}H-DA, sl = slices, syn = synaptosomes.

cannot be done or it is only a weak effect after in-vivo drug administration which is followed by in-vitro incubation of striatal preparations with ^{3}H-DA. A careful review of the literature clearly demonstrates that this is, in fact, not the case as is shown by the results in Table III which are taken from Ross and Renyi (1975) who used striatal slices and Kannengiesser *et al.* (1976) and Carruba *et al.* (1977) who used striatal synaptosomes. The (+)-isomer of amphetamine was found to be quite potent and it is of interest that one can also demonstrate the differential effect of the amphetamine enantiomers in this in-vivo–in-vitro system (Ross and Renyi, 1975). It is also noteworthy that two of these groups (Ross and Renyi, 1975; Kannengiesser *et al.*, 1976) were able to demonstrate an effect with benztropine whilst others using even higher doses were not able to do this (Bauman and Maitre, 1976). Other drugs which have been studied by this in-vivo–in-vitro technique include amphetamine analogues (Ross and Renyi, 1975) fluorinated amphetamines (Kouyoumdtjian *et al.*, 1976) fenfluramine (Kannengiesser *et al.*, 1976) mazindol (Carruba *et al.*, 1977) lithium (Stefanini *et al.*, 1976) and certain steroids (Wirz-Justice *et al.*, 1974).

β-PHENYLETHYLAMINE ANALOGUES AS INHIBITORS OF ^{3}H-DA UPTAKE

In a study of the structure–activity relationships for β-phenylethylamine derivatives as inhibitors of ^{3}H-DA uptake (Horn, 1973) it was shown that these are very similar to those described for inhibition of NA uptake into the perfused rat heart (Iversen, 1967) i.e.:

1. Phenolic hydroxyl groups in the *p* and/or *m* positions enhance uptake site affinity.
2. α-Methylation increases affinity.
3. β-Hydroxylation decreases affinity.
4. Mono- or di-*N*-methylation decreases affinity.
5. *O*-Methylation of phenolic hydroxyl groups produces a marked decrease in affinity.

It must be pointed out, however, that these studies do not distinguish between molecules that are, and those that are not, transported by the DA uptake system. As well as acting as inhibitors of DA uptake a number of these drugs will have an effect on DA release and Raiteri *et al.* (1977) have made a comparison of the effects of structural variation in these molecules and the resulting influence on DA release and the inhibition of its uptake.

INHIBITION OF DA UPTAKE BY OTHER DRUGS

Various other groups of drugs have been tested as inhibitors of radioactive DA uptake in in-vitro preparations of brain tissue, they include:

1. Dopaminergic agonists, ADTN and apomorphine (Horn, 1974; Ferris *et al.*, 1975).
2. 6-Hydroxydopamine and analogues (Kostrzewa and Jacobowitz, 1974; Borchardt *et al.*, 1977).

3. Anticholinergics (Coyle and Snyder, 1969b; Horn *et al.*, 1971; Orlansky and Heikkila, 1974; Palfreyman *et al.*, 1974).
4. Antihistamines (Horn *et al.*, 1971; Orlansky and Heikkila, 1974).
5. Neuroleptics (Horn *et al.*, 1971; Horn, 1976b; Iversen *et al.*, 1976).
6. Narcotic analgesics (Clouet and Williams, 1974).
7. Tryptamine analogues (Heikkila and Cohen, 1973, 1974; Horn *et al.*, 1973).
8. Tetrahydroisoquinoline compounds (Heikkila *et al.*, 1971; Tuomisto and Tuomisto, 1973; Cohen *et al.*, 1974; Alpers *et al.*, 1975).
9. Amphetamine analogues (Tseng *et al.*, 1974; Cho *et al.*, 1975; Ross and Renyi, 1975; Koe, 1976).
10. Tetrahydrocannabinols (Howes and Osgood, 1974; Banerjee *et al.*, 1975).
11. Anorectic drugs (Kruk and Zarrindast, 1976; Kannengiesser *et al.*, 1976; Carruba *et al.*, 1977).
12. Prostaglandin synthesis inhibitors (Clarenbach *et al.*, 1974).
13. Lead and calcium salts (Silbergeld, 1977).

Due to the ability of certain anti-parkinsonian drugs to inhibit uptake into dopaminergic nerve terminals *in vitro* (Coyle and Snyder, 1969b; Ohashi *et al.*, 1972; Palfreyman *et al.*, 1974; Orlansky and Heikkila, 1974) and *in vivo* (Fuxe *et al.*, 1970; Farnebo *et al.*, 1970) it has been suggested that this could possibly account, in part, for their clinical action.

DOPAMINE UPTAKE INTO SYNAPTIC VESICLES

Studies by Philippu and co-workers have shown that the DA uptake mechanisms in synaptosomes and synaptic vesicles from the corpus striatum have some distinct differences (Philippu and Beyer, 1973; Philippu *et al.*, 1975). In synaptic vesicles isolated from pig brain corpus striatum it has been shown that these are two DA uptake processes, i.e. one dependent on ATP, magnesium ion and temperature and a second process that is independent of ATP and magnesium ion and only partially dependent on temperature. It was shown that DA uptake into synaptosomes was not enhanced by the presence of ATP and magnesium ion. Further it was demonstrated that the presence of ouabain did not affect the uptake of ^{14}C-DA either in the presence or absence of ATP and magnesium ion. Ouabain is known to inhibit the uptake of DA into synaptosomes (Holz and Coyle, 1974). The K_m for ^{14}C-DA uptake into the vesicles in the presence of ATP and magnesium ion is 1.52×10^{-6} M, i.e. about a tenth of the affinity for DA displayed by synaptosomes from the corpus striatum (Holz and Coyle, 1974).

The drugs reserpine and prenylamine were potent competitive inhibitors of the ATP–magnesium ion-dependent uptake of DA having K_i's of 0.96 $\times 10^{-8}$ M and 1.74 $\times 10^{-8}$ M, respectively. (+) Amphetamine was also found to be a strong competitive inhibitor ($K_i = 2.7 \times 10^{-6}$ M) whilst desipramine and amantadine were ineffective.

THE UPTAKE OF ^{3}H-DA INTO THE ISOLATED PERFUSED RAT HEART

The isolated perfused rat heart has been shown to accumulate ^{3}H-DA by two distinct uptake mechanisms; these are designated as uptake's 1 and 2 (Hellman *et al.*, 1971a). Uptake 1 occurs at a low perfusion concentration and has a $K_m = 0.69 \times 10^{-6}$ M and $V_{max} = 1.45 \times 10^{-9}$ mol g^{-1} min^{-1}. Uptake 2 is prominent at high perfusion concentrations and has a $K_m = 5.9 \times 10^{-4}$ M and $V_{max} = 0.14 \times 10^{-6}$ mol g^{-1} min^{-1}. In the same system Iversen (1967) has described the existence of two distinct uptake systems for NA. The K_m values of NA for uptake 1 are, (−) NA (0.27 μM), (±) NA (0.67 μM) and (+) NA (1.39 μM) (Iversen, 1967). Studies with 6-hydroxydopamine pretreated rat hearts showed that uptake 1 was greatly impaired but extraneuronal uptake 2 remained intact (Hellman *et al.*, 1971b).

DOPAMINE UPTAKE INTO BLOOD PLATELETS

It is well known that the blood platelet has a serotonin (5-HT) uptake system that is in many ways similar to the neuronal 5-HT transport system (Sneddon, 1973). Recent studies have shown that DA can also be apparently transported by the platelet 5-HT uptake system (Solomon *et al.*, 1970; Boullin and O'Brien, 1970). The DA uptake system in platelets was found to have the following properties:

1. The uptake is linear for 30 minutes.
2. It is saturable.
3. It is temperature sensitive.
4. Dopamine uptake is reduced in the presence of metabolic inhibitors such as iodoacetate, PCMB dinitrophenol, sodium cyanide and also by *N*-ethylmaleimide and ouabain.
5. A Lineweaver–Burk analysis of ^{14}C-DA uptake showed that the system has a $K_m = 6.7 \times 10^{-5}$ M. This is much a higher value (lower affinity) than that of DA for the neuronal uptake site, i.e. 1.3×10^{-7} M. In contrast the K_m for 5-HT uptake into platelets is much lower, i.e. 0.1–0.5 μM (Born and Gillson, 1959; Sneddon, 1969).

6. Competitive kinetic experiments showed that 5-HT and DA can inhibit each others uptake but that 5-HT is a more potent inhibitor of DA uptake ($K_i = 5.8 \times 10^{-6}$ M), than DA is of 5-HT uptake ($K_i = 9.6 \times 10^{-5}$ M).
7. Relatively high concentrations (10^{-4} M) of the following psychotropic drugs potently inhibited ^{14}C-DA uptake, desipramine, benztropine, trihexyphenidyl, haloperidol, diphenhydramine Ro5-9110/1, chlorpheniramine, tripelennamine.

In summary, it can be said that although the blood platelet has an uptake system that can transport DA, it appears to differ in some respects from the neuronal uptake system occurring in the corpus striatum, with regard to the affinity of the amine for the uptake system and also apparently with regard to its sensitivity to inhibition by certain drugs. However, to date there are only reports in the literature on the effects of rather high single concentrations of psychotropic drugs (10^{-4} M) and clearly a detailed investigation with the determination of IC_{50} or K_i values for these drugs is warranted. On the evidence to date it would appear to be true that the blood platelet is a better model of neuronal 5-HT uptake than of neuronal DA uptake.

Platelet DA Uptake in Human Disease States

It has been reported that patients suffering from Huntington's chorea exhibit an increased platelet uptake of DA (Aminoff *et al.*, 1974; McLean and Nihei, 1977).

Barbeau *et al.* (1975) have shown that DA uptake into platelets from Parkinsonian patients is significantly lower than in normal control subjects. The interpretations in this study did not entirely agree with the earlier work of Boullin and O'Brien (1970).

DA UPTAKE IN OTHER SYSTEMS

The accumulation of ^{3}H-DA into rat brain mitochondria has been shown not to be due to an active transport system but is explicable in terms of a simple binding process (Shah, 1972).

The uptake of ^{3}H-DA into chick embryo ganglia in tissue culture has been examined by autoradiography. It was shown that the neurones of sympathetic ganglia, unlike those of other ganglia, accumulated ^{3}H-DA and that this occurred in all parts of the neurone (Burdman, 1968).

Glial cell fractions from rabbit brain have been shown to possess an uptake system for ^{3}H-DA. However, this system apparently has a greater affinity for GABA than for DA (Henn and Hamberger, 1971).

Studies *in vitro* in the pigeon optic tectum (Henke *et al.*, 1976) and in murine neuroblastoma cells (Breakfield, 1975) showed that these two systems possessed only a low affinity uptake system for ^{3}H-DA.

High affinity DA uptake systems in snail neurone tissue and auricle have been studied in detail by Osborne and co-workers (Osborne *et al.*, 1975; Hiripi *et al.*, 1976). Other studies in invertebrates include those of Carpenter *et al.* (1971), who demonstrated an accumulation of DA in *Aplysia* visceral ganglia and auricle, and Myers and Sweeney (1973a, b) who studied the uptake of DA in the pedal ganglion of the freshwater clam, *Quadrula pustulosa*.

HORMONAL INFLUENCES ON DA UPTAKE

It is known that DA uptake can be affected by gonadohormone administration but there are inconsistencies in the reported effects (Meyerson, 1966; Hackmann *et al.*, 1973; Endersby and Wilson, 1974). Recently, however, it has been shown that significant variations in the uptake of ^{3}H-DA occur during the rat oestrous cycle (Davis *et al.*, 1977).

It has been further suggested that the DA uptake system may play a role in the modulation of the neuroendocrine and behavioural events of the rat oestrous cycle.

THE CONFORMATION OF DA AT ITS UPTAKE SITE

In an earlier chapter (Grol, Chapter 2) it has been mentioned that DA is a flexible molecule capable of existing in a number of conformations the most important of which are the *trans* or *anti* form and the two *gauche* forms. Using several rigid and semi-rigid analogues of amphetamine, a competitive inhibitor of catecholamine uptake (Horn *et al.*, 1971), it has been suggested that the preferred conformation for a non-catecholamine as an inhibitor of ^{3}H-DA uptake is probably the fully extended *trans* form (Horn and Snyder, 1972). It was further suggested that the *trans* form of DA might be the uptake site preferred form (Horn and Snyder, 1972). Other workers using catecholamine derivatives of *trans* decalin have reasoned, however, that the preferred conformation for DA at its uptake site is *gauche* (Tuomisto *et al.*, 1974). A major difficulty in the use of these *trans* decalin derivatives is that a large hydrophobic moiety has been added to the catecholamine molecule in

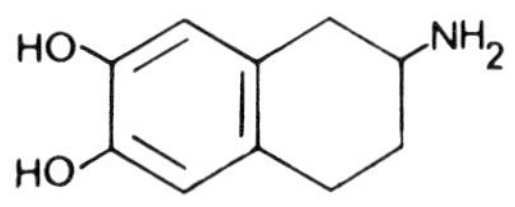

Fig. 3 ADTN.

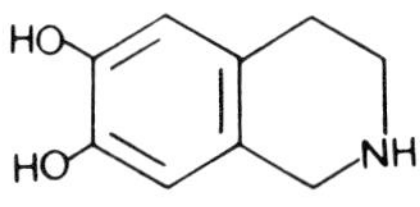

Fig. 4 Norsalsolinol.

order to reduce flexibility. This may produce interactions at the uptake site that are quite different from the natural ligand. An additional complication due to the presence of the bulky decalin group is that the molecule can only approach the uptake site from one side, rather than having, as dopamine does, a more or less unrestricted access. In general, in order to obtain reliable information from the "rigid-analogue" approach one must add as little to the molecule as possible in order to produce a more rigid structure.

In order to circumvent some of these problems two other more rigid analogues of dopamine have recently been examined, i.e. 2-amino-6,7-dihydroxy-1,2,3,4-tetrahydronaphthalene, ADTN (Fig. 3) and 6,7-dihydroxytetrahydroisoquinoline, norsalsolinol (Horn, 1974) (Fig. 4). ADTN was the more potent of the two compounds having an IC_{50} of 8.5×10^{-7} M, whereas norsalsolinol was some 200 times weaker with an IC_{50} of 1.8×10^{-4} M. Kinetic studies showed that the inhibition was of a competitive nature with a K_i of 7.0×10^{-7} M. ADTN is conformationally similar to the *trans* form of DA whilst norsalsolinol may be regarded as an approximate analogue of the *gauche* form of DA with respect to the distance of the amino group from the catechol function. It would thus appear that the *trans* conformation of dopamine is the preferred form for interaction with the uptake site. The *trans* conformation of dopamine has been shown to be the preferred form in the solid state by X-ray crystallography (Bergin and Carlstrom, 1968) in solution by nuclear magnetic resonance analysis and by theoretical calculations (Bustard and Egan, 1971).

CONCLUSION

It is clear that the properties of the neuronal DA uptake system differ significantly from those of the system responsible for the transport of its close structural analogue NA and also from those of the 5-HT system. It is apparently possible with certain drugs to inhibit DA uptake without at the same time causing a release of the amine. Although various classes of drugs inhibit DA transport there is at the moment, however, a real need for a more selective inhibitor of DA uptake, i.e. one that has less effect on the NA and 5-HT uptake systems. Such an agent might be of both academic and clinical interest. Further effort should now also be directed to obtaining a better understanding of the biochemical composition of not only the DA, but also the various other biogenic amine transport systems. A most useful system for these studies is the blood platelet, which is apparently an excellent model for neuronal 5-HT uptake. Isolation of the platelet amine transport system may be possible through the use of radioactive affinity labels or by affinity chromatography.

REFERENCES

Alpers, H. S., McLaughlin, B. R., Nix, W. M. and Davis, V. E. (1975). *Biochem. Pharmacol.* **24,** 1391–1396.

Aminoff, M. J., Trenchard, A., Turner, P., Wood, W. G. and Hills, M. (1974). *Lancet* **ii,** 1115–1116.

Axelrod, J., Weil-Malherbe, H. and Tomchick, R. (1959). *J. Pharmacol. Expl. Ther.* **127,** 252–256.

Banerjee, S. P., Snyder, S. H. and Mechoulam, R. (1975). *J. Pharmacol. Expl. Ther.* **194,** 74–81.

Barbeau, A., Campanella, G., Butterworth, R. F. and Yamada, K. (1975). *Neurology* **25,** 1–9.

Baumann, P. A. and Maitre, L. (1976). *Nature (London)* **264,** 789–790.

Bergin, R. and Carlstrom, D. (1968). *Acta Crystallogr.* **B24,** 1506–1510.

Borchardt, R. T. Burgess, S. K. Reid, J. R. Liang, Y. O. and Adams, R. N. (1977). *Molec. Pharmacol.* **3,** 805–818.

Born, G. V. R. and Gillson, R. E. (1959). *J. Physiol. (London)* **146,** 472–491.

Boullin, D. J. and O'Brien, R. A. (1970). *Br. J. Pharmacol.* **39,** 779–788.

Breakfield, X. O. (1975). *J. Neurochem.* **25,** 877–882.

Burdman, J. A. (1968). *J. Neurochem.* **15,** 1321–1323.

Burn, J. H. (1932). *J. Pharmacol. Expl. Ther.* **46,** 75–95.

Bustard, T. M. and Egan, R. S. (1971). *Tetrahedron* **27,** 4457–4469.

Carlsson, A., Fuxe, K., Hamberger, B. and Lindqvist, M. (1966). *Acta Physiol. Scand.* **67,** 481–497.

Carpenter, D., Breese, G., Schanberg, S. and Kopin, I. (1971). *Int. J. Neurosci* **2,** 49–56.

Carruba, M. O., Picotti, G. B., Zambotti, F. and Mantegazza, P. (1977). *Arch. Pharmacol.* **298,** 1–5.

Cho, A. K., Schaeffer, J. C. and Fischer, J. F. (1975). *Biochem. Pharmacol.* **24,** 1540–1542.

Clarenbach, Raffel, G., Meyer, D. and Hertting, G. (1974). *Naun. Schmied. Arch. Pharmacol.* **284,** R.9.

Clouet, D. H. and Williams, N. (1974). *J. Pharmacol. Expl. Ther.* **188,** 419–428.

Cohen, G., Heikkila, R. E., Dembiec, D., Sang, D., Teitel, S. and Brossi, A. (1974). *Eur. J. Pharmacol.* **29,** 292–297.

Coyle, J. T. and Campochiaro, P. (1976). *J. Neurochem.* **27,** 673–678.

Coyle, J. T. and Snyder, S. H. (1969a). *J. Pharmacol. Expl. Ther.* **170,** 221–231.

Coyle, J. T. and Snyder, S. H. (1969b). *Science* **166,** 899–901.

Cuello, A. C., Horn, A. S., Mackay, A. V. P. and Iversen, L. L. (1973). *Nature (London)* **243,** 465–467.
Cuello, A. C. and Iversen, L. L. (1973). *Brain Res.* **63,** 474–478.
Davis, C. F., Davis, B. F. and Halaris, A. E. (1977). *Life Sci.* **20,** 1319–1332.
De Belleroche, J. S., Bradford, H. F. and Jones, D. G. (1976). *J. Neurochem.* **26,** 561–571.
De Robertis, E. (1967). *Science* **156,** 907–914.
Ehringer, B. and Falck, B. (1971). *Brain Res.* **33,** 157–172.
Endersby, C. A. and Wilson, C. A. (1974). *Brain Res.* **73,** 321–331.
Farnebo, L. O., Fuxe, K., Hamberger, B. and Ljungdahl, H. (1970). *J. Pharm. Pharmacol.* **22,** 733–737.
Ferris, R. M., Tang, F. L. M. and Maxwell, R. A. (1972). *J. Pharmacol. Expl. Ther.* **181,** 407–416.
Ferris, R. M., Tang, F. L. M. and Russell, A. V. (1975). *Biochem. Pharmacol.* **24,** 1523–1527.
Friedman, E., Fung, F. and Gershon, S. (1977). *Eur. J. Pharmacol.* **42,** 47–51.
Fuxe, K., Goldstein, M. and Ljungdahl, A. (1970). *Life Sci.* **9,** 811–824.
Fuxe, K., Hamberger, B. and Malmfors, T. (1967). *Eur. J. Pharmacol.* **1,** 334–341.
Fuxe, K. and Ungerstedt, U. (1968). *Eur. J. Pharmacol.* **4,** 135–144.
Garey, R. E. (1976). *Life Sci.* **18,** 411–418.
Garey, R. E. and Heath, R. G. (1974). *Brain Res.* **79,** 520–523.
Geffen, L. B., Jessell, T. M., Cuello, A. C. and Iversen, L. L. (1976). *Nature (London)* **260,** 258–260.
Glowinski, J., Axelrod, J. and Iversen, L. L. (1966). *J. Pharmacol. Expl. Ther.* **153,** 30–41.
Glowinski, J. and Iversen, L. L. (1966). *J. Neurochem.* **13,** 655–669.
Goodale, D. B. and Moore, K. E. (1975). *Neuropharmacology* **14,** 585–589.
Hackmann, E., Wirz-Justice, A. and Lichtsteiner, M. (1973). *Psychopharmacology* **32,** 183–191.
Halaris, A. E., Belendiuk, K. T. and Freedman, D. X. (1975). *Biochem. Pharmacol.* **24,** 1896–1898.
Hamberger, B. (1967). *Acta Physiol. Scand.* **71,** *Suppl. 295.*
Harris, J. E., and Baldessarini, R. J. (1973a). *Life Sci.* **13,** 303–312.
Harris, J. E. and Baldessarini, R. J. (1973b). *Neuropharmacology* **12,** 669–679.
Heikkila, R. E. and Cohen, G. (1973). *Eur. J. Pharmacol.* **21,** 66–69.
Heikkila, R. and Cohen, G. (1974). *Res. Commun. Chem. Pathol. Pharmacol.* **7,** 539–547.
Heikkila, R. Cohen, G. and Dembiec, D. (1971). *J. Pharmacol. Expl. Ther.* **179,** 250–258.
Heikkila, R. E., Orlansky, H. and Cohen, G. (1975). *Biochem. Pharmacol.* **24,** 847–852.
Hellmann, G., Hertting, G. and Peskar, B. (1971a). *Br. J. Pharmacol.* **41,** 256–269.
Hellmann, G., Hertting, G. and Peskar, B. (1971b). *Br. J. Pharmacol.* **41,** 270–277.
Hendley, E. D., Snyder, S. H., Fawley, J. J. and Lapidus, J. B. (1972). *J. Pharmacol. Expl. Ther.* **183,** 103–116.
Henke, H., Schenker, T. M. and Cuenod, M. (1976). *J. Neurochem.* **26,** 125–130.
Henn, F. A. and Hamberger, A. (1971). *Proc. Natl. Acad. Sci.* **68,** 2686–2690.
Hiripi, L., Neuhoff, V. and Osborne, N. N. (1976). *J. Neurochem.* **26,** 335–343.
Holmes, J. C. and Rutledge, C. O. (1974). *Fed. Proc.* **33,** 523, 1761.
Holz, R. W. and Coyle, J. T. (1974). *Molec. Pharmacol.* **10,** 746–758.

Horn, A. S. (1973). *Br. J. Pharmacol.* **47,** 332–338.
Horn, A. S. (1974). *J. Pharm. Pharmacol.* **26,** 735–737.
Horn, A. S. (1976a). "The Mechanism of Neuronal and Extraneuronal Transport of Catecholamines" (Ed. D. M. Paton), 195–214. Raven Press, New York.
Horn, A. S. (1976b). *Postgrad. Med. J.* **52** (*Suppl. 3*), 25–30.
Horn, A. S., Coyle, J. T. and Snyder, S. H. (1971). *Molec. Pharmacol.* **7,** 66–80.
Horn, A. S., Cuello, A. C. and Miller, R. J. (1974). *J. Neurochem.* **22,** 265–270.
Horn, A. S. and Snyder, S. H. (1972). *J. Pharmacol. Expl Ther.* **180,** 523–530.
Howes, J. and Osgood, P. (1974). *Neuropharmacology* **13,** 1109–1114.
Hunt, P. Kannengiesser, M. H. and Raynaud, J. P. (1974). *J. Pharm. Pharmacol.* **26,** 370–371.
Hyttel, J. (1977). *Psychopharmacology* **51,** 225–233.
Iversen, L. L. (1967). "The Uptake and Storage of Noradrenaline in Sympathetic Nerves", Cambridge University Press, London.
Iversen, L. L. (1972). *In* "The Thyroid and Biogenic Amines" (Eds. T. Rall and I. Kopin), 569–603. North Holland Publishing Co., Amsterdam.
Iversen, L. L. (1975). *In* "Handbook of Psychopharmacology", Vol. 3 (Eds. L. Iversen, S. Iversen and S. Snyder), 381–442, Plenum Press, New York.
Iversen, L. L., Jarrot, B. and Simmonds, M. A. (1971). *Br. J. Pharmacol.* **43,** 845–855.
Iversen, L. L., Rogawski, M. A. and Miller, R. J. (1976). *Molec. Pharmacol.* **12,** 251–262.
Jonason, J. and Rutledge, C. O. (1968). *Acta Physiol. Scand.* **73,** 411–417.
Kannengiesser, M. H., Hunt, P. F. and Raynaud, J. P. (1976). *Eur. J. Pharmacol.* **35,** 35–43.
Koe, B. K. (1976). *J. Pharmacol. Expl. Ther.* **199,** 649–661.
Kostrzewa, R. M. and Jacobowitz, D. M. (1974). *Pharm. Rev.* **26,** 199–288.
Kouyoumdjian, J. C., Belin, M. F., Barakdjian, J. and Gonnard, P. (1976). *J. Neurochem.* **27,** 817–819.
Kramer, S. G., Potts, A. M. and Mangnall, Y. (1971). *Invest. Ophtalmol.* **10,** 617–624.
Kruk, Z. L. and Zarrindast, M. R. (1976). *Br. J. Pharmacol.* **58,** 272–273.
Lichtensteiger, W. and Langemann, J. (1966). *J. Pharmacol. Expl. Ther.* **151,** 400–408.
McLean, D. R. and Nihei, T. (1977). *Lancet* **i,** 249–250.
Meyerson, B. J. (1966). *Acta Physiol. Scand.* **67,** 411–422.
Myers, P. R. and Sweeney, D. C. (1973a). *Int. J. Neurosci.* **5,** 231–234.
Myers, P. R., and Sweeney, D. C. (1973b). *Comp. Gen. Pharmacol.* **4,** 321–325.
Ohashi, T., Akita, H., Tamura, T., Noda, K. and Honda, F. (1972). *Arzneimit. Forsch.* **22,** 966–972.
Orlansky, H. and Heikkila, R. E. (1974). *Eur. J. Pharmacol.* **29,** 284–291.
Osborne, N. N., Hiripi, L. and Neuhoff, V. (1975). *Biochem. Pharmacol.* **24,** 2141–2148.
Palfreyman, M. G., Palfreyman, E. S. and Clark, M. S. G. (1974). *Eur. J. Pharmacol.* **28,** 379–383.
Philippu, A. and Beyer, J. (1973). *Naun. Schmied. Arch. Pharmacol.* **278,** 387–402.
Philippu, A., Matthaei, H. and Lentzen, H. (1975). *Naun. Schmied. Arch. Pharmacol.* **287,** 181–190.
Raiteri, M., Bertollini, A., Angelini, F. and Levi, G. (1975). *Eur. J. Pharmacol.* **34,** 189–245.
Raiteri, M., Bertollini, A., del Carmine, R. and Levi, G. (1976). *In* "Transport

Phenomena in the Nervous System" (Eds. G. Levi, L. Battistin and A. Lajtha), 319–335. Plenum Press, New York.
Raiteri, M., Cerrito, F., Cervoni, A. M., del Carmine, R., Ribera, M. T. and Levi, G. (1978). *In* "Dopamine", Advances in Biochemical Psychopharmacology, Vol. 19 (Eds P. Roberts, G. Woodruff and L. L. Iversen), 35–56. Raven Press, New York.
Randrup, A. and Braestrup, C. (1977). *Psychopharmacology* **53,** 309–314.
Ross, S. B. (1976). *In* "The Mechanism of Neuronal and Extraneuronal Transport of Catecholamines" (Ed. D. M. Paton), 67–93. Raven Press, New York.
Ross, S. B. and Renyi, A. L. (1967). *Eur. J. Pharmacol.* **2,** 181–186.
Ross, S. B. and Renyi, A. L. (1975). *Acta Pharmacol. Toxicol.* **36,** 56–66.
Shah, N. S. (1972). *Brain Res.* **38,** 391–398.
Silbergeld, E. K. (1977). *Life Sci.* **20,** 309–318.
Sneddon, J. M. (1969). *Br. J. Pharmacol.* **37,** 680–688.
Sneddon, J. M. (1973). *In* "Progress in Neurobiology", Vol. 1 (Eds G. Kerkut and W. Phillis), 151–198. Pergamon, Oxford.
Snyder, S. H. and Coyle, J. T. (1969). *J. Pharmacol. Expl. Ther.* **165,** 78–86.
Solomon, H. M., Spirt, N. M. and Abrams, W. B. (1970). *Clin. Pharmacol. Ther.* **11,** 838–845.
Stefanini, E., Argiolas, A., Gessa, G. L. and Fadda, F. (1976). *J. Neurochem.* **27,** 1237–1239.
Tassin, J. P., Thierry, A. M., Blanc, G. and Glowinski, J. (1974). *Naun. Schmied. Arch. Pharmacol.* **282,** 239–244.
Thornburg, J. E. and Moore, K. E. (1973). *Res. Comm. Chem. Pathol. Pharmacol.* **5,** 81–89.
Tseng, L. F., Hitzemann, R. J. and Loh, H. H. (1974). *J. Pharmacol. Expl. Ther.* **189,** 708–716.
Tuomisto, L. and Tuomisto, J. (1973). *Naun. Schmied. Arch. Pharmacol.* **279,** 371–380.
Tuomisto, I., Tuomisto, J. and Smissman, E. E. (1974). *Eur. J. Pharmacol.* **25,** 351–361.
Wang, Y. J., Gurd, J. W. and Mahler, H. R. (1975). *Life Sci.* **17,** 725–734.
Whitby, L. G., Axelrod, J. and Weil-Malherbe, H. (1961). *J. Pharmacol. Expl. Ther.* **132,** 193–201.
Whittaker, V. P. (1966). *Pharmacol. Rev.* **18,** 401–412.
Wirz-Justice, A., Hackmann, E. and Lichtsteiner, M. (1974). *J. Neurochem.* **22,** 187–189.

Notes added in proof

1. The effects of various neuroleptics, antidepressants, CNS stimulants and anticholinergics as inhibitors of ^{3}H-DA uptake has been studied in detail, Hyttel, J. (1978). *Biochem. Pharmacol.* **27,** 1063–1068.
2. ^{3}H-ADTN has been shown to be a good substrate for the DA uptake system, Davis A., Roberts, P. J. and Woodruff, G. N. (1978). *Br. J. Pharmacol.* **63,** 183–190.

CHAPTER 17

Electrical Stimulation as a Tool for the Study of Biochemical Aspects of Dopamine Neurotransmission

Jakob Korf

Department of Biological Psychiatry, Groningen, The Netherlands.

Electrical stimulation in the central nervous system is often used for electrophysiological and behavioural studies. For biochemical studies the technique has not been extensively applied, but I hope this text will indicate its usefulness for the investigation of the biochemical events which a neurotransmitter may undergo during nerve activity.

In this review I have summarized the relevant data from the literature on cerebral dopaminergic systems and compared these results with those obtained in other cerebral nervous systems.

ON THE TECHNIQUE

Several investigators have used a variety of stimulation techniques to evoke biochemical alterations in nerve structures *in vivo*. For this purpose stimulation techniques can only be applied when the neuroanatomy of a system is well established. Thus the localization of the cell bodies, the course of the axons and the distribution of nerve endings have to be documented. Moreover the distance between the relevant nerve terminal area and the cell bodies should be large enough for a selective stimulation of the axons only.

These axons should be localized in a well defined bundle, to ensure that a significant proportion can be stimulated by a single electrode. In several animals the ascending monoaminergic pathways fullfill these requirements (Andén *et al.*, 1964; Ungerstedt, 1971; Lindvall and Björklund, 1974).

The distance between the cell bodies of the substantia nigra or the ventral tegmental area and their nerve terminals in the striatum, the nucleus accumbens, the olfactory tubercle, the cingular or the frontal cortex are between 6 mm (in the adult rat) to at least 12 mm (in the cat). The diameter of the ascending dopaminergic pathway is relatively small, so under maximal stimulation conditions a substantial proportion of the pathway is being stimulated.

The ascending dopaminergic pathway probably consists of thin unmyelinated fibres (Ibata *et al.*, 1973), of which the conduction properties are not well established. In his overview on the use of electrical stimulation in the central nervous system Ranck (1975) emphasizes that no data are available on the current–distance or the current–duration relationships for central finely myelinated or unmyelinated fibres.

As important variables, which determine the efficacy of stimulation, differences can be expected in stimulation of the cell bodies compared to the ascending fibres. The flow of current through brain tissue depends on the composition of the tissue (white or grey matter, proximity to cerebrospinal fluid compartments, etc.) and may differ at different placements of the electrode. For myelinated fibres or cell bodies Ranck (1975) described a relation between the applied current (monophasic, monopolar electrode) with a given pulse length (200 μs) and the maximal distance from the electrode which produces excitation of neural elements. A rough estimate is, that when a current of 50 μA is applied, such distances are about 0.5–1.8 mm.

By using shorter pulse duration more current has to be applied to achieve a given level of effect. In a study of Barry *et al.* (1974) strength–duration functions were determined for rats in which the median forebrain bundle was stimulated in self-stimulation experiments. The same response was obtained when applying a pulse duration of 5 ms and a current of 50 μA or a combination of 0.1 ms and 500 μA. These results may be indicative for the possibly involved dopaminergic fibres, although to date no systematic study has been reported for dopaminergic structures.

The excitability of axons also depends on the inward current. When a fibre is depolarized near a monopolar cathode by an outward current, it can be hyperpolarized on either side by an inward current ("anodal surround"). Such hyperpolarizing effects may attenuate any propagation of the evoked nerve impulses near the electrode.

In the non-stimulated animal the firing activity of the dopaminergic cells and the subsequent activity in the ascending pathway are not all in the same phase of activity, but are more or less randomly depolarized. During its electrical stimulation, the pattern of activity of the nigrostriatal pathway is completely disrupted and the activity of the axons and the nerve terminals to some extent synchronized. Under such conditions it is not unlikely that some naturally occurring regulation mechanisms are affected. Such synchronization of nerve activity should also be taken into account, when behavioural responses following electrical stimulation of axon bundles are studied.

The dopaminergic neurones are highly branched in their nerve terminal areas (Andén *et al.*, 1966). The total length of all branches of the terminal system of one neurone has been estimated to be about 60 cm in the rat. Each nerve cell possesses about 500 000 varicosities (nerve endings). With electrophysiological techniques it has been observed that the evoked action potentials resulting from stimulation do not equally distribute over all branches of the peripheral neurone (see, for example, Chung, 1976). If this is also the case in the central dopaminergic neurones, then any biochemical effects in the striatum seen after electrical stimulation of the dopaminergic axon or cell bodies do not necessarily reflect changes in the whole system, but may be the result of an unknown proportion of it.

During electrical stimulation the brain tissue around the electrode tip will inevitably be destroyed to some extent, depending on the size of the electrodes, the current used, the duration of stimulation and whether mono- or biphasic stimuli were applied. In many instances stimulation in the brain evokes some muscular reactions, and whether the stimulation is still effective can be easily judged in each animal. Electrical stimulation in any area of the brain will never be entirely selective and it may not always be possible to attribute solely to a single neuronal system the consequent biochemical, electrophysiological or behavioural responses.

Despite many uncertainties of the technique, some of which are mentioned here, electrical stimulation has been applied by a number of investigators for the study of aspects of central dopaminergic neurotransmission.

The studies on the use of electrical stimulation in the cerebral dopaminergic systems are summarized in Table I. The stimulus parameters used by the various authors differ in many respects. Both mono- and biphasic stimuli with different durations were used (from 0.1–3 ms). Current intensities varied from 5–600 μA. Electrodes used in the various studies were of the monopolar, bipolar or coaxial type. In some studies the duration of the stimulation was as short as 2 min, while others stimulated for more than 60 min. In all studies square wave pulses were applied. Frequencies of about 1 Hz to about 60 Hz have been used for dopaminergic

Table I Survey of stimulus parameters used in various studies on cerebral dopaminergic processes.

Investigators	Animal	Location[a]	Electrode	Bi/monophasic	Pulse duration (ms)	Intensity	Frequency (Hz) (train per length s^{-1})	Stimulation time (min)	Biochemical Study
Arbuthnott *et al.* (1970)	Rat	MFB	Monopolar	Monophasic	0.1	200 μA	100 Hz of 200 ms per 3 train s^{-1}	60–120	DA depletion after synthesis inhibition
Chiueh and Moore (1974)	Cat	MFB	Bipolar	Monophasic	0.5–1	400 μA	30 Hz	5–90	Release endogenously synthesized ^{3}H-DA into CSF
Korf *et al.* (1976)	Rat	MFB	Monopolar	Biphasic	2.0	50–600 μA	10–75 Hz	15–90	Formation endogenous DA-metabolites in striatum
McLennan (1964)	Cat	Thalamic nucleus centro-medicalis	Monopolar	Monophasic	0.1	—	3–10 Hz	15	Release endogenous DA from caudate
Murrin and Roth (1976)	Rat	MFB	Coaxial	Biphasic	3.0	400 μA	0–30 Hz	10–30	Synthesis of DA in striatum

Nieoullon *et al.* (1977a)	Cat	SN	Coaxial	Monophasic	0.5	100 μA	0.1 train of 100 ms s^{-1}	15	Release of endogenously synthesized ^{3}H-DA from caudate
Portig and Vogt (1969)	Cat	SN	Coaxial	Biphasic	0.5–1	4 V	train of 3 pulses 3 s^{-1}	3–4	Release of endogenous HVA into CSF
Riddell and Szerb (1971)	Cat	SN MFB	Coaxial	Monophasic	0.5	20–50 V	train of 10 pulses 10 s^{-1}	50	Release of endogenously synthesized ^{14}C-DA into CSF
Roth *et al.* (1976)	Rat	MFB	Coaxial	Monophasic	1.5	200 μA	5–60	5–60	Formation of DOPAC in striatum
Von Voigtlander (1971)	Cat	MFB	Bipolar	Monophasic	1.0	100–400 μA	3–100	2	Release of ^{3}H-DA into CSF

[a] MFB = median forebrain bundle, SN = substantia nigra.

neurones. In most studies the ascending bundle was stimulated, but in some the substantia nigra was activated. Although the effects of different frequencies, duration of stimulation and intensity of the applied current have been compared, no systematic study has been done to establish the strength–duration function in the dopaminergic systems.

Endogenous Compounds

McLennan (1964–65) was the first investigator who reported the release of striatal DA in the rat. He used a push-pull cannula system to collect the released amine, and a manual fluorimetric assay was used for the detection of the amine. A release of about 1 ng DA min^{-1} was found. Considering the volume of tissue perfused (0.5–1 mm^3), in a collection period of 10 min a substantial portion of DA present in the perfused system was washed out. Despite these methodological drawbacks, McLennan was able to evoke an increase in the release of DA after electrical stimulation of various brain areas. The methods to detect catecholamines have since been substantially improved, and it should be possible to perform a more systematic study on the effects of electrical stimulation on the release of DA into push-pull cannula perfusates (Bartholini *et al.*, 1976). No such study, however, has yet appeared.

Using a ventricular perfusion system in the cat, Portig and Vogt (1969) detected the formation of HVA and observed an increased output of this metabolite following electrical stimulation of the substantia nigra. In these perfusates it was not possible to detect DA consistently, indicating that released DA is rapidly metabolized. The increase of HVA was rather low (10–20% over the baseline levels) although occasionally increases as high as 80% were found (Vogt, 1969).

Arbuthnott *et al.* (1970) used histofluorescence for the detection of changes in the levels of DA after electrical stimulation of dopaminergic fibres in rats, pretreated with α-methyl-*p*-tyrosine. The depletion of DA following electrical stimulation was *inter alia* dependent on the site of the electrode tip. Thus when the electrode was placed in the medial forebrain bundle changes were only detected in the striatum, while more medially placed DA was found to be depleted in the olfactory tubercle and the nucleus accumbens. A selective depletion of DA in mesolimbic areas was obtained by electrodes placed in or near the dopaminergic cell bodies of the ventral tegmental area. The stimulus parameters used by Arbuthnott *et al.* (1970) were similar to those used in rats which produce maximum self-stimulation.

Increased tissue levels of dopamine metabolites following electrical stimulation of the median forebrain bundle have been reported by a number of authors. The striatal levels of DOPAC, its sulphate conjugate and of HVA were found to depend on the frequency applied. Maximally enhanced DOPAC levels were found when a frequency of 15 Hz and a current of 200 μA were applied for at least 20 min (Roth *et al.*, 1973, 1976; Elchisak *et al.*, 1976). Virtually the same conditions were found for producing maximal HVA formation, although a stimulation time of 60 min was required in probenecid pretreated rats (Korf *et al.*, 1976a). A study on the relation between the current intensity and the HVA-formation revealed that at least 300 μA is necessary for a maximal formation of HVA, although in some cases 100 μA was as effective as 300 μA. Probably at the lower applied current the placement of the electrode in the brain is more critical in order to obtain consistent results.

Exogenous Labelling

Measurements of the release of DA require highly sensitive analytical methods for the amine. To overcome this problem in a number of studies the cerebral DA stores were prelabelled with the radioactive amine or a precursor. Von Voigtlander and Moore (1971a) studied the release of DA into the cerebroventricular perfusate of the cat, when the brain was prelabelled with ^{3}H-DA. The radioactive amine was released after electrical stimulation of the substantia nigra. In addition a metabolite of DA, ^{3}H-3-methoxytyramine, was found to be released in this preparation. Stimulation of the substantia nigra did not produce an enhancement of the release of labelled material, when the brain was prelabelled with ^{14}C-urea. The release of ^{3}H-DA was frequency dependent: maximal efflux of the amine was found at a frequency of 30 Hz (stimulation time 2 min). Maximal output of ^{3}H-DA was found at a current of 300 μA. When the stimulation electrode was placed in the medial forebrain bundle the output of ^{3}H-DA was much higher (Von Voigtlander and Moore, 1971b). The dependency of the release upon the frequency was the same for both the substantia nigra and the medial forebrain bundle. These studies have been criticised, because it is not entirely certain whether exogenously applied ^{3}H-DA is exclusively accumulated in dopaminergic neurones.

The enzyme tyrosine hydroxylase is confined to catecholaminergic neurones and the labelling of cerebral catecholamine stores with radioactive tyrosine is rather selective (Moore *et al.*, 1971). This labelling technique has been used to investigate the release of radioactive DA. Early attempts to measure the release of radioactive DA, synthesized endogenously, in a push-pull cannula perfusate obtained from the cat caudate nucleus did not give consistent results, although in some cases an increase of DA-release was observed during electrical stimulation (Riddell and Szerb, 1971). In this study no relation was found between the formation of radioactive HVA and DOPAC and electrical stimulation.

In their cerebroventricular perfusates of the cat Chiueh and Moore (1973, 1974) observed an increase in the release of ^{14}C-DA, when electrical stimulation (30 Hz, 400 μA, pulse length 0.5 ms, stimulation time 10 min) was applied in the ipsilateral hypothalamus. No relation between various stimulus parameters and the release of DA was established.

The major problem of obtaining sufficient release from the striatum of synthesized DA was solved with the introduction of the cup-technique. In the cat, part of the overlying cortex is removed and a cup is placed on the surface of the caudate nucleus. Such dramatic surgery, however, undoubtedly affects the viability of the striatum, and possibly also of the rest of the brain. Moreover, removal of a substantial part of the cortex destroys the massive cortical innervation of the striatum. Despite these drawbacks Besson *et al.* (1974) observed significant increases in the release of ^{3}H-DA synthesized from ^{3}H-tyrosine following electrical stimulation of the substantia nigra (20 Hz, 0.2 ms, 5 V for 10 min) or of an area close to the ascending dopaminergic pathways, indicating that the striatum was still able to respond to activation of neurones in the substantia nigra.

The technique of measuring release of endogenously synthesized DA has since been improved by the introduction of a permanently installed push-pull cannula system (Nieoullon *et al.*, 1977a). This system allows one to measure the release of DA in no less than four locations in the cat brain (Nieoullon *et al.*, 1977b).

With these push-pull cannula systems, electrical stimulation of the substantia nigra also produced an increased striatal DA-release (Nieoullon, 1977a). In addition it was observed that the mandarin, introduced in the outer cannula of a push-pull cannula system, placed in the substantia nigra, produces DA release in the ipsilateral striatum. This observation demonstrates how easy it is to activate neurones in the dopaminergic system and one should be aware of this by interpreting results obtained with electrical stimulation or after intranigral injection of drugs.

Introduction of perfusion systems, such as the push-pull cannulas and the cup technique will always destroy nervous tissue. Thus nerve terminals, normally closely enveloped by surrounding tissue, will now line a relatively large extracellular space. Under this condition it is likely that processes depending on uptake mechanisms are to some extent impaired. Therefore in the intact tissue, released DA is presumably taken up by nerve terminals and other neural elements. During perfusion, however, such uptake processes are likely to be diminished. In no study thus far published has the fate of DA released into the perfusate been established. On the other hand

Bartholini *et al.* (1976) have emphasized the usefulness of the push-pull cannula system for the determination of DA-release in drug studies.

Dopamine Synthesis

Murrin and Roth (1976) studied the effect of electrical stimulation on the synthesis of ^{3}H-DA from intravenously administered ^{3}H-tyrosine. They found no differences in the levels of endogenous DA during electrical stimulation of the median forebrain bundle, but the radioactive DA increased depending upon the frequency of stimulation. Maximal specific activities of striatal DA were found at 15 and 30 Hz. The time course of the specific activity of DA revealed maximal levels at 20 min of stimulation. A lower specific activity of tyrosine was observed in the striatum as a result of electrical stimulation of the median forebrain bundle.

After 10 min of stimulation a higher incorporation of radioactivity in DA was still found when ^{3}H-tyrosine was administered when the stimulation was ended. By 30 min after the termination of the stimulation the specific activity of DA of the control and stimulated striatum were not significantly different (Murrin and Roth, 1976). In the latter study the rate of DA synthesis was also estimated by the measurement of the formation of endogenous DOPA in the rat striatum. Also with this method electrical stimulation was found to enhance the synthesis of DA.

After blockade of the synthesis of DA with α-methyl-*p*-tyrosine a reduced stimulus-induced HVA-formation was observed (Korf *et al.*, 1976a). All these results point to the induction of an increase in synthesis of DA during electrical stimulation of the ascending dopaminergic fibres.

Wurtman *et al.* (1974) observed an increase in the synthesis of striatal DOPA, following the administration of tyrosine. The question arises whether the access of tyrosine for the stimuli-induced activation of the synthesis of DA would be rate limiting. During high-frequency electrical stimulation, however, no increase in the rate of HVA-formation was seen after loading with tyrosine (Korf *et al.*, 1976a). The mechanism for the increase of the activity of tyrosine hydroxylase following electrical stimulation has been investigated by Murrin *et al.* (1976). They reported that electrical stimulation caused a decrease of the K_m values for both tyrosine and the cofactors 6,7-dimethyl-2-amino-4-hydroxy-5,6,7,8-tetrahydropteridine- and tetrahydrobiopterine. Under these conditions the authors suggest that the K_i for DA of neostriatal tyrosine hydroxylase was increased. Similar changes were observed for tyrosine hydroxylase in the olfactory tubercle. In no case was an increase of the V_{max} for the enzyme observed. Considering the levels of tyrosine in the striatum it is likely that tyrosine hydroxylase is saturated with the precursor aminoacid and Murrin *et al.*

(1976) conclude that a change of the K_m for tyrosine has no great physiological implications. As the level in dopaminergic nerve endings of the natural cofactor is not known, it is difficult to indicate precisely the implication of the changes of the K_m value for these cofactors in the regulation of dopamine synthesis. Nevertheless the changes in the K_m value for the cofactor and the K_i value for DA are likely to be related to the physiological regulation mechanism for tyrosine hydroxylase. Some of these changes may be mediated by cyclic AMP. The biochemical aspects of tyrosine hydroxylase are extensively described by Roth in Chapter 9 of this volume.

In other neuronal systems an increase in the synthesis of the neurotransmitter has been observed during electrical stimulation. Examples are the increase in the synthesis of noradrenaline in sympathetic neurones (Gordon *et al.*, 1966; Roth *et al.*, 1966; Sedvall and Kopin, 1967; Weiner and Rabadjija, 1968) and of serotonin in the brain (Shields and Eccleston, 1972).

In conclusion: the synthesis of DA is activated after electrical stimulation, possibly to meet the higher demands of the transmitter for maintaining sufficient release. A change of the conformation of the enzyme tyrosine hydroxylase appears to be involved in such an increase in DA synthesis.

DOPAMINE POOLS

The compartmentation of DA in striatal nerve terminals is a controversial issue (see Costa *et al.*, 1974; Glowinski, 1974). Costa *et al.* (1974) assumed a single open compartment for the synthesis of DA from intravenously administered radioactive tyrosine. Using this assumption a turnover rate for DA was calculated which was close to values obtained with various other techniques (see, for example, Westerink and Korf, 1976; Carlsson and Lindqvist, 1973). The time course of the specific activity of DOPA is the same as that of tyrosine in striatal tissue, but this, however, does not permit any conclusion as to whether or not synthesis of DA occurs in a specific compartment (as Costa *et al.* conclude). The main argument of Glowinski (1974) for the possible existence of two striatal DA-pools is the observation of a biphasic decline of the amine following an intraperitoneal administration of the DA-synthesis inhibitor, α-methyl-*p*-tyrosine. Such a biphasic decline however, has been attributed to the rapid formation of metabolites of α-methyl-*p*-tyrosine, which displace the amine from the nerve terminals (Costa *et al.*, 1974). If this is so, it is of interest to note here, that only a relatively small amount of the amine, and not all of it, is being displaced after α-methyl-*p*-tyrosine treatment. Glowinski (1974), as a consequence of the compartmentation postulated by him, has presented relatively high turnover rates for striatal DA.

Using electrical stimulation of the median forebrain bundle, Korf *et al.* (1976a, b) observed a biphasic decline of striatal DA. The first rapid phase was concerned with a pool of DA of about 20% and this might represent a releasable pool. The Glowinski-group also obtained evidence for a DA-pool of the same size in their α-methyl-*p*-tyrosine study. After intravenous administration of a high dose of reserpine ($5\,mg\,kg^{-1}$) a rapid, virtually monophasic exponential decline of striatal DA was observed.

An apparently reserpine-resistent pool of DA which was much smaller than the releasable pool was also found in the stimulation studies. The reserpine experiment suggests that the two pools revealed with electrical stimulation may be in close connection with each other. If this is the case than it is conceivable that in the study of Costa *et al.* (1974) striatal DA appears as if it is confined to a single open compartment.

The occurrence of rapidly depletable DA-pools in nerve endings may have consequences for the interpretation of experiments in which electrical stimulation is being applied. The net result of electrical stimulation of

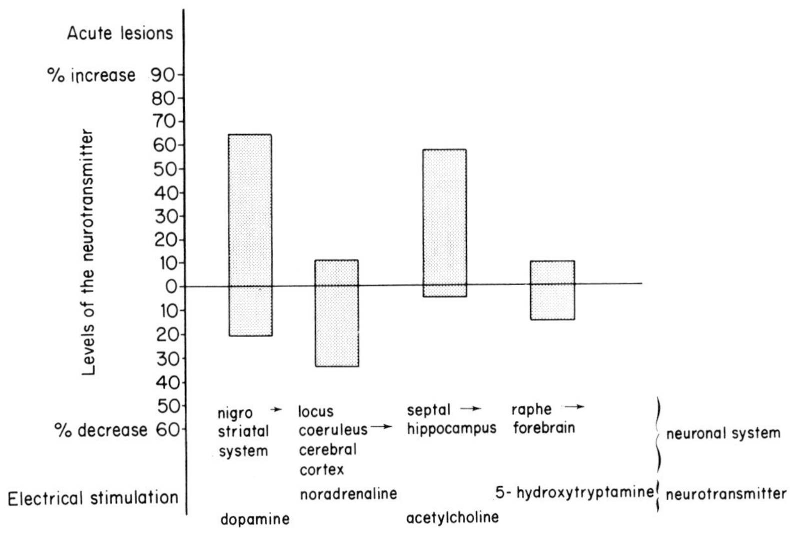

Fig. 1 Estimation of the pool sizes of various neurotransmitters involved in the process of neurotransmission. The increases indicate the steady-state levels of the mentioned neurotransmitter after acute cessation of nerve activity (usually determined after acute axotomy). The decreases of the levels are measured following maximal activation of the neurones by electrical stimulation at high frequency.

dopaminergic pathways may not necessarily induce a hyper-function of striatal DA, but may, conversely lead to a hypo-function of dopaminergic transmission.

Removal of the nerve-impulse flow in striatal dopaminergic nerve terminals produces an enhancement of the levels of the amine (Andén *et al.*, 1972). Apparently under physiological conditions the site in the nerve endings, where the releasable DA is stored, is not entirely filled (see Fig. 1). Both electrical stimulation and acute lesioning have been applied to various cerebral neurotransmitter systems, and in all cases significant changes in the levels of the transmitters have been observed. These results have been summarized in Fig. 1. The following studies were consulted: noradrenaline—Korf *et al.* (1973); acetylcholine—Rommelspacker and Kuhar (1974); serotonin—Sheard and Aghajanian (1968) and Herr and Roth (1976). Taking these results together it appears that all neuronal systems thus far investigated, possess rapidly releasable pools.

RELEASE AND METABOLISM

Both release of DA and the formation of DA metabolites are evoked by electrical stimulation of the median forebrain bundle. The question arises as to where these metabolites are formed, i.e. inside or outside the dopaminergic nerve terminals.

The striatal levels of the two DA-catabolizing enzymes, monoamine oxidase (MAO) and catechol-*O*-methyltransferase (COMT) have been determined after a variety of lesion procedures of the nigrostriatal pathway. After chronic electrolytic lesions of the substantia nigra the activity of MAO, measured with either DA or other substrates, was found not to be decreased (Van der Krogt, 1974; Goldstein *et al.*, 1969; Duncan *et al.*, 1972; Marsden *et al.*, 1972). Using the neurotoxin 6-hydroxy-dopamine some authors reported no alterations of the striatal MAO-activity after the dopaminergic nerve terminals had degenerated (Breese and Traylor, 1971; Uretsky and Iversen, 1970). Agid *et al.* (1973), however, reported decreased MAO-levels after 6-hydroxydopamine treatment, but this observation was not confirmed in a study by Van der Krogt (1974). Thus far only evidence for a localization of COMT outside dopaminergic nerve terminals has been reported (Uretsky and Iversen, 1970; Marsden *et al.*, 1972).

After electrical stimulation during short periods (5–10 min) of the nigro-striatal pathway much more DOPAC than HVA is formed (Korf *et al.*, 1976b). The formation of HVA during a period of stimulation was found not to be influenced by the administration of the potent DA-uptake inhibitor nomifensine (Korf *et al.*, 1976a). Roth *et al.* (1976) used the uptake inhibitor benztropine and they observed decreased formation of DOPAC during electrical stimulation. They concluded that released DA was taken up by the dopaminergic nerve terminals and that the amine was then catabolized. However, it should be realized that benztropine may decrease the release of DA because of its anticholinergic properties (Giorguieff *et al.*, 1977). As MAO is predominantly localized outside the dopaminergic nerve terminals and DOPAC is the most prominent metabolite of DA formed during electrical stimulation (Korf *et al.*, 1976b) we conclude that there is no reason to believe that striatal levels of DOPAC reflect exclusively intraneuronal DA metabolism.

Uptake and Presynaptic Receptors

A synaptosomal preparation of the striatum accumulates DA rapidly from the medium. The high affinity uptake of ^{3}H-DA was found to be proportional to the concentration of Na^+ between 35 and 154 mM (Holz and Coyle, 1974). Optimal uptake of ^{3}H-DA was obtained at 3.9 mM K^+. From these observations we deduce that the optimal uptake of neuronally released DA occurs predominantly when the nerve endings are at rest. This conclusion implies that during high-frequency stimulation released DA cannot be accumulated by the same nerve terminal from which the amine is liberated. Assuming that during supermaximal stimulation of the median forebrain bundle most if not all dopaminergic nerve terminals of the striatum are activated the released DA is likely to be taken up and degraded by non-dopaminergic neurones or gliacells. The lack of effect of nomifensine on the stimulus-induced formation of HVA (Korf *et al.*, 1976a) substantiates such an assumption.

The important issue arises as to what is the function of the (high affinity) uptake of DA into dopaminergic nerve terminals. From the above mentioned considerations it is possible that the high affinity Na^+-dependent uptake mechanism mainly serves to maintain low levels of DA in the extracellular space when the dopaminergic neurone is at rest.

Evidence for the existence of DA-sensitive receptors localized at dopaminergic neurones has been obtained from both electrophysiological studies (Aghajanian and Bunney, 1974) and biochemical studies (Kehr *et al.*, 1972). An important argument for the role of these so-called autoreceptors was obtained by using the DA-mimeticum apomorphine. Apomorphine decreased the synthesis of striatal DA when the nerve impulse was blocked, either after transection of the nigrostriatal bundle (Kehr *et al.*, 1972; Walters and Roth, 1974) or after the administration of drugs (Walters and Roth, 1976). After destruction of presumably post-synaptic DA-receptors by striatal injections of kainic acid, apomorphine is still able to decrease striatal DA-metabolism (DiChiara *et al.*, 1977). All these results together favour the possibility of a direct action of apomorphine on the autoreceptors of dopaminergic nerve terminals. In experiments with relatively low- and high-frequency stimulation of the median forebrain bundle the effect of high dosages of apomorphine was studied on the striatal formation of HVA (Korf *et al.*, 1976a). At high frequency (25 Hz) apomorphine did not change the stimulus-induced formation of HVA. When a submaximal stimulation (10 Hz) was applied there was a small, but significant reduction of impulse-generated HVA formation. From these experiments it appears that the sensitivity of the autoreceptor for DA is decreased during nerve activity. Such conclusions were also drawn from experiments with other, non-dopaminergic neural systems. Both in cholinergic and adrenergic neurones the presumably presynaptic effects of agonists and antagonists on transmitter release diminish when higher frequencies of stimulation are applied (see e.g. Dubocovick and Langer, 1974; Vizi *et al.*, 1973; Starke and Montel, 1973; Bourdois *et al.*, 1974; Langer *et al.*, 1977).

Dendritic Release of Dopamine

Electrical stimulation of cerebral nervous pathways can be used for the study of biochemical effects in both orthodromically and antidromically localized brain areas. Björklund and Lindvall (1975) reported the occurrence of an extensive dendritic network of dopaminergic neurones in the substantia nigra and they suggested a possible physiological role of dendritic DA. Electrical stimulation of the median forebrain bundle increased the levels of DOPAC and HVA both in the striatum and the substantia nigra in a frequency-dependent manner (Korf *et al.*, 1976b and 1977a). These findings suggest release of DA in the substantia nigra. Using entirely different techniques evidence for dendritic release of DA was also observed both *in vitro* (after prelabelling of nigral slices with ^{3}H-DA, Geffen *et al.*, 1976) and *in vivo* (release into push-pull cannula perfusates of ^{3}H-DA, synthesized from ^{3}H-tyr, Nieoullon *et al.*, 1977c). Two other aspects add to the evidence for a physiological role of nigral DA, namely the occurrence of a DA-stimulated adenylate cyclase in the substantia nigra (Kebabian and Saavreda, 1976; Gale *et al.*, 1977; Phillipson and Horn, 1976; Premont *et al.*, 1976 and Spano *et al.*, 1977) and the DA-sensitive receptors on nigral dopaminergic cells (autoreceptors, Aghajanian and Bunney, 1977). The role of dendritic DA is not very clear, but several possibilities have to be considered. Dendritic DA may regulate the neural firing of DA neurones either by lateral inhibition (decreasing the firing activity of adjacent cells) or auto-inhibition (decreasing the firing activity of the same dopaminergic cells from which it is released). Moreover the results with the adenylate cyclase suggest that dendritic DA may have the same function as the DA of the nerve terminals, implying that the dopaminergic cells have an output both in the striatum and the substantia nigra. In this respect it is of interest to view the recent findings of Reubi *et al.* (1977) that addition of DA enhances the release of ^{3}H-GABA previously accumulated in slices of the substantia nigra.

CONCLUSIONS

With electrical stimulation various aspects of cerebral dopaminergic neurotransmission have been studied. These aspects include: release of DA; synthesis and metabolism of the transmitter, presynaptic DA-receptors, neuronal pools and transmitter release from dendrites. The application of electrical stimulation has contributed to a better understanding of neurobiochemical processes associated with neurotransmission of cerebral DA.

As suggested in the first section of this chapter, it is difficult to indicate to what extent application of electrical stimulation induces physiological processes in neurons or whether the technique produces artefacts.

REFERENCES

Aghajanian, G. K., and Bunney, B. S. (1973). *In* "Frontiers in Catecholamine Research" (Eds. S. H. Snyder and E. Usdin), 643–648, Pergamon, New York.

Aghajanian, G. K., and Bunney, B. S. (1977). *Naun. Schmied. Arch. Pharmacol.* **27,** 1–7.

Agid, Y., Javoy, F., and Youdim, M. B. H. (1973). *Br. J. Pharmacol.* **48,** 175–178.

Andén, N. E., Carlsson, A., Dahlström, A., Fuxe, K., Hallarp, N. Å., and Larsson, K. (1964). *Life Sci.* **3,** 523–530.

Andén, N. E., Fuxe, B., Hamberger, B., and Hökfelt, T. (1966). *Acta Physiol. Scand.* **67,** 306–312.

Andén, N. E., Bédard, P., Fuxe, K., and Ungerstedt, U. (1972). *Experientia* **28,** 300–302.

Arbuthnott, G. W., Grow, T. J., Fuxe, K., Olson, L., and Ungerstedt, U. (1970). *Brain Res.* **24,** 471–483.

Barry, F. E., Walters, M. S., and Gallistel, C. R. (1974). *Physiol. Behav.* **12,** 749–754.

Bartholini, G., Stadler, H., Gadea Ciria, M., and Lloyd, K. G. (1976). *Neuropharmacology* **15,** 515–519.

Besson, M. J., Cheramy, A., Glowinsky, J., and Gauchy, C. (1974). *Biochem. Pharmacol. Suppl.* 449–451.

Björklund, A., and Lindvall, O. (1975). *Brain Res.* **83,** 531–537.

Bourdois, P. S., Mitchel, J. F., Somogyi, G. T., and Szerb, J. C. (1974). *Br. J. Pharmacol. Chemother.* **51,** 187–195.

Breese, G. R., and Traylor, T. D. (1971). *J. Pharmacol. Exptl. Ther.* **174,** 413–420.
Carlsson, A., and Lindqvist, M. (1973). *J. Neur. Trans.* **34,** 79–91.
Chiueh, C. C., and Moore, K. E. (1973). *Brain Res.* **50,** 221–225.
Chiueh, C. C., and Moore, K. E. (1974). *J. Neurochem.* **23,** 159–168.
Chung, S. H. (1976). *Nature (London)* **264,** 313–314.
Costa, E., Carenzi, A., Cheney, D., Guidotti, A., Racagni, G., and Zivkovic, B. (1974). *In* "Metabolic Compartmentation and Neurotransmission" (Eds. S. Berl, D. D. Clarke and D. Schneider), 167–186. Plenum Press, New York and London.
DiChiara, G., Porceddu, M. L., Fratta, M., and Gessa, G. L. (1977). *Nature (London)* **267,** 270–272.
Dubocovick, M. L., and Langer, S. Z. (1974). *J. Physiol. (London)* **237,** 505–519.
Duncan, R. J. S., Sourkes, T. L., Boucher, R., Poirier, L. J., and Roberge, A. (1972). *J. Neurochem.* **19,** 2007–2010.
Elchisak, M. A., Murrin, L. C., Roth, R. H., and Maas, J. (1976). *Psychopharmacol. Comm.* **2,** 411–420.
Gale, K., Guidotti, A., and Costa, E. (1977). *Science* **195,** 503–505.
Geffen, L. B., Jessell, T. M., Cuello, A. C., and Iversen, L. L. (1976). *Nature (London)* **260,** 258–260.
Giorguieff, M. F., Le Floc'h, M. L., Glowinski, J., and Besson, M. J. (1977). *J. Pharmacol. Exptl. Ther.* **200,** 535–544.
Glowinski, J. (1974). *In* "Metabolic Compartmentation and Neurotransmission (Eds. S. Berl, D. D. Clarke and D. D. Schneider), 187–203. Plenum Press, New York and London.
Goldstein, M., Anagnoste, B., Battista, A. F., Owen, W. S., and Nakatani, S. (1969). *J. Neurochem.* **16,** 645–653.
Gordon, R., Reid, J. V. O., Sjoerdsma, A., and Udenfriend, S. (1966). *Molec. Pharmacol.* **2,** 610–613.
Herr, B., and Roth, R. H. (1976). *Brain Res.* **110,** 189–193.
Holz, R. W., and Coyle, J. (1974). *Molec. Pharmacol.* **10,** 746–758.
Ibata, Y., Nojivo, Y., Matsuura, T., and Sano, Y. (1973). *Z. Zellforsch. Mikrosk. Anat.* **138,** 333–344.
Kebabian, J. W., and Saavedra, J. M. (1976). *Science* **193,** 683–685.
Kehr, W., Carlsson, A., Lindqvist, M., Magnusson, T., and Atack, C. V. (1972). *J. Pharm. Pharmacol.* **24,** 744–747.
Korf, J., Roth, R. H., and Aghajanian, G. K. (1973). *Eur. J. Pharmacol.* **23,** 276–282.
Korf, J., Grasdijk, L., and Westerink, B. H. C. (1976a). *J. Neurochem.* **26,** 579–584.
Korf, J., Zieleman, M., and Westerink, B. H. C. (1976b). *Nature (London)* **260,** 257–258.
Korf, J., Zieleman, M., and Westerink, B. H. C. (1977a). *Brain Res.* **120,** 184–187.
Korf, J., Van der Heyden, J. A. M., and Westerink, B. H. C. (1977b). *In* "Psychobiology of the Striatum" (Eds. A. R. Cools, A. H. M. Lohman and J. H. L. Van der Bercken), 51–59. North Holland Publishing Co., Amsterdam.
Langer, S. Z., Adler-Graschinsky, E. and Giorgi, O. (1977). *Nature (London)* **265,** 648–650.
Lindvall, O. and Björklund, A. (1974). *Acta Physiol. Scand. Suppl. 412* **92,** 1–47.
Marsden, C. A., Broch, O. J., and Guldberg, H. C. (1972). *Eur. J. Pharmacol.* **19,** 35–42.
McLennan, H. (1964). *J. Physiol. (London)* **174,** 152–161.
McLennan, H. (1965). *Experientia* **21,** 725–726.

Moore, R. Y., Bhatnagar, R. K., and Heller, A. (1971). *Brain Res.* **30,** 119–135.
Murrin, L. C., and Roth, R. H. (1976). *Molec. Pharmacol.* **12,** 463–475.
Murrin, L. C., Morgenroth III, V. M., and Roth, R. H. (1976). *Molec. Pharmacol.* **12,** 1070–1081.
Nieoullon, A., Cheramy, A., and Glowinski, J. (1977a). *J. Neurochem.* **28,** 819–828.
Nieoullon, A., Cheramy, A., and Glowinski, J. (1977b). *Nature (London)* **266,** 375–377.
Nieoullon, A., Cheramy, A., and Glowinski, J. (1977c). *Nature (London)* **269,** 340–342.
Phillipson, O., and Horn, A. S. (1976). *Nature (London)* **261,** 481–482.
Portig, P. J., and Vogt, M. (1969). *J. Physiol. (London)* **204,** 687–715.
Premont, J., Thierry, A. M., Tassin, J. P., Glowinski, J., Blanc, G., and Bockaert, J. (1976). *FEBS Lett.* **68,** 99–104.
Ranck, J. B. (1975). *Brain Res.* **98,** 417–440.
Reubi, J. C., Iversen, L. L., and Jessell, T. M. (1977). *Nature (London)* **268,** 652–654.
Riddell, D., and Szerb, J. C. (1971). *J. Neurochem.* **18,** 989–1006.
Rommelspacher, H., and Kuhar, M. J. (1974). *Brain Res.* **81,** 243–251.
Roth, R. H., Stjärne, L., and Von Euler, U.S. (1966). *Life Sci.* **5,** 1071–1075.
Roth, R. H., Walters, J. R., and Aghajanian, G. K. (1973). *In* "Frontiers in Catecholamine Research" (Eds. E. Usdin and S. Snyder), 567–574. Pergamon, New York and Oxford.
Roth, R. H., Murrin, L. C., and Walters, J. R. (1976). *Eur. J. Pharmacol.* **36,** 163–171.
Sedvall, G. C., and Kopin, I. (1967). *Life Sci.* **6,** 45–51.
Sheard, M. H., and Aghajanian, G. K. (1968). *J. Pharmacol. Exptl. Ther.* **163,** 425–430.
Shields, P. J., and Eccleston, D. (1972). *J. Neurochem.* **19,** 265–272.
Spano, P. F., DiChiara, G., Loddo, P., Tonon, G. C., and Trabucchi, M. (1977). *In* "Nonstriatal Dopaminergic Neurons"(Eds. E. Costa and G. L. Gessa), 613–616. Advances in Biochemical Pharmacology, Vol. 16. Raven Press, New York.
Starke, K., and Montel, H. (1973). *Neuropharmacology* **12,** 1073–1080.
Ungerstedt, U. (1971). *Acta Physiol. Scand. Suppl.* 367 **82,** 1–48.
Uretsky, N. J., and Iversen, L. L. (1970). *J. Neurochem.* **17,** 269–278.
Van der Krogt, J. A. (1974). Thesis, Leiden University, 1–194.
Vizi, E. S., Somogyi, G. T., Hadházy, P., and Knoll, J. (1973). *Naun. Schmied. Arch. Expl. Pathol. Pharmacol.* **280,** 79–91.
Vogt, M. (1969). *Br. J. Pharmacol.* **37,** 325–337.
Von Voigtlander, P. F., and Moore, K. E. (1971a). *Neuropharmacology* **10,** 733–741.
Von Voigtlander, P. F., and Moore, K. E. (1971b). *Brain Res.* **35,** 580–583.
Walters, J. R., and Roth, R. H. (1974). *J. Pharmacol. Expl. Ther.* **191,** 82–91.
Walters, J. R., and Roth, R. H. (1976). *Naun. Schmied. Arch. Pharmacol.* **296,** 5–14.
Weiner, N., and Rabadjija, M. (1968). *J. Pharmacol. Expl. Ther.* **160,** 61–71.
Westerink, B. H. C., and Korf, J. (1976). *Eur. J. Pharmacol.* **37,** 249–255.
Wurtman, R. J., Larin, F., Mostafapour, S., and Fernstrom, J. D. (1974). *Science* **185,** 183–184.

CHAPTER 18

The Effects of Drugs on Dopamine Biosynthesis and Metabolism in the Brain

B. H. C. Westerink

Department of Pharmacy, University of Groningen, Groningen, The Netherlands.

THE SEQUENCE OF DOPAMINE SYNTHESIS AND METABOLISM

The synthesis and metabolism of dopamine (DA) is depicted in Fig. 1. Tyrosine hydroxylase (TH) which is probably the rate-limiting enzyme, catalyses the formation of L-3,4-dihydroxyphenylalanine (L-DOPA) from tyrosine (Levitt *et al.*, 1965). The formation of DA from L-DOPA by the enzyme L-DOPA-decarboxylase is efficient and proceeds at a considerable rate (Holtz, 1959). Dopamine is metabolized by the action of the enzyme monoamine oxidase (MAO), catechol-*O*-methyltransferase (COMT), and aldehyde dehydrogenase (ADH) (Agid *et al.*, 1973; Sharman, 1973; Duncan and Sourkes, 1974). 3,4-Dihydroxyphenylacetic acid (DOPAC), homovanillic acid (HVA) and 3-methoxytyramine (3-MT) are considered as the three main products of DA degradation. Although 3-methoxy-4-hydroxyphenylethanol is a possible metabolite of DA, Wilk and Zimmerberg (1973) using a sensitive gas chromatographic method could not detect this compound in the rat brain.

Catechol-*O*-methyltransferase is often proposed to be extraneuronally localized (Potter *et al.*, 1965; Broch and Fonnum, 1972) and the metabolism of DA was therefore believed to be compartmentalized. Following administration of reserpine, which impairs the storage of

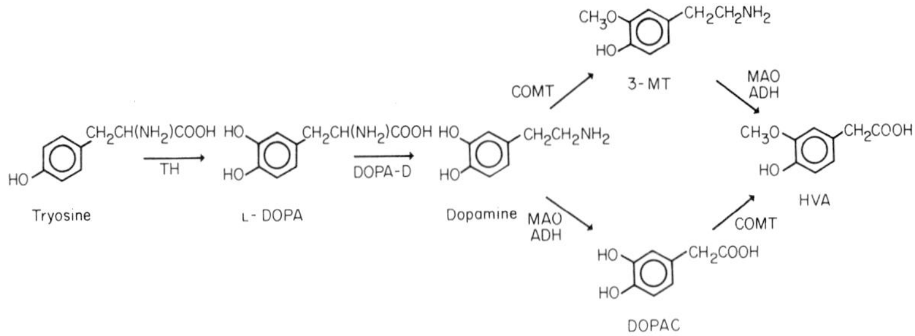

Fig. 1 Synthesis and possible degradation of DA by the action of tyrosine hydroxylase (TH), L-DOPA-decarboxylase (DOPA-D), monoamine oxidase (MAO), aldehyde-dehydrogenase (ADH), and catechol-*O*-methyltransferase (COMT).

intraneuronal DA, DOPAC levels increased before those of HVA, indicating an intraneuronal formation of DOPAC (Guldberg and Broch, 1971; Roffler-Tarlov *et al.*, 1971). Roffler-Tarlov *et al.* (1971) therefore considered DOPAC to be an index of intraneuronal metabolism. The most recent studies, however, indicate an intraneuronal as well as an extraneuronal localization of COMT (Guldberg and Marsden, 1975), while unpublished observations in our laboratory showed that an early rise of DOPAC levels followed by a secondary rise of HVA levels is not only seen after reserpine but also after neuroleptic or cholinomimetic treatment and after electrical stimulation (Korf *et al.*, 1976a) or partial lesioning (Westerink *et al.*, 1978) of the dopaminergic pathway.

Westerink and Korf (1976a) showed that under control conditions DOPAC is only partially methylated to HVA, while the formation of HVA via the 3-MT pathway is quantitatively unimportant in the rat. That DOPAC indeed may leave the brain unmethylated was supported by the occurrence of conjugated DOPAC and HVA (Gordon *et al.*, 1976; Swahn and Wiesel, 1976; Elchisak *et al.*, 1977). In contrast with the data of Westerink and Korf (1976a) Gropetti *et al.* (1977) reported a high specific activity of 3-MT after intraventricular injection of labelled tyrosine. The route of administration of the precursor, however, may have introduced artefacts in the synthesis and metabolism of DA. Although the conclusions of Westerink and Korf (1976a) imply a relatively simple metabolic sequence of DA during control conditions, it is possible, however, that DA metabolism is compartmentalized and that the different metabolic pathways only become apparent after certain drug treatments. In fact, it is well known that treatment with drugs like amphetamine causes a rapid and pronounced decrease of DOPAC levels, while HVA levels do not change, or even increase (e.g. Braestrup, 1977).

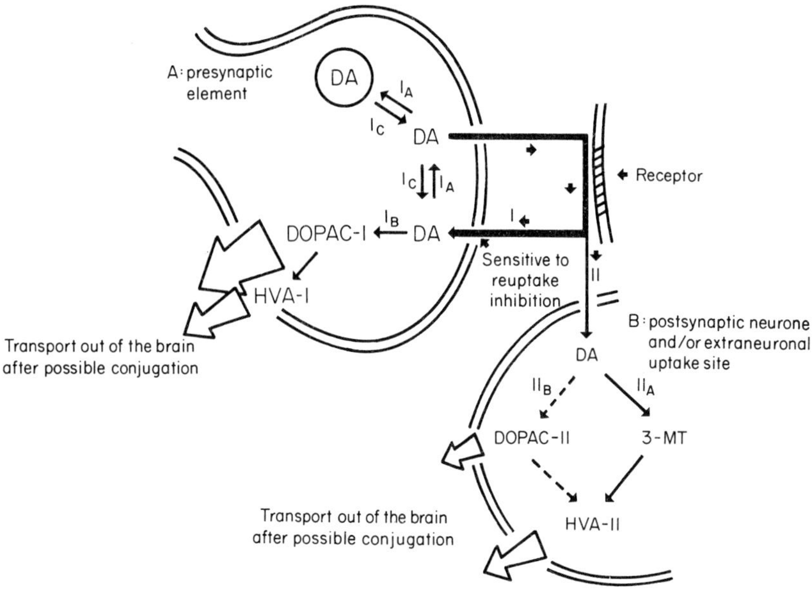

Fig. 2 Hypothetical model of the routes of degradation of DA. The size of the open arrows indicates which of the two main DA metabolites (DOPAC and HVA) is possibly of greater importance in the compartments A and B. For abbreviations see text.

Figure 2 represents a hypothetical model of the routes of degradation of DA, which is consistent with most of the data available in the literature. After release and interaction with the DA receptor most of the neurotransmitter is believed to be taken up again in the presynaptic terminal (compartment A), where it is restored (I_A) or metabolized to DOPAC-I (I_B) by intraneuronally localized MAO. This DOPAC-I is partly removed from the brain (after possible conjugation) and the remaining part is transformed into HVA-I by intra- and/or extraneuronally localized COMT. 3-Methoxytyramine is not thought to be formed in compartment A. This is concluded from experiments in which 3-MT did not seem to be involved in the strong rise in DA catabolism which occurs after reserpine treatment (unpublished observations). The second possible metabolic degradation of DA (pathway II) occurs in compartment B, which represents the postsynaptic neurone and/or extraneuronal uptake sites (glia cells, blood capillaries). As MAO and COMT have a wide distribution, DOPAC-II, HVA-II and 3-MT must be considered as the three products of DA metabolism in this compartment. Homovanillic acid-II, and to a lesser extent DOPAC-II, are supposed to be the main metabolites in compartment B. The small amount of 3-MT found in the rat brain (about 5% of the DOPAC and HVA levels) is derived, in this model, from compartment B. In this respect it is of interest to note that 3-MT is not able to leave the brain, while conjugates could not be detected (Westerink, unpublished observations).

In connection with Fig. 2, the following neurochemical effects can be considered:

Control conditions Pathway I predominates (Westerink and Korf, 1976a), and route II is quantitatively unimportant.

Neuroleptics, electrical stimulation Proportional to the increased release, DA is metabolized at a faster rate in Pathway I. The DOPAC increase precedes the HVA increase.

Reserpine Excessive catabolism of stored DA (I_C). The DOPAC increase precedes HVA increase.

Agonists Proportional to the decreased release, DA is metabolized at a slower rate in pathway I. The DOPAC decrease precedes the HVA decrease.

Amphetamine-like drugs Amphetamine is taken up in the presynaptic terminal and inhibits intraneuronal MAO (Braestrup, 1977), this results in a pronounced and rapid DOPAC decrease (independent of the nerve impulse flow), and a moderate HVA decrease. Pathway II becomes more important as the re-uptake of DA is inhibited at the presynaptic terminal. Higher doses of amphetamine induce DA release and the net result of pathway I and II may result in a moderate rise of HVA levels. In a similar manner 3-MT levels rise at higher doses.

Nomifensine and methylphenidate-like drugs These drugs inhibit the re-uptake of DA at the presynaptic terminal. Increased metabolism of released DA (no restorage) via pathway II. The HVA rises are more pronounced than those of DOPAC (Braestrup, 1977).

Morphine, antidepressants and anaesthetics It is striking that this heterogeneous group of drugs preferentially increases HVA levels in the rat brain (Westerink and Korf, 1976b; Westerink *et al.*, 1977). It is tempting to speculate that the increased DA metabolism caused by these drugs occurs mainly in compartment B.

Benztropine and cocaine Complete block of reuptake of DA at the presynaptic element. Pathway II is the main route of degradation. Homovanillic acid levels are unchanged while DOPAC levels tend to decrease (Gessa and Tagliamonte, 1975; Braestrup, 1977; Scheel-Krüger *et al.*, 1977), probably as the result of a slight decrease of DA biosynthesis (Scheel-Krüger *et al.*, 1977).

The compartmentation of DA metabolism will undoubtedly be more complex than is presented in Fig. 2. Much of our information is still incomplete (especially with respect to the role of 3-MT), and species differences may complicate the total picture. It should be noted that during the various treatments mentioned, there persists a continuous "background degradation of DA" (I_C) which is independent of the state of activity of the dopaminergic neurone. Except after complete MAO-inhibition, no treatment resulted in a complete disappearance of DOPAC and/or HVA levels from the brain.

As DOPAC and HVA can be formed at different sites, and both metabolites are sensitive to efflux inhibition, while in addition DOPAC formation is influenced by re-uptake inhibition (especially by amphetamine-like drugs) it has been suggested that 3-MT levels are a better reflection of impulse-dependent DA release than are DOPAC or HVA levels (Kehr, 1976; Racagni *et al.*, 1977; Westerink, 1979).

DOPAMINE TURNOVER

Effects of drugs on dopaminergic neurotransmission are usually evaluated by their interaction with the turnover rate of DA. The rate of biosynthesis of DA, which under steady state conditions, equals the rate of utilization or catabolism, is usually defined as the turnover of DA. Dopamine release is often considered synonymous with utilization or catabolism of the amine, which is not strictly true as DA can be re-used after re-uptake into the presynaptic terminals. The importance of the re-uptake and re-use in dopaminergic neurotransmission, however, has not been fully evaluated. Moreover, DA release into the synaptic cleft is generally assumed to be a functional process as it is followed by interaction with the DA receptor, whereas DA catabolism may occur at non-functional sites (e.g. after reserpine).

Various authors, using different experimental approaches, have provided evidence that DA is compartmentalized in the brain. A small and rapidly releasable pool is often considered as the "functional pool" (Javoy and Glowinski, 1971; Shore and Dorris, 1975; Korf *et al.*, 1976b), while a larger DA-pool or "granular storage pool" with a relatively small turnover may be involved only indirectly with dopaminergic neurotransmission (Shore, 1976). It is generally assumed that the "granular storage pool" is rapidly depleted by reserpine-like drugs (Andén *et al.*, 1964). The turnover methods which are usually used to evaluate the effects of drugs on dopaminergic neurotransmission, yield values for the average turnover rate of the amine, and do not take into account that DA may be compartmentalized in different pools. The results of the average DA turnover, measured by various methods which were applied to the striatum, are summarized in Table I. The methods used to study DA turnover are described below.

TABLE I Estimations of DA turnover in the striatum of the rat

Method	Turnover ($nmol\,g^{-1}\,h^{-1}$)	References
Use of labelled tyrosine	14.8	Nybäck (1971)
	23.0	Costa *et al.* (1975)
DA decline after AMPT	17.0	Costa and Neff (1966)
	86.0	Javoy and Glowinski (1971)
	21.7	Doteuchi *et al.* (1974)
DA accumulation after MAO-inhibition	148–190	Javoy *et al.* (1973)
L-DOPA accumulation after DOPA-decarboxylase-inhibition	22.0	Calculated from Walters and Roth (1974)
Summation of DOPAC and HVA turnover after MAO-inhibition	30.6	Wilk *et al.* (1975)
DOPAC turnover after MAO-inhibition	23.3	Westerink and Korf (1976c)
	36.0	Karoum *et al.* (1977)
HVA accumulation following probenecid	2.0	Westerink and Korf (1976c)
	5.5	Karoum *et al.* (1977)

The Labelled Precursor Method

The accumulation of labelled DA after intravenous administration of labelled tyrosine, can be used as an index of DA biosynthesis (Nybäck and Sedvall, 1968; Neff *et al.*, 1971). The method is based on steady state kinetics. This implies that, in order to calculate the correct specific activity of tyrosine, the labelled precursor is transported to the tyrosine pool, which is relevant for DA biosynthesis. Compartmentalization of the tyrosine pools in the brain may seriously complicate the assessment of the DA biosynthesis rate. Moreover the method lacks the necessary sensitivity to detect small changes in DA turnover.

The DA Synthesis Inhibition Method

Dopamine decline after α-methyl-*p*-tyrosine (AMPT, which is a potent inhibitor of tyrosine hydroxylase) is often used to measure the rate of DA utilization (e.g. Brodie *et al.*, 1966; Andén *et al.*, 1971). A serious drawback of the method is the incomplete blockade of tyrosine hydroxylase during the first 30 min after AMPT administration (Doteuchi *et al.*, 1974; Moleman and Bruinvels, 1976). When drugs are administered at a time point during which the tyrosine hydroxylase blockade is complete (Papeschi, 1975; Moleman and Bruinvels, 1976) it is quite likely that the "functional

DA-pool", as the result of its high turnover rate, is already exhausted (Shore and Dorris, 1975). This implies that the drug effects studied are mainly concerned with the slow turnover "storage pool". Javoy and Glowinski (1971) observed a biphasic decline in striatal DA after AMPT treatment. Doteuchi *et al.* (1974), who also observed a biphasic decline, maintain that the initial fast component is an artefact due to the formation of the metabolites of AMPT which change the kinetics of striatal DA decay.

The Accumulation of DA after MAO-Inhibition

Javoy *et al.* (1973) observed a rapid initial accumulation of DA in the first 10 min after administration of MAO-inhibitors (pargyline, pheniprazine or tranylcypromine) to rats which resulted in a very high rate of DA synthesis (148–190 nmol g^{-1} h^{-1}). Such a rapid initial accumulation of DA was however, not observed in comparable experiments (Kehr, 1976).

The Accumulation of L-DOPA after DOPA-Decarboxylase Inhibition

The rate of DA biosynthesis can be calculated from the L-DOPA accumulation after a rapid and complete blockade of DOPA-decarboxylase (Carlsson *et al.*, 1972). A drawback of the method is the possible change in tyrosine hydroxylase activity as an indirect result of the decarboxylase inhibition (Costa *et al.*, 1974; Javoy *et al.*, 1974). However, the method has proved to be reliable in the evaluation of the effects of neuroleptics, DA agonists, as well as of hemisection experiments, on DA biosynthesis (e.g. Roth *et al.*, 1973; Walters and Roth, 1976).

The Disappearance of DOPAC and/or HVA after MAO-inhibition

The rate of DA catabolism can be calculated from the turnover of DOPAC and HVA in the brain, assuming that the catabolic sequence of DA is known. The turnover of DOPAC and HVA can be calculated from steady state concentrations and fractional rate constants of the metabolites, assuming steady state conditions. The fractional rate constants of both DOPAC and HVA have been computed from the rate of disappearance of these metabolites from striatal tissue after administration of a rapidly acting irreversible inhibitor of MAO (Wiesel *et al.*, 1973; Wilk *et al.*, 1975b). Although this approach is reliable for DOPAC its use is questionable for HVA, because it does not calculate the possible formation of HVA from the amount of DOPAC still present shortly after administration of the MAO-inhibitor. Westerink and Korf (1976a) have shown that the HVA decline after MAO-inhibition only becomes a first order elimination when, in addition to MAO, COMT is also inhibited. Their data show that under control conditions DOPAC is only partially *O*-methylated to HVA, while the formation of HVA via the 3-MT pathway is unlikely. The turnover of DOPAC, at least in rats, therefore probably approximates DA turnover. The effects of drugs on DA catabolism may be evaluated by this method only, when any possible interference of the drug with the removal of DOPAC and HVA from the brain can be excluded.

The HVA Accumulation Following Probenecid Treatment

Probenecid inhibits the active transport from the brain of HVA and (to a lesser extent) DOPAC. The subsequent accumulation of HVA in the brain is frequently used as an indication of changes in DA turnover (Werdinius, 1966; Sharman, 1967). Calculation of HVA turnover in the rat brain (Table I) indicated that this method reflects only a minor part of DA turnover, and the method is therefore not suitable to estimate drug-modified changes in the turnover of DA. Moreover, the transport inhibition of HVA and DOPAC may differ in various species (Spano and Neff, 1972) and brain regions (Bartholini, 1976; Westerink and Korf, 1975; Karoum *et al.*, 1977).

The determination of DOPAC and HVA

Changes in DA biosynthesis and/or catabolism cause, within minutes, pronounced changes in the levels of the two main metabolites DOPAC and HVA. Determination of DOPAC and HVA levels appeared to be a very sensitive method to detect changes in the turnover of DA. Various drugs, however, were reported to interact with the transport of the metabolites out of the brain (Werdinius, 1966; Sharman, 1967; Horton *et al.*, 1977), and the measurement of drug-induced DOPAC and HVA changes can therefore not be directly related to changes in DA turnover, without additional information. Roth and co-workers found (Roth *et al.*, 1973, 1976) a good correlation between the effects of various drugs on the firing rate of dopaminergic neurones and their ability to increase DOPAC levels in terminal areas. Nevertheless hemisection experiments revealed (Braestrup, 1977; Westerink, 1977) that drugs like amphetamine and apomorphine are still able to decrease DOPAC levels when the impulse flow was prevented. Baclofen, another drug known to inhibit impulse flow of dopaminergic neurones (Olpe *et al.*, 1977) increased DOPAC and HVA levels (DA Prada and Keller, 1976; Braestrup, 1977). Moreover the time–response curves of hemisection experiments (Westerink *et al.*, 1978) as well as those observed after γ-hydroxybutyric acid (GHB) treatment (Spano *et al.*, 1971) revealed complex changes of DOPAC and HVA levels. It seems therefore somewhat hazardous to postulate a correlation between DA metabolite levels and nigral cell activity as was suggested by Roth and co-workers (Roth *et al.*, 1973, 1976).

In summary, apart from the probenecid method, most of the data on the average DA turnover rate observed by different techniques (Table I) agree rather well and indicate a DA turnover rate of 20–30 nmol g^{-1} h^{-1} in rat striatum. Every turnover method is characterized by certain disadvantages, which implies that the interaction of drugs with the functional activity of the dopaminergic neurones can only be evaluated when various methods are studied comparatively.

NEUROLEPTICS

Among the various pharmacological effects of neuroleptic agents, their antagonistic behaviour at DA receptors in the brain has been extensively investigated. Considerable biochemical, neuropharmacological, behavioural and electrophysiological evidence, which is beyond the scope of this chapter, is consistent with the assumption that neuroleptics block DA receptors.

Evidence for an increased DA turnover in various brain areas after neuroleptic treatment has been reported for different species (Table II). Changes in DA biosynthesis and catabolism after neuroleptic treatment have been attributed to blockade of DA receptors (Carlsson and Lindqvist, 1963; Van Rossum *et al.*, 1970). At the molecular level, the induced increase in DA synthesis is maintained by changed kinetic parameters (K_i for DA and K_m for pteridine cofactors) of striatal tyrosine hydroxylase (Roth *et al.*, 1975; Zivkovic *et al.*, 1974).

Various models have been proposed to explain the increased DA turnover by feedback mechanisms based on a hypothetical system of interneurones and/or presynaptic receptors (e.g. Carlsson and Lindqvist, 1963; Kehr *et al.*, 1972; Walters and Roth, 1976; Groves *et al.*, 1975). The "neuronal feedback" model proposed by Carlsson and Lindqvist in 1963, was used for many years to explain drug-induced biochemical and electrophysiological changes in the nigrostriatal pathway. According to this model a striatonigral loop acts as a feedback mechanism for postsynaptic receptor activation and blockade. The neuronal feedback hypothesis was recently challenged by experiments in which postsynaptic structures were damaged by intrastriatal kainic acid injections (Di Chiara *et al.*, 1977). The persistent effects of neuroleptics on DA turnover in the kainic acid treated rats, strongly suggests that postsynaptic structures are not

TABLE II Effect of neuroleptics on DA biosynthesis and metabolism in brain.

Carlsson and Lindqvist (1963)
Increased 3-MT levels were observed in brain of mice pretreated with MAO-inhibitors, after chlorpromazine or haloperidol administration.

Andén *et al.* (1964)
Increased HVA and DOPAC levels were observed in rabbit striatum after chlorpromazine or haloperidol treatment.

Laverty and Sharman (1965)
Increased HVA levels in the basal ganglia of the cat were reported after acute and chronic treatment of chlorpromazine, thioridazine, trifluperazine or thioproperazine.

Andén *et al.* (1966)
An increased DA turnover (AMPT-method demonstrated with histochemical techniques) was observed in nerve terminal areas of the rat, after acute and chronic treatment with chlorpromazine or haloperidol.

Da Prada and Pletscher (1966)
Increased HVA levels were observed in striatum of rat and guinea-pig after chlorpromazine treatment.

Juorio *et al.* (1966)
Increased striatal HVA levels were reported after chlorpromazine treatment of rat, guinea-pig, coypu and meriones.

Sharman (1967)
Increased HVA levels were seen in striatum of mouse brain after acute spiroperidol treatment.

Burkard *et al.* (1967)
Cerebral formation of ^{3}H-catechols from ^{3}H-tyrosine was increased in rats after chlorpromazine treatment.

Guldberg and Yates (1968)
Chlorpromazine ($5\,mg\,kg^{-1}$) increased DOPAC and HVA levels in caudate and globus pallidus of beagle dogs. No effects were seen at a dose of $15\,mg\,kg^{-1}$.

Bartholini and Pletscher (1969)
An increased conversion of intraventriculary injected ^{14}C-tyrosine to ^{14}DA, was observed during chlorpromazine treatment.

Nybäck and Sedvall (1969)
Increased DA biosynthesis (labelled precursor method) during chlorpromazine treatment, was selectively observed in striatum.

Javoy *et al.* (1970)
An increased DA turnover (labelled precursor method) during thioproperazine treatment, was observed in striatum and substantia nigra of rat brain.

O'Keeffe *et al.* (1970)
Chronic treatment with chlorpromazine—but not with thioridazine—increased HVA levels in caudate nucleus of dogs. Chronic treatment with chlorpromazine did not influence HVA levels in different regions of monkey brain. In mice acute treatment with chlorpromazine, haloperidol or thioridazine increased DA turnover (HVA increase; AMPT-method).

Besson *et al.* (1971)
Thioproperazine increased striatal DA biosynthesis (in-vitro labelled precursor method) of rats.

Bartholini *et al.* (1972)
Clozapine increased DA turnover (AMPT-method) and HVA levels in rat brain.

Dailey *et al.* (1972)
Increased HVA levels were observed after acute treatment with chlorpromazine or four of its metabolites in brain of mice. Chlorpromazine sulphoxide had no effect.

Andén and Stock (1973)
Clozapine increased HVA levels in pooled limbic structures somewhat more than in striatum of rabbits.

Green (1973)
Chlorpromazine and trifluperazine increased DA turnover (AMPT-method) in brain of mice. Closely related analogues with the substituent in the 2-position or with a tetramethylene instead of a trimethylene chain were inactive.

Bowers and Rozitis (1974)
Limbic and hypothalamic but not caudate areas of rabbit brain maintained their increased levels of HVA during chronic administration of thioridazine, chlorpromazine or fluphenazine enanthate. Acute treatment with clozapine resulted in a larger rise of HVA levels in limbic and hypothalamic areas, compared with caudate.

Bürki *et al.* (1974)
Dose–response and time–response curves for DOPAC and HVA levels after clozapine, haloperidol, perlapine or thioridazine treatment, were reported. Clozapine elevated striatal DA levels.

Hyttel (1974)
Dose-response curves for the influence of nine neuroleptics on DA biosynthesis (labelled precursor method) were determined in mice brain.

Stille and Lauener (1974)
Time-response curves for rat striatal HVA levels were reported for six neuroleptics.

Bürki *et al.* (1975)
No tolerance to increased DA turnover was observed after chronic clozapine treatment, in contrast with haloperidol, loxapine or chlorpromazine.

Restelli *et al.* (1975)
Chlorpromazine, haloperidol and sulpiride induced an identical percentage HVA increase in striatum and cortex of rat brain.

Scatton *et al.* (1975)
Thioproperazine increased DA biosynthesis (in-vitro labelled precursor method).

Stawartz *et al.* (1975)
The ED100% for the increase in HVA levels after clozapine, chlorpromazine, haloperidol, pimozide and thioridazine, showed the same rank of potency in striatum and limbic forebrain.

Tagliamonte *et al.* (1975)
Sulpiride increased DA turnover (AMPT-method). Dose-response curves for DOPAC and HVA levels were given.

Westerink and Korf (1975)
Clozapine, haloperidol and thioridazine induced a similar percentage HVA increase in striatum, nucleus accumbens and tuberculum olfactorium.

Wiesel and Sedvall (1975)
Similar dose-response curves for HVA levels were observed in striatum and tuberculum olfactorium after chlorpromazine, clozapine, haloperidol or thioridazine treatment.

Wilk *et al.* (1975c)
The effect of haloperidol on DOPAC and HVA levels was more persistent in striatum than in tuberculum olfactorium.

Zivkovic *et al.* (1975b)
Clozapine was more active in increasing the turnover rate of DA (labelled precursor method) in nucleus accumbens than in striatum; the activity of chlorpromazine and haloperidol in these two structures was equal.

Bartholini (1976)
Following probenecid administration, the net effect of clozapine, sulpiride and thioridazine on HVA accumulation was more marked in the limbic system than in striatum, whereas haloperidol and chlorpromazine had a similar effect in the two regions of the rat brain.

TABLE II Continued.

Rollema *et al.* (1976)
Dose–response curves for 14 neuroleptics for rat striatal HVA levels were determined. The antagonism to amphetamine- or apomorphine-induced stereotypy in rats correlated highly with the ED50% HVA increase.

Scatton *et al.* (1976)
A marked decrease of striatal DA biosynthesis (in-vitro labelled precursor method) was measured 24 h after the last injection of a repeated treatment with thioproperazine. At this time point no change in DA biosynthesis was observed in mesolimbic and mesocortical areas.

Von Stralendorff *et al.* (1976)
Chronic treatment with clozapine—in contrast to haloperidol—induced a higher percentage increase in HVA levels the limbic system than in striatum of rats.

Waldmeier and Maître (1976b)
A reduction of the initial DOPAC and HVA increase was observed by repeated treatment with chlorpromazine, haloperidol or thioridazine both in striatum and mesolimbic system. With clozapine this tolerance phenomenon occurred (in both structures studied) only with a dose of 100 mg kg^{-1}.

Waldmeier and Maître (1976a)
Chlorpromazine, perlapine and sulpiride enhanced DA disappearance (AMPT-method) in the mesolimbic area more than in striatum. Haloperidol was slightly more active in striatum than in mesolimbic area. None of the drugs was more efficient in elevating HVA levels in mesolimbic area than in striatum.

Westerink and Korf (1976b)
Clozapine and sulpiride induced a more pronounced rise in DOPAC and HVA levels in nucleus accumbens in comparison with striatum and tuberculum olfactorium.

Westerink and Korf (1976d)
Haloperidol, sulpiride or thioridazine induced identical percentage HVA increase in frontal cortex and striatum; a more pronounced increase in HVA levels was observed after clozapine treatment in frontal cortex.

Westerink and Korf (1976c)
In contrast with a strong increase in DOPAC and HVA levels observed after haloperidol treatment in striatum, a slight rise in metabolite levels was seen in substantia nigra of rat brain.

Westerink and Korf (1976e)
Clozapine, *cis*-flupenthixol or haloperidol induced comparable increases in HVA levels in striatum and retina of rats.

Wilk and Glick (1976)
Dose–response curves for DOPAC and HVA levels after clozapine treatment were similar in striatum and nucleus accumbens.

Elliot *et al.* (1977)
Five substituted benzamides: metoclopramide, sulpiride, tigan, clebopride and its debenzylated metabolite elevated both striatal and mesolimbic HVA levels.

Fadda *et al.* (1977)
Haloperidol produced a strong and equal rise of DOPAC levels in both substantia nigra and striatum of rats.

Gudelsky and Moore (1977)
The rate of DA turnover (AMPT-method) between 16 and 24 h after administration of clozapine, haloperidol or thioridazine was enhanced in the median eminence.

Kebabian *et al.* (1977)
Chlorpromazine caused a slight increase in DOPAC levels in substantia nigra, in contrast with the strong rise in striatum of rats.

Laduron *et al.* (1977)
After chronic treatment with haloperidol, tolerance to the effects on HVA was observed only in striatum and limbic areas, but not in parietal cortex of rabbit brain.

Scatton *et al.* (1977a)
A single injection of the palmitic ester of pipotiazine had a biphasic effect on DA biosynthesis (in-vitro labelled precursor method) in striatum; and acceleration was followed by a marked reduction which lasted for several weeks. A biphasic effect was also seen in the nucleus accumbens+tuberculum olfactorium, although the first phase was of longer duration. Only one phase of increased DA synthesis was seen in the cerebral cortex.

Scatton *et al.* (1977b)
Chlorpromazine, haloperidol and thioridazine caused a more marked increase in DOPAC and HVA levels in striatum than in mesolimbic areas, while the reverse was true for sulpiride.

Scatton *et al.* (1977c)
No tolerance to the effect of chronic neuroleptic treatment on HVA levels, was observed in the retina of rats.

Scatton (1977)
Repeated treatment with haloperidol or sulpiride induced tolerance to the increases of HVA and DOPAC levels in striatum, nucleus accumbens, tuberculum olfactorium and frontal cortex of rats.

Stanley and Wilk (1977)
Haloperidol, chlorpromazine and clozapine produced a dose-dependent increase in DOPAC levels in striatum and tuberculum olfactorium. Non-antipsychotic butyrophenones produced atypically flat curves.

Westerink *et al.* (1977)
Atypical neuroleptics like clozapine, sulpiride, and thioridazine could be differentiated from classical neuroleptics, by their ability to produce a relatively large percentage increase of DOPAC and HVA levels in nucleus accumbens and tuberculum olfactorium in comparison with striatum. A heterogeneous group of 14 non-neuroleptic drugs, however, produced regional changes which were very similar to those of the atypical neuroleptics.

Wilk and Stanley (1977)
Perlapine produced a dose-dependent increase in DOPAC levels in striatum and tuberculum olfactorium. The potency of perlapine as a possible neuroleptic, was similar to that of chlorpromazine.

---

necessarily involved in the biochemical changes induced by these drugs.

Besides the "neuronal feedback" hypothesis, an alternative model was developed to explain the biochemical effects of DA agonists and antagonists. Hemisection experiments suggested that the amount of transmitter released per nerve impulse flow is regulated via a local mechanism involving presynaptic receptors: "the autoreceptor model" (Kehr *et al.*, 1972; Roth *et al.*, 1973; Walters and Roth, 1974, 1976). These DA-autoreceptors are believed to be localized on nerve terminals, cellbodies and/or dendrites of dopaminergic neurones. Electrophysiological studies have provided evidence that neuroleptics are able to block presynaptic receptors (e.g. Bunney *et al.*, 1973). Hemisection experiments revealed that compared with haloperidol the phenothiazines and pimozide appeared less potent in reversing the presynaptic effects of apomorphine than in blocking postsynaptic DA-receptors (Walters and Roth, 1976). The significance of these findings, which were in accordance with electrophysiological studies (Aghajanian and Bunney, 1974) needs further elucidation.

It has been suggested that the capability of neuroleptics to enhance DA turnover correlates well with their antipsychotic potency and their extrapyramidal side-effects (Roos, 1965; Andén *et al.*, 1970; Stawartz *et al.*, 1975; Rollema *et al.*, 1976; Stanley and Wilk, 1977). In this respect it is of interest to note that a significant positive correlation was found between the striatal HVA concentration and the chlorpromazine levels in the brain in acute as well as chronic experiments (Wiesel *et al.*, 1975; Wiesel and Alfredsson, 1976).

Rollema *et al.* (1976) calculated the ED50% HVA increase of 14 neuroleptics. The antagonism to amphetamine- or apomorphine-induced stereotypy in rats—one of the most common screening tests for neuroleptic activity—correlated highly with the ED50% HVA increase in the striatum.

It is questionable whether the screening of new compounds based on DA-receptor blockade will provide new antipsychotic drugs with clinical properties which will differ fundamentally from the hundred or so traditional neuroleptics which have been developed by the pharmaceutical industry. Significantly, some newer drugs like sulpiride and propranolol, despite traditional animal screening tests have antipsychotic activity, yet do not seem to block DA receptors (Trabucchi *et al.*, 1976); these compounds are also devoid of extrapyramidal side-effects. In this respect it should be noted that metoclopramide, a drug known to provoke certain types of extrapyramidal effects in man, to induce catalepsy and to increase DA turnover in rats (Ahtee and Bunçombe, 1974; Peringer *et al.*, 1976; Westerink *et al.*, 1977), is devoid of antipsychotic actions. These data raise questions as to the basic mechanisms by which neuroleptic action is produced. Further clinical evaluation of metoclopramide and a recently synthesized group of related substituted benzamides (Elliot *et al.*, 1977) may be required to resolve this question with certainty.

In the search for interactions between dopaminergic neurones and other putative neurotransmitter systems, various authors have investigated the effects of combined treatment of neuroleptics with different drugs. Apomorphine prevents most of the increased DA turnover produced by neuroleptics (Lahti *et al.*, 1972; Di Chiara *et al.*, 1976; Carlsson *et al.*, 1977), as a consequence of the agonistic behaviour of apomorphine at DA-receptors. The partial suppression of the neuroleptic-induced increased DA turnover by aminooxyacetic acid—a compound known to increase

γ-aminobutyric acid (GABA) levels in the brain by inhibition of GABA-transaminase—indicates an interaction between gabaminergic systems and DA turnover in forebrain structures (Andén, 1974; Lahti and Losey, 1974; Westerink and Korf, 1975). Combined treatment with neuroleptics and benzodiazepines have led to similar speculations about a gabaminergic involvement in the interaction of neuroleptics (Keller *et al.*, 1976). The high dose of benzodiazepines used in this study, however, complicates the interpretation of these findings. Likewise no great value can be attached to the small suppression of the neuroleptic-induced increased DA turnover by anticholinergic drugs applied in a dose which exceeds the clinical dose 1000-fold (O'Keeffe *et al.*, 1970; Corrodi *et al.*, 1972a; Andén, 1972; Westerink and Korf, 1975). The anticataleptic effects of serotonin depletion by *p*-chlorophenylalanine treatment in haloperidol-treated rats and the partial suppression of the haloperidol-induced HVA elevation, suggest an interaction between dopaminergic and sertonergic pathways (Gumulka *et al.*, 1973; Westerink and Korf, 1975).

RESERPINE

Although reserpine has been used as antipsychotic agent, its mechanism of interaction with DA biosynthesis and catabolism differs from that of neuroleptics. Good evidence exists that reserpine and related compounds like benzoquinolines interfere with the granular storage of monoamines in nerve terminals (e.g. Shore, 1962; Saner and Pletscher, 1977). The rapid depletion of DA ("the reserpine-sensitive DA-pool") is followed by a rise in DOPAC and HVA levels (Andén *et al.*, 1964; Juorio *et al.*, 1966; Werdinius, 1967; Guldberg and Broch, 1971; Roffler-Tarlov *et al.*, 1971), while the biosynthesis of DA is enhanced (Koe, 1974; Carlsson, 1975). Although no evidence exists that reserpine blocks DA receptors, its depleting effect on DA storage is often explained as an indirect DA receptor blockade.

DA AGONISTS

Directly Acting Agonists

Apomorphine produces stereotyped behaviour in rodents and birds. It is assumed that this effect is mediated by a direct stimulation of DA receptors in the brain. The apomorphine molecule contains the elements of DA itself, although rigidly bound into a cyclic structure. In support of the view that the turnover of DA is regulated by DA receptors, DA agonists like apomorphine have been found to inhibit the activity of dopaminergic neurones (Roth *et al.*, 1973) and to reduce the synthesis and utilization of DA in the brain (Andén *et al.*, 1967; Roos, 1969). The dose–response curves for DOPA formation (Carlsson *et al.*, 1977) as well as HVA levels (Martres *et al.*, 1977) during apomorphine treatment showed a double S-shape. Apomorphine has also a biphasic action on behaviour, as very low doses ($>30\,\mu g\,kg^{-1}$, i.p.) have been found to induce sedation (Carlsson *et al.*, 1974; Di Chiara *et al.*, 1976), while stereotyped behaviour becomes predominant at dosages $>0.5\,mg\,kg^{-1}$. The biphasic shape of the dose–response curve for HVA and DOPA-accumulation shows similarities with the biphasic actions of apomorphine on behaviour (Carlsson *et al.*, 1977; Martres *et al.*, 1977). Carlsson *et al.* (1974) and Di Chiara *et al.* (1976) have suggested that the sedation induced by low doses of apomorphine might result from stimulation of presynaptic DA-receptors, while the stereotyped behaviour is caused by stimulation of postsynaptic receptors. Both biphasic behavioural and biochemical properties of apomorphine are antagonized by neuroleptic treatment (Carlsson *et al.*, 1977; Di Chiara *et al.*, 1976).

Recently numerous analogues of DA, apomorphine and 2-amino-tetralines have been synthesized and various compounds were found to be potent DA agonists in behavioural tests (e.g. Cannon *et al.*, 1977). Apart from 10-*O*-methylapomorphine (Apocodeine, Lal *et al.*, 1972) and the dipivaloylester of apomorphine (Worms and Scatton, 1977), the effects of these newly synthesized compounds on DA metabolism have not yet been reported.

Various authors have shown that 1-(3,4-methylenedioxybenzyl-4-(2-pyrimidyl)-piperazine (Piribidil, ET495) and its active metabolite 1-(3,4-dihydroxybenzyl)-4-(2-pyrimidyl)-piperazine (S584) induces behavioural and biochemical changes which are comparable with apomorphine (Corrodi *et al.*, 1972b; Jori *et al.*, 1974; Costall and Naylor, 1974).

Ergolines

Another group of drugs which have potent stimulating actions on DA receptors are found among the alkaloids (ergolines) present in ergot (*Claviceps purpurea*). Ergocornine and 2-Br-α-ergocryptine (CB154, Bromocryptine) caused a reduction of the DA turnover in the forebrain, as measured by DA depletion after AMPT treatment (Corrodi *et al.*, 1973; Fuxe *et al.*, 1974). A study of Silbergeld and Pfeiffer (1977) suggested that ergolines enhance dopaminergic function by an action on presynaptic dopaminergic sites in addition to postsynaptic receptor activation. Evidence for stimulation of DA-receptors by relatively low doses of lysergic acid diethylamide (LSD) as well as lisuride (*N*-(D-6-methyl-8-isoergonyl)-*N*′,*N*-

diethylcarbamide) (Da Prada *et al.*, 1975; Kehr, 1977), have been provided by the observation of decreased DA metabolism in the rat brain.

Amantadine

Amantadine—an antiviral agent—has been found to have some therapeutic effect in Parkinson's syndrome (in doses of 3–5 $mg\,kg^{-1}$). Cox (1975) reported a small increase in HVA levels after a dose of 100 $mg\,kg^{-1}$ amantadine, while Vaatstra and Eigeman (1974) could not detect an effect of this drug ($2 \times 40\,mg\,kg^{-1}$) on in-vivo DA turnover, as measured by the labelled precursor method. These data together with the high doses required in animal tests (30–150 $mg\,kg^{-1}$, Baldessarini *et al.*, 1972) suggest that the antiparkinson action of amantadine cannot simply be explained by an interaction with central DA receptors.

Indirectly Acting DA Agonists

Besides the agents directly acting on DA receptors, a number of indirectly acting DA agonists are known. Such compounds of which amphetamine is a classical example, are believed to induce functional DA release from nerve terminals into the synaptic cleft. The behavioural effects seen after amphetamine-like drugs are very similar to those observed after treatment with directly acting agonists like apomorphine.

The amphetamine-like agents have been divided into two subgroups: one in which the stereotyped behaviour is not antagonized by reserpine and short-term AMPT treatment: amphetamine, methamphetamine, phenmetrazine and phenylethylamine; and a second group in which the behavioural effects are prevented by reserpine: amfonelic acid, cocaine, methylphenidate, nomifensine and pipradrol (Scheel-Krüger, 1971; Braestrup, 1977).

As in the case of apomorphine, a decreased activity of dopaminergic cell activity has been observed after amphetamine treatment (Roth *et al.*, 1973). Although the decreased impulse flow leads to a reduction of impulse-flow-dependent DA release, the net amount of the transmitter released is increased (e.g. Lloyd and Bartholini, 1975) probably as a result of re-uptake inhibition and induced DA release. Combined treatment with haloperidol—a drug known to increase the impulse flow in dopaminergic neurones—resulted in a strong additive increase in striatal 3-MT levels (Speckenbach and Kehr, 1976) suggesting that the amount of DA released per nerve impulse is potentiated by amphetamine.

The methylphenidate-like agents induce a moderate increase in both HVA and DOPAC levels (see the "sequence of DA synthesis and metabolism"). For amphetamine-like agents a pronounced decrease of

DOPAC levels have been reported, whereas HVA levels showed a tendency to decrease 15–30 min after treatment, followed by an increase at doses 5 mg kg^{-1} (Sayers and Handley, 1973; Braestrup, 1977). Hemisection experiments revealed that as with apomorphine, the DOPAC decrease seen after amphetamine was independent of the impulse flow (Braestrup, 1977; Westerink, 1977). Braestrup (1977) hypothesized that amphetamine and related compounds act as intraneuronal MAO-inhibitors which may explain, in addition to re-uptake inhibition and induced DA release, the decreased DOPAC levels and the paradoxical effects reported on HVA levels (see above). The effects of amphetamine on DA biosynthesis may be either increased or decreased depending on the dose and time of administration (Kehr *et al.*, 1977; Kuczenski, 1977). Koe (1974) proposed that the interaction of amphetamine with DA biosynthesis and metabolism involves at least five events: two accelerating DA syntheses and three decreasing DA syntheses.

Anorectic Drugs

Anorectic drugs like fenfluramine, phentermine and mazindol are structurally closely related to the amphetamines, but differ in several pharmacological and clinical aspects. Many of them, given in high doses to rats, produce an increase in striatal HVA levels (Jori and Dolfini, 1974). The anorectic activity of amphetamine-like compounds, however, appeared not to be related to the alteration of DA metabolism in the striatum (Jori and Bernardi, 1969). When rats are submitted to repeated treatments with anorectic drugs, tolerance develops to the increase of HVA after amphetamine or mazindol, but not after fenfluramine treatment. As far as the HVA increase is concerned, fenfluramine does not produce cross-tolerance with amphetamine (Jori and Dolfini, 1977). It has been suggested that fenfluramine, unlike amphetamine, increases HVA levels by a neuroleptic-like mechanism or by a serotoninergically mediated process (Jori and Dolfini, 1977).

ANALGESICS

Several lines of evidence suggest that an increased turnover of DA results from the administration of morphine and related compounds like dextromoramide, methadon, pentazocine, and viminol-R_2. An increased DA biosynthesis using the labelled precursor method was reported by various authors (Clouet and Ratner, 1970; Smith *et al.*, 1972; Gauchy *et al.*, 1973; Sugrue, 1974; Costa *et al.*, 1975; Gessa and Tagliamonte, 1975). In addition increased levels of DOPAC and HVA have been observed (Sasame *et al.*,

1972; Ahtee, 1974; Kuschinsky and Hornykiewicz, 1974; Gessa and Tagliamonte, 1975; Kääriäinen and Ahtee, 1976; Westerink and Korf, 1976b, c, d).

A discrepancy appears in the literature when the effects of morphine on the AMPT-induced DA decline are compared. Gunne *et al.* (1969) and Sugrue (1974) reported an increased AMPT-induced DA depletion during morphine treatment, when morphine was administered to rats prior to AMPT. Papeschi *et al.* (1975) reported however, an increased rate of DA depletion by morphine when given before AMPT, but not when given after AMPT. These findings were confirmed by Moleman and Bruinvels (1976), who emphasized the incomplete blockade of tyrosine hydroxylase during the first 30 min after AMPT administration. Furthermore, Moleman and Bruinvels (1976) concluded that the effect of morphine on DA release (as measured by the AMPT-method) depends on the state of activity of the dopaminergic neurones. A clear effect of increased DA release was observed when the impulse flow was prevented by a lesion of the substantia nigra or by administration of apomorphine. No such proposed dependence of the activity state of dopaminergic neurones appeared, however, from a recent electrophysiological study (Iwatsubo and Clouet, 1977). In this study systemic administration of morphine caused a pronounced increase in the firing rate of nigral cells, which could be reversed by naloxon, apomorphine or L-DOPA.

Several lines of evidence suggest that analgesics do not, like the neuroleptics, directly act on postsynaptic DA receptors (Kuschinsky and Hornykiewicz, 1974; Carenzi *et al.*, 1975; Gessa and Tagliamonte, 1975; Westerink *et al.*, 1977). Various authors have suggested that the main site of morphine's action may be presynaptic by increasing the non-functional catabolism of newly synthesized DA, which implies a deficiency of the amine at DA-receptors (Sharman, 1966; Kuschinsky and Hornykiewicz, 1974; Papeschi *et al.*, 1975).

Finally, it should be noted that there is no strict correlation between the catalepsy-inducing potency of narcotic analgesics and their potency in increasing DA turnover (Gessa and Tagliamonte, 1975).

CHOLINOMIMETIC AND ANTICHOLINERGIC DRUGS

Centrally acting cholinomimetic agents increase, in relatively low doses, the turnover of DA in the brain. Arecoline, eserine, nicotine, physostigmine, pilocarpnine and oxotremorine have been found to raise HVA and DOPAC levels (Laverty and Sharman, 1965; Perez-Cruet *et al.*, 1971; Nose and Takemoto, 1974; Bartholini *et al.*, 1975; Haubrich and Goldberg, 1975;

Oelzner *et al.*, 1975; Westerink and Korf, 1976b, c, d). Oxotremorine and physostigmine can also accelerate the AMPT-induced disappearance of brain DA (Corrodi *et al.*, 1967; Bhatnagar, 1973; Paalzov and Paalzov, 1975). An increased DOPA accumulation during DOPA-decarboxylase-inhibition was reported for cholinomimetic drugs (Javoy *et al.*, 1975).

The increased DA biosynthesis and catabolism can be explained by an increased DA release. Evidence for this hypothesis was provided by the experiments of Lloyd and Bartholini (1975) who observed, using a push-pull cannula technique, a dramatically increased release of endogenous DA after oxotremorine injection. Andén and Wachtel (1977) provided evidence that the increased biosynthesis of DA by oxotremorine or physostigmine is dependent on nigral activity. These authors speculate that stimulation of the muscarinic receptors outside the dopaminergic nerve terminals might increase the nerve impulse flow in the dopaminergic neurones. That the interaction between oxotermorine and DA metabolism depends on the neuronal organization of the dopaminergic neurones, was also illustrated by the absence of any effect of this drug on HVA formation in the retina (Westerink and Korf, 1976e).

The effects of the cholinomimetic drugs are probably due to stimulation of muscarinic receptors since the effects on DA metabolism are prevented by centrally active anticholinergic agents like atropine.

It has been shown that administration of anticholinergic agents like atropine and trihexyphenidyl induce a decrease in HVA levels (Bartholini and Pletscher, 1971; O'Keeffe *et al.*, 1970). Andén and Bédard (1971) and Bhatnagar (1973) have shown that anticholinergic drugs cause a slight deceleration of the AMPT-induced DA decline. The doses of anticholinergic drugs used (5–60 $mg\,kg^{-1}$) however, are extremely high in comparison with clinical doses, and this seriously complicates the analysis of the effect of anticholinergics in terms of dopaminergic neurotransmission. The paradoxical effects of atropine on HVA formation (an HVA rise was reported after intraventricular injection and an HVA decrease was seen after systemic application, Bartholini and Pletscher, 1971) illustrate the complexity of the effect of high doses of anticholinergic agents and dopaminergic neurotransmission.

γ-HYDROXYBUTYRATE AND RELATED COMPOUNDS

Application of high doses of γ-hydroxybutyrate (GHB, 750–1500 $mg\,kg^{-1}$) or its precursor, γ-butyrolactone (GBL), induces a dramatic increase in the DA levels in the neuronal terminal network (Gessa *et al.*, 1966; Roth and Surh, 1970; Spano *et al.*, 1971; Walters and Roth, 1972). These changes—of which the mechanisms of action are not known—are probably rather specific for DA since the levels of other amines in the brain were not influenced by these drugs (Andén and Stock, 1973b). The increased DA levels are associated with an enhanced rate of synthesis (Roth *et al.*, 1973; L-DOPA accumulation method) and a decreased DA utilization (Roth and Surh, 1970; Stock *et al.*, 1973; Walters *et al.*, 1973). The maximal DA rise was observed 90 min after GHB treatment (Spano *et al.*, 1971; Roth, 1971), at this time point DA synthesis was normal and not decreased as might be expected as the consequence of a possible end-product inhibition of tyrosine hydroxylase (Roth *et al.*, 1973). Roth *et al.* (1973) observed a slight decrease of DOPAC levels 30 min after GBL treatment. Spano *et al.* (1971) and Roth (1971) showed that HVA levels clearly rose after DA levels were maximally increased by GHB treatment. The newly synthesized DA is apparently stored in a compartment which is relatively invulnerable to MAO-degradation.

The biochemical effects observed after GHB or GBL treatment are mimicked by an acute lesion of the nigrostriatal pathway (Andén *et al.*, 1971; Kehr *et al.*, 1972; Walters *et al.*, 1973). A number of observations suggest that presynaptic receptor activation is able to modify the increase in DA synthesis, observed after hemisection experiments (Kehr *et al.*, 1972) as well as after GHB or GBL treatment (Walters and Roth, 1974, 1976).

Electrophysiological studies have revealed that systemic application of GHB or GBL completely inhibited the spontaneous activity of dopaminergic neurones (Roth *et al.*, 1973). That 3-MT levels indeed reflect impulse-dependent DA release, was confirmed by the pronounced and long lasting (in contrast with DOPAC levels) decrease of the 3-MT concentration observed after GHB treatment (Kehr, 1976; Westerink, 1979).

At the molecular level, Morgenroth *et al.* (1976) demonstrated an increased affinity of both the pteridine cofactor and substrate for tyrosine hydroxylase, while the end-product inhibition was reduced after GHB treatment. Zivkovic *et al.* (1975a) failed to reproduce these findings.

Other types of drugs, which showed, at lower doses, biochemical and electrophysiological (Olpe *et al.*, 1977) similarities to GHB and GBL, are 1-hydroxy-3-amino-pyrrolidone-2 (HA-966; Hillen and Noach, 1971; Van Zwieten-Boot and Noach, 1975) and β-chlorophenyl-GABA (Baclofen; Da Prada and Keller, 1976; Andén and Wachtel, 1977).

ANAESTHETICS

Various authors have described an increased DA turnover (measured with different methods) after anaesthetic drugs like chloral hydrate, ether, etoxadrol, halothane, ketamine and pentobarbital (Roth *et al.*, 1973; Sung *et al.*, 1973; Andén *et al.*, 1974; Lindqvist *et al.*, 1974; Westerink *et al.*, 1977). Mechanisms underlying the interaction of anaesthetics with DA turnover have not been proposed in the literature. An increased firing of dopaminergic neurones has been recorded during chloral hydrate or halothane anaesthesia (Roth *et al.*, 1973).

TABLE III Effect of various drugs on DA biosynthesis and metabolism in brain.

Drug[a]	Effect	Reference
Amino-oxyacetic acid	Decreased DA utilization (AMPT-method)	Andén (1974)
Arginine-vasopressine (intraventricular)	Increased DA utilization (AMPT-method) in various brain areas	Tanaka *et al.* (1977)
Cholera enterotoxin (intranigral)	Increased DOPAC and HVA levels	Quenzer *et al.* (1977)
Contraceptive steroids (oral treatment)	Increased striatal DA turnover (AMPT-method; labelled precursor method)	Algeri *et al.* (1976)
—	HVA maximal in oestrous phase	Jori *et al.* (1976)
β-Endorphin (intraventricular)	Increased HVA levels	Van Loon and Kim (1977)
Ethanol	Increased DOPAC and HVA levels	Karoum *et al.* (1976)
	Increased DA biosynthesis (labelled precursor method)	Carlsson *et al.* (1973)
Histamine (intraventricular)	Increased HVA levels abolished by pretreatment with atropine	Nowak and Maślinški (1977)
Hypoxia	Decreased DA turnover (DOPA-accumulation method and AMPT-method)	Brown *et al.* (1975)
α-Methyldopa	Depletion DA; decreased DOPAC and HVA levels	Uretsky (1974)
α-Methyl-*m*-tyrosine	Depletion DA; increased DOPAC and HVA levels followed by decreased levels	Uretsky *et al.* (1975)
Muscimol	Initial brief fall followed by an increase in DOPAC and HVA levels, without changes in DA concentrations	Biggio *et al.* (1977)
Prolactin in hypophysectomized rats	Increased DA utilization (AMPT-method) in the nucleus accumbens	Fuxe *et al.* (1977)
Substance P (intranigral)	Increased striatal HVA levels	James and Starr (1977)
Tetrahydropapaveroline	Initial increased HVA levels followed by a decrease	Livrea *et al.* (1976)

[a] Drugs were administered systemically if not otherwise indicated.

ANTIDEPRESSANTS

The tricyclic antidepressants amitriptyline, chlorimipramine, doxepine and nortriptyline were found (in relatively high dosages) to induce a moderate increase in DOPAC and HVA levels in various brain areas (Westerink *et al.*, 1977). The structural relationship between the antidepressants and the tricyclic neuroleptics might explain these effects on DA metabolism, as these drugs display weak inhibitory actions in DA-sensitive adenylate cyclase system (Karobath *et al.*, 1975). A similar rise in DOPAC and HVA levels was reported for trazodone, a drug showing antidepressant potency (Stefanini *et al.*, 1976). A moderate increase of DA turnover—as measured with the labelled precursor method—was observed after administration of the tetracyclic antidepressant compound ORG GB 94 (Kafoe and Leonard, 1973).

ANTIEPILEPTICS

The antiepileptics dipropylacetate and ethosuximide were found to increase DOPAC and HVA levels (Westerink *et al.*, 1977). Increased levels of HVA were also observed by Horton *et al.* (1977) after dipropylacetate treatment. Estimation of the DA turnover by means of the AMPT method however, did not indicate any influence of dipropylacetate on the DA utilization. Combined treatment with probenecid suggested that the observed rise in HVA levels after dipropylacetate may result from efflux inhibition of the acidic metabolite from the brain (Horton *et al.*, 1977).

OTHER DRUGS

A heterogeneous group of compounds, which has been reported to interact with DA metabolism in the brain, is summarized in Table III.

THE SIGNIFICANCE OF REGIONAL DOPAMINE METABOLISM

The Nerve Terminal Areas

It is still an open question whether the neuroleptic-induced increase in DA turnover in brain is related to the antipsychotic effects, to the extra-pyramidal effects or to both. As neuroleptics are known to increase DA turnover in various nerve terminal areas of the brain (see Table II), different hypotheses have been concerned with the regional aspects of neuroleptic action in relation to clinically beneficial actions and side-effects.

Andén (1972) found that trihexyphenidyl, one of the anticholinergics often combined with neuroleptics to treat extrapyramidal disorders, selectively suppressed the neuroleptic-induced HVA increase in the striatum as compared to the pooled limbic structures of rabbit brain. In addition it was found that neuroleptics devoid of extrapyramidal side-effects like clozapine, sulpiride and thioridazine (the atypical or non-cataleptogenic neuroleptics), increase DA turnover—as measured with various methods—in the mesolimbic structures (Andén and Stock, 1973a; Bowers and Rozitis, 1974; Zivkovic *et al.*, 1975b; Bartholini, 1976; Crow *et al.*, 1976; Waldmeier and Maître, 1976a; Westerink *et al.*, 1977) and mesocortical (Scatton *et al.*, 1975; Scatton *et al.*, 1976; Westerink and Korf, 1976d; Laduron *et al.*, 1977; Scatton *et al.*, 1977b) somewhat more than in the striatum. Various authors therefore have suggested that the extrapyramidal and antipsychotic actions of neuroleptics might be due to blockade of dopaminergic neurotransmission in the striatum and mesolimbic-mesocortical system respectively.

Westerink and Korf (1975) repeated the study of Andén (1972) and reported opposite findings: a significant suppression was found of haloperidol-induced HVA increase by the anticholinergics atropine and trihexyphenidyl in the mesolimbic structures comprising the nucleus accumbens and tuberculum olfactorium of the rat, while no interaction was observed in the striatum. Similar results for pooled mesolimbic structures of the rat were observed by Bartholini *et al.* (1976). It should be noted that no great value can be attached to these rather small changes in HVA levels because the amount of haloperidol used approximates (in $mg\,kg^{-1}$) the clinical dose, while the dosages of the anticholinergics exceed the clinical dose 1000-fold. Westerink *et al.* (1977) confirmed that the atypical neuroleptics clozapine, sulpiride and thioridazine differ from the classical neuroleptics by their ability to produce a relatively large increase in DOPAC and HVA levels in the mesolimbic structures nucleus accumbens and tuberculum olfactorium. It appeared, however, that a heterogeneous group of 14 non-neuroleptic drugs investigated (analgesics, anaesthetics, antidepressants, antiepileptics and cholinomimetics) all produced a relatively large increase in DOPAC and HVA levels in the two limbic regions studied. It was concluded that, although there is a clear distinction between the pattern of regional DA metabolism for clozapine, sulpiride and thioridazine on the one hand and the classical neuroleptics on the other, such an action is not specific and therefore not necessarily related to the antipsychotic actions of these neuroleptics. In this respect it is of interest to note that the inhibition of the DA-sensitive adenylate cyclase by atypical neuroleptics, showed no regional preference in forebrain dopaminergic structures (Bockaert *et al.*, 1977; Scatton *et al.*, 1977b).

Dopamine metabolism in mesolimbic regions, in contrast to striatal tissue, seems to respond more to atypical neuroleptics and to non-neuroleptics than to classical neuroleptics. Similar suggestions were made by Waldmeier and Maître (1976a). According to this conclusion the findings of Fuxe and co-workers, showing a specific response in mesolimbic structures for drugs like prolactin, propranolol and combined treatment with neuroleptics and "gabaminergic" compounds (Fuxe *et al.*, 1975, 1976, 1977), may belong to this general, and therefore probably non-specific, response of mesolimbic dopaminergic structures.

A possible explanation of the earlier rise in DA metabolites in mesolimbic areas can be found in the more pronounced responsiveness of certain DA-pools in these areas in comparison with the striatum (Westerink, 1978). It should be noted that that atypical neuroleptics are administered in relatively high dosages, which increases the possibility of complex interactions with different DA-pools or transport mechanisms of DA catabolites out of the brain. In this respect it is of interest to note that probenecid blocks the efflux of HVA and DOPAC much more efficiently in the nucleus accumbens and tuberculum olfactorium than in the striatum (Westerink and Korf, 1975, 1976b). It can therefore be concluded that a percentage change of DOPAC and/or HVA levels in a given brain structure after neuroleptic treatment, cannot be considered directly as an index of DA-receptor blockade.

Crow *et al.* (1976) interpreted the larger rise of HVA levels in the nucleus accumbens observed after thioridazine treatment to be a result of a more effective blockade of DA-receptors in this brain region. The suggestion that DA-receptor blockade in the nucleus accumbens—a structure innervated by some hundred dopaminergic neurones—might be responsible for the clinical efficacy of neuroleptic drugs (Stevens, 1973), may lead to an oversimplification of the biological mechanisms underlying antischizophrenic pharmacotherapy. The relation between schizophrenia and DA-receptor blockade is further complicated by the clinical and pharmacological properties of metoclopramide and propranolol (see above). Metoclopramide is known to provoke certain types of extrapyramidal side-effects in man, to induce catalepsy and increase DA turnover in rats but it is devoid of antipsychotic actions (Peringer *et al.*, 1976). It is interesting to note that metoclopramide is similar to the classical neuroleptics in displaying an exceptional pattern of regional DA metabolism (Westerink *et al.*, 1977). In addition the influence of a high dose of propranolol, recently found to have antipsychotic potency (Yorkston *et al.*, 1977), on DA metabolism was reported (Westerink *et al.*, 1977). No statistically significant effect was found in any of the dopaminergic regions studied (Westerink *et al.*, 1977), although Wiesel (1976) recently reported a slight increase of HVA levels in the tuberculum olfactorium. The data on metoclopramide and propranolol suggest that the increased DA turnover observed in the striatum as well as in the limbic areas after neuroleptic treatment, is more likely to be correlated with the extrapyramidal side-effects of these drugs. Stawartz *et al.* (1975) studied various neuroleptics and reached similar conclusions based on the ED100% for HVA increase in the striatum and limbic forebrain and the occurrence of extrapyramidal side effects in man.

With respect to clozapine a fundamental difference between this drug and classical neuroleptics was recently reported by Gianutsos and Moore (1977). These authors observed elevated DA levels in brains of mice which occurred at lower dosages (5–20 $mg\,kg^{-1}$; while 40 $mg\,kg^{-1}$ had no effect), than those required to rise DOPAC and HVA levels. Gianutsos and Moore speculated that clozapine, at lower dosages, may resemble GHB-like compounds in reducing the release of DA. Consistent with this hypothesis are the findings that low dosages of clozapine accelerate DA synthesis (labelled precursor method) without an increase in the disappearance of labelled DA (Hyttel, 1976). In addition, decreased HVA levels were found after a dose of 2.5 $mg\,kg^{-1}$ clozapine (Westerink, unpublished observations).

The Retina

In addition to the DA-containing nigrostriatal and mesolimbic pathways, dopaminergic processes have been described in the retina. Westerink and Korf (1976e) compared the effects of drugs on HVA formation in the rat retina and striatum. Neuroleptic agents such as clozapine, *cis*-flupenthixol and haloperidol induced comparable HVA rises in the retina and striatum. Apomorphine and LSD decreased HVA levels in both retina and striatum (Da Prada *et al.*, 1975; Westerink and Korf, 1976e). Probenecid caused a similar percentage increase of HVA in both structures studied. Morphine and oxotremorine induced a rise in HVA levels in the striatum but not in the retinal samples. These results suggest that a comparison of DA metabolism in the striatum and retina, may differentiate between drugs whose action is dependent on or independent of the connections of dopaminergic neurones with other neuronal systems.

The Substantia Nigra

Several lines of evidence indicate functional DA release from dendrites and/or cellbodies of dopaminergic neurones in the substantia nigra (Björklund and Lindvall, 1975; Korf *et al.*, 1976a; Geffen *et al.*, 1976; Nieoullon *et al.*, 1977). The possible existence of dendro-dendritic dopaminergic transmission in the substantia nigra is of significance for a better understanding of feedback mechanisms and of the coordination between dopaminergic neurones.

A detailed analysis of the effect of various drugs on DA metabolism in both the substantia nigra (including the ventral tegmental area) and striatum, was reported by Westerink and Korf (1976c). Chloral hydrate, morphine, probenecid and oxotremorine induced a similar percentage increase in DOPAC and HVA levels in the substantia nigra as well as in the striatum. Haloperidol, however, caused a small percentage change in the catabolite levels in the substantia nigra when compared to the pronounced rise seen in the striatum. The latter finding was confirmed by Kebabian *et al.* (1977) for chlorpromazine; in contrast with these data Fadda *et al.* (1977) reported a pronounced percentage increase in DOPAC levels in both striatum and substantia nigra. Results obtained with apomorphine were also controversial, as Westerink and Korf (1976c) reported decreased HVA levels in both structures, while Fadda *et al.* (1977) did not notice any change in DOPAC levels in the substantia nigra. In addition, pargyline-induced 3-MT levels measured in the substantia nigra did not decrease significantly after apomorphine treatment (Westerink, 1979).

Evidence for DA release by amphetamine in the substantia nigra, as measured by 3-MT accumulation, could only be obtained using a relatively high dose of (+)-amphetamine.SO_4($10\,mg\,kg^{-1}$). The findings on the different effects of amphetamine and haloperidol on the 3-MT accumulation in the substantia nigra in comparison with the striatum are supported by the small percentage 3-MT increase seen in the substantia nigra after the combined treatment of the two drugs, in contrast with the strong additive increase observed in the striatum (Kehr *et al.*, 1977; Westerink, 1979).

The presence of functional DA release from the substantia nigra is also questioned by the absence of decreased 3-MT formation in this area after GHB treatment (Westerink, 1979). Other evidence that GHB treatment does not influence DA metabolism in the substantia nigra was provided by the reported absence of an increase in DA levels (Hefti *et al.*, 1976; Pericic and Walters, 1976), which is characteristic for terminal areas (Gessa *et al.*, 1966; Roth and Suhr, 1970).

In conclusion, data on the effect of drugs on regional DA metabolism, form a complex picture and reveal important differences between the regulation of DA metabolism in the substantia nigra and terminal areas.

The Median Eminence

Haloperidol lacks an acute effect on DA turnover (AMPT method) in the median eminence (Fuxe *et al.*, 1975; Gudelsky and Moore, 1976). The rate of DA turnover, however, was increased in the median eminence 16 and

24 h after the administration of haloperidol, thioridazine or clozapine, an effect which was blocked by hypophysectomy (Gudelsky and Moore, 1977). These results suggest that the action of neuroleptics on tuberoinfundibular dopaminergic neurones may be mediated by a "sluggish hormonal–neuronal feedback process".

The Superior Cervical Ganglion

Dopamine is present in a special type of small intensely fluorescent (SIF) cell (Björklund *et al.*, 1970) of the superior cervical ganglion of the rat. Dopamine metabolism in this structure of the rat was recently evaluated by Karoum *et al.* (1977). The major metabolite of DA, appeared to be DOPAC, and its rate of formation (estimated after MAO-inhibition) has been found to be about 15 times faster than the rate of formation of DA found for the striatum. Treatment with carbachol alone or in combination with atropine or hexamethonium revealed that DA metabolism in the superior cervival ganglion is modulated by nicotinic and muscarinic mechanisms. The ganglion has a probenecid-sensitive transport system for DOPAC and HVA.

THE EFFECTS OF CHRONIC DRUG TREATMENTS

Neuroleptics

The strong rise in striatal DA metabolism and biosynthesis observed after a single injection of neuroleptics is markedly reduced when laboratory animals are chronically treated with these drugs (Asper *et al.*, 1973; Bürki *et al.*, 1974; Sayers *et al.*, 1975; Hyttel, 1977). In regional studies on HVA levels this tolerance phenomena was found not to occur in mesolimbic (Bowers and Rozitis, 1974) or mesocortical dopaminergic structures (Laduron *et al.*, 1977). Scatton *et al.* (1976) observed a decreased biosynthesis of striatal DA (as measured with an in-vivo labelled precursor method) 24 h after the last injection of the chronic treatment with various neuroleptics, while DA synthesis remained unchanged in mesocortical and mesolimbic tissue. Similar regional differences were observed during the time-course of a single injection of the palmitinic ester of pipotiazine (Scatton *et al.*, 1977a). Other studies however, described tolerance to the increase in HVA and DOPAC levels caused by various neuroleptics (including sulpiride) in striatal as well as in mesolimbic and mesocortical tissue (Waldmeier and Maître, 1976b; Bowers and Rozitis, 1976; Peringer *et al.*, 1976; Scatton, 1977). That clozapine interacts with dopaminergic neurotransmission in a different way from classical neuroleptics (see above) was confirmed by the absence of tolerance to increased DA metabolism in striatal as well as mesolimbic tissue (Bürki *et al.*, 1974; Bürki *et al.*, 1975; Von Stralendorff *et al.*, 1976), although Waldmeier and Maître (1976b) reported a reduction of the initial DA turnover increase by repeated treatment with a high dose (100 mg kg^{-1}) of clozapine.

It should be noted that no tolerance to the effects of neuroleptics on HVA levels was observed in the retina (Scatton *et al.*, 1977c), which indicates that the neuronal organization plays an important role in the manifestation of this phenomenon.

Several authors have speculated about the mesolimbic and/or frontal cortex as the preferential target for neuroleptics in the treatment of schizophrenia (Stevens, 1973; Bowers and Rozitis, 1974: Scatton *et al.*, 1976; Laduron *et al.*, 1977). The earlier and/or more pronounced tolerance to increased DA metabolism developed in the striatum has been related to clinical observations that certain extrapyramidal side-effects show a tendency to diminish during chronic treatment with neuroleptics (Ayd, 1971). It should however, be emphasized that the clinical literature, in this respect, is far from clear (Mattke *et al.*, 1975; Korczyn and Goldberg, 1976). Care should be taken when attempts are made to relate subtle regional and time-dependent differences in neuroleptic-induced DOPAC and HVA levels in the brains of laboratory animals, to the complex behaviour changes which these drugs cause in patients.

Other Drugs

Rats chronically treated with morphine show tolerance to the increase in striatal DA biosynthesis (AMPT-method; Puri and Lal, 1974).

Several anorectic drugs like amphetamine and mazindol administered chronically show tolerance to the normal HVA increase (Jori and Dolfini, 1977). The effect of fenfluramine however, was not reduced by chronic treatment.

CONCLUDING REMARKS

Although it is realized that this chapter is not an exhaustive review, some general conclusions may be drawn. The various data on the interaction of drugs with DA turnover, suggests that most centrally acting drugs interfere with DA biosynthesis and metabolism. Table III clearly illustrates the wide variety of compounds which interfere with DA turnover in the brain.

It should be noted however, that the most pronounced changes in DA turnover are caused by DA-agonists and antagonists (the neuroleptics). The mechanisms of action of most of the other drugs may be non-specific, as is illustrated by the fact that a fall in body temperature may result in an increase in the concentration of HVA in the brain (Hutchins *et al.*, 1975). The relative ease by which a change in DA biosynthesis or metabolism is manifested by a shift in the levels of DOPAC and HVA, could explain why it is in fact rather difficult to find a drug which does not interact with DA turnover in the brain. It may be suggested that other neuronal systems, for which a change in neurotransmitter turnover is less easily detectable, may also show a broad response to psychotropic drug treatment.

REFERENCES

Aghajanian, G. K. and Bunney, B. S. (1974). *In* "Frontiers in Neurology and Neuroscience Research" (Eds. P. Seeman and G. M. Brown), 4–11. Neuroscience Institute, University of Toronto.
Agid, Y., Javoy, F. and Youdim, M. B. H. (1973). *Br. J. Pharmacol.* **48,** 179–181.
Ahtee, L. (1974). *Eur. J. Pharmacol.* **27,** 221–230.
Ahtee, L. and Buncombe, G. (1974). *Acta Pharmacol. Toxicol.* **35,** 429–432.
Algeri, S., Ponzio, F., Dolfini, E. and Jori, A. (1976). *Neuroendocrinology* **22,** 343–351.
Andén, N.-E. (1972). *J. Pharm. Pharmacol.* **24,** 905–906.
Andén, N.-E. (1974). *Naun. Schmied. Arch. Pharmacol.* **283,** 419–424.
Andén, N.-E. and Bédard, P. (1971). *J. Pharm. Pharmacol.* **23,** 460–462.
Andén, N.-E., Butcher, S. G., Corrodi, H., Fuxe, K. and Ungerstedt, U. (1970). *Eur. J. Pharmacol.* **11,** 303–314.
Andén, N.-E., Corrodi, H., Fuxe, K. and Ungerstedt, U. (1971). *Eur. J. Pharmacol.* **15,** 193–199.
Andén, N.-E., Dahlstrom, A. and Fuxe, K. (1966). *Acta Physiol. Scand.* **68,** 419–420.
Andén, N.-E., Magnusson, T. and Stock, G. (1974). *Naun. Schmied. Arch. Pharmacol.* **283,** 409–418.
Andén, N.-E., Roos, B.-E. and Werdinius, B. (1964). *Life Sci.* **3,** 149–158.
Andén, N.-E., Rubenson, A., Fuxe, K. and Hökfelt, T. (1967). *J. Pharm. Pharmacol.* **19,** 627–629.
Andén, N.-E. and Stock, G. (1973a). *J. Pharm. Pharmacol.* **25,** 346–348.
Andén, N.-E. and Stock, G. (1973b). *Naun. Schmied. Arch. Pharmacol.* **279,** 89–92.
Andén, N.-E. and Stromborn, U. (1974). *Psychopharmacologia* **38,** 91–103.
Andén, N.-E. and Wachtel, H. (1977). *Acta Pharmacol. Toxicol.* **40,** 310–320.
Asper, H., Baggliolini, M., Burki, H. R., Lauener, H. and Stille, G. (1973). *Eur. J. Pharmacol.* **22,** 287–294.
Ayd, F. J. (1971). *Int. Drug Ther. Newslett.* **6,** 33–36.
Bartholini, G. (1976). *J. Pharm. Pharmacol.* **28,** 429–433.
Bartholini, G., Haefely, W., Jalfre, M., Keller, H. H. and Pletscher, A. (1972). *Br. J. Pharmacol.* **46,** 736–740.
Bartholini, G., Keller, H. H. and Pletscher, A. (1975). *J. Pharm. Pharmacol.* **27,** 439–442.
Baldessarini, R. J., Lipinski, J. F. and Chace, K. V. (1972). *Biochem. Pharmacol.* **21,** 77–87.

Bartholini, G. and Pletscher, A. (1969). *Experientia* **25,** 919–920.
Bartholini, G. and Pletscher, A. (1971). *Experientia* **27,** 1302–1303.
Besson, M. J., Chéramy, A. and Glowinski, J. (1971). *J. Pharm. Expl. Ther.* **177,** 196–205.
Bhatnagar, S. P. (1973). *Can. J. Physiol. Pharmacol.* **51,** 893–899.
Biggio, G., Casu, M., Corda, M. G., Vernaleone, F. and Gessa, G. L. (1977). *Life Sci.* **21,** 525–531.
Björklund, A., Cegrell, L., Falck, B., Ritzen, M. and Rosengren, E. (1970). *Acta Physiol. Scand.* **73,** 334–338.
Björklund, A. and Lindvall, O. (1975). *Brain. Res.* **83,** 531–537.
Bockaert, J., Tassin, J. P., Thierry, A. M., Glowinski, J. and Premont, J. (1977). *Brain. Res.* **122,** 71–86.
Bowers, M. B. and Rozitis, A. (1974). *J. Pharm. Pharmacol.* **26,** 743–745.
Bowers, M. B. and Rozitis, A. (1976). *Eur. J. Pharmacol.* **39,** 109–115.
Braestrup, C. (1977). *J. Pharm. Pharmacol.* **29,** 463–470.
Broch, O. J. and Fonnum, F. (1972). *J. Neurochem.* **19,** 2049–2055.
Brodie, B. B., Costa, E., Dlabac, A., Neff, N. H. and Smookler, H. H. (1966). *J. Pharmacol. Expl. Ther.* **154,** 493–498.
Brown, R. M., Kehr, W. and Carlsson, A. (1975). *Brain Res.* **85,** 491–509.
Bunney, B. S., Walters, J. R., Roth, R. H. and Aghajanian, G. K. (1973). J. *Pharmacol. Expl. Ther.* **185,** 560–571.
Burkard, W. P., Grey, K. F. and Pletscher, A. (1967). *Nature* (*London*) **213,** 732–733.
Bürki, H. R., Ruch, W., Asper, H., Baggiolini, M. and Stille, G. (1974). *Eur. J. Pharmacol.* **27,** 180–190.
Bürki, H. R., Ruch, W. and Asper, H. (1975). *Psychopharmacologia* **41,** 27–33.
Cannon, J. G., Lee, T. and Goldman, H. D. (1977). *J. Med. Chem.* **20,** 1111–1116.
Carenzi, A., Guidotti, A., Revuelta, A. and Costa, E. (1975). *J. Pharmacol. Expl. Ther.* **194,** 311–318.
Carlsson, A. (1975). *In* "Pre- and Postsynaptic Receptors" (Eds. E. Usdin and W. E. Bunney), 49–65. Marcel Dekker, New York.
Carlsson, A., Davis, J. N., Kehr, W., Lindqvist, M. and Atack, C. V. (1972). *Naun. Schmied. Arch. Pharmacol.* **275,** 153–168.
Carlsson, A., Engel, J., Strömbom, U., Svenson, T. H. and Waldeck, B. (1974). *Naun. Schmied. Arch. Pharmacol.* **283,** 117–128.
Carlsson, A., Kehr, W. and Lindqvist, M. (1977). *J. Neur. Trans.* **40,** 99–113.
Carlsson, A. and Lindqvist, M. (1963). *Acta Pharmacol. Toxicol.* **20,** 140–144.
Carlsson, A., Magnusson, T., Svensson, T. H. and Waldeck, B. (1973). *Psychopharmacologia* **30,** 27–36.
Clouet, D. H. and Ratner, M. (1970). *Science* **168,** 854–856.
Corrodi, H., Farnebo, L.-O., Fuxe, K., Hamberger, B. and Ungerstedt, U. (1972b). *Eur. J. Pharmacol.* **20,** 195–204.
Corrodi, H., Fuxe, K., Hammer, W., Sjöquist, F. and Ungerstedt, U. (1967). *Life Sci.* **6,** 2557–2566.
Corrodi, H., Fuxe, K. and Lidbrink, P. (1972a). *Brain Res.* **43,** 397–416.
Corrodi, H., Fuxe, K., Hokfelt, T., Lidbrink, P. and Ungerstedt, U. (1973). *J. Pharm. Pharmacol.* **25,** 409–412.
Costa, E., Cheney, D. L., Racagni, G. and Zsilla, G. (1975). *Life Sci.* **17,** 1–8.
Costa, E., Guidotti, A. and Zivkovic, B. (1974). *In* "Neuropsychopharmacology of Monoamines and their Regulatory Enzymes" (Ed. E. Usdin), 161–175. Raven Press, New York.

Costa, E. and Neff, N. H. (1966). *In* "Biochemistry and Pharmacology of the Basal Ganglia" (Eds. E. Costa, L. Côté and M. D. Yahr), 141. Raven Press, New York.
Costall, B. and Naylor, R. J. (1974). *Naun. Schmied. Arch. Pharmacol.* **285,** 71–81.
Cox, B. (1975). *Proc. West. Pharmacol. Soc.* **18,** 162–165.
Crow, T. J., Deakin, J. F. W., Johnstone, E. C. and Longden, A. (1976). *Lancet* **ii,** 563–566.
Dailey, J., Sedvall, G. and Sjöqvist, B. (1972). *J. Pharm. Pharmacol.* **24,** 580–581.
Da Prada, M. and Keller, H. H. (1976). *Life Sci.* **19,** 1253–1264.
Da Prada, M. and Pletscher, A. (1966). *Experientia* **22,** 465–466.
Da Prada, M., Sane, A., Burkard, W. P., Bartholini, G. and Pletscher, A. (1975). *Brain Res.* **95,** 67–73.
Di Chiara, G., Porceddu, M. L., Vargin, L., Argiolas, A. and Gessa, G. L. (1976). *Nature (London)* **264,** 564–567.
Di Chiara, G., Porceddu, M. L., Fratta, W. and Gessa, G. L. (1977). *Nature (London)* **267,** 270–272.
Doteuchi, M., Wang, C. and Costa, E. (1974). *Molec. Pharmacol.* **10,** 225–234.
Duncan, R. J. S. and Sourkes, T. L. (1974). *J. Neurochem.* **22,** 663–669.
Elchisak, M. A., Maas, J. W. and Roth, R. H. (1977). *Eur. J. Pharmacol.* **41,** 369–378.
Elliot, P. N. C., Jenner, P., Huizing, G., Marsden, C. D. and Miller, R. (1977). *Neuropharmacology* **16,** 333–342.
Fadda, F., Argiolas, A., Steffanini, E. and Gessa, G. L. (1977). *Life Sci.* **21,** 411–418.
Fuxe, K., Corrodi, H., Hökfelt, T., Lidbrink, P. and Ungerstedt, U. (1974). *Med. Biol.* **52,** 121–132.
Fuxe, K., Hökfelt, T., Ljungdahl, A., Agnati, L., Johansson, O. and Perez de la Mora, M. (1975). *Med. Biol.* **53,** 177–183.
Fuxe, K., Bolme, P., Agnati, L. and Everitt, B. J. (1976). *Neurosci. Lett.* **3,** 45–52.
Fuxe, K., Eneroth, P., Gustafson, J.-A., Löfström, A. and Skett, P. (1977). *Brain Res.* **122,** 177–182.
Gauchy, C., Agid, Y., Glowinski, J. and Cheramy, A. (1973). *Eur. J. Pharmacol.* **22,** 311–319.
Geffen, L. B., Jessel, T. M., Cuello, A. C. and Iversen, L. L. (1976). *Nature (London)* **260,** 258–260.
Gessa, G. L., Crabai, F., Vargiu, L. and Spano, P. F. (1968). *J. Neurochem.* **15,** 377–381.
Gessa, G. L. and Tagliamonte, A. (1975). *Neuropharmacology* **14,** 913–920.
Gessa, G. L., Vargiu, L., Crabai, F., Boezo, G. F., Caboni, F. and Camba, R. (1966). *Life Sci.* **5,** 1921–1930.
Gianutsos, G. and Moore, K. E. (1977). *Res. Commun. Chem. Pathol. Pharmacol.* **17,** 29–39.
Gordon, E. K., Markey, S. P., Sherman, R. L. and Kopin, I. J. (1976). *Life Sci.* **18,** 1285–1292.
Green, A. L. (1973). *J. Pharm. Pharmacol.* **25,** 267–269.
Gropetti, A., Algeri, S., Cattabeni, F., Di Giulio, A. M., Galli, C. L., Ponzio, F. and Spano, P. F. (1977). *J. Neurochem.* **28,** 193–197.
Groves, P. M., Wilson, C. J., Young, S. J. and Rebec, G. Y. (1975). *Science* **190,** 522–529.
Gudelsky, G. A. and Moore, K. E. (1976). *J. Neur. Trans.* **38,** 95–105.
Gudelsky, G. A. and Moore, K. E. (1977). *J. Pharm. Expl. Ther.* **202,** 149–156.
Guldberg, H. C. and Broch, O. J. (1971). *Eur. J. Pharmacol.* **13,** 155–167.

Guldberg, H. C. and Marsden, C. A. (1975). *Pharmacol. Rev.* **27,** 137–204.
Guldberg, H. C. and Yates, C. M. (1968). *Br. J. Pharmacol.* **34,** 220.
Gumulka, W., Kostowski, W. and Czlonkowski, A. (1973). *Pharmacology* **10,** 363–372.
Gunne, L.-M., Jonsson, L. and Fuxe, K. (1969). *Eur. J. Pharmacol.* **5,** 338–342.
Haubrich, D. R. and Goldberg, M. E. (1975). *Neuropharmacology* **14,** 211–214.
Hefti, F., Lienhart, R. and Lichtensteiger, W. (1976). *Nature (London)* **263,** 341–342.
Hillen, F. C. and Noach, E. L. (1971). *Eur. J. Pharmacol.* **16,** 222–224.
Holta, P. (1959). *Pharmacol. Rev.* **11,** 317–329.
Horton, R. W., Anlezark, G. M., Sawaya, M. C. B. and Meldrum, B. S. (1977). *Eur. J. Pharmacol.* **41,** 387–397.
Hutchins, D. A., Pearson, J. D. M. and Sharman, D. F. (1975). *J. Neurochem.* **24,** 1151–1154.
Hyttel, J. (1974). *J. Pharm. Pharmacol.* **26,** 588–596.
Hyttel, J. (1976). *Acta Pharmacol. Toxicol.* **38,** 358–365.
Hyttel, J. (1977). *J. Neurochem.* **28,** 227–228.
Iwatsubo, K. and Clouet, D. H. (1977). *J. Pharmacol. Expl. Ther.* **202,** 429–436.
James, T. A. and Starr, M. S. (1977). *J. Pharm. Pharmacol.* **29,** 181–182.
Javoy, F., Agid, Y., Bouvet, D. and Glowinski, J. (1974). *J. Pharm. Pharmacol.* **26,** 179–185.
Javoy, F., Agid, Y. and Glowinski, J. (1975). *J. Pharm. Pharmacol.* **27,** 707–710.
Javoy, F. and Glowinski, J. (1971). *J. Neurochem.* **18,** 1305–1311.
Javoy, F., Hamon, M. and Glowinski, J. (1970). *Eur. J. Pharmacol.* **10,** 178–188.
Javoy, F., Youdim, M. B. H., Agid, Y. and Glowinski, J. (1973). *J. Neur. Trans.* **34,** 279–289.
Jori, A. and Bernardi, D. (1969). *J. Pharm. Pharmacol.* **21,** 694–697.
Jori, A., Cecchetti, G., Dolfini, E., Monti, E. and Garattini, S. (1974). *Eur. J. Pharmacol.* **27,** 245–248.
Jori, A., Colturani, F., Dolfini, E. and Rutczynski, M. (1976). *Neuroendocrinology* **21,** 262–266.
Jori, A. and Dolfini, E. (1974). *Pharmacol. Res. Comm.* **6,** 175–178.
Jori, A. and Dolfini, E. (1977). *Eur. J. Pharmacol.* **41,** 443–445.
Juorio, A. V., Sharman, D. F. and Trajkov, T. (1966). *Br. J. Pharmacol.* **26,** 385–392.
Kääriäinen, I. and Ahtee, L. (1976). *Med. Biol.* **54,** 56–61.
Kebabian, J. W., Saavedra, J. M. and Axelrod, J. (1977). *J. Neurochem.* **28,** 795–801.
Kafoe, W. F. and Leonard, B. E. (1973). *Arch. Int. Pharmacodyn. Ther.* **206,** 389–391.
Karobath, M. E. (1975). *Eur. J. Pharmacol.* **30,** 159–163.
Karoum, F., Neff, N. H. and Wyatt, R. J. (1977). *Eur. J. Pharmacol.* **44,** 311–318.
Karoum, F., Wyatt, R. J. and Majchrowicz, E. (1976). *Br. J. Pharmacol.* **56,** 403–411.
Kehr, W., Carlsson, A., Lindqvist, M., Magnusson, T. and Atack, C. (1972). *J. Pharm. Pharmacol.* **24,** 744–747.
Kehr, W. (1976). *Naun. Schmied. Arch. Pharmacol.* **293,** 209–215.
Kehr, W. (1977). *Eur. J. Pharmacol.* **41,** 261–273.
Kehr, W., Speckenbach, W. and Zimmerman, R. (1977). *J. Neur. Trans.* **40,** 129–147.
Keller, H. H., Schaffner, R. and Haefely (1976). *Naun. Schmied. Arch. Pharmacol.* **294,** 1–7.
Koe, B. K. (1974). *In* "Neuroleptics" (Eds. S. Fielding and H. Lal), 131–172. Futura, New York.

Korczyn, A. D. and Goldberg, G. J. (1976). *J. Neurol. Neurosurg. Psychiat.* **39,** 866–869.
Korf, J., Zieleman, M. and Westerink, B. H. C. (1976a). *Nature* (*London*) **260,** 257–258.
Korf, J., Grasdijk, L. and Westerink, B. H. C. (1976b). *J. Neurochem.* **26,** 579–584.
Kuschinsky, K. and Hornykiewicz, O. (1974). *Eur. J. Pharmacol.* **26,** 41–50.
Kuczenski, R. (1977). *Eur. J. Pharmacol.* **46,** 249–257.
Laduron, P., De Bie, K. and Leysen, J. (1977). *Naun. Schmied. Arch. Pharmacol.* **296,** 183–185.
Lahti, R. A. and Losey, E. G (1974). *Res. Commun. Chem. Pathol. Pharmacol.* **7,** 31–40.
Lahti, R. A., McAllister, B. and Wozniak, J. (1972). *Life Sci.* **11,** 605–613.
Lal, S., Sourkes, T. L., Missala, K. and Belendiuk, G. (1972). *Eur. J. Pharmacol.* **20,** 71–79.
Laverty, R. and Sharman, D. F. (1965). *Br. J. Pharmacol.* **24,** 538–548.
Levitt, M., Spector, S., Sjoerdsma, A. and Udenfriend, S. (1965). *J. Pharmacol. Expl. Ther.* **148,** 1–8.
Lindqvist, M., Kehr, W. and Carlsson, A. (1974). *Naun. Schmied. Arch. Pharmacol.* **284,** 263–277.
Livrea, P., Di Reda, L., Giovino, A. and Bertolino, A. (1976). *Pharmacology* **14,** 20–26.
Lloyd, K. G. and Bartholini, G. (1975). *Experientia* **31,** 560–562.
Martres, M. P., Consentin, J., Bandry, M., Marcais, H., Protais, P. and Schwartz, J. C. (1977). *Brain Res.* **136,** 319–337.
Mattke, D. J., Mombour, W. and Glötzner, F. L. (1975). *Pharmakopsych. Neuro-Psychopharmakologie* **8,** 36–44.
Moleman, R. and Bruinvels, J. (1976). *Life Sci.* **19,** 1277–1282.
Morgenroth, V. H., Walters, J. R. and Roth, R. H. (1976). *Biochem. Pharmacol.* **25,** 655–661.
Neff, N. H., Spano, P. F., Gropetti, A., Wong, C. T., Costa, E. (1971). *J. Pharmacol. Expl. Ther.* **176,** 701–710.
Nieoullon, A., Cheramy, A. and Glowinski, J. (1977). *Nature* (*London*) **269,** 340–342.
Nose, T. and Takemoto, H. (1974). *Eur. J. Pharmacol.* **25,** 51–55.
Nowak, J. Z. and Maślinśki, Cz. (1977). *Agents Actions* **7,** 27–30.
Nybäck, H. (1971). *Acta Pharmacol. Toxicol.* **30,** 372–384.
Nybäck, H. and Sedvall, G. (1968). *J. Pharmacol. Expl. Ther.* **162,** 294–301.
Nybäck, H. and Sedvall, G. (1969). *Eur. J. Pharmacol.* **5,** 245–252.
Nybäck, H. and Sedvall, G. (1970). *Eur. J. Pharmacol.* **10,** 193–205.
Oelszner, W., Funk, K. F., V. Schwarzenfeld, I., Staib, A. H. and Westermann, K. H. (1975). *Acta Biol. Med. Germ.* **34,** 647–653.
O'Keeffe, R., Sharman, D. F. and Vogt, M. (1970). *Br. J. Pharmacol.* **38,** 287–304.
Olpe, H.-R., Koella, W. P., Wolf, P. and Haas, H. L. (1977). *Brain Res.* **134,** 577–580.
Paalzov, G. and Paalzov, L. (1975). *Eur. J. Pharmacol.* **31,** 261–272.
Papeschi, R. (1975). *Psychopharmacology* **45,** 21–28.
Papeschi, R., Theiss, P. and Herz, A. (1975). *Eur. J. Pharmacol.* **34,** 253–261.
Perez-Cruet, J., Gessa, G. L., Tagliamonte, A. and Tagliamonte, P. (1971). *Fed. Proc. Fed. Am. Soc. Exp. Biol.* **30,** 216.
Pericic, D. and Walters, J. R. (1976). *J. Pharm. Pharmacol.* **28,** 527–529.
Peringer, E., Jenner, P., Donaldson, I. M. and Marsden, C. D. (1976). *Neuropharmacology* **15,** 463–469.

Persson, S. A. (1977). *Eur. J. Pharmacol.* **43,** 73–83.
Potter, L. T., Cooper, T., Willman, V. L. and Wolfe, D. E. (1965). *Circulation Res.* **16,** 468–481.
Puri, S. K. and Lal, H. (1974). *Naun. Schmied. Arch. Pharmacol.* **282,** 155–170.
Quenzer, L. F., Galli, C. L. and Neff, N. H. (1977). *Science* **195,** 78–80.
Racagni, G., Bruno, F., Maggi, A., Cattabeni, F. and Gropetti, A. (1977). *In* "Dopamine" Advances in Biochemical Psychopharmacology, Vol. 19 (Eds. P. J. Roberts, G. N. Woodruff and L. L. Iversen), 363–372, Raven Press, New York.
Restelli, A., Lucchini and Glässer, A. (1975). *Pharmacol. Res. Comm.* **7,** 409–419.
Roffler-Tarlov, S., Sharman, D. F. and Tegerdine, P. (1971). *Br. J. Pharmacol.* **42,** 343–351.
Rollema, H., Westerink, B. H. C. and Grol, C. J. (1976). *J. Pharm. Pharmacol.* **28,** 321–323.
Roos, B. E. (1965). *J. Pharm. Pharmacol.* **17,** 820–821.
Roos, B. E. (1969). *J. Pharm. Pharmacol.* **21,** 263–264.
Roth, R. (1971). *Eur. J. Pharmacol.* **15,** 52–59.
Roth, R. H., Murrin, L. C. and Walters, J. R. (1976). *Eur. J. Pharmacol.* **36,** 163–171.
Roth, R. H. and Surh, Y. (1970). *Biochem. Pharmacol.* **19,** 3001–3012.
Roth, R. H., Walters, J. and Aghajanian, G. (1973). *In* "Frontiers in Catecholamine Research" (Eds. Usdin and S. Snyder), 567–574. Pergamon, New York.
Roth, R. H., Walters, J. R., Murrin, L. C. and Morgenroth, V. H., III (1975). *In* "Pre- and Postsynaptic Receptors" (Eds. E. Usdin and W. E. Bunney, Jr.), 5–48, Marcel Dekker, New York.
Sasame, H. A., Perez-Cruet, J., Di Chiara, G., Tagliamonte and Gessa, G. L. (1972). *J. Neurochem.* **19,** 1953–1957.
Saner, A. and Pletscher, A. (1977). *Eur. J. Pharmacol.* **42,** 155–160.
Sayers, A. C., Bürki, H. R., Ruch, W. and Asper, H. (1975). *Psychopharmacology* **41,** 97–104.
Sayers, A. C. and Handley, S. L. (1973). *Eur. J. Pharmacol.* **23,** 47–55.
Scatton, B. (1977). *Eur. J. Pharmacol.* **46,** 363–369.
Scatton, B., Bischoff, S., Dedek, J. and Korf, J. (1977b). *Eur. J. Pharmacol.* **44,** 287–292.
Scatton, B., Boireau, A., Garret, C., Glowinski, J. and Julou, L. (1977a). *Naun. Schmied. Arch. Pharmacol.* **296,** 169–175.
Scatton, B., Dedek, J. and Korf, J. (1977c). *Brain Res.* **135,** 374–377.
Scatton, B., Glowinski, J. and Julou, L. (1976). *Brain Res.* **109,** 184–189.
Scatton, B., Thierry, A. M., Glowinski, J. and Julou, L. (1975). *Brain Res.* **88,** 389–393.
Scheel-Krüger, J. (1971). *Eur. J. Pharmacol.* **14,** 47–59.
Scheel-Krüger, J., Braestrup, C., Nielson, M., Golembiowska, K. and Mogilnicka, E. (1977). *In* "Cocaine and Other Stimulants" (Eds. E. H. Ellinhood and M. M. Killey), 373–407. Plenum Press, New York.
Sharman, D. F. (1966). *Br. J. Pharmacol.* **28,** 153–163.
Sharman, D. F. (1967). *Br. J. Pharmacol.* **30,** 620–626.
Sharman, D. F. (1973). *Br. Med. Bull.* **20,** 110–115.
Shore, P. A. (1962). *In* "Pharmacological Reviews", vol. 14, 531–550..Williams and Wilkins, Baltimore.
Shore, P. A. (1976). *J. Neur. Trans.* **39,** 131–138.
Shore, P. A. and Dorris, R. L. (1975). *Eur. J. Pharmacol.* **30,** 315–318.

Silbergeld, E. K. and Pfeiffer, R. F. (1977). *J. Neurochem.* **28,** 1323–1326.
Smith, C. B., Sheldon, M. I., Bednarczyk, J. H. and Villareal, J. E. (1972). *J. Pharmacol. Expl. Ther.* **180,** 547–557.
Speckenbach, W. and Kehr, W. (1976). *Naun. Schmied. Arch. Pharmacol.* **296,** 25–30.
Spano, P. F. and Neff, H. (1972). *Brain Res.* **42,** 139–145.
Spano, P. F., Tagliamonte, P., Tagliamonte and Gessa, G. L. (1971). *J. Neurochem.* **18,** 1831–1836.
Stanley, M. and Wilk, S. (1977). *Eur. J. Pharmacol.* **44,** 293–302.
Stawarz, R. J., Hill, H., Robinson, S. E., Setler, P., Dingell, J. V. and Sulser, F. (1975). *Psychopharmacology* **43,** 125–130.
Stefanini, E., Fadda, F., Porceddu, M. L. and Gessa, G. L. (1976). *J. Pharm. Pharmacol.* **28,** 925–926.
Stevens, J. R. (1973). *Arch. Gen. Psychiatry* **29,** 177–189.
Stille, G. and Lauener, H. (1974). *Arzneim. Forsch.* **21,** 252–255.
Stock, G., Magnusson, T. and Andén, N.-E. (1973). *Naun. Schmied. Arch. Pharmacol.* **278,** 347–361.
Sugrue, M. F. (1974). *Br. J. Pharmacol.* **52,** 159–165.
Sung, Y.-F., Frederickson, E. L. and Holtzman, G. (1973). *Anesthesiology* **39,** 478–487.
Swahn, G.-G. and Wiesel, F.-A. (1976). *J. Neur. Trans.* **39,** 281–290.
Tagliamonte, A., Demontis, G., Olianas, M., Vargiu, L., Corsini, G. U. and Gessa, G. L. (1975). *J. Neurochem.* **24,** 707–710.
Tanaka, M., De Kloet, E. R., De Wied, D. and Versteeg, D. H. G. (1977). *Life Sci.* **20,** 1799–1808.
Trabucchi, M., Longoni, R., Fresia, P. and Spano, P. F. (1976). *Life Sci.* **17,** 1551–1556.
Uretsky, N. J. (1974). *J. Pharmacol. Expl. Ther.* **189,** 359–369.
Uretsky, N. J., Snodgrass, S. R. and Lorenzo, A. V. (1975). *J. Pharmacol. Expl. Ther.* **195,** 465–479.
Vaatstra, W. J. and Eigeman, L. (1974). *Eur. J. Pharmacol.* **25,** 185–190.
Van Loon, G. R. and Kim, C. (1977). *Res. Comm. Chem. Path. Pharmacol.* **18,** 171–174.
Van Rossum, J. M., Boissier, J. R., Janssen, P. A. J., Juloi, L., Loew, D. M., Nielsen, I. M., Munkvad, I., Randrup, A., Stille, G. and Tedeschi, B. H. (1970). *In* "The Neuroleptics" (Eds. D. P. Bobon, P. A. J. Jansen and J. Bobon), 23–33. Kargev, Basel.
Van Zwieten-Boot, B. J. and Noach, E. L. (1975). *Eur. J. Pharmacol.* **33,** 247–254.
Von Stralendorff, B., Ackenheil, M. and Zimmerman, J. (1976). *Arzneim-Forsch.* **26,** 1096–1098.
Waldmeier, P. C. and Maître, L. (1976a). *J. Neurochem.* **27,** 589–597.
Waldmeier, P. C. and Maître, L. (1976b). *Eur. J. Pharmacol.* **38,** 197–203.
Waldmeier, P. C. and Maître, L. (1977). *Psychopharmacology* **52,** 137–144.
Walters, J. and Roth, R. (1972). *Biochem. Pharmacol.* **21,** 2111–2121.
Walters, J. R., Roth, R. H. and Aghajanian, G. K. (1973). *J. Pharmacol. Expl. Ther.* **186,** 630–639.
Walters, J. and Roth, R. H. (1974). *J. Pharmacol. Expl. Ther.* **191,** 82–91.
Walters, J. and Roth, R. H. (1976). *Naun. Schmied. Arch. Pharmacol.* **296,** 5–14.
Werdinius, B. (1966). *J. Pharm. Pharmacol.* **18,** 546–547.
Werdinius, B. (1967). *Acta Pharmacol. Toxicol.* **25,** 18–23.
Westerink, B. H. C. (1977). Thesis, Groningen University.

Westerink, B. H. C. (1979). *J. Pharm. Pharmacol.* **31,** 94–99.
Westerink, B. H. C. and Korf, J. (1975). *Eur. J. Pharmacol.* **33,** 31–40.
Westerink, B. H. C. and Korf, J. (1976a). *Eur. J. Pharmacol.* **37,** 249–255.
Westerink, B. H. C. and Korf, J. (1976b). *Eur. J. Pharmacol.* **38,** 281–291.
Westerink, B. H. C. and Korf, J. (1976c). *Eur. J. Pharmacol.* **40,** 131–136.
Westerink, B. H. C. and Korf, J. (1976d). *Brain Res.* **113,** 429–434.
Westerink, B. H. C. and Korf, J. (1976e). *Eur. J. Pharmacol.* **40,** 175–178.
Westerink, B. H. C., Lejeune, B., Korf, J. and Van Praag, H. M. (1977). *Eur. J. Pharmacol.* **42,** 179–190.
Westerink, B. H. C., Van der Heyden, J. A. M. and Korf, J. (1978). *Life Sci.* **22,** 749–756.
Wiesel, F.-A. (1976). *Neurosci. Lett.* **2,** 35–38.
Wiesel, F.-A. and Alfredsson, G. (1976). *Eur. J. Pharmacol.* **40,** 263–272.
Wiesel, F.-A., Alfredsson, G., Likwornik, V. and Sedvall, G. (1975). *Life Sci.* **16,** 1145–1156.
Wiesel, F.-A., Fri, G.-G. and Sedvall, G. (1973). *Eur. J. Pharmacol.* **23,** 104–106.
Wiesel, F.-A. and Sedvall, G. (1975). *Eur. J. Pharmacol.* **30,** 364–367.
Wilk, S. and Glick, S. D. (1976). *Eur. J. Pharmacol.* **37,** 203–206.
Wilk, S. and Stanley, M. (1977). *Eur. J. Pharmacol.* **41,** 65–72.
Wilk, S., Watson, E. and Glick, S. D. (1975a). *Eur. J. Pharmacol.* **30,** 117–120.
Wilk, S., Watson, E. and Stanley, M. E. (1975c). *J. Pharmacol. Expl. Ther.* **195,** 265–270.
Wilk, S., Watson, E. and Travis, B. (1975b). *Eur. J. Pharmacol.* **30,** 238–243.
Wilk, S. and Zimmerberg, B. (1973). *Biochem. Pharmacol.* **22,** 623–628.
Worms, P. and Scatton, B. (1977). *Eur. J. Pharmacol.* **45,** 395–396.
Yorkston, N., Zaki, S. A., Pitcher, D. R., Gruzelier, J. H., Hollander, D. and Sergeant, H. G. S. (1977). *Lancetii* 575–578.
Zivkovic, B., Guidotti, A. and Costa, E. (1974). *Molec. Pharmacol.* **10,** 727–735.
Zivkovic, B., Guidotti, A. and Costa, E. (1975a). *Naun. Schmied. Arch. Pharmacol.* **291,** 193–200.
Zivkovic, B., Guidotti, A., Revuelta, A. and Costa, E. (1975b). *J. Pharmacol. Expl. Ther.* **194,** 37–46.

Neuroanatomy of Dopamine Systems

CHAPTER 19

Fluorescence Histochemistry of Dopamine in Mammalian Tissues

Brigitte Berger and Jeanine Nguyen-Legros

INSERM U-134, Laboratoire de Neuropathologie Charles Foix, Hôpital Salpêtrière, Paris, France.

The demonstration of dopamine by histofluorescence began with the introduction of the Falck–Hillarp technique (Falck *et al.*, 1962) which opened a new era in monoamine research. Another very productive period for the histochemical approach to the dopaminergic systems was initiated ten years later with the introduction of a more sensitive fluorescence technique: the vibratome glyoxylic acid method (Lindvall *et al.*, 1973). Since 1973, a great number of modifications of these fluorescence techniques have been proposed, sometimes in combination with other histochemical methods, to allow an easier or faster demonstration of the dopamine-containing cells and processes. Most of these studies are devoted to the nervous system with which this paper will, therefore, be primarily concerned.

Before reviewing the advantages, the disadvantages and respective field of application of these techniques, one fact must be underlined: neither of the most frequently used fluorescence histochemical methods reveals dopamine alone. Its immediate precursor, DOPA, as well as the next primary catecholamine, noradrenaline, are always visualized at the same time. Moreover, serotonin-containing cells or processes are also demonstrated under particular conditions. Therefore, from the beginning of the "histofluorescence era", other techniques have constantly been used in combination with the histochemical ones to distinguish the dopamine-containing neurones: microspectrofluorometric studies, neurochemical macro- or microanalyses, mechanical or neurotoxic destruction of the other aminergic systems and pharmacological manipulations.

THE FALCK–HILLARP TECHNIQUE

Principles and Methodology

Following condensation with formaldehyde, the amines are converted into high intensity fluorophores. This property was originally shown by Eränkö (1955) in adrenal medulla, using aqueous formaldehyde; however, the technique was not sensitive enough to be successfully applied to the nervous system. In attempts to introduce this technique into neurochemistry, Falck (1962), Falck and Owman (1965), found it to be of major importance "to perform all procedures including the histochemical reactions in a dry milieu". They proposed a method which included: rapid freezing of the tissue in liquid propane cooled by liquid nitrogen, freeze-drying for 3–5 days, reaction with hot paraformaldehyde vapours of a controlled humidity, embedding in paraffin, subsequent sectioning, and mounting with immersion oil. Sections were viewed under a fluorescence microscope with filter combinations optimal for fluorescence excitation and emission of the reaction product (see p. 314).

This fluorescence histochemical procedure has been applied with minor modifications in almost all the histochemical studies on catecholamines until 1972. Its specificity is well established providing that some histochemical criteria were respected. Its chemical background as well as histological methodology have been reviewed many times (Björklund *et al.*, 1972b; Corrodi and Jonsson, 1967; Fuxe and Jonsson, 1973; Fuxe *et al.*, 1970).

The fluorescence formaldehyde reaction takes place with catecholamines and 5-hydroxytryptamine in a two-step reaction, involving a ring closure (Pictet–Spengler reaction) followed by a dehydrogenation step resulting in the formation, in the presence of tissue protein, of strongly fluorescent dihydroisoquinolines from catecholamines and dihydrocarbolines from 5-hydroxytryptamine.

The most important parameters of the reaction are the humidity of the formaldehyde gas, since the histochemical reaction requires a certain amount of water to proceed, the incubation time and the temperature. The standardization of the paraformaldehyde water content for obtaining optimal fluorescence yield without diffusion was performed by Hamberger

(1967) and Hamberger *et al.* (1965). A relative hydration of 50–70% is usually recommended depending on the nature of the tissues under study and on the climatic conditions. The standard duration of the gaseous treatment is 1 h at 80°C; however, formaldehyde condensation at reduced temperature and prolonged reaction time has been proposed to increase the sensitivity and the specificity of the reaction (Håkanson, 1974). Finally, the paraffin embedding can be replaced by araldite embedding to allow better definition of the fluorescent structures on the semi-thin sections (Hökfelt, 1965).

DISTINCTION BETWEEN DOPAMINE AND SEROTONIN

Under standard conditions, the Falck–Hillarp technique when applied to mammalian tissues demonstrates mainly dopamine, noradrenaline, 5-hydroxytryptamine and their immediate precursors, DOPA and 5-hydroxytryptophan. The catecholamine and 5-hydroxytryptamine fluorophores have the same activation peak; however their fluorescence spectrum differs. The two types of fluorophores can therefore be distinguished in the fluorescence microscope by their different colours—green and yellow respectively. However, when high concentrations of catecholamines are present, the formaldehyde-induced fluorescence colour appears more yellowish; the colour has thus to be interpreted very carefully especially in areas or tissues not previously studied. Spectral analysis should be carried out in doubtful cases (Jonsson, 1967).

DISTINCTION BETWEEN DOPA, DOPAMINE AND NORADRENALINE

Initially it was not possible to differentiate directly with the fluorescence microscope between DOPA, dopamine and noradrenaline under standard conditions. Later, a microspectrofluorometric procedure was devised by Björklund *et al.* (1968, 1972a) in which spectral differences namely in the excitation spectra, were readily obtained between the noradrenaline and dopamine fluorophores in tissue sections upon acidification with hydrochloric acid vapours. This analytical histofluorescence technique has been applied to the identification of dopamine in the tubero-hypophyseal neuronal systems as well as in other cell groups in the rat diencephalon (Björklund *et al.*, 1970b; Björklund and Nobin, 1973), in the SIF cells present in the sympathetic ganglia and in the heart (Björklund *et al.*, 1970a; Ehinger *et al.*, 1968), and in the pancreas (Cegrell, 1968).

The distinction between the fluorophores derived from DOPA and dopamine is more difficult. Dowson (1973) has observed a difference in their rate of photodecomposition during irradiation with UV light. Angelakos and King have developed a method based on the trihydroxyindole reaction, which reveals DOPA and noradrenaline but not dopamine (Chiocchio *et al.*, 1971). This technique which has been used to differentiate the various types of SIF cells (Libet and Owman, 1974) does not seem to have been applied to the study of the central nervous system. Finally, a more sophisticated method was proposed, including a combined condensation with glyoxylic acid and formaldehyde, and subsequent microspectrofluorometric analysis (Lindvall *et al.*, 1975). (See p. 303.)

QUANTITATION

The quantitation of cellular stores of monoamines by means of fluorescence intensity measurements presents a number of problems. The first to be resolved was the fluorescence–concentration relationship of formaldehyde-induced catecholamine fluorescence. In many noradrenergic systems, which were the first to be investigated, the local intraneuronal monoamine concentration is so high that there is no linear relationship between fluorescence intensity and monoamine concentration because of a concentration-dependent quenching effect (Jonsson, 1969, 1971). The situation, however, appeared more satisfactory when quantitative microfluorometric studies were carried out on the formaldehyde-induced fluorescence of dopamine. In the striatum of the rat, for instance, a linear relationship was obtained between the fluorescence intensities determined microfluorometrically and the dopamine content measured biochemically (Bacopoulos *et al.*, 1975; Einarson *et al.*, 1975). This result was rather surprising owing to the high dopamine level in this nucleus.

Similar results have been obtained in combined microspectrofluorometric and biochemical studies performed on the lateral part of the median eminence which receives a dense dopaminergic projection (Bacopoulos *et al.*, 1975; Lofstrom *et al.*, 1975). This precise method has also been applied to the study of the relationship between neuronal activity and cellular fluorescence intensity in single cells of the substantia nigra (Lichtensteiger *et al.*, 1976).

It has been suggested that the low affinity of dopamine for the amine storage granules, leading to diffusion of dopamine and its fluorophore into the extragranular space, produced a diminished level of dopamine in each varicosity and thus prevented the quenching phenomenon (Einarson *et al.*, 1975; Fuxe and Jonsson, 1973). The microspectrographic technique which furnishes such precise data unfortunately still remains a difficult, laborious and expensive method although several technical improvements have recently been introduced (Klig *et al.*, 1976; Mazzini *et al.*, 1975; Rost, 1974; Wreford and Schofield, 1975). Therefore, semi-quantitative visual estimations have also been frequently used especially on the noradrenergic systems. The results obtained showed that in an area containing a low level of noradrenaline, such as the cerebral cortex, there was a close relationship between the noradrenaline content recorded biochemically and the fluorescence intensity estimated visually.

However, in densely innervated areas, such as the hypothalamus, a correlation was observed only when the nerves contained less than 50% of the endogenous level. No further proportional increase in fluorescence intensity was observed above this value (Lidbrink and Jonsson, 1971). In the caudate nucleus, Kopin *et al.* (1974) observed that normal fluorescence began to diminish when the dopamine content was about 60% of normal and disappeared when the dopamine levels dropped to about 20% of control. Thus it seems reasonable to conclude, along with several investigators, that if a change in fluorescence intensity is observed compared to the control, this could reflect a true change in monoamine concentration. However, if no change is observed, no definite conclusions can be drawn; moreover, it is safe to compare results of visual estimations with biochemical and/or microspectrofluorometric data.

PRACTICAL CONSIDERATIONS

The Falck–Hillarp technique has great practical advantages for the visualization of dopamine-containing structures. During the freeze-drying procedure, if the pieces cannot be dried immediately, they can be stored on dry ice or in liquid nitrogen for several weeks (Fuxe *et al.*, 1970). Many tissue blocks from the same or from different animals can be handled together. The use of paraffin embedding makes possible the sectioning of all sorts of tissues; serial sections can be easily obtained, on which classical histological techniques can be performed (see below).

However, the Falck–Hillarp technique has also some drawbacks: the time-consuming procedure, the necessity of handling small blocks to prevent their breaking during the freezing, variations in the outcome of the method during different times of the year or during different weather conditions and, above all, its failure to reveal certain dopaminergic systems. It was well known that the non-terminal part of the monoamine neurones were hardly visible with the Falck–Hillarp technique, unless some procedures were used to elevate their amine content: for example, experimental interruption of the axons which induced a piling up of the fluorescent amines rostral to the section, enhancement of the fluorescent content by performing in-vitro incubation in the presence of exogenous amines, or in-vivo intracerebral (intraventricular) injections of exogenous amines. However, in the densely dopaminergic areas like the striatum, it was shown that one varicosity of the dopaminergic terminals could still be detected when the dopamine content had been reduced by about 90% (Anden *et al.*, 1966). Finally, it became obvious that this method was not sensitive enough to visualize the dopaminergic innervation of quite important parts of the central nervous system, such as the cerebral cortex for instance, but this became apparent only after biochemical analysis and other fluorescence techniques had proved the existence of this innervation.

In practice, the Falck–Hillarp method is more sensitive for the visualization of the catecholamine cell bodies which are seen with great detail, than for the visualization of cell processes. Thus, with the help of this method the chief groups of dopaminergic neurones were discovered in the peripheral nervous system (SIF cells of the sympathetic ganglia and elsewhere), and in the central nervous system (the nigrostriatal, mesolimbic, tuberoinfundibular and retinal systems). The dopaminergic terminals and preterminals were identified in densely innervated areas such as the striatum, the nucleus accumbens, olfactory tubercle, amygdala and hypothalamus.

It must be added that some species variations exist. In the monkey, for instance, catecholamines can be demonstrated in the dendrites of the dopaminergic neurones, a fact which was hardly possible in the rat with the Falck–Hillarp technique (Felten, 1977; Sladek and Parnavelas, 1975). Outside the nervous system, dopamine-containing cells were identified in the stomach mucosa of cats and rabbits (Håkanson *et al.*, 1970; Lefranc *et al.*, 1972) and in the mammalian endocrine pancreas (Cegrell, 1968).

THE VIBRATOME GLYOXYLIC ACID METHOD

Principles and Methodology

A vibratome is a simple piece of apparatus which allows one to obtain thin sections, 20–30 μm thick, of fixed or even fresh material without any previous embedding or freezing. The block of tissue is just glued at its bottom to the holder, and sectioning takes place in a liquid cold bath. Hökfelt and Ljungdähl in 1972 were the first to introduce the vibratome as a tissue sectioner for fluorescence histochemical purposes. Sections of fresh or formaldehyde-perfused brain tissue were mounted on glass slides, dried, stored in a desiccator over night and exposed to paraformaldehyde gas obtained as usual. This technique became highly sensitive for the localization of dopamine-containing structures when Lindvall *et al.*, in 1973, replaced paraformaldehyde by glyoxylic acid.

In the original technique, applied to the study of the central nervous system, the brain was rapidly perfused with ice-cold glyoxylic acid. Sections 30 μm thick obtained with the vibratome in an ice-cold bath were further immersed for 1–3 min in ice-cold glyoxylic acid, then dried for 15 min, stored in a desiccator with P_2O_5 and exposed for 2 min at 100°C to glyoxylic acid vapours. However, after 15 min warm-drying, a good fluorescence picture could already be observed (Lindvall and Björklund, 1974).

This glyoxylic acid method can be used in two ways: as far as endogenous dopamine is concerned, the best results are obtained with glyoxylic acid-perfused material. However, when no perfusion is possible, as in the case of human brain biopsies for instance (Berger *et al.*, 1974a, 1976a), or for correlated biochemical and fluorescence analysis (Berger *et al.*, 1974b), an immersion procedure can be used efficaciously. In this case, the rate of penetration into the tissue of the glyoxylic acid is of course a limiting step and only in the first 10 or 15 sections (300–450 μm) are the aminergic structures well preserved.

The glyoxylic acid method is relatively fast and highly reproducible. The two critical points seem to be the time elapsing before the condensation with the glyoxylic acid, and the temperature. Indeed, the specimen must be perfused *in vivo*, or immersed in glyoxylic acid as soon as possible after the dissection to minimize the extraneuronal diffusion of catecholamines induced by an anoxic state. The temperature must be kept very carefully around 0–2°C during all the fixation and sectioning procedure, again to limit the catecholamine diffusion.

Sensitivity

After condensation with glyoxylic acid, the catecholamines become intensely blue-green fluorescent and this fluorescence is very stable upon irradiation with ultraviolet light. In contrast, the indolamine-containing neurones exhibit a weak and unstable brownish yellow fluorescence or are not visible at all. As described for the formaldehyde-induced catecholamine fluorophores, exposure to HCl vapour provides a possibility for the spectral differentiation between noradrenaline on one hand and dopamine and DOPA on the other.

The vibratome glyoxylic acid method permitted the visualization of entire dopaminergic neurones including the non-terminal portion of the axon and

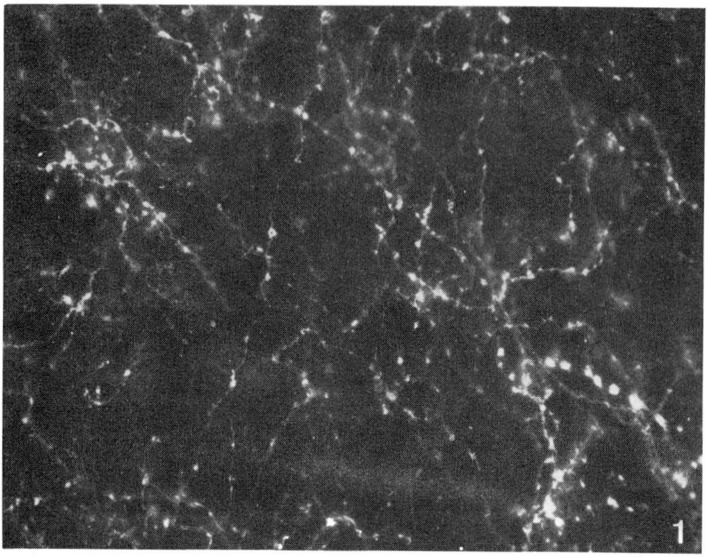

Fig. 1. Dopaminergic fibres in the rat medial frontal cortex, visualized after destruction of the noradrenergic ascending pathways. Perfused material. Vibratome glyoxylic acid method. Formaldehyde vapours treatment. (x550).

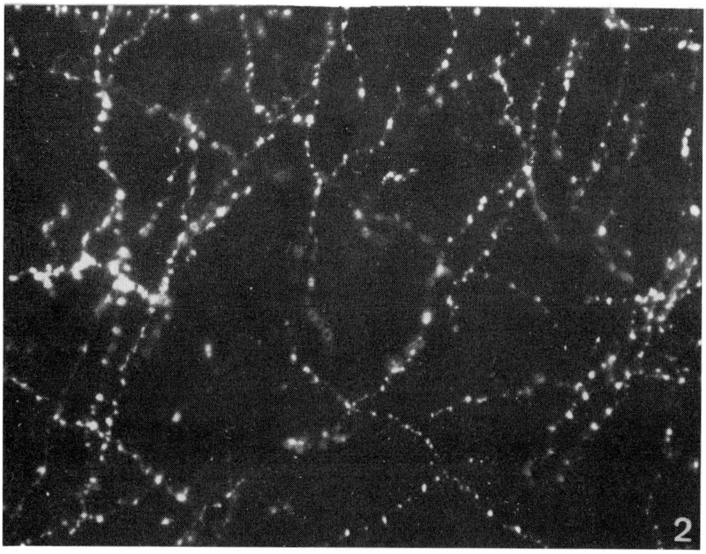

Fig. 2. Noradrenergic fibres in the molecular layer of the rat frontal cerebral cortex. Perfused material. Vibratome glyoxylic acid method. Formaldehyde vapours treatment. (x 550).

sometimes also the dendrites (Björklund and Lindvall, 1975), and revealed several previously unknown axonal systems (Björklund *et al.*, 1975; Lindvall, 1975). In addition, the morphology of the individual fibres could be visualized with precision, and consequently dopaminergic axons (Fig. 1) could be differentiated from noradrenergic axons (Fig. 2) on morphological bases in certain areas and under precise conditions (Berger *et al.*, 1974b, 1976b). In contrast, the morphology of the dopamine-containing cell bodies was of variable quality.

When considering the methodology of the Falck–Hillarp technique, based on the absolute necessity of performing all the reactions in a dry milieu in order to prevent the diffusion of the catecholamines in the tissue, one can wonder how such a wet technique as the glyoxylic acid vibratome method could be so successful and so sensitive; the more so since it has been shown that the reactions underlying the fluorophore formation in the glyoxylic acid method are principally the same as in the formaldehyde method (Axelsson *et al.*, 1973; Lindvall *et al.*, 1974*b*; Svensson *et al.*, 1975). The preservation of larger amounts of internal water resulting in a high fluorescence yield, and the possibility of less diffusion during the drying procedure have been discussed. Moreover, it seems that glyoxylic acid exhibits a considerably higher capacity than formaldehyde to form fluorophores from catecholamines. Indeed, when the glyoxylic acid reaction was performed in histochemical model experiments, under conditions simulating the perfusion immersion procedure used on tissue, the yield of the fluorophore from labelled dopamine was greatly enhanced, 85% of the total recovered radioactivity being found with the strongly fluorescent dihydroisoquinoline (Lindvall *et al.*, 1974b).

Combined Glyoxylic Acid and Paraformaldehyde Treatment

It soon appeared thereafter that formaldehyde and glyoxylic acid could both be used in combination. Formaldehyde vapour treatment instead of glyoxylic acid gas treatment seemed to us to give more reliable results; indeed, the hydration of the fomaldehyde has been standardized but not that of glyoxylic acid which in addition is supplied in more or less hydrated samples. We also used a modified Krebs Ringer bicarbonate buffer in which $CaCl_2$ was replaced by $MgCl_2$ (Berger *et al.*, 1976b) (Figs 1, 2, 6, 10, 11) to minimize the release of catecholamines for which the presence of calcium ions is necessary and which is antagonized by magnesium (Rubin, 1970).

A two-step gaseous treatment including the use of a more or less hydrated paraformaldehyde gas at lower or higher temperature before the glyoxylic acid vapour treatment, has been proposed to allow the differentiation between DOPA, dopamine and noradrenaline fluorophores.

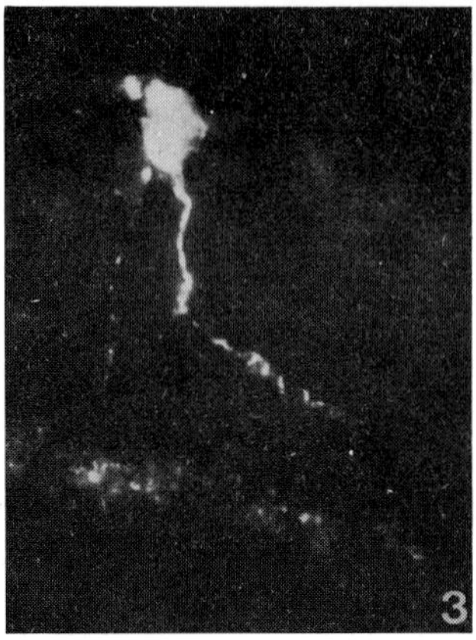

Fig. 3. An alloganglion cell of the rat retina as visualized after consecutive perfusions of glyoxylic acid and paraformaldehyde solution. Cryostat sectioning. Formaldehyde vapours treatment. (x550).

This method would have the great advantage of allying the sensitivity of the glyoxylic acid method with the precision and the possibility of quantitation allowed by the microspectroflurometry; but this technique has so far been applied to only histochemical protein droplet models (Lindvall *et al.*, 1975).

Disadvantage of the Technique

Besides having great advantages, it is obvious that this glyoxylic acid method also has some limitations: the sectioning procedure is rather time consuming, in addition, it is not recommended that the specimens be stored for too long a time in the glyoxylic acid or in the buffer (0.5–1 h and 3–4 h, respectively) before and during sectioning. Consequently, only a limited number of sections can be obtained from each individual specimen, and from each brain, usually only one piece can be sectioned. Moreover, some kinds of tissue like embryonic material or brain regions of a heterogeneous nature are very difficult to cut with the vibratome. Retina cannot be cut at all. To obviate these drawbacks, Loren *et al.* introduced in 1976 a new method combining the advantages of the Falck–Hillarp technique and the sensitivity of the glyoxylic acid method.

MODIFIED GLYOXYLIC ACID/FALCK–HILLARP METHOD

In this technique the animals were perfused with a mixture of glyoxylic acid and paraformaldehyde, at low pH with a high concentration of magnesium salts and the usual Falck–Hillarp technique was then applied. This method seems to have a sensitivity for dopamine-containing structures that was comparable to that of glyoxylic acid vibratome method. It has been applied to the study of the aminergic systems in the developing rat brain (Loren *et al.*, 1976a, b).

Besides its great practical importance, this new technique exhibits two interesting features, the mechanism of which is not yet clear: (1) The effect of the high concentration of magnesium salt which does not seem to depend on the hypertonicity of the solution as the results obtained with similar concentration of sodium chloride or sucrose are much inferior; (2) The fluorescence-promoting effect of the acid pH in the perfusion fluid seems to be contradictory to the inhibiting effect on the fluorescence of low pH glyoxylic acid when used in the vibratome method.

Acid catalysis has been shown to be of importance both in the glyoxylic acid and formaldehyde reactions for the development of the fluorophores (Lindvall *et al.*, 1974b; Svensson *et al.*, 1975). Thus, the acid pH of the perfusion fluid could be thought to create optimal reaction conditions in the tissue. Indeed the highest fluorescence yield from dopamine was obtained with glyoxylic acid at the lowest pHs (the fluorescence intensity at pH 3 was about 70% higher than at pH 7) at least in protein models. However, when an acid solution (pH 3) of glyoxylic acid was tested by the initial glyoxylic acid vibratome method, very few fluorescent structures could be observed, and a neutral glyoxylic acid solution was recommended for obtaining optimal results by dopamine fluorescence histochemistry in tissue sections.

This new fixative has also been assayed in combination with the vibratome method (Loren *et al.*, 1977). The effect of magnesium ions was much less prominent in the glyoxylic acid than in the formaldehyde vibratome technique. Perfusion with a solution of low pH containing formaldehyde, glyoxylic acid and a high concentration of magnesium, before vibratome sectioning, did not seem to have been tried; but, nevertheless, the addition of a large amount of magnesium salt to the perfusion solution whether containing glyoxylic acid or formaldehyde made the tissue difficult to cut.

CRYOSTAT SECTIONING AND OTHER RAPID TECHNIQUES

Special mention must be made of the attempts to obtain satisfactory visualization of the dopaminergic system on cryostat sections as this cutting technique is the least time consuming of all the sectioning techniques used so far.

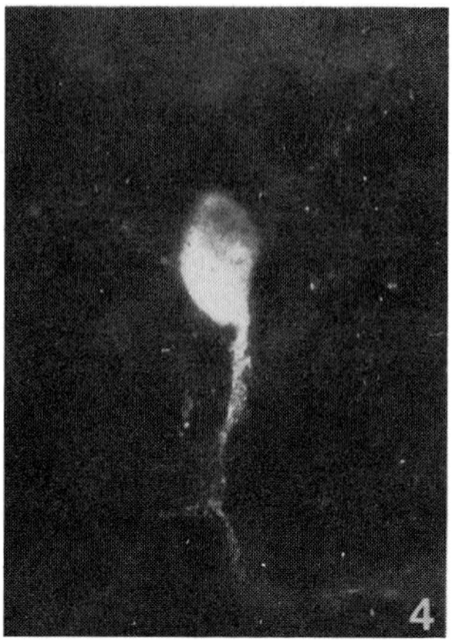

Fig. 4. Dopamine-containing neurone of the rat retina. An intravitreal injection of α-methyl-DOPA was performed after depletion of the endogenous amine content by α-methyl-paratyrosine. Same histofluorescence technique as in Fig. 5. (x 450).

The first assays were devoted to the study of the peripheral catecholamine innervation, using cryostat sections of fresh frozen tissues (El Badawi and Schenk, 1967; Hamberger and Norberg, 1964; Heene, 1968; Laties *et al.*, 1967; Spriggs *et al.*, 1966). Attempts to apply this technique to formaldehyde-fixed brain tissue were unsuccessful (Sakharova and Sakharov, 1971) or were limited by the diffusion of the fluorescent catecholamines (Csillik and Erulkar, 1964). A small improvement was made by freeze-drying of the cryostat sections (Placidi and Masuoka, 1968), but this technique remained quite unsatisfactory for the demonstration of dopamine. During the past three years, several methods were developed based on the high sensitivity resulting from the glyoxylic acid method for the fluorescence histochemical visualization of dopamine. In all these methods, several parameters have to be considered.

Use of Fresh or Fixed Tissues

In the SPG method (De La Torre and Surgeon 1976a, b) cryostat sections of fresh frozen tissue are briefly immersed in glyoxylic acid before air-drying and heating for 5 min at 80°C. This technique presents the advantage of being applicable to unfixed tissues. However, its sensitivity for the overall dopaminergic systems has not yet been tested. Furthermore, the resolution of the fine dopaminergic fibres is poor following this treatment.

Watson and Barchas (1975, 1977), similarly used fresh-frozen tissues and glyoxylic acid fixation of the cryostat sections. A difference was introduced

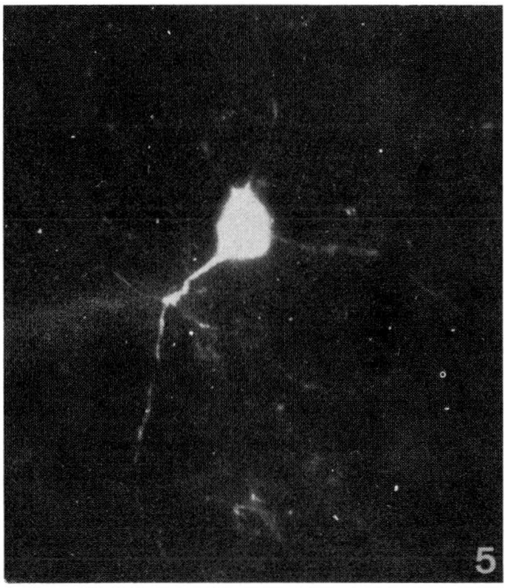

Fig. 5. Fluorescent neurone of the rat retina. Glyoxylic acid perfusion followed by paraformaldehyde fixation. Smears technique. Formaldehyde vapours treatment. (x 640).

by the exposure of the dried sections to glyoxylic acid vapours at 100°C for 2–5 min to reveal the fluorophore. This additional step clearly enhanced the efficiency of the procedure.

In other methods, the animals are perfused before cryostat sectioning with a glyoxylic/paraformaldehyde mixture (Bloom and Battenberg, 1976; Loren *et al.*, 1977; Nygren, 1976) or with a glyoxylic acid solution followed by a paraformaldehyde/glutaraldehyde mixture (Chiba *et al.*, 1976). The pH of the fixative is generally 7–7.4 except the fixative of Loren *et al.* (1977) which contains a high level of magnesium salt. The improvement due to intracardiac perfusion of ice-cold fixative seems to be evident. In our attempts to visualize the dopamine-containing cell bodies in cryostat sections of the retina (Fig. 3), better results were obtained when glyoxylic acid and formaldehyde were perfused successively rather than simultaneously (Nguyen-Legros, 1977). Glyoxylic acid perfusion followed by immersion in paraformaldehyde solution seemed to give similar results. The small percentage of glutaraldehyde added to the second perfusion or immersion solution allowed the observation of fluorescence and ultrastructure on the same processed tissue (Chiba *et al.*, 1976).

Handling of Blocks and Sections in the Cryostat

A rapid freezing of the tissue block obtained by immersion in isopentane cooled by liquid nitrogen, seems to be equivalent to a slow freezing

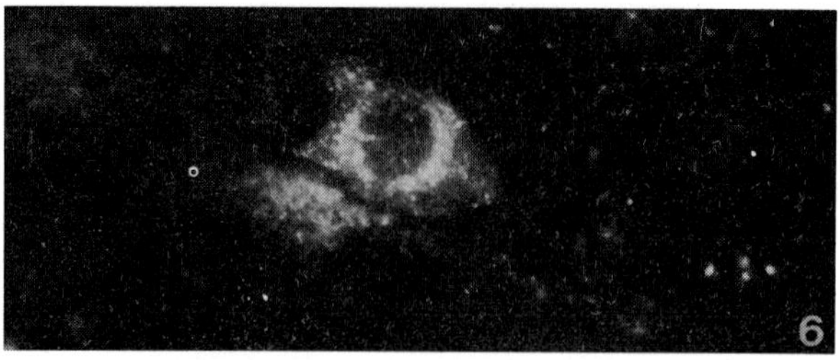

Fig. 6. Dopaminergic neurones in the rat substantia nigra. Perfused material. Vibratome glyoxylic acid method. Formaldehyde vapours treatment. (x 450).

performed in the cryostat chamber at −25°, −30°C with or without the addition of dry ice.

The sections are picked up on slides kept at room temperature and the melting of the tissue occurs immediately. This step is critical because it can induce the diffusion of catecholamines. It must be reduced to a minimum by rapid immersion of the slides in a post-fixative solution or quick air-drying of the sections. The choice between these two steps seems to depend on the technique and the tissue under investigation. We preferred a direct drying of retina sections to allow the demonstration of the fine dopaminergic fibres. However, for the visualization of the dopaminergic cell bodies in the mesencephalon, we found that an immersion of the mounted sections in cold glyoxylic acid or in another solution according to the selected technique (Loren or Chiba) gave better results.

Development of the Fluorophore

Two methods can be used to develop the fluorophore. In the rapid one, the fluorescence is revealed by the warm 15 min air-drying of the sections (Chiba *et al.*, 1976) or by their heating in an oven at 80°–100°C for a few minutes (Bloom and Battenberg, 1976; De La Torre and Surgeon, 1976a, b) or by gassing them with glyoxylic vapours (Watson and Barchas, 1975, 1977). In the slower method, the air-dried sections, whether or not they are post-fixed, are further dehydrated overnight in a desiccator and then exposed to formaldehyde vapours as usual (Loren *et al.*, 1977; Nygren, 1976).

On average, most of the histochemical fluorescent techniques which have been adapted to cryostat sectioning more or less allow the demonstration of the main dopaminergic cell groups (Figs 3, 7, 8). However, following the methods which include a perfusion step and a classical gaseous treatment like those of Loren *et al.* (1977) and Nygren (1976), the cell bodies have a

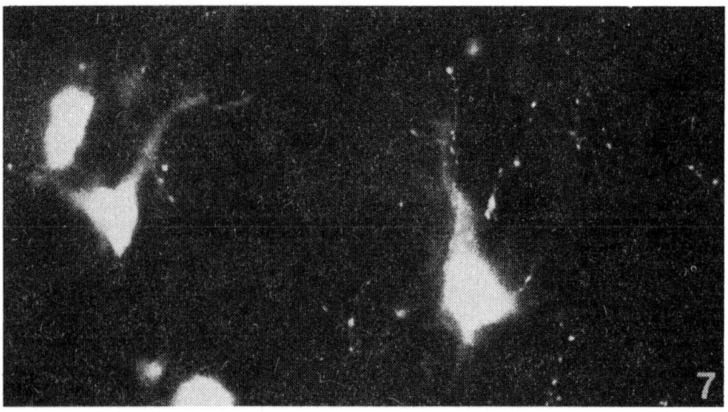

Fig. 7. Dopaminergic neurones in the rat substantia nigra. Cryostat method according to Chiba *et al.*, 1976. (x 450).

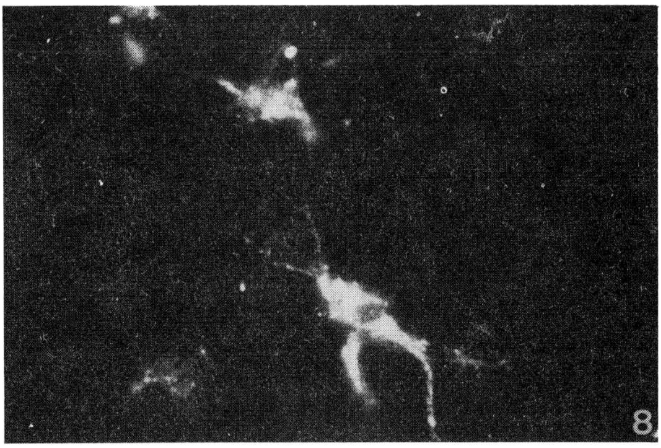

Fig. 8. Dopaminergic meurones in the rat substantia nigra. Cryostat method according to Loren *et al.*, 1977. (x 450).

sharper outline, the fluorophore is distributed in a granulated way in the cytoplasm and the cell processes are more completely visualized (Fig. 8). The presence of dopaminergic axon terminals is readily demonstrated in the densely innervated areas, like the striatum which is used as a test by many workers (De La Torre, Nygren, Watson). The finer dopaminergic systems, in the frontal supragenual cortex or in the incerto-hypothalamic pathway, are however much more difficult to demonstrate in cryostat sections (Fig. 9). It is therefore necessary, whenever possible, to compare the cryostat techniques proposed with the glyoxylic acid vibratome method or the freeze-drying method of Loren *et al.* (1976a) which currently seem to be the most sensitive for the demonstration of dopamine containing structures. Many techniques which provide attractive results when considered alone, appear insufficient when compared with the two previous ones.

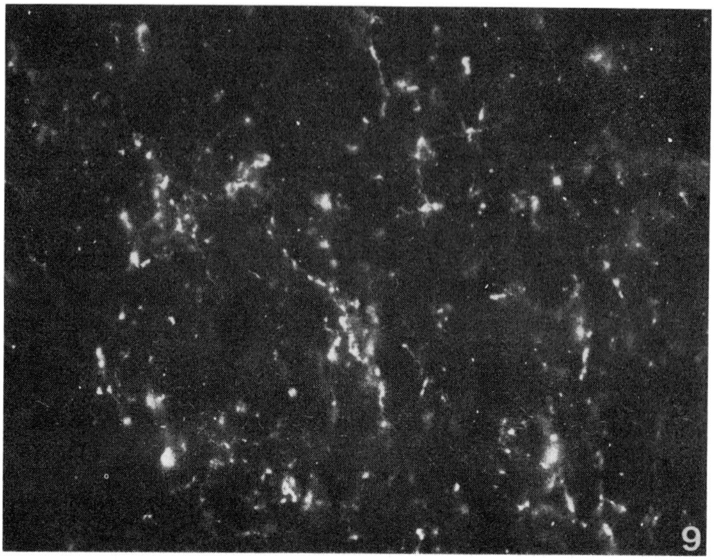

Fig. 9. Dopaminergic fibres in the deep medial frontal cortex of the rat. Cryostat technique according to Loren *et al.*, 1977. (x 500).

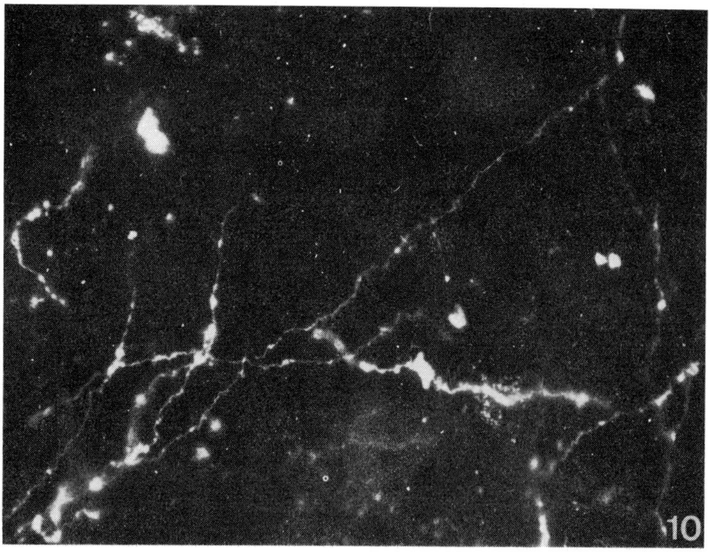

Fig. 10. Fine dopaminergic-like fibres in human cerebral frontal cortex. Biopsy material. Incubation of vibratome sections in the presence of α-methyl-noradrenaline 10^{-6} M. Glyoxylic acid fixation. Formaldehyde vapours treatment. (x 540).

On the other hand, these rapid cryostat techniques might be of interest for obtaining serial sections allowing more precise topographical studies of fluorescent cell bodies. They could also become a valuable tool in combination with other histochemical reactions that could be performed on the same tissue blocks or even the same sections.

Another rapid technique has been proposed, the smear technique. Introduced by Olson and Ungerstedt (1970), it allowed the demonstration of the monoamine nerve terminals in the rat (Fuxe *et al.*, 1974, Lidbrink *et al.*, 1974) and human (Nyström *et al.*, 1972) brain and the dopaminergic cells and processes in the retina (Fig. 5) (Nguyen-Legros and Berger, 1977). A major drawback of this technique is that it prevents extensive topographical studies.

CONCOMITANT USE OF FLUORESCENCE HISTOCHEMICAL AND OTHER MICROSCOPIC TECHNIQUES

As soon as the fluorescence histochemistry of catecholamines had been introduced, attempts were made to combine it with other histological or histochemical techniques. Routine histological stainings have been performed on paraffin (Falck–Hillarp), cryostat or even vibratome sections (Collier and Routtenberg, 1977). Radioautography was mostly combined with the usual Falck–Hillarp method (Fuxe *et al.*, 1968; Hervonen, 1974; Hökfelt and Ljungdähl, 1971; Roth *et al.*, 1974; Winkler and Hempel, 1971), sometimes after embedding in araldite instead of paraffin (Hökfelt, 1965). However, it has also been used on cryostat sections, special care being taken to avoid the diffusion of water-soluble drugs (Masuoka *et al.*, 1971).

Histochemical demonstration of acetylcholine esterase (AChE) combined with histochemical fluorescence method for catecholamines on the same or consecutive section have generally been carried out on cryostat sections of the peripheral nervous system (Ehinger *et al.*, 1968; El Badawi and Schenk, 1967; Eränkö, 1966; Tervo, 1977). Recently, AChE demonstration has been successfully combined with the visualization of the dopaminergic nerve terminals on the same vibratome section of the rat brain septal area (Lindvall, 1977). Lindvall used the glyoxylic acid method but the hot glyoxylic acid vapour step was omitted, the sections just being dried for 15 min with a hair dryer.

We have succeeded in demonstrating catecholamines and exogenous peroxidase reaction products in the same dopaminergic neurone, on vibratome sections of the rat mesencephalon (Berger *et al.*, 1978a, b)

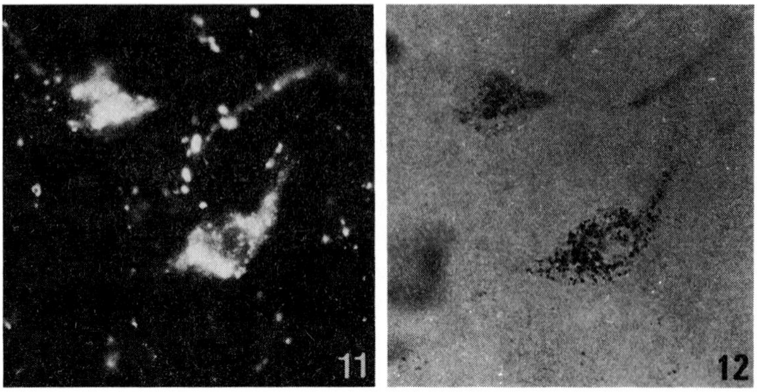

Figs 11–12. Rat substantia nigra. Demonstration of dopamine fluorescence and horseradish peroxidase reaction products in the same neurone. (HRP injection in the striatum.) Histochemical fluorescence demonstration of catecholamines followed by standard HRP reaction applied to the same vibratome section. (x 500).

(Figs 11 and 12). This allowed the precise identification of dopaminergic neurones in neuroanatomical studies on dopaminergic pathways using the retrograde axonal transport of peroxidase.

PHARMACOLOGICAL APPROACHES

Pharmacological tests have long been used to improve the histochemical fluorescence approach of the dopaminergic systems. They were mainly directed to allow a better visualization of dopamine-containing structures by increasing the dopamine levels. They have also provided some possibilities in distinguishing dopamine neurones from other amine-containing structures.

Pharmacological Means Used to Increase Dopamine Levels

The systemic administration of monoamine oxidase inhibitors, such as nialamide or pargyline, is currently used to diminish the catabolism of catecholamines before performing the histochemical fluorescence technique (Fuxe *et al.*, 1970; Javoy *et al.*, 1973). Other procedures are based on the uptake properties of the catecholaminergic neurones. In-vitro incubation, in the presence of exogenous dopamine, of thick (500 μm) unfixed brain sections, further treated by the Falck–Hillarp technique (Fuxe *et al.*, 1970) or of thin (30 μm) unfixed vibratome sections (Fig. 10) further submitted to the glyoxylic acid method, has been shown to enhance markedly the fluorescence of the dopaminergic fibres (Berger *et al.*, 1976). Similar studies have been performed *in vivo* using intracerebral or intraventricular injections; or intravitreous injection for the retina (Fig. 4), of exogenous amines (Ehinger, 1973; Fuxe *et al.*, 1970; Fuxe and Ungerstedt, 1968).

Pharmacological Means Used to Distinguish the Dopamine-Containing Structures

(a) Administration of a selective inhibitor of dopamine-β-hydroxylase has been proposed to reduce brain norepinephrine levels and thus allow the selective visualization of dopaminergic neurones (Hökfelt *et al.*, 1974).

(b) A reserpine-resistant uptake of catecholamines in dopaminergic structures was observed by Lidbrink *et al.* (1974). These authors described, in reserpine pretreated rats, a selective accumulation of dopamine in dopamine nerve terminals, following the parenteral administration of DOPA together with a peripheral decarboxylase inhibitor. When brain sections of animals treated with reserpine (Lidbrink *et al.*, 1974) or with reserpine + nialamide (Lindvall and Björklund, 1974) were incubated *in vitro*, in the presence of exogenous catecholamine, a similar selective uptake in dopaminergic nerve terminals could be observed. These properties have been used to localize in the brain of the rat, the main areas innervated with dopamine (Fuxe *et al.*, 1974; Lindvall *et al.*, 1974a).

(c) A more direct approach could be realized by suppressing the noradrenergic innervation by the use of neurotoxic (6-hydroxy-dopamine) (Berger *et al.*, 1974b, 1976b) or high frequency lesions (Lindvall *et al.*, 1974a).

(d) We proposed another method (Berger and Glowinski, 1978) based on the uptake properties of the aminergic neurones. Unfixed vibratome sections from rats which had been pretreated with α-methyl-*p*-tyrosine to deplete the endogenous catecholamines were incubated in the presence of exogenous dopamine and specific inhibitors of the active transport in noradrenergic, dopaminergic or serotoninergic systems. Dopaminergic and noradrenergic fibres could thus be separately visualized in consecutive sections, after treatment with glyoxylic acid solution or a mixture of glyoxylic acid and paraformaldehyde, and paraformaldehyde gas as seen. above.

EQUIPMENT OF THE FLUORESCENCE MICROSCOPE

For the routine fluorescence microscopic analysis, the instruments equipped to allow incident light illumination (e.g. Leitz–Ploem–Opak system) are mostly used at the present time. The filter selection must take into account the excitation and emission maxima of the dopamine fluorophore which are respectively at 390–410/460–470 nm after formaldehyde induced fluorescence. After glyoxylic acid treatment, the fluorophores are either in the acid or the neutral state, according to the methodology. However, after the standard procedure, most of the fluorophores are in their neutral form, and thus exhibit the excitation and emission maxima indicated above. In this case, the use of a narrow band excitation filter (TAL 405 or S 405) together with a barrier filter at 460–470 nm gives a high fluorescence intensity with a low background. Other filter combinations can be used for special purposes (see in Björklund *et al.*, 1972b; Lindvall and Björklund, 1974).

CONCLUDING REMARKS

A great number of histochemical fluorescence methods are now available which allow the visualization of catecholamine-containing cells and processes in mammalian tissues. Their chemical background which is basically the same whether paraformaldehyde or glyoxylic acid is used, has been thoroughly investigated and their specificity is well established. A number of conditions and criteria permitting the identification of dopamine itself or of dopamine-containing structures have been determined.

From the practical point of view, the following considerations can be of help in the choice of a method as far as the central nervous system is concerned (1) neuronal cell bodies are much easier to visualize than cell processes and the fluorescence is enhanced by the use of formaldehyde in combination with glyoxylic acid solution. Thus, the cryostat techniques, which are currently the fastest, can be used, after it has been verified that the maximum number of cell bodies is demonstrated with the selected technique; (2) the visualization of the neuronal processes—axons, dendrites and nerve terminals—needs techniques of high sensitivity; the glyoxylic acid vibratome method combined with paraformaldehyde vapour treatment, and the glyoxylic acid/paraformaldehyde freezing drying method of Loren *et al.* (1976a) are probably the most sensitive presently available and can be used as references when other methods are tested.

In spite of the wealth of information provided in dopamine research by fluorescence histochemistry, it is obvious that many difficulties still persist. It remains difficult to combine the sensitivity of a technique with an easy routine application. Furthermore, there is presently no equivalent of the dopamine fluorescence method allowing the same histochemical approach at the ultrastructural level. This constitutes a serious limitation for many neuroanatomical investigations which aim at a better knowledge of the intimate relationship between dopaminergic and other systems in the central nervous system.

REFERENCES

Anden, N. E., Fuxe, K., Hamberger, B. and Hökfelt, T. (1966). *Acta Physiol. Scand.* **67,** 306–312.

Axelsson, S., Björklund, A., Falck, B., Lindvall, O. and Svensson, L. A. (1973). *Acta Physiol Scand.* **87,** 57–62.

Bacopoulos, N. G., Bhatnagar, R. K., Schnute, W. J. and Van Orden, L. S. III. (1975.) *Neuropharmacology* **14,** 291–300.

Berger, B., Arluison, M., Escourolle, R. and Moyne, M. A. (1974a). *La Nelle Presse Méd.* **3,** 1081–1082.

Berger, B., Escourolle, R. and Moyne, M. A. (1976a). *Rev. Neurol.* **132,** 183–194.

Berger, B. and Glowinski, J. (1978). *Brain Res.* **147,** 29–45.

Berger, B., Nguyen-Legros, J. and Thierry, A. M. (1978a). *C. R. Acad. Sci., Paris.*

Berger, B., Nguyen-Legros, J. and Thierry, A. M. (1978b). *Neurosc. Lett.* **9,** 297–302.

Berger, B., Tassin, J. P., Blanc, G., Moyne, M. A. and Thierry, A. M. (1974b). *Brain Res.* **81,** 332–337.

Berger, B., Thierry, A. M., Tassin, J. P. and Moyne, M. A. (1976b). *Brain Res.* **106,** 133–145.

Björklund, A. and Lindvall, O. (1975). *Brain Res.* **83,** 531–537.

Björklund, A. and Nobin, A. (1973). *Brain Res.* **51,** 193–206.

Björklund, A., Ehinger, B. and Falck, B. (1968). *J. Histochem. Cytochem.* **16,** 263–270.

Björklund, A., Cegrell, L., Falck, B., Ritzen, M. and Rosengren, E. (1970a). *Acta Physiol. Scand.* **78,** 334–338.

Björklund, A., Falck, B., Hromek, F., Owman, Ch. and West, K. A. (1970b). *Brain Res.* **17,** 1–23.

Björklund, A., Ehinger, B. and Falck, B. (1972a). *J. Histochem. Cytochem.* **20,** 56–64.

Björklund, A., Falck, B. and Owman, Ch. (1972b). *In* "Methods of Investigative and Diagnostic Endocrinology" (Ed. S. A. Berson), Vol. I, 318–368. North Holland Publishing Co., Amsterdam.

Björklund, A., Lindvall, O. and Nobin, A. (1975). *Brain Res.* **89,** 29–42.

Bloom, F. E. and Battenberg, E. L. F. (1976). *J. Histochem. Cytochem.* **24,** 561–571.

Cegrell, L. (1968). *Acta Physiol. Scand. Suppl. 314*, 1–60.

Chiba, T., Hwang, B. H. and Williams, T. H. (1976). *Histochemistry* **49,** 95–106.

Chiocchio, S. R., King, M. P., Carballo, L. and Angelakos, E. T. (1971). *J. Histochem. Cytochem.* **19,** 621–626.

Collier, T. J. and Routtenberg, A. (1977). *Brain Res.* **128,** 354–360.

Corrodi, H. and Jonsson, G. (1967). *J. Histochem. Cytochem.* **15,** 65–78.

Csillik, B. and Erulkar, S. D. (1964). *J. Pharmacol. Exp. Ther.* **146,** 186–193.

De La Torre, J. C. and Surgeon, J. W. (1976a). *Histochemistry* **49,** 81–94.

De La Torre, J. C. and Surgeon, J. W. (1976b). *Neuroscience* **1,** 451–454.

Dowson, J. H. (1973). *Experientia* **29,** 1122–1124.

Ehinger, B. (1973). *Z. Zellforsch.* **139,** 171–178.

Ehinger, B., Falck, B., Persson, H. and Sporrong, B. (1968). *Histochemie* **16,** 197–205.

Einarsson, P., Hallman, H. and Jonsson, G. (1975). *Med. Biol.* **53,** 15–24.
El Badawi, A. and Schenk, E. A. (1967). *J. Histochem. Cytochem.* **15,** 580–588.
Eränkö, O. (1955). *J. Histochem. Cytochem.* **4,** 11.
Eränkö, O. (1966). *Pharmacol. Rev.* **18,** 353–358.
Falck, B. (1962). *Acta Physiol. Scand. Suppl.* 197, 6–25.
Falck, B. and Owman, C. (1965). *Acta Univ. Lund Sect. II* **7,** 5–23.
Falck, B., Hillarp, N. A., Thieme, G. and Torp, A. (1962). *J. Histochem. Cytochem.* **10,** 348–354.
Felten, D. L. (1977). *Brain Res.* **120,** 553–558.
Fuxe, K. and Jonsson, G. (1973). *J. Histochem. Cytochem.* **21,** 293–311.
Fuxe, K. and Ungerstedt, U. (1968). *Histochemie* **13,** 16–28.
Fuxe, K., Hökfelt, T., Ritzen, M. and Ungerstedt, U. (1968). *Histochemie* **16,** 186–194.
Fuxe, K., Hökfelt, T., Jonsson, G. and Ungerstedt, U. (1970). *In* "Contemporary Research Methods in Neuroanatomy" (Eds W. S. H. Nauta and S. O. E. Ebbeson), 275–314. Springer Verlag, New York.
Fuxe, K., Hökfelt, T., Johansson, O., Jonsson, G., Lidbrink, P. and Ljungdähl, A. (1974). *Brain Res.* **82,** 349–355.
Håkanson, R. and Sundler, F. (1974). *J. Histochem. Cytochem.* **22,** 887–897.
Håkanson, R., Owman, Ch., Sjoberg, N. O. and Sporrong, B. (1970). *Histochemie* **21,** 189–220.
Hamberger, B. (1967). *Acta Physiol. Scand Suppl.* 295 . 1–56.
Hamberger, B. and Norberg, K. A. (1964). *J. Histochem. Cytochem.* **12,** 48–49.
Hamberger, B., Malmfors, T. and Sachs, Ch. (1965). *J. Histochem. Cytochem.* **13,** 147.
Heene, R. (1968). *Histochemie* **14,** 324–327.
Hervonen, A., Tuohimaa, P., Kanerva, L. and Lietzen, R. (1974). *Histochemistry* **39,** 371–375.
Hökfelt, T. (1965). *J. Histochem. Cytochem.* **6,** 518–519.
Hökfelt, T. and Ljungdähl, A. (1971). *In* "Progress in Brain Research" Vol. 34, (Ed. O. Eränkö), 87–102. Elsevier, New York.
Hökfelt, T. and Ljungdähl, A. (1972). *Histochemie* **29,** 325–339.
Hökfelt, T., Fuxe, K., Johansson, O. and Ljungdähl, A. (1974). *Eur. J. Pharmacol.* **25,** 108–112.
Javoy, F., Youdim, M. B. H., Agid, Y. and Glowinski, J. (1973). *J. Neur. Trans.* **34,** 279–290.
Jonsson, G. (1967). *Acta Histochem.* **26,** 379–390.
Jonsson, G. (1969). *J. Histochem. Cytochem.* **17,** 714–723.
Jonsson, G. (1971). *In* "Progress in Brain Research", Vol. 34, (Ed. O. Eränkö), 53–62. Elsevier, New York.
Klig, V., Demirjian, C. and Pungaliya, P. (1976). *J.* Microsc. **107,** 173–176.
Kopin, I. J., Palkovits, M., Kobayashi, R. M. and Jacobowitz, D. M. (1974). *Brain Res.* **80,** 229–235.
Laties, A. M., Lund, R. and Jacobowitz, D. (1967). *J. Histochem. Cytochem.* **15,** 535–541.
Lefranc, G., Pradal, G., L'Hermite, A., Dubin, J. C. and Tusques, J. (1972). *C. R. Soc. Biol.* **166,** 1765–1769.
Libet, B. and Owman, C. (1974). *J. Physiol.* **237,** 635–662.
Lichtensteiger, W., Felix, D., Lienhart, R. and Hefti, F. (1976). *Brain Res.* **117,** 85–103.

Lidbrink, P. and Jonsson, G. (1971). *J. Histochem. Cytochem.* **19,** 747–757.
Lidbrink, P., Jonsson, G. and Fuxe, K. (1974). *Brain Res.* **67,** 439–456.
Lindvall, O. (1975). *Brain Res.* **87,** 89–95.
Lindvall, O. (1977). *Histochemistry* **50,** 191–196.
Lindvall, O. and Björklund, A. (1974). *Histochemistry* **39,** 97–127.
Lindvall, O., Björklund, A., Hökfelt, T. and Lungdähl, A. (1973). *Histochemie* **35,** 31–38.
Lindvall, O., Björklund, A., Moore, R. Y. and Stenevi, U. (1974a). *Brain Res.* **81,** 325–331.
Lindvall, O., Björklund, A. and Svensson, L. A. (1974b). *Histochemistry,* **39,** 197–227.
Lindvall, O., Björklund, A., Falck, B. and Svensson, L. A. (1975). *J. Histochem. Cytochem.* **23,** 697–699.
Lofstrom, A., Jonsson, G. and Fuxe, K. (1976). *J. Histochem. Cytochem.* **24,** 415–442.
Loren, I., Björklund, A., Falck, B. and Lindvall, O. (1976a). *Histochemistry* **49,** 177–192.
Loren, I., Björklund, A. and Linvall, O. (1976b). *Brain Res.* **117,** 313–318.
Loren, I., Björklund, A. and Lindvall, O. (1977). *Histochemistry* **52,** 223–240.
Masuoka, D. T., Placidi, G. F. and Gosling, J. A. (1971). *Progr. Brain Res.* **34,** 77–86.
Mazzini, G., Bottiroli, G. and Prenna, G. (1975). *Histochem. J.* **7,** 291–298.
Nguyen-Legros, J. (1977). *C. R. Acad. Sci. D.* **285,** 351–354.
Nguyen-Legros, J. and Berger, B. (1977). *Ophthal. Res.* **9,** 343–350.
Nygren, L. G. (1976). *Med. Biol.* **54,** 278–285.
Nyström, B., Olson, L. and Ungerstedt, U. (1972). *Science* **176,** 924–926.
Olson, L. and Ungerstedt, U. (1970). *Brain Res.* **17,** 343–347.
Placidi, G. F. and Masuoka, D. (1968). *J. Histochem. Cytochem.* **16,** 491–493.
Rost, F. W. D. (1974). *Med. Biol.* **52,** 73–81.
Roth, L. J., Diab, I. M., Watanabe, M. and Dinerstein, R. J. (1974). *Molec. Pharmacol.* **10,** 986–998.
Rubin, R. P. (1970). *Pharmacol. Rev.* **22,** 389–428.
Sakharova, A. V. and Sakharov, D. A. (1971). *In* "Progress in Brain Research", Vol. 34 (Ed. O. Eränkö), 11–25. Elsevier, New York.
Sladek, J. R. Jr. and Parnavelas, J. G. (1975). *Brain Res.* **100,** 657–662.
Spriggs, T. L. B., Lever, J. D., Rees, P. M. and Graham, J. D. P. (1966). *Stain Technol.* **41,** 323–327.
Svensson, L. A., Björklund, A. and Lindvall, O. (1975). *Acta Chem. Scand. B* **29,** 341–348.
Tervo, T. (1977). *Histochemistry* **50,** 291–300.
Watson, S. J. and Barchas, J. D. (1975). *Psychopharmacol. Commun.* **1,** 523–531.
Watson, S. J. and Barchas, J. D. (1977). *Histochem. J.* **9,** 183–196.
Winckler, J. and Hempel, K. (1971). *Histochemie* **25,** 130–142.
Wreford, N. G. M. and Schofield, G. C. (1975). *J. Microsc.* **103,** 127–130.

CHAPTER 20

Dopamine Pathways in the Rat Brain

Olle Lindvall

Departments of Histology and Neurology,
University of Lund, Lund, Sweden

INTRODUCTION

The present knowledge of the organization of central dopamine neurone systems is based mainly on studies carried out with the formaldehyde (Falck, 1962; Falck *et al.*, 1962) and glyoxylic acid (Lindvall and Björklund, 1974a) fluorescence histochemical methods. In addition, during recent years several other techniques have been found highly useful for neuroanatomical studies on these systems. These methods include horseradish peroxidase (HRP) tracing (see, for example, La Vail and La Vail, 1972; Nauta *et al.*, 1974), Fink–Heimer staining (Hedreen and Chalmers, 1972; Maler *et al.*, 1973; Shimizu and Ohnishi, 1973; Simon *et al.*, 1974, 1976), immunohistochemistry (Hökfelt *et al.*, 1975a, 1976; Pickel *et al.*, 1975a, b), autoradiographic tracing after injections of radioactively labelled amino acids (Cowan *et al.*, 1972; Fallon and Moore, 1978b) and determination of dopamine and the dopamine-synthesizing enzyme tyrosine hydroxylase in minute amounts of tissue (Brownstein *et al.*, 1974; Koslow *et al.*, 1974; Ben-Ari *et al.*, 1975; Saavedra and Zivin, 1976). The following will give a brief review of the presently available data on the organization of dopamine neurones in the rat central nervous system. Table I summarizes the major dopaminergic projection systems. Each system will be described separately in the text.

TABLE I Central dopaminergic projection systems

	Cells of origin	Projections
1. The mesostriatal system	Substantia nigra	nc. caudatus-putamen; probably also globus pallidus
	Ventral tegmental area	nc. accumbens
2. The mesocortical system	Substantia nigra and ventral tegmental area	*Allocortical subdivision:* olfactory tubercle, septum, interstitial nucleus of the stria terminalis, amygdala; probably also the piriform cortex *Neocortical subdivision:* suprarhinal cortex, pregenual and supragenual anteromedial cortex, ventral entorhinal cortex
3. The periventricular system	Mesencephalic periaqueductal gray and periventricular gray of caudal thalamus	Periaqueductal gray; medial thalamus and hypothalamus
4. The incertohypothalamic system	Zona incerta and periventricular hypothalamus	Zona incerta; anterior, medial preoptic and periventricular hypothalamus; septum
5. The tuberohypophyseal system	Arcuate and periventricular hypothalamic nuclei	Median eminence; pars nervosa and pars intermedia of the pituitary
6. The periglomerular dopamine neurons	Olfactory bulb	Dendritic processes into olfactory glomeruli
7. The retinal dopamine system	Mainly in the inner nuclear layer of the retina	Local dendritic projections

THE MESOTELENCEPHALIC DOPAMINE SYSTEM

In the present account of the projections of the mesencephalic dopamine neurones (belonging to cell groups A8, A9 and A10 of Dahlström and Fuxe, 1964) the terminology suggested by Björklund *et al.* (1979) has been used. *The meso-telencephalic system*, which refers to the entire ascending forebrain projection of the mesencephalic dopamine neurone system, is divided into two major subsystems (Table I): *the mesostriatal system* (comprising both the classical nigrostriatal system and the projection from the ventral tegmental area to the nc. accumbens (which is part of the striatum)) and *the mesocortical system* which, according to this terminology, includes both the mesolimbic system as defined by Ungerstedt (1971) and the more recently described mesocortical projections. Within the mesocortical dopamine system are distinguished an *allocortical projection* innervating the olfactory tubercle, septum, interstitial nucleus of the stria terminalis, amygdala, and probably also the piriform cortex, and an *iso- or neocortical projection* innervating the pregenual and supragenual parts of the anteromedial cortex, the suprarhinal cortex and the ventral part of the entorhinal cortex.

THE MESOSTRIATAL SYSTEM

The mesostriatal dopamine pathway projecting to the nc. caudatus-putamen originates in the substantia nigra (mainly in the pars compacta) and probably also in the A10 and A8 cell groups in the mesencephalon (Andén *et al.*, 1964; Bertler *et al.*, 1964; Poirier and Sourkes, 1965; Andén *et al.*, 1966a, b; Bédard *et al.*, 1969; Hökfelt and Ungerstedt, 1969; Moore *et al.*, 1971a; Ungerstedt, 1971; Carpenter and Peter, 1972; Nobin and Björklund, 1973; Lindvall and Björklund, 1974b; Nauta *et al.*, 1974; Hökfelt *et al.*, 1976; Fallon and Moore, 1978b). There is a topographical relation between the mesencephalic dopamine cells and the area of termination in the striatum. Thus, the lateral part of the substantia nigra projects to lateral and caudal parts of the neostriatum, the medial part of the nigra to more medial and rostral parts of the neostriatum, and the laterally situated A10 neurones to the most medial and ventral parts of the head of the nc. caudatus-putamen (Ungerstedt, 1971; Domesick *et al.*, 1976; Carter and Fibiger, 1977; Nauta, 1978; Lindvall *et al.*, 1978; Fallon and Moore, 1978b). Moreover, Nauta and Domesick (1978) has suggested from HRP material that the neurones of the A8 cell group give rise to an innervation in the ventral neostriatum.

The nigral axons are, in their initial course, directed medially (Fig. 1; Lindvall and Björklund, 1974b). Medial to the substantia nigra they are

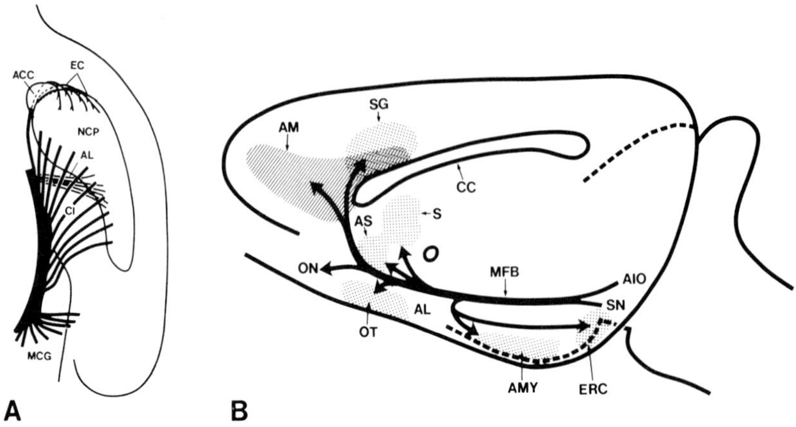

Fig. 1. Schematic representation of the mesostriatal (A) and the mesocortical (B) dopaminergic systems in horizontal and sagittal projections, respectively. Modified from Lindvall and Björklund (1974b) and Lindvall *et al.* (1974a). AL, ansa lenticularis; AM, anteromedial fibre system; AMY, amygdala; AS, nc. accumbens; CC, corpus callosum; CI, internal capsule; EC, external capsule; ERC, entorhinal cortex; MCG, mesencephalic dopamine cell groups; MFB, medial forebrain bundle; NCP, nc. caudatus-putamen; ON, olfactory nuclei; OT, olfactory tubercle; S, septum; SG, supragenual fibre system; SN, substantia nigra; ACC, anterior cingulate cortex (supragenual system) or nucleus accumbens; A10, cell group according to Dahlström and Fuxe (1964).

joined by axons from the A8 group. The axons turn sharply rostrally and assemble into a well-defined bundle that ascends in the H-field of Forel immediately dorsolateral to the MFB system. The most dorsal fibres in the nigrostriatal pathway leave the bundle first (Fig. 1). At the level of the subthalamic nucleus, these dorsal fibres bend first sharply laterally, above the crus cerebri, and then rostrally to turn into the internal capsule from the caudal side towards the caudal parts of the neostriatum. The somewhat more ventrally located fibres deviate less sharply, and run in a rostrolateral direction through the subthalamic region into the internal capsule. The centrally and ventrally located fibres continue rostrally along and partly within the dorsomedial edge of the internal capsule. Along this course, the more dorsally situated fibres in the bundle deviate in a rostrolaterodorsal direction into the internal capsule, towards the central parts of the nc. caudatus-putamen. The ventral portion of the nigrostriatal pathway continues further rostrally up to the level of the globus pallidus. Here, most of the fibres fan out in rostrodorsolateral directions to run along the myelinated fascicles through the globus pallidus into the head of the nc. caudatus-putamen. The most ventral portion, however, continues rostrally, in a position just dorsal to the MFB, up to the anterior commissure; the axons pass ventrally to the commissure, and medially and laterally to its

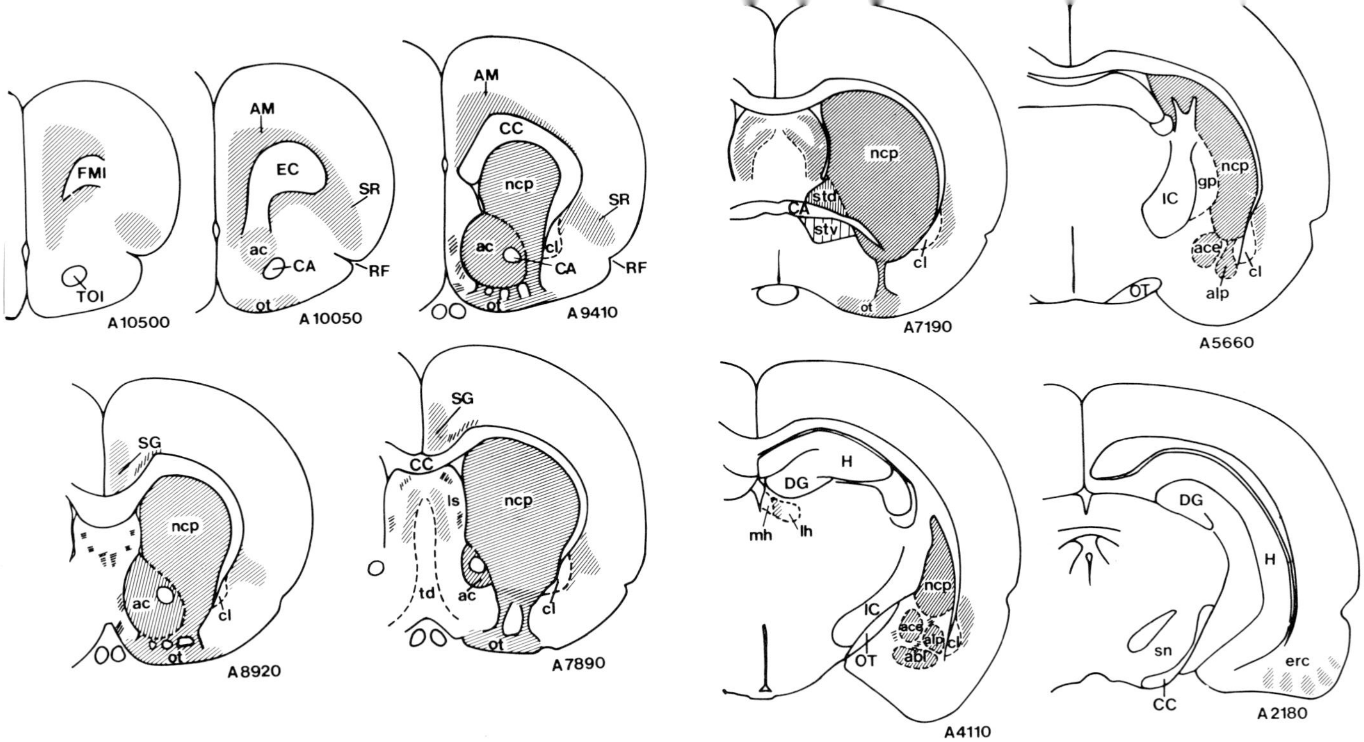

Fig. 2. Schematic representation in nine frontal planes of the terminals of the mesotelencephalic dopamine system. In addition the dopaminergic innervation of the lateral habenular nucleus, probably originating in the mesencephalic cell groups (Herkenham and Nauta, 1977) is represented. Areas of termination are indicated by hatchings. Data compiled from Berger *et al.* (1976), Fuxe (1965), Fuxe *et al.* (1974), Lindvall (1975), Lindvall *et al.* (1974a, b, 1978), Ungerstedt (1971). abl, basal amygdaloid nucleus, lateral part; ac, nc. accumbens; ace, central amygdaloid nucleus; alp, lateral amygdaloid nucleus, posterior part; cl, claustrum; erc, entorhinal cortex; gp, globus pallidus; lh, lateral habenular nucleus; ls, lateral septal nucleus; mh, medial habenular nucleus; ncp, nc. caudatus-putamen; ot, olfactory tubercle; sn, substantia nigra; st, interstitial nucleus of stria terminalis, std = dorsal part, stv = ventral part; td, nc. of diagonal band; AM, anteromedial fibre system; CA, anterior commissure; CC, corpus callosum; DG, dentate gyrus; EC, external capsule; FMI, forceps minor; H, hippocampus; IC, internal capsule; OT, optic tract; RF, rhinal fissure; SG, supragenual fibre system; SR, suprarhinal fibre system; TOI, intermediary olfactory tract.

anterior limb they run into the ventromedial part of the nc. caudatus-putamen and the dorsal part of the interstitial nucleus of the stria terminalis. The ventral portion of the pathway also contributes fibres to the ansa lentricularis, probably innervating the amygdala (e.g. the central amygdaloid nucleus). From this description it is evident that there is a topographic arrangement of the axons within the nigrostriatal pathway. The fibres going to the caudal regions lie more dorsally and laterally, and those going to rostral and ventral regions lie more ventrally.

In the *striatum* the dopaminergic neurones give rise to a dense terminal network that is distributed throughout the nc. caudatus-putamen (Fig. 2). This terminal pattern extends without interruption also into the *nucleus accumbens* (Fuxe, 1965; Andén *et al.*, 1966b). Dopamine fibres of mesencephalic origin which seem to terminate in the *globus pallidus* have recently been demonstrated (Lindvall and Björklund, unpublished observations).

The part of the mesostriatal system projecting to the nc. accumbens originates in the ventral tegmental area, medial to substantia nigra (see below; Andén *et al.*, 1966a; Ungerstedt, 1971; Simon *et al.*, 1976; Carter and Fibiger, 1977). The axons ascend in the dorsal portion of the MFB, in a position immediately ventromedial to the nigrostriatal pathway (Ungerstedt, 1971; Lindvall and Björklund, 1974b). The axons reach nc. accumbens along the MFB. Part of the axons pass dorsally on the rostrolateral aspects of the nucleus to fan out over the external capsule (Fig. 1A). Part of these fibres continue into the frontal cortex; others appear to project into the head of the nucleus caudatus-putamen.

Apart from the axonal processes, projecting to the basal ganglia, the dendrites of the substantia nigra neurones have also been shown to store dopamine (Björklund and Lindvall, 1975). As described originally by Cajal (Cajal and Ramón, 1955) in Golgi preparations, the nigral dendrites form long branching processes that ramify both among the dopamine neurones in the pars compacta as well as into the depth of the pars reticulata. Geffen *et al.* (1976) and Korf *et al.* (1976) have demonstrated a release of dopamine from the pars reticulata upon electrical stimulation. This might suggest a role for dopamine at the dendritic level within the nigra itself. Whether these properties of the dendrites are also valid for other mesencephalic dopamine neurones is so far unknown.

THE MESOCORTICAL SYSTEM

The mesocortical pathways originate in the ventromedial tegmental area (Andén *et al.*, 1966a; Ungerstedt, 1971; Hedreen and Chalmers, 1972; Fuxe *et al.*, 1974; Lindvall and Björklund, 1974b; Lindvall *et al.*, 1974a; Lindvall, 1975; Beckstead, 1976; Simon *et al.*, 1976; Carter and Fibiger, 1977; Lindvall *et al.*, 1978; Lindvall and Stenevi, 1978) with one notable exception: the pathway innervating the outer layers of the supragenual anteromedial frontal cortex originates in the substantia nigra (Lindvall *et al.*, 1974a, 1978; Emson and Koob, 1978).

The different components of the mesocortical system originate in different parts of the mesencephalic dopamine cell system, suggesting that there is a topographical arrangement among these neurones. Figure 3 shows schematically a summary of our own observations obtained from fluorescence histochemical observations in combination with small electrolytic lesions and HRP tracing (Lindvall *et al.*, 1974a; Lindvall, 1975; Lindvall *et al.*, 1978; Lindvall and Stenevi, 1978). Other literature is generally in agreement with this picture (Andén *et al.*, 1966a; Ungerstedt, 1971; Fuxe *et al.*, 1974; Simon *et al.*, 1976; Carter and Fibiger, 1977; Fallon and Moore, 1978b). It should be pointed out that localized HRP injections in the supragenual anteromedial cortex label cells only in the medial part of the substantia nigra, as well as some cells in the lateral A10 area (Beckstead, 1976; Carter and Fibiger, 1977; Lindvall *et al.*, 1978). However, from lesions of varying size and position of the substantia nigra it seems clear that the supragenual dopamine system (see below) originates in cells distributed throughout the mediolateral extent of the substantia nigra (Fig. 3; Lindvall *et al.*, 1978). The reason for the discrepancy between the HRP and the lesion data is unclear (for discussion, see Lindvall *et al.*, 1978).

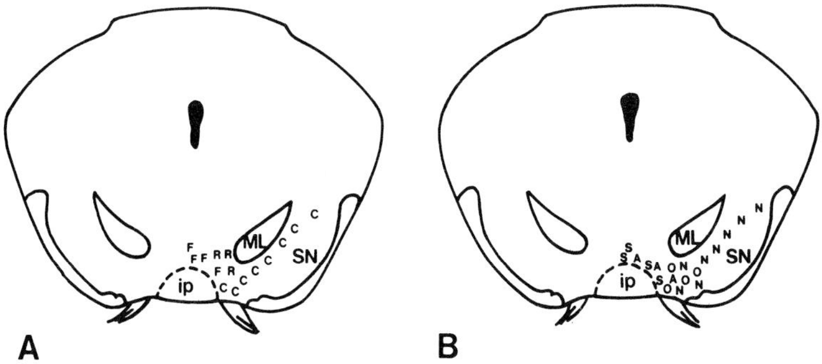

Fig. 3. Relative location of dopamine cell bodies in ventral mesencephalon projecting to neocortical areas (A) and to striatal and allocortical areas (B). F, anteromedial dopamine system; C, supragenual dopamine system; R, suprarhinal dopamine system; S, septum; A, nc. accumbens; O, olfactory tubercle; N, neostriatum. The position of cells projecting to nc. accumbens and the olfactory tubercle rests on preliminary observations with electrolytic lesions and the HRP method (Lindvall and Stenevi, 1978). Abbreviations: ip, interpeduncular nucleus; ML, medial lemniscus; SN, substantia nigra.

It is interesting to note that in our material the cells labelled after HRP injections in the anteromedial frontal cortex and in the lateral septal nucleus had a similar distribution in the medial A10 area (Fig. 3A and B). This would be consistent with the concept that the same dopaminergic cells project both to the frontal cortex and to the septum, an arrangement that finds support in the observation that the axons running into the frontal cortex give off collaterals into the septum (Lindvall and Björklund, 1974b). Interestingly, there seems to be a crude topographic arrangement of the A10 cells in the mediolateral direction, the medial cells projecting to medial telencephalic structures (septum, accumbens, medial frontal cortex) and the lateral cells to areas situated more laterally (olfactory tubercle and neostriatum). As described above (see Lindvall *et al.*, 1977) the rostral cortical areas (such as the pregenual anteromedial cortex) have their cell bodies of origin more medially (in the A10 cell group; Fig. 3A). Going caudally in the medial cortex (i.e. into the supragenual cortex) the cell bodies of origin are found more laterally (in the lateral A10 and substantia nigra), and going caudally along the rhinal sulcus, the cells of origin are again found more laterally in the ventral tegmental area. Such a topographic arrangement is also supported by the recent study of Fallon and Moore (1976, 1978b).

The mesocortical axons ascend along the MFB, in a position ventromedial to the mesostriatal axons (Lindvall and Björklund, 1974b). The axons continue apparently unbranched up to the retrochiasmatic region where fibres (at least partly as collaterals) leave the bundle laterally along the ansa lenticularis towards amygdala and the piriform and entorhinal cortices (see Fig. 1B). After having given off fibres ventrally into the olfactory tubercle and dorsally along the diagonal band into the septum, the remaining portion of the mesocortical pathway leaves the MFB at the level of the rostral septum in a dorsomediorostral direction to run in a position corresponding well to the septohypothalamic tract. The fibres sweep as a broad band along the medial and the medio-ventral aspects of the nc. accumbens. In the region rostromedial to the nc. accumbens, the bundle separates into four main branches: (1) the branch of greatest abundance runs dorsorostrally and laterally into the deep layers of the pregenual part of the anteromedial frontal cortex. These fibres give rise to the extensive anteromedial dopamine terminal system. The branch to the frontal cortex passes along the rostral aspect of the external capsule, and fibres are seen to run into the external capsule; (2) a second branch of the bundle turns more sharply dorsally, to run in a position caudal to the branch to the pregenual cortex. It sweeps caudally above the corpus callosum to ramify into the dense terminal system in the outer layers of the supragenual part of the anteromedial cortex; (3) a third branch runs in a dorsocaudal direction into the septum, contributing to the innervation of this area; (4) fibres in a ventral position in the bundle first turn dorsally for a short distance and then rostrally to join the medial olfactory tract. Presumably, these fibres give rise to terminals in the olfactory nuclei.

The *olfactory tubercle* contains a dense dopaminergic innervation, which is continuous with that of the neostriatum along the cell bridges connecting the two structures. According to Fuxe (1965) the innervation is mainly confined to the medial part of the tubercle, within the lamina pyramidalis. Also the *islands of Calleja* seem to have a dopaminergic input although of lower density (Hökfelt *et al.*, 1977; Fallon and Moore, 1978a; Lindvall, unpublished observations).

The mesencephalic dopamine innervation of the *septal area* is morphologically of two different types (Lindvall and Stenevi, 1978; Moore, 1978). One type is fine-varicose and the other smooth with few varicosities. The latter type of axon often outlines the cell body and the proximal dendrites of septal neurones, forming pericellular basket-like arrangements (Fig. 4A; see also Moore *et al.*, 1971b). At the most rostral levels of the septum scattered pericellular arrangements and smooth fibres are found in the lateral septal nucleus (Fig. 2). Further caudally, the dopamine innervation increases progressively and the highest terminal density is found medially and ventrally in the middle portion of the lateral septal nucleus. The fine-varicose dopamine fibres are aggregated in a dense band outlining the fornix in the medial part of the lateral nucleus. Pericellular arrangements and smooth fibres occur mainly in its lateral and dorsal parts, close to the lateral ventricle. The density of innervation again decreases in the caudal part of the septum (Fig. 2).

The *interstitial nucleus of the stria terminalis* has a dopaminergic innervation that is closely related to that of the septum on one hand (Lindvall, unpublished), and the nc. caudatus-putamen on the other (Fuxe, 1965; Andén *et al.*, 1966a; Ungerstedt, 1971). The dorsal part of the nucleus has the densest innervation, but also the ventral part has a dopaminergic nerve supply, scattered among the dense noradrenergic innervation present in this part of the nucleus (Lindvall, unpublished; see Hökfelt *et al.*, 1977). Dopamine axons can be traced from the interstitial nucleus of the stria terminalis into the dense terminal system in the lateral septal nucleus (Lindvall, unpublished).

In the amygdala a dopamine innervation of high density is found in the central nucleus (Fuxe, 1965; Ungerstedt, 1971) and rich innervations have also been reported in the lateral and basolateral nuclei and the so-called intercalated masses (Fig. 2; Fuxe *et al.*, 1974; Fallon *et al.*, 1978). The innervation in the central nucleus has been described as an extension of that in the nc. caudatus-putamen.

The neocortical projections of the mesocortical dopaminergic system are found in the entorhinal cortex and in the frontal lobe (Berger *et al.*, 1974, 1976; Hökfelt *et al.*, 1974a, b; Lindvall *et al.*, 1974a, 1978; see also Thierry *et al.*, 1973a, b) (Fig. 2). In the *ventral entorhinal area* the dopamine fibres form a series of clusters, which are principally located in the second and third layers but extend into the molecular layer (see also Collier and Routtenberg, 1976; Hökfelt *et al.*, 1977). On the basis of fibre morphology

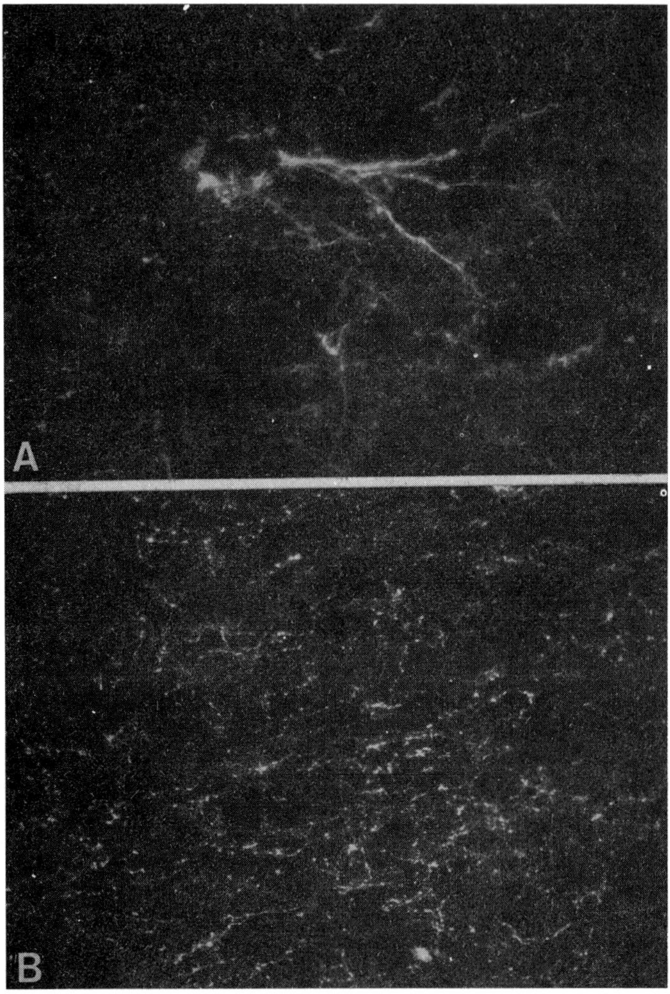

Fig. 4. Examples of dopamine-containing axons in the lateral septal nucleus (A) and the anteromedial frontal cortex (B). In A the dopamine fibres are seen to aggregate around the proximal dendrites and cell body of a non-fluorescent nerve cell.

and localization of cell bodies of origin (see above) three dopaminergic terminal systems have been distinguished in the frontal lobe (Lindvall *et al.*, 1978; Fig. 2). *The anteromedial system* is formed by smooth axons distributed mainly in the pregenual part of the anteromedial cortex and the highest fibre density is found in the basal cortical layers (Fig. 4B). A much more sparse, caudal extension of this system is present in the basal layers in the supragenual part of the anteromedial cortex. *The suprarhinal system* forms the dorsal part of the dopaminergic innervation which is found in the perirhinal cortex, surrounding the rhinal sulcus. It can be regarded as a direct lateral continuation of the anteromedial system, and can be followed from a coronal level just rostral to the nc. accumbens to the level of the most rostral part of the nc. caudatus-putamen. The axons are distributed mainly in the basal cortical layers. *The supragenual system* is formed by very fine-varicose axons distributed in a restricted area of the supragenual anteromedial cortex. The axons are localized in the superficial cortical layers (I–III).

THE PERIVENTRICULAR SYSTEM

A prominent system of catecholamine-containing cell bodies and fibres is distributed in the periventricular and periaqueductal gray of the medulla oblongata, pons, mesencephalon and diencephalon (Lindvall and Björklund, 1974b; Lindvall *et al.*, 1974b). It can be regarded as a catecholamine-containing component of the dorsal longitudinal fasciculus of Schütz. Part of the periventricular fibres originate in the neurones of the midline and periventricular catecholamine cell groups but other catecholamine neurone systems, such as the locus coeruleus and dorsal medullary cell system (group A2), contribute fibres to the periventricular system as well.

It is well established that the major portion of the midline and periventricular cell system is dopaminergic (Björklund and Nobin, 1973; Hökfelt *et al.*, 1976; unpublished data), but some of the neurones might be noradrenergic (Björklund and Nobin, 1973). On the other hand, the fibres coming from other cell groups such as the locus coeruleus, are in all probability noradrenergic. Thus, although the midline and periventricular *cell* system is regarded as primarily dopaminergic, the periventricular *fibre* system is mixed noradrenergic and dopaminergic. The relative contribution and projection patterns of the different components are not known in detail and the description below is therefore by necessity incomplete.

The periventricular fibre system has been distinguished into a dorsal part, the *dorsal periventricular system*, which can be regarded as component of the dorsal longitudinal fasciculus of Schütz, and a ventral part, *the ventral periventricular system*, extending along the periventricular region of the hypothalamus.

Caudally, *the dorsal periventricular system* (DPS) can be observed at the level of the nucleus of the solitary tract. The system extends from this area as a rather sparse bundle running rostrally underneath the ependyma of the fourth ventricle. At the level of the locus coeruleus, the system increases considerably in width and in number of fibres, in that the locus coeruleus and other pontine and medullary catecholamine cell groups contribute fibres to the system. From the region dorsal and medial to the locus coeruleus, the DPS ascends rostrally underneath the ependyma, occupying the lateral part of the periventricular gray.

The DPS gives off fibres to the raphe nuclei in the pons, and to the inferior and superior colliculi in the mesencephalon (Fig. 5). Fibres also run ventrally from the DPS towards the ventral mesencephalic tegmentum. At

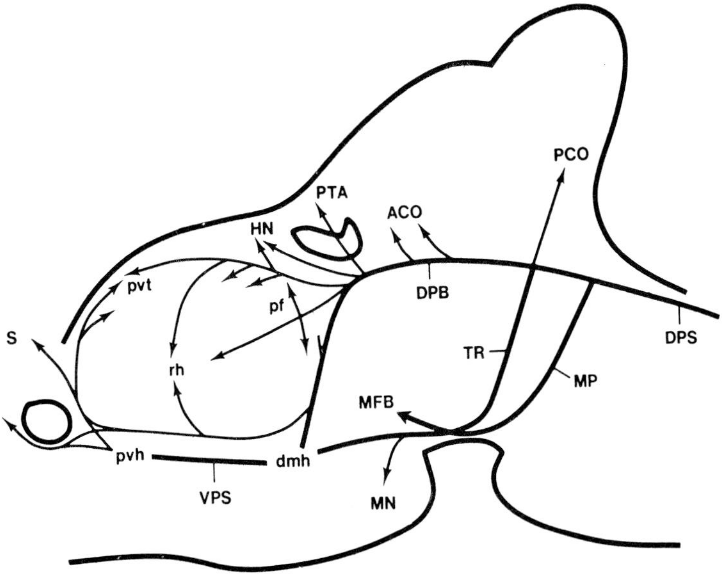

Fig. 5. Schematic representation of the projection routes of the periventricular fibre systems to the tectum, diencephalon and septum (from Lindvall *et al.*, 1974b). Abbreviations: ACO, anterior colliculus; dmh, dorsomedial hypothalamic nucleus; DPB, dorsal periventricular catecholamine bundle; DPS, dorsal periventricular catecholamine system; HN, habenular nuclei; MFB, medial forebrain bundle; MN, mamillary nuclei; MP, mamillary peduncle; PCO, posterior colliculus; pf, parafasciular nucleus; PTA, pretectal area; pvh, paraventricular hypothalamic nucleus; pvt, paraventricular thalamic nucleus; rh, rhomboid nucleus; S, septum; TR, tegmental catecholamine radiations; VPS, ventral periventricular catecholamine system.

the mesodiencephalic junction the DPS splits up into several branches (Fig. 5). One component of the DPS runs across the midline in the posterior commissure to reach *the pretectal area*, and a second component runs immediately in front of the posterior commissure into *the habenula*. A third component runs rostrally into *the paraventricular thalamic nucleus*, and a fourth component runs somewhat more ventral into *the parafascicular nucleus*. The major portion of the DPS turns however ventrally along the periventricular gray of the caudal thalamus through the posterior hypothalamus into *the dorsomedial hypothalamic nucleus*. Some of the axons running in the hypothalamic branch of the DPS turn rostrally, dorsal to the dorsomedial hypothalamic nucleus, to form a distinct bundle in *the medial zona incerta*. At the level of the paraventricular nucleus of the hypothalamus fibres leave the zona incerta bundle in dorsal direction towards *the rhomboid nucleus* of the thalamus. Still further rostrally the zona incerta bundle sweeps dorsally and caudally as periventricular thalamic fibres along the paraventricular thalamic nucleus giving off collaterals into this nucleus (Fig. 5).

The ventral periventricular system (VPS) is observed caudally in the region immediately dorsolateral to the interpeduncular nucleus (Fig. 5). In passing above *the medial mammillary nucleus*, fibres leave the bundle into the nucleus. At the level of the mammillothalamic tract, fibres from the medial part of the MFB leave this bundle, partly as collaterals, in a rostromedial direction to join the VPS. The VPS thus increases in width and fibre number as it passes in a rostral and somewhat dorsal direction through the posterior hypothalamic area into the dense catecholamine-containing terminal system of *the dorsomedial hypothalamic nucleus* (Fig. 5). After having received fibres from the hypothalamic branch of the DPS, the VPS continues as a broad ascending hypothalamic catecholamine fibre system dispersed on the lateral aspect of the periventricular nucleus to innervate *the paraventricular nucleus* and probably *the caudal septum* and *the interstitial nucleus of the stria terminalis*. The VPS also contributes fibres to the innervation of the medial thalamus (Fig. 5).

THE TUBEROHYPOPHYSEAL SYSTEM

The cell bodies of the tuberohypophyseal system are located in the arcuate nucleus (primarily its rostral half) and the portion of the periventricular nucleus just dorsal to the arcuate nucleus (Fuxe, 1964; Björklund *et al*., 1970; Lichtensteiger, 1970; Björklund *et al*., 1973). Their axons seem to project in an orderly manner to all parts of the median eminence, the stalk, the neural lobe, and the pars intermedia of the adenohypophysis (Fig. 6;

Björklund *et al.*, 1970; Jonsson *et al.*, 1972; Smith and Fink, 1972; Björklund *et al.*, 1973; Löfström *et al.*, 1976).

The dopamine innervation of *the median eminence and the stem* originates in cell bodies situated in the acruate nucleus and in the part of the periventricular nucleus lying dorsal to the arcuate nucleus (Fuxe and Hökfelt, 1969; Björklund *et al.*, 1970; Lichtensteiger, 1970; Jonsson *et al.*, 1972; Björklund *et al.*, 1973). These tuberoinfundibular dopamine neurones could be distinguished as two groups on the basis of their mode of projection (Fig. 6; Björklund *et al.*, 1973). One group, situated in the rostral part of the arcuate nuclei, projects more diffusely to all levels of the median eminence and the stem (as well as to the pars intermedia and the neural lobe as mentioned below). The second group, having a more regular dorsoventrally oriented projection, appears to connect each portion of the arcuate nucleus (i.e. the rostral, middle, or caudal portion) with a corresponding part of the ventrally situated median eminence.

The tuberoinfundibular dopamine neurones project to all layers of the median eminence and the stem. The terminals are more abundant in the external layer where they are very densely packed in a pallisade-like manner, close to the capillaries of the portal vessels. In the internal layer it

A

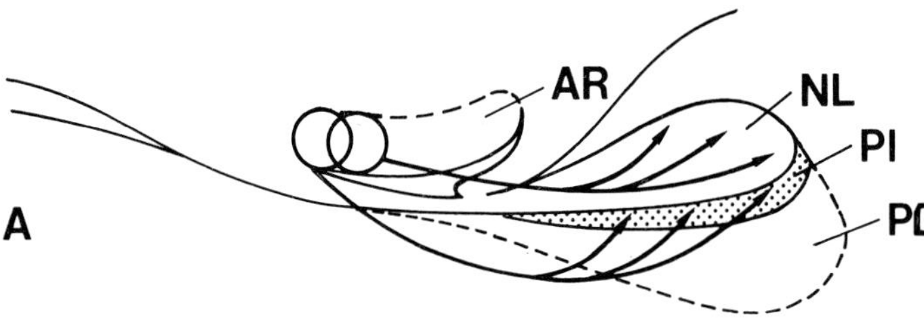

B

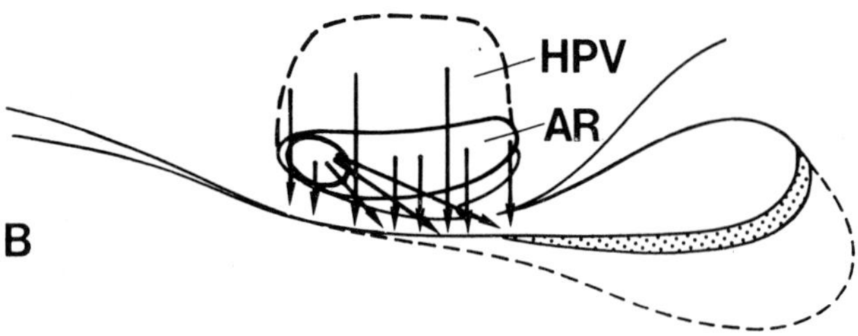

Fig. 6. Organization of the dopamine innervations in the median eminence—pituitary region. A. Arrangement of the dopamine innervation from the arcuate nucleus to the neural lobe and pars intermedia. B. Arrangement of the dopamine innervation from the arcuate and periventricular nuclei to the median eminence. Abbreviations: NL, neural lobe; PI, pars intermedia; PD, pars distalis; AR, arcuate nucleus; HPV, periventricular nucleus.

is likely that part of the fibres are of preterminal nature running parallel to one another in the sagittal and frontal planes. However, part of the fibres appear to ramify together with the noradrenergic fibres in this area to form irregular patterns in the deeper layers, partly in association with the capillary loops of the portal vessels.

Dopamine-containing fibres have been detected by microspectrofluorometric analysis also around the endocrine cells of the pars intermedia, and in the neural lobe and the median eminence (Björklund *et al.*, 1970). In *the neurointermediate lobe* it seems likely that the noradrenaline detected chemically derives from the peripheral sympathetic vascular supply, and that the central innervation is exclusively dopaminergic. Experiments with small, partial lesions of the arcuate nuclei (Björklund *et al.*, 1973) have provided evidence that the dopamine fibres of the neurointermediate lobe—as well as part of the dopamine fibres in the median eminence—originate in the rostral portion of the arcuate nucleus, the cells innervating the pars intermedia lying immediately rostral to those innervating the neural lobe (Fig. 6A).

The dopamine fibres of the neurointermediate lobe are predominantly of a fine varicose type distributed throughout the parenchyma of the neural lobe and around the endocrine cells of the pars intermedia. These fibres form the densest pattern in the vascular border zone between the two lobes and at the outer margins of the lobules formed by the intermedia cells. From here, the fibres penetrate between the intermedia cells to form a plexus inside the lobules.

THE INCERTOHYPOTHALAMIC SYSTEM

This fibre system, which has been demonstrated both by fluorescence histochemistry (Lindvall *et al.*, 1974b) and immunohistochemistry (Hökfelt *et al.*, 1976) can be topographically distinguished into a caudal and a rostral part (Fig. 7; Björklund *et al.*, 1975). *The caudal part* extends from the area of the dopamine cell bodies in the caudal thalamus, the posterior hypothalamic area and the medial zona incerta (groups A11 and A13) into the dorsal part of the dorsomedial nucleus and the dorsal and anterior hypothalamic areas. *The rostral part*, probably originating from the rostral periventricular cell bodies of the A14 cell group, is distributed in the periventricular nucleus of the anterior hypothalamus and in the periventricular, suprachiasmatic and medial preoptic nuclei, up to the level of the anterior commissure. The system probably extends into the most caudal part of the lateral septal nucleus.

The incertohypothalamic system consists of short, locally projecting

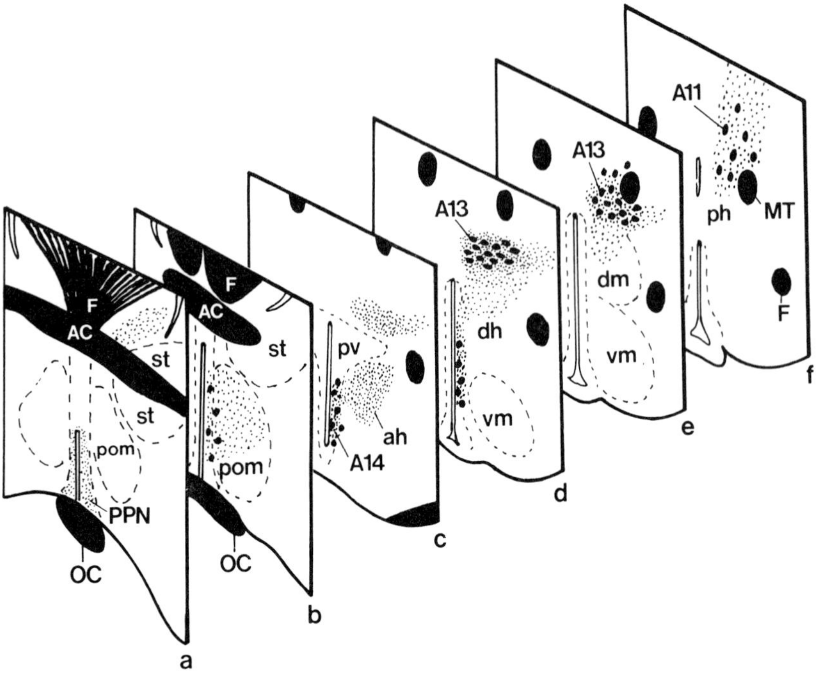

Fig. 7. Distribution of the incertohypothalamic fibre system at six representative frontal levels (a–f). The localization of the catecholamine-containing cell groups A11, A13 and A14 is also indicated. From Björklund *et al.* (1975). Abbreviations: AC, anterior commissure; ah, anterior hypothalamic nucleus; dh, dorsal hypothalamic area; dm, dorsomedial hypothalamic nucleus; F, fornix; MT, mammillothalamic tract; OC, optic chiasm; ph, posterior hypothalamic nucleus; pom, medial preoptic nucleus; PPN, periventricular preoptic nuclei; pv, periventricular hypothalamic nucleus; st, interstitial nucleus of the stria terminalis; vm, ventromedial nuclear group.

neurones whose fibres have an unusual and peculiar fluorescence morphology. The fibres arborize extensively very soon after they have left the cell bodies, into a network of closely and regularly spaced, very fine varicosities with a remarkably low amine content. Interestingly, the incertohypothalamic system has a prominent projection to the preoptic-anterior hypothalamic region, suggesting an involvement in neuroendocrine mechanisms. In the anterior hypothalamic area the fibres of the incertohypothalamic system constitute the predominating type of catecholamine innervation, whereas in the preoptic areas, as well as in the anterior periventricular and the dorsomedial nuclei, they occur intermingled with dense plexuses of noradrenaline-containing fibres. In the anterior hypothalamic and preoptic areas the incertohypothalamic fibres form dense, basket-like patterns around non-fluorescent neuronal perikarya.

THE PERIGLOMERULAR DOPAMINE NEURONES

A system of presumed dopaminergic periglomerular cells have been found in the rat olfactory bulb (Hökfelt *et al.*, 1975b; Halász *et al.*, 1977; see also Dahlström *et al.*, 1965; Lichtensteiger, 1966; and Lidbrink *et al.*, 1974). These cells have been calculated to constitute 8–22% of all periglomerular neurones (Halász *et al.*, 1977). Presumed dopamine-containing fibres have been followed into the glomeruli and it has been assumed that these processes mainly represent dendrites of the periglomerular cells; this is supported by electron microscopic immunohistochemistry. It has been suggested (Halász *et al.*, 1977) that dopamine is released from the periglomerular cells at dendro-dendritic synapses with mitral cells, where dopamine would then act as an inhibitory transmitter (for references, see Halász *et al.*, 1977).

THE RETINAL DOPAMINE SYSTEM

Dopaminergic interneurones have been described in the retina of many submammalian and mammalian species, including the rat (Malmfors, 1963; Ehinger, 1966; Laties and Jacobowitz, 1966; for review see Ehinger, 1976). Most of these neurones have their cell bodies in the inner nuclear layer, among the amacrine cells. In the rat the terminals are located exclusively in a band at the outer border of the inner plexiform layer. In other species, notably New World monkeys and teleost fishes, a second terminal plexus is found in the outer plexiform layer (Laties and Jacobowitz, 1966; Ehinger and Falck, 1969; Ehinger *et al.*, 1969). In these species Dowling and Ehinger (1975) have shown that the dopamine neurones make synapses with amacrine cells in the inner plexiform layer and with bipolar and horizontal cells in the outer plexiform layer. In addition, the dopamine-containing processes were found to be postsynaptic to synapses from processes of amacrine cells in the inner plexiform layer. The retinal dopamine neurone thus seems to be an interesting example of a neurone where the dopamine-containing processes are both pre- and postsynaptic. This arrangement can be regarded as analogous to that observed for the dopamine neurones of the substantia nigra (Björklund and Lindvall, 1975).

CONCLUSIONS

Methodological developments during the last few years have involved a remarkable expansion of our knowledge of the organization of central dopamine neurones. This includes both more data on well-known systems as well as the findings of previously unknown dopamine neuronal systems. Despite this, more work is still required before the organizational features of the central dopamine neurones are clarified in detail. Certain neuronal systems such as the periventricular system, and certain brain regions, above all the lower brain stem, are still incompletely mapped. It will be of particular interest to study under the microscope the relations between the dopamine systems and other neurone systems, e.g. cholinergic and peptidergic. For example, in the lateral septal nucleus there seems to be, at least in some areas, a perfect correlation between the projection area of dopaminergic neurones and acetylcholinesterase-containing neurones (Lindvall, 1977). This result would thus provide a morphological basis for functional studies of dopaminergic-cholinergic nerve interactions in the septum.

An interesting new aspect of the organization of central dopamine neurones is given by the recent finding of an overlap of the neocortical projection fields of the mediodorsal thalamic nucleus and the mesencephalic dopaminergic neurones in the prefrontal cortex of several species (Divac *et al.*, 1975, 1978; Beckstead, 1976; Berger *et al.*, 1976; Björklund *et al.*, 1978. This points to a much more precise and "specific" organization of the dopamine systems than has previously been known. To what extent such a "specificity" exists also for other dopamine systems in the brain remains to be clarified.

ACKNOWLEDGEMENT

This research was supported by a grant from the Swedish Medical Research Council (04X-4493).

REFERENCES

Andén, N.-E., Carlsson, A., Dahlström, A., Fuxe, K., Hillarp, N.-Å., and Larsson, K. (1966a). Ascending monoamine neurons to the telencephalon and diencephalon. *Acta Physiol. Scand.* **67,** 313–326.

Andén, N.-E., Dahlström, A., Fuxe, K., Larsson, K., Olson, L., and Ungerstedt, U. (1966a). Ascending monoamine neurones to the telencephalon and diencephalon. *Acta Physiol. Scand.* **67,** 313–326.

Andén, N.-E., Fuxe, K., Hamberger, B., and Hökfelt, T. (1966b). A quantitative study on the nigro-neostriatal dopamine neuron system in the rat. *Acta Physiol. Scand.* **67,** 306–312.

Beckstead, R. M. (1976). Convergent thalamic and mesencephalic projections to the anterior medial cortex in the rat. *J. Comp. Neurol.* **166,** 403–416.

Bédard, P., Larochelle, L., Parent, A., and Poirier, L. J. (1969). The nigrostriatal pathway: A correlative study based on neuroanatomical and neurochemical criteria in the cat and the monkey. *Exp. Neurol.* **25,** 365–377.

Ben-Ari, Y., Zigmond, R. E., and Moore, K. E. (1975). Regional distribution of tyrosine hydroxylase, norepinephrine and dopamine within the amygdaloid complex of the rat. *Brain Res.* **87,** 96–101.

Berger, B., Tassin, J. P., Blanc, B., Moyne, M. A., and Thierry, A. M. (1974). Histochemical confirmation for dopaminergic innervation of the rat cerebral cortex after destruction of the noradrenergic ascending pathways. *Brain Res.* **81,** 332–337.

Berger, B., Thierry, A. M., Tassin, J. P., and Moyne, M. A. (1976). Dopaminergic innervation of the rat prefrontal cortex: a fluorescence histochemical study. *Brain Res.* **106,** 133–145.

Bertler, Å., Falck, B., Gottfries, C. G., Ljunggren, L., and Rosengren, E. (1964). Some observations on adrenergic connections between mesencephalon and cerebral hemispheres. *Acta Pharmacol. Toxicol.* **21,** 283–289.

Björklund, A., Divac, I., and Lindvall, O. (1978). Regional distribution of catecholamines in monkey cerebral cortex. Evidence for a dopaminergic innervation of the primate prefrontal cortex. *Neurosci. Lett.* **7,** 115–119.

Björklund, A., Falck, B., Hromek, F., Owman, Ch., and West, K. A. (1970). Identification and terminal distribution of the tuberohypophyseal monoamine fibre systems in the rat by means of stereotaxic and microspectrofluorimetric techniques. *Brain Res.* **17,** 1–23.

Björklund, A., and Lindvall, O. (1975). Dopamine in dendrites of substantia nigra neurons: suggestions for a role in dendritic terminals. *Brain Res.* **83,** 531–537.

Björklund, A., Lindvall, O., and Moore, R. Y. (1979). The dopaminergic systems. *In* "The Central Catecholamine Neurone" (Eds. A. Björklund, and R. Y. Moore), Raven Press, New York, to be published.

Björklund, A., Lindvall, O., and Nobin, A. (1975). Evidence of an incertohypothalamic dopamine neurone system in the rat. *Brain Res.* **89,** 29–42.

Björklund, A., Moore, R. Y., Nobin, A., and Stenevi, U. (1973). The organization of tuberohypophyseal and reticuloinfundibular catecholamine neuron systems in the rat brain. *Brain Res.* **51,** 171–191.

Björklund, A., and Nobin, A. (1973). Fluorescence histochemical and microspectrofluorometric mapping of dopamine and noradrenaline cell groups in the rat diencephalon. *Brain Res.* **51,** 193–205.

Brownstein, M., Saavedra, J. M., and Palkovits, M. (1974). Norepinephrine and dopamine in the limbic system of the rat. *Brain Res.* **79,** 431–436.

Cajal, S. and Ramón, Y. (1955). "Histologie du Système Nerveux de l'Homme et des Vertébrés", Vol. 2, 275–278. Consejo Superior de Investigaciones Cientificas, Inst. Ramón y Cajal, Madrid.

Carpenter, M. B., and Peter, P. (1972). Nigrostriatal and nigrothalamic fibers in the rhesus monkey. *J. Comp. Neurol.* **144,** 93–116.

Carter, D. A., and Fibiger, H. C. (1977). Ascending projections of presumed dopamine-containing neurones in the ventral tegmentum of the rat as demonstrated by horseradish peroxidase. *Neuroscience* **2,** 569–576.

Collier, T. J., and Routtenberg, A. (1976). Entorhinal cortex: catecholamine fluorescence and Nissl staining of identical sections. Neuroscience Abstracts, Vol. II, Society for Neuroscience Sixth Annual Meeting, Toronto, Canada, p. 483.

Cowan, W. M., Gottlieb, D. I., Hendrickson, A. E., Price, J. L., and Woolsey, T. A. (1972). The autoradiographic demonstration of connections in the central nervous system. *Brain Res.* **37,** 21–51.

Dahlström, A., and Fuxe, K. (1964). Evidence for the existence of monoamine-containing neurons in the central nervous system. I: Demonstration of monoamines in the cell bodies of brain stem neurones. *Acta physiol. scand.* (Suppl. 232), **62,** 1–55.

Dahlström, A., Fuxe, K., Olson, L., and Ungerstedt, U. (1965). On the distribution and possible function of monoamine nerve terminals in the olfactory bulb of the rabbit. *Life Sci.* **4,** 2071–2074.

Divac, I., Lindvall, O., Björklund, A., and Passingham, R. E. (1975). Converging projections from the mediodorsal thalamic nucleus and mesencephalic dopaminergic neurons to the neocortex in three species. *Exp. Brain Res.* **23,** 58 (Abstract).

Divac, I., Björklund, A., Lindvall, O., and Passingham, R. E. (1978). Converging projections from the mediodorsal thalamic nucleus and mesencephalic dopaminergic neurons to the neocortex in three species. *J. Comp. Neurol.* **180,** 59–72.

Domesick, V. B., Beckstead, R. M., and Nauta, W. J. H. (1976). Some ascending and descending projections of the substantia nigra and ventral tegmental area in the rat. Neuroscience Abstracts, Vol. II, Society for Neuroscience Sixth Annual Meeting, Toronto, Canada, p. 61.

Dowling, J. E., and Ehinger, B. (1975). Synaptic organization of the amine-containing interplexiform cells of the goldfish and cebus monkey retinas. *Science* **188,** 270–273.

Ehinger, B. (1966). Adrenergic nerves to the eye and to related structures in man and in the cynomolgus monkey. *Invest. Opthalmol.* **5,** 42–52.

Ehinger, B. (1976). Biogenic monoamines as transmitters in the retina. *In* "Transmitters of the Visual Process", (Ed. S. L. Bonting). Pergamon Press, Oxford.

Ehinger, B., and Falck, B. (1969). Adrenergic retinal neurons of some New World monkeys. *Z. Zellforsch.* **100,** 364–372.

Ehinger, B., Falck, B. and Laties, A. M. (1969). Adrenergic neurons in teleost retina. *Z. Zellforsch.* **97,** 285–297.

Emson, P. C., and Koob, G. (1978). The origin and distribution of dopamine-containing afferents to the frontal cortex. *Brain Res.* **142,** 249–267.

Falck, B. (1962). Observations on the possibilities of the cellular localization of monoamines by a fluorescence method. *Acta Physiol. Scand. Suppl. 197,* **56,** 1–25.

Falck, B., Hillarp, N.-Å., Thieme, G., and Torp, A. (1962). Fluorescence of catecholamines and related compounds condensed with formaldehyde. *J. Histochem. Cytochem.* **10,** 348–354.

Fallon, J. H., Koziell, D. A., and Moore, R. Y. (1978). Catecholamine innervation of the basal forebrain. II. Amygdala, suprarhinal cortex and entorhinal cortex. *J. Comp. Neurol.* **180,** 509–532.

Fallon, J. H., and Moore, R. Y. (1976). Dopamine innervation of some basal forebrain areas in the rat. Neuroscience Abstracts, Vol. II, Society for Neuroscience Sixth Annual Meeting, Toronto, Canada, p. 486.

Fallon, J. H., and Moore, R. Y. (1978a). Catecholamine innervation of the basal forebrain. III. Olfactory bulb, anterior olfactory nuclei, olfactory tubercle and piriform cortex. *J. Comp. Neurol.* **180,** 533–544.

Fallon, J. H., and Moore, R. Y. (1978b). Catecholamine innervation of the basal forebrain. IV. Topography of the dopamine projection to the basal forebrain and neostriatum. *J. Comp. Neurol.* **180,** 545–580.

Fuxe, K. (1964). Cellular localization of monoamines in the median eminence and infundibular stem of some mammals. *Z. Zellforsch.* **61,** 710–724.

Fuxe, K. (1965). Evidence for the existence of monoamine neurons in the central nervous system. IV: Distribution of monoamine nerve terminals in the central nervous system. *Acta Physiol. Scand. Suppl.* 247, **64,** 39–85.

Fuxe, K., and Hökfelt, T. (1969). Catecholamines in the hypothalamus and the pituitary gland. *In* "Frontiers in Neuroendocrinology" (Eds. W. F. Ganong and L. Martini), 47–96. Oxford University Press, Oxford.

Fuxe, K., Hökfelt, T., Johansson, O., Jonsson, G., Lidbrink, P., and Ljungdahl, Å. (1974). The origin of the dopamine nerve terminals in limbic and frontal cortex. Evidence for meso-cortico dopamine neurons. *Brain Res.* **82,** 349–355.

Geffen, L., Jessell, T. M., Cuello, A. C., and Iversen, L. L. (1976). Release of dopamine from dendrites in rat substantia nigra. *Nature* (*London*) **260,** 258–260.

Halász, N., Hökfelt, T., Ljungdahl, Å., Johansson, O., Ebstein, M., Goldstein, M., Park, D., and Biberfeld, P. (1977). Transmitter histochemistry of the rat olfactory bulb. I. Immunohistochemical localization of monoamine synthesizing enzymes. Support for intrabulbar, periglomerular dopamine neurons. *Brain Res.* **126,** 455–474.

Hedreen, J. C., and Chalmers, J. P. (1972). Neuronal degeneration in rat brain induced by 6-hydroxydopamine; a histological and biochemical study. *Brain Res.* **47,** 1–36.

Herkenham, M., and Nauta, W. J. H. (1977). Afferent connections of the habenular nuclei in the rat. A horseradish peroxidase study with a note on the fiber-of-passage problem. *J. Comp. Neurol.* **173,** 123–146.

Hökfelt, T., Fuxe, K., and Goldstein, M. (1975a). Applications of immunohistochemistry to studies on monoamine cell systems with special references to nervous tissues. *Ann. N.Y. Acad. Sci.* **254,** 407–432.

Hökfelt, T., Fuxe, K., Johansson, O., and Ljungdahl, Å. (1974a). Pharmacohistochemical evidence of the existence of dopamine nerve terminals in the limbic cortex. *Eur. J. Pharmacol.* **25,** 108–112.

Hökfelt, T., Halász, N., Ljungdahl, Å., Johansson, O., Goldstein, M., and Park, D. (1975b). Histochemical support for a dopaminergic mechanism in the dendrites of certain periglomerular cells in the rat olfactory bulb. *Neurosci. Lett.* **1,** 85–90.

Hökfelt, T., Johansson, O., Fuxe, K., Goldstein, M., and Park, D. (1976). Immunohistochemical studies on the localization and distribution of monoamine neuron systems in the rat brain. II. Tyrosine hydroxylase in the mes- and diencephalon. *Med. Biol.* **54,** 427–453.

Hökfelt, T., Johansson, O., Fuxe, K., Goldstein, M., and Park, D. (1977). Immunohistochemical studies on the localization and distribution of monoamine neuron systems in the rat brain. II. Tyrosine hydroxylase in the telencephalon. *Med. Biol.* **55,** 21–40.

Hökfelt, T., Ljungdahl, Å., Fuxe, K., and Johansson, O. (1974b). Dopamine nerve terminals in the rat limbic cortex: aspects of the dopamine hypothesis of schizophrenia. *Science* **184,** 177–179.

Hökfelt, T., and Ungerstedt, U. (1969). Electron and fluorescence microscopical studies on the nucleus caudatus putamen of the rat after unilateral lesions of ascending nigro-neostriatal dopamine neurons. *Acta Physiol. Scand.* **76,** 415–426.

Jonsson, G., Fuxe, K., and Hökfelt, T. (1972). On the catecholamine innervation of the hypothalamus, with special reference to the median eminence. *Brain Res.* **40,** 271–281.

Korf, J., Zieleman, M., and Westerink, B. H. C. (1976). Dopamine release in substantia nigra? *Nature* (*London*) **260,** 257–258.

Koslow, S. H. Racagni, G., and Costa, E. (1974). Mass fragmentographic measurement of norepinephrine, dopamine, serotonin and acetylcholine in seven discrete nuclei of the rat tel-diencephalon. *Neuropharmacology* **13,** 1123–1130.

Laties, A., and Jacobowitz, D. (1966). A comparative study of the autonomic innervation of the eye in monkey, cat and rabbit. *Anat. Rec.* **156,** 383.

La Vail, J. H., and La Vail, M. M. (1972). Retrograde axonal transport in the central nervous system. *Science* **176,** 1416–1417.

Lichtensteiger, W. (1966). Uptake of noradrenaline in periglomerular cells of the olfactory bulb of the mouse. *Nature* (*London*) **210,** 955–956.

Lichtensteiger, W. (1970). Katecholaminhaltige Neurone in der neuroendokrinen Steuerung. *Progr. Histochem. Cytochem.* **1,** 185–276.

Lidbrink, P., Jonsson, G., and Fuxe, K. (1974). Selective reserpine-resistant accumulation of catecholamines in central dopamine neurones after DOPA administration. *Brain Res.* **67,** 439–456.

Lindvall, O. (1975). Mesencephalic dopaminergic afferents to the lateral septal nucleus of the rat. *Brain Res.* **87,** 89–95.

Lindvall, O. (1977). Combined visualization of central catecholamine- and acetylcholinesterase-containing neurons: application of the glyoxylic acid and thiocholine histochemical methods to the same Vibratome section. *Histochemistry* **50,** 191–196.

Lindvall, O., and Björklund, A. (1974a). The glyoxylic acid fluorescence histochemical method: a detailed account of the methodology for the visualization of central catecholamine neurons. *Histochemistry* **39,** 97–127.

Lindvall, O., and Björklund, A. (1974b). The organization of the ascending catecholamine neuron systems in the rat brain as revealed by the glyoxylic acid fluorescence method. *Acta Physiol. Scand. Suppl. 412* . 1–48.

Lindvall, O., Björklund, A., and Divac, I. (1977). Organization of mescencephalic dopamine neurons projecting to neocortex and septum. *In* "Advances in Biochemical Psychopharmacology", Vol. 16 (Eds. E. Costa, and G. L. Gessa), 39–46. Raven Press, New York.

Lindvall, O., Björklund, A., and Divac, I. (1978). Organization of catecholamine neurons projecting to the frontal cortex in the rat. *Brain Res.* **142,** 1–24.

Lindvall, O., Björklund, A., Moore, R. Y., and Stenevi, U. (1974a). Mesencephalic dopamine neurons projecting to neocortex. *Brain Res.* **81,** 325–331.

Lindvall, O., Björklund, A., Nobin, A., and Stenevi, U. (1974b). The adrenergic innervation of the rat thalamus as revealed by the glyoxylic acid fluorescence method. *J. Comp. Neurol.* **154,** 317–348.

Lindvall, O. and Stenevi, U. (1978). Dopamine and noradrenaline neurons projecting to the septal area in the rat. *Cell Tiss. Res.* **190,** 383–407.

Löfström, A., Jonsson, G., and Fuxe, K. (1976). Microfluorometric quantitation of catecholamine fluorescence in rat median eminence. I. Aspects on the distribution of dopamine and noradrenaline nerve terminals. *J. Histochem. Cytochem.* **24,** 415–429.

Maler, L., Fibiger, H. C., and McGeer, P. L. (1973). Demonstration of the nigrostriatal projection by silver staining after nigral injections of 6-hydroxydopamine. *Exp. Neurol.* **40,** 505–515.

Malmfors, T. (1963). Evidence of adrenergic neurons with synaptic terminals in the retina of rats demonstrated with fluorescence and electron microscopy. *Acta Physiol. Scand.* **58,** 99–100.

Moore, R. Y. (1978). Catecholamine innervation of the basal forebrain. I. The septal area. *J. Comp. Neurol.* **177,** 665–684.

Moore, R. Y., Bhatnagar, R. K., and Heller, A. (1971a). Anatomical and chemical studies of a nigro-neostriatal projection in the cat. *Brain Res.* **30,** 119–135.

Moore, R. Y., Björklund, A., and Stenevi, U. (1971b). Plastic changes in the adrenergic innervation of the rat septal area in response to denervation. *Brain Res.* **33,** 13–35.

Nauta, W. J. H. and Domesick, V. B. (1978). *In* "Limbic Mechanisms. The Continuing Evolution of the Limbic System Concept" (Eds. K. Livingston and O. Hornykiewicz), 75–93 Plenum, New York.

Nauta, H. J. W., Pritz, M. B., and Lasek, R. J. (1974). Afferents to the rat caudo-putamen studied with horseradish peroxidase. An evolution of a retrograde neuroanatomical research method. *Brain Res.* **67,** 219–238.

Nobin, A., and Björklund, A. (1973). Topography of the monoamine neuron systems in the human brain as revealed in fetuses. *Acta Physiol. Scand. Suppl. 388*, 1–40.

Pickel, V. M., Joh, T. H., Field, P. M., Becker, C. G., and Reis, D. J. (1975a). Cellular localization of tyrosine hydroxylase by immunohistochemistry. *J. Histochem. Cytochem.* **23,** 1–12.

Pickel, V. M., Joh, T. H., and Reis, D. J. (1975b). Immunohistochemical localization of tyrosine hydroxylase in brain by light and electron microscopy. *Brain Res.* **85,** 295–300.

Poirier, L. J., and Sourkes, T. L. (1965). Influence of the substantia nigra on the catecholamine content of the striatum. *Brain* **88,** 181–192.

Saavedra, J. M., and Zivin, I. (1976). Tyrosine hydroxylase and dopamine-β-hydroxylase: Distribution in discrete areas of the rat limbic system. *Brain Res.* **105,** 517–524.

Shimizu, N., and Ohnishi, S. (1973). Demonstration of nigro-neostriatal tract by degeneration silver method. *Exp. Brain Res.* **17,** 133–138.

Simon, H., LeMoal, M., Galey, D., and Cardo, B. (1974). Selective degeneration of central dopaminergic systems after injection of 6-hydroxydopamine in the ventral mesencephalic tegmentum of the rat. Demonstration by the Fink–Heimer stain. *Exp. Brain Res.* **20,** 375–384.

Simon, H., LeMoal, M., Galey, D. and Cardo, B. (1976). Silver impregnation of dopaminergic systems after radiofrequency and 6-OHDA lesions of the rat ventral tegmentum. *Brain Res.* **115,** 215–231.

Smith, G. C., and Fink, G. (1972). Experimental studies on the origin of monoamine-containing fibres in the hypothalamo-hypophyseal complex of the rat. *Brain Res.* **43,** 37–51.

Thierry, A. M., Stinus, L., Blanc, G., and Glowinski, J. (1973a). Some evidence for the existence of dopaminergic neurons in the rat cortex. *Brain Res.* **50,** 230–234.

Thierry, A. M., Blanc, G., Sobel, A., Stinus, L., and Glowinski, J. (1973b). Dopamine terminals in the rat cortex. *Science* **182,** 499–501.

Ungerstedt, U. (1971). Stereotaxic mapping of the monoamine pathways in the rat brain. *Acta Physiol. Scand. Suppl. 367*, 1–48.

CHAPTER 21

Dopamine Levels of Individual Brain Regions: Biochemical Aspects of DA Distribution in the Central Nervous System

M. Palkovits

Semmelweis University Medical School,
Budapest, Hungary.

Biochemical measurements have verified that the central nervous system contains a relatively high amount of dopamine (DA) that is not evenly distributed, most of it being discernible in the relatively bulky striatum (Glowinski and Iversen, 1966; Popov *et al.*, 1967). By the refinement of the biochemical micromethods and the application of enzymatic isotopic techniques it became possible to measure DA concentrations not only in more extended brain regions, but also in rather small discrete areas. By the application of different radioenzymatic methods (Coyle and Henry, 1973; Cuello *et al.*, 1973; Palkovits *et al.*, 1974; Henry *et al.*, 1975; van der Gugten *et al.*, 1976; Versteeg *et al.*, 1976) the DA content of more than a hundred individual brain regions is now known. By comparing these biochemical data with those obtained from the formaldehyde (Falck–Hillarp, 1959) and glyoxylic acid (Axelsson *et al.*, 1973; Lindvall and Björklund, 1974) induced fluorescence techniques together with the results of the immunofluorescence method (Hökfelt *et al.*, 1973a, b) information may be obtained on the topographical distribution of dopamine within the central nervous system. It is well known that the dopamine-containing cell bodies are found in well-defined areas of the central nervous system (Dahlström and Fuxe, 1964; Björklund and Nobin, 1973) from where hundreds of thousands of their axons and axoncollaterals encompass almost all areas of the central nervous system.

The distribution and concentration of dopamine in the different cell groups of the central nervous system have been subjected to further examination. The DA content of brain areas in this review refers to a concrete time and physiological situation. Therefore, under other measuring parameters the DA concentrations may essentially differ. For neurophysiological examinations, the determination of the actual DA level has to be completed by measuring the activity of the enzymes involved in the biosynthesis and turnover of DA in order to obtain functional conclusions from the results.

The Tables list the DA content from 101 areas of the brain indicated in ng per mg protein. In cases in which data were obtained by different methods, these are indicated. While the measured and calculated values obtained by the modified method (Palkovits *et al.*, 1974) of Coyle and Henry (1973) might differ to a relatively higher percentage from the true DA value, the method of Cuello *et al.* (1973) and its modification (van der Gugten *et al.*, 1976; Versteeg *et al.*, 1976), yielded results in which the cross-interference (the contamination of noradrenaline in the DA assay) was less than 1%. Upon comparing the results obtained with the two methods (in two different laboratories) the differences proved to be minimal—they might have been due to differences of species and environment.

Apart from the absolute values of the DA concentrations relative values are also indicated. Upon comparing the different brain areas whose rank order (1–101) is expressed as a percentage of the DA content of the richest area, the olfactory tubercle, it is perhaps surprising that only one quarter of the examined brain regions reaches or exceeds 10% of the DA value of the olfactory tubercle.

In the Tables the DA concentration is indicated in percentages of noradrenaline in the same brain region. Noradrenaline, like DA, can be revealed in all areas of the central nervous system in measurable quantities. Its concentration is generally (about 80% of the cases in our material) higher than that of dopamine. The question therefore arises whether dopamine elicits its effect everywhere as an independent neurotransmitter, or only as a precursor of noradrenaline. From their simple percentage this question cannot be decided in all brain areas because the noradrenaline–dopamine ratio might change according to the speed of their biosynthesis, the noradrenaline turnover or to the actual functional activity of the system. No brain area exists, from which it could be convincingly ascertained that the neurones or nerve terminals situated there are exclusively of a noradrenergic nature, hence showing that the DA there could only be a precursor. Without any experimental evidence we assumed that where the DA concentration is higher than that of noradrenaline, or happens to exceed its half-value, there DA may well be an independent neurotransmitter. If the DA content exceeds that of NA by between 20 and 50% it is always possible that DA has an independent role, while this is doubtful if the difference is beneath 20%. This purely theoretical assumption still needs evidence but we are convinced that this index is not self-existent but rather yields useful information.

Generally, but not in every case, the concentration of DA agrees with the observations made by histofluorescence studies. This can be assigned to the efficiency and specificity of the histofluorescent methods. The disadvantage of the method, i.e. the subjectivity of its evaluation, does not readily permit quantitative application, and thus comparative estimates with other methods are not possible. Conversely, as the numerical values of the biochemical methods do not rely on any cytological information, only morphological methods can help to determine which structures contain the low, or high DA concentration measured in a given area. It is clear that care should be taken in the choice of the method for an experiment and the choice will, of course, be influenced by the type of experiment performed. It is beyond doubt that their joint application yields the most valuable information.

Among the areas rich in dopamine, the olfactory tubercle, the nucleus accumbens and the nucleus caudatus-putamen (Table 1) play an outstanding role. Owing to a large volume, the caudatus-putamen contains a considerable amount of DA that, however, is not distributed evenly within the nucleus but decreases caudally. The highest DA content is found in the rostral part (nucleus caudatus). In all three mentioned areas DA is located in the nerve terminals, the perikarya of which are found in the ventral tegmental area and in that of the substantia nigra (Dahlström and Fuxe, 1964; Dahlström *et al.*, 1964; Andén *et al.*, 1964, 1965, 1966; Hökfelt and Ungerstedt, 1969, 1974; Ungerstedt, 1971; Fuxe *et al.*, 1974; Lindvall *et al.*, 1974; Koob *et al.*, 1975).

The DA concentration is fairly high in the median eminence (Table 2). Dopamine in the axons and presumably in the nerve terminals originates from the arcuate and periventricular DA cells which are called the A12 catecholaminergic cell-group (Fuxe, 1964, 1965; Fuxe and Hökfelt, 1966; Björklund *et al.*, 1970, 1974). Recently, data were presented that the cell-groups A9 and A10 are also involved in the dopaminergic innervation of the median eminence (Kizer *et al.*, 1976; Palkovits *et al.*, 1977).

A high DA content is found in the globus pallidus (Table 1). The giant

TABLE I Concentrations of dopamine in the telencephalon

	ng per mg protein		Rank	Percentage of the highest DA in the CNS		Percentage of the noradrenaline (100%) in the same area	
	A[a]	B[b]	(1–101)	A[a]	B[b]	A[a]	B[b]
Pallium							
Olfactory bulb	2.8		61–70	3.0		59.2	
Olfactory tubercle	94.6	114.4	1	100.0	100.0	1398.8	1588.9
Nucleus of diagonal band	32.6	35.2	7	34.5	30.8	113.1	138.0
Hippocampus	0.4		91–101	0.4		8.6	
Cortical areas							
Frontal cortex	1.0	0.2[c]	81–90	1.1		45.6	12.1[c]
Parietal cortex	0.8		81–90	0.9		23.7	
Insular cortex		0.3[c]	91–101				15.5[c]
Temporal cortex		0.3[c]	91–101				15.1[c]
Occipital cortex		0.3[c]	91–101				22.3[c]
Retrohippocampal cortex		0.3[c]	91–101				19.9[c]
Cingulate cortex	0.9	0.5[c]	81–90	1.0		20.1	25.1[c]
Pyriform cortex	2.9	2.1[c]	61–70	3.0		63.3	70.6[c]
Entorhinal cortex	2.4	1.1[c]	61–70	2.5		41.6	36.1[c]
Basal ganglia							
Caudate nucleus	70.5	96.6	3	74.5	84.4		8781.8
Caudate-putamen		88.4	4		77.3		6800.0
Nucleus accumbens	87.2		2	92.2			933.3
Globus pallidus	39.6		6	41.9			747.5

Septum							
Dorsal. nucleus	8.1	13.2	31–40	8.6	11.5	123.4	203.1
Intermediate nucleus	14.5	19.3	18	15.3	16.9	175.0	321.7
Medial nucleus	8.6	14.0	30	9.1	12.2	86.1	140.0
Lateral nucleus	21.4	21.2	9	22.6	18.5	173.4	192.7
Fimbrial nucleus	4.3	2.6	51–60	4.6	2.3	49.9	23.9
Triangular nucleus		4.1	51–60		3.6		23.6
Subfornical organ		7.0[d]	31–40		6.1[d]		70.0[d]
Amygdala							
Nucl. tract. olf. lat.	8.2	6.6	31–40	9.7	5.8	81.7	51.6
Lateral nucleus	17.5	11.5	13	18.5	10.1	161.1	95.8
Medial nucleus	7.5	3.6	31–40	7.9	3.1	59.6	31.6
Central nucleus	15.0	16.9	17	15.9	14.8	84.4	126.1
Cortical nucleus	4.8	5.1	51–60	5.1	4.5	56.6	68.0
Basal nucleus	7.6	8.6	31–40	8.1	7.5	78.9	94.5
Medial posterior nucleus	1.4	1.6	71–80	1.5	1.4	13.7	37.2
Basal posterior nucleus		5.4	41–50		4.7		85.7
Posterior nucleus	1.6	1.9	71–80	1.7	1.7	18.7	76.0

[a] A—Versteeg *et al.* (1976).
[b] B—Brownstein *et al.* (1974).
[c] Palkovits, Brownstein, Záborszky, Léránth (in preparation).
[d] Saavedra *et al.* (1976).

TABLE II Concentrations of dopamine in the Diencephalon

	ng per mg protein		rank	Percentage of the highest DA in the CNS		Percentage of the noradrenaline (100%) in the same area	
	A[a]	B[b]	(1–101)	A[a]	B[b]	A[a]	B[b]
Thalamus							
Anterior ventral nucleus	1.2		81–90	1.3		9.4	
Ventral nucleus	0.2		91–101	0.3		6.5	
Posterior nucleus	2.3		71–80	2.4		16.5	
Medial dorsal nucleus	1.0		81–90	1.1		10.3	
Periventricular nucleus		18.6[c]	12				118.5[c]
Midline nuclei	4.0		51–60	4.2		34.3	
Epithalamus							
Habenular nuclei		3.9[c]	51–60				
Metathalamus							
Lateral geniculate body	1.0		81–90	1.0		9.5	
Medial geniculate body	2.2		71–80	2.4		28.6	
Subthalamus							
Subthalamic nucleus	5.3		41–50	5.6		49.8	
Zona incerta	5.5		41–50	5.8		39.8	
Preoptic region							
Medial preoptic nucleus	12.3	10.4[c]	21	13.0		44.0	103.0[c]
Preoptic suprachiasmatic nucleus		11.7	22		10.2		29.6

Nucleus interstitialis striae terminalis							
pars ventralis		13.9	19		12.2		17.3
pars dorsalis		20.1	11		17.6		125.6
Nucleus interstitialis striae medullaris	13.4	13.9	20	14.1	12.2	46.8	36.9
Organum vasculosum laminae terminalis		21.0[d]	10		18.4[d]		190.9[d]
Hypothalamus							
Medial forebrain bundle	7.6	7.5	31–40	8.0	6.6	19.9	46.1
Median eminence	58.0	65.0	5	61.3	56.8	239.9	220.3
Periventricular nucleus	10.9	7.1	23	11.5	6.2	18.9	21.2
Suprachiasmatic nucleus	7.6	9.0	31–40	8.0	7.9	30.5	36.2
Supraoptic nucleus	6.8	3.7	41–50	7.2	3.2	19.9	15.7
Paraventricular nucleus	8.8	10.0	29	9.3	8.7	16.0	19.6
Anterior hypothalamic nucleus	4.5	5.0	51–60	4.8	4.4	18.7	30.9
Retrochiasmatic area		15.1	16		13.2		31.5
Arcuate nucleus	17.1		14	18.1		62.5	
Ventromedial nucleus	5.1		41–50	5.4		18.8	
Dorsomedial nucleus	10.0		25	10.6		11.3	
Perifornical nucleus	6.2	6.0	41–50	6.6	5.2	26.5	33.5
Dorsal premammillary nucleus	4.0	3.9	51–60	4.3	3.4	18.2	27.5
Ventral premammillary nucleus	3.8	3.4	61–70	4.0	3.0	12.6	24.5
Posterior hypothalamic nucleus	7.2	4.3	31–40	7.6	3.8	21.9	30.9
Mammillary body							
Lateral nucleus	2.7	2.5	61–70	2.9	2.2	13.4	14.7
Medial nuclei	4.3	2.6	51–60	4.5	2.3	19.0	20.0

[a] A—Versteeg *et al.* (1976).
[b] B—Brownstein *et al.* (1974) and Palkovits *et al.* (1974).
[c] Kizer *et al.* (1976).
[d] Saavedra *et al.* (1976).

neurones situated here (the dendrites of which terminate in the striatum and most of their axons in the mesencephalon) also receive their innervation from the dopaminergic cells of the mesencephalon. A similar high DA concentration is also perceivable in the nucleus tractus diagonalis (Table I). It is of interest that with regard to acetylcholinesterase content, the above-mentioned seven areas are among the highest reported (Jacobowitz and Palkovits, 1974).

In the above-mentioned regions DA may be found in axons, or nerve terminals. This means that DA (similarly to other neurotransmitters) is found in higher concentrations in the terminals of the neurones than in their perikarya. The brain region richest in dopaminergic cells, the ventral tegmental area (cell group A10), rates only eight in regard to DA concentration having only one-quarter of the highest DA concentration (Table 3). The DA level is relatively high in the arcuate and periventricular nuclei (A12 DA cell-group), yet they rank only 14 and 20, respectively, in the order of DA concentrations.

One or two nuclei of higher DA concentration are occasionally found within major brain areas, hence in this respect the lateral septal nucleus is prominent among the septal nuclei, the lateral and central amygdaloid nuclei are noteworthy within the amygdala (Table 1) and this is also true of the periventricular thalamic nucleus (Table 2) in the thalamus.

The nucleus interstitialis striae terminalis, which is fairly rich in noradrenaline (Palkovits *et al.*, 1974) also contains dopamine in a relatively high concentration. In contrast to the distribution of noradrenaline, DA is found mainly in the pars dorsalis of the nucleus, where it occurs in a higher concentration than noradrenaline (Table 2), while in the pars ventralis it reaches only 17.3% of its noradrenaline concentration (Brownstein *et al.*, 1974).

Dopamine concentrations are high in the circumventricular organs (Saavedra *et al.*, 1976). Among these the highest is the organon vasculosum laminae terminalis, where DA concentration proved to be twice that of noradrenaline (Table 2).

The locus coeruleus which has noradrenaline-containing perikarya is also rich in DA, it ranks 15 (Table 3). It is of interest that among the concentrations obtained by means of different measurements the greatest differences were found in this region (Kobayashi *et al.*, 1974; Versteeg *et al.*, 1976; Thoa *et al.*, 1976). The concentration of DA is 17.6–31.5% of that of noradrenaline. Similar ratios have been described from experiments utilizing mass fragmentography (Koslow *et al.*, 1972) and in studies of in-vivo catecholamines synthesis in the locus coeruleus from ^{3}H-tyrosine (Kuhar *et al.*, 1972). It may be that the DA present constitutes a precursor for noradrenaline, but the relatively high percentage also suggests that dopaminergic cell bodies or varicose nerve terminals may be present (Kobayashi *et al.*, 1974).

Dopamine content is relatively high in the retrochiasmatic area (Table 2), which presumably emerges from the DA axons converging onto the median eminence. The nuclei of the hypothalamus and the preoptic region are generally rich in dopamine: 11 of these are found within the highest 25 values (Table 2).

The DA concentration of the cortical regions—comprising also the hippocampus—is generally low (Table 1). The pyriform, entorhinal and also the cingulate cortical areas have a higher DA-content than the others. In the same areas—compared to noradrenaline—the percentage ratio of DA is high (Table 1). These data provide further support for the histofluorescence observations that there is a dopaminergic innervation in cortical regions (Thierry *et al.*, 1973a,b,c; Hökfelt *et al.*, 1974b; Berger *et al.*, 1974, 1976). The DA found here is of mesencephalic (cell-groups 19–A10) origin (Fuxe *et al.*, 1974; Lindvall *et al.*, 1974).

In the septum the DA content is in the medium range and on average manifests an uneven distribution but with the exception of the septal triangular nucleus, the noradrenaline–dopamine ratio is high (Table 1). This probably means there is an independent dopaminergic innervation of this area. Although only few data are available, it is still probable that DA present in the septum also originates from the neurones of the cell groups A9 and A10 (Lindvall, 1975; Kizer *et al.*, 1976).

Apart from the lateral and central amygdaloid nuclei the DA concentration is medium in the amygdala (Table 1). With the exception of the caudal neclei the percentage ratio of DA—compared to noradrenaline—is high, which probably means that DA is present not only as a precursor.

In the case of the thalamic nuclei, for which only limited data are available, the DA content is low except for the periventricular nucleus (Table 2). This is also apparent from microscopical examination (Fuxe, 1964, 1965; Jacobowitz and Palkovits, 1974). The body of the nucleus is possibly traversed by an ascending DA-pathway and numerous axon varicosities are observable in the nucleus. The fibers are supposed to emerge from the A10 dopaminergic cells (Kizer *et al.*, 1976). Compared to noradrenaline the percentage ratio of DA of medium quantity present in the nucleus subthalamicus and in the zona incerta is higher than would be expected if it was merely serving a precursor role (Table 2).

In the mesencephalon, except for the ventral tegmental area (cell-group A10) DA content is medium-low, yet in the colliculi it is significantly low (Table 3). Apart from the cell-groups A8 (substantia nigra, pars lateralis), A9 and A10, the noradrenaline–dopamine ratio is relatively high in the

TABLE III Concentrations of dopamine in the lower brain stem

	ng per mg protein		Rank	Percentage of the highest DA in the CNS		Percentage of the noradrenaline (100%) in the same area	
	A[a]		(1–101)	A[a]		A[a]	
Mesencephalon							
Central grey matter							
pars dorsalis	4.7		51–60	4.9		44.8	
pars ventralis	3.9		51–60	4.2		16.1	
Superior collicle	1.1		81–90	0.4		19.0	
Inferior collicle	0.4		91–101	0.4		15.9	
Interped. nucleus	8.9		28	9.4		68.0	
Substantia nigra							
zona reticularis	4.9		41–50	5.2		119.6	
zona compacta	5.4		41–50	5.7		45.8	
pars lateralis	7.8		31–40	8.2		94.7	
Ventral tegmental area	22.4		8	23.7		147.9	
Red nucleus	5.8		41–50	6.1		45.5	
Nucleus cuneiformis	2.9		61–70	3.1		12.7	
Dorsal raphe nucl.	5.7		41–50	6.1		28.1	
Subcommissural organ		9.3[b]	26		8.1[b]		143.1[b]
Pons							
Nuclei pontis	1.4		71–80	1.4		12.0	
Reticular tegmental nucleus	1.4		71–80	1.4		20.1	

Pons							
Parabrachial nuclei	1.6		71–80	1.7		6.2	
Locus coeruleus	16.7	5.5[c]	15	17.7		31.5	17.6[c]
Superior olive	1.1		81–90	1.2		11.2	
Cerebellum							
Cortex	0.2		91–101	0.2		6.4	
Nuclei	0.8		91–101	0.9		23.6	
Medulla oblongata							
Area postrema		8.9[b]	27		7.8[b]		103.5[b]
Nucl. retic. gigantocell.	1.1		81–90	1.2		9.8	
Nucl. retic. parvocell.	1.6		71–80	1.7		12.4	
Nucl. retic. med. obl.	1.9		71–80	2.0		13.1	
Nucl. retic. lateralis	3.3		61–70	3.5		21.5	
Nucl. tract. spin. of Vth	2.6		61–70	2.8		31.1	
Motor nucl. of VIIth	1.6		71–80	1.7		10.9	
Motor nucl. of XIIth	2.5		61–70	2.7		16.1	
Inferior olive	0.8	3.4[d]	91–101	0.6		7.8	
Nucl. tractus solitarii							
pars rostralis	3.2		61–70	3.4		22.7	
pars medialis	7.3		31–40	7.7		20.0	
pars commissuralis	10.1		24	10.7		26.3	

A[a]—Versteeg *et al.* (1976).
[b] Saavedra *et al.* (1976).
[c] Kobayashi *et al.* (1975).
[d] Kizer *et al.* (1976).

interpeduncular, the red nuclei and in the dorsal part of the central grey matter. Among the examined areas of the pons, the locus coeruleus is prominent with regard to DA concentration, which is rather low elsewhere (Table 3).

A low DA concentration is also perceivable in the cerebellar nuclei, which do not exceed one-quarter of the value measured in the cortex (Versteeg *et al.*, 1976) and in spite of this its range is among the lowest values (Table 3).

With the only exception of the nucleus tractus solitarii the DA level is rather low in the medulla oblongata. It is low also in the rostral part of this nucleus and relatively high only in the middle and caudal (commissural) parts of it (Table 3). The latter two areas correspond to the A2—catecholaminergic—cell-groups (Dahlström and Fuxe, 1964). It may be assumed that in this area, which is important with regard to blood-pressure regulation, apart from noradrenaline and adrenaline (Hökfelt *et al.*, 1974a; Saavedra *et al.*, 1974; van der Gugten *et al.*, 1976; Palkovits *et al.*, 1977) dopamine also occurs in a considerable quantity.

In the spinal cord the concentrations of DA are low compared with noradrenaline (only one-fifth of it), its presence might merely reflect its role as a precursor of noradrenaline (Zivin *et al.*, 1975).

According to the results obtained so far, dopamine can be revealed in all cell-groups of the central nervous system. Our data, are not complete, however, since measurements were carried out only on about half of the more than 200 cell-groups that can be dissected. The DA content in the other areas ought also to be determined in order to obtain a more complete picture of the distribution of dopamine in the central nervous system.

REFERENCES

Andén, N.-E., Carlsson, A., Dahlström, A., Fuxe, K., Hillarp, N.-L., and Larsson, K. (1964). *Life Sci.* **3,** 523–530.

Andén, N.-E., Dahlström, A., Fuxe, K. and Larsson, K. (1965). *Am. J. Anat.* **116,** 329–333.

Andén, N.-E., Dahlström, A., Fuxe, K., Larsson, K., Olson, L. and Ungerstedt, U. (1966). *Acta Physiol. Scand.* **67,** 313–326.

Axelsson, S., Björklund, A., Falck, B., Lindvall, O. and Svensson, L.-Å. (1973). *Acta Physiol. Scand.* **87,** 57–62.

Berger, B., Tassin, J. P., Blanc, G., Moyne, M. A. and Thierry, A. M. (1974). *Brain Res.* **81,** 332–337.

Berger, B., Thierry, A. M., Tassin, J. P. and Moyne, A. M. (1976). *Brain Res.* **106,** 133–145.

Björklund, A., Falck, B., Hromek, F., Owman, C. and West, K. A. (1970). *Brain Res.* **17,** 1–23.

Björklund, A., Falck, B., Nobin, A. and Stenevi, U. (1974). *In* "Neurosecretion—The Final Neuroendocrine Pathway" (Eds F. Knowles and L. Vollrath), 209–222. Springer Verlag, Berlin.
Björklund, A. and Nobin, A. (1973). *Brain Res.* **51,** 193–205.
Brownstein, M., Saavedra, J. M. and Palkovits, M. (1974). *Brain Res.* **79,** 431–436.
Coyle, J. T. and Henry, D. (1973). *J. Neurochem.* **21,** 61–67.
Cuello, A. C., Hiley, R. and Iversen, L. L. (1973). *J. Neurochem.* **21,** 1337–1340.
Dahlström, A. and Fuxe, K. (1964). *Acta Physiol. Scand. Suppl. 233*, **62,** 1–55.
Dahlström, A., Fuxe, K., Olson, L. and Ungerstedt, U. (1964). *Acta Physiol. Scand.* **62,** 485–486.
Falck, B. and Hillarp, N.-Å. (1959). *Acta Anat.* (*Basel*) **38,** 277–279.
Fuxe, K. (1964). *Z. Zellforsch.* **61,** 710–724.
Fuxe, K. (1965). *Acta Physiol. Scand. Suppl. 247* . **64,** 39–85.
Fuxe, K. and Hökfelt, T. (1966). *Acta Physiol. Scand.* **66,** 245–246.
Fuxe, K., Hökfelt, T., Johansson, O., Lonsson, G., Lidbrink, P. and Ljungdahl, Å (1974). *Brain Res.* **82,** 349–355.
Glowinski, J. and Iversen, L. L. (1966). *J. Neurochem.* **13,** 655–669.
van der Gugten, J., Palkovits, M., Wijnen, H. L. J. M. and Versteeg, D. H. G. (1976). *Brain Res.* **107,** 171–175.
Henry, D. P., Starman, B. J., Johnson, D. C. and Williams, R. H. (1975). *Life Sci.* **16,** 375–384.
Hökfelt, T., Fuxe, K. and Goldstein, M. (1973a). *Brain Res.* **53,** 175–180.
Hökfelt, T., Fuxe, K. and Goldstein, M. (1973b). *Brain Res.* **62,** 461–469.
Hökfelt, T., Fuxe, K., Goldstein, M. and Johansson, O. (1974a). *Brain Res.* **66,** 235–252.
Hökfelt, T., Fuxe, K., Johansson, O. and Ljungdahl, Å. (1974b). *Eur. J. Pharmacol.* **25,** 108–112.
Hökfelt, T. and Ungerstedt, U. (1969). *Acta Physiol. Scand.* **76,** 415–426.
Hökfelt, T. and Ungerstedt, U. (1974). *In* "Dynamics of Degeneration and Growth in Neurons" (Eds K. Fuxe, L. Olson and Y. Zotterman), 141–149. Pergamon, Oxford.
Jacobowitz, D. M. and Palkovits, M. (1974). *J. Comp. Neurol.* **157,** 13–28.
Kizer, J. S., Palkovits, M. and Brownstein, M. J. (1976). *Brain Res.* **108,** 363–370.
Kobayashi, R. M., Palkovits, M., Kopin, I. J. and Jacobowitz, D. M. (1974). *Brain Res.* **77,** 269–279.
Koob, G. F., Balcom, G. J. and Meyerhoff, J. L. (1975). *Brain Res.* **94,** 45–55.
Koslow, S., Cattabeni, F. and Costa, E. (1972). *Science* **176,** 177–180.
Kuhar, M., Roth, R. and Aghajanian, G. (1972). *Biochem. Pharmacol.* **21,** 2280–2282.
Lindvall, O. (1975). *Brain Res.* **87,** 89–95.
Lindvall, O. and Björklund, A. (1974). *Histochemistry* **39,** 97–127.
Lindvall, O., Björklund, A., Moore, R. Y. and Stenevi, U. (1974). *Brain Res.* **81,** 325–331.
Palkovits, M., Brownstein, M. J., Saavedra, J. M. and Axelrod, J. (1974). *Brain Res.* **77,** 137–149.
Palkovits, M., de Jong, W., Zandberg, P., Versteeg, D. H. G., van der Gugten, J. and Léránth, Cs. (1977). *Brain Res.* **127,** 307–312.
Palkovits, M., Léránth, Cs., Záborszky, L. and Brownstein, M. J. (1977). *Brain Res.* **136,** 339–344.
Popov, N., Pohle, W., Rosler, W. und Matthies, H. (1967). *Acta Biol. Med. Germ.* **18,** 695–702.

Saavedra, J. M., Brownstein, M., Kizer, J. S. and Palkovits, M. (1976). *Brain Res.* **107,** 412–417.

Saavedra, J. M., Palkovits, M., Brownstein, M. J. and Axelrod, J. (1974). *Nature (London)* **248,** 695–696.

Thierry, A. M., Blanc, G., Sobel, A., Stinus, L. and Glowinski, J. (1973a). *Science* **182,** 499–501.

Thierry, A. M., Stinus, L., Blanc, G. and Glowinski, J. (1973b). *Brain Res.* **50,** 230–234.

Thoa, N. B., Tizabi, Y. and Jacobowitz, D. M. (1976). *In* "Catecholamines and Stress" (Eds E. Usdin, R. Kvetnansky and I. J. Kopin), 61–67. Pergamon, Oxford.

Ungerstedt, U. (1971). *Acta Physiol. Scand.* **82,** Suppl. 367, 1–48.

Versteeg, D. H. G., van der Gugten, J., de Jong, W. and Palkovits, M. (1976). *Brain Res.* **113,** 563–574.

Zivin, J. A., Reid, J. L., Saavedra, J. M. and Kopin, I. J. (1975). *Brain Res.* **99,** 293–301.

CHAPTER 22

The Distribution of Dopamine in Vertebrates

Margarethe Holzbauer and D. F. Sharman

Agricultural Research Council Institute of Animal Physiology, Babraham, Cambridge, U.K.

INTRODUCTION

This chapter is intended to summarize the information available on the distribution of dopamine (DA) in vertebrates. It is focused on the mammalian brain but examples are also given for DA concentrations in lower vertebrates and in peripheral organs and body fluids of mammals. In a previous, more detailed, review on the distribution of catecholamines in vertebrates (Holzbauer and Sharman, 1972) the data obtained before 1972 were collected and the methods used for identification and quantitative estimation of catecholamines were discussed.

Since then progress in our knowledge of the distribution of DA has been based mainly on a further increase in the sensitivity of the fluorimetric assay methods (see Atack, 1977), the development of an assay method in which gas chromatography was combined with mass fragmentography (see Costa *et al.*, 1975) and of radioenzymatic assay methods (Coyle and Henry, 1973; Saavedra *et al.*, 1973). Thus it has become possible to measure the quantities of catecholamines in a single nucleus of the brain of the rat (e.g. Versteeg *et al.*, 1976, Table II in the present paper). The improvement of the histofluorescence techniques (Vibratome cutting procedure: Hökfelt and Ljungdahl, 1972; glyoxylic acid reaction: Axelsson *et al.*, 1973) has made it possible to obtain a very detailed picture of the pathways of aminergic neurones in the brain.

In the present review the values included in the Tables and text are mean values as given in the references. In most cases standard errors are omitted as they do not contribute to the overall picture of DA-distribution. The space allotted does not allow us to refer to every publication that is available on the subject. We have, therefore, selected representative examples. For human material the mean values obtained by a larger number of authors are given, because the number of samples was often limited. The methods used are indicated in text and Tables: FL for fluorimetry; GLC–MF for gas-liquid chromatography–mass fragmentography; REA for radioenzymatic assay.

LOWER VERTEBRATES

Dopamine has been identified in the brain and eyes of a number of lower vertebrates. Concentrations of between 3.7 and 1.4 $\mu g\, g^{-1}$ tissue were found in the nucleus basalis of the tortoises *Geochelone chilensis* (Juorio, 1969; FL), *Testudo graeca* and *Testudo hermanni* (Juorio, 1973; FL). The DA concentration in the hypothalamus of *Rana temporaria* was 1.24 $\mu g\, g^{-1}$ tissue and that of the hypothalamus of *Carassius auratus* (goldfish) 1.8 $\mu g g^{-1}$ (Jurio, 1973). Baumgarten (1972) measured DA (FL) in various brain areas of *Lampetra fluviatilis* (lamprey) (0.57 $\mu g\, g^{-1}$, total brain without hypophysis), *Acanthias vulgaris* (spiny dogfish) (0.7 $\mu g\, g^{-1}$, forebrain), *Carassius auratus* (0.28 $\mu g\, g^{-1}$, diencephalon) and *Lacerta viridis* (green lizard) (0.36 $\mu g\, g^{-1}$, forebrain). A dopaminergic innervation of the pituitary gland, originating in the hypothalamus, has also been observed in several species of lower vertebrates, for example in the eel (*Anguilla anguilla*) (Fremberg and Meurling, 1975).

In the retina of the eye of the toad and the frog the DA concentrations were lower after exposure to light (0.17 and 0.86 $\mu g\, g^{-1}$ tissue respectively) than when the animals were kept in the dark (0.31 or 1.37 $\mu g\, g^{-1}$) (Drujan *et al.*, 1965; FL).

BIRDS

Examples of DA concentrations in the brain of several species of birds are listed in Table I. The highest concentrations were found in the anterior part of the nucleus basalis. Dopamine is also present in peripheral tissues of chickens. Manukhin *et al.* (1969) for example found 3.9 μg DA g^{-1} tissue in the heart, 5.3 $\mu g\, g^{-1}$ in the spleen and 2.7 $\mu g\, g^{-1}$ in the intestine (FL);

TABLE I Dopamine in the avian brain

Order and species	Strain or age	Brain region	Dopamine (μg per g tissue)	Ref.[a]
Galliformes				
Gallus domesticus (chicken)	Chick	Whole brain	0.13, 0.19	(1)
	Adult White Leghorn	Pallium	0.81	(2)
		Striatum	1.31	
		Diencephalon and mesencephalon	0.19	
		Pons and medulla	0.24	
		Cerebellum	0.09	
		Spinal cord	<0.08	
	Adult Rhode Island Red and Plymouth Rock	Nucleus basalis:		(3)
		anterior part	2.80	
		posterior part	0.40	
		Other regions	<0.05	
	"Ross 1" cocks 6 months	Cerebral hemispheres	0.44	(4)
		Diencephalon and mesencephalon	0.10	
		Other regions	<0.03	
	7 days	Hypothalamus	0.30	
Ansiformes				
Cairina moschata (muscovy duck)		Nucleus basalis:		(3)
		anterior part	3.61	
		posterior part	0.55	
		Other regions	<0.05	
Passeriformes				
Lonchura punctulata (Spice finch)		Nucleus basalis:		(3)
		anterior part	7.5	
		posterior part	1.5	
Columbiformes				
Columba livia (pigeon)		Nucleus basalis:		(3)
		anterior part	3.0	
		posterior part	0.6	
		Hypothalamus	0.15	
		Other regions	<0.1	

All estimations were carried out fluorimetrically.
[a] References: (1) Montagu, 1957; (2) Pscheidt and Haber, 1965; (3) Juorio and Vogt, 1967; (4) Callingham and Sharman, 1970.

Anton and Sayre (1964) reported the presence of 0.09 μg DA g^{-1} in the heart and 0.13 μg g^{-1} in the liver of the pigeon (FL). In the retina of the eye of the pigeon and chicken, Stoeckel *et al.* (1976) observed an intensely fluorescing narrow marginal zone and gave values of 0.97 pmol DA per mg retina (or more) for this outer zone whereas the DA concentration in the rest of the retina was 0.16 pmol per mg retina (FL).

MAMMALS

Nervous System

Brain

Examples of the distribution of DA in the mammalian brain are given in Tables II, IIIa, b, IV, V and VI. In all species the highest concentrations were found in the corpus striatum and the nucleus accumbens septi. Table II gives data obtained in an analysis of individual nuclei of the brain of the rat with the use of a micropunch technique and a radioenzymatic assay (REA) to measure DA (Versteeg *et al.*, 1976).

Gas-liquid chromatography combined with mass fragmentography (GLC–MF) was used by Koslow *et al.* (1974) to measure DA in various regions of the rat brain. There was a good agreement between the average DA concentrations in the various brain regions estimated by this method and the values obtained with the REA by Versteeg *et al.* (1976). For example, the respective figures (expressed as pg per μg protein) for the caudate nucleus, were 87 (GLC–MF) and 71 (REA), the nucleus accumbens 93 and 87, the nucleus amygdaloideus centralis 37 and 15 and the pyriform cortex 1.04 and 2.86. When the caudate-putamen complex of the rat brain was divided, transversely, from rostral to caudal into 400 μm sections it was found that the concentration of DA (GLC–MF) had a bimodal type of distribution along the sagittal axis (Koslow *et al.*, 1974; Costa *et al.*, 1975). Taking the level of the decussation of the anterior commissure as "0", the highest values were found 2 mm (600 pmol DA per mg protein $\pm$ 50), 1.6 mm (720 $\pm$ 65) and 1.2 mm (620 $\pm$ 37) rostral from it. The lowest value in the rostral sections was found in the section 0.8 mm from "0" (205 $\pm$ 18). At "0" it was 470 $\pm$ 26 and fell gradually to 93 $\pm$ 12 at -1.2 mm caudal from "0".

Dopamine concentrations in the brains of some domestic species are listed in Tables IIIa, b, of the monkey in Table IV and of man in Table V.

The brain dopamine concentrations vary with age: For example, Fry (1977) measured (REA) the dopamine concentrations in six regions of the brain of the pig between birth and 8 weeks. In the caudate nucleus there was a steady rise during this period from about 1.5 μg DA per g tissue to 5 μg per g whereas in the putamen, the nucleus accumbens and the olfactory tubercle it remained constant (about 2 $\mu g\,g^{-1}$). In the substantia nigra and the hypothalamus the figures were lower than 0.5 $\mu g\,g^{-1}$ during the first eight weeks of life. Coyle and Henry (1973) (REA) found a rise in the dopamine content of the whole brain of the rat from 13 $ng\,g^{-1}$ (15th day of gestation) to 193 $ng\,g^{-1}$ at birth; a value of 613 $ng\,g^{-1}$ was obtained for the adult rat (REA). In the corpus striatum alone there was an eight-fold increase in the DA concentration between birth and adulthood. In the

TABLE II Dopamine in the brain of the rat (pg per μg protein)[a]

Telencephalon	
Frontal cortex	1.04
Cingulate cortex	0.91
Parietal cortex	0.84
Piriform cortex	2.86
Entorhinal cortex	2.41
Hippocampus	0.36
Nucleus caudatus	70.53
Globus pallidus	39.62
Nucleus accumbens	87.17
Olfactory lobe	2.79
Olfactory tubercle	94.56
Nucleus tractus diagonalis	32.57
Medial forebrain bundle (rostral)	51.34
Nucleus interstitialis striae terminalis	19.09
Nucleus septalis dorsalis	8.13
Nucleus septalis lateralis	21.35
Intermediate septal nucleus	14.54
Nucleus septalis medialis	8.57
Nucleus septalis fimbrialis	4.34
Nucleus tractus olfactorius lateralis	8.19
Nucleus amygdaloideus medialis	7.51
Nucleus amygdaloideus centralis	15.01
Nucleus amygdaloideus lateralis	17.48
Nucleus amygdaloideus corticalis	4.82
Nucleus amygdaloideus basalis	7.65
Nucleus amygdaloideus medialis posterior	1.37
Nucleus amygdaloideus posterior	1.56
Mesencephalon	
Nucleus ruber	5.76
Substantia nigra zona compacta (A9)	5.36
Substantia nigra zona reticularis	4.94
Substantia nigra pars lateralis (A8)	7.76
Colliculus superior	1.12
Colliculus inferior	0.40
Substantia grisea centralis, pars dorsalis	4.68
Substantia grisea centralis, pars ventralis	3.95
Nucleus raphe dorsalis	5.75
Nucleus interpeduncularis	8.87
Area tegmentalis ventralis (Tsai) (A10)	22.43
Nucleus cuneiformis (A7)	2.91
The pons	
Nuclei pontis	1.37
Nucleus reticularis pontis	1.39
Nucleus tegmentalis dorsalis	2.99
Locus coeruleus (A6)	16.72
Nuclei parabrachiales	1.65
A5 region	1.79
Superior olive	1.12
Diencephalon	
Habenula	4.97
Nucleus praeopticus medialis	12.33
Zona incerta	5.50
Nucleus subthalamicus	5.28
Nucleus interstitialis striae medullaris	13.38
Nucleus suprachiasmatis	7.57
Nucleus supraopticus	6.84
Nucleus paraventricularis	8.79
Nucleus periventricularis	10.92
Nucleus hypothalamicus anterior	4.55
Nucleus arcuatus	17.14
Eminentia mediana	57.99
Nucleus ventromedialis	5.10
Nucleus dorsomedialis	10.01
Nucleus perifornicalis	6.21
Medial forebrain bundle (caudal)	7.59
Nucleus hypothalamicus posterior	7.18
Nucleus premammillaris ventralis	3.80
Nucleus premammillaris dorsalis	4.04
Nucleus mammillaris medialis	4.27
Nucleus mammillaris lateralis	2.74
Nucleus anterior ventralis thalami	1.20
Nucleus medialis thalami	1.02
Nucleus ventralis thalami	0.24
Nucleus reticularis thalami	n.d.
Nucleus posterior thalami	2.27
Midline thalamic nuclei	4.00
Corpus geniculatum mediale	2.25
Corpus geniculatum laterale	0.98
Medulla oblongata	
Nucleus raphe magnus	1.18
Nucleus motorius n. facialis	1.64
Nucleus tractus spinalis n. trigemini	2.63
Nucleus vestibulares (lateral and medial)	0.94
Nucleus cochleares (dorsal and ventral)	0.58
Nucleus reticularis gigantocellularis	1.11
Nucleus reticularis parvocellularis	1.62
Nucleus reticularis medullae oblongatae (ventral)	1.92
Nucleus reticularis lateralis (A1)	3.27
Nucleus tractus solitarii (rostral)	3.20
A2 region	7.28
Nucleus commissuralis	10.08
Nucleus motorius n. hypoglossi	2.53
Inferior olive	0.85
Cerebellum	
Cortex	0.17
Nuclei	0.82
Spinal cord	
"Lateral spinal cord"	2.40

[a] From Versteeg *et al.* (1976) (REA). n.d.: not detectable.

TABLE IIIa Dopamine in the brain of the cat and dog (μg per g tissues).

Brain region	*Felis catus* DA ($\mu g\,g^{-1}$)	Ref.[a]	*Canis familiaris* DA ($\mu g\,g^{-1}$)	Ref.[a]
Whole brain	0.28	(1)	0.19	(1)
Cerebral hemispheres[b]				
rostral part	0.08		0.07	
caudal part	0.10		0.08	
Caudate nucleus	8.00		5.90	
Lentiform nucleus	1.90		1.63	
Hippocampus	0.08		0.13	
Hypothalamus	0.75		0.26	
Diencephalon[c]	0.16		0.09	
Mesencephalon	0.19		0.20	
Pons	0.11		0.10	
Medulla oblongata	0.08		0.13	
Cerebellum	0.02		0.03	
Association cortex	0.30	(2)		
Sensory cortex	0.13			
Orbital cortex	0.30			
Septal region	1.62			
Caudate nucleus	3.12			
Midbrain	0.53			
Hypothalamus	0.70			
Thalamus	0.50			
Amygdala + fornix + hippocampus	0.21			
Cingulate gyrus	0.58			
Quadrigeminal plate	1.59			
Pons	0.32			
Medulla	0.32			
Neocerebellum	0.20			
Archicerebellum	0.14			
Cervical cord	0.45			
Caudate nucleus	9.9	(3)	9.9	(3)
Hypothalamus	0.20		0.25	
Massa intermedia of thalamus	0.05		0.05	
Midbrain	0.15		0.33	
Cerebral cortex	0.07		0.01	
Palaeocerebellum	<0.003		0.003	

[a] References for Tables IIIa and IIIb: (1) Bertler and Rosengren, 1959; (2) McGeer *et al.*, 1963; (3) Laverty and Sharman, 1965.

[b] Without corpus striatum and hippocampus.

[c] Excluding hypothalamus.

TABLE IIIb Dopamine in the brain of farm animals (μg per g tissue).

Brain region	*Sus scrofa domesticus* (pig) DA ($\mu g\,g^{-1}$)	Ref.	*Ovis aries* (sheep) DA ($\mu g\,g^{-1}$)	Ref.	*Capra hircus* (goat) DA ($\mu g\,g^{-1}$)	Ref.
Whole brain	0.22	(1)	0.29	(1)		
Cerebral hemispheres[b]						
rostral part	0.08		0.12			
caudal part	0.05		0.02			
Caudate nucleus	6.1		6.68			
Lentiform nucleus	3.6		4.78			
Hippocampus	0.04		0.06			
Hypothalamus	0.88		0.19			
Diencephalon[c]	0.11		0.16			
Mesencephalon	0.14		0.24			
Pons	0.06		0.04			
Medulla oblongata	0.05		0.17			
Cerebellum	0.01		0.03			
Caudate nucleus			11.78	(3)	10.9	(3)
Hypothalamus						
superior			0.27		0.15	
inferior			0.21		0.11	
medial			—		0.10	
Massa intermedia of thalamus			0.34		0.36	
Anterior midbrain			—		0.41	
Posterior midbrain			—		0.04	
Pituitary stalk and median eminence			5.05		2.00	

[b] Without corpus striatum and hippocampus.
[c] Excluding hypothalamus.

caudate nucleus and putamen of the rabbit a rise from about 0.3 μg DA g^{-1} on day 20 of gestation to about 2.0 μg g^{-1} at birth was observed. The values for the adult rabbits were about 10 μg g^{-1} (Tennyson *et al.*, 1975; FL). Hyyppä (1972) measured DA in the hypothalamus of the human foetus (FL). The values varied between 0.26 and 0.98 μg g^{-1} between the 11th and 15th week (adult: 0.34 μg g^{-1}).

Dopamine was also found in the pineal gland. Pellegrino de Iraldi and Zieher (1966) estimated 34.3 μg DA g^{-1} in the rat pineal gland (FL). Unlike noradrenaline, the DA concentration did not change after cervical sympathectomy. Hyyppä (1971) found 4 ng DA per pineal gland in 10 day old rats and 12 ng per gland in 60 day old rats (FL). A value of 0.5 μg DA per g tissue was given for the human pineal gland (Sano *et al.*, 1959; FL).

Of special interest to the endocrinologist are the high concentrations of dopamine in the median eminence and the pituitary stalk. Thus, Chiocchio

TABLE IV Dopamine in the brain of the monkey (*Macaca mulatta*) ($\mu g\, g^{-1}$)

	(1)[a]	(2)	(3)
Telencephalon			
Basal ganglia	1.59	—	—
Caudate nucleus	3.50	10.3	6.12
Putamen	—	12.4	—
Diencephalon			
Thalamus	0.11	—	0.42
Hypothalamus	1.69	—	0.48
Mesencephalon			
Midbrain (total)	0.39	—	0.13
Substantia nigra	—	0.8	—
Quadrigeminal plate	0.76	—	—
Rhombencephalon			
Medulla	0.25	—	0.17
Cerebellum	0.08	—	0.05
Pons	0.27	—	—
Rhinencephalon			
Amygdala + Hippocampus + Fornix	0.18	—	—
Septal region	0.28	—	—
Cortex			
Cerebral cortex (unspec.)	—	—	0.10
Cingulate gyrus	0.18	—	—

All estimations were carried out fluorimetrically.

[a] References: (1) Wada and McGeer (1966); (2) O'Keeffe *et al.* (1970); (3) Segal *et al.* (1972).

et al. (1976) found in the median eminence of rats (a piece of the floor of third ventricle weighing on average 65 μg) 28.98 ± 2.93 μg DA per g tissue ($n = 10$) (REA). Even if the region analysed included most of the arcuate nucleus and small portions of the periventricular and the ventromedial nucleus in addition to the median eminence (weighing on average 2.6 mg) the dopamine concentrations were still as high as 2.2 μg per g tissue (Brown *et al.*, 1972; FL). With the aid of the micropunch method Versteeg *et al.* (1976) found 58 ± 3.6 ng DA per mg protein ($n = 5$) (REA) in a circumscribed section of the median eminence of the rat. Laverty and Sharman (1965) found 5.1 μg DA per g tissue in the combined median eminence and

TABLE V Dopamine in human brain (μg per g tissue)

	(a)[a]	(2)	(3)	(4)	(5)	(6)
Telencephalon						
Caudate nucleus (CN)	3.40	5.74	3.12	—	—	1.16
Head of CN	—	—	—	4.948	—	—
Putamen	4.14	8.25	5.27	5.163	—	2.46
Globus pallidus	0.41	1.01	0.32	0.213	—	1.24
Claustrum	0.31	—	0.44	—	—	—
Centrum semiovale	—	—	—	0.009	—	—
Internal capsule	—	0.38	0.42	—	—	—
Bed nucleus of striae terminalis	—	—	—	—	0.65	—
Diencephalon						
Thalamus (total)	0.13	—	—	—	—	0.94
medial	0.22	0.46	0.03	0.006	—	—
rostral	—	0.11	0.07	—	—	—
lateral	—	0.30	0.01	0.021	—	—
Hypothalamus (total)	1.25	1.12	—	—	—	—
anterior	< 0.05	—	0.18	—	—	—
posterior	—	—	0.22	—	—	—
medial	—	—	0.14	—	—	—
lateral	—	—	—	—	0.20	—
Hypothalamus						
AMD[b]	—	—	—	0.368	—	—
AMV	—	—	—	0.064	—	—
ALD	—	—	—	0.277	—	—
ALV	—	—	—	0.175	—	—
PMD	—	—	—	0.195	—	—
PMV	—	—	—	0.181	—	—

TABLE V (continued)

	(1)[a]	(2)	(3)	(4)	(5)	(6)
Hypothalamus						
PLD	—	—	—	0.394	—	—
PLV	—	—	—	0.345	—	—
Median eminence + infundibular stem (a)	—	—	—	—	—	0.24
Medial basal hypothalamus (b)	—	—	—	—	—	0.18
Hypothalamus—(a + b)	—	—	—	—	—	0.34
Pituitary gland, total	—	—	0.02	—	—	—
Mesencephalon						
Nucleus ruber	0.30	1.17	0.19	0.207	—	—
Substantia nigra (total)	0.60	0.38	0.40	0.950	0.68	—
pars compacta	0.76	—	—	—	—	—
pars reticularis	0.34	—	—	—	—	—
Corpora quadrigemina (total)	0.12	0.07	—	0.011	—	—
Superior colliculus	—	—	0.13	—	—	—
Inferior colliculus	—	—	0.10	—	—	—
Rhombencephalon						
Grey matter on floor of IVth ventricle	0.35	—	—	—	—	—
Medulla oblongata (total)	—	0.17	—	—	—	—
Dorsal part of medulla including floor of IVth ventricle	—	—	< 0.01	—	—	—
Ventral part	—	—	0.01	—	—	—

Rhombencephalon						
Pons (total)	0.13	—	< 0.01	—	—	—
Dorsal part	—	0.08	—	—	—	—
Ventral part	0.07	0.07	—	0.006	—	—
Tegmentum pontis	—	—	—	0.119	—	—
Cerebellum (total)	0.01	—	—	—	—	—
Cerebellar cortex	—	—	0.02	—	0.09	—
Nucleus dentatus	< 0.006	0.03	0.08	—	—	—
Rhinencephalon						
Parolfactory gyrus	—	—	—	—	0.35	—
Nucleus accumbens (septi)	—	—	—	—	3.38	—
Septum	0.31	—	—	—	0.23[c]	—
Medial olfactory area	—	—	—	—	0.79	—
Lateral olfactory area[d]	—	—	—	—	0.45	—
Hippocampus						
caudal	—	—	—	0.025	—	—
rostral	—	—	—	0.006	—	—
Amygdaloid						
nucleus	0.30	0.13	—	—	—	—
body	—	—	—	0.051	—	—

[a] References: (1) Hornykiewicz *et al.*; FL: overall means calculated from mean values given in: Ehringer and Hornykiewicz, 1960; Hornykiewicz, 1973; Lloyd, Davidson and Hornykiewicz, 1975. (2) Sano *et al.*, 1959; FL; (3) Bertler, 1961; FL; (4) Moses and Robins, 1975; REA; (5) Farley *et al.*, 1977; FL, REA; (6) Rinne and Sonninen, 1968; FL; Rinne *et al.*, 1971; FL.

[b] The hypothalamus was divided into 12 pieces which were analysed individually as: AMD, anterior medial dorsal; AMV, anterior medial ventral; ALD, anterior lateral dorsal; ALV, anterior lateral ventral; PMD, posterior medial dorsal; PMV, posterior medial ventral; PLD, posterior lateral dorsal; PLV, posterior lateral ventral.

[c] Ventral septum + nucleus diagonal band of Broca.

[d] Olfactory tubercle, substantia innominata, anterior perforate substance.

TABLE VI Dopamine in the pituitary gland and median eminence

Species	Median eminence[a] + pituitary stalk	Neurointermediate lobe	Pituitary gland Anterior lobe	Zona tuberalis	Ref.[d]
Rattus norvegicus (rat)	—	neural lobe: 7.2 ng per mg protein intermediate lobe: 1.8 ng per mg protein	0.2 ng per mg protein	—	(1)
		2.88 ng per lobe	0.25 ng per lobe	—	(2)
Cavia porcellis (guinea-pig)	—	1.20 ng per lobe	0.1–3.2 ng per lobe[b]	—	(2)
Oryctolagus cuniculus (rabbit)	—	3.6 ng per lobe	1.8–8.6 ng per lobe[c]	—	(2)
Sus scrofa (pig)	—	0.41 $\mu g\ g^{-1}$	0.01 $\mu g\ g^{-1}$	—	(3)
	10.0 ng per mg protein	38.1 ng per mg protein	0.9 ng per mg protein	3.7 ng per mg protein	(2)
Capra hircus (goat)	50.0 ng per mg protein	14.5 ng per mg protein	0.6 ng per mg protein	2.1 ng per mg protein	(2)
Ovis aries (sheep)	10.0 ng per mg protein	13.0 ng per mg protein	1.0 ng per mg protein	13.7 ng per mg protein	(2)
Mus musculus (mouse)		Total pituitary gland: 1.43 ng protein mg protein			(4)

[a] Including some basal-hypothalamic tissue.
[b] In some guinea-pigs the stalk was fused to the anterior lobe.
[c] Anterior lobe contains disseminated islets of zone tuberalis.
With exception of reference (1) the intermediate and posterior lobe were homogenized together = neurointermediate lobe. A radioenzymatic method was used for all assays with the exception of reference (3) in which case a larger number of glands was pooled and a fluorimetric assay was used.
[d] References: (1) Saavedra, Palkovits, Kizer, Brownstein and Zivin, 1975; (2) Holzbauer, Sharman and Godden, 1978; (3) Björklund, Falck and Rosengren, 1967; (4) Lorden and Oltmans, 1977.

pituitary stalk of the sheep, 2.0 μg DA g^{-1} in the same structure of a goat and 8.7 μg g^{-1} in that of a cat (FL).

Dopamine concentrations in the posterior, intermediate and anterior lobes of the pituitary glands of several mammalian species are given in Table VI. In the rat the posterior and intermediate lobe of the pituitary gland contains dopaminergic nerve fibres which have their cell bodies in the most rostral part of the arcuate nucleus and reach the gland via the median eminence and the pituitary stalk (Björklund *et al.*, 1973). An involvement of DA in the control of the secretion of pituitary hormones can be assumed.

The presence of DA in the cerebral cortex has recently attracted attention in connection with a theory which suggests that DA might play a role in schizophrenia. Thierry *et al.* (1973) estimated endogenous DA concentrations in the rat brain cortex to be between 0.08 and 0.2 μg g^{-1} (FL). Using specific drug treatment and histofluorescence methods the presence of dopaminergic nerve terminals was demonstrated in the gyrus cinguli, the entorhinal cortex, the amygdaloid cortex, and the frontal cortex of the rat (Hökfelt *et al.*, 1974; Berger *et al.*, 1974). Dopaminergic fibres were also visualized in the suprasylvian gyrus of the cortex of the cat brain and a DA concentration of 230 ng per g tissue (S.E. ± 30, $n = 3$) (FL) was estimated (Thierry *et al.*, 1974). The quantities of DA (between 1 and 3 ng per mg protein, REA) estimated in four cortical regions of the rat brain by Versteeg *et al.* (1976) are included in Table II. Kehr *et al.* (1976) divided the rat cortex from rostral to caudal into 4 regions and found mean values of 97 (most rostral), 68, 57 and 33 (most caudal) ng DA per g tissue (FL). The differences between the regions were statistically significant. In the neocortex of the rhesus monkey (two years old) DA concentrations were highest in the orbital prefrontal cortex (71 ng per g tissue ± 8, $n = 7$) and the dorsolateral prefrontal cortex (81 ng g^{-1} ± 8, $n = 7$). In the parietal cortex the values were 26 ng g^{-1} ± 5 ($n = 7$) and in the visual cortex 8 ng g^{-1} ± 2 ($n = 2$) (McBrown and Goldman, 1977; FL). The quantities of DA found in the human cortex are small. For example, Bertler (1961) could not detect DA (< 10 ng g^{-1}, FL) in the frontal lobe, the anterior and posterior central gyrus and the occipital pole; Moses and Robins (1975) found between 2 and 11 ng DA per g tissue (REA) in the frontal, temporal, parietal, occipital and cerebellar grey of the human cortex; Farley *et al.* (1977) could not detect DA (< 30 ng g^{-1}, FL, REA) in the cingulate, hippocampal or olfactory gyrus, or in the isthmus of the human brain. The paraolfactory gyrus (area 25) however contained 350 ± 90 ng DA per g tissue ($n = 4$). During postnatal development, Coyle and Molliver (1977) estimated the concentration of dopamine in the rat cortex on day one to be 37 ng DA per g tissue, on day six 41 ng g^{-1} and in the adult rat 104 ng g^{-1} (REA).

So far no dopamine-producing cell bodies have been seen in cortical tissue. In the rat brain, projections of dopaminergic fibres to the cortex were described from cell bodies in A10, and the substantia nigra (Fuxe *et al.*, 1974; Lindvall *et al.*, 1974; Collier and Routtenberg, 1977).

Spinal Cord

The quantities of DA reported before 1971 to be present in the spinal cord were between 8 and 50 ng per g tissue, with the exception of observations by McGeer and McGeer (1962) which were somewhat higher (FL) (see Holzbauer and Sharman, 1972, Table 23). With the improvement in the chemical assay techniques it has become possible to analyse restricted sections of the spinal cord. Thus, Zivin *et al.* (1975) obtained the figures shown in Table VII for the spinal cord of the rabbit (REA). Magnusson

TABLE VII Dopamine in the spinal cord of the rabbit. (From Zivin *et al.* (1975) (REA).)

Region	Cervical cord	Thoracic cord dopamine concentration (ng g^{-1} wet weight, $n = 15$)	Lumbar cord
Ventral horn	40 ± 9	113 ± 29	55 ± 10
Central region[a]	92 ± 15	141 ± 25	76 ± 23
Dorsal horn	68 ± 16	120 ± 27	52 ± 8
White matter	9 ± 9	17 ± 8	21 ± 11

[a] In the cervical and lumbar cord a region was punched out around the central canal, in the thoracic cord a piece was punched out from the lateral horn.

(1973) found in the rat spinal cord about 23 ng DA per g tissue (FL). A value of 2.4 ng per mg protein (REA) was obtained for the rat spinal cord by Versteeg *et al.* (1976) (see Table II). Observations were also made on dogs and monkeys in which the spinal cord was removed under anaesthesia in order to maintain a satisfactory blood flow during this rather lengthy dissection procedure. Bingham *et al.* (1975) (FL) observed a rise in the DA concentrations towards the lumbosacral cord and showed a positive correlation between the amount of grey matter present and the catecholamine concentrations in any section of the cord. For the dog their values lay between 100 ng DA per g tissue (cervical cord) and 300 ng g^{-1} (S_3). Similarly, in the monkey, the concentrations lay between 150 and 400 ng g^{-1}, increasing towards the sacral end. Stanton *et al.* (1975) obtained much lower values for the spinal cord of the dog: 1–10 ng DA per g tissue for the segments from T_1 to L_3, 14 ng g^{-1} for L_4–L_6 and 24 ng g^{-1} for

L_7–S_3 (FL). In the human lumbar cord dissected 4 hours after death between 60 and 190 ng DA per g tissue were found (Commissiong and Sedgwick, 1975) (FL).

Peripheral nervous tissue

A list of DA concentrations in peripheral nervous tissue is given in Table VIII. More recently Koslow *et al.* (1975) reported a concentration of 94 ± 8.1 pmol DA per mg protein ($n = 9$) to be present in sympathetic ganglia of newborn rats and of 141 ± 18 pmol per mg protein ($n = 8$) in sympathetic ganglia of adult rats (GLC–MF). In a study of the subcellular localization of DA in the bovine superior cervical ganglion 25.1 ng DA per mg protein were found in the P_2 fraction, 12.1 and 10.2 ng per mg protein in the P_1 and P_3 fraction and 22.8 ng per mg protein in the S_3 supernatant; the concentration in the whole ganglion was 1.6 µg per g tissue (Wilson *et al.*, 1973; FL).

TABLE VIII Dopamine in peripheral nervous tissue (µg per g tissue)

Tissue	*Oryctolagus cuniculus* (rabbit)	*Felis catus* (cat)	*Canis familiaris* (dog)	*Ovis aries* (sheep)	*Capra hircus* (goat)	*Bos taurus* (cattle)
Superior cervical ganglion	0.85	0.82	2.00	—	—	—
Inferior cervical ganglion	—	—	0.80	—	—	—
Stellate ganglion	4.64	1.32	1.53	1.48	0.85	—
Sympathetic chain	—	—	0.17, 0.1[a]	—	0.60[a]	—
Cardiac nerve	—	—	0.78	—	—	—
Splenic nerve	—	—	0.45	—	—	2.5[b]
Sciatic nerve	—	—	—	—	0.4[a]	—

From Laverty and Sharman, 1965; except [a] Carlsson, 1959; [b] v. Euler and Lishajko, 1957.
All values were obtained by fluorimetric methods.

Peripheral Organs

Examples of the occurrence of DA in peripheral organs are given in Table IX. The high concentrations of DA in some of the tissues of ruminants is probably due to the presence of mast cells which contain DA in these species (Carlsson, 1959; Bertler *et al.*, 1959; Falck *et al.*, 1959; Falck *et al.*, 1964). Large quantities of DA (up to 30 µg per g tissue) were also isolated from certain sections of the intestinal tract of ruminants (Schümann, 1959; Bertler *et al.*, 1959; FL).

Relatively high dopamine concentrations were found in the sino-atrial node of the heart of the dog and the rabbit. Snider (1974) measured DA in

TABLE IX Dopamine in peripheral tissues of mammals (μg per g tissue) tissue).

Species	Heart		Ref.	Organ and reference (Ref.)[a] Lung	Ref.	Kidney	Ref.	Spleen	Ref.	Liver	Ref.	Pancreas	Ref.
Oryctolagus cuniculus (rabbit)	Sino-atrial node (SAN)	3.42	(1)	0.2	(6)	0.02	(2)	0.04	(2)	0.02	(2)	—	
	R. atrium (-SAN)	0.29		0.3	(8)	n.d.	(6)			n.d.	(6)		
	L. atrium	0.07											
	R. ventricle	0.08											
	L. ventricle	0.03											
	Whole heart	0.04	(2)										
Rattus norvegicus (rat)	Whole heart	0.03	(2)	n.d.	(6)	0.03	(2)	0.02	(2)	0.03	(2)	—	
	Whole heart	0.014	(3)	0.18	(8)	n.d.	(6)	0.1	(6)	n.d.	(6)		
Mus musculus (mouse)	Whole heart	0.01	(4)	0.15	(8)	—		—		—		—	
Cavia porcellus (guinea-pig)	R. atrium	1.39	(1)	0.22	(8)	0.04	(2)	0.05	(2)	n.d.	(6)	0.03	(12)
	L. atrium	0.68				n.d.	(6)						
	R. ventricle	0.08											
	L. ventricle	0.06											
Canis familiaris (dog)	R. atrium	0.78	(5)	n.d.	(6)	0.11	(11)	0.24	(11)	<0.06	(11)	—	
	L. atrium	0.68		0.27	(9)	0.03	(2)	0.03	(2)	0.02	(2)		
	R. ventricle	0.44		0.24	(8)	n.d.	(6)			n.d.	(6)		
	L. ventricle	0.43											
	Whole heart	0.05	(2)										
	SAN	2.00	(1)										
Felis catus (cat)	Whole heart	0.07	(2)	n.d.	(6)	0.07	(2)	0.12	(2)	0.07	(2)	—	
	Whole heart			0.44	(8)	n.d.	(6)			n.d.	(6)		

Species	Heart		Ref.	Lung	Ref.	Kidney	Ref.	Spleen	Ref.	Liver	Ref.	Pancreas	Ref.
				Organ and reference (Ref.)[a]									
Sus scrofa (pig)		—		n.d.	(6)	—		—		—		0.47	(13)
Ovis aries (sheep)	Whole heart	0.3	(6)	3.6–9.8	(7)	n.d.	(6)	0.4	(9)	Whole liver 0.4	(9)	0.20	(9)
	Whole heart	0.3	(7)	18.6	(9)			0.9	(7)	Whole liver 0.3 Capsule 1.4 Parenchyma 0.1	(7)		
										Whole liver 0.3	(6)		
Capra hircus (goat)	Whole heart	0.5	(6)	5.3	(6)	2.0	(6)	1.0	(7)	Capsule 1.6 7.3 Parenchyma 1.6	(7)	—	
	Whole heart	0.6	(7)	6.45	(8)								
										Whole liver 1.6	(6)		
Bos taurus (cattle)	Whole heart	1.5	(6)	9.5	(6)	0.2	(6)	Whole spleen 0.88	(6)	Whole liver 2.5	(6)	0.2	(9)
	Whole heart	1.5	(7)	3.5–15.0	(7)			Whole spleen 0.8	(7)	Whole liver			
				1.0	(10)			Capsule 0.5		Capsule 7–17 Parenchyma 0.6–1.8	(7)		
										Whole liver 1.5	(9)		

All estimations were carried out by fluorimetric methods. The large variations in the DA content of some organs of ruminants are due to DA-containing mast cells (Carlsson, 1959; Bertler *et al.*, 1959; Falck *et al.*, 1959, 1964). n.d.: not detected.

[a] References: (1) Angelakos *et al.*, 1963; (2) Anton and Sayre, 1964; (3) Snider *et al.*, 1973; (4) Laverty *et al.*, 1965; (5) Miyahara, 1962; (6) Carlsson, 1959; (7) Bertler *et al.*, 1959; (8) Aviado and Sadavongvivad, 1970; (9) Schümann, 1959; (10) v. Euler and Lishajko, 1957; (11) Wegmann, 1963; (12) Cegrell and Falck, 1968; (13) Cegrell *et al.*, 1964.

five human hearts and found in the atrium 4.4 ng per g tissue (range 0.45–9.2), in the ventricle 13.3 ng g^{-1} (range 7.9–27.5) and in the aorta 5.7 ng g^{-1} (range 4.4–7.8) (FL). Concentrations of DA between 10 and 50 ng per g tissue were found in different arteries of two human subjects (Neubauer and Christensen, 1976; REA). In the carotid body of the rabbit Dearnaley *et al.* (1968) found 20–40 µg DA per g tissue (FL); Chioccio *et al.* (1966) found 60–200 ng DA per carotid body in the cat (FL) and Hellström and Koslow (1976) estimated 30 pmol DA per pair of carotid bodies in the rat (GLC–MF).

Dopamine is also present in the retina of the mammalian eye. Its possible significance in retinal neurotransmission has been discussed by Kramer (1976) and Ehinger (1976). Whereas in the toad and frog (see p. 358) the DA concentration was reported to decrease on light exposure, in mammals it appears to increase. Thus, in the posterior segment of the eyeball of the rat about 3.3 ng DA per eye were found when the animals were kept in the dark, 5.2 when they were kept in a light surrounding. For the posterior segment of the eyeball of the guinea-pig the values were about 3.5 ng DA per eye (dark) and 4.8 ng (light), for that of the rabbit about 8.8 ng DA per eye (dark) and 13.7 ng (light) (Nichols *et al.*, 1967; FL). Häggendal and Malmfors (1965) quoted a mean value of 9.2 ng DA per eye for the rat and could not detect DA in the choroid (FL). Ruffolo and Patil (1974) compared the DA content of the eyes of albino rabbits with that of the eyes of pigmented rabbits. In the retina and iris of the albinos there were <2 µg g^{-1}, in the respective tissues of the pigmented rabbits there were >2 µg g^{-1} and 6.5 µg g^{-1} (FL).

The presence of DA in the adrenal glands was first reported for the gland of the sheep by Goodall (1951). Shepherd and West (1951) measured DA in the adrenal medulla of ox (35 µg DA g^{-1}), cow (17 µg g^{-1}) and sheep (350 µg g^{-1}) with a colorimetric technique on a paper chromatogram. Vandermeulen and Cession-Fossion (1968) found in the goat 7.8 µg DA per g tissue (whole adrenal gland) (FL). More recently Carlsson *et al.* (1973) estimated in the rat adrenal a quantity of 1 µg DA per kg body weight (FL); DA accounted for about 1% of the total catecholamines in the gland. They presented good evidence that adrenal DA serves mainly as a precursor for noradrenaline and adrenaline. Snider *et al.* (1977) reported a DA content of the rat adrenal of 1.01 ± 0.06 nmol per pair of glands (FL).

Dopamine in Body Fluids

Blood

Free dopamine was estimated in blood plasma of the cat by Street and Roberts (1969) (two groups: 0.69–1.52 and 3.85–14.46 ng ml^{-1}) and of the sheep by Kelly *et al.* (1970) (0.1–0.4 ng ml^{-1}) using fluorimetric methods. The development of radioenzymatic methods for the estimation of DA has enabled it to be determined in small volumes of human peripheral blood plasma. The following values have been reported for the concentration of free DA: Christensen (1973): 0.11–0.27 ng ml^{-1}; Callingham and Barrand (1976): male 0.20, female 0.19 ng ml^{-1}; Da Prada and Zürcher (1976): 0.127, 0.113 ng ml^{-1}; Weise and Kopin (1976): 0.033 ng ml^{-1} and Peuler and Passon (1977): 0.034, 0.030 ng ml^{-1}. Estimates of the concentration of DA in human plasma obtained by gas chromatography and gas chromatography–mass spectrometry have given values corresponding to the higher values in the above group (Wang *et al.*, 1975: 0.12–0.58 ng ml^{-1}). Wang *et al.* (1976) have also measured the total DA in human blood plasma and found that approximately 90% of the DA was in a conjugated form. The presence of a large proportion of conjugated DA was also observed by Tyce *et al.* (1974) in the blood plasma of patients treated with L-DOPA.

Da Prada and Zürcher (1976) estimated the concentration of DA in rat blood plasma to be 0.274 ± 0.028 ng ml^{-1} (REA). In plasma obtained from rat hypophyseal portal blood, Ben-Jonathan *et al.* (1977) have found DA in a concentration of 1.32 ng ml^{-1} during pro-oestrus rising to 3.87 ng ml^{-1} at oestrus. Values as high as 18 ng ml^{-1} were observed during pregnancy. These authors reported values of 0.4–0.8 ng DA ml^{-1} in arterial plasma of female rats (REA).

Urine

The presence of DA in urine was first reported by Holtz *et al.* (1947). They estimated the amounts of "Urosympathin" excreted in man in 24 hours to be equivalent to 2–3 mg DA or 100–150 μg adrenaline or noradrenaline using the cat's blood pressure. Wong *et al.* (1973), using a gas-liquid chromatographic method, found a mean excretion of free DA of 402.4 ± 86.3 μg 24 h^{-1} in normal human subjects. They estimated that 50–75% of the total DA excreted was in a conjugated form. These authors also reviewed the results obtained by other workers.

Gastric juice

Dopamine was found in the gastric juice of man in concentrations of 2–100 ng ml^{-1} after the injection of histamine or insulin (Häggendal, 1967; FL). Christensen and Brandsborg (1974) found a basal concentration of 5.4 ng ml^{-1} (range: 3.6–8.2) in human gastric juice (REA).

REFERENCES

Angelakos, E. T., Fuxe, K. and Torchiana, M. L. (1963). *Acta Physiol. Scand.* **59,** 184–192.

Anton, A. H. and Sayre, D. F. (1964). *J. Pharmac. Exp. Ther.* **145,** 326–336.

Atack, C. (1977). *Acta Physiol. Scand. Suppl. 451* .

Aviado, D. M. and Sadavongvivad, C. (1970). *Br. J. Pharmacol.* **38,** 374–385.

Axelsson, S., Björklund, A., Falck, B., Lindvall, O. and Svensson, L.-Å. (1973). *Acta Physiol. Scand.* **87,** 57–62.

Baumgarten, H. G. (1972). *Prog. Histochem. Cytochem.* **4,** 1–90.

Ben-Jonathan, N., Oliver, C., Weiner, H. J., Mical, R. S. and Porter, J. C. (1977). *Endocrinology* **100,** 452–458.

Berger, B., Tassin, J. P., Blanc, G., Moyne, M. A. and Thierry, A. M. (1974). *Brain Res.* **81,** 332–337.

Bertler, Å. (1961). *Acta Physiol. Scand.* **51,** 97–107.

Bertler, Å., Falck, B., Hillarp, N.-Å., Rosengren, E. and Torp, A. (1959). *Acta Physiol. Scand.* **47,** 251–258.

Bertler, Å. and Rosengren, E. (1959). *Acta Physiol. Scand.* **47,** 350–361.

Bingham, W. G., Ruffolo, R., Goodman, J. H., Knofel, J. and Friedman, S. (1975). *Life Sci.* **16,** 1521–1526.

Björklund, A., Falck, B. and Rosengren, E. (1967). *Life Sci.* **6,** 2103–2110.

Björklund, A., Moore, R. Y., Nobin, A. and Stenevi, U. (1973). *Brain Res.* **51,** 171–191.

Brown, G. M., Krigstein, E., Dankova, J. and Hornykiewicz, O. (1972). *Neuroendocrinology* **10,** 207–217.

Callingham, B. A. and Barrand, M. A. (1976). *J. Pharm. Pharmacol.* **28,** 356–360.

Callingham, B. A. and Sharman, D. F. (1970). *Br. J. Pharmacol.* **40,** 1–5.

Carlsson, A. (1959). *Pharmacol. Rev.* **11,** 490–493.

Carlsson, A., Snider, S. R., Almgren, O. and Lindqvist, M. (1973). *In* "Frontiers in Catecholamine Research" (Eds. E. Usdin and S. H. Snyder), 551–556. Pergamon, Oxford.

Cegrell, L. and Falck, B. (1968). *Acta Physiol. Scand. Suppl. 314*, 24–34.

Cegrell, L., Falck, B. and Rosengren, E. (1968). *Acta Physiol. Scand. Suppl. 314*, 8–13.

Chiocchio, S. R., Biscardi, A. M. and Tramezzani, J. H. (1966). *Nature* (*London*) **212,** 834–835.

Chiocchio, S. R., Cannata, M. A. and Tramezzani, J. H. (1976). *Brain Res.* **110,** 612–618.

Christensen, N. J. (1973). *Scand. J. Clin. Lab. Invest.* **31,** 343–346.

Christensen, N. J. and Brandsborg, O. (1974). *Scand. J. Clin. Lab. Invest.* **34,** 315–320.

Collier, T. J. and Routtenberg, A. (1977). *Brain Res.* **128,** 354–360.

Commissiong, J. W. and Sedgwick, E. M. (1975). *Lancet* **i** (7902) 347.

Costa, E., Koslow, S. H. and Le Fevre, H. F. (1975). *In* "Handbook of Psychopharmacology", Vol. I, (Eds L. L. Iversen, S. D. Iversen and S. H. Snyder), 1–24. Plenum, New York.

Coyle, J. T. and Henry, D. (1973). *J. Neurochem.* **21,** 61–67.

Coyle, J. T. and Molliver, M. E. (1977). *Science* **196,** 444–447.

Da Prada, M. and Zürcher, G. (1976). *Life Sci.* **19,** 1161–1174.

Dearnaley, D. P., Fillenz, M. and Woods, R. I. (1968). *Proc. R. Soc. B.* **170,** 195–203.

Drujan, B. D., Diaz Borges, J. M. and Alvarez, N. (1965). *Life Sci.* **4,** 473–477.

Ehinger, B. (1976). *In* "Transmitters in the Visual Process" (Ed. S. L. Bonting), 145–163. Pergamon, Oxford.
Ehringer, H. and Hornykiewicz, O. (1960). *Klin. Wschr.* **38,** 1236–1239.
v. Euler, U. S. and Lishajko, F. (1957). *Acta Physiol. Pharmacol. Neerl.* **6,** 295–303.
Falck, B., Hillarp, N.-Å. and Torp, A. (1959). *J. Histochem. Cytochem.* **7,** 323–328.
Falck, B., Nystedt, T., Rosengren, E. and Stenflo, J. (1964). *Acta Pharmacol. (Kbh.)* **21,** 51–58.
Farley, I. J., Price, K. S. and Hornykiewicz, O. (1977). *Adv. Biochem. Psychopharmacol.* **16,** 57–64.
Fremberg, M. and Meurling, P. (1975). *Cell Tiss. Res.* **157,** 53–72.
Fry, J. P. (1977). Cerebral Dopamine Metabolism and Animal Behaviour. Ph.D. Thesis, University of Cambridge.
Fuxe, K., Hökfelt, O., Johansson, O., Jonsson, G., Lidbrink, P. and Ljungdahl, Å. (1974). *Brain Res.* **82,** 349–355.
Goodall, McC. (1951). *Acta Physiol. Scand. Suppl. 85*, 24.
Häggendal, J. (1967). *Acta Physiol. Scand.* **71,** 127–128.
Häggendal, J. and Malmfors, T. (1965). *Acta Physiol. Scand.* **64,** 58–66.
Hellström, S. and Koslow, S. H. (1976). *Brain Res.* **102,** 245–254.
Hökfelt, T., Fuxe, K., Johansson, O. and Ljungdahl, Å. (1974). *Eur. J. Pharmacol.* **25,** 108–112.
Hökfelt, T. and Ljungdahl, Å. (1972). *Histochemie* **29,** 325–339.
Holtz, P., Credner, K. and Kroneberg, G. (1947). *Naunyn-Schmiedebergs Arch. Exp. Path. Pharmak.* **204,** 228–243.
Holzbauer, M. and Sharman, D. F. (1972). *In* "Handbuch der Experimentellen Pharmakologie". Heffter Heubner New Series XXXIII. Catecholamines. (Eds. H. Blaschko and E. Muscholl), 110–185. Springer-Verlag, Berlin.
Holzbauer, M., Sharman, D. F. and Godden, U. (1978). *Neuroscience* **3,** 1251–1262.
Hornykiewicz, O. (1973). *Br. Med. Bull.* **29,** 172–178.
Hyyppä, M. (1971). *Experientia* **27,** 336–337.
Hyyppä, M. (1972). *Neuroendocrinology* **9,** 257–266.
Juorio, A. V. (1969). *J. Physiol. (London)* **204,** 503–509.
Juorio, A. V. (1973). *J. Neurochem.* **20,** 641–645.
Juorio, A. V. and Vogt, M. (1967). *J. Physiol (London)* **189,** 489–518.
Kehr, W., Lindqvist, M. and Carlsson, A. (1976). *J. Neur. Trans.* **38,** 173–180.
Kelly, M., Sharman, D. F. and Tegerdine, P. (1970). *J. Physiol. (London)* **210,** 130P.
Koslow, S. H., Bjegovic, M. and Costa, E. (1975). *J. Neurochem.* **24,** 277–281.
Koslow, S. H., Racagni, G. and Costa, E. (1974). *Neuropharmacology* **13,** 1123–1130.
Kramer, S. G. (1976). *In* "Transmitters in the Visual Process," (Ed. S. L. Bonting), 166–198. Pergamon, Oxford.
Laverty, R. and Sharman, D. F. (1965). *Br. J. Pharmacol. Chemother.* **24,** 538–548.
Laverty, R., Sharman, D. F. and Vogt, M. (1965). *Br. J. Pharmacol. Chemother.* **24,** 549–560.
Lindvall, O., Björklund, A., Moore, R. Y. and Stenevi, U. (1974). *Brain Res.* **81,** 325–331.
Lloyd ,K. G., Davidson, L. and Hornykiewicz, O. (1975). *J. Pharmacol. Exp. Ther.* **195,** 453–464.
Lorden, J. F. and Oltmans, G. A. (1977). *Brain Res.* **131,** 162–166.
McBrown, R. and Goldman, P. S. (1977). *Brain Res.* **124,** 576–580.
McGeer, E. G. and McGeer, P. L. (1962). *Can. J. Biochem.* **40,** 1141–1151.
McGeer, P. L., McGeer, E. G. and Wada, J. A. (1963). *Archs Neurol., Chicago* **9,** 81–89.

Magnusson, T. (1973). *Naunyn-Schmiedebergs Arch. Exp. Path. Pharmak.* **278,** 13–22.

Manukhin, B. N., Pustovoĭtova, Z. E. and Vyaz'mina, N. M. (1969). *Zh.* Evol. Biol. *Fiziol.* **V,** 42–48.

Miyahara, M. (1962). *Jap. Circulat. J.* (*Ni*) **26,** 1–7.

Montagu, K. A. (1957). *Nature* (*London*) **180,** 244–245.

Moses, S. G. and Robins, E. (1975). *Psychopharmacol. Commun.* **1,** 327–337.

Neubauer, B. and Christensen, N. J. (1976). *Diabetes* **25,** 6–10.

Nichols, C. W., Jacobowitz, D. and Hottenstein, M. (1967). *Invest. Ophthal.* **6,** 642–646.

O'Keeffe, R., Sharman, D. F. and Vogt, M. (1970). *Br. J. Pharmacol.* **38,** 287–304.

Pellegrino de Iraldi, A. and Zieher, L. M. (1966). *Life Sci.* **5,** 149–154.

Peuler, J. D. and Passon, G. A. (1977). *Life Sci.* **21,** 625–635.

Pscheidt, G. R. and Haber, B. (1965). *J. Neurochem.* **12,** 613–618.

Rinne, U. K. and Sonninen, V. (1968). *Experientia* **24,** 177–178.

Rinne, U. K., Sonninen, V. and Hyyppä, M. (1971). *Life Sci.* **10,** 549–557.

Ruffolo, R. R. Jr. and Patil, P. N. (1974). *Eur. J. Pharmacol.* **25,** 255–258.

Saavedra, J. M., Brownstein, M. and Axelrod, J. (1973). *J. Pharmac. Exp. Ther.* **186,** 508–515.

Saavedra, J. M., Palkovits, M., Kizer, J. S., Brownstein, M. and Zivin, J. A. (1975). *J. Neurochem.* **25,** 257–260.

Sano, I., Gamo, T., Kakimoto, Y., Taniguchi, K., Takesada, M. and Nishinuma, K. (1959). *Biochim. Biophys. Acta.* **32,** 586–587.

Schümann, H.-J. (1959). *Naunyn-Schmiedebergs Arch. Exp. Path. Pharmak.* **236,** 474–482.

Segal, M., Deneau, G. A. and Seevers, M. H. (1972). *Neuropharmacology* **11,** 211–222.

Shepherd, D. M. and West, G. B. (1951). *Br. J. Pharmacol.* **6,** 665–674.

Snider, S. R. (1974). *J. Neur. Trans.* **35,** 87–90.

Snider, S. R., Almgren, O. and Carlsson, A. (1973). *Naunyn-Schmiedebergs Arch. Exp. Path. Pharmak.* **278,** 1–12.

Snider, S. R., Sahar, D., Prasad, A. L. N. and Fahn, S. (1977). *Life Sci.* **20,** 1077–1086.

Stanton, E. S., Smolen, P. M., Nashold, B. S. Jr., Dreyer, D. A. and Davis, J. N. (1975). *Brain Res.* **89,** 93–98.

Stoeckel, M. E., Roussel, G., Zwiller, J., Madarasz, B. and Porte, A. (1976). *Cell Tiss. Res.* **173,** 335–341.

Street, D. M. and Roberts, D. J. (1969). *J. Pharm. Pharmacol.* **21,** 199–201.

Tennyson, V. M., Mytilineou, C., Heikkila, R., Barrett, R. E., Cohen, G., Côté, L., Duffy, P. E. and Marco, L. (1975). *In* "Brain Mechanism in Mental Retardation" (Eds. N. A. Buchwald and M. A. B. Brazier), 227–264. Academic Press, New York and London.

Thierry, A. M., Hirsch, J. C., Tassin, J. P., Blanc, G. and Glowinski, J. (1974). *Brain Res.* **79,** 77–88.

Thierry, A. M., Stinus, L., Blanc, G. and Glowinski, J. (1973). *Brain Res.* **50,** 230–234.

Tyce, G. M., Sharpless, N. S. and Muenter, M. D. (1974). *Clin. Pharmacol. Ther.* **16,** 782–788.

Vandermeulen, R. and Cession-Fossion, A. (1968). *Arch. Int. Physiol. Biochim.* **76,** 916–919.

Versteeg, D. H. G., van der Gugten, J., de Jong, W. and Palkovits, M. (1976). *Brain Res.* **113,** 563–574.
Wada, J. A. and McGeer, E. G. (1966). *Archs Neurol. (Chicago)* **14,** 129–142.
Wang, M.-T., Imai, K., Yoshioka, M. and Tamura, Z. (1975). *Clinica Chim. Acta* **63,** 13–19.
Wegmann, A. (1963). *Naunyn-Schmiedebergs Arch. Exp. Path. Pharmak.* **246,** 184–190.
Weise, V. K. and Kopin, I. J. (1976). *Life Sci.* **19,** 1673–1686.
Wilson, W. S., Roth, R. H. and Cooper, J. R. (1973). *Life Sci.* **13,** 281–285.
Wong, K. P., Ruthven, C. R. J. and Sandler, M. (1973). *Clinica Chim. Acta* **47,** 215–222.
Zivin, J. A., Reid, J. L., Saavedra, T. M. and Kopin, I. J. (1975). *Brain Res.* **99,** 293–301.

CHAPTER 23

Interconnections of Dopamine Systems

P. L. McGeer and E. G. McGeer

University of British Columbia,
Vancouver, B. C., Canada.

Previous chapters in this review have discussed the location of dopaminergic cell bodies as well as the distribution of their axons and terminals. There has also been a description of dopaminergic receptors and the second messenger adenylate cyclase system which is linked to these postsnyaptic elements. This chapter deals with a related aspect, namely the transmitters used by neurones making afferent or efferent connections with some of the dopaminergic cells.

The information to be gained from knowledge of these interconnections is of more than just theoretical interest. Drugs which interact with dopamine have proved to be of considerable clinical value in a wide variety of conditions, including behavioural disorders, extrapyrimidal movement disorders such as Parkinson's disease and Huntington's chorea, and circumstances producing persistent nausea. The individual drugs are many and the interactions complex, but the prospects for more precise drug action in the future are fairly bright. One obvious possibility for sharpening the action of dopaminergic drugs is to look beyond the dopaminergic cell itself to the differing neurones which interact with it in its various anatomical locations. If these secondary neurones differ, then appropriate drug combinations should be capable of modifying selectively specific dopaminergic functions. In some locations where reliable double labelling techniques have been used at the electron microscopic level, the interaction can be described with a considerable degree of certainty. In other areas, where double labelling techniques are less specific, or where the less reliable techniques of pharmacology have been employed, the information is only speculative. Nevertheless, an indication may be gained from such tentative work which can form the basis of precisely designed but technically difficult experiments.

6-Hydroxydopamine (6-OHDA) is a valuable tool for investigating these interconnections because it can be used to mark dopaminergic structures at the electron microscopic level. It has been particularly useful in demonstrating linkages of the nigrostriatal tract.

Input neurones synapsing with degenerating dopaminergic cells in the nigra have been marked autoradiographically by axoplasmic flow of labelled protein from the basal ganglia. Cholinergic cells in the neostriatum receiving output from degenerating dopaminergic terminals have been marked by immunohistochemical staining for choline acetyltransferase (CAT).

THE NATURE OF NEOSTRIATAL DOPAMINERGIC CONNECTIONS

The morphology of dopaminergic nerve endings in the neostriatum can be demonstrated both by 6-OHDA degeneration and by labelling with radioactive materials transported by axoplasmic flow from dopaminergic cell bodies in the substantia nigra. Figure 1 shows a molecule of labelled

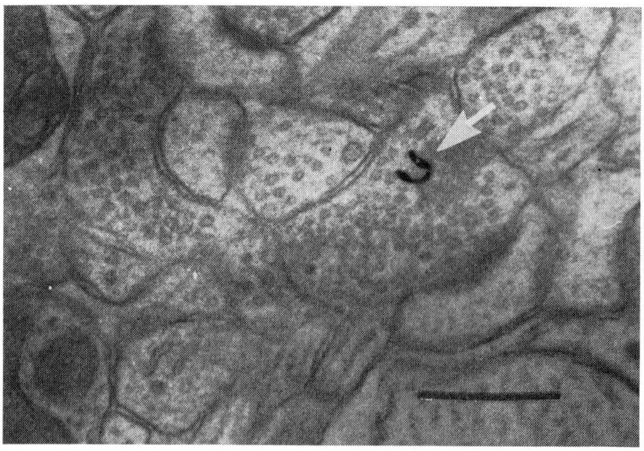

Fig. 1. Labelled bouton (→) in caudate-putamen following injection of ^{3}H-norepinephrine into the rat substantia nigra. Note asymmetrical axospinous contact and mildly pleomorphic vesicles. Bar indicates 0.5 μm.

norepinephrine residing in a small terminal in the neostriatum 24 h after the label had been injected into the substantia nigra. Only terminals of this particular morphology are labelled and chemical evidence has indicated that this transport of labelled norepinephrine is specific to the dopaminergic tract (McGeer *et al.*, 1975).

The bouton is small, makes an asymmetrical axospinous contact, and contains mildly pleomorphic vesicles of 45–55 nm diameter. Similar nerve terminals are labelled following the injection of ^{3}H-leucine into the substantia nigra (Hattori *et al.*, 1973). Following the injection of 6-OHDA degeneration of nerve terminals containing such labelled protein has been observed (Hattori *et al.*, 1973).

To test whether these dopaminergic nerve endings make contact with cholinergic neurones, neostriatal tissue was immunohistochemically stained for CAT. Following such staining, positive cell bodies, dendrites and nerve endings can be observed in striatal slices. The stained neuronal somata are 7–15 μm in diameter with smooth nuclei and few Golgi bodies. The electron-dense stain in dendrites is particularly associated with neurotubules, but expands into the cytoplasm of spines. Some of the stained dendrites receive asymmetrical boutons containing mildly pleomorphic vesicles similar to those seen in dopaminergic nerve endings (Fig. 2). Following the intraventricular application of 6-OHDA, dendritic spines staining for CAT can be found to receive degenerating boutons (Fig. 3). This is direct evidence of a dopaminergic-cholinergic neural connection in the neostriatum. It strongly supports previous pharmacological data suggesting that at least some dopaminergic nigrostriatal neurones relate to cholinergic

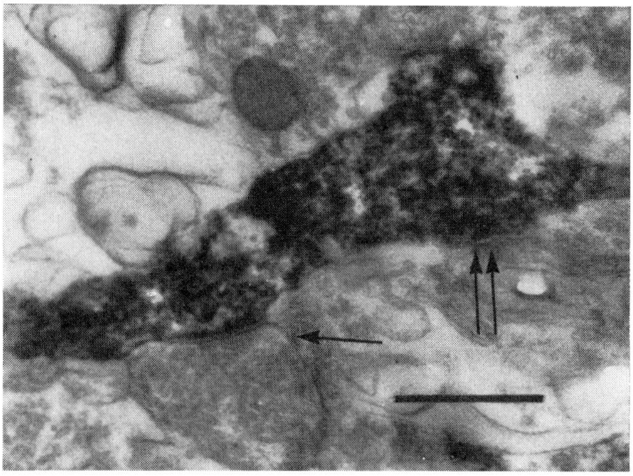

Fig. 2. Guinea-pig neostriatal dendrite (⇉) stained for CAT. Note unstained nerve ending (→) containing mildly pleomorphic vesicles making asymmetric contact. Bar indicates 0.5 μm.

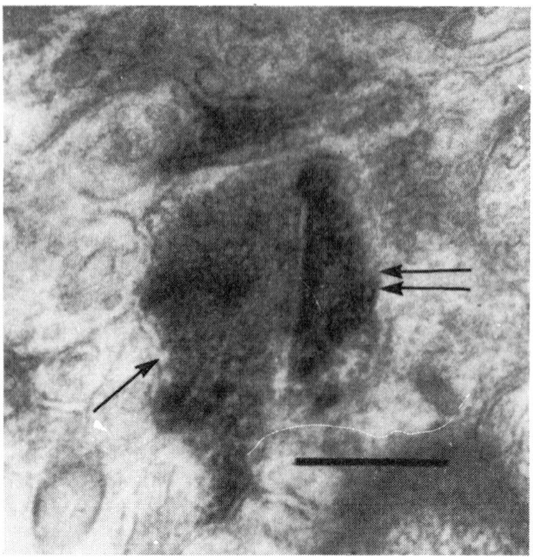

Fig. 3. Degenerating nerve ending (→) in guinea-pig neostriatum following 6-OHDA administration making asymmetrical synaptic contact with dendritic spine ($\overrightarrow{\underrightarrow{}}$). Bar indicates 0.5 μm.

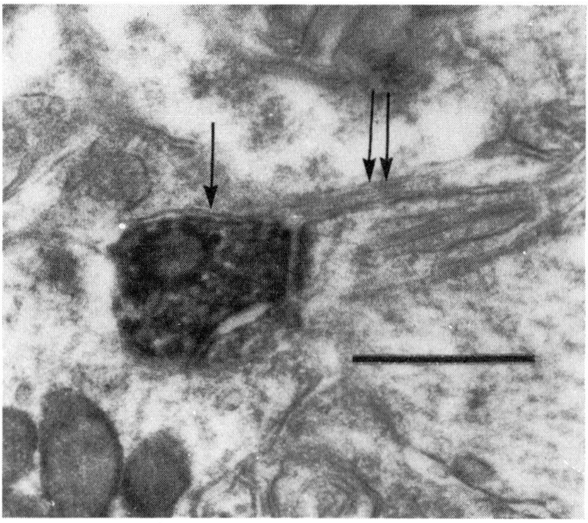

Fig. 4. Guinea-pig neostriatal nerve ending (→) stained for CAT making assymetrical synaptic contact with unstained dendritic spine ($\overrightarrow{\underrightarrow{}}$). Bar indicates 0.5 μm.

neurones in the corpus striatum of rodents (McGeer *et al.*, 1975; McGeer *et al.*, 1974).

Synaptic boutons are also stained, particularly when a part of their outer membrane structure is ruptured by poor fixation. Occasionally, an intact asymmetrical synaptic bouton is stained (Fig. 4). The stained boutons make, without exception, asymmetric synaptic contacts with negative dendrites or dendritic spines. The boutons contain strongly positive round vesicles (25.5–40 nm). These morphological characteristics correspond to those previously ascribed to cholinergic nerve endings in the rat neostriatum on the basis of postnatal developments (Hattori and McGeer, 1973) and ^{3}H-DFP uptake (Hattori and McGeer, 1975).

THE NATURE OF NIGRAL DOPAMINERGIC CONNECTIONS

Gamma-aminobutyric acid neurones play an important role in extrapyramidal function at several levels. Descending GABA neurones were traced by anterograde axoplasmic flow of ^{3}H-protein (following ^{3}H-leucine injection), or ^{3}H-GABA from pallidal cell bodies to substantia nigra nerve endings (Hattori *et al.*, 1975) and other neurones, possibly also GABA containing, were traced by flow of ^{3}H-protein from the neostriatum to the substantia nigra. When the injection site of ^{3}H-leucine was clearly localized to the caudate-putamen by light microscopic autoradiography, there was always considerable labelling of the globus pallidus and the entopeduncular nucleus, but only trace labelling of the subthalamic nucleus. The zona reticulata of the substantia nigra was always labelled diffusely while the zona compacta showed only slight labelling.

When the injection site was clearly localized to the globus pallidus, however, there was only slight labelling of the caudate-putamen near the injection site, but there was heavy labelling of the entopenduncular nucleus, intense labelling of the subthalamic nucleus, moderately heavy labelling of the zona compacta of the substantia nigra, and less labelling of the zona reticulata.

The dopaminergic cell bodies in the substantia nigra were marked by 6-OHDA degeneration. The distribution of silver grains in the substantia nigra following neostriatal or pallidal injections of ^{3}H-leucine was then considered in relation to their proximity to normal or degenerating dendrites. It was found that following pallidal injections, there was a five-fold greater frequency of association with degenerating dopaminergic dendrites than following neostriatal injections (Fig. 4).

These results indicate that the main descending input to dopaminergic

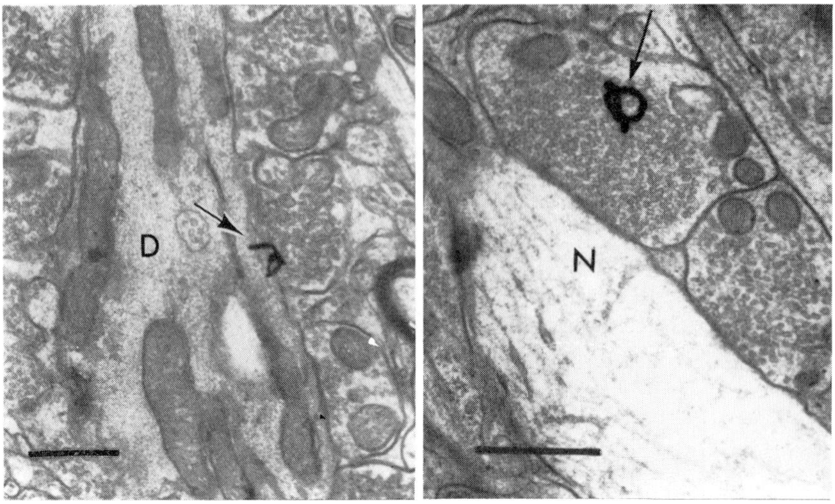

Fig. 5. Boutons with labelled protein (→) in the substantia nigra following injection of ^{3}H-leucine into the globus pallidus (5A, left) or neostriatum (5B, right) of 6-OHDA treated rats. Note symmetrical contacts with normal (N) or degenerating (D) dendrite. Bars indicate 0.5 µm.

cells of the substantia nigra comes from the pallidum. There is a lesser input from the neostriatum. The caudato-nigral tract would appear to project principally to cells of the pars reticulata. The sharp drop in GAD following pallidal lesions would suggest that the pallido-nigral tract is GABA-containing (Hattori *et al.*, 1973; McGeer *et al.*, 1974). There is also evidence that the caudato-nigral tract may be GABA-containing as well (Kim *et al.*, 1971; Fonnum *et al.*, 1974).

THE NATURE OF A10 DOPAMINERGIC CONNECTIONS

Differences seem to exist between the afferent connections to the A9 dopaminergic neurones of the substantia nigra, which feed the neostriatum, and the A10 dopaminergic cells of the midbrain which feed the accumbens, septum, and cortex.

The A9 cells receive a major GABAnergic feedback from the basal ganglia. They do not appear to receive a cholinergic feedback as evidenced by choline acetyltransferase levels which are low and which are unaffected by hemitransections between the nigra and the basal ganglia (McGeer *et al.*, 1973).

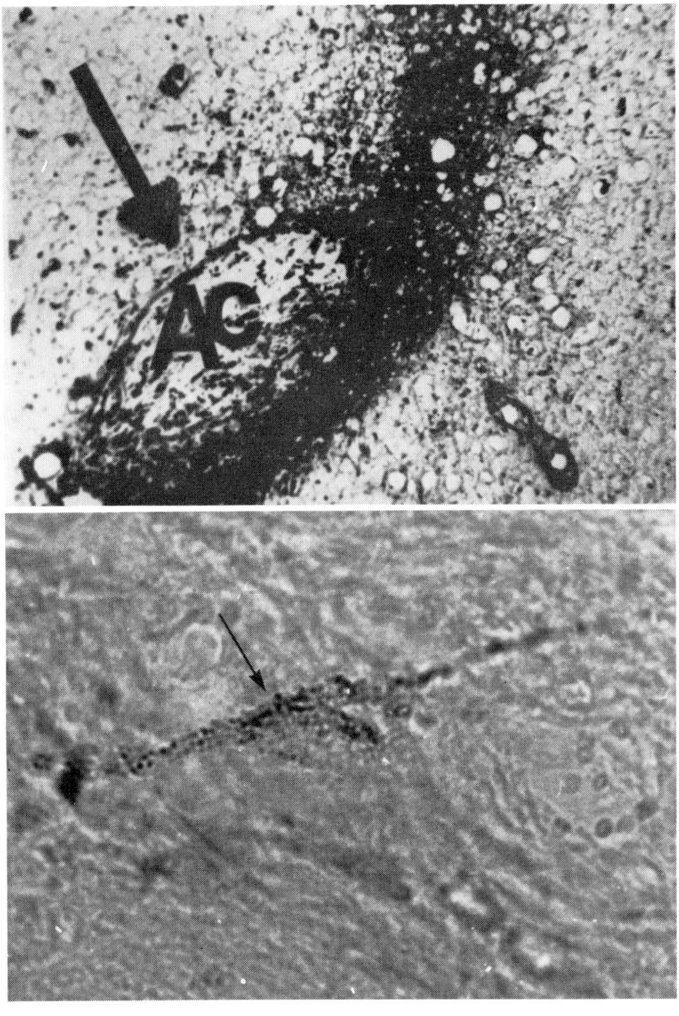

(A)

(B)

Fig. 6. Horseradish peroxidase (HRP) staining of the injected nucleus accumbens (Fig. 6A, × 84) and of dopaminergic cell body in the A10 region (→, Figure 6B, × 625) to which the HRP is transported.

The A10 cells, on the other hand, do not appear to receive any major GABAnergic feedback from the cortex and limbic system, but do appear to have some feedback, possibly cholinergic, from the habenula.

Lesions of the medial forebrain bundle at the level of the anterior commissure, which would interrupt descending pathways from the septum

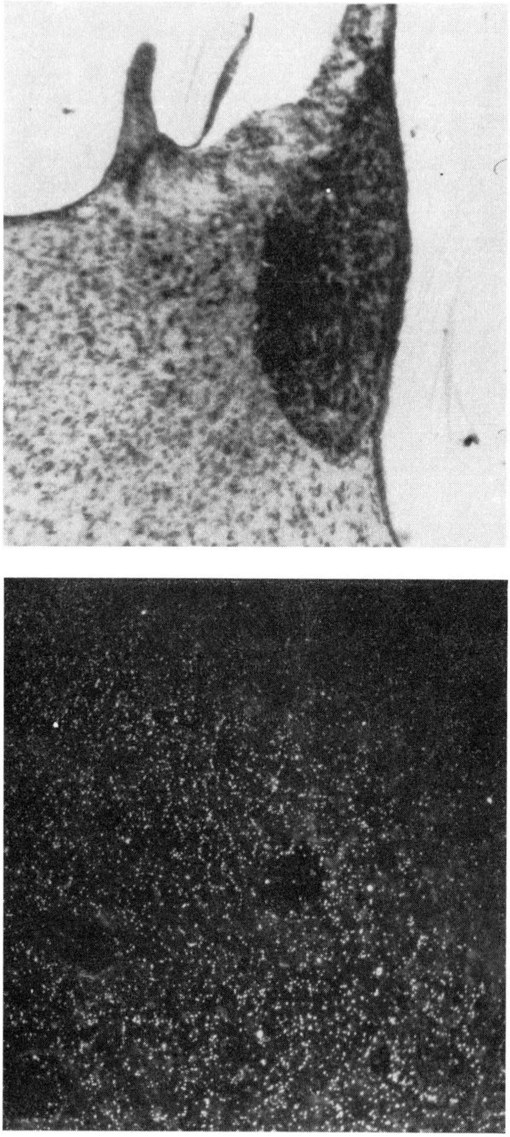

(A)

(B)

Fig. 7. Autoradiographic labelling of medial habenula (Fig. 7A, × 160), of the interpeduncular nucleus (by dark field; Fig. 7B, × 300), and of the A10 region (Fig. 7C, × 625) two days after injection of ^{3}H-leucine into the medial habenula. Figure 7C is an overlay of two pictures of the identical area but taken at different focal depths to show the peroxidase staining in the tissue (outlined cell) and the autoradiographic grains in the emulsion layer.

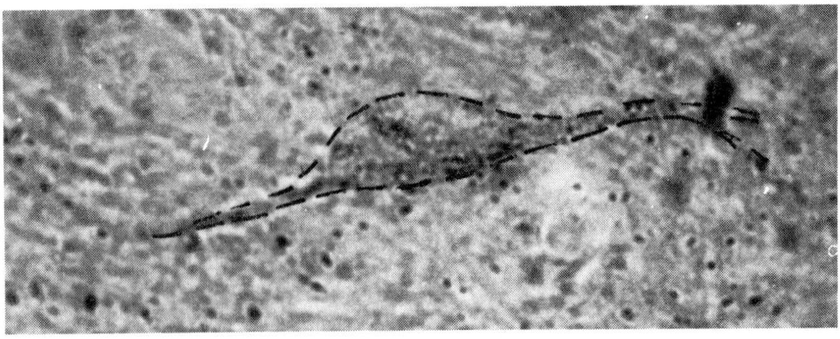
(C)

and accumbens, do not cause a decrease in GAD levels in the region of the A10 cells. This would seem to rule out any direct GABA feedback analogous to that from the basal ganglia to the substantia nigra (McGeer *et al.*, 1977).

An indirect feedback from the limbic system to the midbrain is via the habenula. The habenulo-interpeduncular tract appears to innervate at least some A10 cells. These A10 cells were marked by retrograde axoplasmic flow using horseradish peroxidase. Figure 6A shows the injection site in the accumbens and 6B shows the outline of presumed dopaminergic cell bodies in the midbrain. Particularly prominent staining was seen in those A10 cells which form a cap over the interpeduncular nucleus. These cells may be a special subgroup since they are somewhat separated from those A10 cells that lie dorsally and spread towards the midline, and those that lie laterally between the interpeduncular nucleus and the substantia nigra.

The tract from the habenula to the interpeduncular nucleus was labelled by anterograde axoplasmic flow using injections of ^{3}H-leucine into the medial habenula (Fig. 7A). Heavy labelling of the interpeduncular nucleus was observed two days following the injection (Fig. 7B). This labelling also extended to the region of dendrites of the presumed A10 cells outlined by horseradish peroxidase (Fig. 7C). That actual contact was being made by these nerve endings with dendrites of cells outlined by horseradish peroxidase, is indicated by electron microscopic autoradiography (Fig. 8).

The biochemical nature of these nerve endings has not yet been conclusively demonstrated but it is well known that the habenulo-interpeduncular tract has a massive cholinergic component, as well as substance-P-containing fibers. Present evidence suggests that these nerve endings on dopaminergic dendrites in the A10 region may be cholinergic.

Figure 9 is a summarizing diagram of the descending connections to midbrain dopaminergic cells as they are so far understood. Figure 9A shows descending GABA neurones from the pallidum to nigral dopaminergic cells, and also indicates a descending GABA pathway from the neostriatum to

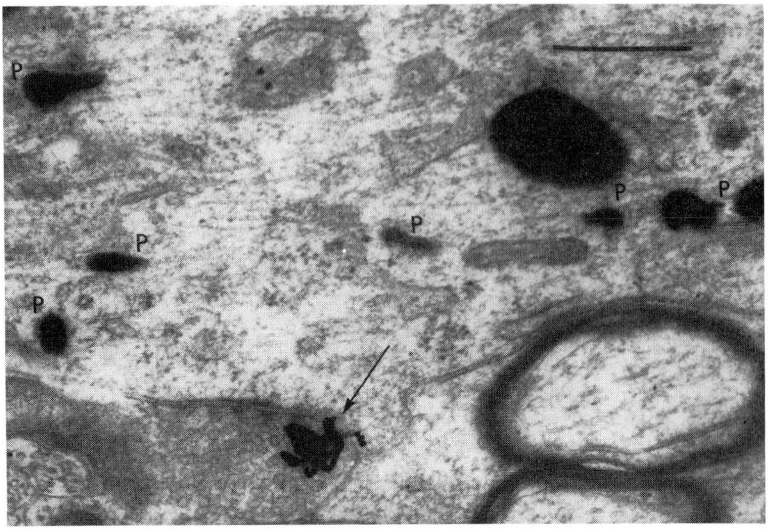

Fig. 8. An electron microscopic detail from the area shown in Fig. 7C; P(s) indicate HRP-produced pigment granules in the dendrite and → indicates the radioactive label transported from the habenula. Bar indicates 0.5 μm.

the pars reticulata of the substantia nigra. These endings may also make occasional contact with dopaminergic dendrites. Figure 9B shows descending cholinergic neurones from the habenula to A10 cells just dorsal to the interpeduncular nucleus. Thus, the feedback innervation of these two major sources of dopaminergic pathways, one to the basal ganglia, and one to the limbic system and neocortex, is probably significantly different. They march, as it were, to different neurotransmitter tunes.

There may be some input of cells using a neurotransmitter common to these two dopaminergic sources. There is a descending substance P pathway from the neostriatum to the nigra (Hong *et al.*, 1977) and from the habenula to the interpeduncular nucleus (Kataoka, 1973). Both could innervate dopaminergic structures, but so far the appropriate double labelling experiments have not been done.

THE NATURE OF SMALL INTENSELY FLUORESCING (SIF CELL) DOPAMINERGIC CONNECTIONS

Double labelling experiments have not been done in sympathetic ganglia, but enough is known about the rather simple anatomical arrangements in these structures to make assumptions about the anatomical connections of these cells. The presynaptic input to ganglion cells of all types is cholinergic. Small intensely fluorescing cells make contact with cell bodies in the superior cervical ganglion (Libet, 1977) which are overwhelmingly noradrenergic. Thus, SIF cells undoubtedly rcccive cholinergic input and make noradrenergic output contacts (Fig. 9C).

Mention has been made of a few of the probable contacts of dopaminergic neurones in the nigra, midbrain, tegmentum and sympathetic ganglia where they affect extrapyramidal, limbic and sympathetic activity,

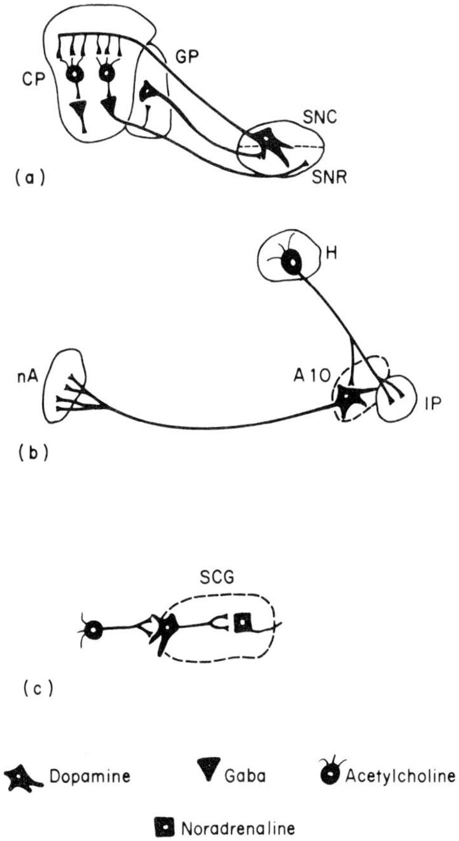

Fig. 9. Hypothesized interconnections of some of the neurones in (a) extrapyramidal, (b) limbic and (c) ganglionic systems. GP = globus pallidus; CP = caudate-putamen; SNR = zona reticularis; SNC = zona compacta; H = habenular; IP = interpeduncular nucleus; A10 = ventral tegmental area of Tsai; nA = nucleus accumbens; SCG = superior cervical ganglion.

respectively. In each case, the connections are different, suggesting the possibility of a selective neurochemical approach to modifying each of these functions based on the secondary neurones involved. Little is known about other contacts in areas of dopaminergic cell population such as the hypothalamus and retina where the functions are presumably related to neuroendocrine function and vision.

Obviously, there is much more to be learned, even about those areas where some information already exists. There are undoubtedly several important inputs to each of these dopaminergic cell populations with only one or none so far understood. A similar situation applies to the output side, providing considerable scope for future research.

REFERENCES

Fonnum, F., Grofova, I., Rinvik, E., Storm-Mathisen, J. and Walberg, G., (1974). *Brain Res.* **71,** 77–92.

Hattori, T., Fibiger, H. C., McGeer, P. L. and Maler, L. (1973). *Exp. Neurol.* **41,** 599–611.

Hattori, T., Fibiger, H. C. and McGeer, P. L. (1975). *J. Comp. Neurol.* **162,** 487–504.

Hattori, T. and McGeer, P. L. (1973). *Exp. Neurol.* **38,** 70–79.

Hattori, T. and McGeer, P. L. (1975). *Exp. Neurol.* **45,** 541–548.

Hattori, T., McGeer, P. L., Fibiger, H. C. and McGeer, E. G. (1973). *Brain Res.* **54,** 103–104.

Hong, J. S., Yang, H. Y. T., Racagni, G. and Costa, E. (1977). *Brain Res.* **122,** 541–544.

Kataoka, K., Nakamura, Y. and Hassler, R. (1973). *Brain Res.* **62,** 264–267.

Kim, J. S., Bak, I. J., Hassler, R. and Okado, Y. (1971). *Exp. Brain Res.* **14,** 95–104.

Libet, B., *In* "Nonstriatal Dopaminergic Neurons", (1977). (Eds E. Costa and G. L. Gessa), 541. Raven Press, New York.

McGeer, E. G., Fibiger, H. C., McGeer, P. L. and Brooke, S. (1973). *Brain Res.* **52,** 289–300.

McGeer, P. L., Fibiger, H. C., Hattori, T., Singh, V. K., McGeer, E. G. and Maler, L. (1974). *In* "Advances in Behavioural Biology", Vol. 10, (Eds R. D. Myers and R. R. Drucker-Colin), 27. Plenum, New York.

McGeer, P. L., Grewaal, D. S. and McGeer, E. G. (1974). *Brain Res.* **80,** 211–217.

McGeer, E. G., Hattori, T. and McGeer, P. L. (1975). *Brain Res.* **86,** 478–482.

McGeer, E. G., McGeer, P. L., Grewaal, D. S. and Singh, V. K. (1975). *J. Pharmacol. (Paris)* **6,** 143–152.

McGeer, P. L., McGeer, E. G. and Hattori, T. (1977). *In* "Nonstriatal Dopaminergic Neurons", (Eds E. Costa and G. L. Gessa), 397. Raven Press, New York.

Neurophysiology of Dopamine Systems

CHAPTER 24

The Neurophysiology of Dopamine Receptors

Donald H. York

University of Missouri, Columbia, U.S.A.

INTRODUCTION

Many studies have now demonstrated the existence of axonal projections and terminations of the large cell bodies localized in the pars compacta of the substantia nigra (SN). These fibres course through the median forebrain bundle to terminate in the caudate nucleus and putamen. They are relatively small unmyelinated fibres and contain dopamine concentrated in their end terminals (Andén *et al.*, 1964; Fuxe *et al.*, 1964; Andén *et al.*, 1966; Hornykiewicz, 1966; Poirier *et al.*, 1967; Ungerstedt, 1968; Bedard *et al.*, 1969; Hökfelt and Ungerstedt, 1969; Moore and Heller, 1969; Moore *et al.*, 1971; Siegel *et al.*, 1971; Ungerstedt, 1971; Carpenter and Peter, 1972; Hedreen and Chalmers, 1972; Fibiger *et al.*, 1973; Hökfelt and Ungerstedt, 1973; Ibata *et al.*, 1973; Maler *et al.*, 1973; Shimizu and Ohnishi, 1973; Kuypers *et al.*, 1974; Nauta *et al.*, 1974; Simon *et al.*, 1974). The release of dopamine from terminals of the nigrostriatal pathway by stimulation of the substantia nigra has been demonstrated by collecting either dopamine (Portig *et al.*, 1968) or labelled dopamine in ventricular cerebrospinal fluid (von Voigtlander and Moore, 1971; Chiueh and Moore, 1973) or by employing a push-pull cannula stereotaxically inserted into the striatum (McLennan, 1964, 1965). Other studies have utilized measurements of the dopamine metabolites HVA and DOPAC as indicators of dopamine having been released, as a result of neuronal activity presumably in dopaminergic neurones (Portig *et al.*, 1968; Vogt, 1969). Several of these studies have also used drug-induced techniques to evaluate dopamine release from the striatum *in vivo* (McKenzie and Szerb, 1968; Riddell and Szerb, 1971; Besson *et al.*, 1971; Chiueh and Moore, 1973; von Voigtlander and Moore, 1973).

Neither the demonstration of dopamine release by activation of the nigrostriatal pathway nor the presence of dopamine in the striatum, by itself, proves dopaminergic transmission. Each criterion for synaptic transmission must be evaluated together with all of the evidence satisfying other criteria (Werman, 1966; McLennan, 1970) before it is certain that dopamine is functioning as a "classic neurotransmitter".

Thus, there is no question of a direct synaptic linkage from the substantia nigra to the striatum, nor that dopamine is released in the striatum. The question remaining centres on what is the action of dopamine when it is released from the terminals of the nigrostriatal tract? Is dopamine an excitatory or an inhibitory neurotransmitter? The answer to this question will have considerable implications for the pathophysiology of diseases affecting the basal ganglia and especially for their subsequent pharmacological management. In particular, degeneration of the nigrostriatal dopamine pathway is known to be directly involved with Parkinson's disease (for reviews see Hornykiewicz, 1966, 1972; Barbeau, 1972). Clarification of the basic synaptic mechanisms involving this pathway and its inter-relationships with the basal ganglia circuitry is thus of crucial importance. Why the action of dopamine has continued to be so ill defined and controversial at the present time is the subject of this Chapter.

Neurophysiological studies have demonstrated the monosynaptic nature of the nigrocaudate pathway (Frigyesi and Purpura, 1967; Hull *et al.*, 1970; Feltz and Albe-Fessard, 1972; Kitai *et al.*, 1975) and the nigroputamen pathway (York, 1970). The long latency of monosynaptic responses has been demonstrated by several of these studies and suggests a relatively slow conduction velocity of approximately one metre per second in this pathway (Frigyesi and Purpura, 1967; McLennan and York, 1967; Connor, 1970; Feltz and McKenzie, 1969; Hull *et al.*, 1970; York, 1970; Feltz and Albe-Fessard, 1972; Gonzales-Vegas, 1974; Liles, 1974; Kitai *et al.*, 1975; 1976; Zarzecki *et al.*, 1976). These long latency responses have in general been attributed to the small poorly myelinated fibres which characterize this pathway (Verhaart, 1950; Fuxe *et al.*, 1964; Mori, 1966; Adinolfi and Pappas, 1968; Kemp, 1968).

Stimulation of the substantia nigra and its resultant effects upon single caudate unit discharge have been examined in several studies utilizing extracellular recording techniques (McLennan and York, 1967; Connor, 1970; Feltz and McKenzie, 1969; Feltz and Albe-Fessard, 1972; Gonzales-

Vegas, 1974; Liles, 1974; Dafny, 1975; Zarzecki *et al.*, 1976; Richardson, *et al.*, 1977). More in-depth analysis of the synaptic responses of striatal cells to electrical stimulation of the SN has been evaluated with intracellular recording studies, such as those undertaken by Frigyesi and Purpura (1967); Hull *et al.* (1970); Buchwald *et al.* (1973); Frigyesi and Szabo (1974); Kitai *et al.* (1975, 1976).

STUDIES INVOLVING SPONTANEOUSLY FIRING STRIATAL NEURONES—A SELECT POPULATION?

In a study of spontaneously active units in the caudate nucleus of the rat, Dafny (1975) found that cells which exhibited spontaneous activity could not be driven by stimulation of SN. However, some 55 units (71%) responded to stimulation of SN by alteration of their spontaneous activity. Twenty-three units (30%) responded by increasing their discharge frequency and 32 units (41%) responded by decreasing their activity. Two cells which did not fire spontaneously could be driven by stimulation of SN to elicit spikes at latencies of 18–22 ms. Of the spontaneously firing caudate units that were studied, most fired relatively slowly (1–3 spikes s^{-1}), although several exhibited firing rates of more than 10 spikes s^{-1}.

Many studies which have used spontaneously firing cells to evaluate nigral effects have presumed homogeneity within the nigrostriatal tract with regard to there being only one neurotransmitter in this pathway, namely dopamine. Further, this one neurotransmitter is presumed to have only one postsynaptic action. However, such predicted homogeneity may not be the case, as demonstrated anatomically and electrophysiologically (Hedreen, 1971; Feltz and DeChamplain, 1972a, b).

After the injection of 6-hydroxydopamine (6-OHDA) into the SN which resulted in marked decreases in striatal dopamine levels ($\sim$80% of control), a monosynaptic nigrostriatal excitatory and inhibitory functional pathway could still be demonstrated (Feltz and DeChamplain, 1972a, b). There are several possible conclusions from this study. It may be that the 20% dopamine remaining in the striatum reflects an incomplete lesioning of the nigrostriatal dopamine pathway and enough undisrupted fibres remain intact to allow functional effects to be obtained, or it may be the case that there is a non-dopaminergic nigrostriatal pathway. This latter possibility has also been suggested from anatomical studies involving 6-OHDA (Hedreen, 1971).

Other studies which examined only spontaneously firing cells have attempted to correlate changes in the discharge frequency of striatal units following lesioning of the SN, with the nature of synaptic action of the nigrostriatal pathway.

It is also presumed in such a study that spontaneously firing striatal cells are a representative sample of striatal cells which receive nigral dopaminergic input. However, since few striatal cells (~10%) demonstrate spontaneous activity (McLennan and York, 1967; Connor 1970; York, 1970, 1972; Feltz and Albe-Fessard, 1972; Zarzecki *et al.*, 1976), the population being sampled is a very small proportion of the total. Since there is a very diffuse innervation of cells by dopamine terminals in the striatum, it is more probable that the 90% of the striatal cell population that is normally quiescent will receive a larger portion of dopaminergic nigrostriatal input.

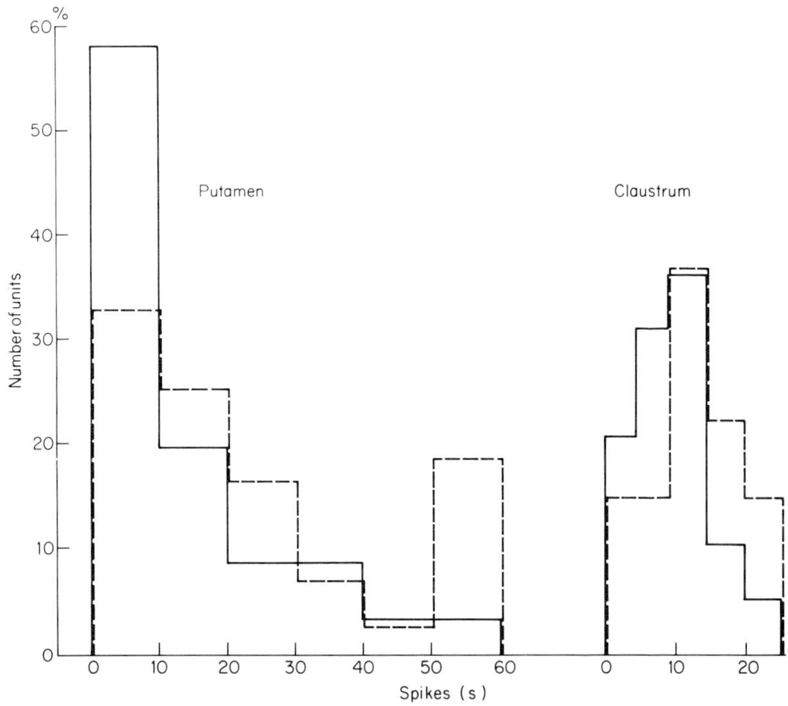

Fig. 1 Frequency (maximal spontaneous firing) histogram of extracellular unit activities of cells of the putamen and claustrum recorded on the intact (unbroken line) and lesioned (broken line) side, respectively, in different animals. It includes 44 and 36 cells in the putamen of the lesioned and intact side, respectively and 14 and 20 cells in the claustrum of the lesioned and intact side, respectively. All cells of the putamen on the lesioned side that fired at a frequency exceeding 50 spikes s^{-1} are plotted in the column of 50 spikes s^{-1}. (From Ohye, Bouchard, Boucher and Poirier, 1970; with permission of Williams and Wilkins, Co.)

With this background in mind, we will review the data which has been used to classify dopamine as an inhibitory neurotransmitter. It has been claimed that there is an increase in spontaneous firing frequency in 30 out of 44 cells recorded from in the putamen of cats deprived of their incoming nigrostriatal fibres, compared with the discharge rate of cells in the putamen of the intact side (Ohye *et al.*, 1970). In this latter study, 36 cells were examined in the putamen of the intact side and 44 cells in the putamen of the lesioned side. The histogram computed from neuronal discharges is shown in Fig. 1. This histogram clearly shows that of the cells in the putamen discharging spontaneously (of which the slow-firing cells, i.e. $0–10\,s^{-1}$ comprise some 60–80%) there is a marked decrease in the proportion of slowly firing cells and an increase in the proportion of cells firing over $50\,s^{-1}$. It is difficult to see how the overall conclusion was reached that there was a net increase in firing frequency of neurones after SN lesioning. Although this was observed with cells which tended to fire above $10\,s^{-1}$, with these particular data, it could also be argued that the most dramatic effect of the lesion was the removal of an excitatory input to the slow-firing population of neurones, while causing facilitation of another population of cells which already tended to be firing faster before lesioning.

In a most comprehensive analysis of caudate units spontaneously discharging before and following lesions of the median forebrain bundle and SN, together with a correlative neurochemical determination of striatal dopamine levels following such lesions, it was concluded, that in cats with a significant decrease in ipsilateral dopamine levels produced by such lesions there was no statistically significant concomitant change in striatal firing rate (Hull *et al.*, 1974).

In the studies to be discussed, it will also be demonstrated that spontaneously active cells do not tend to be affected by nigral stimulation (Feltz and Albe-Fessard, 1972; Dafny, 1975). Thus the effects of lesions or denervation on spontaneously firing cells may not be relevant to studies of the nigral input to the caudate nucleus.

NIGRAL ACTIVATION LEADS TO EXCITATION, OR INHIBITION OR BOTH?

There is clearly a substantial population of neurones in the striatum whose response to electrical stimulation of SN is excitation. These neurones are mainly localized in the medial two-thirds of the caudate nucleus (McLennan and York, 1967; Feltz and Albe-Fessard, 1972; Kitai *et al.*, 1976; Zarzecki *et al.*, 1976). This is in contrast to caudate neurones receiving mainly inhibitory input from SN which would appear to be part of a different neuronal population scattered throughout the whole caudate nucleus (Albe-Fessard *et al.*, 1967; Connor, 1970; Feltz and Albe-Fessard, 1972). This latter group of nigral-inhibited neurones are more easily activated by iontophoretically applied amino acids (L-glutamic acid or D,L-homocysteic acid) than are striatal cells which are excited by nigral stimulation (Connor, 1970; Feltz, 1970). On the other hand nigral-excited cells are seldom spontaneously active (Dafny, 1975), even in decerebrate or spinal-local analgesic cat preparations (Feltz and Albe-Fessard, 1972), although some discharge may be apparent after prolonged repetitive stimulation (Feltz and MacKenzie, 1969).

An analysis of the intracellular postsynaptic potentials (PSPs) produced by nigral stimulation in some 51 impaled caudate neurones demonstrated that excitation was the most common effect observed (Fig. 2), either as pure EPSPs (16% of cells) or EPSP–IPSP sequences (43% of cells). Only 1% of caudate neurones showed a pure IPSP response. The remainder of cells (40%) showed no response to nigral stimulation (Hull *et al.*, 1970). There was some degree of variability in the distribution of excitatory PSPs obtained depending on whether a lateral or medial nigral stimulation site was chosen, i.e. pure EPSPs were seen in 21% of cells given lateral versus 11% given medial nigral activation. Also, the EPSP–IPSPs evoked

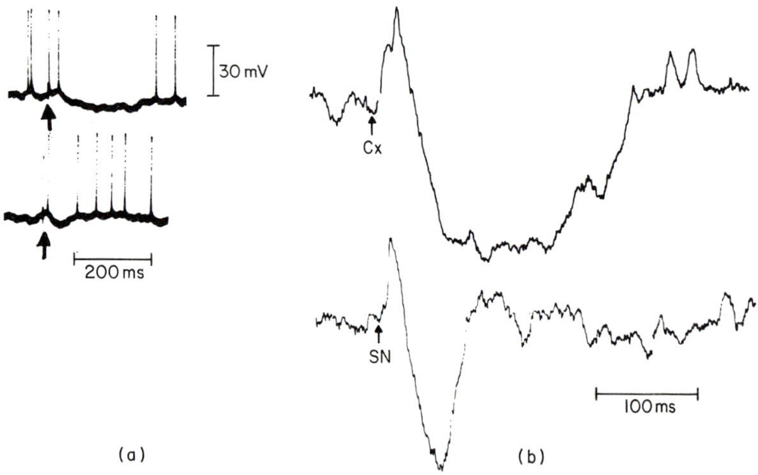

Fig. 2 Comparison of intracellular responses of a caudate neurone to anterior sigmoid gyrus and nigral stimulations. (a) Oscillographic tracing of intracellular responses to cortical (upper trace) and nigral (lower trace) stimuli. (b) Averages of 10 consecutive responses to cortical (upper trace) and nigral (lower trace) stimuli. Note increased length of IPSP to cortical as compared with nigral stimulus. Arrow indicates stimulus shock artifact. (From Hull *et al.*, 1970, with permission of Elsevier Publishing Co.)

sequence was observed in 37% of cells given lateral, versus 50% of cells driven by a medial nigral stimulation site.

The above studies clearly indicated that nigral activiation produced predominantly excitatory effects on striatal cells. However, "muddying of the waters" arose when iontophoretic data were now evaluated with respect to the above studies.

ACTION OF DOPAMINE ON SINGLE STRIATAL NEURONES

Studies in which the activity of striatal neurones was examined by iontophoresis, a technique which also allows non-spontaneously firing cells to be evaluated by applying amino acids which excite otherwise quiescent cells (Curtis, 1964) have demonstrated results quite different to those obtained from studies of spontaneously firing cells. The effects of dopamine applied in the local vicinity of a single neurone also receiving nigral input may also be evaluated with this technique. Early studies by Bloom, *et al.* (1965); McLennan and York (1967); Herz and Zieglgänsberger (1968) found that iontophoretically applied dopamine had predominantly depressant actions on caudate neurones. Although one of these studies (McLennan and York, 1967) also demonstrated nigral-driven cells, the effects of dopamine never appeared to facilitate such nigral-evoked response. A later, more in-depth examination of the effects of iontophoretically applied dopamine and stimulation of the SN by Connor (1970) revealed that nigral activation produced several types of responses. Pure depression of cell firing was seen in 64 cells (24%), a depression followed by a late facilitation was seen in 57 (22%), a "pure" facilitation response in 42 cells (16%). In an examination of the effects of dopamine applied iontophoretically on 101 caudate neurones whose responses to nigral stimuli were "identified" on the basis of such poststimulus histograms, 31 units which were depressed by SN were also depressed by iontophoretic dopamine and no instances of dopamine facilitation were observed on neurones depressed by SN stimuli. The iontophoretic administration of α-methyldopamine blocked the depressant responses elicited by both iontophoretic dopamine and nigral stimulation in six cells. A large number of caudate neurones were facilitated by nigral stimulation (42 out of 260 cells), or were initially depressed followed by late facilitation (57 out of 260 cells) suggesting that nigral-induced facilitation was also a significant effect, since it was observed on 38% of the total cell population compared to 24% of the cells showing pure depression. The problem with the nigral-induced facilitation was that it was only seen on seven cells which were also facilitated by iontophoretically applied dopamine, versus 31 cells which were depressed by both dopamine and SN.

Thus there is a major problem in the interpretation of the iontophoretic data. Either there are two nigrocaudate dopaminergic tracts, one being excitatory and the other inhibitory, or the technique itself had biased the sampling of nigral input on striatal cells, i.e. possibly, only the larger cell spikes may be sampled by the micropipette, thus producing a skewed sample which omits smaller cells, which are in fact in greater actual number in the striatum and may receive more of the SN input.

The resolution of the physiological action of dopamine released from nigrostriatal fibres became more perplexing when a similar study was undertaken in the putamen (York, 1970). In this study dopamine was tested iontophoretically on 120 cells; it was found to cause a depression of discharge frequency in 37 (31%) and excited 53 (44%). However, other types of cells, some 22 in number in the putamen, also demonstrated a mixed action consisting of an initial depressant action which was followed by an excitant action. The late excitation was apparent from several seconds, up to one minute, after application had ceased. When the results of dopamine iontophoresis onto single cells were compared with the cells' responsiveness to stimulation of the SN, by repetitive trains of pulses delivered to the nigra, it was observed that 43 (77%) of some 56 cells which were excited by dopamine given iontophoretically were also excited by stimulation of SN. Cells were also encountered which were excited by SN stimulation, but were either unaffected or depressed by dopamine, or alternatively were excited by dopamine but unaffected by SN activation. The probability that cells excited by stimulation of the substantia nigra would also be excited by iontophoretically applied dopamine was high, considering that many cells not responsive to SN stimulation were also unresponsive to dopamine application (17% or 125 cells) (York, 1970).

A more recent study examining the effects of dopamine on single neurones in the putamen and amygdala of the cat was undertaken by Ben-Ari and Kelly (1976). In this study, although no synaptic inputs were studied, the action of dopamine applied iontophoretically was compared to that of γ-aminobutyric acid as a method of assessing the relative potency of dopamine on single neurones. Of 24 cats studied, dopamine was tested on the spontaneous or glutamate activity of 386 units and found to have a powerful inhibitory action on all but 41 cells (11%). Of these uninhibited cells, 27 (7%) were considered to be unaffected by dopamine. No dopamine-excited cells were encountered in pentobarbitone-anaesthetized cats ($n = 12$), although 12 cells were excited by dopamine in halothane-anaesthetized cats ($n = 12$). In nine of the 14 excited cells relatively high currents (greater than 100 nA) were apparently required to produce the excitation, which was then often followed by a variable period of reduced activity. These excitations were described as similar to those reported by York (1970, 1972), in that "the cells rapidly became refractory to dopamine when the same dose was repeatedly tested at intervals of 40 seconds, although the inhibitory response to dopamine was considered to never exhibit tachyphylaxis".

It should be pointed out that in the previous two studies of York (1970, 1972), in which more dopamine excitations were observed, the dopamine current never exceeded 80 nA. Also, in the latter studies dopamine was specifically being evaluated on putamen cells which responded to electrical stimulation of the SN rather than "a random survey" approach utilized by Ben-Ari and Kelly (1976). In the neostriatum of the rat, a rather large proportion of dopamine-excited cells (36%) have also been reported by Spencer and Havlicek (1974).

Other studies (Gonzalez-Vegas, 1974; Zarzecki *et al.*, 1976) utilizing extracellular recording which have reported nigral-induced excitation followed by inhibition have found that a pure inhibition similar to that described by Connor (1970) is a more predominant occurrence of nigral activation. These studies, however, did not find any correlation between the effects of iontophoretically applied dopamine and synaptically mediated effects of the nigrostriatal pathway (Gonzalez-Vegas, 1974; Zarzecki *et al.*, 1976).

RESOLUTION OF NIGROSTRIATAL DOPAMINERGIC ACTION

Many studies have shown that excitation of striatal neurones by means of electrical stimulation of the SN is a consistent reproducible finding with extracellular recording (McLennan and York, 1967; Connor, 1968, 1970; Feltz and MacKenzie, 1969; York, 1970; Feltz and Albe-Fessard, 1972; Gonzalez-Vegas, 1974; York, 1976; Zarzecki *et al.*, 1976). However, whether these typical long latency-induced responses were monosynaptic or not was of some concern and not resolved satisfactorily until the intracellular studies of Frigyesi and Purpura (1967), Hull, *et al.* (1970), and Buchwald *et al.* (1973) were viewed collectively. Further, whether such monosynaptic evoked responses were due to the release of dopamine also remained to be proven.

A major advance in the resolution of the monosynaptic nature of the excitatory dopaminergic nigrostriatal pathway came with the studies of Kitai *et al.* (1975, 1976) who recorded intracellularly from caudate units which were excited monosynaptically by nigral stimulation and then evaluated the effects of iontophoretically administered dopamine applied extracellularly from an adjacent multibarrelled microelectrode. The technical difficulty involved with the undertaking of such experiments cannot be overemphasized and these experiments truly herald a major advance in understanding the nigrostriatal dopaminergic pathway.

Dopamine was found to produce depolarization of caudate neurones which were monosynaptically excited by substantia nigra stimulation (Fig. 3). This study offers evidence that the depolarizing potentials produced in caudate neurones by nigral stimulation are in fact EPSPs and not disinhibitory phenomena. This is based on the observation that the amplitude of the depolarizing potentials could be increased by hyperpolarizing current injected into the cell or decreased and reversed to the

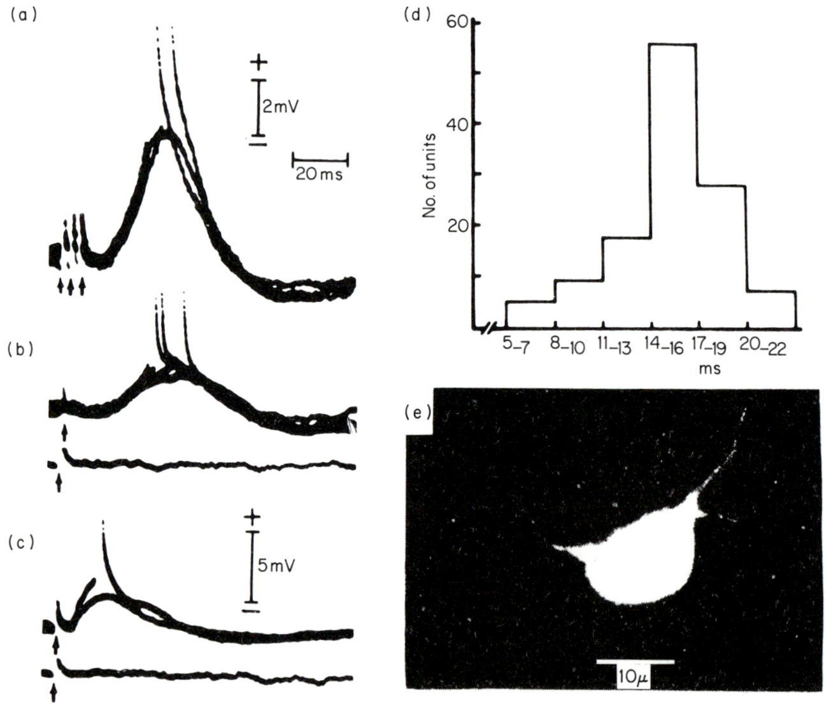

Fig. 3 EPSPs and action potentials recorded from a Cd neurone activated from the SN ((a) 3 stimuli, (b) 1 stimulus) and the MFB ((c) 1 stimulus). The bottom traces in B and C are extracellular records. Calibrations: 2 mV for (a) and (b), 5 mV for (c), 20 ms, (a)–(c). (d) Latency histogram of the SN-induced EPSPs in Cd neurones. Abscissa: latencies in milliseconds. Ordinate: number of cells. (e) A photomicrograph of a procion yellow-injected Cd neurone. (From Kitai *et al.*, 1976; with permission of Springer-Verlag Publishing Co.)

negative potential by depolarizing current (Fig. 4) (Coombs *et al.*, 1955). In contrast to other studies involving the iontophoresis of dopamine, Kitai *et al.* (1976) utilized current pulses of short-duration currents. The dopamine effect commenced some time after the cessation of the current. A current of longer duration resulted in the depolarizing action starting some time after

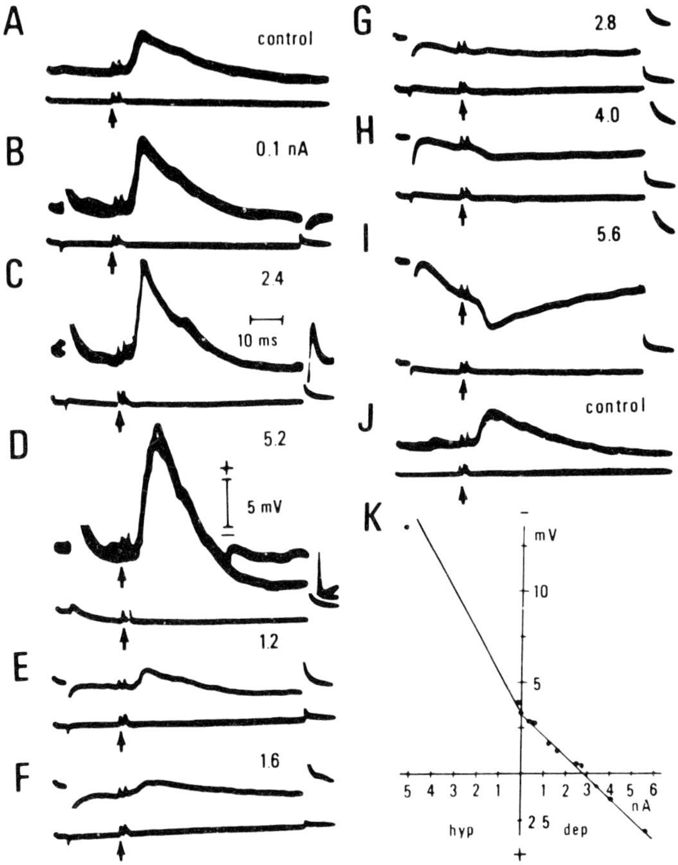

Fig. 4 Effects of hyperpolarizing (B–D) and depolarizing (E–I) current on MFB-induced EPSPs evoked by a double stimulus and recorded from a Cd neurone. A and J show control EPSPs prior to and immediately after the termination of current respectively. The bottom traces in A–J are extracellular records. Numbers on right-hand corner of records in B–I indicate the intensity of intracellular current in nA; Calibrations, 5 mV and 10 ms, A–J. (K) Changes in amplitude and polarity of EPSPs with hyperpolarizing (hyp) and depolarizing (dep) current. Abscissa: current in nA. Ordinate; amplitude of EPSPs in mV. Upgoing arrows indicate the onset of stimulation in this and all of the subsequent figures. (From Kitai *et al.*, 1976; with permission of Springer-Verlag Publishing Co.)

the onset of the current and continuing for a considerable amount of time after the offset of the current. Controls were also undertaken to make sure the depolarization recorded was not due to a current effect caused by the application of current to the dopamine barrel, rather than to the action of dopamine (Fig. 5). This was evaluated by passing anodal and cathodal currents through the sodium chloride containing barrels of the multibarrel microelectrode. Such controls clearly indicated that there were no current effects at the level of intensity at which dopamine was applied (maximal current used was 60 nA). Together with the short duration of the ejection current, a relatively low intensity current was also used (maximum 60 nA), so that the compound applied would be relatively localized to the area of application. The proximity of the drug application site and the receptor site is indicated by the fact that drug effects were seen in this study within 20 or 30 ms after the application. Repeated applications of dopamine were evaluated whenever the depolarizing effect was initially observed and subsequent applications often were found to be additive, leading to a larger depolarization or occasionally to firing of the cell. This depolarizing action

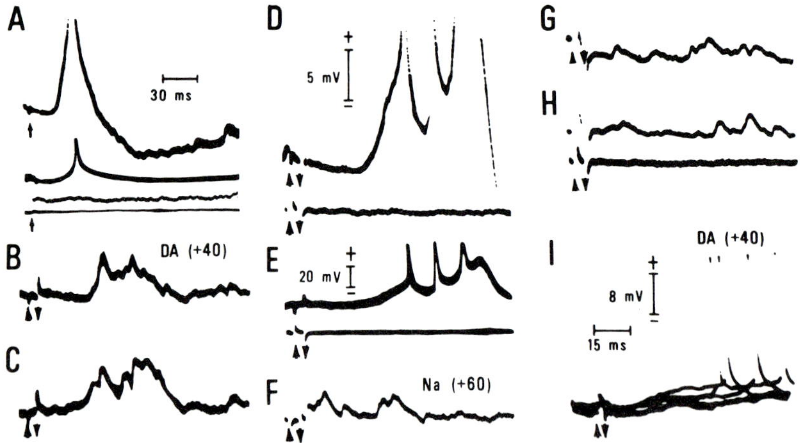

Fig. 5 (A) Action potential evoked by SN stimulation. Top two traces high gain AC and low gain DC record respectively. Bottom two traces, respective extracellular records. (B) DA, 40 nA, for 10 ms. An enhanced effect of successive doses of dopamine on the same cell is shown in C (2nd dose) and D (high gain AC records) and E (low gain DC records) (4th dose). (F–H) Effects of electrophoretic Na on the same neurone. (F) First Na administration, (G) second, (H) fourth. Bottom traces in D, E and H are the extracellular control records. Calibration; 5 mV for high gain AC records in A–D and F–H. 20 mV for low gain DC records in A and E. 30 ms for A–H. (I) Superimposed records of DA-induced depolarization and spike potentials from another Cd neurone. DA was ejected with a current of 40 nA for 5 ms for each dose. (From Kitai *et al.*, 1976; with permission of Springer-Verlag Publishing Co.)

of dopamine was found in approximately 30% of the neurones tested. Hyperpolarizing potentials following dopamine application were *never* observed.

Iontophoretic application of chlorpromazine was also examined in this study since earlier work had indicated that chlorpromazine may have dopamine-blocking properties on striatal neurones (York, 1972). Also, Feltz (1971) had shown that haloperidol blocked the excitant effects of SN stimulation, whereas nigrostriatal inhibition of caudate neurones was unaffected by superfusion of the caudate nucleus with haloperidol. Chlorpromazine was found to suppress the nigral or median forebrain bundle-evoked EPSPs and spikes evoked in caudate cells (Fig. 6), indicating that the synaptic evoked response could be mediated by dopamine. The identification anatomically, of which neurones were excited by nigral stimulation and by iontophoretically applied dopamine, was also under-

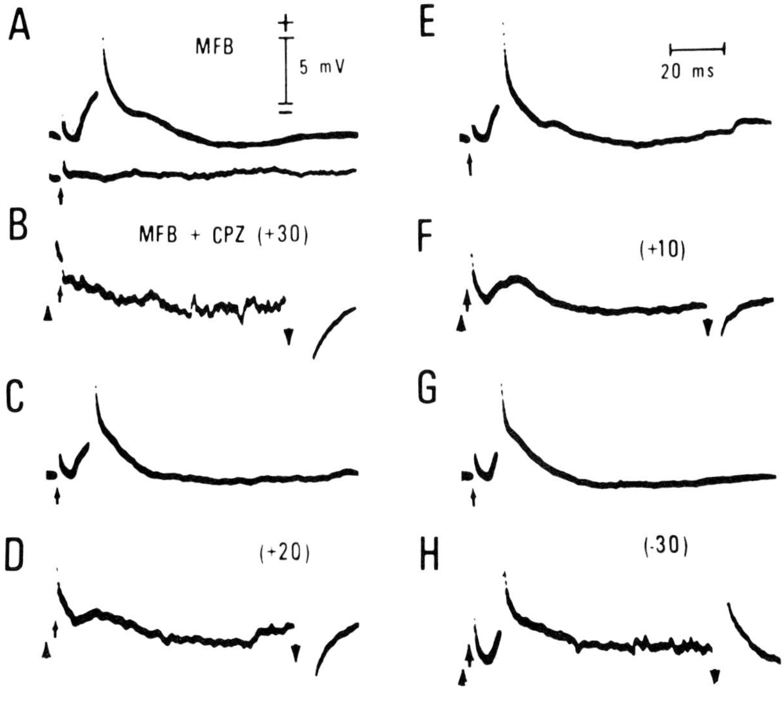

Fig. 6 Effects of chlorpromazine (CPZ) on MFB-induced EPSPs and action potentials recorded from a Cd neurone. (A) Control records; bottom trace, extracellular record. CPZ was administered with currents of 30 nA in B, 20 nA in D and 10 Na in F. (C, E, G) Control records in absence of CPZ. (H) CPZ ejection current was reversed to negative 30 nA. Calibrations: 5 mV and 20 ms. (From Kitai *et al.*, 1976; with permission of Springer-Verlag Publishing Co.)

taken in this study. Some of the impaled caudate neurones were marked by the injections of the dye procion yellow through the intracellular recording electrode (Fig. 3). Analysis of the morphology of the cells indicated that they resembled medium-sized intrinsic spiny cells described by Fox *et al.* (1971) and Kemp and Powell (1971) which account for over 95% of the caudate neurone population. No caudate neurones recorded from could be activated antidromically by stimulation of the SN, cerebral cortex or cerebral peduncle, further suggesting the intrinsic nature of these neurones. The conclusion from this study then indicates that the nigrocaudate pathway is monosynaptically excitatory on a population of caudate neurones which are also excited by dopamine. These cells are intrinsic interneurones within the caudate nucleus.

MECHANISM OF NIGRAL-INDUCED INHIBITION

A recent study of nigrostriatal mechanisms of excitation and inhibition utilizing extracellular recording techniques have been undertaken in the rat,

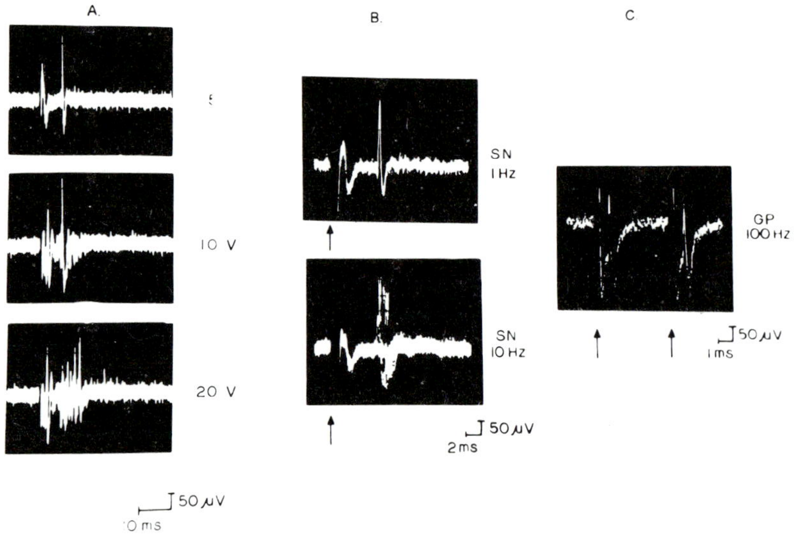

Fig. 7 Activation of large amplitude neurones in neostriatum following stimulation of SN. (A) effects of increasing stimulus intensity from 5 to 20 V. Note that the large amplitude spike is blocked by the burst response at higher intensities. (B) activation of large spike following SN stimulation at 1 and 10 Hz. The response failed to follow above 50 Hz. (C) the same cell as in (B) evoked antidromically from globus pallidus (GP) stimulation and indicating ability of this response to follow 100 Hz stimulation. Arrows indicate shock artefacts. (From Richardson *et al.*, 1977; with permission of Elsevier Publishing Co.)

anaesthetized with urethane (Richardson *et al.*, 1977). In this study, low-intensity stimulation of the substantia nigra was found to evoke single large-amplitude spikes, while higher intensities (10–20 V) elicited a high frequency burst of small-amplitude spikes or waves (Fig. 7). When spontaneous or glutamate-induced large spikes were being recorded, it was

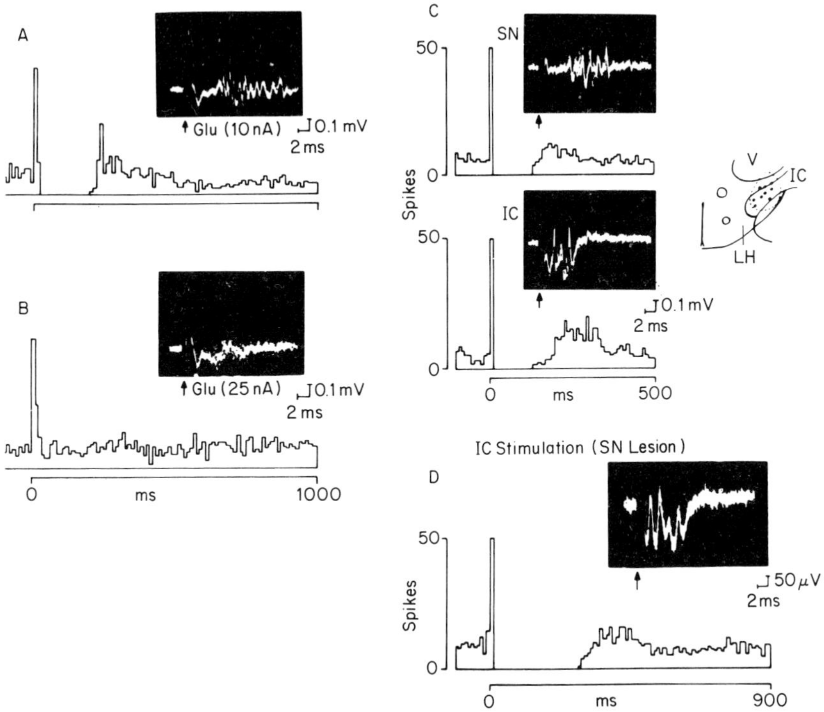

Fig. 8 The effects of iontophoretic application of glutamate (Glu), substantia nigra (SN) and internal capsule (IC) stimulation on the response pattern of large amplitude neostriatal neurones. (A) post-stimulus histogram (PSH) showing inhibition of spontaneously firing neurone following single pulse stimulation of SN. Inset record shows prolongation of burst response following application of 10 nA of Glu. (B) elimination of inhibitory period following depolarization block of burst response (inset) by 25 nA Glu. Note that spontaneous discharge of large amplitude spike was only slightly altered by Glu. (C) PSH indicating inhibition from both SN and IC sites. Inset records of fast sweep speeds illustrate comparable burst responses evoked from both sites. Diagrammatic section of the rat brain at the level of the hypothalamus indicates effective sites (o) in region of the IC which evoked burst response. Dotted line indicates extent of electrolytic lesions which eliminate the SN evoked responses in striatum. (D) PSH showing inhibition of large spike produced by IC stimulation three weeks following electrolytic lesion of SN. Inset record shows IC elicited burst response recorded at same site. Arrows on inset recordings indicate shock artifact. LH, lateral hypothalamus; V, ventral thalamus. (From Richardson *et al.*, 1977; with permission of Elsevier Publishing Co.)

found that nigral stimulation would depress the spikes coincidental with the development of a burst response (Fig. 8). If the burst was prevented, the inhibitory response disappeared. Both the nigral-evoked inhibition and burst response were unaffected by iontophoretically or systemically administered antagonists of dopamine or by chemical lesions of dopamine-containing nigral neurones. Monosynaptic activation of large-amplitude spike potentials in striatal neurones, which were antidromically identified by stimulation of globus pallidus could be reversibly blocked by dopamine antagonists such as haloperidol (Fig. 9). Thus it was concluded that the burst responses are induced through the antidromic excitation of striatonigral axons within the striatum. Striatal neurones activated in this manner are in fact inhibitory neurones. The dopamine-containing neurones

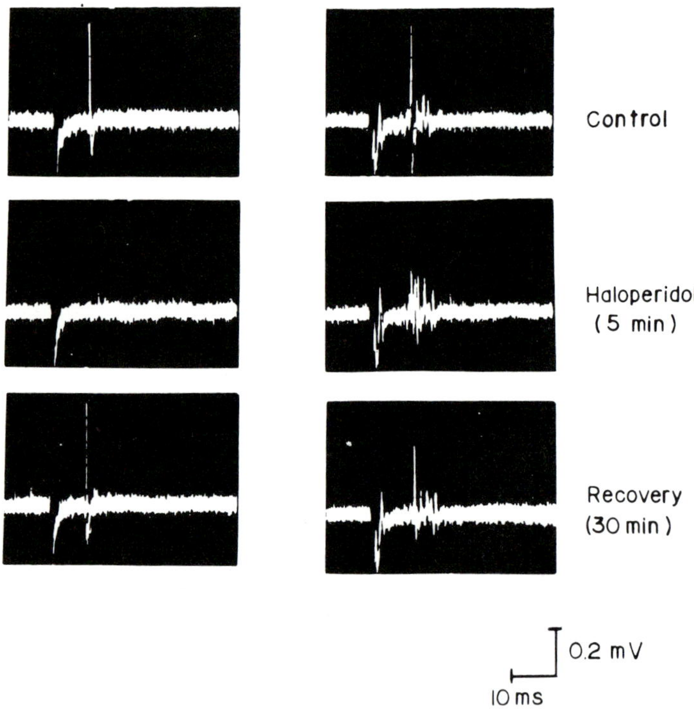

Fig. 9 The effects of intravenous administration of haloperidol (0.5 mg kg) on SN-evoked responses. Left and right panels are recordings from two different cells. On the right simultaneous recording of burst response with large amplitude spike. Haloperidol blocks synaptically evoked spikes but not burst responses 5 min following injection. Recovery of responses occurs after 30 min. (From Richardson *et al.*, 1977; with permission of Elsevier Publishing Co.)

of the nigra make an excitatory contact with a population of striatal output cells, some of which at least project to the globus pallidus.

A scheme illustrating this neuronal arrangement is shown in Fig. 10. Thus striatal target cells which receive nigra input show two types of responses to SN stimulation. First, a monosynaptic excitation which is frequently followed by inhibition (see also Frigyesi and Purpura, 1967; Hull *et al.*, 1970; York, 1976) and secondly, a pure inhibition. However, the important contribution made by the study of Richardson *et al.* (1977) is that it clearly shows that the nigrostriatal dopaminergic pathway evokes the excitatory response, whereas the inhibitory response is mediated in both cases (i.e. pure inhibition and excitation followed by inhibition) by a recurrent collateral system which drives inhibitory interneurones. The neurones giving rise to the two efferent systems are located in different

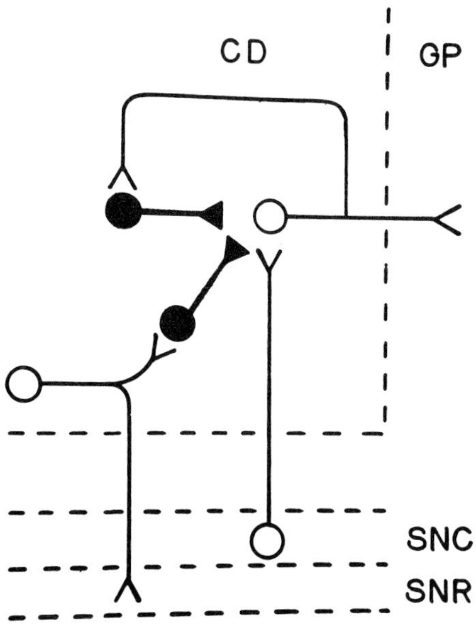

Fig. 10 A scheme illustration of the proposed synaptic arrangements of neostriatal neurones (CD) with the substantia nigra and globus pallidus (GP). Stimulation of the nigral compacta (SNC) region produces an activation–inhibition sequence of striatal target cells, the inhibitory component being mediated by recurrent axon collaterals impinging on inhibitory interneurones (shown in black). GP stimulation elicits antidromic spike in the same target cell and inhibition of spontaneous firing by the same axon collateral system. Stimulation of the reticulata region of SN (SNR) evokes antidromic responses mediated by striatonigral pathway and inhibition of CD target cells by recurrent collaterals impinging on inhibitory interneurones. (From Richardson *et al.*, 1977; with permission of Elsevier Publishing Co.)

regions of the striatum. The nigral-evoked, antidromically activated interneurones were localized to ventral and peripheral circumferential sites in the striatum, thus identifying the source of striatonigral output cells. The pallidal-evoked antidromically activated neurones were localized mainly to the central core region of the striatum.

An earlier study reporting the lack of effect of haloperidol on nigrostriatal inhibition of caudate neurones (Feltz, 1971) now fits into place, since the inhibition was being evoked through non-dopaminergic interneurones (see Fig. 10, from Richardson *et al.*, 1977).

CONCLUSIONS— DOPAMINE, AN EXCITATORY NEUROTRANSMITTER

There is thus a recent consolidation of opinion as to the mechanism of action of the nigrostriatal dopamine pathway. The recent studies of Kitai *et al.* (1976) and Richardson *et al.* (1977) coupled with earlier observations (Frigyesi and Purpura, 1967; McLennan and York, 1967; Feltz and MacKenzie, 1969; Connor, 1970; Hull *et al.*, 1970; York, 1970; Feltz, 1970; Feltz and Albe-Fessard, 1972; Buchwald *et al.*, 1973; Hull *et al.*, 1974) have clearly delineated that dopamine is an excitatory neurotransmitter released at terminals of the nigrostriatal tract. Why did it take 10 years to sort out this relatively straightforward problem? It is clear that iontophoresis and extracellular recording of single cells cannot be expected to yield a definitive answer when looking at complex neuronal populations which consist of cells of different sizes, some of which exhibit extensive collateral branching onto smaller intrinsic cells. To assume that the diffusion from the electrode tip, the distance from the tip to the active receptor site and the location of the tip in relation to the cell geometry will be similar with each cell recorded from is obviously a fallacy. It would appear that intracellular recording has remedied some of these problems to some extent, by the fact that the cell is impaled and the distance of the drug applying barrel is fixed relative to the specific site of impalement (presumably the cell soma). Thus drug effects with short-duration pulsed applications should be localized to the immediate vicinity surrounding the micropipette tip and evidenced by localized effects on the neurone impaled, rather than a more diffuse spread which will affect interneurones nearby causing alternations in the cell under observation.

Certain questions still remain unanswered at the present time regarding certain functional aspects of the nigrostriatal pathway. Why do only one-third of the striatal cells sampled receive an excitatory dopaminergic input when fluorescent dopamine terminals are very densely scattered throughout the striatum? In addition to the dopaminergic nigrostriatal excitatory tract, are there other components to this tract which are non-dopaminergic? Furthermore, it would also appear that one cannot generalize on what the action of dopamine will be at a particular synapse, based on its actions elsewhere. For example, in the snail (*Aplysia*) dopamine has both excitatory and inhibitory actions which can be distinguished pharmacologically (see Chapter 29); whereas in the sympathetic ganglia, dopamine is localized in an inhibitory interneurone producing a slow IPSP (Libet and Tosaka, 1970), and also has a modulatory action which may enhance a cholinergic slow EPSP for several hours. Evidence for a long lasting modulatory action of nigral stimulation in the striatum has also been recently reported (York and Lentz, 1976).

It is important that the above conclusions of dopamine as an excitatory transmitter not be misinterpreted in view of what the net effect of nigrostriatal activation may be several hundred milliseconds, or seconds later. It is clear that the inhibitory axon collateral network will cause marked inhibition of striatal cell firing, following nigral stimulation. However, the initial event in the sequence is the dopaminergic excitation, when then activates a much longer duration inhibition, as is apparent from intracellular recordings (Frigyesi and Purpura, 1967; Hull *et al.*, 1970). Considerations of a more global picture of interactions between the basal ganglia nuclei in terms of a feedback loop hypothesis (Groves *et al.*, 1975), in which the dopaminergic cell interacts with a cholinergic striatal neurone to activate ultimately a gabaminergic output cell will also have to take these recent findings into consideration.

REFERENCES

Adinolfi, A. M. and Pappas, G. D. (1968). *J. Comp. Neurol.* **133,** 167–184.

Albe-Fessard, D., Raieva, S. and Santiago, W. (1967). *J. Physiol.* (*Paris*) **59,** 324–325.

Andén, N. E., Carlsson, A., Dahlström, A., Fuxe, L., Hillarp, N. A. and Larsson, K. (1964). *Life Sci.* **3,** 523–530.

Andén, N. E., Dahlström, A., Fuxe, K., Larsson, K., Olsson, L. and Ungerstedt, U. (1966). *Acta Physiol. Scand.* **67,** 313–326.

Barbeau, A. (1972). Monogr Human Genet. **6,** 114–136.

Bedard, P., Larochelle, L., Parent, A. and Poirier, L. (1969). *Expl. Neurol.* **25,** 365–377.

Ben-Ari, Y. and Kelly, J. S. (1976). *J. Physiol.* (*London*) **256,** 1–21.

Besson, M., Cheramy, A., Feltz, P., and Glowinski, J. (1971). *Brain Res.* **32,** 407–424.

Bloom, F. E., Costa, E. and Salmoiraghi, G. C. (1965). *J. Pharmacol. Expl. Ther.* **150,** 244–252.

Buchwald, N. A., Price, D. D., Vernon, L. and Hull, C. D. (1973). *Expl. Neurol.* **38,** 311–323.

Carpenter, M. B. and Peter, P. (1972). *J. Comp. Neurol.* **144,** 93–116.
Chiueh, C. C. and Moore, K. E. (1973). *Brain Res.* **50,** 221–225.
Connor, J. D. (1970). *J. Physiol. (London)* **208,** 691–703.
Coombs, J. S., Eccles, J. C. and Fatt, P. (1955). *J. Physiol. (London)* **130,** 374–395.
Curtis, D. R. (1964). *In* "Physical Techniques in Biological Research", Vol. 5, Electrophysiological Methods (Ed. W. L. Nastuk), Part A, 144–190. Academic Press, London and New York.
Dafny, N. (1975). *Expl. Neurol.* **47,** 503–508.
Feltz, P. and MacKenzie, J. S. (1969). *Brain Res.* **13,** 612–616.
Feltz, P. (1970). *J. Physiol. (Paris)* **62,** 151.
Feltz, P. and Albe-Fessard, D. (1972). *Electroenceph. Clin. Neurophysiol.* **33,** 179–193.
Feltz, P. and DeChamplain, J. (1972a). *Brain Res.* **43,** 595–600.
Feltz, P. and DeChamplain, J. (1972b). *Brain Res.* **43,** 601–605.
Fibiger, H. C., McGeer, E. G. and Atmadja, S. (1973). *J. Neurochem.* **21,** 373–385.
Fox, C. A., Andrade, A. M., Hillman, D. E. and Schwyn, R. C. (1971). *J. Hirnforsch* **13,** 181–201.
Frigyesi, T. L. and Purpura, D. P. (1967). *Brain Res.* **6,** 440–456.
Frigyesi, T. L. and Szabo, J. (1974). *IRCS* **2,** 1692.
Fuxe, K., Hökfelt, T. and Nilsson, O. (1964). *Z. Zellforsch* **63,** 701–706.
Gonzalez-Vegas, J. A. (1974). *Brain Res.* **80,** 219–228.
Groves, P. M., Wilson, C. J., Young, S. J. and Rebec, G. V. (1975). *Science* **190,** 522–529.
Hedreen, J. C. (1971). *Anat. Rec.* **169,** 338.
Hedreen, J. C. and Chalmers, J. P. (1972). *Brain Res.* **47,** 1–36.
Herz, A. and Zieglgänsberger, W. (1968). *Int. J. Neuropharmacol.* **7,** 221–230.
Hökfelt, T. and Ungerstedt, U. (1969). *Acta Physiol. Scand.* **76,** 415–426.
Hökfelt, T. and Ungerstedt, U. (1973). *Brain Res.* **60,** 269–297.
Hornykiewicz, O. (1966). *Pharmacol. Rev.* **18,** 925–964.
Hornykiewicz, O. (1972). *In* "Handbook of Neurochemistry" (Ed. A. Lajtha), Chapter 19. Plenum Press, New York.
Hull, C. D., Bernardi, G. and Buchwald, N. A. (1970). *Brain Res.* **22,** 163–179.
Hull, C. D., Levine, M. S., Buchwald, N. A., Heller, A. and Browning, R. A. (1974). *Brain Res.* **73,** 241–262.
Ibata, Y., Nojyo, Y., Matsuura, T. and Sano, Y. (1973). *Z. Zellforsch.* **138,** 333–344.
Kemp, J. (1968). *Brain Res.* **11,** 464–467.
Kemp, J. and Powell, T. P. S. (1971). *Phil. Trans. B.* **262,** 383–401.
Kitai, S. T., Wagner, A., Precht, W. and Ohno, T. (1975). *Brain Res.* **85,** 44–48.
Kitai, S. T., Sugimori, M. and Kocsis, J. D. (1976). *Expl. Brain Res.* **24,** 351–363.
Kuypers, H. G. J. M., Kievit, J. and Groen-Klevant, A. C. (1974). *Brain Res.* **67,** 211–218.
Libet, B. and Tosaka, T. (1970). *Proc. Natl. Acad. Sci.* **67,** 667–673.
Liles, S. L. (1974). *J. Neurophysiol.* **37,** 254–265.
Maler, L., Fibiger, H. C., and McGeer, P. L. (1973). *Expl. Neurol* **40,** 505–515.
McKenzie, G. M. and Szerb, J. C. (1968). *J. Pharmacol. Expl. Ther.* **162,** 302–308.
McLennan, H. (1964). *J. Physiol. (London)* **174,** 152–161.
McLennan, H. (1965). *Experientia* **21,** 725.
McLennan, H. and York, D. H. (1967). *J. Physiol. (London)* **189,** 393–402.
McLennan, H. (1970). *In* "Synaptic Transmission", 23. W. B. Saunders Publishing Co., Philadelphia.

Moore, R. and Heller, A. (1969). *In* "Progress in Neurogenetics" (Eds. A Barbeau and J. R. Brunette), 276–282. Elsevier, Amsterdam.
Moore, R. Y., Bhatnager, R. K. and Heller, A. (1971). *Brain Res.* **30,** 119–136.
Mori, S. (1966). *Z. Zelforsch* **70,** 461–488.
Nauta, H. Kaiserman-Abramof, I. and Lasek, R. J. (1974). *Brain Res.* **85,** 373–384.
Ohye, C., Bouchard, R., Boucher, R. and Poirier, L. J. (1970). *J. Pharmacol. Expl. Ther.* **175,** 700–708.
Poirier, L. J., Singh, P., Sourkes, T. L. and Boucher, R. (1967). *Brain Res.* **6,** 654–666.
Portig, P. J., Sharman, D. F. and Vogt, M. (1968). *J. Physiol.* (*London*) **194,** 565–572
Richardson, T. L., Miller, J. J., and McLennan, H. (1977). *Brain Res.* **127,** 219–234.
Riddell, D. and Szerb, J. C. (1971). *J. Neurochem.* **18,** 989–1006.
Shimizu, N. and Ohnishi, S. (1973). *Expl. Brain Res.* **17,** 133–138.
Siegel, A., Sasso, L. and Tassoni, J. P. (1971). *Expl. Neurol.* **33,** 130–146.
Simon, H., LeMoal, M., Galey, D. and Cardo, B. (1974). *Expl. Brain Res.* **20,** 375–384.
Spencer, H. J. and Havlicek, V. (1974). *Can. J. Physiol. Pharmacol.* **52,** 808–813.
Ungerstedt, U. (1968). *Eur. J. Pharmacol.* **5,** 107–110.
Ungerstedt, U. (1971). *Acta Physiol. Scand.* **367,** 1–48.
Verhaart, W. J. C. (1950). *J. Comp. Neurol.* **93,** 425–440.
Vogt, M. (1969). *Br. J. Pharmacol.* **37,** 325–337.
Voigtlander, P. F. von and Moore, K. E. (1973). *J. Pharmacol. Expl. Ther.* **184,** 542–552.
Werman, R. (1966). *Comp. Biochem. Physiol.* **18,** 745–766.
York, D. H. (1970). *Brain Res.* **20,** 233–249.
York, D. H. (1972). *Brain Res.* **37,** 91–99.
York, D. H. (1976). *J. Pharmacol. Ther. B* **2,** 1–7.
York, D. H. and Lentz, S. (1976). *Proc. Soc. Neurosci.* **2,** 70.
Zarzecki, P., Blake, D. J. and Somjen, G. G. (1976). *Brain Res.* **115,** 257–272.

CHAPTER 25

The Electrophysiological Pharmacology of Midbrain Dopaminergic Systems

Benjamin S. Bunney

Yale University School of Medicine, New Haven, USA.

INTRODUCTION

Numerous pharmacological agents have been found to have effects on central dopaminergic (DA) systems. Biochemical, electrophysiological, and behavioural methods have been used for determining these effects. This section is concerned with the effects of drugs on central DA systems, primarily as measured by electrophysiological techniques. Except where specifically stated otherwise the results described were obtained using extracellular single unit recording techniques (i.e. recording from outside a cell and one cell at a time). Drugs studied were administered either parenterally or by means of microiontophoresis. (Studies involving direct infusion of drugs into the brain and studies involving drug effects on stimulation-induced changes in the activity of recorded cells will not be reviewed unless the area of diffusion of the drug and the identity of the pathway stimulated were determined.) The use of microiontophoretic techniques permits the direct application of drugs to the cell recorded from by means of a multibarrel micropipette. As with most techniques, both microiontophoresis and single unit recording have limitations and difficulties associated with them which can make interpretation of results difficult if careful attention is not paid to the design and execution of the experiments. These problems will not be described in detail here (for detailed reviews see Bloom, 1974; York, Chapter 24), but will be briefly mentioned as they pertain to a particular experiment being discussed.

Notwithstanding these limitations, valuable information can be derived from the use of such techniques regarding the direct effects of pharmacological agents on neurochemically and anatomically defined neuronal systems within the brain. Central dopamine-containing cells form such a system. This review will be concerned primarily with the three midbrain subdivisions of the dopamine system—the nigrostriatal, the mesolimbic and the mesocortical.

MIDBRAIN DOPAMINERGIC NEURONES

Anatomy

The cells of origin for the DA nigrostriatal, mesolimbic and mesocortical systems are found in the midbrain. Cells forming the nigrostriatal dopamine system (A9) are located in the zona compacta of the substantia nigra whereas the DA cell bodies making up the other two systems (A10) primarily surround the interpeduncular nucleus in the ventral tegmental area (Dahlström and Fuxe, 1964; Ungerstedt, 1971; Lindvall and Björklund, 1974; see Lindvall, Chapter 20 for detailed description).

Electrophysiological Studies of A9 and A10 Neurones

Identification

Identification of the specific type of cell from which recordings are made is essential if one is to study the effects of drugs on a particular neuronal system. In many of the studies concerning the effects of drugs on A9 and A10 neurones, specific criteria for their identity were used. A9 and A10 neurones are similar in their neurophysiological properties and are clearly distinguishable from neighbouring cells in the zona reticulata of the substantia nigra, red nucleus, reticular formation, and interpeduncular nucleus (Bunney *et al.*, 1973b; Aghajanian and Bunney, 1973, 1974; Guyenet and Aghajanian, 1978). A9 and A10 cells typically have an unusually wide action potential (4–5 ms) with a distinct initial segment and a late positive component (Fig. 1A). The unusual shape of these cells is not due to injury as they can be recorded from for several hours without any change in shape or firing rate. Many of these cells exhibit a characteristic slow (2–8 s^{-1}) bursting complex spike pattern in which there is a progressive decrease in spike amplitude within each cluster of action

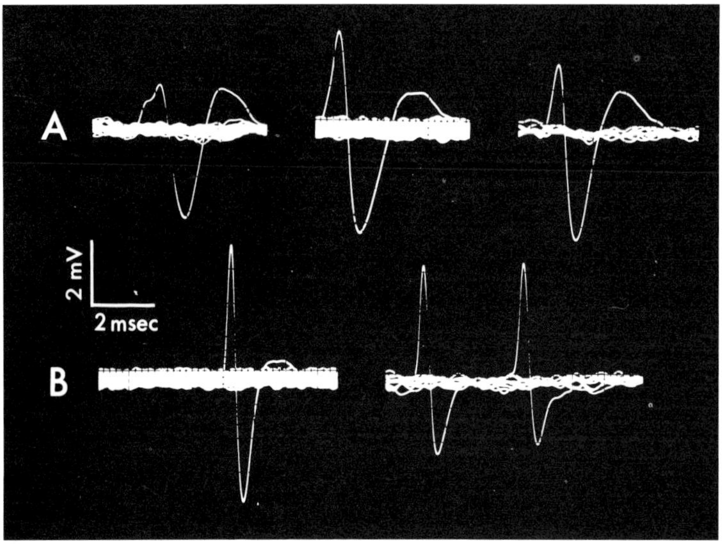

Fig. 1 Characteristic action potential shape of nigral type I (DA) and type II (non-DA) cells. A. Three different type I cells: Notice the unusual width of the action potential, the presence of a marked initial segment component (left) and the large size of the late positive component (positivity upwards). B. Two type II cells: These action potentials are much narrower, do not present a marked initial segment component. The later positive component of the spike is very small when it exists. (From Guyenet and Aghajanian, 1978.)

potentials (Bunney *et al.*, 1973b) (Fig. 2). These cells also possess a unique pharmacological profile of responses to dopamine agonists (e.g. inhibition by extremely low doses of apomorphine; described in following Sections).

Two criteria were used originally to establish the identity of these neurones as DA (Bunney and Aghajanian, 1973b). First, after recording from cells with the above characteristics a small amount of L-dihydroxyphenylalanine (L-DOPA) was ejected from the recording electrode by means of microiontophoresis. This led to a markedly enhanced histochemical fluorescence of DA neurones immediately surrounding the electrode tip. The enhanced fluorescence was specific for DA neurones in that L-DOPA, ejected into cell areas (e.g. red nucleus, reticular formation) that do not contain DA neurones, failed to produce any fluorescent cell bodies. Secondly, small amounts of 6-hydroxydopamine were injected unilaterally into the vicinity of A9 or A10 cells (a treatment which selectively destroys DA neurones; Ungerstedt, 1968; Agid *et al.*, 1973; Hökfelt and Ungerstedt, 1973; Sotelo *et al.*, 1973). When examined one and six days following this procedure there was a loss, on the lesioned side, solely of those cells which

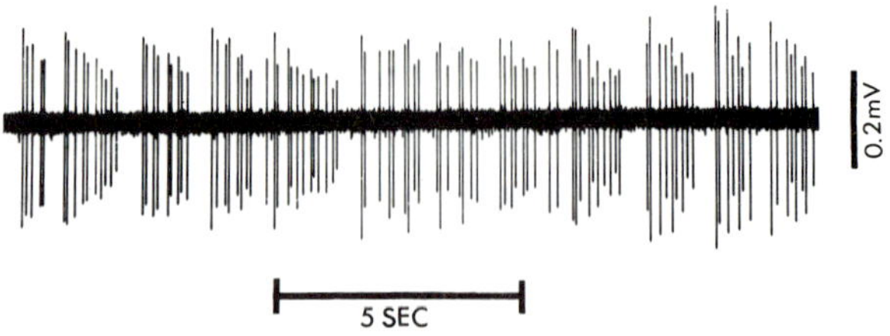

Fig. 2 Oscillograph recording of DA cell firing pattern in chloral hydrate-anaesthetized animal. Note repeated bursts with progressively decreasing spike amplitude. (From Bunney *et al.*, 1973b.)

had the electrophysiological characteristics (e.g. duration, wave form and firing pattern) described above (Bunney *et al.*, 1973b). Cells with these characteristics, however, were still readily found on the non-lesioned side. Histochemical studies of the same animals showed that there was a selective loss of dopamine-containing cells on the side ipsilateral to the injection site.

Recently Guyenet and Aghajanian (1978) identified two types of cells in the substantia nigra using antidromic stimulation of various areas thought to be innervated by nigral efferents. Type I cells were found mainly in the substantia nigra zona compacta and were activated by stimulating points along the course of the nigrostriatal pathway (caudate-putamen, globus pallidus and medial forebrain bundle). These cells possessed the characteristic action potential duration and shape, firing pattern and response to DA agonists described above. Type I cells were not activated by stimulating the ventromedial thalamus, motor cortex or pontine reticular formation. In addition, they were found to have a slow conduction velocity ($0.58\,m\,s^{-1}$) which is compatible with available data describing axons of DA cells as fine and non-myelinated (Hökfelt and Ungerstedt, 1973). This slow conduction velocity was in marked contrast to the $2.8\,m\,s^{-1}$ conduction velocity (Fig. 3) of the second type of antidromically activated cells (e.g. by stimulation in the ventromedial thalamus and medial forebrain bundle) that were found scattered throughout the substantia nigra. These type II cells had a much narrower action potential (2–3 ms) and their wave form was entirely different from that of the type I neurones (Fig. 1). Microiontophoretic injection of 6-hydroxydopamine into the vicinity of the medial forebrain bundle selectively blocked propagation of antidromic spikes in the presumed DA neurones (type I). Contrastingly, type II cell activation by antidromic stimulation of the medial forebrain bundle was unaffected by 6-hydroxydopamine pretreatment (Guyenet and Aghajanian, 1978) Type I cells were inhibited by extremely low doses (ID50: $9.3\,\mu g\,kg^{-1}$) of i.v. apomorphine, whereas type II cells were unresponsive to $20\,\mu g\,kg^{-1}$ of apomorphine.

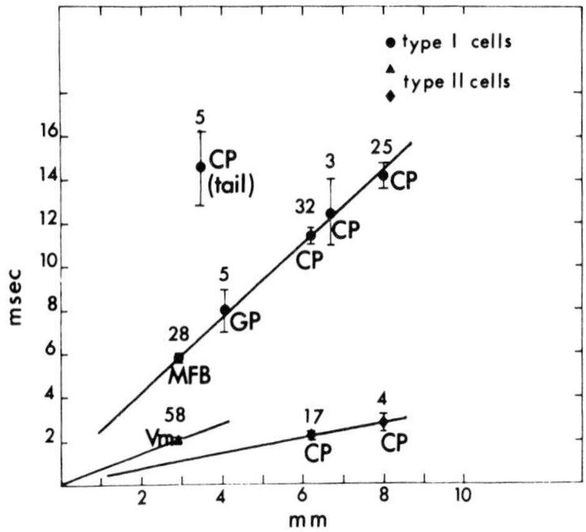

Fig. 3 Conduction velocity of type I and type II cells. The abscissa indicates the straight line distance between the stimulating electrode and the recording site in 250 g rats. The conduction time (mean ± S.E.M.) is plotted on the ordinate. Data for type I cells fit a straight line except for the point corresponding to the tail of the caudate for obvious anatomical reasons. The numbers refer to the size of the sampling. Abbreviations: CP, caudate-putamen; GP, globus pallidus; MFB, median forebrain bundle; Vm, nucleus medialis thalami. (From Guyenet and Aghajanian, 1978.)

Taken together, the above evidence strongly suggests that midbrain DA neurones are identifiable by the shape and duration of their extracellular action potential, firing characteristics and response to low doses of apomorphine. Consequently, if the above criteria are used, the effects of pharmacological agents on the activity of these neurones can be studied with confidence in their dopaminergic identity.

Acute effects of parenterally administered drugs on midbrain dopaminergic neurones

Dopamine Agonists

All agonists tested, whether acting directly or indirectly, inhibit A9 and A10 cells (Table I). *d*-Amphetamine, which is thought to act indirectly by releasing DA and blocking its reuptake (Besson *et al.*, 1971; Costa and Groppetti, 1970; Glowinski *et al.*, 1966; Von Voigtlander and Moore, 1973), sometimes initially increases (Rebec and Groves, 1975) but then always decreases the activity of A9 and A10 units (Bunney *et al.*, 1973a, b; Rebec and Groves, 1975). The indirect nature of *d*-amphetamine's action is supported by the finding that its inhibitory effects are blocked and reversed by α-methyl-*p*-tyrosine (a selective inhibitor of tyrosine hydroxylase;

TABLE I Acute effects of systemically administered drugs upon the activity of midbrain dopaminergic neurones.

Type of drug	Cells tested	Effect on cell activity	Comment	Reference
Agonist				
d-Amphetamine	A9, A10	Decrease	Effect blocked by pretreatment with α-methylparatyrosine and high doses of Ro4-4602	Bunney *et al* (1973a, b); Rebec and Groves (1975); Bunney and Aghajanian (1976d); Groves *et al.* (1978)
l-Amphetamine	A9	None or decrease	50% of cells treated were not inhibited at any dose. For those inhibited, potency was 5–20 times less than *d*-amphetamine	Bunney *et al.* (1975b)
Apomorphine	A9, A10	Decrease	Unaffected by pretreatment with α-methylparatyrosine or high doses of Ro4-4602	Bunney *et al.* (1973a); Iwatsubo and Clouet (1977); Guyenet and Aghajanian (1978)
L-DOPA	A9, A10	Decrease	Unaffected by pretreatment with α-methylparatyrosine. Blocked by pretreatment with high doses of Ro4-4602	Bunney *et al.* (1973a); Iwatsubo and Clouet (1977)
Methylphenidate	A9, A10	Decrease		Bunney and Aghajanian, unpublished data
Piribedil (ET 495)	A9	Decrease		Walters *et al.* (1975)
Antagonist				
Chlorpromazine	A9, A10	Increase		Bunney *et al.* (1973b); Bunney and Aghajanian (1974)
Chlorpromazine sulphoxide	A9, A10	None	Inactive metabolite of chlorpromazine	Bunney and Aghajanian (1974)
7-OH-Chlorpromazine	A9, A10	None	Active metabolite of chlorpromazine	Bunney and Aghajanian (1974)
Clozapine	A9, A10	None		Bunney and Aghajanian (1975a, b)
Diethazine	A9, A10	None	Phenothiazine lacking antipsychotic activity and extrapyramidal side effects	Bunney and Aghajanian (1975a, b)
Fluphenazine	A9	Increase		Bunney *et al.* (1973b)
Haloperidol	A9, A10	Increase		Bunney *et al.* (1973b); Iwatsubo and Clouet (1977)

Mepazine	A9, A10	None	Phenothiazine lacking antipsychotic properties	Bunney and Aghajanian (1975a)
Molindone	A9	None		Bunney *et al.* (1975a)
Perphenazine	A9	Increase		Bunney *et al.* (1973b)
Promethazine	A9, A10	None	Phenothiazine lacking antipsychotic properties and extrapyramidal side effects	Bunney *et al.* (1973b)
Thioridazine	A9	None		Bunney and Aghajanian (1975a); Bunney *et al* (1973b)
Trifluoperazine	A9	Increase		Bunney *et al.* (1973b)
Miscellaneous				
Baclofen	A10	Decrease	Inhibition not reversed by fluphenazine	Bernard *et al* (1975)
Chloral hydrate	A9, A10	Increase	Does not block effects of DA agonists and antagonists	Bunney and Aghajanian (1975b); Bunney *et al.* (1973b)
Diazepam	A9, A10	None		Bunney and Aghajanian (1975b)
γ-Hydroxybutyrate	A9	Decrease		Roth *et al.* (1973)
Halothane	A9	Increase	Does not block effects of DA agonists and antagonists	Bunney *et al.* (1973b)
LSD	A9	See comment	78% decrease, 22% no effect or slight excitation	Christoph *et al.* (1977)
Brom LSD	A9	See comment	29% slightly inhibited, 71% unaffected	Christoph *et al.* (1977)
5-Methoxy *N*,*N*-dimethyltryptamine	A9	Increase		Christoph *et al.* (1977)
Morphine	A9	Increase		Iwatsubo and Clouet, 1977 Nowycky *et al.* (1978)
Nicotine	A9	See comment	Initial transient decrease followed by sustained increase	Lichtensteiger *et al.* (1976)
Pentobarbital	A9, A10	Decrease	Only occurred in anesthetic dose range 20–40 $mg\ kg^{-1}$, i.v.	Bunney and Aghajanian (1975b)
Physostigmine	A9	See comment	Initial increase followed by decrease	Lichtensteiger *et al* (1976)
Picrotoxin	A9	See comment	Slowly firing cells—increase Fast cells—no effect	Bunney and Aghajanian, (1978a)
Scopolamine	A9	None		Bunney and Aghajanian (1975b)

Spector *et al.*, 1965) and by high doses ($800\,mg\,kg^{-1}$ i.p.) of the decarboxylase inhibitor Ro4-4602 (Bunney and Aghajanian, 1973; Bunney *et al.*, 1973a) which completely inhibits DOPA decarboxylase in the CNS (Weiss *et al.*, 1972). In contrast to *d*-amphetamine, *l*-amphetamine has only a slight inhibitory effect on approximately 50% of A9 cells even when given intravenously in doses as high as $25\,mg\,kg^{-1}$ (ID50 for *d*-amphetamine is $1.60\,mg\,kg^{-1}$). When *l*-amphetamine does depress A9 cell activity it is 5–20 times less potent than *d*-amphetamine (Bunney and Aghajanian, 1973; Bunney *et al.*, 1975b).

Intravenously administered methylphenidate, another indirectly acting dopamine receptor stimulator (Chiueh and Moore, 1975; Scheel-Krüger, 1971), also inhibits these cells (Bunney and Aghajanian, unpublished data) as does parenterally administered L-DOPA, the immediate precursor of DA (Bunney *et al.*, 1973a; Iwatsubo and Clouet, 1977). L-DOPA probably exerts its inhibitory action by being converted to dopamine as high doses ($800\,mg\,kg^{-1}$, i.p.) of Ro4-4602, abolish L-DOPA's inhibitory effect (Bunney *et al.*, 1973a). Contrastingly, the inhibitory effect of apomorphine (Bunney *et al.*, 1973a; Iwatsubo and Clouet, 1977) and piribedil (Walters *et al.*, 1975), directly acting dopamine receptor stimulators (Andén *et al.*, 1967; Ernst, 1967), are unaffected by pretreatment of animals with α-methyl-*p*-tyrosine (Bunney *et al.*, 1973*a;* Walters *et al.*, 1975). Thus, their action is not dependent upon ongoing DA synthesis.

Dopamine Antagonists

Many antipsychotic drugs with diverse chemical structures increase the firing rate of A9 neurones (see Table I). Despite this wide variation in chemical structure, all antipsychotic drugs tested have been found to block and reverse the inhibitory effects of dopamine agonists on A9 and A10 neurones (Bunney and Aghajanian, 1975a, b; Bunney *et al.*, 1973b, 1975a) (Table II). In addition, one metabolite of the neuroleptic chlorpromazine, 7-hydroxychlorpromazine, has been found to have the same blocking effect. However, another metabolite, chlorpromazine sulphoxide, which lacks biochemical effects on DA metabolism, is totally inactive in this regard (Bunney and Aghajanian, 1974), as are phenothiazine analogues lacking antipsychotic efficacy (promethazine and diethazine) (Bunney *et al.*, 1973b; Bunney and Aghajanian, 1975b).

Whether direct or indirect in their action all DA agonists are thought to decrease the activity of DA cells through an *activation* of DA receptors. The interactions of neuroleptics with DA agonists, as well as the increase in DA cell activity and turnover induced by most antipsychotic drugs (Andén, 1972; Andén *et al.*, 1964; Bartholini, 1976; Carlsson and Lindqvist, 1963;

TABLE II Acute effect of systemically administered drugs upon agonist-induced depression of midbrain dopaminergic neurones.

Type of drug	Cells tested	Effect on depressed activity	Comment	Reference
Chlorpromazine	A9, A10	Reversal		Bunney *et al.* (1973b)
Chlorpromazine sulphoxide	A9, A10	None	Inactive metabolite of chlorpromazine	Bunney and Aghajanian (1974)
7-OH-Chlorpromazine	A9, A10	Reversal	Active metabolite of chlorpromazine	Bunney and Aghajanian (1974)
Clozapine	A9, A10	Reversal		Bunney and Aghajanian (1975a, b)
Diethazine	A9, A10	None	Phenothiazine lacking antipsychotic activity and extrapyramidal side-effects	Bunney and Aghajanian (1975a)
Fluphenazine	A9, A10	Reversal		Bunney *et al.* (1973b)
Haloperidol	A9, A10	Reversal		Bunney *et al.* (1973b)
Mepazine	A9, A10	See comment	A9—reversed, A10—no effect. Phenothiazine lacking antipsychotic activity	Bunney and Aghajanian (1975a)
Molindone	A9, A10	Reversal		Bunney *et al.* (1975a)
Pimozide	A9, A10	Reversal		Bunney and Aghajanian, unpublished data
Promethazine	A9, A10	None	Phenothiazine lacking antipsychotic activity and extrapyramidal side-effects	Bunney *et al.* (1973b)
Miscellaneous				
Baclofen	A10	None		Bernard *et al.* (1975)
Diazepam	A9, A10	Reversal	Possible mechanism—effect on GABA efferents	Bunney and Aghajanian (1975b)
Pentobarbital	A9, A10	Reversal	Only decreased in anaesthetic dose range—20–40 mg kg^{-1} i.v.	Bunney and Aghajanian (1975b)
Picrotoxin	A9	Reversal		Bunney and Aghajanian (1975b)
Scopolamine	A9, A10	None		Bunney and Aghajanian (1975b)

Glowinski *et al.*, 1966; Nyback and Sedvall, 1968; Persson, 1970), is presumed to be mediated through their ability to *block* dopamine receptors (Carlsson and Lindqvist, 1963; Matthysse, 1973; Meltzer and Stahl, 1976; Snyder *et al.*, 1974). These receptors are thought to be on neurones in A9 and A10 innervated areas which form a link in feedback pathways that modulate A9 and A10 cell activity (Carlsson and Lindqvist, 1963; Corrodi *et al.*, 1967) and/or on DA receptors located on DA cells themselves (Kehr *et al.*, 1972) (see Section on DA autoreceptors below).

Miscellaneous Drugs

There is an ever-increasing number of pharmacological agents which appear to affect A9 and A10 cell activity but which are not classified as classic DA agonists or antagonists. The mode of action of these compounds is often unknown. However, in some cases they would appear to act through an effect on non-DA neurones which directly or indirectly modulate DA cell activity. Thus the putative γ-aminobutyric acid (GABA) antagonist picrotoxin, in low subconvulsant doses, reverses *d*-amphetamine-induced inhibition of A9 cells. This appears to be a specific effect as the glycine antagonist, strychnine (even in near convulsant doses) has no effect in this regard (Bunney and Aghajanian, 1976a). When given by itself picrotoxin appears to increase the firing rate of slow DA cells while having little effect on faster-firing neurones (Bunney, unpublished data). Presumably these effects of picrotoxin are due to its interaction with GABAergic neurones or receptors in the striatonigral feedback pathway thought to modulate DA cell activity (McGeer *et al.*, 1971; Kim *et al.*, 1971; Feltz, 1971; Precht and Yoshida, 1971; McGeer *et al.*, 1974).

Diazepam and high doses of pentobarbital also reverse *d*-amphetamine-induced inhibition of A9 and A10 cells (Bunney and Aghajanian, 1975b). Diazepam given alone appears to have very little effect on baseline activity whereas high doses of pentobarbital depresses DA cell firing rate. Conceivably these effects could be mediated through an effect on some part of the feedback pathway. Whether diazepam acts on the GABAergic link while pentobarbital's effect is non-specific has not yet been determined, nor is it clear why the anaesthetics chloral hydrate and halothane (Bunney *et al.*, 1973b) have an effect opposite to that of pentobarbital—i.e. increase the activity of A9 and A10 neurones.

In addition to a GABAergic innervation, some investigators feel that there may be a cholinergic input to A9 cells (Javoy *et al.*, 1974; Jacobowitz and Goldberg, 1977). In urethane-anaesthetized rats, nicotine administered subcutaneously has been reported to cause a transient decrease in unit activity followed by a sustained increase in the activity of presumed A9 DA neurones (Lichtensteiger *et al.*, 1976). The exact converse of these results was found for i.p. physostigmine (Lichtensteiger *et al.*, 1976). On the other hand, in chloral hydrate-anaesthetized rats, scopolamine was found neither to affect A9 cell firing rate nor to modify the usual effects of apomorphine and haloperidol on A9 cell activity (Bunney and Aghajanian, 1975b). The reason for the different results in these two studies is unresolved at the present time.

Recent biochemical (Von Hungen *et al.*, 1974, 1975; Bockaert *et al.*, 1976) and binding (Creese *et al.*, 1975) studies have suggested that D-lysergic acid diethylamide (LSD) is a mixed agonist–antagonist of dopamine. Recording from presumed A9 DA neurones in chloral hydrate-anaesthetized rats, Christoph *et al.* (1977) found that LSD (25–50 $\mu g\,kg^{-1}$, i.v.) transiently depressed the activity of most A9 cells studied. Attenuation of the effect was observed upon repeated administration. This inhibitory effect was both blocked and reversed by small i.v. doses of haloperidol. Furthermore, after pretreatment with LSD the amount of *d*-amphetamine needed to produce 50% inhibition of the presumed A9 cells was significantly increased (Christoph *et al.*, 1977). These electrophysiological results, therefore, appear to correlate well with the biochemical ones, suggesting that LSD is a mixed DA agonist–antagonist. However, in two other studies, LSD was reported to have no effect on the activity of DA-containing neurones (Bunney *et al.*, 1973b; Svensson *et al.*, 1975). Additional experiments in chloral hydrate-anaesthetized animals yielded mixed results (Bunney and Aghajanian, unpublished data). Intravenous LSD, in a dose range of 2–500 μg, failed to affect the activity of the majority of A9 cells tested. Some DA cells showed a transient (15–30 s) 50% decrease in activity. Rarely a DA cell would exhibit a dose dependent progressive decrease in activity to approximately 10% of baseline activity. However, even after doses of 500 $\mu g\,kg^{-1}$, LSD appeared neither to have an affect on apomorphine's ability to inhibit A9 cell activity, nor to have an effect on the reversal of that inhibition by low doses of haloperidol ($<0.1\,mg\,kg^{-1}$, i.v.). These latter findings would suggest that despite the presence of LSD, DA receptors were available to interact with a direct acting DA receptor stimulator and blocker. The reason for discrepancies between these findings and those of Christoph *et al.* (1977) remains to be explained.

Numerous biochemical studies have demonstrated the presence of enkephalin (Pasternak *et al.*, 1976; Hong *et al.*, 1977) as well as opiate receptors (Pert and Snyder, 1975) in the basal ganglia. However, whether there are neurones in the caudate nucleus or globus pallidus that use enkephalin as their neurotransmitter is still unclear, as is the way that enkephalin systems might interact with midbrain dopaminergic systems. Electrophysiological experiments have shown that morphine ($5\,mg\,kg^{-1}$), when given intravenously (Iwatsubo and Clouet, 1977; Bunney and Aghajanian, unpublished data) and when given intraperitoneally (Nowycky *et al.*, 1978), increases the firing rate of A9 cells. This effect is reversed by naloxone and levallorphan. The inhibitory effect of morphine does not appear to be due to an action on DA receptors as, in their usual low doses, apomorphine inhibits DA cells and haloperidol reverses that inhibition in the presence of morphine and/or naloxone (Iwatsubo and Clouet, 1977; Bunney and Aghajanian, unpublished data).

Chronic effects of peripherally administered drugs on midbrain dopaminergic neurones

All of the experiments described above have been concerned with the acute effects of parenterally administered drugs on the activity of midbrain DA neurones. However, as many of the clinical effects of psychotropic drugs (e.g. the antipsychotic and neurological effects of neuroleptics) often take days to weeks to appear, one could seriously question the relevance of acute studies for determining the mechanism of action of the clinical effects of these drugs. Therefore, chronic studies, although time consuming and often technically more difficult, must be carried out. Although the importance of such studies is widely recognized, there are, to date, very few electrophysiological studies of the effects of chronic drug treatment on A9 and A10 cell activity. The first such study was performed by Nowycky *et al.* (1978) in an effort to determine the effects of chronic morphine treatment on midbrain DA neurones. In that study, seven days of morphine injections (i.p.) were found to result in a significantly higher average firing rate of A9 DA cells. An acute challenge dose of morphine given at the end of chronic treatment produced a further increase in rate. Small doses of naloxone rapidly slowed neurones activated by chronic morphine treatment (Nowycky *et al.*, 1978).

The chronic effects of two antipsychotic drugs have also been studied. In one study (Nowycky, 1976), rats received two subcutaneous injections of fluphenazine decanoate ($10\,mg\,kg^{-1}$ initially; $5\,mg\,kg^{-1}$ 5–6 days later). Recordings from A9 cells were made in gallamine-paralysed animals on the day following the second injection and compared to findings in acutely treated ($10\,mg\,kg^{-1}$ s.c.) animals and to control animals. Although differences in the mean and median firing rate of A9 cells between the three groups were minimal there was a notable change in the distribution of firing rates in the acutely treated animals when compared to chronically treated and control animals. Thus, in non-treated (control) animals a normal distribution was found. However, in acutely treated animals the firing rates were found to be approximately equal at all frequencies. Contrastingly, in chronically treated animals, the distribution of DA cell firing rates was almost identical to those in control animals.

In another study, haloperidol (0.5 mg kg^{-1} s.c.) was administered to rats for 22 days and the effects on A9 cell activity determined on the last day of injection and one week later (Bunney and Grace, 1978). Recordings were made in both chloral hydrate-anaesthetized and gallamine-paralysed rats. Results obtained from these two preparations were qualitatively similar. When the A9 area was recorded from on the day of the last injection, significantly fewer DA cells were found to be firing when compared to saline- or promethazine-treated controls. After 0.08 mg kg^{-1} of apomorphine was administered intravenously to these animals there was a marked increase in the number of DA cells found to be firing. Recordings made from animals in which the haloperidol had been discontinued one week before found the average number of DA cells firing to be nearly back to that of control animals.

Using microiontophoretic techniques, the reason for decreased numbers of spontaneously firing DA cells in three-week treated animals was explored. In these chronically treated animals, many "silent" A9 cells were found by pulsing GABA from one barrel of multibarrel micropipettes. Microiontophoretically applied glutamic acid did not activate these cells. However, cells made to appear by low ejection currents of GABA could be inhibited by higher ejection currents and under these conditions glutamic acid would activate these cells. The above results suggest that chronic treatment with low doses of haloperidol causes most DA cells to go into depolarization block. The parenteral administration of the directly acting dopamine agonist apomorphine, which in control animals inhibits A9 neurones, appears to overcome this block as does local application of the inhibitory neurotransmitter GABA. The mechanisms by which chronic treatment with haloperidol causes this change in DA cell activity and the consequences of this change in terms of its effect on the activity of cells innervated by the A9 system have still to be determined.

Effect of microiontophoretically administered neurotransmitters and drugs on midbrain dopaminergic neurones—interaction with presynaptic receptors (autoreceptors)*

Biochemical Evidence for Dopaminergic Autoreceptors in the Central Nervous System

Paralleling findings in the periphery (Brown and Gillespie, 1957; Enero *et al.*, 1972; Farnebo and Hamberger, 1970, 1971; Farnebo and Malmfors,

*The term autoreceptor was first used in place of "presynaptic receptor" by A. Carlsson in 1974 (see Carlsson, 1975) to describe receptors for dopamine on central DA cell bodies, dendrites and terminals. The term presynaptic receptor is still often used in the literature to

1971; Häggendal, 1970; Kirpekar and Puig, 1971; Langer *et al.*, 1971; Langer, 1973; Starke, 1971, 1972, 1973; Thoenen *et al.*, 1964), recent biochemical evidence has suggested the presence of a presynaptic feedback mechanism in the central nervous system (for reviews see Carlsson, 1975; Nowycky and Roth, 1978). Under certain conditions dopamine receptor blocking agents (e.g. antipsychotic drugs) have been shown to enhance the release of dopamine from terminals in neostriatal slices (Cheramy *et al.*, 1970; Farnebo and Hamberger, 1971) whereas a dopamine agonist, apomorphine, blocks dopamine release (Farnebo and Hamberger, 1971). Thus, there is evidence in the CNS suggesting the existence of either a local *trans*synaptic feedback mechanism involving the postsynaptic cell or a direct action of catecholamine agonists and antagonists on presynaptic receptors. All of the above CNS studies, however, were done *in vitro*. The question as to whether these mechanisms obtain *in vivo* is more difficult to answer. Both DA agonists and antagonists are thought to interact with postsynaptic receptors on neurones innervated by A9 and A10 cells (for reviews see Carlsson, 1978; Meltzer, 1977; Sulser and Robinson, 1978). Some of these cells, as mentioned earlier, are thought to form links in feedback pathways that modulate the neuronal activity of DA neurones in the midbrain (Carlsson and Lindqvist, 1963; Corrodi *et al.*, 1967)—pathways which would be inactivated in in-vitro preparations such as brain slices. How then can one differentiate, *in vivo*, a presynaptic action of dopamine agonists or antagonists from a postsynaptic one?

Inhibition of Impulse Flow In order to distinguish *in vivo* between pre- and postsynaptic effects a method is needed which eliminates the effect of possible neuronal feedback loops but which still provides a means for measuring the effects of any presynaptic receptor that might be present. The discovery of an unexpected biochemical change seen within 30 min following blockade of impulse flow in the nigrostriatal DA pathway has provided the in-vivo model needed to study the hypothesized presynaptic receptor (Carlsson, 1972; Kehr, 1972). It has been found that interruption of impulse flow by any means (e.g. electrothermic lesions, mechanical transection, injection of local anaesthetics, or the administration of γ-butyrolactone, a compound which has been shown to stop dopamine cell firing (Roth *et al.*, 1973), produces a marked increase in neostriatal dopamine levels and synthesis (Andén *et al.*, 1971; Carlsson *et al.*, 1972; Roth *et al.*, 1973, 1974; Walters and Roth, 1972; Walters *et al.*, 1973). With

refer to DA receptors located on DA neurones. However, with the discovery of DA receptors on DA cell dendrites and bodies as well as terminals, the term autoreceptor would appear more appropriate and less confusing.

† For a detailed review of some of the biochemical aspects of dopamine autoreceptors, see Nowycky and Roth, 1978.

impulse flow non-existent in this preparation the effects of a neuronal feedback loop are eliminated and any drug-induced alterations in the above effects must be mediated through a local synaptic effect. Thus it has been shown that the administration of directly acting DA agonists (e.g., apomorphine and ET495—trivastal) or pretreatment with *d*-amphetamine, a dopamine releaser, prevent the increase in dopamine levels and synthesis induced by cessation of impulse flow (Kehr, *et al.*, 1972, 1977; Peričić and Walters, 1976; Roth *et al.*, 1973, 1974; Walters and Roth, 1974, 1976b). Many antipsychotic drugs (hypothesized postsynaptic DA receptor blockers) have been found to block the ability of dopamine agonists to inhibit the γ-butyrolactone-induced increase in dopamine levels and synthesis (Kehr *et al.*, 1972; Kehr *et al.*, 1977; Walters and Roth, 1976a). However, one antipsychotic drug, clozapine, was found to be ineffective in this regard (Walters and Roth, 1974). A variety of drugs which lack antipsychotic activity (e.g. promethazine, ethopropazine, mepazine, (−) butaclamol, phenoxybenzamine and propranalol) also have no effect (Walters and Roth, 1975, 1976a, b). A biochemical feedback mechanism involving end-product inhibition, seen under some circumstances (Costa and Neff, 1966; Javoy *et al.*, 1972) could not explain the change in DA synthesis as dopamine levels actually increase not decrease when DA neuronal impulse flow ceases. Consequently, it has been hypothesized that interruption of impulse flow leads to a decreased release of dopamine into the synaptic cleft resulting in a depletion of dopamine at receptor sites located on the DA terminal membrane. The resultant decrease in DA autoreceptor activation is then thought to cause an activation of tyrosine hydroxylase leading to an increase in intraneuronal dopamine levels (Carlsson *et al.*, 1972; Kehr *et al.*, 1972; Walters and Roth, 1972; Walters and Roth, 1974). The fact that known postsynpatic dopamine receptor agonists and antagonists interact in an identical manner in a system lacking the influence of a neuronal feedback loop provides further evidence to support the concept of a DA autoreceptor.

Stimulation-induced Acceleration of Impulse Flow Further in-vivo biochemical evidence for the existence of a presynaptic receptor has recently been provided in experiments analogous to the in-vitro field-stimulated brain slice experiments described earlier. It has been found that electrical stimulation of the nigrostriatal pathway results in a frequency-dependent increase in tyrosine hydroxylase activity which reaches a maximum (at 15 Hz for 10 min) that persists for 15 min following cessation of stimulation (Roth *et al.*, 1975a, b; Murrin *et al.*, 1976). Pretreatment with a dopamine agonist (trivastal) attentuated the post stimulus increase of tyrosine hydroxylase activity, whereas the administration of a potent DA receptor blocker, haloperidol, produced a marked augmentation of the stimulus induced activation of this enzyme (Roth *et al.*, 1975a). It is hypothesized that this may be due to the removal of an inhibitory effect on tyrosine hydroxylase activity mediated by increased quantities of dopamine in the synaptic cleft. The increase in dopamine concentration in the synaptic cleft, by stimulating DA autoreceptors on DA terminals, would exert a modulating effect on the stimulus-induced activity of tyrosine hydroxylase. Dopamine-receptor blockers are hypothesized to remove this breaking effect and therefore to potentiate the stimulation-induced activation of tyrosine hydroxylase. Since stimulation is performed at a frequency and time period which produced a maximal increase in tyrosine hydroxylase activity, any drug effect mediated by a postsynaptic feedback loop should be cancelled out.

Thus, a large body of biochemical evidenced suggests the existence of presynpatic DA receptors (autoreceptors) in addition to the long known postsynaptic receptors. Does electrophysiological evidence exist to support these biochemical findings in the central nervous system?

Electrophysiological evidence for dopaminergic autoreceptors in the central nervous system

In trying to establish the presence or absence of DA autoreceptors the problem of differentiating between the pre- and postsynaptic actions of drugs which effect the dopamine system must be confronted by the electrophysiologist, just as it was by the biochemist. For the electrophysiologist two techniques are helpful in this regard. The possible presynaptic action of parenterally administered drugs on A9 DA neurones can be studied by selectively lesioning the anatomically identified (McGeer *et al.*, 1974; Grofova, 1975; Hattori *et al.*, 1975; Bunney and Aghajanian, 1976c) striatonigral feedback pathways. Whether the lesions are made electrothermically or using a fine wire retractable knife, the direct-acting DA receptor stimulator, apomorphine still inhibits dopamine cells when given intravenously in very low doses. This inhibition is reversed by low doses of intravenous haloperidol (a presumed DA-receptor blocker) (Aghajanian and Bunney, 1974a; Bunney and Aghajanian, 1976d).

Although recent studies using autoradiographic (Conrad and Pfaff, 1976) and horseradish peroxidase (Dray and Oakley, 1978) techniques support the existence of a pathway from the accumbens nucleus to unidentified cells in the ventral tegmental area, it has not yet been determined whether this pathway can affect the activity of A10 cells. Consequently, lesioning techniques have not yet been used for studying A10 DA autoreceptors. However, there is another method which circumvents the possible effects of a feedback pathway. Neurotransmitters and drugs can be applied locally to

A9 and A10 cells by means of a cannula or by microiontophoresis while simultaneously recording from them.

Microiontophoretically applied DA and apomorphine rapidly decrease DA neurone activity (Aghajanian and Bunney, 1973) at low ejection currents, whereas *d*-amphetamine ejected with high currents over long periods of time, either around A9 cell bodies (Bunney and Aghajanian, 1973) or around A9 dendrites (Bunney and Aghajanian, 1976e), rarely causes more than a 50% decrease in rate (Bunney and Aghajanian, 1973, 1976e). Following i.v. haloperidol, the microiontophoresis of dopamine no longer produced an inhibitory effect. Similar results were obtained with i.v. chlorpromazine except that a total blockade of dopamine could not be achieved even at doses 20 times that of haloperidol (Aghajanian and Bunney, 1974a). Microiontophoretic administration of the antipsychotic drug, trifluoperazine, has also been shown to block dopamine-induced inhibition of these cells (Aghajanian and Bunney, 1977). Although i.v. trifluoperazine, like haloperidol and chlorpromazine, increases the firing rate of A9 cells (Bunney *et al.*, 1973) it has very little effect on cell activity when administered microiontophoretically (Aghajanian and Bunney, 1976).

Clozapine at doses twice those of chlorpromazine was ineffective in blocking DA inhibition of these cells (Bunney and Aghajanian, 1975c). Thus there is a direct parallel between the relative potency of these three drugs (haloperidol, chlorpromazine and clozapine) in blocking dopamine depression of DA neurone activity and their potency in blocking the ability of dopamine agonists to inhibit impulse flow cessation-induced activation of tyrosine hydroxylase (Walters and Roth, 1974).

The local infusion of drugs into the substantia nigra by means of a cannula has sometimes yielded results different from those obtained with microiontophoresis. Local infusion of *d*-amphetamine was found to markedly inhibit A9 cells and this effect was significantly reduced by pretreatment with α-methyl-*p*-tyrosine (Groves *et al.*, 1976). Haloperidol, on the other hand, increased the firing rate of A9 cells (Groves *et al.*, 1975). The reasons for the different results obtained with these two techniques remain unclear at the present time.

Can the DA autoreceptor be classified as an α or β receptor? Microiontophoretically applied norepinephrine inhibits dopamine cells as does isoproterenol, although it has a potency only about one-fifth that of dopamine (Aghajanian and Bunney, 1977). The α agonist, clonidine, as well as the α antagonist, piperoxane, and the β antagonist, sotol, are entirely without blocking activity (Aghajanian and Bunney, 1977) (Table III). Thus, from the profile of microiontophoretically administered drugs affecting the activity of A9 and A10 cells, it would appear that the DA autoreceptor is neither α nor β but uniquely dopaminergic.

TABLE III Effects of catecholamine agonists and antagonists upon the activity of A9 and A10 dopaminergic neurones[a].

Agonists	Inhibitory effect on dopaminergic neurones[b]
Dopamine	+ + to + + +
Norepinephrine	+ + to + + +
Apomorphine	+ +
Isoproterenol	+
d-Amphetamine	0 to +
Clonidine	0
Antagonists	**Blockade of inhibition**
Trifluoperazine	+ + +
Haloperidol	+ + +[c]
Chlorpromazine	+[c,d]
Piperoxane	0
Sotalol	0

[a] All drugs administered by means of microiontophoresis except where otherwise noted. Ratings: + + +, 75–100% inhibition (or antagonism); + +, 50%–75% inhibition; +, 25%–50% inhibition.

[b] All data from reference (Aghajanian and Bunney, 1977) except where otherwise noted.

[c] Administered intravenously. Data from reference (Aghajanian and Bunney, 1974a).

[d] Produced only partial blockade even at dose 20 times that of haloperidol. (Table modified from Aghajanian and Bunney, 1977.)

Dopaminergic Autoreceptors—Physiological and Pharmacological Controversies

The above evidence strongly suggests the presence of DA autoreceptors on both dopaminergic cell terminals and bodies or dendrites. Biochemical studies suggest that the function of the receptors on DA terminals is to modulate dopamine synthesis and release. But what is the purpose of these receptors on DA cell bodies and dendrites? It has recently been suggested that dopamine cells can exert a tonic inhibitory effect on each other through DA cell dendrodendritic synapses (Groves *et al.*, 1975). Supporting this contention is both anatomical evidence (Hajdu *et al.*, 1973; Björklund and Lindvall, 1975; Wilson *et al.*, 1977) and electrophysiological evidence (Wilson *et al.*, 1977; Groves *et al.*, 1977) for dendrodendritic synapses within the substantia nigra. Furthermore, there is evidence of dopamine release in the substantia nigra induced by potassium (Geffen *et al.*, 1976), the indirectly acting DA agonists—amphetamine and benztropine (Paden *et al.*, 1976; Nieoullon *et al.*, 1977a), electrical stimulation (Korf *et al.*, 1976) and sensory stimuli (Nieoullon *et al.*, 1977b). As described above, in addition to the microiontophoretic electrophysiological evidence for the presence of DA autoreceptors on DA cells in the zona compacta, it has been reported that *d*-amphetamine introduced locally into the substantia nigra by means of a cannula inhibits DA cell firing (Groves *et al.*, 1975; Groves *et al.*, 1976). (This latter finding not only adds further support to the proposed concept of autoregulation of DA cell activity but also obviates the need for a striato-nigral feedback pathway to mediate *d*-amphetamine's mechanism of action.) Thus, the evidence is mounting that DA cell dendrites can release dopamine which may in turn affect the activity of neighbouring DA and/or non-DA cells. However, not all data fit this concept.

Recently, a dopamine-sensitive adenylate cyclase has been described in the substantia nigra (Kebabian and Saavedra, 1976; Phillipson and Horn, 1976; Spano *et al.*, 1976) which is apparently located presynaptically on terminals of striato-nigral afferents and *not* on DA cell dendrites (Gale *et al.*, 1977; Spano *et al.*, 1977). Compatible with this data, dopamine has been shown to release GABA from slices of rat substantia nigra (Reubi *et al.*, 1977). It has been suggested, therefore, that the inhibition of ZC cells caused by microiontophoretic application of DA directly onto them may not be due to the interaction of dopamine on autoreceptors but rather may be due to dopamine interacting with DA receptors on afferent GABAergic terminals to cause release of GABA (Gale *et al.*, 1977; Reubi *et al.*, 1977).

These findings, on the other hand, are not compatible with the following data. Dopamine applied microiontophoretically has been found to inhibit zona compacta DA cells with its usual potency in animals in which afferent striatal nigral pathways were destroyed by injecting kainic acid (kainic acid destroys neurones without affecting nerve terminals or fibres of passage; Coyle and Schwarcz, 1976; Schwarcz and Coyle, 1977; Hattori and McGeer, 1977), into the rat caudate putamen (Bunney and Aghajanian, 1978a). Dopamine cells in these kainic acid lesioned animals were also totally inhibited by the usual low dose of apomorphine (0.1 mg kg^{-1} i.v.) and this inhibition was reversed by 0.1 mg kg^{-1} of i.v. haloperidol. Thus, despite a 60% decrease in nigral GABA concentration in these lesioned animals no attenuation of the inhibitory effects of DA or apomorphine was observed. These results suggest that the ability of at least some dopamine agonists and antagonists to affect the activity of A9 neurones can occur due to an interaction on DA autoreceptors and is not totally dependent upon interaction with an afferent GABA system.

Contrary to kainic acid lesion effects on apomorphine-induced inhibition, these lesions markedly attenuated the ability of low i.v. doses of *d*-amphetamine to inhibit DA cells (although higher doses still had an inhibitory effect; Bunney and Aghajanian, 1978a). Diencephalic lesions anterior to the substantia nigra have also been reported disrupt the inhibition of A9 cells induced by local infusion of *d*-amphetamine through a cannula (Groves *et al.*, 1978) suggesting that the block by kainic acid lesions of i.v. amphetamine-induced inhibition was a non-specific effect. However, in the diencephalic lesion studies the axons of DA neurones were cut as well as those of any striatonigral feedback neurones. Although not everyone would agree, one might conclude from the evidence available that an intact striato-nigral feedback pathway is necessary for the marked inhibition of DA cells induced by *low i.v.* doses of *d*-amphetamine and that at higher doses *d*-amphetamine can cause inhibition through local effects in the substantia nigra.

DA autoreceptors—epilogue

The existence of dopamine receptors on DA cell terminals, dendrites and/or bodies, although still controversial, is supported by an ever-increasing body of evidence. However, a great deal more information is necessary before we shall have a clear understanding of their physiological purpose and the manner in which they function. The majority of evidence suggests that pharmacological agents can interact with these receptors to change the activity of DA neurones as well as to regulate dopamine synthesis and release. Therefore, it would now appear essential to assess the effect and relative potency of drugs on both pre- and postsynaptic receptors in order to understand fully the basis of their biochemical, physiological and behavioural actions.

With the discovery of autoreceptors has come a new site for possible malfunction in the brain which could be important in the aetiology of various diseases of the central nervous system. If autoreceptors differ from postsynaptic receptors either in character or sensitivity they may also provide a new method of treatment for central nervous system diseases. Thus, for example, if autoreceptors were more sensitive than postsynaptic receptors to a given pharmacological agent, low doses of the drug might be expected to interact preferentially with them and therefore have effects directly opposite to those obtained with higher doses (e.g. preferential action of a DA agonist on DA autoreceptors would cause a decrease in the amount of DA seen by DA receptors on cells innervated by DA neurones and therefore decrease the influence of the DA system on the rest of the brain). This line of reasoning already has begun an intensive search for new drugs with preferential effects on autoreceptors as well as a reinvestigation of known drugs for possible preferential autoreceptor effects at low doses. Differential behavioural effects between low and high doses of DA agonists have already been found in animals (Carlsson, 1976; Baudry *et al.*, 1977;

Strömbom, 1977). Recent studies have also suggested such effects may exist in man. In four diseases believed to be due, at least in part, to an increased influence of dopamine systems in the brain low doses of apomorphine, a directly acting DA-receptor stimulator, have been found to have therapeutic effects. Thus, low doses of apomorphine temporarily decreased symptoms of tardive dyskinesia (Carroll *et al.*, 1977; Smith *et al.*, 1977b), Gilles de la Tourette's disease (Feinberg and Carroll, personal communication), Huntington's chorea (Tolosa and Sparber, 1975) and schizophrenia (Smith *et al.*, 1977a). Although these studies need to be repeated and the mechanisms involved clarified, they suggest an exciting, rational new approach to the treatment of these disorders unknown before the discovery of DA autoreceptors.

Pharmacological differences between A9 and A10 neurones in their electrophysiological response to drugs

Numerous acute and chronic biochemical studies have suggested that the A9 and A10 cell systems are not identical in their response to drugs (for a brief review see Bunney and Aghajanian, 1978; Nowycky and Roth, 1978). Although systematic comparative studies have not yet been carried out, several findings suggest that such differences may also be found using electrophysiological techniques.

Agonists

Although the effects of apomorphine on the firing rate of A9 and A10 cells appear the same, marked differences can be seen in the recovery of the two cell populations after inhibition by apomorphine. A9 cells demonstrate a biphasic recovery curve—recovery to approximately half the basal rate occurs very rapidly (within 2 min) whereas the second half of the recovery is extremely slow, taking as much as an hour after a single i.v. dose of 0.1 mg kg^{-1} (Bunney *et al.*, 1973a). Cells in the A10 area respond differently in that they lack an initial rapid phase, the entire recovery being a slow continuous process (Bunney *et al.*, 1973a).

Antagonists

Only one antagonist has been found that appears clearly to differentiate between the two systems. After *d*-amphetamine-induced inhibition of A9 cells low doses of the phenothiazine analog, mepazine, reverse the inhibition of these cells. Contrastingly, mepazine has no effect on *d*-amphetamine-induced inhibition of A10 cells even when administered in doses 10 times that necessary to produce an effect on A9 cells (Bunney and Aghajanian,

1975a). The mechanism responsible for these differences is at present unknown.

Thus, it would appear that A9 and A10 systems may be different in terms of their response to drugs. A difference in postsynaptic DA receptors in the two areas would be one explanation for this difference. However, against the concept of different DA receptors in different parts of the brain are the findings that dopamine-stimulated adenylate cyclase (Clement-Cormier *et al.*, 1974; Horn *et al.*, 1974) and dopamine and haloperidol binding (Creese *et al.*, 1976; Snyder *et al.*, 1975) do not appear to differ in the various areas innervated by the two systems. Other possible explanations such as different neuronal feedback circuits, presence or absence of dendro-dendritic synapses (a dopamine-sensitive adenylate cyclase has not been found in the A10 area; Bockaert *et al.*, 1977), and differences in the DA autoreceptors in the two areas must also be considered. Obviously, any differences that can be determined and understood are important as they may permit the design of drugs which will act selectively on one system or the other as well as provide the potential for furthering our understanding of the therapeutic action of neuroleptics and thus, ultimately, our understanding of the pathogenesis of schizophrenia.

EFFECTS OF DRUGS ON NEURONES INNERVATED BY MIDBRAIN DOPAMINERGIC NEURONES

The Nigrostriatal DA System—A9 Innervated Neurones in the Caudate-Putamen

By far the most intensively studied of all the areas receiving a DA innervation, the caudate-putamen is also one of the most difficult to study electrophysiologically because the great majority of cells in the anaesthetized or immobilized animal are either silent or fire extremely slowly. For this reason the great majority of studies have been carried out on cells activated by the microiontophoretic application of an excitatory substance (e.g. glutamic acid). The use of such a technique unfortunately introduces a variable whose relevance to the normal physiological function of cells in the caudate nucleus is unknown. On the other hand, studies which only record from spontaneously firing cells obviously suffer from a problem of bias. Thus, spontaneously firing cells may form a very special and select population of cells within the caudate nucleus. Consequently, the effect of dopamine agonists and antagonists upon them may not represent the sole or even the major effect of these substances within the caudate nucleus.

Further complicating the problem of studying the effects of DA agonists and antagonists on cells in this area is the basic question, intensively debated at the moment, of whether dopamine is an inhibitory or excitatory neurotransmitter (for a detailed description and analysis of this debate, see Chapter 24). Most studies examining the effects of dopamine agonists and antagonists on dopamine-innervated striatal cells use extracellular micro-iontophoretic techniques. As most extracellular iontophoretic studies find dopamine to be predominantly an inhibitory neurotransmitter (see Siggins, 1978, for a review), the effects of putative DA antagonists have been mainly tested against this inhibitory effect.

DA agonists

Acute Studies

d-Amphetamine, supposedly acting indirectly through its ability to increase dopamine release and block dopamine uptake, inhibits caudate neurones when applied either iontophoretically (Bunney and Aghajanian, 1973; Feltz and De Champlain, 1973; Spencer and Havlicek, 1974; Stone, 1976; Stone and Bailey, 1975) or by direct infusion through a cannula (Groves *et al.*, 1975). In some studies the indirect action of amphetamine has been questioned. Thus destruction of the dopamine innervation to the caudate nucleus by 6-hydroxydopamine, depletion of dopamine by reserpine, and blockade of dopamine synthesis by α-methyl-*p*-tyrosine have been reported to have no effect on *d*-amphetamine's inhibitory action (Feltz and De Champlain, 1972, 1973).

The direct acting dopamine agonist apomorphine, applied microiontophoretically, also inhibits caudate cell activity (Siggins *et al.*, 1974; Yarbrough, 1975).

Numerous biochemical studies have demonstrated the presence of a dopamine-sensitive adenylate cyclase in the rat caudate nucleus (Kebabian *et al.*, 1972; Miller *et al.*, 1974; for review see Iversen, 1977). It has been suggested that the electrophysiological effects of dopamine might be mediated through its ability to generate cyclic adenosine monophosphate in postsynaptic cells (Siggins *et al.*, 1974). In support of this hypothesis it has been found that microiontophoretically applied cyclic AMP inhibits the great majority of rat caudate neurones studied (Siggins *et al.*, 1974).

Chronic Studies

Only one study, to date, has attempted to examine the effects of chronic DA agonist treatment on the activity of caudate neurones (Rebec and Groves, 1976). In this study *d*-amphetamine (2.5 mg kg^{-1} day, i.p.) or saline was administered to rats for up to 36 days. On the day following the last injection the response of spontaneously firing caudate neurones to i.p. *d*-amphetamine ($2.5\,mg\,kg^{-1}$) or apomorphine ($0.25\,mg\,kg^{-1}$) was determined. It was found that compared to saline-treated controls the inhibitory effects of both DA agonists was greatly prolonged. In addition, the degree of inhibition produced by apomorphine was increased. No difference was found between *d*-amphetamine- and saline-treated controls in the spontaneous activity of caudate neurones. The authors suggest that the effects observed may be due to the development of DA receptor supersensitivity secondary to decreased central catecholamines (caused by long-term *d*-amphetamine administration). However, DA levels in the caudate nucleus were not measured. Interestingly, Klawans *et al.* (1977) have recently reported behavioural evidence suggesting the development of supersensitivity after chronic treatment with a variety of DA agonists—*d*-amphetamine, L-DOPA, bromocriptine and methylphenidate.

DA antagonists

Many antipsychotic drugs are insoluble at the high concentrations needed for iontophoretic studies. Consequently, the number of these drugs studied has been relatively small and problems with ejection of the drug have often been encountered by the investigator.

Acute Studies

The first electrophysiological evidence that antipsychotic drugs might block DA receptors was obtained by York (1970). He found that microiontophoretically applied chlorpromazine could block the dopamine induced inhibition and excitation of striatal cells in anaesthetized cats (York, 1970) and monkeys (York, 1972). The majority of cells studied were activated with either L-glutamic acid or D,L-homocysteic acid. Subsequently microiontophoretically applied α-flupenthixol has also been shown to block dopamine inhibition of spontaneous and activated caudate cells in the anesthetized rat (Siggins *et al.*, 1974) and cat (Ben-Ari and Kelly, 1976). Fluphenazine (Siggins *et al.*, 1974) and trifluoperazine (Bunney and Aghajanian, unpublished data) have been found to have the same effect when administered iontophoretically to anaesthetized rats.

Chronic Studies

To date, despite numerous biochemical studies examining the effects of chronic neuroleptic treatment on dopamine systems, only one chronic electrophysiological study has been published (Yarbrough, 1975). Albino rats were treated for nine days with haloperidol (2 mg kg^{-1} per day, i.p.) or saline. Four to eight days after the last injection the response of glutamate activated caudate neurones to microiontophoretically applied DA, apomorphine and GABA was determined in chloral hydrate-anaesthetized animals. The number of cells responsive to DA and apomorphine in haloperidol-pretreated animals was over twice that found in saline controls. There was no difference between groups in number of cells responsive to GABA. These findings were interpreted as suggesting the development of DA receptor supersensitivity.

In another study (Skirboll and Bunney, in preparation) haloperidol was administered subcutaneously to rats in a dose of 0.5 mg kg^{-1} per day for 21 days. Using single unit recording and microiontophoretic techniques, spontaneously firing cells in the caudate nucleus were studied in gallamine paralyzed preparations on the last day of chronic treatment (day 21) and one week after haloperidol was discontinued. The number of spontaneously firing cells was found to be greatly increased at the end of the treatment period as compared to saline-or promethazine-treated animals. Dopamine applied microiontophoretically, inhibited caudate neurones with the same potency in control and chronically treated (21-day) animals. One week after chronic treatment had ended fewer cells were firing spontaneously than in control animals. However, those few cells found to be firing were much more sensitive to dopamine-induced inhibition when compared to controls. The latter results could be interpreted as suggesting that dopamine receptor supersensitivity had developed due to chronic haloperidol treatment. It is unclear at this time why chronic treatment with haloperidol did not block dopamine-induced inhibition in the 21-day animals. One possible explanation would be that the development of dopamine receptor supersensitivity allowed one to see an inhibitory effect of dopamine despite the presence of the haloperidol.

The Mesolimbic DA System—A10 Innervated Cells in Limbic Areas

Compared to the nigrostriatal dopamine system very little information exists concerning the effects of drugs on postsynpatic DA receptors in areas comprising the mesolimbic dopamine system. Almost all the available studies involve cells which are presumed to have a dopamine input based on fluorescence histochemical evidence but whose identity has never been firmly established. The following is a summary of the most relevant studies arranged according to area studied.

Accumbens nucleus and olfactory tubercle

Spontaneously firing cells in both the accumbens nucleus and olfactory tubercle have been shown to be inhibited by microiontophoretic dopamine, *d*-amphetamine and adenosine-3′,5′-monophosphate (Bunney and Aghajanian, 1973). Inhibitory effects of dopamine in the accumbens nucleus are blocked by microiontophoretic trifluoperazine and intravenous haloperidol and clozapine (Bunney and Aghajanian, 1975b). In another study dopamine and also the dopamine analogue 2-amino-6,7-dihydroxy-1,2,3,4-tetrahydronapthalene (ADTN) and the ergot alkaloid ergometrine (applied by means of microiontophoresis), have been shown to inhibit cells in the accumbens nucleus activated by continuous iontophoretic application of D,L-homocysteic acid (Woodruff *et al.*, 1976). Many of the cells inhibited by dopamine were also inhibited by direct application of dibutyryl cyclic AMP (Woodruff *et al.*, 1976).

Septum

Dopamine innervation of the septum has been demonstrated to be confined mainly to the lateral septal nucleus (Lindvall, 1975). Recent studies have demonstrated norepinephrine (NE) depression of cells in this area (Segal, 1974). In preliminary studies, cells in the lateral septal nucleus have been found to be depressed by microiontophoretic dopamine as well as NE. Trifluoperazine administered microiontophoretically was found to block the dopamine-induced depression while having no effect on NE inhibition (Bunney, unpublished data). Electrical stimulation of A10 neurones has been found to cause an excitation or excitation–inhibition of lateral septal neurones which was blocked by haloperidol and flupenthixol (dopamine-receptor blockers) and absent after 6-hydroxydopamine pretreatment (Assaf and Miller, 1976). The discrepancy between the excitatory effects of A10 cell stimulation and inhibitory effects of microiontophoretic dopamine remains to be explained.

Amygdala

The greatest concentration of dopamine in the amygdaloid complex appears to be in the central nucleus of the amygdala (Ben-Ari *et al.*, 1975). Measurements of the ratio of tyrosine hydroxylase to dopamine-β-hydroxylase in various regions of the amygdala have suggested a dopamine innervation of the lateral and basal posterior portions of the amygdaloid complex (Saavedra and Zivin, 1976). Fluorescence histochemical studies have demonstrated DA nerve terminals in both areas (Fuxe, 1965; Hökfelt *et al.*, 1974).

One of the first pharmacological studies of the amygdala using electrophysiological techniques was performed in cats in 1965 (Straughan and Legge). Most of the cells studied were in the lateral and basal medial complex. In this study NE and dopamine, applied microiontophoretically, were found to depress synaptically activated and glutamate-excited cells. Other studies have demonstrated dopamine inhibition of glutamate-excited cells in the central amygdaloid complex (Ben-Ari and Kelly, 1974). Furthermore, this dopamine-induced inhibition was blocked by microiontophoretic and i.v. α-fluphenthixol (Ben-Ari and Kelly, 1974). In a follow-up electrophysiological study of the basal lateral amygdala of the cat, Ben-Ari and Kelly (1976) obtained the same results as in the previous study with the notable exception that this time i.v. α-flupenthixol and pimozide did not block dopamine-induced inhibition. In preliminary studies DA has been found to inhibit spontaneously firing cells (rat, low-cerveau isolé preparation) in the region of the central nucleus and this inhibition is blocked by trifluoperazine applied microiontophoretically (Bunney, unpublished data).

The Mesocortical Dopamine System—A10 Innervated Cells in the Cortex

One of the first studies of the effects of iontophoretic dopamine on cortical neuronal activity was in the pericruciate region of the cat on primarily glutamate-activated cells (Krnjevic and Phillis, 1963). In this study dopamine was found to have a greater inhibitory effect than NE but neither were inhibited by chlorpromazine. Since that time various other studies have appeared in the literature, most of them reporting the effect of dopamine or dopamine agonists or antagonists on cell activity in the parietal cortex (Stone, 1973a, b, 1974, 1976; Stone and Bailey, 1975; Yarbrough *et al.*, 1973). All of the above studies suffer from the problem that they were carried out in areas of the cortex now believed not to receive a dopamine innervation. However, in 1976, two reports appeared in which the cells studied were located in a dopamine-innervated area—the prefrontal cortex (Bunney and Aghajanian, 1976b; Mora *et al.*, 1976). In one study, spontaneously firing cells in anaesthetized rats were recorded simultaneously from the prefrontal cortex and an area known not to receive a dopamine innervation (somatosensory cortex or hippocampus) (Mora *et al.*, 1976). Forty-two per cent of the cells recorded from in the prefrontal cortex were inhibited by i.v. apomorphine, whereas no cells were inhibited by apomorphine in control areas. Pretreatment with spiroperidol blocked the apomorphine-induced inhibition as well as inhibition of prefrontal cells by *d*-amphetamine. *d*-Amphetamine increased the firing rate of most cells recorded from control areas. L-DOPA when combined with a peripheral decarboxylase inhibitor (Ro4-4602) was also found to inhibit some cells in the prefrontal cortex. No attempt was made in this study to correlate dopamine agonist effects on the activity of a given cell with the location of that cell within the various layers of the prefrontal cortex. However, the distribution of DA and NE terminals in this area is not random. Thus, the heaviest dopamine innervation of this area has been shown to occur in layers V and VI of a region which is rostral to the genu of the corpus callosum and medial to the corpus callosum and forceps minor (Berger *et al.*, 1974; Berger *et al.*, 1976). The superficial layers (I–III) of this same area receive a prominent NE projection as well as a light dopamine input (Dahlström and Fuxe, 1964). Thus, the anatomical location of dopamine and NE nerve terminals within the prefrontal cortex is known.

Recently a series of experiments was published in which an attempt was made to use this anatomical information to identify pharmacologically the catecholamine input to a given cell within the prefrontal cortex (Bunney and Aghajanian, 1976b). Cells in each layer of the frontal cortex surrounding the dorsal and medial aspects of the forceps minor were tested for their response to dopamine and NE. A clear correspondence was found between the response of a cell to DA or NE and the layer in which that cell resided. Thus, most cells in layers II and III were found to be depressed by low ejection currents of NE and minimally responsive to dopamine, whereas cells in layers V and VI were depressed primarily by dopamine and not NE. A small percentage of cells in layers III, IV and V were equally responsive to both. Scattered throughout all layers were occasional cells which did not respond to either catecholamine. Cells lateral to the foreceps minor did not respond to DA although some were depressed by NE.

Desmethylimipramine, an NE uptake blocker (Horn *et al.*, 1971), has recently been shown to be a poor blocker of DA uptake into cerebral cortex homogenates after 6-OH-dopamine destruction of NE ascending pathways (Tassin *et al.*, 1974). Conversely, benztropine (a drug which has previously been shown to block DA uptake in the corpus striatum; Coyle and Snyder, 1969), was found to be a very good blocker of dopamine uptake in the cerebral cortex (Tassin *et al.*, 1974). Therefore, the ability of desmethylimipramine and benztropine to potentiate the depressant effect of microiontophoretically applied NE and dopamine on cells in the various layers of the frontal cortex was tested. Microiontophoretically applied desmethylimipramine potentiated NE depression of cells in layers II and III although it either had no effect or tended to block DA depression of dopamine-sensitive cells in layers V and VI. Benztropine, on the other hand, did not potentiate NE depression of NE sensitive cells but did potentiate DA depression of cells in layers V and VI.

Trifluoperazine (a potent antipsychotic agent; Klein and Davis, 1969) whose mechanism of action is thought to be its ability to block DA receptors (Snyder *et al.*, 1974; Meltzer and Stahl, 1976), when applied microiontophoretically to cells in the frontal cortex, was found to block DA depression of DA-sensitive cells in layers V and VI but was ineffective in blocking NE depression of cells in layers II and III.

In further studies it was found that pyramidal cells in the hippocampus, which receive only an NE input, and cells in the nucleus accumbens, which receive only a dopamine innervation, responded to dopamine, NE, benztropine, desmethylimipramine, and trifluoperazine in a manner identical to that found in cortical layers II to IV and V and VI, respectively (Bunney and Aghajanian, 1976b).

The above findings demonstrate that, using extracellular single unit recording techniques, it is possible to differentiate pharmacologically between cells in the frontal cortex primarily innervated by the NE system vs those primarily innervated by the dopamine system. In addition, NE innervated cells, whether in the hippocampus or frontal cortex, respond in a similar manner to the drugs tested. The ability of trifluoperazine to block dopamine depression of cells in the frontal cortex, while having no effect on NE depression of NE sensitive cells, suggests a specific dopamine-receptor blocking action and adds further support to the concept that dopamine cortical systems may play an important role in mediating the antipsychotic effects of neuroleptics.

CONCLUSION

In this chapter we have seen that many drugs affect the activity of midbrain dopaminergic systems. For some of these drugs (dopamine agonists and antagonists) we have some idea of their cellular mechanism of action. On the other hand, we are totally ignorant about how they produce their behavioural effects. For most of the drugs under the heading "Miscellaneous" even the cellular mechanism of action is at best conjecture. The primary reason for these gaps in our knowledge is that the neuronal systems interacting with the DA system have not been neurochemically identified and the neural circuitry responsible for the drug-induced behaviours is unknown. We know that a specific DA system innervates a given area of the brain but we do not know the type of cell it innervates (interneurone or efferent) nor do we know the neurotransmitters used by these neurones. To study a drug's effect on behaviour adequately, we must know the location of all of the neurones responsible for mediating the behaviour as well as the neurotransmitters they use. In addition, we must know the role that the drug-affected system plays in that circuit. Only when we have this information for all of the neurones responsible for a particular drug-induced behaviour will we begin to be able to understand how a drug's action at a cellular level can alter an animal's behaviour.

The anatomical and histochemical techniques (autoradiography, horse-radish peroxidase retrograde tracing, immunofluorescence histochemistry) are available for obtaining part of this information and are being used extensively. However, it is really only *after* this information has been obtained that the electrophysiologist can use his techniques to obtain reliable and easily interpretable information concerning the functioning of these circuits and the effects of drugs upon them. Because of the complexity of the brain, examining one cell at a time will probably not enable the electrophysiologist fully to understand a circuit's function. New techniques will have to be developed in which concomitant multiple single unit recordings can be made from neurones throughout a given circuit. Ultimately, this will have to be done in freely moving animals as only then can correlations be made with behaviour.

The obstacles in the way of achieving these goals are formidable; however, there is no lack of talented motivated investigators interested in these problems. Many of them, for the first time, bring knowledge and expertise from traditionally non-biological disciplines (e.g. physics, electrical engineering) to bear on problems concerning the actions of psychotropic drugs. Provided funding is made available, it would now appear possible, in time, to understand fully the way in which psychotropic drugs alter behaviour. The potential benefit to mankind derived from achieving this goal is obvious.

ACKNOWLEDGEMENT

This work was supported in part by United States Public Health Service Grants MH-28849 and MH-25642.

REFERENCES

Aghajanian, G. K. and Bunney, B. S. (1973). *In* "Frontiers in Catecholamine Research" (Eds. S. H. Snyder and E. Usdin), 643–648. Pergamon, New York.

Aghajanian, G. K. and Bunney, B. S. (1974a). *In* "Frontiers of Neurology and Neuroscience Research" (Eds. P. Seeman and G. M. Brown), 4–11. University of Toronto Press, Toronto.

Aghajanian, G. K. and Bunney, B. S. (1974b). *In* "Proceedings of the 9th international Congress of the Collegium Internationale Neuropsycho-

pharmacologicum" (Eds. J. R. Boisser, H. Hippius, and P. Pichot), 444–452. Excerpta Medica Foundation.
Aghajanian, G. K. and Bunney, B. S. (1977). *Naun. Schmied. Arch. Pharmacol.* **297,** 1–7.
Agid, Y., Javoy, F., Glowinski, J., Bouvet, D., and Sotelo, C. (1973). *Brain Res.* **58,** 291–301.
Andén, N.-E. (1972). *J. Pharm. Pharmacol.* **24,** 905–906.
Andén, N.-E., Roos, B. E. and Werdinius, B. (1964). *Life Sci.* **3,** 149–158.
Andén, N.-E., Rubenson, A., Fuxe, K., and Hökfelt, T. (1967). *J. Pharmacol. Pharm.* **19,** 627–629.
Andén, N.-E., Corrodi, H., Fuxe, K., and Ungerstedt, U. (1971). *Eur. J. Pharmacol.* **15,** 193–199.
Assaf, S. Y. and Miller, J. J. (1976). *Neurosci. Abst.* Vol. II, Part I, No. 677.
Bartholini, G. (1976). *J. Pharm. Pharmacol.* **28,** 429–433.
Baudry, M., Costentin, J., Marcois, H., Martres, M. P., Protais, P., and Schwartz, J. C. (1977). *Neurosci. Lett.* **4,** 203–207.
Ben-Ari, Y. and Kelly, J. S. (1974). *J. Physiol.* **242,** 66–67.
Ben-Ari, Y. and Kelly, J. S. (1976). *J. Physiol.* **256,** 1–21.
Ben-Ari, Y., Zigmond, R. E., and Moore, K. E. (1975). *Brain Res.* **87,** 96–101.
Berger, B., Tassin, J. P., Blanc, G., Moyne, M. A., and Thierry, A. M. (1974). *Brain Res.* **81,** 332–337.
Berger, B., Thierry, A. M., Tassin, J. P., and Moyne, M. A. (1976). *Brain Res.* **106,** 133–145.
Bernard, P., Edwards, S. J., Fielding, S., Robson, R. D., Saelens, J. K., Simke, J. P., and Welch, J. (1975). *Pharmacologist* **17,** 447.
Besson, M., Cheramy, A., Feltz, P., and Glowinski, J. (1971). *Brain Res.* **32,** 407–424.
Björklund, A. and Lindvall, O. (1975). *Brain Res.* **83,** 531–537.
Bloom, F. E. (1974). *Life Sci.* **14,** 1819–1834.
Bockaert, J., Premont, J., Glowinski, J., Thierry, A. M., and Tassin, J. P. (1976). *Brain Res.* **197,** 303–315.
Bockaert, J., Premont, J., Glowinski, J., Tassin, J. P., and Thierry, A: M. (1977). *In* "Advances in Biochemical Psychopharmacology", Vol. 16 (Eds. E. Costa and G. L. Gessa), 29–37. Raven Press, New York.
Brown, G. L. and Gillespie, J. S. (1957). *J. Physiol.* (*London*) **138,** 81–102.
Bunney, B. S. and Aghajanian, G. K. (1973). *In* "Frontiers in Catecholamine Research" (Eds. S. H. Snyder and E. Usdin), 957–963. Pergamon, New York.
Bunney, B. S. and Aghajanian, G. K. (1974). *Life Sci.* **15,** 309–318.
Bunney, B. S. and Aghajanian, G. K. (1975a). *In* "Predictability in Psychopharmacology: Preclinical and Clinical Correlations" (Eds. A. Sudilovsky, S. Gershon and B. Beers), 225–245. Raven Press, New York.
Bunney, B. S. and Aghajanian, G. K. (1975b). *In* "Antipsychotic Drugs, Pharmacodynamics and Pharmacokinetics", Wenner-Gren Center International Symposium Series, 305–318. Pergamon Press, New York.
Bunney, B. S. and Aghajanian, G. K. (1975c). *In* "Pre- and Postsynaptic Receptors" (Eds. E. Usdin and W. E. Bunney, Jr.), 89–123. Marcel Dekker, New York.
Bunney, B. S. and Aghajanian, G. K. (1976a). *In* "The Basal Ganglia" (Eds. M. D. Yahr), 249–267. Raven Press, New York.
Bunney, B. S. and Aghajanian, G. K. (1976b). *Life Sci.* **19,** 1783–1792.
Bunney, B. S. and Aghajanian, G. K. (1976c). *Brain Res.* **117,** 423–435.
Bunney, B. S. and Aghajanian, G. K. (1976d). *Science* **192,** 391–393.

Bunney, B. S. and Aghajanian, G. K. (1976e). *In* "Non-Striatal Dopaminergic Neurons, Advances in Biochemical Psychopharmacology Series" (Eds. G. L. Gessa and E. Costa), 65–70. Raven Press, New York.

Bünney, B. S. and Aghajanian, G. K. (1978a). *Naun. Schmied. Arch. Pharmacol.* **304,** 255–261.

Bunney, B. S. and Aghajanian, G. K. (1978b). *In* "Psychopharmacology: A Generation of Progress" (Eds. M. A. Lipton and A. DiMascio), 159–171. Raven Press, New York.

Bunney, B. S., Aghajanian, G. K., and Roth, R. H. (1973a). *Nature (London)* **245,** 123–125.

Bunney, B. S. and Grace, A. A. (1978). *Life Sci.* **23,** 1715–1728.

Bunney, B. S., Walters, J. R., Roth, R. H., and Aghajanian, G. K. (1973b). *J. Pharmacol. Expl. Ther.* **185,** 560–571.

Bunney, B. S., Roth, R. H., and Aghajanian, G. K. (1975a). *Psychopharmacol. Commun.* **1**(4), 349–358.

Bunney, B. S., Walters, J. R., Kuhar, M. J., Roth, R. H., and Aghajanian, G. K. (1975b). *Psychopharmacol. Commun.* **1**(2), 177–190.

Carlsson, A. (1975). *In* "Pre- and Postsynaptic Receptors" (Eds. E. Usdin and W. E. Bunney, Jr.), 49–65. Marcel Dekker, New York.

Carlsson, A. (1976). *In* "The Basal Ganglia" (Ed. M. D. Yahr), 181–189. Raven Press, New York.

Carlsson, A. (1978). *In* "Psychopharmacology: A Generation of Progress" (Eds. M. A. Lipton, A. DiMascio and K. F. Killam), 1057–1070. Raven Press, New York.

Carlsson, A. and Lindqvist, M. (1963). *Acta Pharmacol. Toxicol.* **20,** 140–144.

Carlsson, A., Kehr, W., Lindqvist, M., Magnusson, T., and Atack, C. V. (1972). *Pharmacol. Rev.* **24,** 371–384.

Carroll, B. J., Curtis, G. C., Kokmen, E. (1977). *Am. J. Psychiatr.* **134,** 785–789.

Cheramy, A., Besson, M. J. and Glowinski, J. (1970). *Eur. J. Pharmacol.* **10,** 206–214.

Chiueh, C. C. and Moore, K. E. (1975). *J. Pharmacol. Expl. Ther.* **193,** 559–563.

Christoph, G. R., Kuhn, D. M., and Jacobs, B. L. (1977). *Life Sci.* **21,** 1585–1596.

Clement-Cormier, Y. C., Kebabian, J. W. and Greengard, P. (1974). *Proc. Natl. Acad. Sci., U.S.A.* **71,** 113–117.

Conrad, L. C. A. and Pfaff, D. W. (1976). *Brain Res.* **113,** 589–596.

Corrodi, H., Fuxe, K., and Hökfelt, T. (1967). *Eur. J. Pharmacol.* **1,** 363–368.

Costa, E. and Groppetti, A. (1970). *In* "Amphetamines and Related Compounds" (Eds. E. Costa and S. Garattini), 231–255. Raven Press, New York.

Costa, E. and Neff, N. H. (1966). *In* "Biochemistry and Pharmacology of the Basal Ganglia" (Eds. E. Costa, L. Cote and M. D. Yahr), 141. Raven Press, New York.

Coyle, J. T. and Schwarcz, R. (1976). *Nature (London)* **263,** 244–246.

Coyle, J. T. and Snyder, S. H. (1969). *Science* **166,** 899–901.

Creese, I., Burt, D. R., and Snyder, S. H. (1975). *Life Sci.* **17,** 1715–1720.

Creese, I., Burt, D. R., and Snyder, S. H. (1976). *Science* **192,** 481–483.

Dahlström, A. and Fuxe, K. (1964). *Acta Physiol. Scand. Suppl. 232*, **62,** 1–55.

Dray, A. and Oakley, N. R. (1978). *Experientia* **4,** 68–70.

Enero, M. A., Langer, S. Z., Rothlin, R. P., and Stefano, F. J. E. (1972). *Br. J. Pharmacol.* **44,** 672–688.

Ernst, A. M. (1967). *Psychopharmacologia* **10,** 316–323.

Farnebo, L. O. and Hamberger, B. (1970). *J. Pharm. Pharmacol.* **22,** 855–857.

Farnebo, L. O. and Hamberger, B. (1971). *Acta Physiol. Scand.* **371,** 35–44.

Farnebo, L. O. and Malmfors, T. (1971). *Acta Physiol. Scand.* **371,** 1–18.

Feltz, P. (1971). *Can. J. Physiol. Pharmacol.* **49,** 1113–1115.
Feltz, P. and De Champlain, J. (1972). *Brain Res.* **43,** 595–600.
Feltz, P. and De Champlain, J. (1973). *In* "Frontiers in Catecholamine Research" (Eds. S. H. Snyder and E. Usdin), 951–956. Pergamon, New York.
Fuxe, K. (1965). *Acta Physiol. Scand.* **247,** 41–85.
Gale, K., Guidotti, A. and Costa, E. (1977). *Science* **195,** 503–505.
Geffen, L. B., Jessell, T. M., Cuello, A. C., and Iversen, L. L. (1976). *Nature (London)* **260,** 258–260.
Glowinski, J., Axelrod, J., and Iversen, L. L. (1966). *J. Pharmacol. Expl. Ther.* **153,** 30–41.
Grofová, I. (1975). *Brain Res.* **91,** 286–291.
Groves, P. M., Wilson, C. J., Young, S. J., and Rebec, G. V. (1975). *Science* **190,** 522–529.
Groves, P. M., Young, S. J., and Wilson, C. J. (1976). *Neuropharmacology* **15,** 755–762.
Groves, P. M., Wilson, C. J., and MacGregor, R. J. (1977). *In* "Interactions Among Putative Neurotransmitters in the Brain" (Ed. S. Garattini), Raven Press, New York.
Groves, P. M., Young, S. J., and Wilson, C. J. (1978). *In* "Cholinergic-Monoaminergic Interactions in the Brain" (Ed. L. L. Butcher), in press. Academic Press, New York and London.
Guyenet, P. G. and Aghajanian, G. K. (1978). *Brain Res.*, **150,** 69–84.
Häggendal, J. (1970). *In* "Bayer-Symposium II", 100. Springer-Verlag, Heidelberg.
Hajdu, F., Hassler, R., and Bak, I. J. (1973). *Z. Zellforsch.* **146,** 207–221.
Hattori, T. and McGeer, E. G. (1977). *Brain Res.* **129,** 174–180.
Hattori, T., Fibiger, H. C., and McGeer, P. L. (1975). *J. Comp. Neur.* **162,** 487–504.
Hökfelt, T., and Ungerstedt, U. (1973). *Brain Res.* **60,** 269–297.
Hökfelt, T., Ljungdahl, A., Fuxe, K., and Johansson, O. (1974). *Science* **184,** 177.
Hong, J. S., Yang, H.-Y. T., Fratta, W., and Costa, E. (1977). *Brain Res.* **134,** 383–386.
Horn, A. S., Coyle, J. T., and Snyder, S. H. (1971). *Molec. Pharmacol.* **7,** 66–80
Horn, A. S., Cuello, A. C. and Miller, R. J. (1974). *J. Neurochem.* **22,** 265–270.
Iwatsubo, K. and Clouet, D. H. (1977). *J. Pharmacol. Expl. Ther.* **202,** 429–436.
Iwatsubo, K. and Clouet, D. H. (1977). *J. Pharmacol. Expl. Ther.* **202,** 429.
Jacobowitz, D. M. and Goldberg, A. M. (1977). *Brain Res.* **122,** 575–577.
Javoy, F., Agid, Y., Bouvet, D., and Glowinski, J. (1972). *J. Pharmacol. Expl. Ther.* **182,** 454–563.
Javoy, F., Agid, Y., Bouvet, D., and Glowinski, J. (1974). *Brain Res.* **68,** 253–260.
Kebabian, J. W. and Saavedra, J. M. (1976). *Science* **193,** 683–685.
Kebabian, J. W., Petzold, G. L., and Greengard, P. (1972). *Proc. Natl. Acad. Sci.* **69,** 2145–2149.
Kehr, W., Carlsson, A., Lindqvist, M., Magnusson, T., and Atack, C. V. (1972). *J. Pharm. Pharmacol.* **24,** 744–747.
Kehr, W., Carlsson, A., and Lindqvist, M. (1977). *Naun. Schmied. Arch. Pharmacol.* **297,** 111–117.
Kim, J. S., Bak, I. J., Hassler, R., and Okada, Y. (1971). *Expl. Brain Res.* **14,** 95–104.
Kirpekar, S. M. and Puig, M. (1971). *Br. J. Pharmacol.* **43,** 359–369.
Klawans, H. L., Hitri, A., Nausieda, P. A., and Weiner, W. J. (1977). *In* "Animal Models in Psychiatry and Neurology" (Eds. I. Hanin and E. Usdin), 351–365. Pergamon, New York.

Klein, D. E. and Davis, J. M. (1969). "Diagnosis and Drug Treatment of Psychiatric Disorders". The Williams and Wilkins Company, Baltimore.

Korf, J., Zieleman, M., and Westerink, B. H. C. (1976). *Nature* (*London*) **260,** 257–258.

Krnjevic, K. and Phillis, J. W. (1963). *Br. J. Pharmacol.* **20,** 471–490.

Langer, F. Z., Adler, E., Enero, M. A., and Stefano, J. F. E. (1971). *Proc. Int. Union Physiol. Sci.* **9,** 335.

Langer, F. Z. (1973). *In* "Frontiers in Catecholamine Research" (Eds. S. H. Snyder and E. Usdin), 543. Pergamon, New York.

Lichtensteiger, W., Felix, D., Lienhart, R., and Heft, F. (1976). *Brain Res.* **117,** 85–103.

Lindvall, O. (1975). *Brain Res.* **87,** 89–95.

Lindvall, O. and Björklund, A. (1974). *Acta Physiol. Scand. Suppl. 412.*

Matthysee, S. (1973). *Fed. Proc.* **32,** 200–205.

McGeer, P. L., McGeer, E. G., Wada, J. A., and Jung, E. (1971). *Brain Res.* **32,** 425–431.

McGeer, P. L., Fibiger, H. C., Hattori, T., Singh, V. K., McGeer, E. G., and Maler, L. (1974). *In* "Advances in Behavioural Biology", Vol. 10 (Eds. R. D. Myers and R. R. Drucker-Colin), 27–47. Plenum Press, New York.

Meltzer, H. Y. and Stahl, S. M. (1976). *Schizophrenia Bull.* **2,** 19–76.

Miller, R. J., Horn, A. S., and Iversen, L. L. (1974). *Molec. Pharmacol.* **10,** 759–766.

Mora, F., Sweeney, K. F., Rolls, E. T., and Sanguinetti, A. M. (1976). *Brain Res.* **116,** 516–522.

Murrin, L. C., Morgenroth, V. H., III, and Roth, R. H. (1976). *Molec. Pharmacol.* **12,** 1070–1081.

Nieoullon, A., Cheramy, A., and Glowinski, J. (1977a). *Nature* (*London*) **266,** 375–377.

Nieoullon, A., Cheramy, A., and Glowinski, J. (1977b). *Nature* (*London*) **269,** 340–342.

Nowycky, M. C. (1976). "Chronic Drug Effects on Dopamine Neurons", Ph.D. Dissertation, Yale University.

Nowycky, M. C. and Roth, R. H. (1977). *Naun. Schmied. Arch. Pharmacol.* **300,** 247–254.

Nowycky, M. C. and Roth, R. H. (1978). *Prog. Neuro-Psychopharmacol.*, **2,** 139–158.

Nowycky, M. C., Walters, J. R., and Roth, R. H. (1978). *J. Neur. Trans.* **42,** 46–55.

Nyback, H. and Sedvall, G. (1968). *J. Pharmacol. Expl. Ther.* **162,** 294–301.

Paden, C., Wilson, C. J., and Groves, P. M. (1976). *Life Sci.* **19,** 1499–1506.

Pasternak, G. W., Simantov, R., and Snyder, S. H. (1975). *Molec. Pharmacol.* **12,** 504–513.

Peričić, D. and Walters, J. R. (1976). *J. Pharm. Pharmacol.* **28,** 527–529.

Persson, T. (1970). *Acta Pharmacol. Toxicol.* **28,** 378–390.

Pert, C. B. and Snyder, S. H. (1975). *Life Sci.* **16,** 1623–1634.

Phillipson, O. T. and Horn, A. S. (1976). *Nature* (*London*) **261,** 418–420.

Precht, W. and Yoshida, M. (1971). *Brain Res.* **32,** 229–233.

Rebec, G. V. and Groves, P. M. (1975). *Neuropharmacology* **14,** 275–282.

Rebec, G. V. and Groves, P. M. (1976). *Pharmacol. Biochem. Behav.* **5,** 349–357.

Reubi, J. C., Iversen, L. L., and Jessell, T. M. (1977). *Nature* (*London*) **268,** 652–654.

Roth, R. H., Walters, J. R., and Aghajanian, G. K. (1973). *In* "Frontiers in Catecholamine Research" (Eds. S. H. Snyder and E. Usdin), 567–574. Pergamon, New York.

Roth, R. H., Walters, J. R., and Morgenroth, V. H. III (1974). *In* "Neuropsychopharmacology of Monoamines and Their Regulatory Enzymes" (Ed. Usdin), 369. Raven Press, New York.

Roth, R. H., Morgenroth, V. H., III, and Murrin, L. C. (1975a). *In* "Antipsychotic Drugs, Pharmacodynamics and Pharmacokinetics", Wenner Gren International Symposium Series, 133–145. Pergamon, New York.

Roth, R. H., Walters, J. R., Murrin, L. C., and Morgenroth, V. H., III (1975b). *In* "Pre- and Postsynaptic Receptors" (Eds. E. Usdin and W. E. Bunney, Jr.), 5–48. Marcel Dekker, New York.

Saavedra, J. M. and Zivin, J. (1976). *Brain Res.* **105,** 517–524.

Scheel-Krüger, J. (1971). *Eur. J. Pharmacol.* **14,** 47–59.

Schwarcz, R. and Coyle, J. T. (1977). *Brain Res.* **127,** 235–249.

Segal, M. (1974). *Life Sci.* **14,** 1345–1351.

Siggins, G. R. (1978). *In* "Psychopharmacology: A Generation of Progress" (Eds. M. A. Lipton, A. Di Mascio and K. F. Killam), 143–157. Raven Press, New York.

Siggins, G. R., Hoffer, B. J., and Ungerstedt, U. (1974). *Life Sci.* **16,** 779–792.

Smith, R. Ç., Tamminga, C., and Davis, J. M. (1977a). *J. Neur. Trans.* **40,** 171–176.

Smith, R. C., Tamminga, C. A., Harasziti, J., Pandey, G. N., and Davis, J. M. (1977b). *Am. J. Psychiatr.* **134,** 763–768.

Snyder, S. H., Banerjee, S. P., Yamamura, H. I., and Greenberg, D. (1974). *Science* **184,** 1243–1253.

Snyder, S. H., Creese, I., and Burt, D. R. (1975). *Psychopharmacol. Commun.* **1,** 663–673.

Sotelo, C., Javoy, F., Agid, Y., and Glowinski, J. (1973). *Brain Res.* **58,** 269–290.

Spano, P. F., Trabucchi, M., and Di Chiara, G. (1977). *Science* **196,** 1343–1345.

Spector, S., Sjoerdsma, A., and Udenfriend, S. (1965). *J. Pharmacol. Expl. Ther.* **147,** 86–95.

Spencer, H. J. and Havlicek, V. (1974). *Can. J. Physiol. Pharmacol.* **52,** 808–813.

Starke, K. (1971). *Naturwissenschaften* **58,** 420.

Starke, K. (1972). *Naun. Schmied. Arch. Pharmacol.* **274,** 18–45.

Starke, K. (1973). *In* "Frontiers in Catecholamine Research" (Eds. E. Usdin and S. H. Snyder), 561–565. Pergamon, New York.

Stone, T. W. (1973). *Naun. Schmied. Arch. Pharmacol.* **278,** 333–346.

Stone, T. W. (1973). *Arch. Int. Pharm. Ther.* **205,** 29–39.

Stone, T. W. (1974). *Brain Res.* **72,** 177–180.

Stone, T. W. (1976). *Br. J. Pharmacol.* **56,** 101–110.

Stone, T. W. and Bailey, E. V. (1975). *Brain Res.* **85,** 126–129.

Straughan, D. W. and Legge, K. F. (1965). *J. Pharm. Pharmacol.* **17,** 675–677.

Strömbom, U. (1977). *J. Neur. Trans.* **40,** 191–194.

Sulser, F. and Robinson, S. E. (1978). *In* "Psychopharmacology: A Generation of Progress" (Eds. M. A. Lipton, A. DiMascio, and K. F. Killam), 943–954. Raven Press, New York.

Svensson, T. H., Bunney, B. S., and Aghajanian, G. K. (1975). *Brain Res.* **92,** 291–306.

Tassin, J. P., Thierry, A. M., Blanc, G., and Glowinski, J. (1974). *Naun. Schmied. Arch. Pharmacol.* **282,** 239–244.

Thoenen, H., Hürlimann, A., and Haefely, W. (1964). *Helv. Physiol. Pharmacol. Acta* **22,** 148.

Tolosa, E. S. and Sparber, S. B. (1975). *Life Sci.* **15,** 1371–1380.

Ungerstedt, U. (1968). *Eur. J. Pharmacol.* **5,** 107–110.

Ungerstedt, U. (1971). *Acta Physiol. Scand. Suppl.* **367,** 1–48.
Von Hugen, K., Roberts, S., and Hill, D. F. (1974). *Nature (London)* **252,** 588–589.
Von Hugen, K., Roberts, S., and Hill, D. F. (1975). *Brain Res.* **94,** 57–66.
Von Voigtlander, P. F. and Moore, K. E. (1973). *J. Pharmacol. Expl. Ther.* **184,** 542–552.
Walters, J. R. and Roth, R. H. (1972). *Biochem. Pharmacol.* **21,** 2111–2121.
Walters, J. R. and Roth, R. H. (1974). *J. Pharmacol. Expl. Ther.* **191,** 82–91.
Walters, J. R. and Roth, R. H. (1975). *In* "Pre- and Postsynaptic Actions of Antipsychotic Drugs", Wenner-Gren Center International Symposium Series. Pergamon, New York.
Walters, J. R. and Roth, R. H. (1976a). *Naun. Schmied. Arch. Pharmacol.* **296,** 5–14.
Walters, J. R. and Roth, R. H. (1976b). *Biochem. Pharmacol.* **25,** 649–654.
Walters, J. R., Roth, R. H., and Aghajanian, G. K. (1973). *J. Pharmacol. Expl. Ther.* **186,** 630–639.
Walters, J. R., Bunney, B. S., and Roth, R. H. (1975). *In* "Advances in Neurology" Vol. 9 (Eds. D. Calne, T. N. Chase, and A. Barbeau), 273–285. Raven Press, New York.
Weiss, B. F., Munroe, L. A., Ordonez, R. J., and Wurtman, R. J. (1972). *Science* **177,** 613–616.
Wilson, C. J., Groves, P. M., and Fifkova, E. (1977a). *Expl. Brain Res.* **30,** 161–174.
Wilson, C. J., Young, S. J., and Groves, P. M. (1977b). *Brain Res.* **136,** 243–260.
Woodruff, G. N., McCarthy, P. S., and Walker, R. J. (1976). *Brain Res.* **115,** 233–242.
Yarbrough, G. G. (1975). *Eur. J. Pharmacol.* **31,** 367–369.
Yarbrough, G. G., Lake, N., and Phillis, J. W. (1973). *Life Sci.* 703–711.
York, D. H. (1970). *Brain Res.* **20,** 233–249.
York, D. H. (1972). *Brain Res.* **37,** 91–99.

CHAPTER 26

Dopaminergic Synaptic Processes in the Superior Cervical Ganglion: Models for Synaptic Actions

Benjamin Libet

University of California, San Francisco, U.S.A.

At least two types of physiological synaptic actions appear to have been established for dopamine (DA) in the mammalian superior cervical ganglion (SCG), particularly of the rabbit. (1) Dopamine acts as a direct synaptic transmitter for the slow inhibitory postsynaptic potential, s-IPSP (Libet and Tosaka, 1970; Libet, 1970; Libet and Owman, 1974). (2) Dopamine produces a long-lasting and selective modulatory change in subsequent postsynaptic responses to another transmitter, namely, acetylcholine (ACh) acting muscarinically (Libet and Tosaka, 1970; Libet *et al.*, 1975). These postsynaptic responses to DA, as well as the slow muscarinic depolarization by ACh (slow excitatory postsynaptic potential, s-EPSP), are strikingly different from the classically known (fast) EPSPs and IPSPs, in their temporal characteristics and associated membrane changes (Libet, 1970). In addition, evidence has been developed for participation by cyclic nucleotides in the intracellular mediation of these slow responses (see review by Greengard, 1976; Libet *et al.*, 1975); and intraganglionic monoaminergic interneurones or their equivalent, which can be activated by the preganglionic transmitter ACh and in turn provide the presynaptic source of DA (or other monoamines), have been established as morphological entities (see below). These and other features endow the ganglionic slow synaptic actions with a potential significance as novel models of synaptic processes especially in the brain.

SLOW POSTSYNAPTIC POTENTIALS

In addition to the (fast) EPSP, virtually all the ganglion cells can also exhibit a slow IPSP and slow EPSP, at least in rabbit SCG—see Fig. 1 (Libet and Tosaka, 1969; Libet, 1970). Synaptic delay is about 25 ms for the s-IPSP and about 200–300 ms for the s-EPSP (Libet, 1967), instead of the less than 1 ms for the fast EPSP. Some repetition of orthodromic inputs is required in order to develop maximal slow PSP's. Durations of each slow PSP can be in the range of seconds to minutes. Classical (fast) PSP's show an increase in membrane conductance (see Eccles, 1964). By contrast, both the s-IPSP and s-EPSP are not accompanied by any change in membrane conductance, at least in cells with physiological resting potentials of −60 to −70 mV (Kobayashi and Libet, 1968; Kobayashi and Libet, 1970; kobayashi and Libet, 1974; Hashiguchi *et al.*, 1978). The electrogenic mechanisms for the slow PSP's are yet to be established; however, there is definitive evidence *against* proposals that involve activation of the Na–K electrogenic pump for the s-IPSP (see Libet, 1970; Libet *et al.*, 1977) or a decrease (inactivation) of ionic conductances (see Kobayashi and Libet, 1968; Kobayashi and Libet, 1974; Libet and Kobayashi, 1974). The selective sensitivity of the s-EPSP to loss or inhibition of oxidative metabolism (Kobayashi and Libet, 1968), a feature also of many brain processes, suggests a more immediate dependence upon intracellular processes or products that require sustained energy inputs greater than those needed for general maintenance of ionic imbalances (see below, on role of guanylate cyclase).

DOPAMINERGIC INTERNEURONES

The existence of a type of intraganglionic adrenergic interneurone in rabbit SCG was postulated by R. M. Eccles and Libet (1961) as part of the synaptic pathway leading to the s-IPSP response. It was later proved that a second transmitter must in fact be released, by a muscarinic action of ACh, in order to obtain the hyperpolarizing (s-IPSP) response to ACh (Libet and Tosaka, 1970; Libet, 1970). Subsequently, small monoaminergic neurones, "small-intensely-fluorescent (SIF)" cells, uniquely distinct from the principal neurones (ganglion cells), were histologically demonstrated in many autonomic ganglia (Eränkö and Härkönen, 1965; Norberg *et al.*, 1966; see

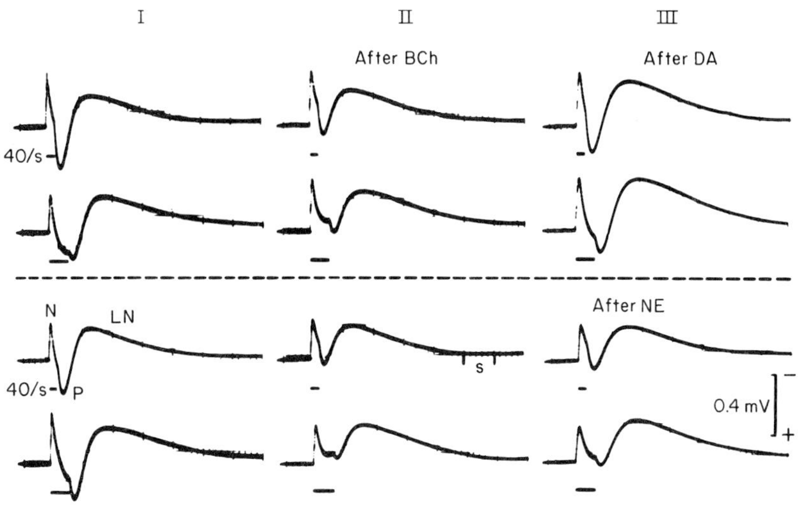

Fig. 1 Postsynaptic potentials of ganglion cells; effects of depletion and re-uptake of DA on the s-IPSP response. Extracellular recordings of postsynaptic potentials of the ganglion cell population in the curarized rabbit SCG at 37°C. Each tracing is a response to a train of supramaximal stimulus pulses at $40\,s^{-1}$, applied to the preganglionic nerve (during stimulus bar below each) for 0.25 s in the upper horizontal row and for 0.7 s in the lower row of each vertical pair. (I) Each response shows: (a) Initial, surface-negative or N component during the stimulus train. N consists of (fast) EPSP's that are summated during the repetitive train; EPSP's have been depressed by the D-tubocurarine to amplitudes well below the firing threshold. (b) A surface-positive (P) component, the s-IPSP, that starts during the train (see superimposed on the summated EPSP's) but continues on after the stimulus. (c) A later-negative (LN) component, the s-EPSP, that continues for some seconds here. P and LN can both be eliminated by atropine (Eccles and Libet, 1961). In an uncurarized ganglion, however, the surface-positive potential follows actual discharges during the stimulus period; P in such a case would contain a large post-tetanic after-hyperpolarization that is not atropine-sensitive and that overlaps in parallel with the s-IPSP component (see Libet, 1970). (II) Responses "after BCh" were obtained following a 30 min incubation with the muscarinic agonist bethanechol (0.6 mM) and a subsequent 60 min recovery period without BCh present. This treatment has been shown to deplete the SIF cells of about 50% of their DA content (Libet and Owman, 1974). Note that the s-IPSP components are now greatly reduced or absent. (III) Responses "after DA" (upper pair rows) or "after NE" (lower pair) were obtained following a 30 min exposure of each of the ganglia in part II to DA or NE respectively (each at $10\,\mu g\,ml^{-1}$, with an MAO inhibitor and ascorbic acid also present), and a subsequent washing-out of these additives for 20 min before testing in part III. Such a procedure has been shown, for the case of DA, to result in a re-uptake of DA by the depleted SIF cells (Libet and Owman, 1974). Note that the s-IPSP is now largely restored after exposure to DA, but that it is only slightly restored after exposure to NE. (From Libet and Tosaka, 1970; and Libet, 1970.)

Eränkö, 1976); additionally, adrenergic-type synaptic terminal structures of intraganglionic origin were described in rat SCG (Grillo, 1966, and others; see Dail, 1977 for rabbit SCG). It was then definitively established that SIF cells can qualify morphologically as interneurones; at least some of them were found to make efferent-type synaptic contacts with principal neurones in rat SCG, in addition to their receiving afferent-type contacts made by preganglionic cholinergic fibre terminals (Williams, 1967; Siegrist *et al.*, 1968; Matthews and Raisman, 1969; Williams and Palay, 1969; Matthews and Nash, 1970; Matthews, 1971). Small intensely fluorescent cells may be grouped according to the specific catecholamines they contain: Thus far DA has been identified as the monoamine in SIF cells of rat (Björklund *et al.*, 1970; Baker *et al.*, 1974) and rabbit SCG's (Libet and Owman, 1974) and, by indirect evidence, in SCG's of monkey (Chiba *et al.*, 1977), cat and cow (Black *et al.*, 1977); but norepinephrine (NE) is the monoamine in SIF cells of the guinea-pig's SCG and NE plus epinephrine (E) in those of its inferior mesenteric ganglion (Elfvin *et al.*, 1975; Dail, 1976). Small intensely fluorescent cells may also be grouped by their structural characteristics (Williams *et al.*, 1975): "Type I" cells are relatively isolated and few in number but give rise to long processes that ramify extensively among the principal ganglion cells (Libet and Owman, 1974); they are regarded as the interneurones (Williams, 1967; Williams *et al.*, 1975). "Type II" cells occur in larger clusters, near fenestrated blood vessels and lack conspicuous processes; they have been termed "paraneurones", since they do not make efferent synaptic contacts with other neurones (Chiba *et al.*, 1977).

It has been often assumed that the specialized efferent synaptic contact

---

Fig. 2 Schema of pathways for synaptic inputs and intracellular mediators, involved in responses to DA and ACh in the rabbit SCG. Acetylcholine liberated by terminals of preganglionic fibres onto (principal) ganglion cells acts on nicotinic receptors for (fast) EPSP, and on muscarinic ones for s-EPSP. Acetylcholine liberated at SIF cell acts on muscarinic receptor to induce release of DA by the SIF cell terminations on the ganglion cell. Dopamine may act on an α-type receptor to elicit the s-IPSP response, and on a "modulator" (M) receptor. The M receptor couples to adenylate cyclase; the resulting rise in cAMP in the ganglion cell induces, via a series of unknown reactions, an enduring molecular change in the neurone. This altered molecular state or product acts in such a manner as to enhance the s-EPSP response to the same ACh input (see dashed s-EPSP response "after DA"). The general possible sites for this action are indicated by dashed lines leading from the "modulator neuronal change"; they are currently under investigation by us. The process for developing the modulatory molecular change, but not the latter's final nature or effectiveness on the s-EPSP, can be antagonized or "disrupted" by cyclic GMP. (The proposal for a guanylate cyclase-cyclic GMP role in mediation of the s-EPSP response was made previously by the Greengard group, see 1976; it has gained further support from evidence obtained by Hashiguchi *et al.*, 1978, on the postsynaptic membrane changes involved in the s-EPSP and cyclic GMP depolarizations.) (This schema is modified from those in Libet, 1970, and Libet *et al.*, 1975.)

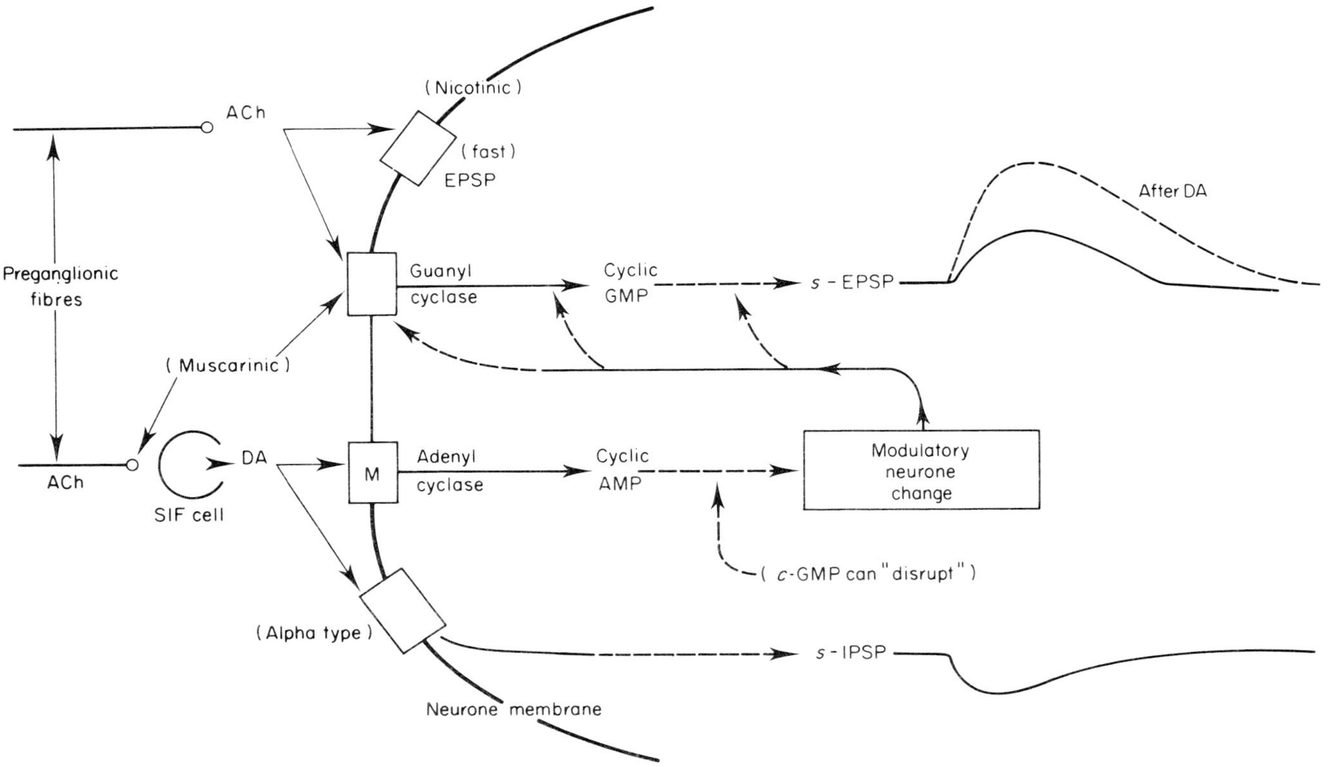

(Nicotinic)
ACh
(fast)
EPSP
After DA
Preganglionic
fibres
Guanyl
cyclase
Cyclic
GMP
s - EPSP
(Muscarinic)
ACh
SIF cell
DA
M
Adenyl
cyclase
Cyclic
AMP
Modulatory
neurone
change
(c-GMP can "disrupt")
(Alpha type)
s - IPSP
Neurone membrane

(20 nm cleft, etc.) that is characteristic of classical chemical synapses, is essential for producing a transmitter-mediated synaptic action; such efferent synapses from SIF cells have actually been demonstrated in SCG's of rat (see above), of rabbit (Dail, personal communication), and probably of monkey (Chiba *et al.*, 1977). However, the synaptic delays for the slow PSP's make it possible to argue that such close contacts may not necessarily be required for all modes of chemical transmission. The synaptic delay of 25 ms or more for the s-IPSP (Libet, 1967) could accommodate transmitter diffusion distances in "looser" synaptic arrangements (see Libet, 1965), e.g. with release from beaded varicosities of the fine fibre networks found to be closely apposed to and surrounding the ganglion cells; in rabbit SCG these fine networks have been shown to contain DA, not NE (Libet and Owman, 1974). The much greater delay for the onset of the DA modulatory action (on the s-EPSP response to ACh) has not yet been definitively quantitated, but it would appear to be no less than 30 s and could be as long as 2–4 min (Libet *et al.*, 1975). Such a delay could even accommodate delivery of a neurosecretory type from the axonless type II SIF cells (Libet, 1976; Williams *et al.*, 1976), especially if an intraganglionic portal circulation were to be involved (Williams *et al.*, 1976; Chiba *et al.*, 1977). Indeed, three kinds of actual sites of release of catecholamine by SIF cells, made apparent by either preganglionic or muscarinic stimulation, have recently been demonstrated histologically in the rat SCG by Grillo (1977); extruded material related to dense core vesicles appeared (1) at classical type chemical synapses, (2) at non-junctional membrane specializations called "synaptoid", and (3) at "sheathless portions of a SIF cell surface bordering on the interstitial space and, in some cases, a fenestrated blood vessel".

Intraganglionic pathways for the responses are shown in Fig. 2 (see also Libet, 1970; Libet, 1976; Libet, 1977a). Preganglionic fibres release ACh which acts directly on two types of postsynaptic receptors on ganglion cells, a muscarinic (atropine sensitive) receptor for the s-EPSP and a nicotinic receptor for the (fast) EPSP. They also release ACh at recognized synaptic endings on SIF cells, activating the latter muscarinically; the monoaminergic interneurone or SIF cell in turn releases DA which can act directly on two types of postsynaptic receptors, an alpha-adrenergic type for the s-IPSP* and a modulatory (M) receptor discussed below (Eccles and Libet, 1961; Libet, 1970; Libet *et al.*, 1975).

* Actually, evidence has now been developed for the existence in the rabbit SCG of a second and parallel s-IPSP component that is mediated by a nicotinic type of ACh-action (Libet, 1977b). This s-IPSP appears to be mediated by ACh activation of NE release at dendrodendritic junctions between ganglion cells, rather than by activation of DA release by SIF cells onto ganglion cells. This second, NE-type of s-IPSP will not be considered here further.

DOPAMINE AS THE TRANSMITTER FOR s-IPSP

Initial clues that the non-cholinergic direct transmitter for the s-IPSP is a catecholamine came from pharmacological evidence. The s-IPSP could be selectively blocked by α (but not by β) adrenergic antagonists (Eccles and Libet, 1961; Libet, 1970). Inhibitors of catechol-*O*-methyltransferase (COMT), such as tropolone or U-0521 (3′,4′-dihydroxy-2-methyl-propriophenone), increased the amplitude and especially the duration of the s-IPSP (Libet and Tosaka, 1971; Libet and Kobayashi, 1974). Incidentally, the latter finding suggests that COMT located in the vicinity of such synaptic regions can limit the physiological duration and even intensity of the monoamine transmitter action, in a manner analogous to that of acetylcholinesterase for ACh junctions.

To establish the identity of the specific physiological transmitter three criteria are essential, and these have been fulfilled by DA in at least the rabbit SCG. (1) Postsynaptic membrane changes must be identical with those found in the orthodromically produced s-IPSP. DA does hyperpolarize the ganglion cells (Libet and Tosaka, 1970; Libet, 1970); and, more critically, no detectable change in membrane conductance accompanies the hyperpolarization by DA (Dun and Nishi, 1974) or by NE (Libet and Kobayashi, 1969; Kobayashi and Libet, 1970). (2) The putative transmitter, and the enzymes for its synthesis, should be adequately available in appropriate presynaptic sites. This requirement is fulfilled by the DA-containing SIF cells and their extensive ramifications around the ganglion cells in at least the rabbit SCG (Libet and Owman, 1974) and in the rat SCG (Björklund *et al*., 1970; Baker *et al*., 1974) which also exhibits the s-IPSP response (Dunant and Dolivo, 1967). (3) Release of the transmitter from its presynaptic sites should be producible either by stimulation of the preganglionic nerve that normally elicits the s-IPSP response, or by the application of a muscarinic agonist (for example, bethanechol) that elicits a response equivalent to the s-IPSP. Either preganglionic stimulation or bethanechol were in fact found to be capable of producing a depletion of DA content of the SIF cells, accompanied by a corresponding loss in the ability to elicit an s-IPSP response to orthodromic input, as in Fig. 1 (Libet, 1970; Libet and Owman, 1974). In addition, suitable exposure of such a depleted ganglion to exogenous DA for re-uptake could restore both the DA content of SIF cells and the s-IPSP response (see Fig. 1). Dopamine content of SIF cells is thus causally and functionally related to production of s-IPSP.

Norepinephrine also meets criteria (1) and perhaps (2), the latter by virtue of its presence in the presynaptic side of a kind of dendro-dendritic junction between ganglion cells which has been described for inferior mesenteric ganglion of cat (Elfvin, 1971) and for rat SCG (Van Orden *et al.*, 1970). (Norepinephrine probably does mediate a parallel component of s-IPSP that involves a nicotinic-ACh action for its release, see footnote above, p. 458, and Libet, 1977.) However, NE can be excluded as the transmitter for that s-IPSP which includes a muscarinic ACh step in its mediation, on the basis of the following evidence: (a) after depletion of DA content of SIF cells, this s-IPSP can be substantially restored by exposure to (and re-uptake) of exogenous DA, but not at all or very little by exogenous NE or epinephrine—see Fig. 1 (Libet and Tosaka, 1970; Libet, 1970; Libet and Owman, 1974); (b) pretreatment with reserpine can produce a depletion of total NE in the rabbit SCG (Muscholl and Vogt, 1958); this depletion includes the catecholamine in the dendro-dendritic junctions but not that in the (DA-containing) SIF cells of the rat SCG (Van Orden *et al.*, 1970). Similar pretreatment of rabbits with reserpine has been found not to affect significantly the ability to elicit the s-IPSP response (Libet, 1964; Libet, *et al.*, in preparation). (c) Blockade of DβH (DA-β-hydroxylase), by diethyldithiocarbamate or fusaric acid or FLA-63, did not depress the s-IPSP (Libet and Tosaka, 1970; Libet, unpublished). In further accord with this, DβH cannot be detected in the SIF cells of rat SCG (Fuxe *et al.*, 1971). (d) Desmethylimipramine has been found to be relatively ineffective in blocking presynaptic re-uptake of DA, although it is of course highly effective against NE re-uptake (Carlsson *et al.*, 1966). In accord with this distinction, the drug had no effect on the s-IPSP at concentrations well above those effective for NE-mediated responses (Libet *et al.*, in preparation).

Other Synaptic Depressive Actions by DA

Other slow IPSP's

DA appears to be the probable transmitter for a type of s-IPSP response found in ganglion cells of the myenteric plexus in the guinea-pig ileum (Hirst and Silinsky, 1975). However, this s-IPSP differs from the one in the rabbit SCG, in that (i) it is accompanied by an increase in K-conductance and (ii) it is blocked by methysergide (Hirst and Silinsky, 1975), a drug that does not antagonize the s-IPSP of rabbit SCG (Libet, unpublished). It would appear that even for a postsynaptic inhibitory response there may be more than one kind of DA-receptor within the mammalian nervous system.

Presynaptic function

Exogenously applied DA, like NE and epinephrine, has been found capable of producing a potent depression of the release of ACh by the terminals of preganglionic fibres in rabbit SCG; this effect is also antagonized by *alpha*-adrenergic blockers (Dun and Nishi, 1974). The presynaptic action probably accounts for much of the observed depression of "ganglionic transmission" produced by any of the catecholamines when applied exogenously, a depression first described by Marrazzi (1939); it cannot, of course, account for the postsynaptic hyperpolarizing response to the catecholamines (Libet and Tosaka, 1970; Libet, 1970), a response first described by Lundberg (1952). However, one should distinguish actions which may be purely pharmacological from those that are part of a physiological process. In contrast to the case for the s-IPSP, there is as yet no evidence that presynaptic depression by catecholamines is actually producible physiologically by preganglionic nerve impulses. Also, there is no morphological evidence for adrenergic terminations onto cholinergic presynaptic terminals; i.e., there are no axo-axonal synapses in which presynaptic terminals contain dense core vesicles (Elfvin, Grillo; personal communications).

DOPAMINE MODULATORY ACTION ON THE s-EPSP-TYPE OF RESPONSE

The slow depolarizing responses to a muscarinic agonist like methacholine (MCh) can be employed as tests for the postsynaptic equivalent of the s-EPSP response (see Libet and Tosaka, 1970; Libet *et al.*, 1975). After a brief single exposure to a small dose of DA, subsequent MCh tests can be considerably enhanced (by 50–100%) for at least some hours; see Fig. 3. This DA-induced modulatory change in the response to MCh is not directly correlated with or tied to the processes involved in the hyperpolarizing response to DA; it can be induced by a dose of DA too small to produce any hyperpolarization, and it is sensitive to different blocking agents (see below). The action is selective for the slow muscarinic depolarizing response; there is no effect on the nicotinic depolarization by ACh. While the other catecholamines are as or more effective than DA in eliciting the direct hyperpolarizing response, both NE and E were relatively ineffective in producing any modulatory change in the MCh response.

This type of modulatory synaptic action by DA can also be demonstrated via physiological routes. (1) The s-EPSP response to the usual brief testing train of preganglionic impulses was found to be considerably and consistently enlarged after applying an inhibitor of the enzyme COMT (Libet and Tosaka, 1971). Presumably the significant action of the drug was to antagonize destruction of endogenous DA after its release from SIF cells. (2) The s-EPSP response was also enhanced following a conditioning period of preganglionic volleys (Kobayashi *et al.*, 1978); the s-EPSP in this study was recorded intracellularly, and so the enhanced state could be shown to occur with no significant shifts in either membrane potential or membrane resistance. In ganglia showing evidence of depletion of the DA content of their SIF cells, similar preganglionic conditioning did not induce modulatory enhancement of s-EPSP (except for the much briefer, non-specific post-tetanic potentiation of all PSPs that is a well-known function of presynaptic terminals). (3) The discharge of ganglion cells gives rise to action potentials recordable in the postganglionic nerve. A long-lasting,

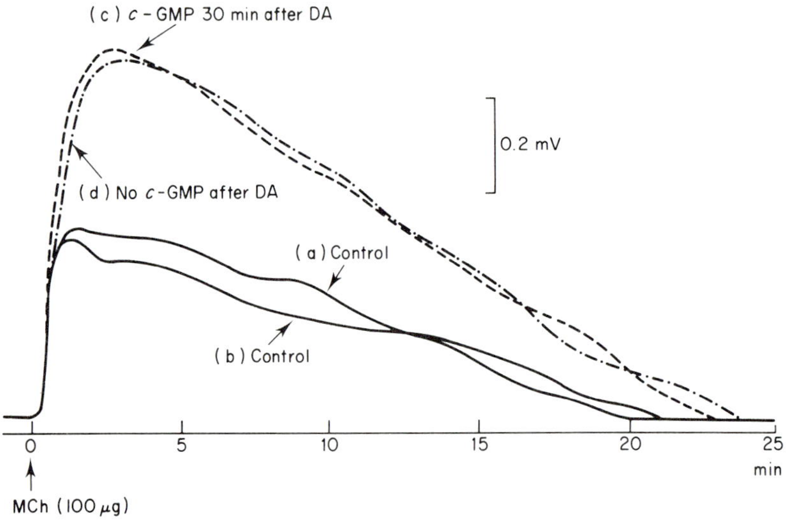

Fig. 3 Dopamine modulatory enhancement of slow muscarinic depolarizations. Recording of ganglionic potentials in a sucrose-gap perfusion chamber, rabbit SCG at 22°C. Each tracing is a depolarizing response to a single dose of acetyl-β-methylcholine, MCh (100 μg), injected within one second into the superfusate. Tracings (a) and (b) are control MCh responses before DA. Tracing (c) is MCh response tested 60 min after a brief exposure to DA. (A single dose of DA (50 μg) was injected during 1 s, after completing control response (b).) Note that the MCh response is now larger in amplitude and duration. (Later test responses to MCh in other experiments generally remain larger without decrement for at least some hours, until general deterioration of function of the isolated SCG develops (Libet and Tosaka, 1970).) In the present experiment, dibutyryl cyclic GMP (50 μM) was added to the ganglionic superfusate during the period between 30 and 38 min after DA. The usual enhancement of MCh response was seen in tracing (c), tested 60 min after DA or 22 min after end of the cGMP treatment. Acetyl-β-methylcholine test (d) was made 60 min after a second dose of DA; the latter DA was injected after test (c) but was not followed by a second cGMP treatment. The similarity of the (c) and (d) test responses indicates that the usually maximal enhancement following the first DA dose had not been affected by the cGMP treatment in this case; compare to Fig. 4. (From Libet *et al.*, 1975.)

asynchronous discharge of postganglionic nerve can appear *in vivo* either after an intra-arterial injection of a test dose of a muscarinic agonist (Volle, 1962), or as an atropine-sensitive after-discharge following a brief test train of stimuli to the preganglionic nerve (Brown, 1967; McIsaac, 1977). This muscarinically induced after-discharge undoubtedly reflects firing of cells by the s-EPSP (Libet, 1970) and may be regarded as an indirect but not linear indicator of the amplitude and duration of s-EPSP. It was found that repetitive conditioning stimulation of the preganglionic nerve for a period of 15 s or so was followed by a prolonged enhancement of the "s-EPSP-based" test discharge of postganglionic nerve for many hours, whether the test response was MCh induced in the cat SCG (Volle, 1962) or neurally induced by a brief preganglionic test train in rat SCG (McIsaac, 1977).

We have then a novel type of synaptic action, in which one transmitter (DA) can selectively induce a long-lasting change in the postsynaptic responses to another transmitter (ACh, acting muscarinically here). The molecular nature of the enduring neuronal change whose development is induced initially by a brief exposure to DA, and the cellular site at which it controls the amplitude and duration of s-EPSP responses, are yet to be discovered. However, some clues to the possible kinds of neuronal changes are available, since cAMP appears to mediate this DA action and cGMP can antagonize it in a time-dependent way (see below).

In any case, the DA-modulatory action and the time-dependent "disruptability" of its retention, show striking formal similarities to long-lasting functional changes in reactivities of the brain. The action has been proposed as at least the partial basis of a possible model for synaptic coupling into the production of a neuronal memory trace (Libet *et al.*, 1975).

POSTSYNAPTIC RECEPTORS FOR DA ACTIONS

Present evidence indicates that the two postsynaptic actions of DA, as direct transmitter for the hyperpolarizing s-IPSP response and as enduring modulator of the s-EPSP, are mediated independently by two different types of receptors. (1) Relative effectiveness of DA and NE is not very dissimilar for the hyperpolarization, but is strikingly different for the modulatory one (Libet and Tosaka, 1970). For the hyperpolarizing response, modified dose–response tests showed NE (and E) actually distinctly more effective than DA, but the maximum producible response was similar for all three monoamines (Kobayashi *et al.*, unpublished). On the other hand, the modulatory action is hardly detectable with NE or E (Libet and Tosaka, 1970). (2) The concentration range in which DA can induce a threshold to maximal degree of modulatory change is about five-fold or more below the analogous range for the DA-hyperpolarizing response (Libet and Tosaka, 1970). (3) Apomorphine was found to be relatively weakly effective for eliciting any hyperpolarization, but it was about as effective as DA for the modulatory action (on the depolarizing response to methacholine; Libet *et al.*, in preparation). Because of its greater differential effectiveness for the two actions, apomorphine could induce a maximal (not merely threshold) modulatory change with a dose that produced very little or no hyperpolarization. (4) Alpha-adrenergic antagonists are effective against the s-IPSP response (Eccles and Libet, 1961; Libet, 1970; Libet and Kobayashi, 1974), and even more especially so against the hyperpolarizing response to exogenously applied DA or NE (Libet and Tosaka, 1970). We had previously reported that α-blockers also antagonized the DA-modulatory action, but this conclusion was based on the use of phenoxybenzamine (Libet and Tosaka, 1970). We realized later that, in the concentrations employed, the somewhat variable atropinic action of this agent could have been sufficient to obscure any enhancement, by DA, of the slow muscarinic depolarization produced by a test dose of methacholine. We have since found that the relatively specific, competitive α-antagonist dihydroergotamine, at concentrations that completely eliminated the DA-hyperpolarizing response but had no detectable atropinic actions in the rabbit SCG, did *not* block the DA modulatory effect (Libet *et al.*, in preparation). (5) On the other hand, spiroperidol, shown to be a potent inhibitor of certain "specific" DA-receptors (Andén *et al.*, 1970; Iversen *et al.*, 1975) did appear to antagonize the DA-modulatory action in a concentration that did *not* block the DA-hyperpolarizing response (Libet *et al.*, in preparation). It has been suggested that the related drug haloperidol may be a selective antagonist of the hyperpolarizing response to DA, because haloperidol at 10^{-6}–10^{-7} M was found to block this response (as recorded intracellularly, with extracellular iontophoretic application of DA; Dun *et al.*, 1977). However, the significance of this haloperidol effect remains to be established, since (a) a depression of depolarizing responses to ACh was also observed with not greatly different haloperidol treatment, and (b) this modest selectivity of haloperidol was not compared to that possible with a competitive α-adrenergic antagonist (e.g. dihydroergotamine) under the same experimental conditions.

It would appear, then, that the DA-receptor mediating the hyperpolarizing response is more closely related to the classical α-receptor than it is to so-called specific DA receptors; it has somewhat greater affinity for the other catecholamines and is selectively blocked by α-antagonists. But the DA-receptor mediating the modulatory change (of the muscarinic-ACh response) may be more related to the so-called specific DA receptors studied in the brain—in its greater responsiveness to DA and to apomorphine, and in its selective blockade by spiroperidol.

POSSIBLE INTRACELLULAR MEDIATION OF DA ACTIONS BY CYCLIC AMP

After the appearance of our evidence indicating that DA is the specific synaptic transmitter for the s-IPSP in the rabbit SCG (Libet, 1970; Libet and Tosaka, 1970), the Greengard group investigated the possible role of cAMP in mediating this response to DA (see Greengard, 1976). They and others established that exogenous DA can stimulate adenylate cyclase activity in the SCGs of at least some animals (Kebabian and Greengard, 1971; Roch and Kalix, 1975; Black *et al.*, 1977, for cow; Kalix *et al.*, 1974, for rabbit; Shinnick-Gallagher *et al.*, 1976, for rat; Chiba *et al.*, 1977, for monkey), and that preganglionic stimulation or application of a muscarinic agonist could also raise cAMP levels in the rabbit SCG itself (Kalix *et al.*, 1974; Aleman *et al.*, 1974). (The relatively negative results with exogenous DA on adenyl cyclase in SCG of rat, reported by Cramer *et al.*, 1973 and by Lindl and Cramer, 1975, could be a function of either prolonged isolation of ganglia or insufficient phosphodiesterase inhibitor, etc.) These findings appear to fulfill one of the essential criteria for a role by cAMP in mediating a DA synaptic action, i.e., the stimulation of adenylate cyclase by suitable inputs.

Two other criteria that should be met are (i) duplication by cAMP of the postsynaptic response to DA, and (ii) suitability of the pharmacological actions on both the relevant synaptic actions and the associated processes involving production or actions of cAMP. In analysing these criteria one must remember that DA can exert *two* different postsynaptic actions, a direct hyperpolarizing one (s-IPSP) and an enduring modulatory change in the s-EPSP response. Therefore, the possible role of cAMP in mediating a response to DA must be examined with respect to both of these postsynaptic actions (Libet, 1979).

Does cAMP Elicit a Postsynaptic Hyperpolarization like DA?

It was reported by McAfee and Greengard (1972) that superfusion with monobutyryl cAMP (1–2.5 mM) elicited a hyperpolarizing response in eight of 11 rabbit SCG's, as studied with the sucrose gap recording chamber. However, in our own extensive tests the hyperpolarizing response was either not obtainable (see also Dun and Karczmar, 1977) or, using the same technique and sensitivity as McAfee and Greengard (1972), a similar hyperpolarization could appear in a variable manner that was non-specific and was probably an insidiously deceptive artefact (Libet 1979).

With the intracellular recording technique, cyclic AMP has thus far never elicited any hyperpolarizing response in ganglion cells, whether the cAMP was applied extracellularly (as butyryl-compounds) (Kobayashi *et al.*, 1977; Dun *et al.*, 1977; Kuba and Nishi, 1976, in frog ganglia), or intracellularly (Gallagher and Shinnick-Gallagher, 1977; Kobayashi *et al.*, 1977, 78); only a depolarizing, apparently non-specific response has been obtainable as the amount injected intracellularly was raised (Gallagher and Shinnick-Gallagher, 1977).

Does cAMP Produce Postsynaptic Modulation like DA?

The modulatory action of DA on the s-EPSP, or on the equivalent slow depolarizing response to metacholine, is fully mimicked by cAMP in either the butyryl- or non-butyryl form (Libet *et al.*, 1975; Libet, 1977b). The effects of DA and cAMP on test depolarizing responses to metacholine are similar in magnitude, duration and consistency of occurrence (Libet *et al.*, 1975), as well as in the time-dependent antagonism of them by cGMP (see below). A maximal modulatory effect is obtainable by superfusion for 5 min with as little as 0.1 mM cAMP. Cyclic AMP could induce the modulatory change even when injected intracellularly into the postsynaptic unit; following iontophoretic injection of a relatively small amount of cAMP into a ganglion cell, there was a prolonged and selective enhancement of the s-EPSP response, with no accompanying changes in membrane resistance or in membrane potential (Kobayashi *et al.*, 1977, 1978). Extracellularly applied cAMP can still induce the modulatory change in the presence of spiroperidol concentrations which eliminate the modulatory action by DA (Libet *et al.*, in preparation); such a result would of course be expected if cAMP mediates the modulatory action at an intracellular site that is at least one step beyond the DA-receptor. It is also noteworthy that apomorphine, which selectively mimics DA in its modulatory action in the ganglion, has been found to be similarly effective for stimulating adenylate cyclase activity in DA-containing brain structures (Kebabian *et al.*, 1972; its effectiveness for adenylate cyclase in ganglia is yet to be tested). Indirect evidence for a modulatory action by cAMP is also seen in the prolonged augmentation, of the post-tetanic facilitatory phase in rat SCG (test spike responses of postganglionic nerve), following an exposure to dibutyryl-cAMP (Hsu and McIsaac, 1978).

Pharmacological Evidence

Phosphodiesterase inhibitors should tend to enhance responses mediated by cAMP. In apparent support of the postulated mediation of the s-IPSP (Greengard, 1976), theophylline was found to enhance the s-IPSP as well as the hyperpolarizing response to applied DA (McAfee and Greengard, 1972). However, it now appears that this effect is due to an action other than that of inhibiting phosphodiesterase. The more potent and selective phosphodiesterase inhibitor RO-20-1724 [4-(3-butoxy-4-methoxybenzyl)-2-imidazolidinone] produced no significant change in s-IPSP (Libet, 1979) or in DA-hyperpolarization (Ushiyama, unpublished), even when this drug was applied under conditions of concentration (0.18 mM), temperature, etc., similar to those shown by Kalix *et al.* (1974) and by Roch and Kalix (1975) to be highly effective for accumulation of cAMP within rabbit SCG. Theophylline has in fact been found to have a more general action; it augments also another type of hyperpolarizing response, the positive after-potential of the action potential (Dun and Karczmar, 1977; in frog ganglia, Busis and Weight, 1976), while RO-20-1724 and even MIX (methyl-isobutyl-xanthine at 0.5 mM) have no such action (Libet, 1979). On the other hand, RO-20-1724 or MIX could, at least under certain conditions, induce some enhancement of the s-EPSP response (Libet and Tanaka, unpublished). Theophylline generally depresses the recorded s-EPSP (Dun and Karczmar, 1977), probably not directly but as a result of its non-specific augmentation of the hyperpolarizing s-IPSP that overlaps electrically with the depolarizing s-EPSP. However, when spike responses of postganglionic nerve to a preganglionic stimulus were tested after a conditioning volley or train, they were found to be potentiated in the presence of theophylline in lower concentrations (2 mM); since the extra postconditioning potentiation in theophylline was abolished by atropine, it may be inferred that it actually reflected an enhanced s-EPSP that was produced by each testing stimulus following the conditioning period (Hsu and McIsaac, 1978).

Prostaglandin E_1 was found to depress both the s-IPSP and the hyperpolarizing responses to applied DA (McAfee and Greengard, 1972). On the assumption that prostaglandin E_1 would antagonize DA-activation of adenylate cyclase, this finding appeared to support the proposed role in the s-IPSP. However, direct tests later showed the assumption to be incorrect (McAfee, 1976). The interesting depressive effect of prostaglandin E_1 on the s-IPSP must therefore be mediated at some other intracellular site; it cannot be regarded as support for the proposed role of cAMP in the s-IPSP.

Conclusions on Possible Roles of cAMP in the Ganglionic Actions by DA

Present evidence does provide strong support for cAMP mediation of the DA-modulatory action (the prolonged enhancement of the s-EPSP response to ACh), as proposed by Libet *et al.*, 1975. On the other hand, evidence to support the prior proposal by the Greengard group (see Greengard, 1976), that cAMP mediates the direct hyperpolarizing response to DA (the s-IPSP response), must presently be regarded as grossly inadequate. In addition, at least some evidence appears to be contradictory to a proposed role in mediating the s-IPSP, as follows. (1) Timing of physiological and chemical events: after a single preganglionic volley, synaptic delay for the onset of s-IPSP is approximately 25–35 ms (Libet, 1967) and peak amplitude is achieved by about 150 ms at 37°C (Libet and Tosaka, 1969) or by about 500–600 ms at 22°C. But the time required for even a detectable rise of ganglionic cAMP appears to be in the range of minutes, following either preganglionic stimulation (Aleman *et al.*, 1974; Kalix *et al.*, 1974) or incubation with DA (Roch and Kalix, 1975), with the peak increase not achieved for 5 to 10 minutes. The further putative sequence of enzymatic reactions in a cAMP mediated response (Greengard, 1976) would have to consume additional time for completion; the time constants for one or more of these reactions are apparently in the range of at least hundreds of msec. With the DA-modulatory action, however, such delays in developing the adequate series of biochemical changes are readily accommodated. (2) NE vs DA actions: NE can elicit a hyperpolarization having membrane characteristics similar to those in the DA-hyperpolarization (Libet and Kobayashi, 1969; Dun and Nishi, 1974), and in the s-IPSP itself (Kobayashi and Libet, 1970). As noted above, both NE and E (epinephrine) are more effective than DA in the hyperpolarizing action but NE and E are very much less effective for the modulatory action. Alpha-adrenergic antagonists are selectively effective against the hyperpolarizing responses to NE, E or DA, or against the s-IPSP itself (see Libet, 1970). Yet, in those ganglia where the actions of DA, NE and E on adenylate cyclase have been studied, those of NE and E (but not of DA) are blocked predominantly if not exclusively by β-adrenergic antagonists (Kebabian and Greengard, 1971; Cramer *et al.*, 1973; Roch and Kalix, 1975). The claim that alpha-adrenergic antagonists depress DA-activation of adenylate cyclase might appear to offer some partial support for a cAMP role in the s-IPSP response, if one assumes that a different mechanism mediates the otherwise similar NE- or E-induced hyperpolarizations. However, that claim was based on the use of phenoxybenzamine and phentolamine (Kebabian and Greengard, 1971) in concentrations that might also have produced a relatively non-specific depression of the DA-receptor involved in the modulatory response. It would be desirable to test an α-adrenergic antagonist that can more selectively depress DA-hyperpolarization (as opposed to the DA-modulatory action), as dihydroergotamine appears to do (see above).

CYCLIC GMP AS ANTAGONIST OF DA/cAMP MODULATORY ACTION

In the course of our investigating the proposal that cGMP mediates the s-EPSP (see Greengard, 1976), an additional capability for antagonizing the modulatory effect of DA or cAMP on the s-EPSP was discovered (Libet *et al*., 1975). Treatment of the ganglion with cyclic GMP (50 μM for 6 min) either during or soon after the exposure to DA or cyclic AMP could prevent the later appearance of the enhancement of the MCh-depolarization that either of these agents can produce (Fig. 4). However, this cGMP effect on the modulatory change shows a striking time-dependency: it can be 100% effective even if onset of cGMP treatment is delayed for up to 4–5 min after the exposure to a single dose of DA, as in Fig. 4; but cGMP is completely ineffective if delayed for more than 10–15 min after DA, as in Fig. 3 (Libet *et al*., 1975). Whether cGMP antagonism applies only to

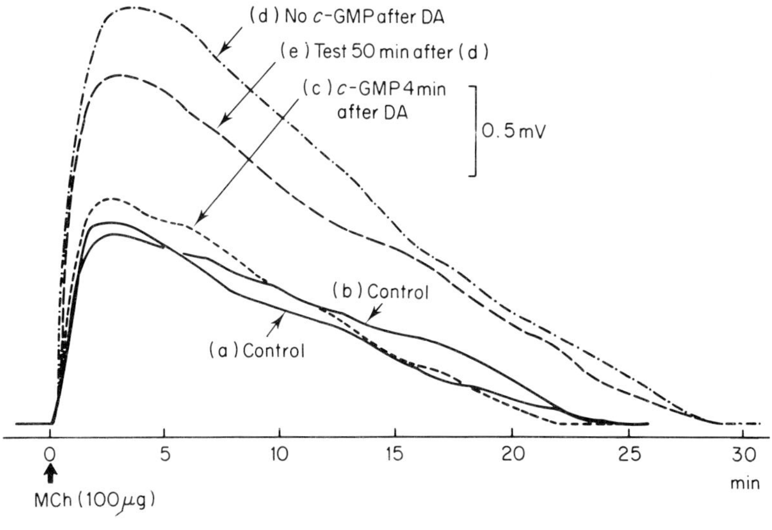

Fig. 4 Cyclic GMP prevents appearance of the DA-modulatory enhancement effect on MCh responses, when cGMP treatment is begun up to 4 min after DA. (a) and (b) Control MCh responses. (c) MCh response at 35 min after the single DA dose (50 μg), with an interposed addition of cyclic GMP (50 μM for 8 min) which began at 4 min after DA. (d) MCh response 35 min after a second dose of DA, with no interposed treatment with cyclic GMP; this MCh response does show enhancement. (e) Additional MCh response at 50 min after that in (d), with no interposed treatments at all. (The bulk of the enhancement effect is still evident in (e); but the smaller response in (e) relative to (d) is an example of some late ganglionic deterioration, not of a fall-off in the effect.) (From Libet *et al*., 1975.)

retention of an already developing modulatory change induced by DA or cAMP, or whether it in fact prevents the *development* of any modulatory change in the first place, is still an unsettled question; the minimal latency for onset of modulatory enhancement of s-EPSP, following application of DA, has not yet been definitively established. It is clear, however, that (a) the enduring form of the modulatory neuronal change, and its capacity to alter the s-EPSP response, are not affected by cyclic GMP; but that (b) at least those intracellular reactions which impart the "non-disruptible" enduring quality to the modulatory change take place chiefly during a time interval of 5 to 15 min after exposure to DA (at 22°C).

SIGNIFICANCE OF THE GANGLIONIC SYNAPTIC ACTIONS OF DA FOR BRAIN FUNCTIONS

The existence of two quite different postsynaptic actions by DA in the rabbit SCG should serve as an example for similar multiple potentialities elsewhere by DA, and perhaps by other monoamines as well (see also Cools, 1977). Additional potentialities exist for actions on presynaptic terminals, although in sympathetic ganglia only a pharmacological depression of ACh release has been demonstrated (see above).

It should be recognized that it is the postsynaptic DA-modulatory receptor, rather than the DA-hyperpolarizing one, that bears the greater resemblance to those relatively specific DA receptors in the brain (a) whose physiological action is mimicked by apomorphine (see Colpaert *et al.*, 1976), (b) which are selectively blocked by spiroperidol and related agents rather than by α-adrenergic antagonists, and (c) which activate adenylate cyclase when either DA or apomorphine acts on them (Kebabian *et al.*, 1972; Iversen *et al.*, 1975). Past experimental investigations of the synaptic mode of DA actions in a cerebral structure have in general not been capable of detecting such a modulatory action by DA (or by other monoamines). For this experimental capability it would be essential that the responses to *another* input (a muscarinic-ACh one in the present ganglionic example) be compared before and for some time after the monoaminergic input. In addition, any previous enduring modulatory action may occlude the change which might be produced by the experimental input. Consequently, it may also be necessary to eliminate or reduce endogenous ongoing release of the monoamine for a period sufficiently lengthy to permit any existing modulatory change to "wear off" down to some submaximal level.

Some investigations of brain function in which at least partially appropriate experimental designs were employed have, in fact, begun to demonstrate that analogous enduring modulatory actions by one transmitter on the responses to other transmitter inputs do exist in the central nervous system (e.g. Yamamoto, 1973; Freedman *et al.*, 1977; Moises and Woodward, 1977; York and Lentz, 1976; Reader *et al.*, 1977). Whether the 30 min increase in locomotor activity, that follows injection of a small amount of DA into nucleus accumbens (Cools, 1977), is also a modulatory rather than a direct excitatory action, remains to be seen.

Probably all enduring brain processes which appear to involve roles for various monoamines (e.g. learning and memory formation, mood changes, sleep states, hypothalamic regulation of neuroendocrine systems, etc.) should be candidates for analogous experimental analysis, of the kind that could detect modulatory mechanisms similar in principle to that found with DA in the rabbit SCG.

Morphological similarities also exist between the monoaminergic systems in the brain and in ganglia. The patterns of innervation of cerebral regions by the monoaminergic nuclei in the brain stem bear some striking resemblances to those by the Type I SIF cells in ganglia. Relatively small numbers of monoaminergic cells "innervate" immense numbers of cerebral or ganglionic cells respectively by means of extensively ramifying processes. Also, the monoaminergic terminations in both the central and peripheral structures often appear as fine, beaded fibres, meandering extensively among and close to the "innervated" neurones. In both cases, "loose" synaptic contacts rather than classical 20 nm separations appear to be common (see above). The slow synaptic actions in ganglia offer physiological models of processes that can accommodate the delays inherent in such morphological arrangements.

ACKNOWLEDGEMENT

This research was supported in part by US Public Health Service Research Grant NS-00884 from the National Institute of Neurological and Communicative Disorders and Stroke.

REFERENCES

Aleman, V., Bayon, A. and Molina, J. (1974). *Adv. Behav. Biol.* **10,** 115–124.

Andén, N.-E., Butcher, S. G., Corrodi, H., Fuxe, K. and Ungerstedt, U. (1970). *Eur. J. Pharmacol.* **11,** 303–314.

Baker, N. A., Redick, J. A., Schnute, W. J. and Van Orden, III, L. L. (1974). (*Soc.*) *Neurosci. Abstr.*, 4th Annual Meeting, 126.

Björklund, A., Cegrell, L., Falck, B., Ritzén, M. and Rosengren, E. (1970). *Acta Physiol. Scand.* **78,** 334–338.

Black, Jr., A. C., Chiba, T., Wamsley, J. K., Bhalla, R. C. and Williams, T. H. (1978). *Brain Res.* **148,** 389–398.
Brown, A. M. (1967). *J. Physiol.* (*London*) **191,** 271–288.
Busis, N. A. and Weight, F. F. (1976). *Nature* (*London*) **263,** 434–436.
Carlsson, A., Fuxe, K., Hamberger, B. and Lindqvist, M. (1966). *Acta Physiol. Scand.* **67,** 481–497.
Chan-Palay, V. (1975). *Anat. Embryol.* **148,** 235–265.
Chiba, T., Black, Jr., A. C. and Williams, T. H. (1977). *J. Neurocytol.* **6,** 441–453.
Colpaert, F. C., Van Bever, W. F. M. and Leysen, J. E. M. F. (1976). *In* "International Review of Neurobiology", Vol. 19 (Eds. C. C. Pfeiffer and J. R. Smythies), 225–268. Academic Press, New York and London.
Cools, A. R. (1977). *In* "Nonstriatal Dopaminergic Neurons" *Advances in Biochemical Psychopharmacology*, Vol. 16 (Eds. E. Costa and G. L. Gessa), 215–225. Raven Press, New York.
Cramer, H., Johnson, D. G., Hanbauer, I., Silberstein, S. D. and Kopin, I. J. (1973). *Brain Res.* **53,** 97–104.
Dail, Jr., W. G. (1976). *In* "SIF Cells: Structure and Function of the Small, Intensely Fluorescent Sympathetic Cells" (Ed. O. Eränkö), 8–18. U.S. Government Printing Office, Washington, D.C.
Dail, Jr., W. G. and Wood, J. (1977). (*Soc.*) *Neurosci. Abstr.* **III,** 248.
Dun, N. J. and Karczmar, A. G. (1977). *J. Pharmacol. Expl. Ther.* **202,** 89–96.
Dun, N. and Nishi, S. (1974). *J. Physiol.* (*London*) **239,** 155–164.
Dun, N. J., Kaibara, K., and Karczmar, A. G. (1977). *Science* **197,** 778–780.
Dunant, Y., and Dolivo, M. (1967). *J. Physiol.* (*Paris*) **59,** 281–294.
Eccles, J. C. (1964). "The Physiology of Synapses". Springer-Verlag, Berlin.
Eccles, R. M. and Libet, B. (1961). *J. Physiol.* (*London*) **157,** 484–503.
Elfvin, L.-G. (1971). *J. Ultrastruct. Res.* **37,** 411–425.
Elfvin, L.-G., Hökfelt, T. and Goldstein, M. (1975). *J. Ultrastruct. Res.* **51,** 377–396.
Eränkö, O. (1976). "SIF Cells: Structure and Function of the Small, Intensely Fluorescent Sympathetic Cells". U.S. Government Printing Office, Washington, D.C.
Eränkö, O. and Härkönen, M. (1965). *Acta Physiol. Scand.* **63,** 511–512.
Freedman, R., Hoffer, B. J., Woodward, D. J. and Puro, D. (1977). *Expl. Neurol.* **55,** 269–288.
Fuxe, K., Goldstein, M., Hökfelt, T. and Joh, T. H. (1971). *Progr. Brain Res.* **34,** 127–138.
Gallagher, J. P. and Shinnick-Gallagher, P. (1977). *Science* **198,** 851–853.
Greengard, P. (1976). *Nature* (*London*) **260,** 101–108.
Grillo, M. A. (1966). *Pharmacol. Rev.* **18,** 387–399.
Grillo, M. A. (1977). *J. Cell Biol.* **75,** 118a.
Hashiguchi, T., Ushiyama, N., Kobayashi, H. and Libet, B. (1978). *Nature* (*London*) **271,** 267–268.
Hirst, G. D. S. and Silinsky, E. M. (1975). *J. Physiol.* (*London*) **251,** 817–832.
Hsu, S. Y. and McIsaac, R. J. (1978). *J. Pharmacol. Expl. Ther.* **205,** 91–103.
Iversen, L. L., Horn, A. S. and Miller, R. J. (1975). *In* "Pre- and Postsynaptic Receptors", *Modern Pharmacological-Toxicology*, Vol. 3 (Eds. E. Usdin and W. E. Bunney, Jr.), 207–243. Marcel Dekker, New York.
Kalix, P., McAfee, D. A., Schorderet, M. and Greengard, P. (1974). *J. Pharmacol. Expl. Ther.* **188,** 676–686.
Kebabian, J. W. and Greengard, P. (1971). *Science* **174,** 1346–1349.

Kebabian, J. W., Petzold, G. L. and Greengard, P. (1972). *Proc. Natl. Acad. Sci., U.S.A.* **69,** 2145–2149.
Kobayashi, H. and Libet, B. (1968). *Proc. Natl. Acad. Sci., U.S.A.* **60,** 1304–1311.
Kobayashi, H. and Libet, B. (1970). *J. Physiol. (London)* **208,** 353–372.
Kobayashi, H. and Libet, B. (1974). *Life Sci.* **14,** 1871–1883.
Kobayashi, H., Ushiyama, N. S. and Hashiguchi, T. (1977). *Proc. Int. Union Physiol. Sci.* **13,** 394.
Kobayashi, H., Hashiguchi, T. and Ushiyama, N. S. (1977/78). *Nature (London)* (in press).
Kuba, K. and Nishi, S. (1976). *J. Neurophysiol.* **39,** 547–563.
Libet, B. (1964). *J. Physiol. (London)* **174,** 1–25.
Libet, B. (1965). *In* "Studies in Physiology" (Eds. D. R. Curtis and A. K. McIntyre), 160–165. Springer-Verlag, Berlin.
Libet, B. (1967). *J. Neurophysiol.* **30,** 494–514.
Libet, B. (1970). *Fed. Proc.* **29,** 1945–1956.
Libet, B. (1976). *In* "SIF Cells: Structure and Function of the Small, Intensely Fluorescent Sympathetic Cells" (Ed. O. Eränkö), 163–177. U.S. Government Printing Office, Washington, D.C.
Libet, B. (1977a). *In* "Nonstriatal Dopaminergic Neurons" *Advances in Biochemical Psychopharmacology*, Vol. 16 (Eds. E. Costa and G. L. Gessa), 541–546. Raven Press, New York.
Libet, B. (1977b). *Proc. Int. Union Physiol. Sci.* **12,** 558.
Libet, B. (1979). *Life Sci.* **24,** 1043–1058.
Libet, B. and Kobayashi, H. (1969). *Science* **164,** 1530–1532.
Libet, B. and Kobayashi, H. (1974). *J. Neurophysiol.* **37,** 805–814.
Libet, B. and Owman, Ch. (1974). *J. Physiol. (London)* **237,** 635–662.
Libet, B. and Tosaka, T. (1969). *J. Neurophysiol.* **32,** 43–50.
Libet, B. and Tosaka, T. (1970). *Proc. Natl. Acad. Sci., U.S.A.* **67,** 667–673.
Libet, B. and Tosaka, T. (1971). *Fed. Proc.* **30,** 323.
Libet, B., Kobayashi, H. and Tanaka, T. (1975). *Nature (London)* **258,** 155–157.
Libet, B., Tanaka, T. and Tosaka, T. (1977). *Life Sci.* **20,** 1863–1870.
Lindl, T. and Cramer, H. (1975). *Biochem. Biophys. Res. Commun.* **65,** 731–739.
Lundberg, A. (1952). *Acta Physiol. Scand.* **26,** 252–263.
McAfee, D. A. (1976). *In* "SIF Cells: Structure and Function of the Small, Intensely Fluorescent Sympathetic Cells" (Ed. O. Eränkö), 132–142. U.S. Government Printing Office, Washington, D.C.
McAfee, D. A. and Greengard, P. (1972). *Science* **178,** 310–312.
McIsaac, R. J. (1977). *J. Pharmacol. Expl. Ther.* **200,** 107–116.
Marrazzi, A. S. (1939). *Am. J. Physiol.* **127,** 738–744.
Matthews, M. R. (1971). *J. Physiol. (London)* **218,** 95–96.
Matthews, M. R. and Nash, J. R. G. (1970). *J. Physiol. (London)* **210,** 11–14.
Matthews, M. R. and Raisman, G. (1969). *J. Anat.* **105,** 255–282.
Moises, H. C. and Woodward, D. J. (1977). *(Soc.) Neurosci. Abstr.* **III,** 255.
Muscholl, E. and Vogt, M. (1958). *J. Physiol. (London)* **141,** 132–155.
Norberg, K.-A., Ritzen, M. and Ungerstedt, U. (1966). *Acta Physiol. Scand.* **67,** 260–270.
Reader, T. A., Ferron, A. and Descarries, L. (1977). *(Soc.) Neurosci. Abstr.* **III,** 257.
Roch, Ph. and Kalix, P. (1975). *Neuropharmacology* **14,** 21–29.
Shinnick-Gallagher, P., Williams, B. J. and Gallagher, J. P. (1976). *(Soc.) Neurosci. Abstr.* **II**(2), 800.

Siegrist, B., Dolivo, M., Dunant, Y., Foroglou-Kerameus, C., DeRibaupierre, Fr. and Rouiller, C. (1968). *J. Ultrastruct. Res.* **25,** 381–407.

Van Orden, L. S. III, Burke, J. P., Geyer, M. and Lodoen, F. V. (1970). *J. Pharmacol. Expl. Ther.* **174,** 56–71.

Volle, R. L. (1962). *J. Pharmacol. Expl. Ther.* **136,** 68–74.

Williams, T. H. (1967). *Nature* (*London*) **214,** 309–310.

Williams, T. H. W. and Palay, S. L. (1969). *Brain Res.* **15,** 17–34.

Williams, T. H., Black, Jr., A. C., Chiba, T. and Bhalla, R. C. (1975). *Nature* (*London*) **256,** 315–317.

Williams, T. H., Chiba, T., Black, Jr., A. C., Bhalla, R. C. and Jew, J. (1976). *In* "SIF Cells: Structure and Function of the Small, Intensely Fluorescent Sympathetic Cells" (Ed. O. Eränkö), 143–162, U.S. Government Printing Office, Washington, D.C.

Williams, T. H., Black, Jr., A. C., Chiba, T. and Jew, J. Y. (1977). *In* "Nonstriatal Dopaminergic Neurons" (Eds. E. Costa and G. L. Gessa), *Advances in Biochemical Psychopharmacology*, Vol. 16, 505–511, Raven Press, New York.

Yamamoto, C. (1973). *Progr. Brain Res.* (*Jpn.*) **17,** 64–70.

York, D. H. and Lentz, S. (1976). (*Soc.*) *Neurosci. Abstr.* **II,** (1), 70.

CHAPTER 27

The Neuronal Dopaminergic System of the Retina

J. S. Wassenaar

Department of Neurophysiology, University of Groningen, Groningen, The Netherlands

INTRODUCTION

Embryologically the retina is an ectopic portion of the primitive forebrain and thus it is understandable that the neuronal network of the retina in its fully developed stages is very similar in its morphology, physiology and pharmacology to other brain regions. However, the retina as a whole subserves two inter-related functions: one is the sensory function of the photoreceptor layer, and the other is the data-processing function of the neuronal network. This double function makes the retina different from other brain areas. In the photoreceptor layer of the retina, *rods* and *cones* (Fig. 1) can be distinguished with light and electron microscopy. These photoreceptors use photopigments, such as rhodopsine, to transduce light signals, within the range of the visible wavelengths and intensities, into a hyperpolarization of the membrane potential of the photoreceptors themselves.

NEURONAL ELEMENTS

In the neuronal network of the retina four different types of neurones, which can be divided into subclasses and which are interconnected by their processes, are laid down in rows in a recognizable way (Fig. 1). Thus

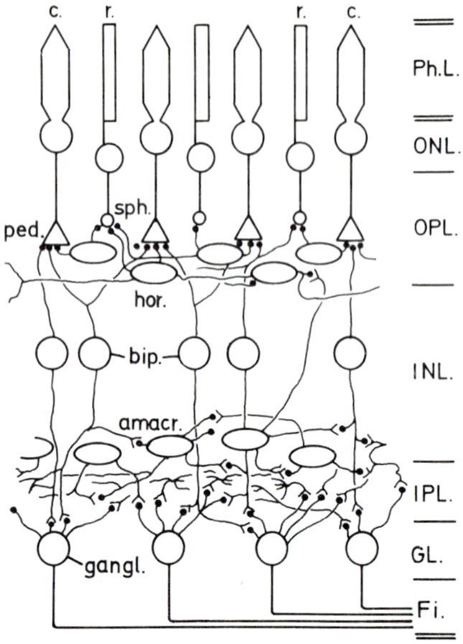

Fig. 1 Schematic drawing of layers and cells in the vertebrate retina. Abbreviations: Ph.L.: photoreceptor layer; ONL.: Outer nuclear layer; OPL.: outer plexiform layer; INL.: inner nuclear layer; IPL.: inner plexiform layer; GL.: ganglion cell layer; Fi.: fibres of ganglion cells to optic nerve; c.: cone; r.: rod; ped.: cone pedicle; sph.: rod spherule; hor.: horizontal cell; bip.: bipolar cell; amacr.: amacrine cell; gangl.: ganglion cell. This figure is a schematic representation of the retina derived from numerous authors.

processes of the three subtypes of horizontal cells, situated in the outer plexiform layer, and the dendritic processes from bipolar cells in this layer, receive their input signals via connections from the cones and rods. The slow graded receptor potentials (cones and rods) induce hyperpolarizing and sometimes depolarizing slow sustained graded potentials in the horizontal and bipolar cells. From morphological and electrophysiological data it is known that horizontal cells and bipolar cells seem to integrate and coordinate the input signals from the photoreceptors. Moreover, bipolar cells appear to conduct the more or less integrated activity state of the rods and cones either directly towards the amacrine cells, situated in the inner nuclear layer, or towards the ganglion cells (Dowling and Boycott, 1966). Among the amacrine cells dopaminergic cells have been demonstrated. In the retinal network many glia cells (Müller cells) are intermingled between the neuronal elements. The ganglion cells send the continuously changing information from the rods and cones, filtered and processed by the retinal network, via axonal fibres in the optic nerve to the lateral geniculate body. This is the second relay station at which incoming action potentials are again processed; connections are then made with the visual cortex and motor nuclei subserving the eye muscles.

In this section it will be mentioned that several types of synapses are found in the retinal network i.e. ribbon type, flat type, conventional type, and serial types of synapses.

Almost all the synapses have synaptic vesicles in their presynaptic structures. The *ribbon type* synapse has a ribbon-like structure in the presynaptic element. This ribbon is perpendicular to the presynaptic membrane. This type of synapse usually has more than one postsynaptic structure. The *flat type* synapse has flat but specialized pre- and postsynaptic structures. No vesicles seem to be present. The *serial synapse* has several pre- and postsynaptic structures, concentrated in a cluster, one presynaptic structure being postsynaptic for another cell. As far as the dopaminergic amacrine cells are concerned, conventional type synaptic contacts have usually been observed. It is beyond the scope of this chapter, however, to describe the several synapses in detail; for a more detailed account see Polyack (1957), Dowling and Boycott (1966) and Werblin (1974).

DOPAMINE AND THE RETINA

Though identification and localization of DA and the characterization of its mode of action in the retinal network is hampered, of course, by technical limitations, the use of histofluorescence microscopy, electron microscopy, autoradiography, electrophysiology, neurochemistry and neuropharmacology has revealed much information on dopaminergic pathways and on the transmitter role of DA in the retina. In the next subdivisions of this Chapter, results obtained by the different approaches will be briefly reviewed. The neurochemical and neuropharmacological aspects will be discussed more extensively in subsequent sections.

Localization

The retinal network contains cell bodies, especially in the inner nuclear layer, 5–10% of which are dopaminergic. Dopaminergic fibres and synaptic terminals, undoubtedly derived from these cell bodies, are located predominantly in the inner plexiform layer, in the inner nuclear layer and sometimes in the outer plexiform layer (Figs. 1, 2 and 3). One of the criteria for a substance being a neurotransmitter is that it should be localized in the cells of the structures under consideration. The use of histofluorescence microscopy for the detection of catecholamines was first reported in 1962

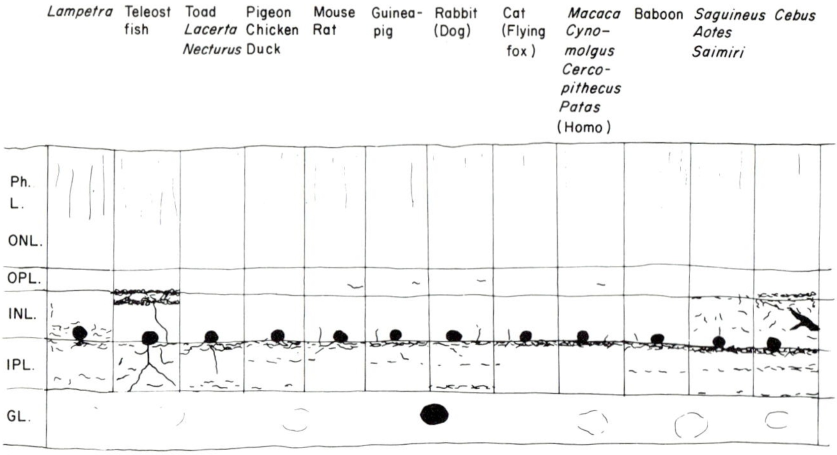

Fig. 2 Catecholamine (DA) containing neurones and fibres (black denoted) in the retina of different species (with kind permission from Ehinger, 1977).

(Falck *et al.*, 1962). A few years later the first studies on the retina appeared (Häggendal and Malmfors, 1965; Nichols *et al.*, 1967). Knowledge on this subject was rapidly advanced by the application of this technique to the retina of many species by Ehinger (1966a, b); Ehinger and Falck (1969a, b); Ehinger *et al.* (1969); Ehinger (1976, 1977). Studies since then have now clearly established that dopamine-containing cell bodies are present in the innermost cell row of the inner nuclear layer in the retinae of fish, amphibians, reptiles, subprimate mammals and primates, including humans (Fig. 2) (Ehinger and Falck, 1969a; Ehinger *et al.*, 1969; Ehinger, 1976, 1977). Other catecholamines such as noradrenaline (NA) and adrenaline (A), seem to play a minor role or are even absent in the retina (Drujan and Diaz Borges, 1968; Drujan *et al.*, 1965; Ehinger and Falck, 1969a).

Extensive studies have revealed five types of DA-cells, existing in the retinae of several species (Fig. 2). Most species contain more than one of these types of DA-cells.

Dopaminergic junctional cells and fibres belonging to them, probably account for 5–10% of amacrine cells. They are located in the inner nuclear layer (Häggendal and Malmfors, 1965; Nichols *et al.*, 1967; Ehinger and Falck, 1969a; Ehinger *et al.*, 1969).

Dopaminergic alloganglion cells, with their fibres sparsely spread among the ganglion cells are found in retinae of many species (Ehinger, 1966a). Their number is limited.

Dopaminergic eremite cells, have a special shape and are located in the midrow of the inner plexiform layer (Ehinger, 1966a).

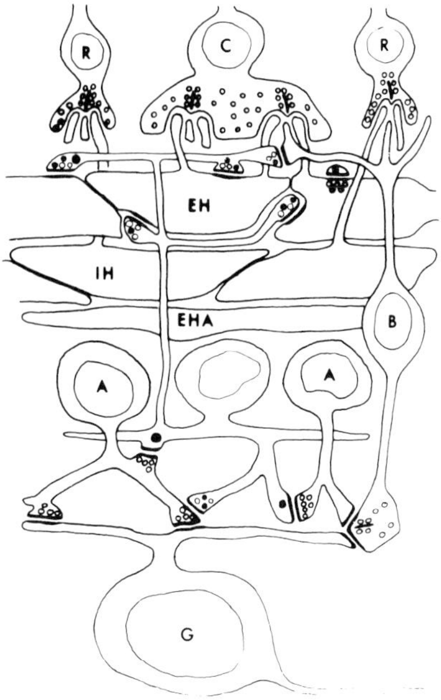

Fig. 3 Synaptic connections of a so-called interplexiform cell of the goldfish retina. The electrical input is from some amacrine cells (A), for which the interplexiform cells are postsynaptic. The interplexiform cell has synapses onto amacrine cell processes in the IPL as well, but the interplexiform cell does not have contacts with ganglion cells (G). There are synaptic connections in the OPL onto horizontal cells (EH) and onto bipolar cell (B)-dendrites; IH: intermediate (rod) horizontal cell; EHA: external horizontal cell axon process (with kind permission from Dowling and Ehinger; see Dowling and Ehinger, 1975).

Dopaminergic pleomorph cells, which have only been found in the Cebus monkeys are situated in the middle of the inner nuclear layer (Ehinger and Falck, 1969b).

Dopaminergic interplexiform neurones have been characterized with the Golgi stain technique (Boycott *et al.*, 1975; Dawson and Perez, 1973) and with histofluorescence and electron miscroscopical techniques (Dowling and Ehinger, 1975). The cell bodies of these cells are located in the inner nuclear layer. Fibres and terminals of these cells branch within both the inner and outer plexiform layer (Fig. 3). The cells have their fibre branches mostly in the inner plexiform layer, but send some dopaminergic centrifugal fibres from the inner nuclear layer to the outer plexiform layer. These fibres make synaptic contacts onto bipolar cell dendrites and onto horizontal cell somata in the outer plexiform layer. In addition they make synaptic contacts with amacrine cells in the inner nuclear layer (Fig. 3). The fibres of these interplexiform cells do not make contacts with the rod horizontal cells, but they do have connections with the cone horizontal cells (see also under Electrophysiology).

Kramer *et al.* (1971b) developed an autoradiographic technique in which ^{3}H-DA-labelled structures could be distinguished in the retinae of cats. They found the DA-localization at the junction of the inner nuclear and inner plexiform layers. This is further evidence for DA being a neurotransmitter in the retina.

Electrophysiology

To obtain more information about the physiological role of catecholamines in the retina, records have been made of membrane potentials of ganglion cells in the retina. Thus changes in spike pattern were correlated with the effects of catecholamines and related drugs. Straschill and Perwein (1969) applied DA and noradrenaline (NA) iontophoretically to retinal ganglion cells of the eyes of cats. In both cases a decrease of spontaneous and light-driven activity was observed, suggesting that the effects of DA or NA might be inhibitory. Ames and Pollen (1969) used microelectrodes and perfused an isolated retinal preparation of the rabbit and found inhibitory effects of DA and NA on ON- and ON-OFF cells. However, on OFF-cells they found an enhancement of the activity state. The responses of several kinds of cells to phentolamine were opposite to the responses to DA. Propranolol inhibited the responses of all the cells. However, none of these drugs are specific DA-antagonists. Moreover the effect of propranolol is due to a blockade of β-adrenergic receptors. Thus from these results one cannot reach a clear conclusion. Dowling *et al.* (1976) investigated interplexiform dopaminergic neurones (Fig. 3) in the goldfish (see above, under Localization). They applied DA in "atomized" solutions of different concentrations to all kinds of cells (Hedden and Dowling, 1977) and found this catecholamine to counteract (to depolarize) the cone (luminosity) type of horizontal cells and to reduce the amplitude of light-evoked responses in these cells. Phentolamine completely abolished the DA-effects on the cone horizontal cells, whilst propranolol was ineffective or inhibited the DA-effect only at high concentrations. Moreover the authors found no effect of DA on the rod horizontal cells. Sometimes a depressant effect was observed on the depolarization by red light on C-type cone horizontal cells during DA-application. An increase of the hyperpolarization of these cells was obtained during exposure to white light in the presence of DA. Moreover, evidence was gathered that cones are involved in a feedback system via L-type horizontal cells; dopamine which is probably released by recurrent processes of interplexiform cells, then influences the cone responses to light, via these L-type horizontal cells. A depressant effect by DA on bipolar cells with surrounding illumination, which induces hyperpolarization, and an enhancement of the DA-effect with centre illumination, which induces depolarization, was clearly shown. Reversed effects were demonstrated with OFF-centre bipolar cells. Finally, a strong depolarizing effect on transient amacrine cells, but hardly any response of sustained types of amacrine cells during DA application was demonstrated. Because of the fact that the authors used "atomized" solutions the real concentrations on the retinal elements could not be given or calculated, however, it may be worth mentioning that the threshold response value was 1–2 mM and the maximal response value was 20–40 mM DA.

LIGHT-INDUCED CHANGES OF DA CONTENT AND SYNTHESIS

To study intrinsic functions of DA in the intact retina several authors have performed experiments using the photosensory properties of the retina. Thus Kramer *et al.* (1971a, b) have developed a perfusion technique by which the cat retina was superfused with physiological saline with and without ^{3}H-DA. They found that retinae previously labelled with ^{3}H-DA showed an increased release of the amine after stimulation with light flashes. The increased release appeared to be flash frequency dependent in dark- and light-adapted retinae. Bilateral optic nerve section prior to perfusion did not alter the general pattern from that found with animals with intact optic nerves. The fact that labelled retinae displayed a variable DA-release on variable changes of the light intensities (see also review: Kramer, 1976) might be indicative that DA is a transmitter with a control function in the neuronal processing of the retina. Da Prada (1977) investigated the changes of DA synthesis and content in dark- and light-adapted rat retinae. He found no differences in the DA content of these retinae. However, DA synthesis was enhanced considerably by light and inhibited in the dark.

Adenylate cyclases, which convert ATP to cAMP and which are specifically sensitive to DA, exist in several dopaminergic structures of the brain. The DA-sensitive receptor coupled to adenylate cyclase is considered to be a model for a postsynaptic neurophysiological process (Brown and Makman, 1972; Kebabian *et al.*, 1972; Greengard, 1976; Makman *et al.*, 1975: Bucher and Schorderet, 1974; Schorderet, 1977a, b; Wassenaar and Korf, 1976; Spano *et al.*, 1977).

Spano *et al.* (1977) found a supersensitive reaction for the DA receptor adenylate cyclase system in retinae of light-deprived adult rats. They also found supersensitivity of this DA-system after reserpine treatment in which the DA-receptors are artificially deprived of DA. These data support the observations of Kramer *et al.* (1971a, b), who observed in cat retinae a light-dependent release of DA and of Da Prada (1977), who found a relatively low DA content in rat retinae studied at or before 14 days postnatally. Spano *et al.* (1977) and Da Prada (1977) are of the opinion that a physiological supersensitivity of the DA-cAMP-system exists during the ontogenesis of the rat. Thus just before the opening of the eyelids, this DA system appears to be supersensitive compared with the normal DA-sensitivity of the system before nine days or after 21 days postnatally or at an adult stage of life. Thus, after 21 days of age DA synthesis is enhanced by light (Da Prada, 1977) and DA-receptor sensitivity seems to have adjusted to a new steady state (Spano *et al.* 1977) comparable with the sensitivity of the DA-receptors before nine days of postnatal life.

RETINAL DOPAMINE AND DRUGS

Dopaminergic agonists and antagonists, which decrease and increase DA turnover respectively, possibly influence both post- and presynaptic DA-receptors. Thus on looking at the mode of action of apomorphine (a DA agonist) on DA-turnover rate, one finds HVA (homovanillic acid) levels to be decreased in several brain structures including the retina (Da Prada *et al.*, 1975; Westerink and Korf, 1976; Da Prada, 1977). The reverse is true for neuroleptics (DA-antagonists) such as haloperidol, flupenthixol, butaclamol etc. (Scatton *et al.*, 1975, 1977; Da Prada, 1977).

The effects of lysergic acid diethylamide (LSD) on DA turnover and on DA-adenylate cyclases in the retina and other dopaminergic brain areas are often explained as being due to a DA-agonistic activity of this drug (Spano *et al.*, 1975). Thus haloperidol inhibits LSD-induced cAMP-formation in addition to its inhibition of DA-induced cAMP-synthesis (Spano *et al.*, 1977). Supporting evidence for this idea is the finding that LSD decreases HVA-production in the striatum and the retina (Da Prada *et al.*, 1975). On the other hand LSD appears to enhance single unit firing in the visual system of cats, indicating that it acts as a DA-antagonist (Brawley and Duffield, 1972).

Effects of Neuroleptic Drugs on Adenylate Cyclases and Phosphodiesterases

It has already been mentioned that adenylate cyclases which convert ATP to cAMP, and which are activated by DA and/or NA and/or A have been demonstrated in several brain areas including the retina (Kebabian *et al.*,

1972; Makman *et al.*, 1975). The DA-sensitive receptor coupled adenylate cyclase system can be inhibited by different neuroleptic drugs (Brown and Makman, 1973; Clement-Cormier *et al.*, 1974; Iversen, 1975). In addition, however, quite a number of reports point to an inhibitory action of various psychotropic drugs on brain phosphodiesterases (PDases) (Moffat *et al.*, 1972; Weiss *et al.*, 1974; Levin and Weiss, 1976; Chasin and Harris, 1976).

With these data in mind, we reasoned that a given cAMP-level, induced by one of the catecholamines, is dependent on the actual adenylate cyclase and PDase activity in a specific tissue. We therefore checked whether neuroleptic drugs do inhibit both synthesis (adenylate cyclases) and breakdown (phosphodiesterases) of cAMP in the living cell of the retina. These actions must be opposite to their effects on the cAMP levels. Thus we investigated the synthesis and breakdown of catecholamine-induced cAMP levels and their susceptibility to neuroleptic drugs in combination with classical phosphodiesterase inhibitors in the intact retinae incubated *in vitro*. The experimental procedures are mainly as described by Brown and Makman (1972) and Wassenaar and Korf (1976).

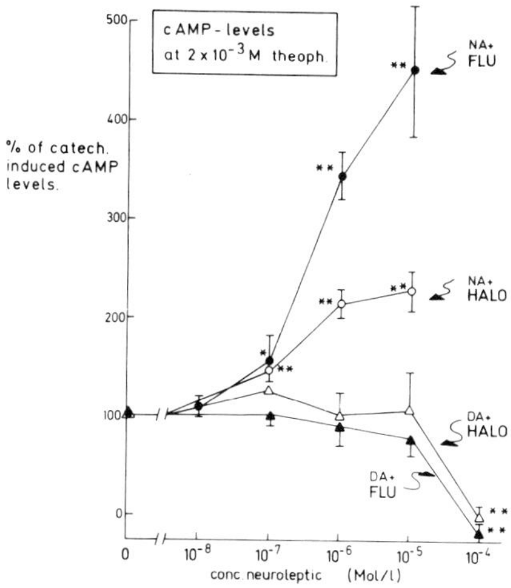

Fig. 4 Enhancement or inhibition by fluphenazine (FLU) and haloperidol (HALO) of the NA- and DA-induced cAMP level in intact retinae incubated *in vitro*. Incubations were performed in the presence of a "critical" theophylline concentration of 2×10^{-3} M (see text). The net cAMP levels were found by subtracting the cAMP levels of non-stimulated retinae (varying from 18–25 pmol cAMP per mg protein) from the cAMP-levels of retinae stimulated by DA and NA. The net values of DA- and NA-induced cAMP-levels without drugs were taken as 100%. These values in pmoles cAMP per mg protein were: NA 5×10^{-5} M (HALO-curve): 58.5 ± 4.5 ($n = 8$); NA 5×10^{-5} M (FLU-curve): 62 ± 2.5 ($n = 12$); DA 5×10^{-5} M (HALO- and FLU-curves): 38 ± 3.5 ($n = 11$), values ± SEM; number of experiments: 4–12; Significance: $P^* < 0.01$ and $^{**}P < 0.005$ (Student's-t-test). As can be seen: the NA-responses are enhanced, the DA-responses are inhibited by haloperidol and fluphenazine.

TABLE I Effects of neuroleptic drugs on catecholamine-induced cAMP levels at sub-maximal theophylline concentration.

10^{-4} M of Neuroleptics	DA	NA	Adr.
Haloperidol	↘	↗	↗
Fluphenazine	↘	↗	↗
d-Butaclamol	↘	↗	↗
l-Butaclamol	↘	↗	↗
cis-Flupenthixol	↘	↗	↗
trans-Flupenthixol	↘	↗	↗

Effects of neuroleptic drugs (10^{-4} M) on catecholamine-induced cAMP-levels at a "critical" theophylline concentration of 2×10^{-3} M (see text). The DA-induced cAMP levels are always decreased, the NA- and A-induced cAMP levels are always enhanced by the neuroleptic compounds. No stereoselectivity is present. Arrows pointing up indicate an increase and vice versa.

We found that at an optimal ("non-critical") concentration of 10^{-2} M theophylline (a classical PDase inhibitor) the DA-induced cAMP-levels are inhibited by the neuroleptics, haloperidol and fluphenazine; the NA-induced cAMP-levels appeared to be only partly inhibited by these psychotropic drugs. However, at 2×10^{-3} M theophylline we always found an enhancement of the NA-response by fluphenazine and haloperidol; we again found an inhibition of the DA-response by these drugs (Fig. 4). The enhancement of the NA-response occurs at a concentration as low as 10^{-6} M for both fluphenazine and haloperidol. It is likely that the addition of fluphenazine or haloperidol induces an extra inhibition of the PDase activity resulting in an enhancement of the NA-induced cAMP-level. We therefore concluded from these results (Fig. 4) that neuroleptic drugs may inhibit the PDase activity at so called "critical" PDase inhibitor concentrations. Low concentrations of RO-20-1724 and papaverine (PDase-inhibitors) showed the same picture as for a "critical" concentration of theophylline (2×10^{-3} M).

In our experiments we did not find a stereoselective action of the active stereoisomers of butaclamol and flupenthixol. In Table I data are listed for catecholaminergic effects at a "critical" theophylline concentration of 2×10^{-3} M. In this table one can clearly see the induced enhancements of the NA-responses and the induced inhibitions of the DA-responses. The adrenaline (A) responses are also enhanced. Once again the stereoselective action of butaclamol and flupenthixol was absent in our experiments. As is

TABLE II Single and additional incubation of NA or DA in a normal incubation medium; $[K^+] = 6.25$ mM

Combination DA; NA; DA+NA	Incubation time (2 min+2 min)		cAMP (pmol per mg protein) Mean ± S.E.M. (*n*)
—	contr.		56.4 ± 5.2 (6)
Single	DA	DA	123.4 ± 2.9 (5)
Single	—	DA	144.9 ± 6.6 (6)
Single	NA	NA	201.5 ± 15.1 (6)
Single	—	NA	224.1 ± 9.1 (9)
Addition	DA —	DA NA	195.5 ± 17.5 (6)
Addition	NA —	NA DA	238.2 ± 20.1 (6)
Addition	DA NA	DA DA	209.8 ± 11.4 (6)

cAMP-levels in the presence of DA; NA; and (DA + NA). Mean values (pmol cAMP per mg protein) ± S.E.M.; (*n*) is number of incubated retinae. The cAMP-levels are determined after 2 or 4 min of incubation time, as indicated. These levels were obtained from retinae incubated *in vitro* in the presence of 5×10^{-5} M DA and/or 5×10^{-5} M NA. The addition value is calculated either by taking the mean value of the single DA and NA values (minus one control value) or by taking each appropriate single value combination for DA and NA (minus one control value). As can be seen, the calculated addition value (about 290 pmol cAMP per mg protein) is not reached by the addition values found experimentally.

pounds whilst their stereoisomers are nearly inactive (Lippmann *et al.*, 1975; Miller *et al.*, 1974, 1975; Iversen, 1975; Robinson and Sulser, 1976; Horn and Philipson, 1976; Skolnic *et al.*, 1976; Schorderet, 1977a).

Effects of the Absence of K^+-ions and the Presence of Li^+-ions on cAMP-Levels

We wanted to elucidate whether it was possible to obtain more evidence for the existence of two different NA-receptors, one being more comparable with DA-receptors than the other (Forn *et al.*, 1974; Harris, 1976). We studied the effects of the absence of K^+-ions in the pre-incubation and incubation medium on the DA- and NA-induced cAMP-levels. The experiments were performed in the presence of a "non-critical" theophylline concentration. Our first striking result was the absence of a complete addition of DA- and NA-responses in a normal K^+-containing medium (Table II, see legend). Performing the same experiments with media without K^+, we found firstly a complete addition of the DA- and NA-responses

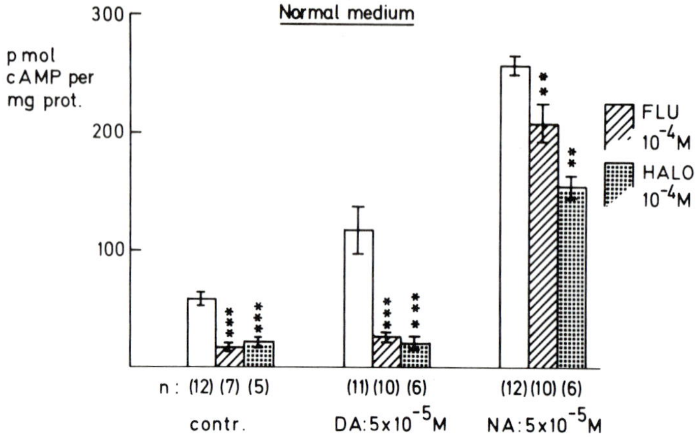

Fig. 5 Noradrenaline- and DA-induced cAMP-levels at a "non-critical" theophylline concentration of 10^{-2} M and in the presence of 6.24 mM K^+-ions (normal medium); the DA-response is reduced to control values and the NA-response is only partly blocked by 10^{-4} M fluphenazine (FLU) and 10^{-4} M haloperidol (HALO); cAMP levels are mean values ± S.E.M. (vertical bars); n = number of experiments. $^{**}P < 0.025$; $^{***}P < 0.0005$ (Student's-t-test) with respect to the corresponding NA- or DA-responses without neuroleptics.

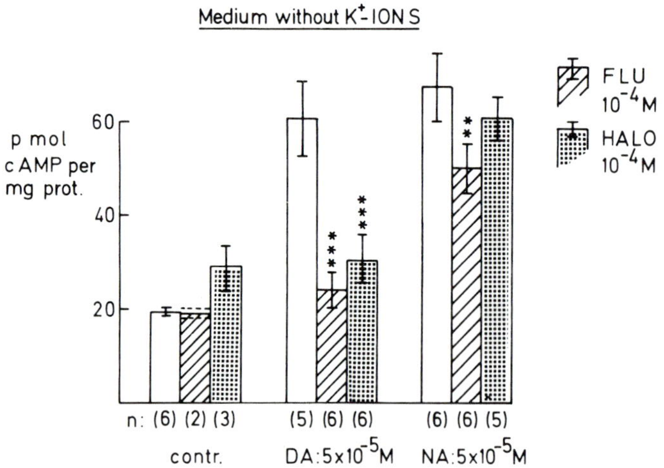

Fig. 6 Legend as for Fig. 5, except that pre-incubations and incubations were in media without K^+ ions; cAMP levels are mean values ± S.E.M. or with the deviation as such (dotted line); n = number of experiments.

TABLE III Single and additional incubation of NA or DA in an incubation medium without potassium-ions.

Combination DA ; NA ; DA + NA	Incubation time (2 min + 2 min)		cAMP (pmol per mg protein) Mean ± S.E.M. (n)
	control		28.7 ± 3.5 (4)
Single	DA	DA	69.5 ± 6.0 (8)
Single	—	DA	66.2 ± 7.5 (4)
Single	NA	NA	68.9 ± 6.2 (7)
Single	—	NA	82.9 ± 9.2 (4)
Addition	DA	DA	115.0 ± 11.5 (4)
	—	NA	
Addition	NA	NA	119.7 ± 15.5 (4)
	—	DA	
Addition	DA	DA	116.2 ± 12.9 (4)
	NA	NA	

cAMP-levels in the presence of DA; NA; and (DA + NA). Mean values (pmol cAMP per mg protein) ± S.E.M.; (n) is number of incubated retinae. The cAMP-levels are determined after 2 or 4 min of incubation time, as indicated. These levels are obtained from retinae incubated *in vitro* in the presence of 5×10^{-5} M DA and/or 5×10^{-5} M NA. The addition value is calculated either by taking the mean value of the single DA- and NA-values (minus one control value) or by taking each appropriate single value combination for DA and NA (minus one control value). As can be seen, the calculated addition value (about 115 pmol cAMP per mg protein) is very near to the addition values found experimentally.

already known (+)-butaclamol and α-flupenthixol are the active com- (Table III) and secondly a more profound decrease of the NA-response than of the DA-response. However, both single responses were reduced by the absence of K^+-ions (Table III, see legend).

The NA- and DA-responses in the presence of haloperidol and fluphenazine, in the absence of K^+-ions and in the presence of a "non-critical" theophylline concentration (Fig. 6) were comparable with the responses obtained in normal media with neuroleptic drugs (Fig. 5). Finally we studied the effects of Li^+-ions on DA- and NA-responses in media with and without K^+-ions (Wassenaar and Korf, 1977). We found that at a Li^+-ion concentration of 20 mM and 10 mM both DA- and NA-responses are inhibited in both types of media. This inhibition indicates a less specific action of Li^+-ions on DA- and NA-receptors than was found with the neuroleptics, haloperidol and fluphenazine.

CONCLUSIONS

To understand more about the role of DA in the retina one should combine the information mentioned in the several subdivisions presented in this chapter. The question whether NA affects specific NA-receptors in the retina or not, is still unsolved (Brown and Makman, 1972; Wassenaar and Korf, 1976). Neither is it clear if NA-receptors are comparable with DA-receptors and if NA is endogenously released onto these receptors.

Our experiments with optimal ("non-critical") and non-optimal ("critical") concentrations (Fig. 4; Table I) of PDase inhibitors combined with neuroleptics, indicate that neuroleptic drugs inhibit PDase activity (Fig. 4; Table I). The sensitivity of this PDase activity to neuroleptics seems to be very high (Fig. 4). Experiments with and without potassium, with and without neuroleptics and with and without lithium point to a complex mechanism of the DA- and NA-induced cAMP levels in different ionic milieux, which might have relevance for postsynaptic events. Further research is needed to elucidate this point (Schorderet, 1977b and 1978).

The knowledge of local circuits of DA-containing fibres and the action of drugs on specific DA- (and NA-) receptors in other brain areas may elucidate the properties and dendritic composition of the retina (Makman *et al.*, 1975; Dowling and Ehinger, 1975; Korf *et al.*, 1976; Geffen *et al.*, 1976). Most of the studies reviewed here are concerned with only one aspect of DA function. Therefore it appears that multidisciplinary studies on retinal DA may reveal more detailed information of the intrinsic function of dopamine in the retina.

ACKNOWLEDGEMENTS

I gratefully acknowledge the help of all the members of the Department of Neurophysiology for critical discussions during the course of this work. I am much indebted to Dr C. Rashbass and Dr F. Veringa of this department for helpful discussions during preparation of the manuscript. I would also like to thank Dr J. Korf (Department of Biological Psychiatry) and Dr A. Horn (Department of Pharmacy) for useful discussions. I am much obliged to Miss G. Popken for bibliographical work, to Mrs G. Wortmann-Even for typing the manuscript and to Miss H. Roelse for expert technical assistance. I also thank Mr Th. Deddens for the excellent drawings, Mr B. Tebbes for the photography of the Figures and Mr J. Mast for the development of various instruments.

REFERENCES

Ames III, A. and Pollen, D. A. (1969). *J. Neurophysiol.* **32,** 424–442.
Blumberg, J. B., Vetulani, J., Stawarz, R. J. and Sulser, F. (1976). *Eur. J. Pharmacol.* **37,** 357–366.
Boycott, B. B., Dowling, J. E., Fisher, S. K., Kolb, H. and Laties, A. M. (1975). *Proc. R. Soc. Lond.* (*Biol.*) **191,** 353–368.
Brawley, P. and Duffield, J. C. (1972). *Pharmacol. Rev.* **24,** 31–66.
Brown, J. H. and Makman, M. H. (1972). *Proc. Natl. Acad. Sci., U.S.A.* **69,** 539–543.
Brown, J. H. and Makman, M. H. (1973). *J. Neurochem.* **21,** 477–479.
Bucher, M. B. and Schorderet, M. (1974). *Biochem. Pharmacol.* **23,** 3079–3082.
Chasin, M. and Harris, D. N. (1976). *In* "Advances in Cyclic Nucleotide Research", Vol. 7 (Eds. P. Greengard and G. A. Robison), 225–264. Raven Press, New York.
Clement-Cormier, Y. C., Kebabian, J. W., Petzold, G. L. and Greengard, P. (1974). *Proc. Natl. Acad. Sci., U.S.A.* **71,** 1113–1117.
Da Prada, M., Saner, A., Burkard, W. P., Bartholoni, G. and Pletscher, A. (1975). *Brain Res.* **94,** 67–73.
Da Prada, M. (1977). *In* "Advances in Biochemical Psychopharmacology", Vol. 16 (Eds. E. Costa and G. L. Gessa), 311–319. Raven Press, New York.
Dawson, W. W. and Perez, J. M. (1973). *Science* **181,** 747–749.
Dowling, J. E. and Boycott, B. B. (1966). *Proc. R. Soc. Lond.* (*Biol.*) **166,** 80–111.
Dowling, J. E. and Ehinger, B. (1975). *Science* **188,** 270–273.
Dowling, J. E., Ehinger, B. and Hedden, W. L. (1976). *Invest. Ophthalmol.* **11,** 916–926.
Drujan, B. D., Diaz Borges, J. M. and Alvares, N. (1965). *Life Sci.* **4,** 473–477.
Drujan, B. D. and Diaz Borges, J. M. (1968). *Experientia* **24,** 676–677.
Ehinger, B. (1966a). *Z. Zellforsch.* **71,** 146–152.
Ehinger, B. (1966b). *Invest. Ophthalmol.* **5,** 42–52.
Ehinger, B. (1976). *In* "Transmitters in the Visual Process" (Ed. S. L. Bonting), 145–163. Pergamon, New York.
Ehinger, B. (1977). *In* "Advances in Biochemical Psychopharmacology", Vol. 16 (Eds. E. Costa and G. L. Gessa), 299–306. Raven Press, New York.
Ehinger, B. and Falck, B. (1969a). *Albrecht v. Graefes Arch. Klin. Expl. Ophthalmol.* **178,** 295–305.
Ehinger, B. and Falck, B. (1969b). *Z. Zellforsch.* **100,** 364–375.
Ehinger, B., Falck, B. and Laties, A. M. (1969). *Z. Zellforsch.* **97,** 285–297.
Falck, B., Hillarp, N. Å., Thieme, G. and Torp, A. (1962). *J. Histochem. Cytochem.* **10,** 348–354.
Forn, J., Krueger, B. K. and Greengard, P. (1974). *Science* **186,** 1118–1120.
Greengard, P. (1976). *Nature* (*London*) **260,** 101–108.
Geffen, L. B., Jessel, T. M., Cuello, A. C. and Iversen, L. L. (1976). *Nature* (*London*) **260,** 258–260.
Häggendal, J. and Malfors, T. (1965). *Acta Physiol. Scand.* **64,** 58–66.
Harris, J. E. (1976). *Molec. Pharmacol.* **12,** 546–558.
Hedden, W. L. and Dowling, J. E. (1977). *Proc. R. Soc. London* (*Biol.*) **201,** 27–55.
Horn, A. S. and Philipson, O. T. (1976). *Eur. J. Pharmacol.* **37,** 1–11.
Iversen, L. L. (1975). *Science* **188,** 1084–1089.
Kebabian, J. W., Petzold, G. L. and Greengard, P. (1972). *Proc. Natl. Acad. Sci., U.S.A.* **69,** 2145–2149.

Korf, J., Zieleman, M. and Westerink, B. H. C. (1976). *Nature* (*London*) **260,** 257–258.

Kramer, G. (1971a). *Invest. Ophthalmol.* **10,** 438–452.

Kramer, S. G. (1976). *In* "Transmitters in the Visual Process" (Ed. S. L. Bonting), 165–198. Pergamon, New York.

Kramer, S. G., Potts, A. M. and Mangnall, Y. (1971b). *Invest. Ophthalmol.* **10,** 617–624.

Levin, R. M. and Weiss, B. (1976). *Molec. Pharmacol.* **12,** 581–589.

Lippmann, W., Pugsley, T. and Merker, J. (1975). *Life Sci.* **16,** 213–224.

Makman, M. H., Brown, J. H. and Mishra, R. K. (1975). *In* "Advances in Cyclic Nucleotide Research", Vol. 5 (Eds. G. I. Drummond, P. Greengard and G. A. Robison), 661–679. Raven Press, New York.

Miller, R. J., Horn, A. S. and Iversen, L. L. (1974). *Molec. Pharmacol.* **10,** 759–766.

Miller, R. J., Horn, A. S. and Iversen, L. L. (1975). *J. Pharm. Pharmacol.* **27,** 212–213.

Moffat, A. C., Patterson, D. A., Curry, A. S. and Owen, P. (1972). *Eur. J. Toxicol.* **5,** 160–162.

Nichols, C. W., Jacobowitz, D. and Hottenstein, M. (1967). *Invest. Ophthalmol.* **6,** 642–646.

Polyack, S. L. (1957). *In* "The Vertebrate Visual System", Chicago University Press, Chicago.

Robinson, S. E. and Sulser, F. (1976). *J. Pharm. Pharmacol.* **28,** 645–646.

Scatton, B., Thierry, A. M., Glowinski, J. and Julou, L. (1975). *Brain Res.* **88,** 389–393.

Scatton, B., Dedek, J. and Korf, J. (1977). *Brain Res.* **135,** 374–377.

Schorderet, M. (1977a). *Life Sci.* **20,** 1741–1748.

Schorderet, M. (1977b). *Biochem. Pharmacol.* **26,** 167–170.

Schorderet, M. (1978). *Gerontology* **24,** 86–93.

Skolnic, P., Daly, J. W., Freedman, R. and Hoffer, B. J. (1976). *J. Pharmacol. Expl. Ther.* **197,** 280–292.

Spano, P. F., Kumakura, K., Tonon, G. C., Govoni, S. and Trabucchi, M. (1975). *Brain Res.* **93,** 164–167.

Spano, P. F., Govoni, S., Hofmann, M., Kumakura, K. and Trabucchi, M. (1977). *In* "Advances in Biochemical Psychopharmacology", Vol. 16 (Eds. E. Costa and G. L. Gessa), 307–310. Raven Press, New York.

Straschill, M. and Perwein, J. (1969). *Pflügers' Arch.* **312,** 45–54.

Wassenaar, J. S. and Korf, J. (1976). *In* "Transmitters of the Visual Process" (Ed. S. L. Bonting), 199–218. Pergamon, New York.

Wassenaar, J. S. and Korf, J. (1977). To be published.

Weiss, B., Fertel, R., Figlin, R. and Uzunov, P. (1974). *Molec. Pharmacol.* **10,** 615–625.

Werblin, F. S. (1974). *In* "The Eye", Vol. 6 (Eds. H. Davson and L. T. Graham Jr.), 257–281. Academic Press, New York and London.

Westerink, B. H. C. and Korf, J. (1976). *Eur. J. Pharmacol.* **40,** 175–178.

CHAPTER 28

The Neuroendocrinology of Dopamine Systems

Walter Lichtensteiger

University of Zürich, Zürich, Switzerland.

Early observations on a possible participation of central catecholaminergic systems in neuroendocrine control included observations on ovulation (Sawyer *et al.*, 1947; Barraclough and Sawyer, 1957), pseudopregnancy (Shelesnyak, 1958) and lactation (Meites, 1957). The interest in the role of catecholamines was considerably revived by the observation of catecholamine-containing nerve terminals in the median eminence (Carlsson *et al.*, 1962; Fuxe, 1964) which were found to originate mainly in the arcuate nucleus of the hypothalamus (Fuxe and Hökfelt, 1966; Lichtensteiger and Langemann, 1966) and were characterized as dopaminergic (DA). It later became evident that several other brain areas involved in neuroendocrine regulation also receive a DA innervation. However, there is still very little precise information available on the relationship between these "extra-infundibular" DA systems and neuroendocrine control.

There can be little doubt that DA, as well as noradrenaline (NA), and possibly also adrenaline, plays a role in neuroendocrine processes. The data presented in this chapter have been selected according to whether they could be related with some certainty to DA. Emphasis is given to findings that lead from a descriptive inventory to an understanding of the underlying mechanisms. Since the majority of these observations were made on the rat and in some cases on amphibians, some predominance of these species ensued. It is quite obvious that different situations may be encountered in other species. Clinical studies could be considered only in cases where they provided additional information on DA mechanisms.

DOPAMINERGIC EFFECTS ON PITUITARY FUNCTION AND THE INVOLVEMENT OF THE TUBEROINFUNDIBULAR DA SYSTEM

Dopamine may interfere with neuroendocrine regulation at various levels of the central nervous system. Although drugs can be expected to reach all these levels, the effects of pharmacological manipulations of DA mechanisms will be discussed primarily with respect to a possible involvement of the tuberoinfundibular DA neurones. These neurones occupy a key position, as they interfere at the last or second-last level of pituitary control. Interactions with DA mechanisms at this level will tend to over-ride DA effects at higher levels, especially in those cases where the secretion of the pituitary hormone is inhibited by DA. The possible involvement of other DA systems is discussed later.

The tuberoinfundibular DA neurones form a very conspicuous system which originates mainly in the arcuate nucleus of the hypothalamus and in some periventricular cells. The system projects to the external layer of the median eminence (with the highest density of terminals in the lateral part) and in addition, sends an important projection to the pars intermedia, while a smaller number of fibres end in the posterior lobe (Fuxe, 1964; Fuxe and Hökfelt, 1966; Lichtensteiger and Langemann, 1966; Björklund *et al.*, 1970, 1973; Löfström *et al.*, 1976a). The notion of an intrahypothalamic origin of this system is supported by the ineffectiveness of total deafferentation of the mediobasal hypothalamus on the DA content of this area (Weiner *et al.*, 1972). More recently, Kizer and co-workers (1976a) have suggested, on the basis of lesion experiments combined with amine determination, that DA neurones of the ventromedial midbrain tegmentum also contribute to the innervation of the median eminence. The idea of extrahypothalamic contributions has been considered to be supported by the incomplete disappearance of DA after arcuate nucleus lesions by neonatal monosodium-L-glutamate administration (Nemeroff *et al.*, 1977). Since the midbrain DA neurones project to the telencephalon in a fan-like way, it seems not impossible that some of the medial fibres might reach the median eminence. Yet, there exists a considerable discrepancy between the results of total deafferentation and lesions at the mesodiencephalic junction. As an alternative explanation of the lesion data, one might think of an alteration in the functional state of the hypothalamic DA system. This question presently remains unsettled.

The topographical situation of the terminals of the tuberoinfundibular DA system suggests that three types of effects may occur: (1) actions on brain constituents (nerve terminals or non-neuronal cells) in the median eminence and posterior lobe, (2) effects on glandular cells by direct innervation in the case of pars intermedia, (3) effects on glandular cells by release of DA into the pituitary portal system. All of these mechanisms appear to be used in the control of one or the other of the various hormone axes.

CONTROL OF PROLACTIN AND MELANOTROPIN SECRETION—DIRECT ACTIONS ON PITUITARY CELLS?

Prolactin—Control by the Portal System

Drug effects *in vivo*

After inhibition of catecholamine synthesis or administration of chlorpromazine, pseudopregnancy and lactation ensue in the rat (Barraclough and Sawyer, 1959; Talwalker *et al*., 1960; Coppola *et al*., 1965). The suckling-induced release of prolactin is potentiated by inhibition of catecholamine synthesis (Voogt and Carr, 1975). Certain drug effects (Donoso *et al*., 1971) as well as the absence of a change in prolactin concentration after total hypothalamic deafferentation (Blake *et al*., 1972) which greatly reduced NA levels in this region (Weiner *et al*., 1972, see above), indicated that the catecholamine was DA rather than NA. In female and male rats, prolactin levels rise after administration of neuroleptics (Lu *et al*., 1970; Dickerman *et al*., 1972; Lawson and Gala, 1975; Horowski and Gräf, 1976); in urethane-anaesthetized males, the response was detected within 10 min after i.v. injection (Iwasaki, 1976). In-vivo blockade of DA receptors also led to increased prolactin synthesis *in vitro* (MacLeod, 1976; Maurer and Gorski, 1977). Whether this is due directly to a DA action or results from increased hormone release, remains to be clarified. Even before they were classified as DA receptor-stimulating agents, certain ergot alkaloids were found to interfere with prolactin secretion (Shelesnyak, 1958). These substances and other DA receptor-stimulating agents such as apomorphine and piribedil have later been shown to inhibit the secretion of this hormone in a variety of functional situations (Wuttke *et al*., 1971; Shaar and Clemens, 1972; Lu *et al*., 1971; Smalstig *et al*., 1974; Lawson and Gala, 1975; Flückiger and Kovacs, 1974; Markó and Flückiger, 1974; Horowski and Gräf, 1976). L-DOPA appears to exert its inhibitory action through DA rather than NA, since its effect persists after inhibition of dopamine-β-hydroxylase (Donoso *et al*., 1971).

More recently, Van der Gugten and co-workers (1976) postulated on the basis of indirect evidence that the fall of prolactin levels after L-DOPA might be due (also) to increased retention of the hormone by peripheral tissues. Although amphetamine should likewise reinforce DA mechanisms, its administration yielded controversial results (Lu and Meites, 1971; Horowski and Gräf, 1976).

Inhibitory control by tuberoinfundibular DA neurones

In the case of prolactin, the pharmacological experiments point to an inhibitory DA control which probably is relieved when an acute prolactin surge occurs, e.g. upon suckling or in pro-oestrus. This inhibitory control appears to be exerted by the tuberoinfundibular DA system, as first suggested by Van Maanen and Smelik (1968) who observed pseudopregnancy after implantation of reserpine into the mediobasal hypothalamus of rats, together with a depletion of catecholamine terminals restricted to this area. Their hypothesis is corroborated by the finding that inhibition of catecholamine synthesis elicits a rise in prolactin in rats with a completely deafferented mediobasal hypothalamus, where the tuberoinfundibular DA system would appear to be the only target left for the synthesis inhibitor (Weiner, 1975).

Prolactin feedback

If prolactin secretion is under the inhibitory control of the tuberoinfundibular DA neurones, a rise in serum levels of the hormone should provoke an activation of these neurones. Evidence for such a feedback effect stems from a number of biochemical and histochemical studies. In pregnancy, pseudopregnancy and lactation, an increased rate of disappearance of catecholamine fluorescence in the external layer of rat median eminence (Fuxe *et al.*, 1967) and an increase in catecholamine fluorescence of nerve cells in the arcuate nucleus was seen. Variations in histofluorescence that are possibly related in part to prolactin secretion, occur in the tuberoinfundibular DA system also during the oestrous cycle and after acute exposure to cold; they will be discussed later. Experimental elevation of prolactin levels by injection of prolactin or by implantation of pituitaries underneath the kidney capsule, affects the functional state of the tuberoinfundibular DA system in the rat. Some uncertainty exists about the time course of the feedback effect and the sensitivity of the system. The rise in prolactin induced by suckling after 4 min was followed by an increase of hypothalamic DA synthesis 11 min later (Voogt and Carr, 1974); the effect on DA could be due to prolactin, but a direct influence of the suckling stimulus cannot be excluded. In studies on the cell bodies (Lichtensteiger and Lienhart, 1975, 1977) or on the terminals of the DA system (Olson *et al.*, 1972; Hökfelt and Fuxe, 1972), a response became apparent between 30 min (cells) and 2–4 h (terminals) after either injection of prolactin (2 or

5 mg kg^{-1} i.p. or i.v.) or transplantation of pituitaries in rats of various hormonal states. Hypophysectomized rats appeared to be more sensitive. A similar time-course is suggested by studies on the suppression of stress-induced prolactin release by prolactin injections (Advis *et al.*, 1977). In contrast, an increase in median eminence DA turnover as detected by biochemical assay, was not seen before 26 h, with s.c. prolactin injections every 8 h (Gudelsky *et al.*, 1976). Moreover, a low dose, 0.1 mg kg^{-1}, was ineffective on DA terminals of male rats at 4 hours; in this case, prolactin persisting in spite of hypophysectomy may have reduced the sensitivity of the system (Fuxe *et al.*, 1977a).

The picture is complicated by the possibility of crossed feedback mechanisms. Removal or transplantation of the pituitary may exert additional effects on the DA system. Thus, hypophysectomy was accompanied by a decreased rate of fluorescence disappearance in the lateral part of the DA terminal area, but prolactin failed to reverse this change (Fuxe *et al.*, 1977a). One of the hormones that should also be considered, is α-melanotropin (α-MSH) which exerts a marked feedback influence on arcuate DA neurones in female and intact or hypophysectomized male rats (Lichtensteiger and Lienhart, 1975, 1977; Lichtensteiger and Monnet, 1978). The reaction extended over the entire arcuate DA neurone group, not only its rostral part from which the projection to the intermediate lobe appears to arise (Björklund *et al.*, 1973), moreover, prolactin levels tended to fall in the females after α-MSH (Lichtensteiger *et al.*, 1977). Conversely, the prolactin feedback on DA neurones may affect the secretion of other hormones. Implantation of pituitaries suppressed the cyclic changes in DA terminals of the median eminence (Olson *et al.*, 1972) and the episodic release of LH in ovariectomized rats, at least for some time (Beck and Wuttke, 1977). In the latter study, chronically elevated prolactin levels finally appeared to decrease the sensitivity of DA receptors involved in the control of LH, but not prolactin secretion. Oestrogen effects on the tuberoinfundibular DA system of male rats also appear to be mediated by prolactin (Eikenburg *et al.*, 1977), yet, as will be discussed later, this is certainly not the only way gonadal steroids influence this DA system.

Direct action on the anterior pituitary

L-DOPA lowers serum prolactin levels in spite of lesions of the mediobasal hypothalamus (Donoso *et al.*, 1973). Similarly, injection of L-DOPA or of DA agonists result in reduced prolactin levels in rats bearing pituitary transplants (Malven and Hodge, 1971; Lu and Meites, 1972; Shaar and Clemens, 1972; Donoso *et al.*, 1974; Horowski and Gräf, 1976). A drug with indirect action, *d*-amphetamine, was ineffective. The pituitary site of action of DA which is suggested by these data, has been demonstrated in a series of in-vitro experiments. Dopamine inhibits prolactin secretion by pituitary glands taken from male or female rats at concentrations of 10^{-7} to 5×10^{-6} M (Koch *et al.*, 1970; MacLeod *et al.*, 1970; MacLeod and Lehmeyer, 1974a; Quijada *et al.*, 1973/74) or less (Weiner *et al.*, 1978: 50% inhibition with 1×10^{-7} M DA). Inhibitory effects are further seen *in vitro* with DA agonists such as apomorphine and certain ergot alkaloids (Nicoll *et al.*, 1970; Lu *et al.*, 1971; Pasteels *et al.*, 1971; MacLeod and Lehmeyer, 1974a, b; Smalstig *et al.*, 1974). Further evidence for a direct control of prolactin secretion by DA stems from infusion experiments: DA inhibited prolactin secretion by the pituitary *in situ* when infused into portal vessels (Takahara *et al.*, 1974a), by pituitaries transplanted underneath the kidney capsule when infused into the renal artery (Chodoroff *et al.*, 1977). The fall of prolactin in the venous outflow from the kidney was rapid, with the minimum 5 min and normalization 15 min after the end of a 12 min infusion period. Also at the level of the pituitary, the stimulatory action of TRH on prolactin release was antagonized by DA (Hill-Samli and MacLeod, 1974; Takahara *et al.*, 1974b). A similar mechanism may underlie the suppressive action of L-DOPA on TRH-stimulated prolactin release in man (Noel *et al.*, 1973).

Neuroleptics applied *in vitro* antagonize the effects of DA or DA agonists; they are also active when given *in vivo* prior to the excision of the pituitary (MacLeod and Lehmeyer, 1974a, b; Smalstig *et al.*, 1974; Quijada *et al.*, 1973/74; MacLeod, 1976; MacLeod and Robyn, 1977). In the absence of DA, the antagonists have mostly been found not to influence prolactin secretion *in vitro* (MacLeod and Lehmeyer, 1974a; Blackwell *et al.*, 1973; MacLeod and Robyn, 1977; but compare Debeljuk *et al.*, 1974, for an opposite finding). Under in-vivo conditions, DA antagonists caused a further rise in prolactin in animals bearing pituitary transplants (Blackwell *et al.*, 1973; Horowski and Gräf, 1976) or with lesions of the mediobasal hypothalamus (Iwasaki *et al.*, 1976). This effect is probably brought about by removal of inhibitory actions of circulating DA and other catecholamines.

Pituitary DA receptor

In support of a physiological significance of direct DA actions on prolactin release, stereospecific, reversible, high affinity–low capacity binding of labelled DA, dihydroergocryptine or haloperidol has been observed on crude membrane preparations from pituitaries or anterior lobes of rat, bovine or monkey (Brown *et al.*, 1976; Calabro and MacLeod, 1977; Caron *et al.*, 1977; Cronin *et al.*, 1977; Weiner *et al.*, 1978). The affinity constants fit reasonably well with the lower range of inhibitory concentrations of DA mentioned above. Other catecholamines and α-blockers in higher concentrations also compete with the binding site; this probably explains the in-vitro action of these substances on prolactin (MacLeod, 1969; Birge *et al.*, 1970; Koch *et al.*, 1970). The binding characteristics of the pituitary DA receptor resemble those of the receptor of a typical DA terminal area in brain, the caudate (Brown *et al.*, 1976; Cronin *et al.*, 1977); also, a correlation exists in man between the antischizophrenic potency of neuroleptics and their capacity to elevate prolactin (Langer *et al.*, 1977). This points to a considerable degree of resemblance between the two receptors. The pituitary receptor has recently been localized in electron micrographs to the surface of mammotrophs (Weiner *et al.*, 1978).

The pituitary DA receptor system appears to undergo changes in sensitivity analogous to those observed in the brain. Supersensitivity to apomorphine—as tested *in vivo* and *in vitro*—developed after destruction of the mediobasal hypothalamus in female rats, i.e. after removal of the DA influence (Cheung and Weiner, 1976; Weiner *et al.*, 1978). Enhanced activity of DA receptors controlling prolactin secretion has also been noted upon withdrawal from prolonged haloperidol treatment (Lal *et al.*, 1977). As mentioned above, chronic increase of DA activity by elevated prolactin levels did not appear to influence the sensitivity of prolactin-secreting cells (Beck and Wuttke, 1977).

Under normal conditions, prolactin appears to be the only hormone of the anterior pituitary that is placed under direct DA control. Experiments with LH, FSH, GH and ACTH have been negative. In very high concentrations, LH, FSH and ACTH are inactivated (Schneider and McCann, 1969; Kamberi and McCann, 1969; Van Loon and Kragt, 1970; Quijada *et al.*, 1973/74). These findings are in keeping with the electron microscopic evidence for a selective localization of DA receptors on mammotrophs (see above). A different situation is encountered with pituitary tumour cells: DA agonists inhibit GH secretion in acromegalia, probably by a direct action on the tumour cells (see Müller *et al.*, 1977; Chiodini *et al.*, 1975). In tumours, the prolactin response to DA can also be atypical (Malarkey *et al.*, 1977).

DA in portal blood

Another prerequisite for a direct control of prolactin secretion by DA is the occurrence of the amine in the blood circulating in the pituitary portal vessels. Dopamine has indeed been found there in concentrations between *c.* 4×10^{-9} M (male rats), 2×10^{-8} M (cycling rats) and 10^{-7} M (pregnancy;

Ben-Jonathan *et al.*, 1977). Since the pentobarbital anaesthesia used in these experiments was found to lower catecholamine levels, the actual values are probably higher. It appears then that DA concentrations in portal blood may well be in the range required for efficient receptor occupation and inhibition of prolactin release.

Hence, a considerable body of evidence presently supports the notion of a direct inhibitory control of prolactin secretion by DA. This does not completely exclude the existence of other factors, however. Recently, several reports on catecholamine-free hypothalamic extracts with PIF activity have appeared (Greibrokk *et al.*, 1974; Schally *et al.*, 1977; Enjalbert *et al.*, 1977). In one of these, GABA has been identified, yet, GABA proved to be about 100 times less potent than DA on cultured pituitary cells. Also, it should be noted that there is no evidence so far for a well-developed GABA projection to the external layer of the median eminence that could be compared with the DA system.

Melanotropin (MSH)—Control by Direct Innervation

The control of melanotropin (melanophore-stimulating hormone, MSH) secretion bears many similarities to that of prolactin, with the exception that the pars intermedia of mammals and many submammalian species receives a direct innervation by catecholaminergic, mainly DA fibres (see Björklund *et al.*, 1970, 1973; Baumgarten *et al.*, 1972). In this respect, the pars intermedia may be compared with the anterior lobe of certain fishes which is penetrated by fibres from the neural lobe part of which appear to contain a catecholamine (Knowles and Vollrath, 1965; see Fremberg and Meurling, 1975). Catecholamines appear to inhibit the secretion of both, MSH and prolactin, in teleosts (Follénius, 1971; Fremberg and Olivereau, 1973; Olivereau, 1975; Fremberg and Meurling, 1975), but interestingly the prolactin axis was found to respond only in freshwater but not in seawater eels (Olivereau, 1975).

In-vivo studies point to an inhibitory action of catecholamines on MSH secretion in amphibians and mammals (Iturriza, 1966; Kastin and Schally, 1966; Platt and Norris, 1974; Taleisnik *et al.*, 1972, 1973/74). Catecholamines or apomorphine maintain their activity on transplanted intermediate lobes (Ito, 1974; Smith, 1975) which suggests a direct action on the gland.

The idea of an inhibitory DA control is further supported by data from in-vitro experiments on frog, rat and mouse neurointermediate lobes. Dopamine or DA agonists consistently inhibited MSH release at concentrations (10^{-7} M) comparable to those active on prolactin secretion; the effect was antagonized by chlorpromazine and dibenamine (Bower *et al.*,

1974; Baker, 1976; Morgan and Hadley, 1976). In a superfusion system of neurointermediate lobe, high potassium induced release of DA and inhibition of MSH secretion; the effect on MSH was blocked by haloperidol (Tilders and Mulder, 1975). In the control of both, MSH and prolactin, the final effect of DA may be one on the membrane potential of the endocrine cell; there is evidence for a link between membrane depolarization and secretion for both hormones (Parsons, 1970; Davis and Hadley, 1976). Dopamine might thus hyperpolarize the cells. In contrast to DA, adrenaline and NA were found to inhibit MSH release only at higher concentrations, while lower concentrations actually facilitated hormone secretion (Bower *et al.*, 1974). The same type of reversal has been noted for prolactin release by rat pituitaries (Koch *et al.*, 1970). In the case of MSH, the facilitatory mechanism exhibited β-receptor characteristics, the inhibitory one was of the α-type. Relatively high concentrations of agonists and antagonists were required which is in keeping with data from affinity studies on the pituitary DA receptor (Caron *et al.*, 1977). Whether this dual action of adrenaline and NA is of any physiological meaning, remains uncertain at the present time; it may, however, influence certain pharmacological experiments. Besides catecholamines, peptides have also been proposed as potential mediators of the inhibitory control of MSH secretion. However, re-investigations of the effect of proline–leucine–glycinamide (PLG, MIF) have yielded negative or uncertain results (Grant *et al.*, 1973; Kastin *et al.*, 1973; Baker, 1976).

In analogy to the situation with prolactin, one would expect a stimulatory effect of MSH on the tuberoinfundibular DA system which would result in increased inhibition of MSH release. Characteristic signs of neuronal activation have indeed been detected in DA neurones of the arcuate nucleus of female and male rats after i.p. injection of α-MSH. The effect persisted after hypophysectomy which indicates that it was due to the peptide itself (Lichtensteiger and Lienhart, 1975, 1977; Lichtensteiger and Monnet, 1978).

CONTROL OF THE SECRETION OF OTHER PITUITARY HORMONES AND THE POSSIBILITY OF DA EFFECTS IN MEDIAN EMINENCE AND NEURAL LOBE

As mentioned earlier, there is no evidence so far for direct effects of DA on the secretion of anterior lobe hormones other than prolactin, except for pathological conditions. Effects of the tuberoinfundibular DA system on these hormones would therefore be expected at a higher level, i.e. the median eminence. Control mechanisms exerted at this level influence the secretory activity of the anterior lobe in an indirect way. This opens up the possibility of interactions with other factors and hence, of variable and even contrasting effects.

Control of Gonadotropin Secretion

Although changes related to the oestrous cycle were among the first to be detected in the tuberoinfundibular DA system (Fuxe *et al.*, 1967; Lichtensteiger, 1967, 1969a), data on a dopaminergic involvement in the regulation of LH and FSH secretion are still conflicting.

Secretion of luteinizing hormone (LH) in gonadectomized animals

The most striking effect of gonadectomy in females is the immediate disappearance of cyclic changes in the tuberoinfundibular DA system (Fuxe *et al.*, 1969, Lichtensteiger *et al.*, 1969). Apart from the loss of cyclicity, data on this DA system do not yet seem to permit a precise comparison of its functional state with that of intact animals. An early report describes a decreased rate of disappearance, after synthesis inhibition, of catecholamine fluorescence in the external layer of the median eminence of ovariectomized rats in comparison with cycling animals, but the stage of the cycle was not specified (Fuxe *et al.*, 1969). More recently, orchidectomy was found not to change the disappearance rate in the lateral palisade zone which mainly contains the LH-RH terminals, while an enhancement was noted in the medial palisade zone (Löfström *et al.*, 1976b). In females, fluorescence intensity of DA nerve cells remained at an intermediate level after gonadectomy as compared to the cycle (Lichtensteiger *et al.*, 1969). In acute conditions, cellular fluorescence intensity reflects firing rate (Lichtensteiger *et al.*, 1976), but since it also depends upon amine synthesis (Lichtensteiger, 1971), some additional influence from the changes in tyrosine hydroxylase activity cannot be ruled out. The activity of this enzyme increased in the hypothalamus after ovariectomy (Beattie *et al.*, 1972) and, more specifically, in the median eminence after orchidectomy. The latter change appears to occur in DA terminals, since it was seen in spite of destruction of the NA projection to the median eminence (Kizer *et al.*, 1974, 1976b). Whether these changes in enzyme activity were due to changes in neuronal activity or to other metabolic influences, remains uncertain. Altogether, the response of the tuberoinfundibular DA system to gonadectomy does not appear to be very drastic except for the loss of cyclicity, which would be in keeping with the idea that this system is not of great importance for the postcastration rise in LH secretion, in contrast to NA neurones (see Ojeda and McCann, 1973, and below). Although DA agonists suppressed the characteristic episodic release of LH in ovariectomized rats (Drouva and Gallo, 1976, 1977; Gnodde and Schuiling, 1976; Beck and Wuttke, 1977), pimozide and *d*-butaclamol which effectively antagonized this action, did not by themselves show any effect on LH release when given alone. This contrasts with the reduction of LH levels and the disappearance of episodic release after selective impairment of NA transmission (Donoso *et al.*, 1971; Drouva and Gallo, 1976, 1977; Gnodde and Schuiling, 1976). These observations could be interpreted as indicating an exclusive role of NA, if there were not the finding that simultaneous impairment of both, NA and DA transmission by inhibition of tyrosine hydroxylase left LH secretion unchanged, quite in contrast to a selective action on NA (Donoso *et al.*, 1971; Carr *et al.*, 1975; Drouva and Gallo, 1976; Gnodde and Schuiling, 1976). Hence, there may exist some kind of interaction between the two catecholamine systems also in gonadectomized animals. As far as FSH secretion is concerned, some reports suggesting a facilitatory action have appeared (Kamberi *et al.*, 1970a; Choudhury *et al.*, 1974) while others failed to see changes in FSH (Lachelin *et al.*, 1977; Beck and Wuttke, 1977; Fuxe *et al.*, 1977b).

Steroid feedback

While the actual role of the tuberoinfundibular DA system in the gonadectomized animal remains somewhat uncertain, these neurones certainly react to administration of gonadal steroids. In histochemical studies, changes were seen in the rat after testosterone, oestrogen, oestrogen +progesterone but not progesterone alone (Fuxe *et al.*, 1967, 1969; Lichtensteiger *et al.*, 1969). The disappearance of catecholamine fluorescence after synthesis inhibition was enhanced in both, the lateral and medial pallisade zone, by administration of oestradiolbenzoate to ovariectomized females and of 5-α-dihydrotestosterone to orchidectomized males (Fuxe *et al.*, 1977b, 1978). An enhanced DA turnover was seen in male rats after oestrogen administration (Eikenburg *et al.*, 1977). Since in-vivo administration of oestrogen alone to gonadectomized animals does not appear to influence DA uptake by the mediobasal hypothalamus (Endersby and Wilson, 1974; Wirz-Justice *et al.*, 1974; Cardinali and Gómez, 1977), the change in the rate of DA disappearance—which possibly might have been influenced by variations in uptake—appears to provide a reliable picture of increased DA release. At variance with this idea of increased neuronal activity is the finding of a reduction of tyrosine hydroxylase activity in the mediobasal hypothalamus after oestrogen and in the median eminence after testosterone (Luine *et al.*, 1977a; Kizer *et al.*, 1974). Such an effect would not be expected together with increased activity, but it may have some other metabolic basis. The oestrogen effect on tyrosine hydroxylase might possibly explain the decrease in cellular fluorescence intensity seen 24 h after steroid injection (Lichtensteiger *et al.*, 1969) which then would not reflect a decrease in neuronal activity as it usually does. Oestrogens affect the activity of COMT and MAO in the hypothalamus (Luine *et al.*, 1975; Luine and McEwen, 1977; Scardapane and Cardinali, 1977); they are transformed into catecholoestrogens which compete for COMT (Breuer *et al.*, 1978). It is not known whether these effects are in any way related to the functional state of the tuberoinfundibular DA system.

The uptake of labelled oestrogen by arcuate DA neurones (Grant and Stumpf, 1975) indicates that these cells are themselves a target for gonadal steroids. This does not preclude additional mechanisms. Thus, elevation of serum prolactin by oestrogen certainly constitutes an additional feedback mechanism and may be even the major one upon prolonged action of the steroid (Eikenburg *et al.*, 1977; Fuxe *et al.*, 1977b). Yet, it should be noted that androgens influenced these DA neurones in spite of unchanged prolactin levels (Fuxe *et al.*, 1978).

Short-term activation of the tuberoinfundibular DA neurones of ovariectomized, oestrogen–progesterone-pretreated rats either by electrical stimulation or by nicotine failed to change LH levels significantly except for certain cases where LH-RH-containing neurones may have been stimulated (Keller and Lichtensteiger, 1971; Lichtensteiger and Keller, 1974; Lichtensteiger *et al.*, 1978). When the time-course of the response to medial preoptic stimulation (Keller and Lichtensteiger, 1971) is considered in the light of more recent data on the characteristics of the response of DA nerve cells to stimulation (Lichtensteiger *et al.*, 1976), it appears that the rise in serum LH took place at a time when the activation of DA neurones had reached its climax and declined. Intraventricular DA was ineffective—in contrast to NA— in inducing LH release in ovariectomized, oestrogen–progesterone-pretreated rats (Krieg and Sawyer, 1976). These observations after acute activation of DA mechanisms in steroid-treated rats would also be compatible with an inhibitory DA action.

As far as DA effects on LH can be visualized in gonadectomized animals, they thus appear to be predominantly inhibitory. In the absence of gonadal steroids, the role of the DA system is not clear, but it appears to be involved in the suppression of LH secretion in response to oestrogen and androgen.

LH secretion in intact males and cycling females

Some findings suggest that in intact animals, certain facilitatory DA effects may occur besides inhibitory actions. Dopamine increased the release of LH and FSH from rat anterior pituitaries coincubated with hypothalami (Schneider and McCann, 1969; Kamberi *et al.*, 1970a); supportive evidence for facilitatory DA actions was obtained with intraventricular injections in anaesthetized rats (Kamberi *et al.*, 1969, 1970b, 1971; Schneider and McCann, 1970). The earlier in-vitro studies used comparatively high DA concentrations (10^{-5}–10^{-6} M) and were hampered by the necessity to measure release of pituitary hormones as an index of a DA action on hypothalamic tissue. More recently, however, DA was found to cause LH-RH release from rostral mediobasal hypothalamus of rat at lower concentrations (3×10^{-8}–10^{-7} M; Rotszteijn *et al.*, 1977). Blockade of DA by pimozide and ineffectiveness of NA point to a DA mechanism. The DA antagonist was ineffective when given alone, which might indicate a low residual activity of the DA mechanism in the tissue slice. While the site of action of DA cannot be specified, some regional specificity is suggested by the lack of a DA action on an area with LH–RH-containing terminals that does not receive a DA innervation, the organon vasculosum laminae terminalis. Moreover, DA-induced LH–RH release was seen with synaptosomal fractions from the median eminence (Bennett *et al.*, 1975; see below). The DA effects appear to depend in some way upon gonadal steroids: they were seen with hypothalami taken from intact males or from females on the second day of dioestrus (D_2) or pro-oestrus, but not with tissue from castrated males (Schneider and McCann, 1969; Rotszteijn *et al.*, 1977). Similar variations were noted with intraventricular DA injections; they yielded uncertain results with ovariectomized rats (Schneider and McCann, 1970). Bennett and co-workers (1975) reported that LH–RH release from hypothalamic synaptosomes was facilitated by oestrogen and reduced by oestrogen + progesterone treatment *in vivo*; the possibility of an interaction with DA was not tested. Even if these in-vitro studies present various problems for interpretation, it is interesting to note that they always failed to provide any evidence for inhibitory effects, in contrast to in-vivo studies.

Cell bodies and terminals of the tuberoinfundibular DA system exhibit characteristic changes during the oestrous cycle of the rat (Fuxe *et al.*, 1967; Lichtensteiger, 1969a) which disappear after ovariectomy (see above) or in persistent oestrus (Fuxe *et al.*, 1972). Changes in histofluorescence of this DA system were also observed in female rabbits after coitus, but their relationship to ovulation has been questioned (Bensch *et al.*, 1975; Fink *et al.*, 1975). While the cyclic variations in the tuberoinfundibular DA system are reviewed here in the context of LH secretion, it should be emphasized that it is not yet clear at all to what extent they are related to the control of LH or to that of other hormones, notably of prolactin. From the first to the second day of dioestrus (D_1 or metoestrus and D_2, respectively), fluorescence intensity of arcuate DA neurones of the rat increased

(Lichtensteiger, 1969a) which is indicative of increased activity unless unknown metabolic factors interfere (see above). Fluorescence disappearance, after synthesis inhibition, remained largely unchanged in the external layer of the median eminence (Löfström, 1977). Uptake of labelled DA by rat hypothalamus was low on D_1 and relatively high on D_2 at 14.00 hours (Davis *et al.*, 1977). Provided these changes occurred in the median eminence, they might possibly have influenced the disappearance rate of DA. This would offer one possible explanation for the discrepancy between observations on nerve cells and terminals. In the early phase of the cycle, i.e. on D_2, inhibition of the synthesis of both DA and NA, or administration of relatively pure DA antagonists, led to prolonged dioestrus with a rise in prolactin as a main feature, whereas impairment of NA transmission delayed ovulation for a shorter period of time and was accompanied by signs of prolonged oestrus (Terasawa *et al.*, 1975; Kun *et al.*, 1977). In women, L-DOPA lowered LH and prolactin in the early phase of the cycle (Lachelin *et al.*, 1977).

Administration of DA agonists in the morning of pro-oestrus was followed by an early rise of LH in the afternoon; this was considered to indicate that the afternoon surge of LH in pro-oestrus may be preceded by increased DA activity (Clemens *et al.*, 1977). During the critical period (studied between 15.00 and 17.00 hours), histochemical signs of reduced catecholamine turnover were observed in the rat in areas with DA terminals, especially in the lateral palisade zone where LH–RH-containing terminals are mainly found (Ahrén *et al.*, 1971; Löfström, 1977). In agreement with this finding, DA concentration in rat portal blood was lowest on pro-oestrus (Ben-Jonathan *et al.*, 1977). At the beginning of the afternoon (14.00–14.15 hours), fluorescence intensity of DA nerve cells in the arcuate nucleus was somewhat higher than on D_2 (Lichtensteiger, 1969a). This apparent discrepancy between cells and terminals could be due either to differences in timing—the cellular data might reflect the state immediately preceding the critical period (see above)—or else, DA release might be affected by other factors in addition to neuronal firing, such as presynaptic influences (see p. 511). As far as DA uptake is concerned, contrasting observations have been reported for pro-oestrus (Endersby and Wilson, 1974; Davis *et al.*, 1977). Drug administration on pro-oestrus led to controversial results. Reduction of the LH surge or blockade of (spontaneous or, in immature rats, PMS-induced) ovulation was observed after DA agonists (Wuttke *et al.*, 1971; Markó and Flückiger, 1974; Fuxe *et al.*, 1975) as well as after treatment with DA antagonists, even with relatively pure antagonists such as pimozide (Barraclough and Sawyer, 1957; Markó and Flückiger, 1974; Choudhury *et al.*, 1974), and after inhibition of tyrosine hydroxylase (Kordon and Glowinski, 1969). Blockade of ovulation was further caused by injection of α-methyl-DOPA into the mediobasal hypothalamus (Kordon, 1971); whether this effect was due to interaction with NA or DA, remains uncertain.

The available data do not seem to permit one to depict a coherent picture of DA involvement in LH control at the present time. It appears that strong dopaminergic activation suppresses LH release also in cycling animals, but the possibility of facilitatory influences occurring under certain conditions should not be disregarded. A dual type of control could be brought about by DA receptors on different target structures, by hormone-dependent changes in the responsiveness of the same receptor mechanism, or by concentration-dependent effects. (The density of the network of DA terminals would seem to allow for a wide range of local DA concentrations.) In addition, DA mechanisms at other brain levels play a role (see p. 509), yet, they could not account for discrepant observations that appear to be related to the basomedial hypothalamus. It also remains an open question whether all the observed effects reflect processes of physiological significance.

Control of the Secretion of Growth Hormone, Thyrotropin and Adrenocorticotropic Hormone

Growth hormone (GH)

Several observations indicate an involvement of DA systems, yet, controversial results were obtained in different species and under different experimental conditions (see Martin, 1976; Müller *et al*., 1977). Effects of DA agonists and antagonists in male and ovariectomized female rats speak in favour of a facilitatory DA influence (MacLeod and Lehmeyer, 1974b; Mueller *et al*., 1976; Maurer and Gorski, 1977). This corresponds to observations in normal man (see Müller *et al*., 1977). The histochemical evidence for reduced DA turnover in the median eminence after GH injections in hypophysectomized rats (Fuxe *et al*., 1978) suggests that the facilitatory influence might be exerted by the tuberoinfundibular DA system. Yet, the role of this DA system is still uncertain. The hypothalamus is capable of some regulation of GH secretion, as shown by the persistence of episodic GH release after total deafferentation (Willoughby *et al*., 1977). When rats with hypothalamic deafferentation were treated with α-methyltyrosine or chlorpromazine, they showed a rise in GH; GH did not rise further upon combination of both drugs (Chihara *et al*., 1975). L-DOPA but not D,L-DOPS reduced GH to some extent after α-methyltyrosine. This suggests the existence of a local inhibitory DA mechanism rather than a facilitatory one. Inhibitory DA actions have been suggested before (Collu *et al.*, 1972; Kato *et al.*, 1973), but the DA nature of the effect was uncertain. The suppression by α-methyltyrosine of GH release by hippocampal or amygdaloid stimulation (Martin *et al.*, 1973) and of episodic GH release (Martin, 1976), could also be explained by impaired NA transmission, since NA appears to facilitate GH release (see Martin, 1976). It should be noted that the evidence for stimulatory DA effects stems mainly from un-anaesthetized rats, whereas inhibitory actions were visualized in urethane anaesthesia which is known to affect GH secretion. It also prevents GH release by stimulation in a hypothalamic area, the ventromedial nucleus (see Martin, 1976). Other inhibitory effects such as that of nicotine, do not seem to involve a DA link, since nicotine reduces GH in rats treated with DA antagonists (Kato *et al.*, 1974; Lichtensteiger *et al.*, 1978). The observation of differences between infant and adult rats introduces an additional dimension: according to Müller and co-workers (1977), DA may facilitate GH secretion in the former and inhibit it in the latter condition. Under physiological conditions, DA appears to act upon GH release at a suprapituitary level (MacLeod and Robyn, 1977), whereas its suppressive effect in acromegalia may occur at the pituitary cell (see Müller *et al.*, 1977).

Thyrotropin (TSH)

Information on a possible involvement of DA in TSH control is still rather fragmentary. Thyroidectomy is accompanied by an increase in tyrosine hydroxylase activity in arcuate nucleus and median eminence where most of the enzyme activity appears to be derived from DA neurones (Kizer *et al.*, 1976b). This effect may be linked with a change in neuronal activity or reflect some metabolic action. Prolonged thyroxine treatment did not change DA concentration in this area (Brown *et al.*, 1972), but since unchanged DA levels have also been reported for other neuroendocrine conditions (Donoso *et al.*, 1967; Kueng *et al.*, 1976), this does not appear to rule out a response of the DA system. The control of THS secretion by the central nervous system becomes especially prominent upon acute environmental changes. Therefore, some information may be expected from acute exposure to cold. This condition elicited a marked stimulatory response in the arcuate DA neurones of rats (Lichtensteiger, 1969b). The response was suppressed by thyroxine pretreatment which is known to abolish the THS reaction to cold, but it is nevertheless not yet certain whether the reaction of the DA system was related specifically to TSH or possibly also to other hormone axes such as prolactin. Reports on the action of DA agonists on cold-induced TSH release are conflicting, showing inhibition (Tuomisto *et al.*, 1975) or no change (Scapagnini *et al.*, 1977). Resting levels of TSH were reduced by DA agonists but also to some extent by haloperidol

(Mueller *et al.*, 1976), which points to a possible interplay of different DA mechanisms.

It is difficult to know which of these would be represented by the tuberoinfundibular system. In contrast to prolactin, TSH did not change after α-methyltyrosine in rats with hypothalamic deafferentation (Scapagnini *et al.*, 1977) which rather speaks against the existence of an inhibitory DA mechanism in the mediobasal hypothalamus. A facilitatory influence of DA at this level would not seem to be excluded, since the necessary input to the DA system may have been interrupted. If there is a DA influence on TSH, the site of action would probably again be at a suprapituitary level. In support of this assumption, apomorphine was found to reduce the TRH-induced rise of prolactin but not that of TSH in man (Cooper and Jacobs, 1977). Dopamine induced release of TRH from hypothalamic slices (Grimm and Reichlin, 1973) and from synaptosomal fractions of sheep median eminence (Bennett *et al.*, 1975). In the former case, this was found to be due to conversion of DA into NA; Bennett and co-workers tested NA only with regard to LH–RH but not TRH release.

Adrenocorticotropic hormone (ACTH)

While an NA involvement in the regulation of ACTH secretion is rather well established, the participation of DA systems remains uncertain. Disappearance of catecholamine fluorescence after synthesis inhibition was enhanced in the external layer of rat median eminence after adrenalectomy (with the midtuberal part and the medial zone being most affected, Gouget *et al.*, 1973; Löfström *et al.*, 1976b), but it is not clear whether this change was related directly to the adrenal axis. No response to ACTH 1–24 injections could be detected in the tuberoinfundibular DA neurones (Lichtensteiger and Lienhart, 1975, 1977; Fuxe *et al.*, 1978), but the possibility of feedback actions of adrenal steroids remains open. Reserpine implants in the mediobasal hypothalamus which interfered with prolactin secretion, did not abolish the adrenal response to stress (Smelik, 1967). Treatment with relatively high doses of pimozide decreased by *c.* 12% the corticosteroid response to ether stress (Upton and Corbin, 1975). In the rhesus monkey, no evidence for an involvement of DA in the control of the adrenal axis was obtained, but the results were also negative with respect to NA (Chambers and Brown, 1976). Release of corticotropin-releasing factor (CRF) from hypothalamic synaptosomes by acetylcholine or electrical stimulation was reduced by DA (Edwardson and Bennett, 1974). The specificity of this effect remains to be established, since a similar action of NA on hypothalamic slices has been characterized as α-adrenergic (Hillhouse *et al.*, 1975).

DA Mechanisms in Median Eminence and Neural Lobe

The failure of DA to influence the secretion of hormones other than prolactin by the anterior pituitary *in vitro* (see above), is suggestive of some action in the median eminence. So far, DA receptors have not been identified at this level (Brown *et al.*, 1976), but a DA-sensitive adenylate cyclase was found in rats (Clement-Cormier *et al.*, 1977). What are the possible targets of DA in median eminence? The terminals of the tuberoinfundibular DA neurones do not form axo-axonal synapses (Ajika and Hökfelt, 1973). Nevertheless, effects on neighbouring non-dopaminergic terminals seem possible. This notion gains support from the stimulatory effect of low concentrations of DA on the release of LH–RH from hypothalamic synaptosomes and, more specifically, of LH–RH and TRH from synaptosomes of sheep median eminence (Bennett *et al.*, 1975). Noradrenaline has only been tested on the hypothalamic synaptosome fraction (rat) where it was ineffective. No case of inhibition of peptide release from synaptosomal fractions by DA has been reported so far, except for the reduction of acetylcholine- or stimulation-induced CRF release (Bennett *et al.*, 1975) whose DA nature is uncertain. These synaptosomal preparations were somewhat peculiar in that peptide release could not be elicited by high potassium but only by electrical field stimulation. Certainly, further experimentation is required until it becomes clear whether these DA effects reflect a physiological mechanism. In the absence of axo-axonal synapses, any effect of DA necessarily becomes rather non-specific. Some degree of specificity may be achieved by topographical subdivision of the DA neurone group, as indicated by anteroposterior differences in intensity changes and by changes in the variance of the nerve cell population (Lichtensteiger *et al.*, 1969; Lichtensteiger and Keller, 1974) as well as by differences in turnover rate and magnitude of response in different sections of the terminal area (Gouget *et al.*, 1973; Löfström *et al.*, 1976b; Fuxe *et al.*, 1978).

Besides nerve terminals, non-neuronal elements of the median eminence are candidates for being targets of DA. Tanycytes, modified ependymal cells that extend from the third ventricle to the basement membrane at the capillaries of the hypophyseal portal system, receive synaptoid contacts. Some of these have been reported to contain catecholamines (Mazzuca and Poulain, 1974) and to take up labelled DA (Scott and Krobisch-Dudley, 1975). They may partly belong to tuberoinfundibular DA neurones (Scott and Krobisch-Dudley, 1975; Sladek *et al.*, 1978), while others are of extrahypothalamic origin (Cuello *et al.*, 1973). Activation of the tubero-infundibular DA system by nicotine (20 min) or by electrical stimulation in medial amygdala or medial preoptic area (15 min) was accompanied by a decrease in the coverage of portal capillaries by tanycyte processes in ovariectomized, steroid-treated rats (Lichtensteiger *et al.*, 1978). Blockade of the nicotine effect by pimozide points to the involvement of a DA mechanism. The response may possibly be influenced by the hormonal state and also may vary in time. Thus, an extension of tanycyte processes was reported after intraventricular injection of 10 μg DA in the absence of gonadal steroids (Hökfelt, 1973) and 30–60 min after electrochemical preoptic stimulation in male rats (Stoeckart *et al.*, 1975). The functional significance of these changes in tanycytes is as yet unknown; no direct correlation with serum hormone levels (LH, GH, prolactin) was found in individual rats (Lichtensteiger *et al.*, 1978).

The tuberoinfundibular DA neurones project also to the posterior lobe of the pituitary (see Björklund and Falck, 1968; Björklund *et al.*, 1973). In consequence, the situation in this part is in some way analogous to that in the median eminence, but with a much smaller proportion of DA terminals as compared to peptidergic ones. Dopamine caused some release of vasopressin and oxytocin from rat neural lobes *in vitro* (Bridges *et al.*, 1976), but more pronounced effects were seen with incubation of hypothalami and also with intraventricular DA application. Whether these more efficient sites of action of DA were localized in the median eminence or at higher levels in the hypothalamus, cannot be decided at the present time.

NEUROENDOCRINE INVOLVEMENT OF OTHER DA SYSTEMS

Intradiencephalic DA Systems

Besides the tuberoinfundibular DA neurones, other DA neurones which belong to the incerto-hypothalamic system described by Björklund and co-workers (1975) also project to regions which participate in neuroendocrine regulation, e.g. the medial preoptic area. Neurones in this area which project to the mediobasal hypothalamus, were found to respond to microiontophoretic DA (mainly inhibition; Dyball *et al.*, 1974), however, the response pattern was very similar to that seen with NA. Biochemical data suggest an involvement of these DA systems in various neuroendocrine processes, but it is often not clear to what extent changes in DA and NA systems contributed to the findings. Gonadectomy and gonadal steroids influence DA concentration or uptake in some but not all anterior diencephalic areas investigated (Donoso and Stefano, 1967; Donoso *et al.*, 1967; Wirz-Justice *et al.*, 1974; Cardinali and Gómez, 1977; Shoemaker and Schlumpf, 1977). Prolonged thyroxine treatment lowered DA concentration in the anterior hypothalamus (Brown *et al.*, 1972). In lactating rats, the DA concentration was low in the anterior periventricular area (Shoemaker and Schlumpf, 1977), while DA turnover increased in the anterior hypothalamus after prolactin injection (Gudelsky *et al.*, 1976).

Mesencephalic DA Systems

Some observations suggest that DA mechanisms in the *amygdala* which receives DA projections from the mesencephalon, may be involved directly in the control of pituitary function: local application of DA agonists and antagonists to the basomedial amygdala of ovariectomized rats affected LH and prolactin secretion (Borrell *et al.*, 1977). The data are difficult to interpret because bromocriptine and pimozide produced responses that were only partly opposed to each other, and with considerable differences in time-course. The DA effect may possibly be inhibitory to prolactin and stimulatory to LH.

Nigrostriatal DA system

In contrast to several other catecholamine neurone groups, the DA nerve cells of substantia nigra do not accumulate oestrogen to any noticeable extent (Grant and Stumpf, 1975; Stumpf *et al.*, 1978). This does not exclude indirect steroid effects. During the oestrous cycle, cellular fluorescence intensity varied little in nigral DA neurones of rats (Lichtensteiger, 1969a); no variations in caudate DA turnover were observed in a histochemical study (Fuxe *et al.*, 1972). Some effect of the cycle is indicated by changes in HVA which showed a maximum in rat caudate on dioestrus (Jori and Cecchetti, 1973), however, this contrasts with data from cycling mice where caudate DA turnover appeared to be highest in oestrus (Jori *et al.*, 1976). While the effect of the oestrous cycle remains to be clarified, signs of increased caudate DA turnover were consistently seen after prolonged treatment of cycling rats with a combination of oestrogen and progestational agent of the type used in contraception (Algeri *et al.*, 1976). In ovariectomized animals, the effects of gonadal steroids on the nigrostriatal DA system vary considerably with the experimental conditions and do not permit one to draw general conclusions (Greengrass and Tonge, 1974; Luine *et al.*, 1977; Hackmann *et al.*, 1973); in male rats, oestrogen was without effect (Eikenburg *et al.*, 1977).

The nigral DA neurone group is further influenced by systemic administration of pituitary peptide hormones and related peptides. In doses comparable to those used in behavioural studies, α-MSH elicited complex response patterns in this neurone group in rats (Lichtensteiger and Lienhart, 1975, 1977; Lichtensteiger and Monnet, 1977, 1978). In contrast to the peptide effect on the arcuate DA neurones, that on the nigral system depended upon the hormonal state of the animal and the presence of the pituitary. In experiments with ACTH4-10 and α-MSH 11-13, differences were also seen with respect to the importance of the two active sites of the α-MSH molecule. The tripeptide proline–leucine–glycinamide (PLG, MIF) which potentiates L-DOPA-induced locomotor activation also in hypophysectomized mice (Plotnikoff *et al.*, 1971), has been reported to stimulate striatal DA synthesis *in vitro* when given *in vivo*; this effect was abolished by hypophysectomy (Friedman *et al.*, 1973). Results with prolactin have been variable: a reaction of the cell bodies occurred 30 min after prolactin ($2\,mg\,kg^{-1}$) in ovariectomized, steroid-treated rats (Lichtensteiger and Lienhart, 1975, 1977), whereas no change in caudate DA turnover was seen four hours after a single dose ($0.1\,mg\,kg^{-1}$) in hypophysectomized males (Fuxe *et al.*, 1977a) or after prolonged treatment in ovariectomized rats (Gudelsky *et al.*, 1976). In contrast, Fuxe and co-workers (1977a) reported an enhanced DA turnover in the nucleus accumbens. In spite of certain discrepancies, the results on anterior hypothalamus (p. 510), nucleus accumbens and nigral DA neurones indicate that extrahypothalamic DA systems can be influenced by prolactin. It seems conceivable that the effects of steroid hormones and peptides on mesencephalic DA systems may be linked with behavioural manifestations of neuroendocrine processes, either under physiological or pharmacological conditions.

REGULATION OF NEURONAL ACTIVITY IN THE TUBEROINFUNDIBULAR DA NEURONES—COMBINATION OF NEURAL AND HORMONAL CIRCUITS

The tuberoinfundibular DA neurones differ from a system such as the nigrostriatal one, in that the regulatory loop which controls their activity is closed by hormonal links. Extrahypothalamic sites appear to interfere in this control, but some regulation probably takes place already on a local basis. This is suggested, among other things, by the observations on prolactin which remains unchanged after total deafferentation of the mediobasal hypothalamus but rises when α-methyltyrosine is given to such animals (Weiner, 1975, cf. p. 494).

Local Circuitry

In the case of the tuberoinfundibular DA system, the activation of "postsynaptic" receptors finally leads to a hormonal response. As already pointed out by Moore and Gudelsky (1977), this may be the reason why haloperidol did not cause an increase in DA turnover of the median eminence initially, but only, in their study, after a delay of 16 (possibly 8) hours, when prolactin levels were high. This late response was abolished by hypophysectomy. In contrast, one can assume that a control of the activity of the DA system should be possible at somatodendritic and presynaptic levels. In rats withdrawn from morphine or haloperidol, an unexpected rise in prolactin after a low dose of apomorphine has been tentatively interpreted as a consequence of supersensitivity of presynaptic receptors (Lal *et al.*, 1977). Opioid peptides have been suggested to exert a presynaptic control on DA release in the caudate (Loh *et al.*, 1976); terminals immunoreactive to enkephalins occur in the median eminence (Simantov *et al.*, 1977). Intraventricular injection of met-enkephalin or β-endorphin was followed by a reduction of the rate of fluorescence disappearance, after α-methyltyrosine, in the area of DA terminals (Ferland *et al.*, 1977; Fuxe *et al.*, 1978), and a rise in prolactin. In caudate, nicotinic presynaptic effects have also been described (Giorguieff *et al.*, 1976). The existence of a cholinergic tuberoinfundibular system has recently been postulated for the rat (Carson *et al.*, 1977), but nothing is known about possible interactions between cholinergic and DA terminals.

Dopamine effects at the level of the nigral neurones may be mediated by GABA (Reubi *et al.*, 1977). GABAergic mechanisms appear to exert an inhibitory control on arcuate neurones projecting to the median eminence (Sawaki and Yagi, 1976), but it is not known whether DA neurones were among the cells studied. Some pharmacological evidence suggests the existence of a GABA-ergic link behind the DA system facilitating MSH secretion and hence, possibly, inhibiting DA neurones (Taleisnik *et al.*, 1973/74). It is not clear whether this could be an interneurone or a link in an afferent neuronal chain. Reinforcement of GABA mechanisms by γ-butyrolactone does not elicit an increase in DA concentration in the terminal area of the tuberoinfundibular neurones as might be expected from observations on caudate (Gudelsky and Moore, 1976; Hefti and Lichtensteiger, unpublished observations). Whether this reflects the absence of a GABAergic influence or a different kind of response of these terminals, remains uncertain, however. Similar findings have been made on dendrites of nigral DA neurones (Hefti and Lichtensteiger, 1978).

Hormonal Feedback

This link has been considered in previous sections; it may be sufficient to point out that so far, the data are most clear-cut with respect to the two systems with a direct inhibitory control, i.e. prolactin and MSH. The interpretation of these findings is nevertheless somewhat complicated by the observation of nerve terminals with prolactin-like immunoreactivity in the arcuate nucleus (Fuxe *et al.*, 1977c) and of immunoreactive α-MSH in synaptosomal fractions of rat hypothalamus (Oliver *et al.*, 1977). Among other possible feedback connections, that of oestrogen appears to be well documented by the accumulation of the steroid in arcuate DA neurones (Grant and Stumpf, 1975).

Extrahypothalamic Influences

Afferent connections to the tuberoinfundibular DA neurones from outside the mediobasal hypothalamus can transmit certain hormonal feedback signals but can be used also in the coordination of pituitary control, autonomic functions and behaviour. Although extrahypothalamic influences on pituitary function are well documented, our understanding of the interplay of various extrahypothalamic regions in certain neuroendocrine processes is still very limited. It becomes difficult, then, to interpret extrahypothalamic influences on the DA system with respect to their possible neuroendocrine significance. The following description aims at pointing out certain interesting parallelisms between effects on the DA system and endocrine changes which would seem to be in keeping with what is presently known of DA effects on certain hormone axes.

The tuberoinfundibular DA neurones are activated by stimulation in the corticomedial amygdala (Lichtensteiger and Keller, 1974; Lichtensteiger *et al.*, 1978), the dorsal hippocampus (Lichtensteiger and Lienhart, 1975) and in some places of the ventromedial midbrain tegmentum (area of Tsai) as shown in ovariectomized, oestrogen–progesterone-treated rats in urethane anaesthesia. The endocrine responses to stimulation in these regions depend to some extent on the hormonal state. Inhibitory effects on LH secretion are known for corticomedial amygdala (in ovariectomized rats, Lawton and Sawyer, 1970; Ellendorff *et al.*, 1973), dorsal hippocampus (Kawakami *et al.*, 1973; Carillo *et al.*, 1977) and ventral tegmental area of Tsai (Carrer and Taleisnik, 1970). The same amygdaloid and hippocampal areas and possibly also certain afferents from midbrain appear to exert an inhibitory influence on prolactin (Kawakami *et al.*, 1973; Flerkó and Bárdos, 1966). Growth hormone decreased slightly after stimulation in the corticomedial amygdala of male rats, whereas the remaining two regions exerted stimulatory actions (Martin *et al.*, 1973; Martin, 1976). In the studies on DA neurones, LH levels remained unchanged after 10–15 min which would be compatible with inhibitory influences; prolactin tended to be slightly elevated in the first but was unchanged in the second investigation (Lichtensteiger and Keller, 1974, and unpublished observations). The connection to the DA system might possibly play a role in proestrus in the sense of a modulatory influence. Stimulation in the corticomedial amygdala *before* the onset of the critical period was followed by a sharpening of the LH peak ("early synchronization", Carillo *et al.*, 1977) which is reminiscent of the "early rise" of LH observed after administration of DA agonists on the morning of pro-oestrus (Clemens *et al.*, 1977).

Stimulation in the posteromedial thalamic region is also followed by a stimulatory response of the tuberoinfundibular DA system (Lichtensteiger and Lienhart, 1975). The secretion of LH and (as indicated by lesion experiments) probably also that of prolactin can be inhibited from this area (Cáceres and Taleisnik, 1976; Flerkó and Bardós, 1966). This region has further been implicated in behavioural actions of MSH (Bohus and de Wied, 1967); it appears to be involved in acquisition and extinction of avoidance behaviour. The posteromedial thalamic region thus seems to constitute yet another relay for the integration of certain aspects of behaviour and several hormone axes.

Little is known about the transmitter systems that pass on extra-hypothalamic influences to the tuberoinfundibular DA neurones, but cholinergic systems appear to play a role (Lichtensteiger and Keller, 1974; Lichtensteiger and Lienhart, 1975). A DA mechanism appears to mediate the reduction of prolactin levels by pilocarpine as well as by nicotine (Grandison and Meites, 1976; Lichtensteiger *et al.*, 1978), whereas the suppressive effect of nicotine on GH does not appear to involve a DA system (Kato *et al.*, 1974; Lichtensteiger *et al.*, 1978, see above).

Afferent influences arriving at the tuberoinfundibular DA neurones are directed on a system with a dual nature: in the control of prolactin and MSH, it is probably analogous to peptidergic systems that act on the pituitary cell, but at the same time, it appears to modulate the release of such peptides. As mentioned earlier, this arrangement seems to allow for a higher degree of complexity in pituitary control than a situation where all DA actions would be directed on the pituitary.

REFERENCES

Advis, J. P., Hall, T. R., Hodson, C. A., Mueller, G. P. and Meites, J. (1977). *Proc. Soc. Expl. Biol. Med.* **155,** 567–570.

Ahrén, K., Fuxe, K., Hamberger, L. and Hökfelt, T. (1971). *Endocrinology* **88,** 1415–1424.

Ajika, K. and Hökfelt, T. (1973). *Brain Res.* **57,** 97–117.

Algeri, S., Ponzio, F., Dolfini, E. and Jori, A. (1976). *Neuroendocrinology* **22,** 343–351.

Baker, B. I. (1976). *J. Endocrinol.* **68,** 283–287.
Barraclough, C. A. and Sawyer, C. H. (1957). *Endocrinology* **61,** 341–351.
Barraclough, C. A. and Sawyer, C. H. (1959). *Endocrinology* **65,** 563–571.
Baumgarten, H. G., Björklund, A., Holstein, A. F. and Nobin, A. (1972). *Z. Zellforsch.* **126,** 483–517.
Beattie, C. W., Rodgers, C. H. and Soyka, L. F. (1972). *Endocrinology* **91,** 276–279.
Beck, W. and Wuttke, W. (1977). *J. Endocrinol.* **74,** 67–74.
Ben-Jonathan, N., Oliver, C., Weiner, H. J., Mical, R. S. and Porter, J. C. (1977). *Endocrinology* **100,** 452–458.
Bennett, G. W., Edwardson, J. A., Holland, D., Jeffcoate, S. L. and White, N. (1975). *Nature (London)* **257,** 323–325.
Bensch, C., Lescure, H., Robert, J. and Faure, J. M. A. (1975). *J. Neur. Trans.* **36,** 1–18.
Birge, C. A., Jacobs, L. S., Hammer, T. and Daughaday, W. H. (1970). *Endocrinology* **86,** 120–130.
Björklund, A. and Falck, B. (1968). *J. Histochem. Cytochem.* **16,** 717–720.
Björklund, A., Falck, B., Hromek, F., Owman, C. and West, K. A. (1970). *Brain Res.* **17,** 1–23.
Björklund, A., Moore, R. Y., Nobin, A. and Stenevi, U. (1973). *Brain Res.* **51,** 171–191.
Björklund, A., Lindvall, O. and Nobin, A. (1975). *Brain Res.* **89,** 29–42.
Blackwell, R., Vale, W., Rivier, C. and Guillemin, R. (1973). *Proc. Soc. Expl. Biol. Med.* **142,** 68–71.
Blake, C. A., Scaramuzzi, R. J., Norman, R. L., Hilliard, J. and Sawyer, C. H. (1973). *Endocrinology* **92,** 1419–1425.
Bohus, B. and de Wied, D. (1967). *Physiol. Behav.* **2,** 221–223.
Borrell, J., Piva, F., Motta, M., Leto, S. and Martini, L. (1977). *In* "Advances in Biochemical Psychopharmacology", Vol. 16 (Eds. E. Costa and G. L. Gessa), 115–126. Raven Press, New York.
Bower, A., Hadley, M. E. and Hruby, V. J. (1974). *Science* **184,** 70–72.
Bridges, T. E., Hillhouse, E. W. and Jones, M. T. (1976). *J. Physiol. (London)* **260,** 647–666.
Brown, G. M., Krigstein, E., Dankova, J. and Hornykiewicz, O. (1972). *Neuroendocrinology* **10,** 207–217.
Brown, G. M., Seeman, P. and Lee, T. (1976). *Endocrinology* **99,** 1407–1410.
Cáceres, A. and Taleisnik, S. (1976). *Neuroendocrinology* **22,** 30–37.
Calabro, M. A. and MacLeod, R. M. (1977). *Fed. Proc.* **36,** 366.
Cardinali, D. P. and Gómez, E. (1977). *J. Endocrinol.* **73,** 181–182.
Carillo, A. J., Rabii, J., Carrer, H. F. and Sawyer, C. H. (1977). *Brain Res.* **128,** 81–92.
Carlsson, A., Falck, B. and Hillarp, N. A. (1962). *Acta Physiol. Scand. Suppl.* 56 **196,** 1–28.
Caron, M. G., Raymond, V., Lefkowitz, R. J. and Labrie, F. (1977). *Fed. Proc.* **36,** 278.
Carr, L. A., Conway, P. M. and Voogt, J. L. (1975). *J. Pharmacol. Expl. Ther.* **192,** 15–21.
Carrer, H. F. and Taleisnik, S. (1970). *J. Endocrinol.* **48,** 527–539.
Carson, K. A., Nemeroff, C. B., Rone, M. S., Youngblood, W. W., Prange, A. J. Jr and Hanker, J. S. (1977). *Brain Res.* **129,** 169–173.
Chambers, J. W. and Brown, G. M. (1976). *Endocrinology* **98,** 420–428.

Cheung, C. Y. and Weiner, R. I. (1976). *Endocrinology* **99,** 914–917.
Chihara, K., Kato, Y., Ohgo, S. and Imura, H. (1975). *Neuroendocrinology* **18,** 192–203.
Chiodini, P. G., Liuzzi, A., Botella, L., Oppizzi, G., Müller, E. E. and Silvestrini, F. (1975). *J. Clin. Endocr. Metab.* **40,** 705–708.
Chodoroff, B., Chodoroff, G. and Gala, R. R. (1977). *Experientia* **33,** 773–774.
Choudhury, S. A. R., Sharpe, R. M. and Brown, P. S. (1974). *J. Reprod. Fert.* **39,** 275–283.
Clemens, J. A., Tinsley, F. C. and Fuller, R. W. (1977). *Acta Endocrinol.* (*Kbh.*) **85,** 18–24.
Clement-Cormier, Y. C. and Robison, G. A. (1977). *Biochem. Pharmacol.* **26,** 1719–1722.
Collu, R., Fraschini, F., Visconti, P. and Martini, L. (1972). *Endocrinology* **90,** 1231–1237.
Cooper, D. S. and Jacobs, L. S. (1977). *J. Clin. Endocrinol. Metab.* **44,** 404–407.
Coppola, J. A., Leonardi, R. G., Lippmann, W., Perrine, J. W. and Ringler, I. (1965). *Endocrinology* **77,** 485–490.
Cronin, M., Roberts, J. and Weiner, R. (1977). *Fed. Proc.* **36,** 278.
Cuello, A. C., Weiner, R. I. and Ganong, W. F. (1973). *Brain Res.* **59,** 191–200.
Davis, M. D. and Hadley, M. E. (1976). *Nature* (*London*) **261,** 422–423.
Davis, C. F., Davis, B. F. and Halaris, A. E. (1977). *Life Sci.* **20,** 1319–1332.
Debeljuk, L., Darkal, H., Rozados, R. and Guitelman, A. (1974). *Experientia* **30,** 1355–1356.
Dickerman, S., Clark, J., Dickerman, E. and Meites, J. (1972). *Neuroendocrinology* **9,** 332–340.
Donoso, A. O. and Stefano, F. J. E. (1967). *Experientia* **23,** 665–666.
Donoso, A. O., Stefano, F. J. E., Biscardi, A. M. and Cukier, J. (1967). *Am. J. Physiol.* **212,** 737–739.
Donoso, A. O., Bishop, W., Fawcett, C. P., Krulich, L. and McCann, S. M. (1971). *Endocrinology* **89,** 774–784.
Donoso, A. O., Bishop, W. and McCann, S. M. (1973). *Proc. Soc. Expl. Biol. Med.* **143,** 360–363.
Donoso, A. O., Banzán, A. M. and Barcaglioni, J. C. (1974). *Neuroendocrinology* **15,** 236–239.
Drouva, S. V. and Gallo, R. V. (1976). *Endocrinology* **99,** 651–658.
Drouva, S. V. and Gallo, R. V. (1977). *Endocrinology* **100,** 792–798.
Dyball, R. E. J., Dyer, R. G. and Drewett, R. F. (1974). *Brain Res.* **71,** 140–143.
Edwardson, J. A. and Bennett, G. W. (1974). *Nature* (*London*) **251,** 425–427.
Eikenburg, D. C., Ravitz, A. J., Gudelsky, G. A. and Moore, K. E. (1977). *J. Neur. Trans.* **40,** 235–244.
Ellendorff, F., Colombo, J. A., Blake, C. A., Whitmoyer, D. I. and Sawyer, C. H. (1973). *Proc. Soc. Expl. Biol. Med.* **142,** 417–420.
Endersby, C. A. and Wilson, C. (1974). *Brain Res.* **73,** 321–331.
Enjalbert, A., Priam, M. and Kordon, C. (1977). *Eur. J. Pharmacol.* **41,** 243–244.
Ferland, L., Fuxe, K., Eneroth, P., Gustafsson, J.-A. and Skett, P. (1977). *Eur. J. Pharmacol.* **43,** 89–90.
Fink, G., Smith, G. C., McMaster, R., Osborne, L. W. and Chiappa, S. A. (1975). *Brain Res.* **89,** 71–80.
Flerkó, B. and Bárdos, V. (1966). *Expl. Brain Res.* **1,** 299–305.
Flückiger, E. and Kovacs, E. (1974). *Experientia* **30,** 1173.

Follénius, E. (1971). *C.r. Acad. Sci. (Paris) Sér. D.* **272,** 733–736.
Fremberg, M. and Meurling, P. (1975). *Cell Tiss. Res.* **157,** 53–72.
Fremberg, M. and Olivereau, M. (1973). *Acta Zool.* **54,** 231–239.
Friedman, E., Friedman, J. and Gershon, S. (1973). *Science* **182,** 831–832.
Fuxe, K. (1964). *Z. Zellforsch.* **61,** 710–724.
Fuxe, K. and Hökfelt, T. (1966). *Acta Physiol. Scand.* **66,** 245–246.
Fuxe, K., Hökfelt, T. and Nilsson, O. (1967). *Life Sci.* **6,** 2057–2061.
Fuxe, K., Hökfelt, T. and Nilsson, O. (1969). *Neuroendocrinology* **5,** 107–120.
Fuxe, K., Hökfelt, T. and Nilsson, O. (1972). *Acta Endocr.* **69,** 625–639.
Fuxe, K., Goldstein, M., Hökfelt, T., Jonsson, G. and Lidbrink, P. (1974). *Adv. Neurol.* **5,** 405–419.
Fuxe, K., Eneroth, P., Gustafsson, J.-A., Löfstöm, A. and Skett, P. (1977a). *Brain Res.* **122,** 177–182.
Fuxe, K., Löfstöm, A., Eneroth, P., Gustafsson, J.-A., Skett, P., Hökfelt, T., Wiesel, F. A. and Agnati, L. (1977b). *Psychoneuroendocrinology* **2,** 203–225.
Fuxe, K., Hökfelt, T., Eneroth, P., Gustafsson, J. A. and Skett, P. (1977c). *Science* **196,** 899–900.
Fuxe, K., Ferland, L., Andersson, K., Eneroth, P., Gustafsson, J.-A., and Skett, P. (1978). *In* "Brain–Endocrine Interaction III" (Eds. D. E. Scott, G. P. Kozlowski and A. Weindl), 172–182. Karger, Basel.
Giorguieff, M. F., Le Floc'h, M. L., Westfall, T. C., Glowinski, J. and Besson, M. J. (1976). *Brain Res.* **106,** 117–131.
Gnodde, H. P. and Schuiling, G. A. (1976). *Neuroendocrinology* **20,** 212–223.
Gouget, A., Duvernoy, J. and Bugnon, C. (1973). *C.r. Soc. Biol.* **167,** 919–923.
Grandison, L. and Meites, J. (1976). *Endocrinology* **99,** 775–779.
Grant, L. D. and Stumpf, W. E. (1975). *In* "Anatomical Neuroendocrinology" (Eds. W. E. Stumpf and L. D. Grant), 445–463. Karger, Basel.
Grant, N. H., Clark, D. E. and Rosanoff, E. I. (1973). *Biochem. Biophys. Res. Comm.* **51,** 100–106.
Greengrass, P. M. and Tonge, S. R. (1974). *Arch. Int. Pharmacodyn.* **211,** 291–304.
Greibrokk, T., Currie, B. L. and Johansson, K. N. G. (1974). *Biochem. Biophys. Res. Comm.* **59,** 704–709.
Grimm, Y. and Reichlin, S. (1973). *Endocrinology* **93,** 626–631.
Gudelsky, G. A. and Moore, K. E. (1976). *J. Neur. Trans.* **38,** 95–105.
Gudelsky, G. A., Simpkins, J., Mueller, G. P., Meites, J. and Moore, K. E. (1976). *Neuroendocrinology* **22,** 206–215.
Hackmann, E., Wirz-Justice, A. and Lichtsteiner, M. (1973). *Psychopharmacologia* **32,** 183–191.
Hefti, F. and Lichtensteiger, W. (1978). *J. Neurochem.* **30,** 1217–1230.
Hill-Samli, M. and MacLeod, R. M. (1974). *Endocrinology* **95,** 1189–1192.
Hillhouse, E. W., Burden, J. L. and Jones, M. T. (1975). *Neuroendocrinology* **17,** 1–11.
Hökfelt, T. (1973). *Acta Physiol. Scand.* **89,** 606–608.
Hökfelt, T. and Fuxe, K. (1972). *Neuroendocrinology* **9,** 100–122.
Horowski, R. and Gräf, K.-J. (1976). *Neuroendocrinology* **22,** 273–286.
Ito, T. (1974). *Neuroendocrinology* **16,** 312–322.
Iturriza, F. C. (1966). *Gen. Comp. Endocrinol.* **6,** 19–25.
Iwasaki, Y., Kato, Y., Chihara, K., Ohgo, S., Maeda, K. and Imura, H. (1976). *Neuroendocrinology* **21,** 267–274.
Jori, A. and Cecchetti, G. (1973). *J. Endocrinol.* **58,** 341–342.

Jori, A., Colturani, F., Dolfini, E. and Rutczynski, M. (1976). *Neuroendocrinology* **21,** 262–266.

Kamberi, I. A. and McCann, S. M. (1969). *Endocrinology* **85,** 815–824.

Kamberi, I. A., Mical, R. S. and Porter, J. C. (1969). *Science* **166,** 388–389.

Kamberi, I. A., Schneider, H. P. G. and McCann, S. M. (1970a). *Endocrinology* **86,** 278–284.

Kamberi, I. A., Mical, R. S. and Porter, J. C. (1970b). *Endocrinology* **87,** 1–12.

Kamberi, I. A., Mical, R. S. and Porter, J. C. (1971). *Endocrinology* **88,** 1003–1011.

Kastin, A. J. and Schally, A. V. (1966). *Endocrinology* **79,** 1018–1020.

Kastin, A. J., Plotnikoff, N. P., Viosca, S., Anderson, M. S. and Schally, A. V. (1973). *Yale J. Biol. Med.* **46,** 617–622.

Kato, Y., Dupré, J. and Beck, J. C. (1973). *Endocrinology* **93,** 135–145.

Kato, Y., Chihara, K., Ohgo, S. and Imura, H. (1974). *Neuroendocrinology* **16,** 237–242.

Kawakami, M., Terasawa, E., Kimura, F. and Wakabayashi, K. (1973). *Neuroendocrinology* **12,** 1–16.

Keller, P. J. and Lichtensteiger, W. (1971). *J. Physiol.* (*London*) **219,** 385–401.

Kizer, J. S., Palkovits, M., Zivin, J., Brownstein, M., Saavedra, J. M. and Kopin, I. J. (1974). *Endocrinology* **95,** 799–812.

Kizer, J. S., Palkovits, M. and Brownstein, M. J. (1976a). *Brain Res.* **108,** 363–370.

Kizer, J. S., Muth, E. and Jacobowitz, D. M. (1976b). *Endocrinology* **98,** 886–893.

Knowles, F. and Vollrath, L. (1965). *Nature* (*London*) **208,** 1343–1344.

Koch, Y., Lu, K. H. and Meites, J. (1970). *Endocrinology* **87,** 673–675.

Kordon, C. (1971). *Neuroendocrinology* **7,** 202–209.

Kordon, C. and Glowinski, J. (1969). *Endocrinology* **85,** 924–931.

Krieg, R. J. and Sawyer, C. H. (1976). *Endocrinology* **99,** 411–419.

Kueng, W., Wirz-Justice, A., Menzi, R. and Chappuis-Arndt, E. (1976). *Neuroendocrinology* **21,** 289–296.

Kun, I., Feszt, G., Cuparencu, B. and Orban, I. (1977). *Pharmacology* **15,** 104–111.

Lachelin, G. C. L., Leblanc, H. and Yen, S. S. C. (1977). *J. Clin. Endocrinol. Metab.* **44,** 728–732.

Lal, H., Brown, W., Drawbaugh, R., Hynes, M. and Brown, G. (1977). *Life Sci.* **20,** 101–106.

Langer, G., Sachar, E. J., Gruen, P. H. and Halpern, F. S. (1977). *Nature* (*London*) **266,** 639–640.

Lawson, D. M. and Gala, R. R. (1975). *Endocrinology* **96,** 313–318.

Lawton, I. E. and Sawyer, C. H. (1970). *Am. J. Physiol.* **218,** 622–626.

Lichtensteiger, W. (1967). *Helv. Physiol. Pharmacol. Acta* **25,** CR423–CR425.

Lichtensteiger, W. (1969a). *J. Pharmacol. Expl. Ther.* **165,** 204–215.

Lichtensteiger, W. (1969b). *J. Physiol.* (*London*) **203,** 675–687.

Lichtensteiger, W. (1971). *J. Physiol.* (*London*) **218,** 63–84.

Lichtensteiger, W. and Keller, P. J. (1974). *Brain Res.* **74,** 279–303.

Lichtensteiger, W. and Langemann, H. (1966). *J. Pharmacol. Expl. Ther.* **151,** 400–408.

Lichtensteigner, W. and Lienhart, R. (1975). *In* "Cellular and Molecular Bases of Neuroendocrine Processes" (Ed. E. Endröczi), 211–221. Akadémiai Kiadó, Budapest.

Lichtensteiger, W. and Lienhart, R. (1977). *Nature* (*London*) **266,** 635–637.

Lichtensteiger, W. and Monnet, F. (1978). *Bull. Schweiz. Akad. Med. Wiss.* **34,** 197–208.

Lichtensteiger, W., Korpela, K., Langemann, H. and Keller, P. J. (1969). *Brain Res.* **16,** 199–214.
Lichtensteiger, W., Felix, D., Lienhart, R. and Hefti, F. (1976). *Brain Res.* **117,** 85–103.
Lichtensteiger, W., Lienhart, R. and Kopp, H. G. (1977). *Psychoneuroendocrinology* **2,** 237–248.
Lichtensteiger, W., Richards, J. G. and Kopp, H. G. (1978). *In* "Brain–Endocrine Interaction III" (Eds. D. E. Scott, G. P. Kozlowski and A. Weindl), 251–262. Karger, Basel.
Löfström, A. (1977). *Brain Res.* **120,** 113–131.
Löfström, A., Jonsson, G. and Fuxe, K. (1976a). *J. Histochem. Cytochem.* **24,** 415–429.
Löfström, A., Jonsson, G., Wiesel, F.-A. and Fuxe, K. (1976b). *J. Histochem. Cytochem.* **24,** 430–442.
Loh, H. H., Brase, D. A., Sampath-Khanna, S., Mar, J. B., Way, E. L. and Li, C. H. (1976). *Nature* (*London*) **264,** 567–568.
Lu, K.-H. and Meites, J. (1971). *Proc. Soc. Expl. Biol. Med.* **137,** 480–483.
Lu, K.-H. and Meites, J. (1972). *Endocrinology* **91,** 868–872.
Lu, K.-H., Amenomori, Y., Chen, C.-L. and Meites, J. (1970). *Endocrinology* **87,** 667–672.
Lu, K.-H., Koch, Y. and Meites, J. (1971). *Endocrinology* **89,** 229–233.
Luine, V. N., Khylchevskaya, R. I. and McEwen, B. S. (1975). *Brain Res.* **86,** 293–306.
Luine, V. N. and McEwen, B. S. (1977). *J. Neurochem.* **28,** 1221–1227.
Luine, V. N., McEwen, B. S. and Black, I. B. (1977). *Brain Res.* **120,** 188–192.
MacLeod, R. M. (1969). *Endocrinology* **85,** 916–923.
MacLeod, R. M. (1976). *In* "Frontiers in Neuroendocrinology", Vol. 4 (Eds. L. Martini and W. F. Ganong), 169–194. Raven Press, New York.
MacLeod, R. M. and Lehmeyer, J. E. (1974a). *Endocrinology* **94,** 1077–1085.
MacLeod, R. M. and Lehmeyer, J. E. (1974b). *Proc. Soc. Expl. Biol. Med.* **145,** 1128–1131.
MacLeod, R. M. and Robyn, C. (1977). *J. Endocrinol.* **72,** 273–277.
MacLeod, R. M., Fontham, E. H. and Lehmeyer, J. E. (1970). *Neuroendocrinology* **6,** 283–294.
Malarkey, W. B., Groshong, J. C. and Milo, G. E. (1977). *Nature* (*London*) **266,** 640–641.
Malven, P. V. and Hodge, W. R. (1971). *Endocrinology* **88,** 445–449.
Markó, M. and Flückiger, E. (1974). *Experientia* **30,** 1174–1176.
Martin, J. B. (1976). *In* "Frontiers in Neuroendocrinology", Vol. 4 (Eds. L. Martini and W. F. Ganong), 129–168. Raven Press, New York.
Martin, J. B., Kontor, J. and Mead, P. (1973). *Endocrinology* **92,** 1354–1361.
Maurer, R. A. and Gorski, J. (1977). *Endocrinology* **101,** 76–84.
Mazzuca, M. and Poulain, P. (1974). *Brain Res.* **68,** 281–295.
Moore, K. E. and Gudelsky, G. A. (1977). *In* "Advances in Biochemical Psychopharmacology", Vol. 16 (Eds. E. Costa and G. L. Gessa), 227–235. Raven Press, New York.
Morgan, C. M. and Hadley, M. E. (1976). *Neuroendocrinology* **21,** 10–19.
Mueller, G. P., Simpkins, J., Meites, J. and Moore, K. E. (1976). *Neuroendocrinology* **20,** 121–135.
Müller, E. E., Liuzzi, A., Cocchi, D., Panerai, A. E., Oppizzi, G., Locatelli, V.,

Mantegazza, P., Silvestrini, F. and Chiodini, P. G. (1977). *In* "Advances in Biochemical Psychopharmacology", Vol. 16 (Eds. E. Costa and G. L. Gessa), 127–138. Raven Press, New York.

Nemeroff, C. B., Konkol, R. J., Bissette, G., Youngblood, W., Martin, J. B., Brazeau, P., Rone, M. S., Prange, A. J. Jr., Breese, G. R. and Kizer, J. S. (1977). *Endocrinology* **101,** 613–622.

Nicoll, C. S., Yaron, Z., Nutt, N. and Daniels, E. (1970). *Biol. Reprod.* **5,** 59–66.

Noel, G. L., Suh, H. K. and Frantz, A. G. (1973). *J. Clin. Endocrinol. Metab.* **36,** 1255–1258.

Ojeda, S. R. and McCann, S. M. (1973). *Neuroendocrinology* **12,** 295–315.

Oliver, C., Barnea, A., Warberg, J., Eskay, R. L. and Porter, J. C. (1977). *In* "Frontiers in Hormone Research", Vol. 4 (Eds. F. J. H. Tilders, D. F. Swaab and Tj. B. van Wimersma Greidanus), 162–166. Karger, Basel.

Olivereau, M. (1975). *Gen. Comp. Endocrinol.* **26,** 550–561.

Olson, L., Fuxe, K. and Hökfelt, T. (1972). *Acta Endocrinol.* (*Kbh.*) **71,** 233–244.

Parsons, J. A. (1970). *J. Physiol.* (*London*) **210,** 973–987.

Pasteels, J. L., Danguy, A., Frerotte, M. and Ectors, F. (1971). *Ann. Endocrinol.* (*Paris*) **32,** 188–192.

Platt, J. E. and Norris, D. O. (1974). *J. Expl. Zool.* **189,** 7–12.

Plotnikoff, N. P., Kastin, A. J., Anderson, M. S. and Schally, A. V. (1971). *Life Sci.* **10,** 1279–1283.

Quijada, M., Illner, P., Krulich, L. and McCann, S. M. (1973/74). *Neuroendocrinology* **13,** 151–163.

Reubi, J.-C., Iversen, L. L. and Jessell, T. M. (1977). *Nature* (*London*) **268,** 652–654.

Rotsztejn, W. H., Charli, J. L., Patton, E. and Kordon, C. (1977). *Endocrinology* **101,** 1475–1483.

Sawaki, Y. and Yagi, K. (1976). *J. Physiol.* (*London*) **260,** 447–460.

Sawyer, C. H., Markee, J. E. and Hollinshead, W. H. (1947). *Endocrinology* **42,** 395–402.

Scapagnini, U., Annunziato, L., Di Renzo, G. F., Schettini, G. and Preziosi, P. (1977). *In* "Advances in Biochemical Pyschopharmacology", Vol. 16 (Eds. E. Costa and G. L. Gessa), 369–375. Raven Press, New York.

Scardapane, L. and Cardinali, D. P. (1977). *J. Neur. Trans.* **40,** 81–86.

Schally, A. V., Redding, T. W., Arimura, A., Dupont, A. and Linthicum, G. L. (1977). *Endocrinology* **100,** 681–691.

Schneider, H. P. G. and McCann, S. M. (1969). *Endocrinology* **85,** 121–132.

Schneider, H. P. G. and McCann, S. M. (1970). *Endocrinology* **86,** 1127–1133.

Scott, D. E. and Krobisch-Dudley, G. (1975). *In* "Brain–Endocrine Interaction II" (Eds. K. M. Knigge, D. E. Scott, H. Kobayashi and S. Ishii), 29–39. Karger, Basel.

Shaar, C. J. and Clemens, J. A. (1972). *Endocrinology* **90,** 285–288.

Shelesnyak, M. C. (1958). *Acta Endocrinol.* (*Kbh.*) 27, 99–109.

Shoemaker, W. J. and Schlumpf, M. (1977). *In* "Advances in Biochemical Psychopharmacology", Vol. 16 (Eds. E. Costa and G. L. Gessa), 359–367. Raven Press, New York.

Simantov, R., Kuhar, M. J., Uhl, G. R. and Snyder, S. H. (1977). *Proc. Natl. Acad. Sci., U.S.A.* **77,** 2167–2171.

Sladek, J. R. Jr., Sladek, C. D., McNeill, T. H. and Wood, J. G. (1978). *In* "Brain–Endocrine Interaction III" (Eds. D. E. Scott, G. P. Kozlowski and A. Weindl), 154–171. Karger, Basel.

Smalstig, E. B., Sawyer, B. D. and Clemens, J. A. (1974). *Endocrinology* **95,** 123–129.
Smelik, P. G. (1967). *Neuroendocrinology* **2,** 247–254.
Smith, A. F. (1975). *Neuroendocrinology* **19,** 363–376.
Stoeckart, R., Kreike, A. J. and Jansen, H. G. (1975). *J. Endocrinol.* **64,** 57P.
Takahara, J., Arimura, A. and Schally, A. V. (1974a). *Endocrinology* **95,** 462–465.
Takahara, J., Arimura, A. and Schally, A. V. (1974b). *Endocrinology* **95,** 1490–1494.
Taleisnik, S., Tomatis, M. E. and Celis, M. E. (1972). *Neuroendocrinology* **10,** 235–245.
Taleisnik, S., Celis, M. E. and Tomatis, M. E. (1973/74). *Neuroendocrinology* **13,** 327–338.
Talwalker, P. K., Meites, J., Nicoll, C. S. and Hopkins, T. F. (1960). *Am. J. Physiol.* **199,** 1073–1076.
Terasawa, E., Bridson, W. E., Davenport, J. W. and Goy, R. W. (1975). *Neuroendocrinology* **18,** 345–359.
Tilders, F. J. H. and Mulder, A. H. (1975). *J. Endocrinol.* **64,** 63P–64P.
Tuomisto, J., Ranta, T., Mannisto, P., Saarinen, A. and Leppaluoto, J. (1975). *Eur. J. Pharmacol.* **30,** 221–229.
Upton, G. V. and Corbin, A. (1975). *Experientia* **31,** 249–250.
Van der Gugten, A. A., Sahuleka, P. C., Van Galen, G. H. and Kwa, H. G. (1976). *J. Endocrinol.* **68,** 369–381.
Van Loon, G. R. and Kragt, C. L. (1970). *Proc. Soc. Expl. Biol. Med.* **133,** 1137–1141.
Van Maanen, J. H. and Smelik, P. G. (1968). *Neuroendocrinology* **3,** 177–186.
Voogt, J. L. and Carr, L. A. (1974). *Neuroendocrinology* **16,** 108–118.
Voogt, J. L. and Carr, L. A. (1975). *Endocrinology* **97,** 891–897.
Weiner, R. I. (1975). *In* "Hypothalamic Hormones" (Eds. M. Motta, P. G. Crosignani and L. Martini), 249–253. Academic Press, London and New York.
Weiner, R. I., Shryne, J. E., Gorski, R. A. and Sawyer, C. H. (1972). *Endocrinology* **90,** 867–873.
Weiner, R. I., Selmanoff, M. K., Cheung, C. Y., Cronin, M. J. and Goldsmith, P. C. (1978). *In* "Cell Biology of Hypothalamic Neurosecretion" (Eds. J. D. Vincent and C. Kordon), 101–115. Editions du C.N.R.S., Paris.
Willoughby, J. O., Terry, L. C., Brazeau, P. and Martin, J. B. (1977). *Brain Res.* **127,** 137–152.
Wirz-Justice, A., Hackmann, E. and Lichtsteiner, M. (1974). *J. Neurochem.* **22,** 187–189.
Wuttke, W., Cassell, E. and Meites, J. (1971). *Endocrinology* **88,** 737–741.

CHAPTER 29

Dopamine Receptors in Invertebrates

G. N. Woodruff

School of Biochemical and Physiological Sciences, University of Southampton, Southampton, U.K.

INTRODUCTION

Dopamine is widely distributed throughout the animal kingdom. In the invertebrates its presence has been detected in most of the major Phyla, for example Platyhelminthes, Coelenterates, Annelids, Molluscs, Arthropods and Echinoderms. The distribution of dopamine in the invertebrate kingdom has been fully covered in recent reviews (Welsh, 1972; Fange and Hanson, 1973; Kerkut, 1972).

There have been relatively few studies on the biosynthesis of dopamine in invertebrates. As far as is known it seems that dopamine is mainly synthesised by a similar route to that occurring in mammals. Details of the biosynthesis of dopamine are included in the comprehensive review of Fange and Hanson (1973).

One of the major physiological roles of dopamine in invertebrates is that of a neurotransmitter. It may, however, have additional roles. For example in insects, dopamine is the precursor of *N*-acetyldopamine, which is involved in cuticular tanning (Sekeris and Karlson, 1962; Karlson and Sekeris, 1962; Murdock, 1971).

The present review will be almost entirely concerned with the receptors for dopamine in invertebrate nervous systems. A comparison will be made between invertebrate dopamine receptors and those in the mammalian system, and the value of invertebrate systems as models of mammalian dopamine receptors will be discussed.

DOPAMINE RECEPTORS IN *Helix aspersa* AND RELATED SPECIES

The isolated snail brain has proved to be a useful preparation for pharmacological studies. The snail brain contains large neurones, which are readily penetrated with microelectrodes. The neurones are easily recognizable from one preparation to another. In studying the effects of potential agonists, the drugs can be applied directly to the experimental bath. In the study of antagonists, it is often more convenient to apply the dopamine by iontophoretic application.

In the study of dopamine or dopamine-like drugs on invertebrate neurones, care should be taken to ascertain the pharmacological identity of the receptor mediating the response. For example, there is evidence indicating that there might be receptors for octopamine and phenylethanolamine on some molluscan neurones (Carpenter and Gaubatz, 1974; Batta *et al.*, 1977) and dopamine might possibly have affinity for these receptors. In the mammalian system dopamine is known to have actions on α-adrenoceptors. The presence of more than one type of receptor on a single neurone would further complicate the situation.

Effect of Agonists on *Helix* Dopamine Receptors

Cells inhibited by dopamine

Effect of phenylethylamine derivatives

Woodruff and Walker (1969) studied the structure–activity requirements for dopamine like activity on neurones in *Helix aspersa* that were inhibited by dopamine. The neurones studied were located mainly in the right parietal ganglion and on these cells dopamine caused a powerful inhibition of firing with hyperpolarization of the resting potential (Fig. 1). Dopamine and dopamine analogues were applied by addition to the bath or by microiontophoresis. The potency of the compounds was expressed as the equipotent molar ratio (EPMR) relative to dopamine. The EPMR is defined as the amount of compound (mol) required to produce the same response as 1 mol dopamine.

It was immediately apparent that the receptor mediating the inhibitory effect of dopamine on these cells was quite unlike the mammalian α- or β-adrenoceptor. Thus the α-stimulant phenylephrine and the β-stimulant

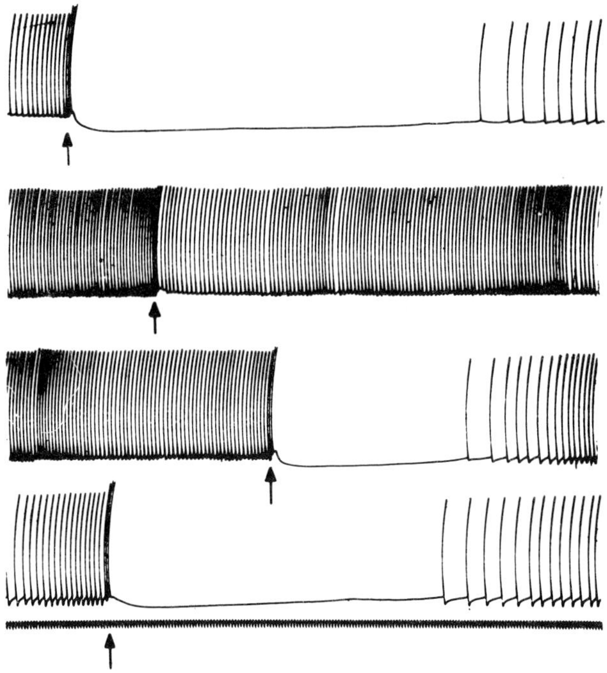

Fig. 1 Intracellular recording from neurone in the isolated brain of *Helix aspersa*. Drugs were applied at the arrows. In A and D, the addition of ADTN (4 nmol) caused a strong hyperpolarization and inhibition. In B, 2-amino-5,6-dihydroxy-1,2,3,4-tetrahydronaphthalene (4 nmol) had little effect. C. The effect of 5 nmol dopamine. Bath volume 20 ml. All traces are consecutive records from the same cell. Time trace 1 s.

isoprenaline are both completely inactive in concentrations of up to 2000 times the effective dose of dopamine (Table I). Epinine, which is *N*-methyldopamine, was equipotent with dopamine in causing inhibition of firing of these cells. The introduction of an ethyl or an isopropyl group on to the terminal nitrogen, to give *N*-ethyldopamine and *N*-isopropyldopamine causes a considerable diminution of dopamine-like activity. Compounds containing a hydroxyl group on the carbon atom β to the terminal nitrogen are similarly less active than dopamine. Thus (−)noradrenaline and (−)adrenaline are about 25 and 92 times less active than dopamine respectively.

The structure activity requirements for dopamine-like activity on *Helix* neurones are fairly precise (Woodruff and Walker, 1969; Woodruff, 1971). For example, the only active phenylethylamine derivatives on this system are those which contain hydroxyl groups in both the 3 and 4 positions of

TABLE I Relative potencies of some phenylethylamine derivatives on dopamine receptors in snail and rat. In the snail, EPMR is expressed relative to dopamine tested on the same cell. On adenylate cyclase, EPMR is expressed relative to the EC_{50} (concentration causing 50% stimulation) for dopamine in striatal homogenates. In = inactive. Data from Woodruff and Walker, 1969 (snail neurones inhibition), Struyker Boudier, 1975 (snail neurones excitation) and Munday *et al.*, 1976, 1977 (adenylate cyclase).

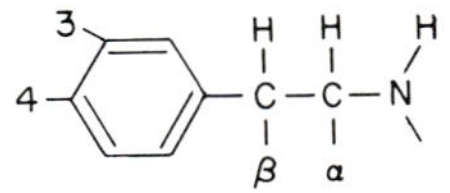

	3	4	β	α	N	Equipotent molar ratio (EPMR): Snail neurones: Inhibition	Snail neurones: Excitation	Rat adenylate cyclase: Striatum	Rat adenylate cyclase: Nucleus accumbens
Dopamine	HO	HO	H	H	H	1	1	1	2
Epinine	HO	HO	H	H	Me	1	1	1	1
(−)-Noradrenaline	HO	HO	HO	H	H	25	4	24	9
(−)-Adrenaline	HO	HO	HO	H	Me	92	13	63	16
(±)-Isoprenaline	HO	HO	HO	H	$CH(Me)_2$	In	300	In	In
(−)-Phenylephrine	HO	H	HO	H	Me	In	In	In	In
Tyramine	H	HO	H	H	H	In	In	In	In

the benzene ring. Compounds containing a single hydroxyl group on either the 3 or the 4 positions of the benzene ring are inactive on this *Helix* dopamine receptor (Table I).

In recent years, the dopamine-sensitive adenylate cyclase, present in several regions of the mammalian brain (Kebabian *et al.*, 1972; Horn *et al.*, 1974; Phillipson and Horn, 1976) has been successfully used to study the actions of dopamine-like drugs. It is of interest that the structural requirements for dopamine-like activity on the adenylate cyclase from rat striatum and rat nucleus accumbens are similar to the structural requirements for dopamine-like inhibitory activity on *Helix* neurones (Table I). This suggests that the dopamine receptors in invertebrates are similar to those in the mammalian brain.

Effect of apomorphine

Apomorphine (Fig. 2) was suggested by Ernst (1965), as a result of behavioural studies, to be an agonist at dopamine receptors. It was therefore somewhat surprising to find that apomorphine had no dopamine-like inhibitory activity in concentrations of up to 1000 times the effective concentration of dopamine (Woodruff and Walker, 1969). It has more recently been shown that apomorphine in fact behaves as an antagonist in this system (Struyker Boudier, 1975; Batta, Woodruff and Walker, unpublished). Recent work has indicated that apomorphine may not be as powerful an agonist on mammalian postsynaptic receptors as was previously imagined. For example, on the dopamine-sensitive adenylate cyclase in rat striatum, apomorphine behaves as a partial agonist, producing a maximum response that is only about 50% of that produced by dopamine (Kebabian *et al.*, 1972; Miller *et al.*, 1974). It is possible that on snail neurones apomorphine does initiate small conductance changes that are not sufficient to trigger membrane hyperpolarizations. Such changes could possibly be detected using voltage clamp techniques. If this were so, then apomorphine could be regarded as a partial agonist on dopamine receptors mediating inhibition of *Helix* neurones.

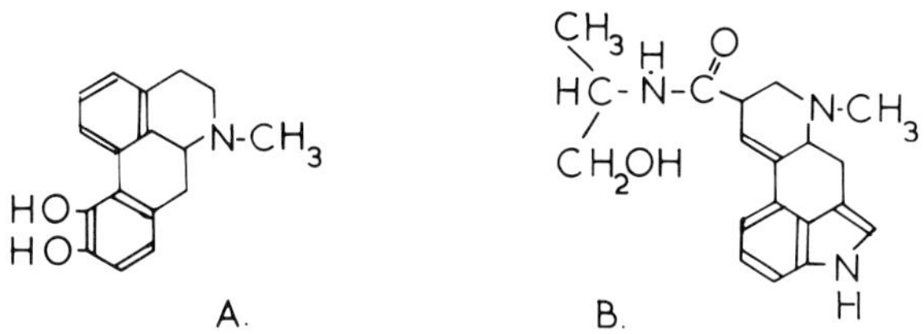

Fig. 2 The structures of A, apomorphine; B, ergometrine.

Effect of 3,4-(*dihydroxyphenylamino*)-2-*imidazoline* (*DPI*)

3,4-Dihydroxyphenylamino-2-imidazoline (DPI) has been reported as an agonist at inhibitory dopamine receptors in *Helix aspersa* brain, whilst having no effect on cells that are excited by dopamine (Struyker Boudier *et al.*, 1975). This indicates that the inhibitory dopamine receptors are different from the receptors on neurones where dopamine has an excitatory action. Further studies on DPI would be of value. We have shown that DPI has only very weak dopamine-like activity on the dopamine-sensitive adenylate cyclase from the rat nucleus accumbens (Watling, Poat and Woodruff, unpublished).

Conformational studies

In solution, dopamine can exist in extended or folded conformations. In the extended conformation, rotation about the phenyl–carbon bond leads to the possible existence of two rotameric extremes α and β (for references see Woodruff *et al.*, 1977). It is possible to obtain information about the active conformation of dopamine at its receptor site by studying the actions of rigid analogues of dopamine which contain the dopamine side chain locked in a more or less fixed conformation. Thus, norsalsolinol contains the dopamine skeleton in a folded conformation. The extended conformation of dopamine (β-rotamer) is contained in the molecule of 2-amino-6, 7-dihydroxy-1,2,3,4-tetrahydronaphthalene (ADTN). The compound 2-amino-5,6-dihydroxy-1,2,3,4-tetrahydronaphthalene contains the dopamine skeleton as it exists in the extended conformation (α-rotamer). These different conformations of dopamine, together with the corresponding rigid analogues are illustrated in Fig. 3. Norsalsolinol and salsolinol (1-methyl-6,7-dihydroxy-1,2,3,4-tetrahydroisoquinoline) have no dopamine like activity on these *Helix* dopamine receptors (Woodruff, 1971; Batta, Woodruff and Walker, unpublished). We have previously shown that ADTN is a potent agonist on these dopamine receptors. More recently we have directly compared the potency of ADTN with that of 2-amino-5,6-dihydroxy-1,2,3,4-tetrahydronaphthalene on the dopamine receptors mediating neuronal inhibition in *Helix*. (Batta, Woodruff and Walker, unpublished). In this latest study ADTN was equipotent with dopamine on many cells. At its least active ADTN was only about five times less active than dopamine. On the other hand 2-amino-5,6-dihydroxy-1,2,3,4-tetrahydronaphthalene was always much less active than dopamine. The potency of ADTN was between 100 and 1000 times greater than that of 2-amino-5,6-dihydroxy-1,2,3,4-tetrahydronaphthalene (Fig. 1). These results strongly suggest that the active conformation of dopamine on these receptors is the extended form, β rotamer. This represents a further point of

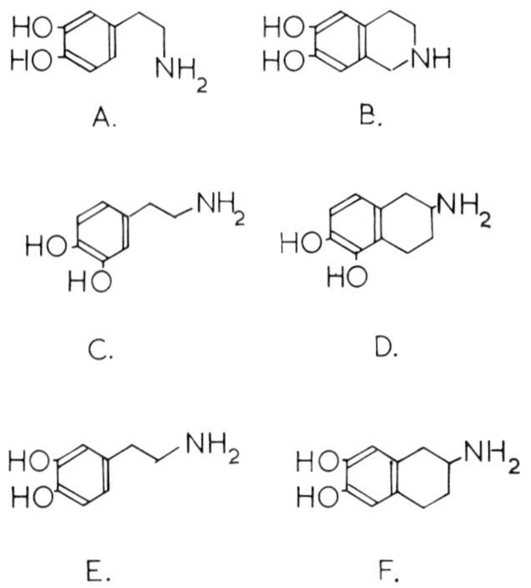

Fig. 3 Conformations of dopamine and the structures of the corresponding rigid analogues. Dopamine in the folded conformation (A) corresponds to norsalsolinol (B). In the extended form, dopamine can exist as the α-rotamer (C), which corresponds to 2-amino-5,6-dihydroxy-1,2,3,4-tetrahydronaphthalene (D), or as the β rotamer (E), which corresponds to ADTN (F).

similarity between the invertebrate dopamine receptors and those in the mammalian brain, since we have previously shown that ADTN is much more active than 2-amino-5,6-dihydroxy-1,2,3,4-tetrahydronaphthalene on dopamine receptors associated with dopamine sensitive adenylate cyclase in rat striatum or nucleus accumbens (Woodruff *et al.*, 1977).

Cells excited by dopamine

In addition to the neurones in *Helix* brain that are inhibited by dopamine, there are a second group of cells on which dopamine has an excitatory action. Other neurones are unaffected by high concentrations of dopamine. Ascher (1972) showed that neurones in *Aplysia* responded to iontophoretically applied dopamine by depolarization, hyperpolarization or, in a third group of cells, the response was of a biphasic nature. He produced evidence that the receptors associated with hyperpolarization differed from those concerned with depolarization.

Effect of phenylethylamine derivatives

The structure activity requirements for dopamine-like activity on cells excited by dopamine has been studied by Struyker Boudier *et al.* (1974). The results are shown in Table I. It can be seen that noradrenaline and adrenaline were less active than dopamine although both were slightly more active, relative to dopamine, on cells excited by dopamine than on cells inhibited.

Struyker Boudier *et al.* (1974) have suggested, partly as a result of structure activity studies on agonists, that the dopamine receptors mediating inhibitions were different from the dopamine receptors mediating excitation of *Helix* neurones. The differences between the relative potencies of the phenylethylamine derivatives on the two types of neurones are, however, slight. Such differences could possibly be explained by differences in the relative affinities for uptake sites or metabolizing enzymes on the two types on cell. Certainly there are great points of similarity between the structure–activity relationships for dopamine-induced inhibitions and excitations. For example epinine is equipotent with dopamine in causing excitation or inhibition of dopamine-sensitive cells, whereas noradrenaline and adrenaline are both less active. The α-stimulant phenylephrine is inactive on both types of cell. The β-stimulant isoprenaline, is inactive on neurones that are inhibited by dopamine and weakly active (300 times less than dopamine) on neurones on which dopamine has a depolarizing action. Tyramine, 3-methoxytyramine, amphetamine and 3,4-dimethoxyphenylethylamine are inactive on both types of neurone. Swann and Carpenter (1975) have investigated three types of dopamine-sensitive cells in *Aplysia.* The first cell type was excited by dopamine, the response being apparently due to an increase in Na^+ conductance. The other two types of neurone were both inhibited by dopamine; the responses in these cells were brought about by either a Cl^- or a K^+ conductance increase. The cells were unaffected by octopamine or phenylethanolamine although other cells have receptors for these substances. It was concluded that the fact that the various phenylethylamines have similar relative potencies in activating the three dopamine responses is consistent with the view that the dopamine binding site is identical irrespective of the conductance change elicited (Swann and Carpenter, 1975).

Effect of apomorphine

Apomorphine, which has no agonist activity on cells inhibited by dopamine is nevertheless capable of stimulating neurones that are excited by dopamine. It is, however, considerably less active than dopamine in this respect. The maximum depolarization produced by apomorphine is less than half that produced by dopamine. Apomorphine thus acts as a partial agonist with an intrinsic activity of between 0.2 and 0.5 (Struyker Boudier, 1975). The effect of apomorphine on cells inhibited by dopamine has already been discussed.

Conformational studies

The strongest point of similarity between the dopamine receptors on cells that are inhibited or excited by dopamine comes from conformational studies. Norsalsolinol has little effect on cells depolarized by dopamine, whereas ADTN is equipotent with dopamine on these neurones. 2-Amino-5,6-dihydroxy-1,2,3,4-tetrahydronaphthalene is a weak agonist on these cells in *Helix aspersa*, being more than a 100 times less active than ADTN. These results suggest that the active conformation of dopamine on cells depolarized by dopamine is in the extended form (β-rotamer). This is identical to the situation on neurones on which dopamine has an inhibitory effect.

Effect of Antagonists on *Helix* Dopamine Receptors

Cells inhibited by dopamine

Neuroleptics

The neuroleptic drugs haloperidol and chlorpromazine were found to have no blocking action on dopamine-evoked inhibitions in *Helix* brain (Woodruff *et al.*, 1970, 1971; Woodruff, 1971). These results were confirmed by Struyker Boudier *et al.* (1974). Heiss and Hoyer (1974) were able to show in *Aplysia*, that the inhibitory action of dopamine could be antagonized by haloperidol (50 μM) or fluphenazine (10 or 100 μM), but only after the neuroleptic had been present in the bath for periods of from 8 to 40 min, conditions where specificity of action must be questioned. Indeed, the authors showed that "long lasting perfusion with 100 μM fluphenazine impaired the acetylcholine response" and that after 35 min perfusion with mepazine 100 μM (a non-antipsychotic phenothiazine with presumably little dopamine blocking activity), the dopamine response "started to decrease".

In recent years it has become apparent that the butyrophenone type of neuroleptics may have considerable presynaptic activity and that their antagonistic activity at postsynaptic dopamine receptors might not be as powerful as would have been predicted by behavioural studies (Iversen, 1975; Seeman and Lee, 1975). The use of the dopamine sensitive adenylate cyclase model has shown that the thioxanthine, α-flupenthixol, the phenothiazine, fluphenazine and (+)-butaclamol are among the most potent dopamine antagonists *in vitro*. Metoclopramide, which behaviourally has dopamine antagonistic activity, is inactive as a dopamine antagonist on the dopamine-sensitive adenylate cyclase (Elliott *et al.*, 1977). We have recently shown (Batta, Woodruff and Walker, unpublished) that fluphenazine, α-flupenthixol and (+)-butaclamol are able to block the inhibitory effect of dopamine on snail neurones, although the compounds are not particularly potent in this respect. Metoclopramide is similarly an antagonist of dopamine-induced inhibitions, although it is not certain that this compound is blocking at the level of the receptor. On the cells on which we have studied the actions of metoclopramide dopamine appears to cause a hyperpolarization by increasing the K^+ conductance. On other neurones glutamate also hyperpolarizes by increasing K^+ conductance. Metoclopramide can antagonize the actions of dopamine and glutamate respectively on the two types of cell. Thus it is possible that the metoclopramide effect is due to a blockade of K^+ conductance rather than to an action at the receptor level (Batta, Woodruff and Walker, unpublished). Further investigations on this are required.

Ergot alkaloids

The most potent known antagonists of dopamine on invertebrate dopamine receptors are certain derivatives of lysergic acid. Ergometrine (Fig. 2) was first shown to be a potent, long lasting, antagonist on the dopamine inhibitory response in *Helix aspersa* (Walker *et al.*, 1968). An example of this blocking action of ergometrine is shown in Fig. 4. Although it is a potent dopamine antagonist, ergometrine has little or no α-blocking activity (Brown and Dale, 1936). Ergometrine has been shown to antagonize the inhibitory actions of dopamine on neurones in *Aplysia* (Ascher, 1972) and *Planorbis* (Berry and Cottrell, 1975). Ergometrine will also antagonize the excitatory actions of dopamine on molluscan neurones, although it is less active on these cells than on neurones where dopamine has an inhibitory action. It is of course well known that ergometrine is a potent antagonist of 5-hydroxytryptamine (Cerletti and Doepfner, 1958). In the snail brain the 5 HT-blocking activity of ergometrine seems to be unrelated to its dopamine-blocking activity (Woodruff, 1971). In the mammalian central nervous system ergometrine acts as a partial agonist, with both agonistic and antagonistic activity (Woodruff, 1978).

Other ergot alkaloids are also dopamine antagonists in molluscs. Lysergic acid diethylamide is particularly potent as an antagonist of dopamine-evoked inhibitions in *Helix* (Woodruff *et al.*, 1971).

6-Hydroxydopamine

Apart from its known actions in depleting catecholamine stores, 6-hydroxydopamine may have postsynaptic actions. It has been shown that 6-hydroxydopamine can block both the inhibitory and excitatory action of dopamine in *Planorbis*, being more active on the former actions (Cottrell *et al.*, 1974).

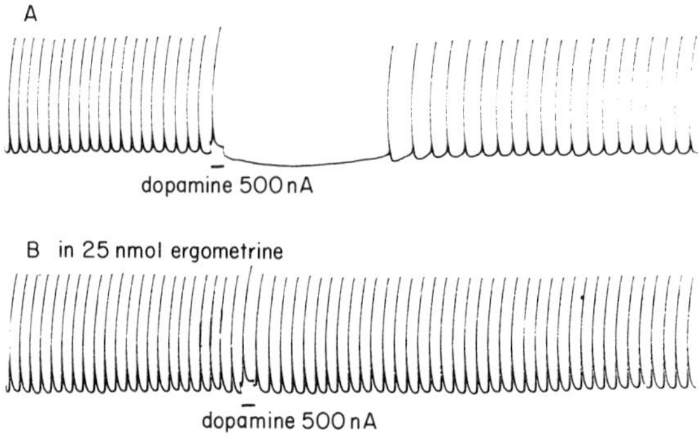

Fig. 4 Effect of ergometrine on the dopamine inhibitory response in *Helix aspersa*. A. Intracellular recording from dopamine-sensitive cell; dopamine (500 nA) applied iontophoretically for 1 s caused hyperpolarization and inhibition. B. In the presence of 25 nmol ergometrine, the dopamine response was completely blocked.

Cells excited by dopamine

Struyker Boudier *et al.* (1974) showed that the neuroleptics haloperidol and chlorpromazine, which as previously mentioned have little blocking activity on dopamine-induced inhibitions in *Helix*, are nevertheless able to block the excitatory actions of dopamine in the snail brain. This was taken as evidence that the receptors mediating dopamine-induced excitations are different from those mediating neuronal inhibitions. Ascher (1972) investigated the actions of a series of what he called noradrenaline antagonists on the inhibitory and excitatory actions of dopamine on *Aplysia* neurones. He found that phentolamine, tolazoline, phenoxybenzamine, chlorpromazine, pronethalol, dichloroisoprenaline and 1-(4-nitrophenyl)-2-isopropylaminoethanol were all found to be "better antagonists of the excitatory DA effects than of the inhibitory ones". In addition most of these compounds reduced the acetylcholine response. Unfortunately the paper was lacking in further details. These results raise the possibility that the excitatory actions of dopamine are more readily blocked, by a non-specific mechanism, than are the inhibitory actions. Since many neuroleptics are known to have non-specific membrane effects, it is possible that at least a component of the apparent blocking action of neuroleptics on dopamine excitations is due to a non-specific mechanism.

Dopamine as a Neurotransmitter in Molluscs

There have been several studies which indicate that the presence of dopamine receptors on molluscan neurones reflects the physiological role of dopamine as a neurotransmitter in these species. A detailed account of this topic is outside the scope of the present article. One particularly convincing piece of work in this field has been carried out in the isolated brain of *Planorbis corneus* by Cottrell and his colleagues. Marsden and Kerkut (1970) first showed, using the technique of of fluorescence histochemistry, that there was a giant dopamine-containing neurone in the left pedal ganglion of *Planorbis corneus*. Subsequent microbiochemical studies (Powell and Cottrell, 1974) have confirmed that this neurone contains dopamine but lacks noradrenaline. The axonal branches of this giant dopamine neurone have been traced and shown to innervate follower neurones in the left parietal and visceral ganglia (Berry and Cottrell, 1975; Cottrell *et al.*, 1974). Action potentials in the giant dopamine cell result in the appearance of EPSPS, IPSPS, or biphasic hyperpolarizing/depolarizing

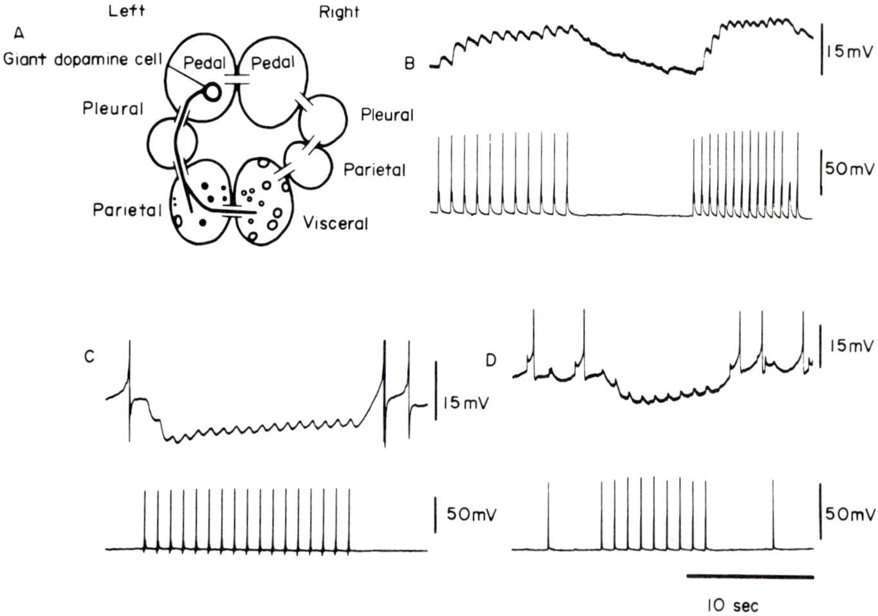

Fig. 5 A. Diagram of the dorsal surfaces of the suboesophageal ganglionic ring of *Planorbis corneus*; the positions of the giant dopamine cell and of its follower neurones are shown. Open and closed circles indicate follower cells receiving inhibitory and excitatory inputs respectively from the giant dopamine cell. B. Shows excitatory responses of a parietal ganglion cell (upper record) to stimulation of the giant dopamine cell (lower record). The ganglion cell was artificially hyperpolarized to −90 mV. C. Inhibitory responses of a visceral ganglion cell (upper record) produced by stimulation of the giant dopamine cell (lower record). D. Biphasic response of a visceral ganglion cell. Figure reproduced from Cottrell *et al.* (1974) by kind permission of Dr G. A. Cottrell and Pergamon Press.

potentials in the follower cells. Examples of this phenomenon are shown in Fig. 5 which is taken from the paper of Cottrell *et al.* (1974). The authors have also shown that ergometrine inhibits both the IPSPS in the follower cell resulting from stimulating the giant cell, and also the hyperpolarization produced by applied dopamine. Thus there is strong evidence that dopamine functions both as an inhibitory transmitter and as an excitatory transmitter in the brain of *Planorbis*.

DOPAMINE RECEPTORS IN SALIVARY GLANDS

Salivary Glands of the Cockroach

Salivary glands of the cockroach (*Nauphoeta cinerea*, Olivier) can be dissected out with an intact nerve input. Ginsborg, House and their co-workers at Edinburgh have shown that the addition of dopamine to the preparation produces a powerful hyperpolarization of the salivary gland cells and have provided strong evidence that dopamine is the transmitter involved in this hyperpolarization (House, 1973; Bland *et al.*, 1973; House *et al.*, 1973; Ginsborg *et al.*, 1974; Bowser-Riley and House, 1976). Ginsborg *et al.* (1976) have studied the actions of a series of dopamine-like compounds on this preparation. The only active compounds tested contained hydroxyl groups in the 3 and 4 positions of the benzene ring. *N*-Acetyldopamine was inactive, whereas epinine was an active agonist. In a more recent study, House and Ginsborg (1976) have shown that ADTN is equipotent with dopamine in hyperpolarizing cockroach salivary gland cells. The same authors have also shown that α-flupenthixol blocks both the effects of dopamine and the hyperpolarizing potentials elicited by nerve stimulation. The dopamine receptors in the cockroach salivary gland thus appear to resemble those in other preparations. The preparation is of great potential value in the study of drugs affecting dopamine receptors.

Salivary Glands of Ixodid Ticks

Salivary gland cells can be isolated from two species of ixodid ticks (*Dermacentor andersoni*, Stiles and *Amblyomma hebraeum*, Koch). Kaufman (1977) has studied the effects of a series of catecholamines and other derivatives for their ability to elicit fluid secretion in these isolated salivary glands. Dopamine was the most potent catechol derivative tested, noradrenaline and adrenaline were less active than dopamine. Phenylephrine and isoprenaline were each weakly active. Dopamine-induced secretions were depressed by a number of drugs such as phenoxybenzamine, α-flupenthixol, phentolamine and propranolol, but only when the antagonist was present in concentrations of from 10 to 1000 times that of dopamine. Ergometrine was itself a potent agonist in causing fluid secretion, but Kaufman (1977) has not yet determined whether ergometrine acts via a catecholamine receptor or by a direct action on the secretory mechanism. Whether or not the effects of dopamine are mediated via a true dopamine receptor is not yet certain. For example the tick salivary gland responds to several phenylethylamine derivatives which are known to be inactive on conventional dopamine receptors. However, as was pointed out by Kaufman (1977), some of these active compounds may exert their effects indirectly by releasing an endogenous transmitter. The mechanism of action of dopamine on this interesting preparation deserves further investigation.

DOPAMINE AND CILIARY ACTIVITY

There is evidence for a physiological role for dopamine in controlling ciliary activity in the gill of *Mytilus edulis* and *Modiolus demissus* (Malanga, 1975). Dopamine produces a stimulation of the transport of particles by the gill of *Mytilus*. It is believed that this is due to a stimulation of frontal ciliary activity. Malanga (1975) has shown that epinine mimics this effect of dopamine, whereas ergometrine is a dopamine antagonist in this system. Noradrenaline, adrenaline and apomorphine were inactive in the concentrations tested. More recently, in a brief report, Malanga *et al.* (1977) have shown that ADTN is a potent dopamine-like agonist of the cilioexcitatory response of *Modiolus demissus*.

DOPAMINE RECEPTORS IN MOLLUSCAN HEART AND GUT

Tapes watlingi

Dougan and McLean (1970) showed that dopamine was a potent inhibitor of electrically induced contractions of the isolated cardiac intestine of the mollusc *Tapes watlingi*. (—)-Noradrenaline, (—)-adrenaline and isoprenaline had a similar action but were considerably less active than dopamine. The effects of dopamine were blocked by ergometrine. The authors produced good evidence for a physiological role for dopamine in this tissue and for the presence of specific dopamine receptors in the gut of *Tapes*.

Metoclopramide antagonizes the inhibitory effects of dopamine on the electrically stimulated gut of *Tapes watlingi* (Dougan *et al*., 1974).

Chione stutchburyi

Gordon and McGregor (1976) have examined the action of dopamine on spontaneously beating ventricle of another lamellibranch mollusc, *Chione stutchburyi*. Low concentrations of dopamine caused a dose-dependent decrease in the amplitude of the spontaneous beats. The inhibitory action of dopamine was antagonized by metoclopramide.

CONCLUSIONS

This review has been concerned with only a small part of the work that has been carried out on the role of dopamine in invertebrates. The molluscan nervous system is of value in the study of drugs believed to act on mammalian dopamine receptors. One of the major disadvantages of this system as a test system is that, in addition to the receptors for dopamine, there may also be receptors for octopamine and possibly other phenylethylamines on molluscan neurones. The presence of more than one receptor type on a single neurone can make studies on dopamine receptors difficult to interpret. As was pointed out by Ascher and Kehoe (1975), "the usefulness of gastropod neurones for the study of drug receptors will depend on the degree to which investigators take into account the special characteristics and limitations of this preparation". Nevertheless there is good evidence that dopamine is a neurotransmitter in gastropods and that some neurones contain specific dopamine receptors. In using this system as a model on which to study dopamine-like drugs, it is preferable to work on identified neurones, and, if possible, to select neurones which are known not to contain receptors for closely related substances such as octopamine or phenylethanolamine. Evidence has been presented that there is more than one type of dopamine receptor in invertebrate nervous system. This evidence cannot, at present, be refuted, although the evidence is not particularly compelling. There are, however, strong points of similarity between the receptors mediating dopamine-induced excitation or inhibition of neurones. For example conformational studies indicate that the active conformation of dopamine is in the extended form (β-rotamer) at both inhibitory and excitatory dopamine receptors in *Helix aspersa*. The study of the effects of agonists on some invertebrate neurones has also revealed remarkable similarities between invertebrate dopamine receptors and those in the mammalian system. The invertebrate dopamine receptors do, however, seem to be less susceptible to blockade by some of the neuroleptic drugs than are mammalian dopamine receptors. The original finding that ergometrine and LSD have effects on invertebrate dopamine receptors has now been extended to include dopamine receptors in the mammalian brain.

Other invertebrate dopamine receptors, such as those in salivary glands and those controlling ciliary activity seem to resemble those in molluscs and may provide very satisfactory model systems on which to study dopaminergic agonists and antagonists.

REFERENCES

Ascher, P. (1972). *J. Physiol.* (*London*) **225,** 173–209.
Ascher, P. and Kehoe, J. S. (1975). *In* "Handbook of Psychopharmacology", Vol. 4 (Eds. L. L. Iversen, S. D. Iversen and S. H. Snyder), 265–310. Plenum Press, New York.
Batta, S., Walker, R. J. and Woodruff, G. N. (1977). *J. Physiol.* (*London*) **270,** 63–64.
Berry, M. S. and Cottrell, G. A. (1975). *J. Physiol.* (*London*) **244,** 589–612.
Bland, K. P., House, C. R., Ginsborg, B. L. and Laszlo, I. (1973). *Nature* (*London*) **244,** 26.
Bowser-Riley, F. and House, C. R. (1976). *J. Expl. Biol.* **64,** 665–676.
Brown, G. L. and Dale, H. H. (1936). *Proc. R. Soc. B.* **118,** 446–477.
Carpenter, D. O. and Gaubatz, G. L. (1974). *Nature* (*London*) **252,** 483–485.
Cerletti, A. and Doepfner, W. (1958). *J. Pharmacol. Expl. Ther.* **122,** 124–136.
Cottrell, G. A., Berry, M. S. and Macon, J. Б. (1974). *Neuropharmacology* **13,** 431–439.
Dougan, D. F. H. and McLean, J. R. (1970). *Comp. Gen. Pharmacol.* **1,** 33–46.
Dougan, D. F. H., Mearrick, P. T. and Wade, D. N. (1974). *Clin. Expl. Pharmacol. Physol.* **1,** 473–478.
Elliott, P. N. C., Jenner, P., Huizing, G., Marsden, C. D. and Miller, R. (1977). *Neuropharmacology* **16,** 333–342.
Ernst, A. M. (1965). *Psychopharmacologia* **7,** 391–399.
Fange, R. and Hanson, A. (1973). *In* "Comparative Pharmacology", Vol. I (Ed. M. J. Michelson), 391–517. Pergamon, Oxford.
Ginsborg, B. L., House, C. R. and Silinsky, E. M. (1974). *J. Physiol.* (*London*) **236,** 723–731.
Ginsborg, B. L., House, C. R. and Turnbull, K. W. (1976). *Br. J. Pharmacol.* **57,** 133–140.
Gordon, A. R. and McGregor, D. D. (1976). *Proc. Univ. Otago Med. Sch.* **54,** 72–74.
Heiss, W. D. and Hoyer, J. (1974). *Experientia* **30,** 1318–1320.
Horn, A. S., Cuello, A. C. and Miller, R. J. (1974). *J. Neurochem.* **22,** 265–270.
House, C. R. (1973). *J. Expl. Biol.* **58,** 29–43.
House, C. R., Ginsborg, B. L. and Silinski, E. M. (1973). *Nature* (*London*) **245,** 63.
Iversen, L. L. (1975). *Science* **188,** 1084–1089.
Karlson, P. and Sekeris, C. E. (1962). *Nature* (*London*) **195,** 183–184.
Kebabian, J. W., Petzhold, G. L. and Greengard, P. (1972). *Proc. Natl. Acad. Sci., U.S.A.* **69,** 2145–2149.
Kerkut, G. A. (1972). *Br. Med. Bull.* **29,** 100–104.
Malanga, C. J. (1975). *Comp. Biochem. Physiol.* **51C,** 25–34.
Malanga, C. J., O'Donnell, J. P. and Poll, K. A. (1977). *Pharmacologist* **19,** 438.
Marsden, C. and Kerkut, G. A. (1970). *Comp. Gen. Pharmacol.* **1,** 101–116.

Miller, R. J., Horn, A. S., Iversen, L. L. and Pinder, R. J. (1974). *Nature (London)* **261,** 238–241.

Munday, K. A., Poat, J. A. and Woodruff, G. N. (1976). *Br. J. Pharmacol.* **57,** 452–453.

Munday, K. A., Poat, J. A., Watling, K. J. and Woodruff, G. N. (1977). *Br. J. Pharmacol.* **61,** 150–151.

Murdock, L. L. (1971). *Comp. Gen. Pharmacol.* **2,** 254–274.

Phillipson, O. T. and Horn, A. S. (1976). *Nature (London)* **261,** 418–420.

Powell, B. and Cottrell, G. A. (1974). *J. Neurochem.* **22,** 605–606.

Seeman, P. and Lee, T. (1975). *Science* **188,** 1217–1219.

Sekeris, C. E. and Karlson, P. (1966). *Pharmacol. Rev.* **18,** 89–94.

Struyker Boudier, H. A. J. (1975). "Catecholamine Receptors in Nervous Tissue", Thesis, University of Nijmegen.

Struyker Boudier, H. A. J., Gielen, W., Cools, A. R. and Rossum, J. A. van. (1974). *Arch. Int. Pharmacodyn.* **209,** 324–331.

Struyker Boudier, H., Teppama, L., Cools, A. and Rossum, J. van. (1975). *J. Pharm. Pharmacol.* **27,** 882–883.

Swann, J. W. and Carpenter, D. O. (1975). *Nature (London)* **258,** 751–754.

Walker, R. J., Woodruff, G. N., Glaizner, B., Sedden, C. B. and Kerkut, G. A. (1968). *Comp. Biochem. Physiol.* **24,** 455–469.

Welsh, J. H. (1972). *In* "Catecholamines" (Eds. H. Blaschko and E. Muscholl), 79–109. Springer, Berlin.

Woodruff, G. N. (1971). *Comp. Gen. Pharmacol.* **2,** 439–455.

Woodruff, G. N. (1978). *In* "Dopamine" (Eds. P. J. Roberts, G. N. Woodruff and L. L. Iversen), 89–118. Raven Press, New York.

Woodruff, G. N. and Walker, R. J. (1969). *Int. J. Neuropharmacol.* **8,** 279–289.

Woodruff, G. N., Walker, R. J. and Kerkut, G. A. (1970). *Comp. Gen. Pharmacol.* **1,** 54–60.

Woodruff, G. N., Walker, R. J. and Kerkut, G. A. (1971). *Eur. J. Pharmacol.* **14,** 77–80.

Woodruff, G. N., Watling, K. J., Andrews, C. D., Poat, J. A. and McDermed, J. D. (1977). *J. Pharm. Pharmacol.* **29,** 422–427.

CHAPTER 30

The Dopamine Vascular Receptor

Leon I. Goldberg

University of Chicago, Chicago, U.S.A.

Increased bloodflow produced by dopamine (DA) in the renal and mesenteric vascular beds of experimental animals and humans is the result of action of the amine on a specific DA vascular receptor (Goldberg, 1972; Goldberg *et al.*, 1978b). This statement now can be made because the following criteria necessary for identifying a new receptor have been met (Furchgott, 1972): (1) structure–activity relationships have been defined and a potency series has been generated (Table I); (2) the same order of potency occurs in the renal and mesenteric vascular beds; (3) vasodilation can be selectively antagonized (Table II); and (4) specific relaxation by DA can be demonstrated in isolated blood vessels.

METHODS

In Vivo

Most studies have been carried out in mongrel dogs anaesthetized with pentobarbital (McNay and Goldberg, 1966; Volkman *et al.*, 1977b). The kidney is exposed by flank incision and retroperitoneal dissection and an electromagnetic flow probe is placed on the renal artery. A 23-gauge hypodermic needle, bent to an angle of approximately 80° and connected to a constant infusion system is inserted into the artery proximal to the flow probe. The system is arranged for injection of drug into the renal artery.

TABLE I Approximate relative potency of active dopamine agonists in the intact canine renal vascular bed.

Agonists	Potency
Dopamine	1.0
Epinine (*n*-methyldopamine)	1.0
A-6,7-DTN	1.0
N-Methyl-A-6,7-DTN	0.5–1
N,*N*-Di-*n*-propyl dopamine	0.05
N,*n*-Di-propyl A-6,7-DTN	0.05
N,*n*-Propyl-*N*-*n*-butyl dopamine	0.05
N,*n*-Propyl-*N*-*n*-pentyl dopamine	0.05
N,*n*-Propyl-*N*-isobutyl dopamine	0.05
6-*N*-*n*-Propylnorapomorphine	0.02
Apomorphine (partial agonist)	0.01
N-*n*-propyl-*N*-isobutyl dopamine	0.003
N-*n*-propyl-*N*-phenethyl dopamine	0.003

TABLE II Inactive compounds.

Norepinephrine
Epinephrine
α-Methyl dopamine
N,*N*-Dimethyl dopamine
N-Propyl dopamine
A-5,6-DTN
N-Methyl-A-5,6-DTN
N,*N*-Dimethyl A-5,6-DTN
N,*N*-Dimethyl A-6,7-DTN
6,7-Dihydroxyisoquinoline
N-Methyl-6,7-dihydroxyisoquinoline
Amphetamine
Bromocryptine
Lergotrile
Piribedil
S-584

Arterial blood pressure is measured from the femoral or carotid artery and blood flow and blood pressure is simultaneously recorded on an appropriate polygraph. In order to eliminate the vasoconstriction produced by DA, phenoxybenzamine (POB) is administered intra-arterially in doses of $5\,mg\,kg^{-1}$ over a 15–30 min period. If the vasoconstricting effect of *l*-norepinephrine (1 μg) injected intra-arterially is not reversed, an additional

$5\,mg\,kg^{-1}$ of POB is administered. After POB, a complete dose-response curve of the increments in renal blood flow caused by DA can be recorded without interference from opposing vasoconstriction. Antagonists are studied by injecting the antagonist into the renal artery immediately before or simultaneously with DA, isoproterenol, and/or bradykinin. Most antagonists have exhibited a very short duration of action and the majority of studies were carried out with simultaneous injections. Specificity is determined by the extent DA-induced vasodilation is attenuated without affecting vasodilation caused by bradykinin or isoproterenol (Yeh *et al.*, 1969; Goldberg and Yeh, 1971; Kohli *et al.*, 1978). The same methods are used for measuring mesenteric and femoral blood flow with the exception that in the femoral experiments both sciatic and femoral nerves are cut and the paw circulation is occluded to minimize neurogenic effects and cutaneous shunts.

AGONISTS

Table I lists the active DA vascular agonists found thus far and their approximate relative potencies in the renal vascular bed. The compounds may be classified into three groups: (1) phenylethylamines; (2) tetralines; and (3) aporphines.

Phenylethylamines

In an initial study of 44 phenylethylamines only one compound, *N*-methyl DA (epinine), was active (Goldberg *et al.*, 1968). *N*-methyl DA is equipotent to DA. Inactive phenylethylamines included agents with substitution on the benzene ring other than OH groups on the 3 and 4 positions, analogues with substitutions on the α or β carbons, catecholamines with 2- and 4-carbon side chains, single nitrogen substituents larger than CH_3 (*N*-ethyl, *N*-propyl, and *N*-isopropyl), disubstitutions on the N atom other than propyl (*N*,*N*-dimethyl, *N*,*N*-diethyl). As shown in Table I, *N*,*N*-dipropyl DA is an active but a considerably weaker agonist than DA (Volkman *et al.*, 1977a). Further studies of *N*,*N*-disubstituted compounds indicated that one of the substituents must be *N*-propyl and the other a group larger than C_2H_5 (Ginos *et al.*, 1977).

Aminotetralins

Two series of aminotetralins were investigated (Fig. 1). The first derivative studied was 2-amino-6,7-dihydroxy-1,2,3,4-tetrahydronaphthalene (A-6,7-

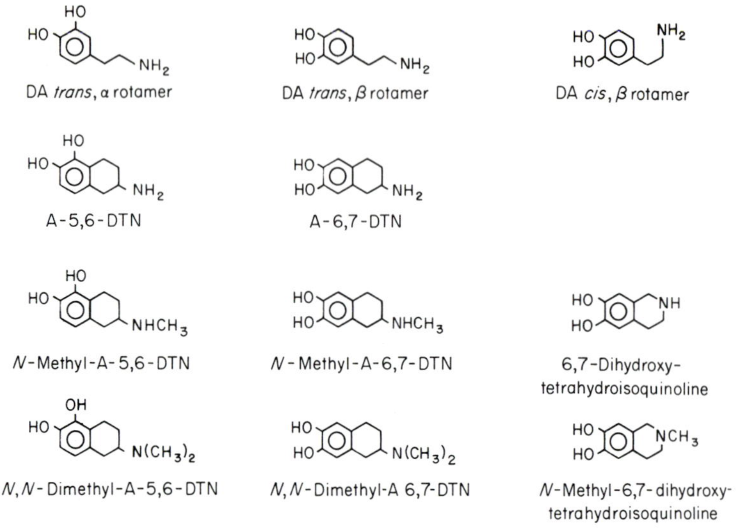

Fig. 1 Semi-rigid analogues of dopamine (see text for details).

DTN) which contains the DA molecule in the β-rotameric *trans* form. This compound was found to be equipotent to DA in both renal and mesenteric vascular beds (Volkman *et al.*, 1977b; Crumly *et al.*, 1976). The structure–activity relationship for the A-6,7-DTN molecule appears to be similar to that for DA. *N*-Methyl A-6,7-DTN is approximately equipotent to A-6,7-DTN (Volkman *et al.*, 1977b) and *N*,*N*-dipropyl A-6,7-DTN is approximately equipotent to *N*,*N*-dipropyl DA (Kohli *et al.*, 1977). *N*,*N*-Dimethyl A-6,7-DTN is inactive (Volkman *et al.*, 1977b). In contrast, the α-rotamer with a *trans* conformation, 2-amino-5,6-dihydroxy-1,2,3,4-tetrahydronaphthalene (A-5,6-DTN) (Fig. 1) and its *N*-methyl analogue are inactive (Volkman *et al.*, 1977b).

Aporphines

Apomorphine contains the structure of *N*-methyl DA with the side chain fully extended and amino group *trans* to the catechol moiety in an α-rotameric conformation. Apomorphine is a much weaker agonist of the DA vascular receptor than DA and appears to be a partial agonist (Goldberg *et al.*, 1968; Goldberg and Musgrave, 1971; Goldberg, 1978). 6-*n*-Propylnoraporphine is more potent than apomorphine and appears to be a full agonist of the vascular DA receptor with 1/30 to 1/50 the potency of DA (Crumly *et al.*, 1976). Study of several aporphines with a single OH group demonstrated that they were all inactive. Isoapomorphine, which contains the structure of *N*-methyl DA in a β-rotameric conformation, is inactive (Volkman *et al.*, 1977b). The partial agonist activity of apomorphine and the inactivity of isoapomorphine appears to contradict the hypothesis that the β-rotameric form of DA is the active rotamer. This assumption is not necessarily true, however, since it is possible to hypothesize a better fit for apomorphine into catechol and nitrogen binding sites on the postulated DA receptor than for isoapomorphine (Goldberg *et al.*, 1978a).

Table II presents a list of compounds which are inactive as DA vascular agonists. Many of these compounds have been considered putative DA agonists in other systems.

ANTAGONISTS

The antagonists studied are listed in Table III. We are limited to the study of soluble compounds and compounds which did not affect renal blood flow. Accordingly, because of insolubility we were unable to investigate adequately pimozide, which has been reported to be inactive as a vascular DA antagonist by Setler *et al.* (1975) and Imbs *et al.* (1976). We have also been unable to confirm the efficacy of ergometrine as an antagonist which was reported by Bell *et al.* (1974).

Table III lists two qualities of the antagonists. First, relative potency, which is the dose required to shift the dose–response curve of DA-induced renal vasodilation by 3–4 log units and; secondly, specificity, which is defined as the minimum dose attenuating vasodilating responses of bradykinin and isoproterenol per minimum dose attenuating vasodilating responses of DA.

Examination of Table III demonstrates that sulpiride is the most potent antagonist and shares with metoclopramide the greatest range of specificity. These data are interesting in view of the fact that sulpiride (Trabucci *et al.*, 1976; Roufogalis *et al.*, 1976) and metoclopramide (Donaldson *et al.*, 1976) are not antagonists of DA-induced increments of caudate cyclic AMP.

Summary

The structure–activity relationships and order of potency of agonists and antagonists of the DA vascular receptor are different from those described for several central nervous system models and other putative DA receptor models. A pertinent question which must be answered is whether these

TABLE III Antagonists of the dopamine vascular receptor.

	Relative potency[a] (mol)	Range of specificity[b]
Sulpiride	2.9×10^{-8}	>10
Bulbocapnine	4.7×10^{-8}	~8
Haloperidol	1.4×10^{-7}	<2
Chlorpromazine	2.5×10^{-7}	<2
Prochlorperazine	2.5×10^{-7}	<2
Trifluoperazine	2.5×10^{-7}	<2
Fluphenazine	2.5×10^{-7}	<2
Thioridazine	5.0×10^{-7}	<2
Metoclopramide	1.5×10^{-6}	>10

[a] Dose to produce 3–4 × shift in dopamine dose–response curves.

[b] $\dfrac{\text{(Minimal dose attenuating vasodilating responses of bradykinin or isoproterenol)}}{\text{Minimal dose attenuating vasodilating responses of dopamine)}}$

discrepancies are related to the existence of different DA receptors or whether some of the effects attributed to DA receptors in the central nervous system and elsewhere are due to action of DA and its analogues on other receptors. This subject has been extensively discussed in a recent review (Goldberg *et al.*, 1978b).

IN-VITRO MODELS

Studies of isolated canine renal, mesenteric, coronary, and cerebral arteries have confirmed the existence of the DA vascular receptor. Dose-related DA-induced relaxation can be demonstrated in these vessels under special conditions (Goldberg and Toda, 1975; Toda, 1976; Toda and Goldberg, 1975). Small (less than 0.5 mm outside diameter) arteries must be used. Vessels are bathed in a strong POB solution (10^{-5} M) for 1 h and are contracted with K^+ or prostaglandin $F_{2\alpha}$ ($PGF_{2\alpha}$). Prostaglandin $F_{2\alpha}$ is a more appropriate contracting agent and most vessels treated by this agent relax with DA. It was difficult to demonstrate the antagonism of DA-induced relaxation in these vessels. However, recently we have demonstrated antagonism by metoclopramide and sulpiride without affecting nitroglycerin relaxation (Ozaki, Kohli, and Goldberg, unpublished data). Toda (personal communication) found that DA-induced relaxation can be specifically antagonized by droperidol without affecting relaxation produced by adenosine.

ACTION OF DA ON OTHER RECEPTORS IN THE CARDIOVASCULAR SYSTEM

Dopamine acts on several receptors in the cardiovascular system in addition to the DA vascular receptor. Accordingly, DA vascular agonists have been screened for action on β_1-, β_2-, and α-adrenergic receptors.

β_1-Adrenergic Receptors

Dopamine increases cardiac contractile force and heart rate by direct action on β_1-adrenergic receptors and by releasing norepinephrine (NE) from sympathetic storage sites in the myocardium (Goldberg, 1972). β_1-Adrenergic activity is measured in the open-chest, vagotomized, anaesthetized dog by means of a Walton–Brodie strain gauge arch sutured to the right ventricle (McDonald and Goldberg, 1963). Separation of the direct and indirect effects are obtained by intravenous administration of methylphenidate, 5 mg kg^{-1} (Meyer and Goldberg, 1966). Methylphenidate potentiates the increments in contractile force of direct acting agonists; whereas, indirect antagonists are completely blocked. In preliminary studies we found that DA and epinine exert approximately the same β_1-adrenergic activity with approximately the same mix of direct and indirect actions. 2-Amino-6,7-dihydroxy-1,2,3,4-tetrahydronaphthalene (A-6,7-DTN) is a weaker β_1-adrenergic agonist than DA and appears to have a greater indirect component than DA (Crumly *et al.*, 1976). Interestingly, *N*,*N*-di-*n*-propyl DA, *N*,*n*-propyl-*N*-*n*-butyl DA, and *N*,*n*-propyl-*N*-*n*-pentyl DA do not increase cardiac contractile force and, thus, lack β_1-adrenergic activity (Volkman *et al.*, 1977a; Ginos *et al.*, 1978). The reasons why a propyl group in a disubstituted compound prevent β_1-adrenergic activities are unknown.

β_2-Adrenergic Receptors

β_2-Adrenergic activity is studied by observing effects of intra-arterial injections of agonists into canine femoral arteries after administration of POB and determining whether agents which cause vasodilation are antagonized by propranolol (Volkman *et al.*, 1977b). Dopamine and A-6,7-DTN do not usually increase bloodflow in this preparation until doses greater than 1000 times the threshold dose of isoproterenol are administered. In contrast, A-5,6-DTN, which is inactive on DA vascular receptors, is approximately 100 times less potent than isoproterenol as a β_2-adrenergic agonist (Volkman *et al.*, 1977b). These data suggest that the conformation for β_2-adrenergic activity in the DA molecule is the α-rotameric form.

Addition of an *N*-methyl group to either the ADTN or DA molecule increases β_2-adrenergic activity.

α-Adrenergic Receptors

α-Adrenergic activity is studied by observation of vasoconstriction in the renal or femoral vascular bed prior to administration of POB and by studies with isolated rabbit aortic strips (Kohli *et al.*, 1978). Slightly different potency series were observed in the two preparations. In the rabbit aortic strip the relative order of potency of DA, epinine, and aminotetralins is epinine > DA > *N*-methyl A-6,7-DTN > A-6,7-DTN. The A-5,6-DTN derivatives were much weaker as α-adrenergic agonists, indicating that the β-rotamer appears to be the preferred conformation. A discrepancy exists between the isolated rabbit aortic strip and the intact dog studies in that A-6,7-DTN and *N*-methyl A-6,7-DTN are more potent vasoconstricting agents than DA and epinine in the intact animal. *N*,*N*-Di-*n*-propyl DA exhibits weaker blood vessel contracting activity in both the intact dog and isolated rabbit aortic strip.

These studies indicate that it is possible to eliminate or greatly attenuate β_1- and β_2-adrenergic activity in active DA vascular agonists. An active DA agonist without α-adrenergic activity has not been found. However, since *N*,*N*-di-*n*-propyl DA is a much weaker alpha-adrenergic agonist than DA, the goal of synthesizing an agonist acting only on DA vascular receptors may be possible. Such a compound would be extremely useful in comparing putative DA receptors in different areas and could have useful clinical indications.

POTENTIAL PHYSIOLOGICAL ROLE OF DA IN THE KIDNEY

Several lines of evidence support the hypothesis that DA may have a physiological role in the kidney, possibly in the regulation of sodium balance. The hypothesis is based on the initial finding that DA-induced renal vasodilation is associated with a natriuresis (McDonald *et al.*, 1964). Because plasma levels of DA are either undetectable or very low, most studies have been concerned with measurement of urinary excretion of DA. Cuche *et al.* (1972) observed that subjects on normal and sodium-restricted diets experienced a reduction in urinary DA excretion, an increase in NE and epinephrine excretion, and a decrease in sodium excretion when assuming the upright position. A positive and significant correlation was found between the urinary excretion of DA and sodium. Alexander *et al.*

(1974) reported that when sodium intake was increased in normal subjects, urinary DA increased and urinary NE decreased. In addition, intravenous infusions of saline increased sodium excretion and urinary DA levels. Faucheux *et al.* (1977) compared the effects of saline and albumin infusions on plasma urinary catecholamines in anaesthetized dogs. Plasma and urinary NE and epinephrine decreased following both saline and albumin infusions. Urinary DA, however, increased in saline-expanded dogs but did not significantly change in control or in albumin-expanded dogs, indicating that the sodium load and not the increase in volume was responsible for the increase in DA excretion. Plasma-free DA in dogs in which this measurement could be made did not significantly change, suggesting that the increase in urinary DA excretion was due to a generation of DA in the kidney. Finally Kuchel *et al.* (1977) have presented evidence that a pathological condition "idiopathic oedema" is associated with reduction in urinary DA excretion. Sixteen women with this condition had lower urinary DA excretions than a series of control women.

A mechanism by which the kidney could generate DA in response to a sodium load has not been determined. However, it is possible that there are dopaminergic nerves in the kidney which may subserve this function. Bell and Lang (1973) stimulated specific areas of the midbrain in anaesthetized dogs, pretreated with guanethidine to prevent NE release, and found that renal blood flow increased. This effect was not attenuated by atropine or the antihistamine, mepiramine, but the DA vascular antagonist, haloperidol, produced complete or nearly complete elimination of the renal vasodilating responses in 8 of 10 dogs tested. These investigators concluded that the renal vasculature is innervated by nerves which release DA as a transmitter and that these nerves can be activated by midbrain stimulation. More recently, Dinerstein *et al.* (1978) reported histofluorescent evidence for dopaminergic innervation in the dog kidney. Kidneys were prepared by a modified Falck–Hillarp procedure and histofluorescence was analysed microspectrofluorometrically for identification of NE and DA. Specific localization of endogenous DA in kidney was found to be associated with blood vessels at the glomerular vascular pole. Innervation of the remaining arteries was noradrenergic. Surgical denervation abolished DA and NE fluorescence.

CONCLUSIONS

1. The chemical structural requirements for activation of DA vascular receptors are extremely specific. Of hundreds of compounds studied only a few have been found to be active. The preferred conformation that will fit

into the receptor appears to be the β-rotameric form of DA, as typified by A-6,7-DTN.

2. Specific antagonists have been found which differ with respect to potency and range of specificity.

3. Preliminary studies of several DA vascular agonists have demonstrated that they differ with respect to their actions on α-, β_1-, and β_2-adrenergic receptors.

4. Evidence has accumulated that DA may have a physiological role in the kidney in the regulation of sodium balance. It is possible that DA may be released in response to changes in sodium load by dopaminergic nerves located in the kidney.

ACKNOWLEDGEMENT

This research was supported in part by NIH Grants GM-22220 and NS-12324.

REFERENCES

Alexander, R. W., Gill, J. R., Jr, Yamabe, H., Lovenberg, W., and Keiser, H. R. (1974). *J. Clin. Invest.* **54,** 194–200.

Bell, C., Conway, E. L., and Lang, W. J. (1974). *Br. J. Pharmacol.* **52,** 591–595.

Bell, C. and Lang, W. J. (1973). *Nature (London)* **246,** 27–29.

Crumly, H., Hinshaw, W., Pinder, R., and Goldberg, L. I. (1976). *Nature (London)* **259,** 584–587.

Cuche, J. L., Kuchel, O., Barbeau, A., Boucher, R., and Genest, J. (1972). *Clin. Sci.* **43,** 481–491.

Dinerstein, R. J., Henderson, R. C., Goldberg, L. I., and Hoffmann, P. C. (1978). *Fed. Proc.* **37,** 346.

Donaldson, I. M., Jenner, P., Marsden, C. D., Miller, R., Peringer, E. (1976). *Br. J. Pharmacol.* **56,** 373P.

Faucheux, B., Buu, N. T., and Kuchel, O. (1977). *Am. J. Physiol.* **232,** F123–F127.

Furchgott, R. F. (1972). *In* "Handbook of Experimental Pharmacology" (Eds. H. J. Blaschko and E. Muscholl), 283–335. Springer, New York.

Ginos, J. Z., Kohli, J. D., and Goldberg, L. I. (1978). *Fed. Proc.* **37,** 683.

Goldberg, L. I. (1972). *Pharmacol. Rev.* **24,** 1–9.

Goldberg, L. I. (1978). *In* "Dopamine". Advances in Biochemical Pharmacology, Vol. 19. (Eds. P. Roberts, G. Woodruff and L. Iversen), 119–129. Raven Press, New York.

Goldberg, L. I., Kohli, J. D., Kotake, A. N., and Volkman, P. H. (1978a). *Fed. Proc.* **37,** 2396–2402.

Goldberg, L. I. and Musgrave, G. (1971). *Pharmacologist* **13,** (2) 227.

Goldberg, L. I., Sonneville, P. F., and McNay, J. L. (1968). *J. Pharmacol. Expl. Ther.* **163,** 188–197.

Goldberg, L. I. and Toda, N. (1975). *Circ. Res.* **36,** 11–197.

Goldberg, L. I., Volkman, P. H., and Kohli, J. D. (1978b). *Ann. Rev. Pharmacol. Toxicol.* **18,** 57–79.

Goldberg, L. I. and Yeh, B. K. (1971). *Eur. J. Pharmacol.* **15,** 36–40.

Imbs, J. L., Schmidt, M., Velly, J., and Schwartz, J. (1976). *Eur. J. Pharmacol.* **38,** 175–178.

Kohli, J. D., Nichols, D. E., Volkman, P. H., and Goldberg, L. I. (1977). *Clin. Res.* **25,** 602A.

Kohli, J. D., Volkman, P. H., Glock, D., and Goldberg, L. I. (1978). *Fed. Proc.* **37,** 792.

Kuchel, O., Cuche, J. L., Buu, N. T., Guthrie, G. P., Jr., Unger, T., Nowaczynski, W., Boucher, R., and Genest, J. (1977). *J. Clin. Endocrin. Metab.* **44,** 639–646.

McDonald, R. H., Jr., and Goldberg, L. I. (1963). *J. Pharmacol. Expl. Ther.* **140, 60–66.**

McDonald, R. H., Jr., Goldberg, L. I., McNay, J. L., and Tuttle, E. P., Jr. (1964). *J. Clin. Invest.* **43,** 1116–1124.

McNay, J. L. and Goldberg, L. I. (1966). *J. Pharmacol. Expl. Ther.* **151,** 23–31.

Meyer, M. B. and Goldberg, L. I. (1966). *Cardiologia* **49,** 1–10.

Roufogalis, B. D., Thornton, M., and Wade, D. N. (1976). *Life Sci.* **19,** 927–934.

Setler, P. E., Pendleton, R. G., and Finlay, E. (1975). *J. Pharmacol. Expl. Ther.* **192,** 702–712.

Toda, N. (1976). *Br. J. Pharmacol.* **58,** 121–126.

Toda, N. and Goldberg, L. I. (1975). *Cardiovasc. Res.* **9,** 384.

Trabucci, M., Longoni, R., Fresia, P. and Spano, P. F. (1976). *Life Sci.* **17,** 1551–1556.

Volkman, P. H., Kohli, J. D., Goldberg, L. I., and Cannon, J. G. (1977a). *Fed. Proc.* **36,** 1049.

Volkman, P. H., Kohli, J. D., Goldberg, L. I., Cannon, J. G., and Lee, T. (1977b). *Proc. Natl. Acad. Sci., U.S.A.* **74,** 3602–3606.

Yeh, B. K., McNay, J. L., and Goldberg, L. I. (1969). *J. Pharmacol. Expl. Ther.* **168,** 303–309.

Dopamine and Behaviour

CHAPTER 31

Behavioural Aspects of Dopamine Agonists and Antagonists

B. Costall and R. J. Naylor

University of Bradford, Bradford, U.K.

Until recently the study of behavioural responses induced by dopamine agonists and antagonists has been limited not only by the nature of the tests available, but by the limited number of drugs which may have been applied to these tests. Thus, for a long time apomorphine, amphetamine and related agents were the sole source for our knowledge of agonist activity, and the typical neuroleptics of the butyrophenone, phenothiazine and other series produced our only knowledge of the behavioural effects of dopamine antagonists. The literature was generally limited to changes in stereotyped behaviour, hyperactivity and catalepsy production as being the major behavioural changes to be readily observed after the peripheral administration of dopamine agonists or antagonists.

Advances have been made in recent years by the introduction of more sophisticated stereotaxic techniques which allow the deposition of agonists/antagonists into specific brain regions, and by the introduction of new agents for study, particularly new dopamine agonists. It is now appreciated that an indiscriminate stimulation or blockade of cerebral dopamine systems is at best a limited approach, since it is now well accepted that the different cerebral dopamine systems have different functional capacities. It has been in the definition of these functional capacities that the most important advances have been made, since such knowledge not only allows the possible prediction of wanted or unwanted effects of a potential agonist/antagonist, but provides the basis for the possible design of drugs having a specificity of action.

The final bases for such predictions are as yet, far from complete, since studies have been bound to intensive, yet limited investigations in selected brain areas. Thus, knowledge has been gained as to the nature of the dopaminergic mechanisms in the mesolimbic and extrapyramidal nuclei, and their role in the control of motor behaviour but, so far, studies on the dopamine receptor mechanisms in other undoubtedly important areas for the mediation of the overall activity of dopamine agonists/antagonists have not been subject to any careful analysis, just a few examples are the mesocortical, tuberoinfundibular and amygdaloid systems, and the area postrema. Further, we are not in a position to extrapolate many of the animal findings to man, although in certain areas it has been rewarding to find that predictions made on the basis of studies in the rodent do apply to man. Advances here have been particularly in the field of dyskinesias.

The behavioural studies which have been most widely applied to the analysis of differential cerebral dopamine function are stereotypy induction/antagonism, hyperactivity induction/antagonism and dyskinesia induction/antagonism. The recent synthesis and development of many novel dopamine agonists from the phenylethylamine, tetrahydroisoquinoline, 2-aminotetralin, ergot and aporphine series, and the use of these agents has allowed a definition of those extrapyramidal and mesolimbic dopamine mechanisms which modulate motor function. This has necessitated a virtual renaissance of our concepts of dopamine receptor mechanisms for, if nothing else, the findings to date question many basic hypotheses and it is hoped that the evidence presented here will lead the reader to our own inevitable conclusions that within one brain area there may be more than one receptor population which may be differentially stimulated or antagonized to induce a particular behavioural response, and that a particular behavioural response may be induced in what appears to be an identical form by differential stimulation of dopamine mechanisms in different brain areas. Many of the pharmacological tools which have been used are still basically experimental but unless one's clinical colleagues can be convinced of the essential need for a clinical evaluation of their activity, the assessment and improvement of existing test procedures will be as impossible as the chances of improving existing therapy. Such advances in the field of dopamine agonist/antagonist therapy are dependent almost entirely on the selection of the correct agent with a specificity of action for the malfunctioning system which may be thus corrected with minimal interference with other dopamine mechanisms.

THE FUNCTIONAL ROLE OF CEREBRAL DOPAMINE SYSTEMS IN MOTOR CONTROL

At the present time there is little evidence to suggest that the dopamine mechanisms within the area postrema may modify motor function, but this lack of evidence relates to lack of experimentation and one cannot lightly regard the findings of Lichtensteiger and Lienhart (1977): we must await further knowledge before judging the possible importance of the dopamine function within this brain area. In contrast, dopamine agonists and antagonists have a clear potential both to influence neuroendocrine function and to cause marked changes in motor performance. However, at this time the available data is considered to be too general in nature to allow the formulation of any hypothesis as to the possible role of hypothalamic-pituitary systems in normal or abnormal motor activities. Therefore, the present chapter is restricted to a review of the roles of striatal, mesolimbic and mesocortical systems in motor control.

Stereotyped responding, hyperactivity and abnormal involuntary movements induced in animals are considered to be good indicators of changes in motor behaviour which depend on changes in extrapyramidal and mesolimbic-cortical dopamine function. An elucidation of the relevance of these systems to stereotyped responding and hyperactivity using the brain lesion techniques has been reviewed elsewhere (Cools, 1977; Costall and Naylor, 1977a): the present analysis will primarily consider the results obtained using the intracerebral injection technique.

THE STRIATUM AND MOTOR CONTROL

The association between a reduced striatal dopamine function and Parkinson's disease and the success of L-DOPA in alleviating Parkinsonian symptoms stimulated numerous studies designed to elucidate the mechanisms of dopamine action and drug treatment with dopamine agonists. Whilst the paleostriatum has been subject to only superficial examination, many authors have demonstrated that neostriatal dopamine systems play an important role in the modulation of "stereotyped" responses to dopamine agonists. In the rat, the intrastriatal injection of dopamine can induce repetitive motor behaviour which is invariably characterized by perioral movements—biting, gnawing or licking (Fog, 1972; Costall *et al.*, 1974) and, depending upon the precise experimental conditions, a more generalized hyperactivity may be induced (Costall and Naylor, 1976b). Of the catecholamine receptors present in the striatum, the evidence indicates that dopamine *is* stimulating dopamine receptors in order to cause these motor changes; from a series of phenylethylamine derivatives dopamine was shown to possess optimal activity, noradrenaline and adrenaline were inactive (Costall *et al.*, 1974, 1975b). The structural requirements for stereotypy induction appeared more precise than those for hyperactivity production, but that the mechanisms involved are in fact dopaminergic is further supported by the specificity of the blockade of these motor responses by the neuroleptic agents (Costall *et al.*, 1975a).

Although both the stereotypy and hyperactivity mechanisms in the striatum may be termed "dopaminergic" these may be further characterized and differentiated by their responses to dopamine agonists. Firstly, the classical dopamine agonist apomorphine induces stereotypy after peripheral or intrastriatal administration, but a hyperactivity component is conspicuously absent in both situations (Costall *et al.*, 1974, 1977b). Secondly, investigations into the dopamine agonist properties of a series of *N*-alkylated 2-aminotetralins revealed that some derivatives clearly dissociated stereotypy and hyperactivity, for example, 2-(*N*,*N*-diethyl)amino-5,6-dihydroxytetralin exhibited a potent stereotypic potential on intrastriatal injection, but did not induce hyperactivity (Costall *et al.*, 1977b; Cannon *et al*, 1977). This contrasts with the actions of intrastriatal amphetamine which generally causes a hyperactivity in the absence of stereotyped peri-oral movements (Costall *et al.*, 1974).

We consider that the striatal dopamine mechanisms involved in hyperactivity production, which may be influenced by peripherally administered drugs, may be important for an "anti-akinetic" action, but whilst it is certainly a possibility that the striatum is a site at which neuroleptics may inhibit stereotyped responses (Costall *et al.*, 1972; Pijnenburg *et al.*, 1975a), recent studies in the guinea-pig have urged caution in the use of such terminology so commonly applied to oral movements, and some revision would certainly appear necessary. In the guinea-pig, intrastriatal dopamine was found to induce "peri-oral" movements and hyperactivity (Costall *et al.*, 1975b) similar to those observed in the rat but, unexpectedly, the peri-oral movements induced in the guinea-pig (in contrast to those induced in the rat and termed stereotypies) were found to be resistant to classical neuroleptic treatment (Costall and Naylor, 1975a) and thus appeared to relate more to the abnormal peri-oral movements observed in the clinic—movements generally termed dyskinesias (Marsden *et al.*, 1975). We have termed the peri-oral movements induced by intrastriatal dopamine in the guinea-pig as dyskinesias. Our experiments to substantiate this terminology have consistently differentiated these "dyskinesias" from stereotyped behaviour (Costall and Naylor, 1976a; Costall *et al.*, 1977d). Thus, in the guinea-pig at least, there appears to be two distinct dopamine mechanisms, stimulation of one causing hyperactivity which is readily amenable to neuroleptic therapy, and stimulation of the other causing dyskinesias which are generally resistant to neuroleptic action. We have termed these dopamine mechanisms DA-1 and DA-2 respectively (Costall and Naylor, 1975a). Of particular interest to a differentiation of these receptor mechanisms is the observation that three agents in particular may be selected from a large number tested as being "antidyskinetic", i.e. DA-2 blockers, pimozide, oxiperomide and tiapride. The actions of tiapride further emphasise the differences in the DA-1 and DA-2 mechanisms, since large pharmacological doses of this agent fail to block the DA-1 hyperactivity mechanisms whilst the DA-2 dyskinetic mechanisms may be inhibited (Costall and Naylor, 1977b).

Cools and colleagues have also reported that intrastriatal dopamine can induce dyskinetic phenomena in the monkey (Cools *et al.*, 1975). In this species, three types of behavioural change occur depending on the dose of dopamine. From low to higher doses the observations are (a) an increase in the number of skilled manipulative movements (dynamic phase), (b) a dystonic torticollis (dystonic phase) and (c) oro-lingual-facial dyskinesia and dyskinesias in the extremities (dyskinetic phase). It is possible that these three phases relate to an increasing stimulation of the same dopamine mechanism, certainly, no evidence was presented to show that these phases may be differentially induced by different dopamine agonists. A single antagonist, haloperidol, was employed and shown to antagonize selectively the first two phases of the dopamine response. The resistance of the dyskinesias correlates with our own findings, but although we are in agreement with the conclusions of Cools and colleagues that hyperfunctioning of a DA-2 system results in motor disorders such as oro-lingual-facial dyskinesias (see Cools, 1975), it would appear essential to confirm the observations in monkey by using a variety of control antagonists and neuroleptics having differing DA-1 : DA-2 blocking capacities, in particular, pimozide, oxiperomide and tiapride. If these differential antagonistic capacities (which have been confirmed in the clinic) can be confirmed in the monkey, then the ability of this species to exhibit a wider spectrum of motor abnormalities than any rodent, and which are more closely related to the abnormalities observed in man, should theoretically render this model the most appropriate for the detection of potential antidyskinetic agents than any test presently available.

There are no reports as to the possible different topographical distribution of dopamine receptors in the striatum of the rat, guinea-pig or monkey, but in the cat where the caudate nucleus is distinct from the putamen, Cools and colleagues have indicated that the caudate nucleus may contain a heterogeneous distribution of two functionally distinct types of dopamine mechanism (Cools, 1973, 1975, 1977; Cools *et al*., 1976). In the rostromedial part of the caput (Cools' terminology), local application of L-DOPA, dopamine, apomorphine or amphetamine results in the contralateral turning syndrome, counteracted by haloperidol. At intracaudate sites outside the rostromedial area, low doses of L-DOPA, dopamine, amphetamine and DPI ((3,4-dihydroxy-phenylamino)-2-imidazoline) are able to elicit the ipsilateral turning syndrome whilst haloperidol and apomorphine are ineffective at this site; the dopamine effects are antagonized by ergometrine, piribedil and noradrenaline. The relevance of this complexity of pharmacological data to either the production of hyperactive movements or abnormal involuntary movements is, at present, far from clear, but from the studies of these authors it is apparent that there are two further potential sites of drug action which may be important to an understanding of dyskinesias. Firstly, within the rostromedial area an injection of L-DOPA, dopamine, 3-methoxytyramine and amphetamine induced athetoid and choreiform hyperkinesias which were restricted to the forelimbs (Cools, 1973). Unfortunately, no attempt was made to analyse the nature of the dopamine mechanism mediating these responses other than to establish that they were ameliorated by α-methylparatyrosine—one requires knowledge on the possible antagonistic effects of at least haloperidol and ergometrine. It is possible that the dyskinetic responses reported are merely due to an excessive stimulation of the dopamine mechanisms mediating contralateral movements. The further use of different types of dopamine agonist may help either to differentiate or to indicate the similarity in the characteristics of these mechanisms mediating contralateral movements or dopamine dyskinesias. Secondly, the bilateral application of DPI into the caudate nucleus outside the rostromedial area was reported to cause orofacial dyskinesias (Cools *et al*., 1976). However, it is unfortunate that again no attempts were made to assess the pharmacological characteristics of this "dyskinetic mechanism" using potential dopamine agonists or antagonists (although it is mentioned in an addendum to another manuscript (Cools and Van Rossum, 1976) that ergometrine and piribedil blocked the dyskinesias—however, one must question the pharmacological relevance of isolated findings: it would be most interesting to determine the specificity and nature of these blockades). It can only be questioned whether the dyskinetic syndrome induced by DPI is due to excessive stimulation of the DPI-sensitive ipsilateral motor mechanism? (see Table I).

Cools and colleagues have reviewed their own data with that of others (Cools and Van Rossum, 1976) and have summarized that, within the mammalian brain, there are at least two distinct types of dopamine receptor. They propose that DA-1 receptors are largely located within the nucleus accumbens (see p. 564) and tuberculum olfactorium, and are scattered throughout the striatum, whilst DA-2 receptors are located in the medial part of the neostriatum and are diffusively found elsewhere. On the basis of electrophysiological evidence these authors have further defined these dopamine receptors as "inhibitory" or "excitatory": it is forwarded that the DA-1 receptors may essentially represent an "inhibition mediating receptor", i.e. DAi, whilst the DA-2 receptors are essentially excitatory, i.e. DAe. Elaborate hypotheses regarding the neuronal localization of these receptors and a balance between DAi and DAe systems in the control of normal and abnormal motor functions are presented and offer scope for an experimental verification. At the present time, however, the available experimental data is limited and it would appear incautious to consider a DAi mechanism as being the only mechanism that may be involved with dyskinesia production, and a DAe mechanism as being involved in the modulation of normal motor activity. For example, it would appear possible that the athetoid and choreiform dyskinesias induced by dopamine in the ventromedial area of the neostriatum may be mediated via a DAe system, whereas "normal" contralateral and ipsilateral movements may be mediated via DAe and DAi systems respectively. Possibly even more pertinent to the present discussion is the absence of data from the above experiments to provide evidence that it is possible to stimulate *selectively* or block a hyperactivity (contralateral or ipsilateral movements) or dyskinetic mechanism. For example, piribedil and ergometrine block DPI-induced ipsilateral movements *and* DPI-induced oro-facial effects; H44/68 blocks both the dopamine-induced contralateral movements *and* dopamine-induced chorea/athetosis. It would be interesting (indeed, it would appear necessary) to assess the actions of a range of dopamine agonists and antagonists (particularly pimozide, oxiperomide and tiapride) in these systems.

THE MESOLIMBIC SYSTEM AND MOTOR CONTROL

Of those mesolimbic brain areas whose concentration of dopamine is sufficiently high to indicate a possible neurotransmitter role, the tuberculum olfactorium, nucleus accumbens, central amygdaloid nucleus, septal nucleus etc. (Versteeg *et al.*, 1976), only the nucleus accumbens has received detailed attention. The primary response to dopamine injected into the nucleus accumbens of rat is the production of hyperactivity with stereotypy contributing only a minor component to the overall effect (Pijnenburg and Van Rossum, 1973; Costall and Naylor, 1975b; Jackson *et al.*, 1975). Whilst a number of phenylethylamines, including noradrenaline and adrenaline,

TABLE I Summary of the findings on which Cools and colleagues base their hypotheses of differing striatal dopamine mechanisms.

Anatomical location	Behaviour	System	Agonists	Antagonists	Ineffective
CRM	Contralateral movements	1	DA, amphetamine, L-DOPA, apomorphine	Haloperidol H44/68	Piribedil Ergometrine
	Athetosis/Chorea	2	DA, amphetamine, L-DOPA	H44/68	?
Extra-CRM	Ipsilateral movements	3	DA, amphetamine, L-DOPA, DPI	Piribedil Ergometrine	Haloperidol, Apomorphine
	Orofacial dyskinesias	4	DPI	Piribedil Ergometrine	?

CRM equivalent to rostromedial caudate (see Cools, 1973, for details of terminology). Full reference to the systems 1–4 is made in the text.
DA—Dopamine.

are also able to induce this hyperactivity response (Jackson *et al.*, 1975; Costall *et al.*, 1976b), it is, nevertheless, specifically antagonized by neuroleptic agents (Jackson *et al.*, 1975; Pijnenburg *et al.*, 1975b; Costall *et al.*, 1976b), and thus it may be postulated that it is a dopamine mechanism which is important for the mediation of hyperactivity. This may, however, be too simplistic a view of the role for accumbens dopamine since recent experiments using the intra-accumbens injection technique and a wide range of dopamine agonists from the *N*-alkylated phenylethylamine (Cannon *et al.*, 1978), 2-aminotetralin (Costall *et al.*, 1977a) and aporphine series (Costall *et al.*, 1977a, b) have clearly shown that the nucleus accumbens can sustain a stereotyped response. The close proximity of the nucleus accumbens to the striatum may initially cause some doubts as to the specificity of the stereotyped responses to the nucleus accumbens, but not only are there agents which exert a greater stereotypic potency from the nucleus accumbens than from the striatum, such as *N*,*N*-dipropyldopamine, but, even more convincing, *N*-propylnorapomorphine has been shown to induce stereotyped responses from the accumbens but *not* from the striatum (Costall *et al.*, 1977b). Even though dopamine itself fails to induce consistent stereotyped responses on intra-accumbens injection, it would appear that the stereotypy induced by other dopamine agonists *is* dopamine dependent since these stereotypies are all specifically blocked by neuroleptic agents. It should be noted here that although we are using the word "stereotypies" to describe the peri-oral movements initiated from this mesolimbic area, we are in no position to comment as to whether these responses are stereotypic or dyskinetic in nature. It still remains to characterize the nature of these accumbens receptor mechanisms and therefore, basically for simplistic reasons, we use the term stereotypies as indicating the repetitive nature of the peri-oral movements. Thus it is apparent that there are mechanisms in *both* the nucleus accumbens and striatum for mediating *both* hyperactivity and stereotyped behaviour, but it is equally as apparent that the structural requirements for the stimulation of these mechanisms are *not* the same for both areas.

The hyperactivity response initiated from the nucleus accumbens by dopamine has assumed a particular importance since it has been considered that this brain region may be involved in the regulation of normal motor activity, and possibly in the control of higher psychomotor/mental functions (see Hornykiewicz, 1977). Thus an overactivity of the system (pathologically or drug induced—as by dopamine) may be related to excessive psychomotor activity. On this basis, the "mesolimbic hyperactivity model" has been forwarded for the detection of drugs having antimanic and/or antipsychotic potential (Costall and Naylor, 1976b). Indeed, this "model" has been shown to detect not only classical neuroleptics but also agents such as sulpiride, clozapine and thioridazine which are reported to possess antipsychotic action whilst failing to give a positive response in the standard tests for neuroleptic activity (Costall and Naylor, 1975c). The sensitivity of the test is worthy of note and even the atypical agents are active at pharmacologically acceptable doses. But, possibly more important, this test eliminates metoclopramide. However, the interpretation of findings with this model are not simple; it would appear that the dopamine hyperactivity mechanism is under a strong modulatory influence from 5-hydroxytryptamine (Costall *et al.*, 1976a); this must certainly be considered when discussing the inhibitory actions of the neuroleptic agents (Costall *et al.*, 1977a), and also when considering the ability of apomorphine to antagonize the dopamine response similarly to the neuroleptics (Costall and Naylor, 1976c). This latter, possibly somewhat surprising observation, indicates that the routine use of apomorphine as a classical dopamine "agonist" may require considerable caution.

Cools and colleagues have also considered the nature of the dopamine receptors in the accumbens, and have extended to this area their hypothesis of the DAi and DAe receptors (Cools, 1977). It is suggested that the DAe system is activated by dopamine to produce hyperactivity, and classical neuroleptics can inhibit this mechanism. The second DAi system is normally considered to oppose the DAe system and cause hypoactivity. This DAi receptor mechanism is thought to be blocked by ergometrine (to induce hyperactivity) and stimulated by DPI (to antagonize ergometrine-induced hyperactivity). If this hypothesis were correct then there are clearly increased "opportunities" for the modification of mesolimbic dopamine function, but the pharmacological bases of this hypothesis are yet to be established and some data which is presented tends to confuse rather than clarify the situation. Thus, it is not understood why the locomotor stimulant effects of ergometrine (which presumably depend on a release of dopamine at the DAe system) are not abolished by a pretreatment with reserpine, α-methylparatyrosine or a combination of reserpine/α-methylparatyrosine (Pijnenburg, 1976), or why the intra-accumbens injection of DPI has no effect on normal locomotor activity (Pijnenburg, 1976). The elucidation of these factors is of some importance.

THE MESOCORTICAL SYSTEM AND MOTOR CONTROL

The motor functions of the mesocortical system have not been the subject of any extensive studies. Data on the fundamental role of the mesocortical system has been obtained somewhat indirectly by lesions of the area A10

(Galey *et al.*, 1977) which largely innervates the forebrain and cortical limbic structures. It has been suggested that activation of this system may exert a general inhibitory effect on behaviour. If this can be substantiated, it may provide a basis for further studies designed to investigate some paradoxical motor effects, for example, akathisia induced by neuroleptics or motor depression caused by dopamine agonists.

CHEMICAL STRUCTURES OF DOPAMINE AGONISTS AND ANTAGONISTS

There is a vast number of dopamine agonists and antagonists available to the pharmacologist and/or clinician, certainly too many to include detailed structure for in this review. We are most interested in forwarding the novel compounds which have been used by ourselves and others to establish the existence of different cerebral dopamine mechanisms. Tables II, III and IV show the structures and major behavioural responses to selected dopamine agonists of the phenylethylamine, aporphine and 2-aminotetralin series. The structures of other agents which have been used to contribute to the above hypotheses are shown in Fig. 1.

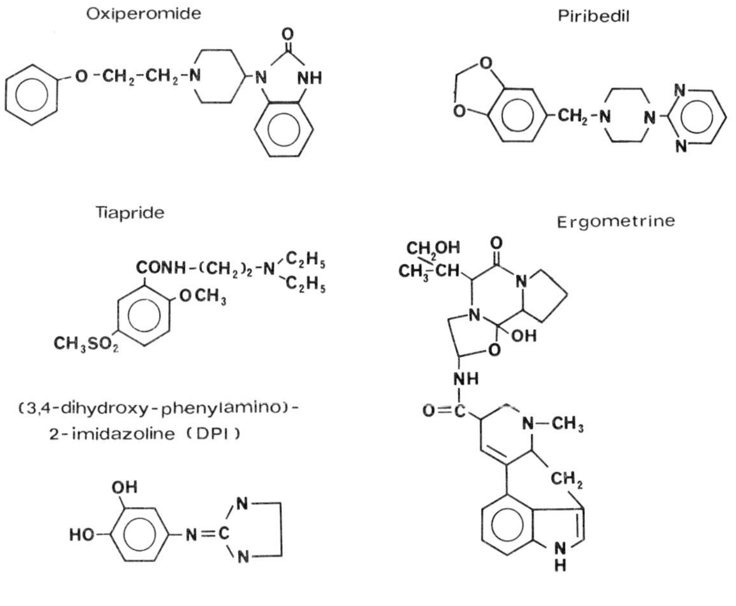

Fig. 1 Structures of putative dopamine agonists and antagonists used as pharmacological tools.

TABLE II Structures and behavioural effects of some *N*,*N*-disubstituted dopamines.

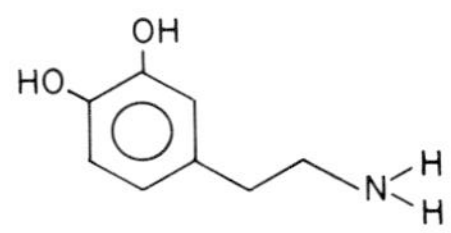

	Behavioural effects					
	Peripheral		Intra-ACB		Intra-CP	
Dopamine	Circling	Stereotypy	Hyperactivity	Stereotypy	Hyperactivity	Stereotypy
Dopamine	0	0	+	0	+	+
N,*N*-Diethyldopamine	+	0	+	+	0	+
N,*N*-Dipropyldopamine	+	+	+	+	0	0
N,*N*-Dibutyldopamine	0	0	0	0	0	0

Only the major responses are indicated as being present (+) or absent (0). Low intensity effects are also categorized as 0. For a discussion of the optimal conformation of dopamine for stimulation of dopamine receptors see Costall *et al.* (1977c), Grol and Rollema (1977), Munday *et al.* (1977) and Woodruff *et al.* (1977).

TABLE III Structures and behavioural effects of some aporphine derivatives.

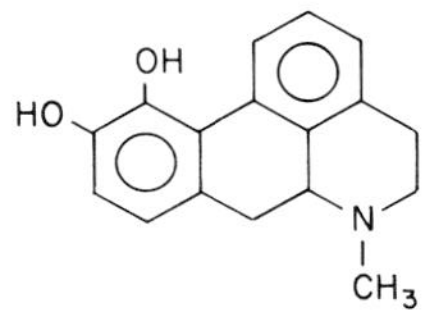

Apomorphine	Behavioural effects					
	Peripheral		Intra-ACB		Intra-CP	
	Circling	Stereotypy	Hyperactivity	Stereotypy	Hyperactivity	Stereotypy
Apomorphine	+	+	0	0	+	+
N-Propylnorapomorphine	+	+	+	+	0	0

Only the major responses are indicated as being present (+) or absent (0). Low intensity effects are also categorized as 0.

TABLE IV Structures and behavioural effects of some 2-aminotetralin derivatives.

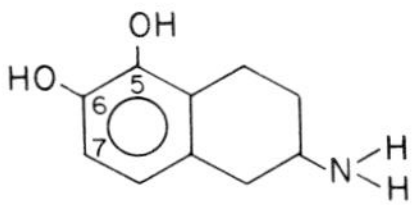 Primary amine (5,6-dihydroxy deriv.)	Behavioural effects					
	Peripheral		Intra-ACB		Intra-CP	
	Circling	Stereotypy	Hyperactivity	Stereotypy	Hyperactivity	Stereotypy
2-(Amino)5,6-dihydroxytetralin	0[a]	0[a]	+	+	+	+
2-(*N*,*N*-Diethyl)amino-5,6-dihydroxytetralin	+	+	+	+	0	+
2-(*N*,*N*-Dipropyl)amino-5,6-dihydroxytetralin	+	+	0	+	0	+
2-(Amino)6,7-dihydroxytetralin	0[a]	0[a]	+	+	0	+
2-(*N*,*N*-Dipropyl)amino-6,7-dihydroxytetralin	+	+	0	0	+	+

[a] These agents fail to pass the blood–brain barrier.

Only the major responses are indicated as being present (+) or absent (0). Low intensity effects are also categorized as 0.

DOPAMINE AGONIST AND ANTAGONIST EFFECTS ON PRESYNAPTIC RECEPTORS

Our interpretation of drug action in all the above studies has tacitly assumed a post-synaptic effect, but both dopamine agonists and antagonists can modify the synthesis and release of dopamine, and many biochemical/electrophysiological studies have indicated a direct drug effect on the dopamine neurone (Kehr *et al.*, 1975; Walters *et al.*, 1975). However, an assessment of presynaptic action at the behavioural level has been considered difficult because of an assumed predominance of postsynaptic actions. However, some interesting experiments have been carried out recently using the mouse: it was shown that very small doses of apomorphine may cause sedation which was antagonized by non-sedative doses of pimozide and the other neuroleptic agents (but not clozapine and trifluperidol) (Chiari *et al.*, 1977; Strombom, 1977). It is suggested that this agonist–antagonist interaction occurs at "autoreceptors", although the precise location of these receptors, for example, at the neuronal terminal and/or cell body, is uncertain. These autoreceptors presumably possess dopaminergic characteristics, although the specificity of the receptor interaction cannot be ascertained in the absence of data using other catecholamine receptor-blocking agents and an extensive range of other dopamine agonists. If, however, the functional and biochemical antagonism observed in the mouse *does* reflect an agonist–antagonist interaction on inhibitory dopamine autoreceptors, then this would provide a suitable substrate(s) for the development of selective agonist–antagonist action on dopamine autoreceptors and, thus, an alternative and important method of modulating dopamine function.

INDIRECTLY ACTING DOPAMINE AGONISTS AND ANTAGONISTS

A number of dopamine agonists are known to cause stimulation of dopamine receptors, at least in part, via indirect mechanisms such as the release of dopamine (for example, amphetamine), blockade of dopamine re-uptake (for example, nomifensine (Tuomisto, 1977)) or theoretically by a partial agonist–antagonist action on presynaptic dopamine receptors. Very little is known of the latter mechanisms which remain an interesting subject of investigation. The dopamine antagonists can, theoretically, act via the same indirect means. However, the major importance of such knowledge is for the selection of agents for specific disease therapy, for example, it has been assumed that for the therapy of disease states characterized by dopamine neurone degeneration, a directly acting dopamine agonist is required, and it becomes important to assess the relative direct and indirect properties of a drug. Such knowledge is generally required for the prediction of possible drug effects in disease states, but for the experimental analysis of cerebral dopamine receptor mechanisms such drugs (once the nature of their indirect action has been defined) have simply been used as manipulative tools.

CLINICAL IMPLICATIONS OF EXPERIMENTAL STUDIES ON DOPAMINE AGONIST AND ANTAGONIST ACTION IN MOTOR CONTROL

In considering the clinical implications of the experimental studies we firstly assume that the dopamine mechanisms which mediate the general locomotor activity following intracerebral injection into mesolimbic or extrapyramidal areas in animals have an analogous substrate(s) in man which mediate an anti-akinetic action. Secondly, we assume that the stereotyped peri-oral movements observed after the injection of dopamine and dopamine agonists into dopamine containing areas of the animal brain may relate to an excessive stimulation of normal or latent dopamine mechanisms which have an analogous substrate(s) involved with the mediation of abnormal peri-oral/oro-bucco-lingual motor behaviour in the clinic. It is not an essential requirement that the dopamine mechanisms in animals and man should be precisely the same, nor is it prerequisite that the mechanisms are located in precisely the same anatomical substrate (indeed, this would not even appear possible when one considers the differences in cerebral anatomy of the rat and man; taking the nucleus accumbens as a pertinent example, this area is anatomically distinct from the striatum in the rat, but probably forms a component of the rostral caudate complex in man—Bird, personal communication). The essential points are that the dopamine receptors and/or mechanisms which mediate locomotor activity or peri-oral dyskinesias are firstly discrete and, secondly, may possess different pharmacological characteristics. The concept of two different dopamine receptor mechanisms, which have been termed DA-1 (for hyperactivity induction) and DA-2 (for dyskinesia induction) (Costall and Naylor, 1975a), considerably expands the traditional concept in which it is tacitly assumed that a single type of dopamine receptor mediates the hyperactivity response on modest stimulation and the dyskinetic manifestations on marked stimulation.

It is appreciated that many neurotransmitter mechanisms are relevant to a control of normal and abnormal motor function, but the ability of L-DOPA and bromocriptine to induce dyskinesias (see review by Marsden *et al.*, 1975; Parkes *et al.*, 1976), and the ability of other dopamine agonists to modify existing dyskinetic phenomena (Gessa *et al.*, 1972; Tolosa and Sparber, 1974; Smith *et al.*, 1976), strongly supports an important involvement of dopamine mechanisms. Dopamine agonists are widely used to alleviate akinesia, but the appearance of dyskinesias constitutes one of the most limiting aspects of dopamine agonist therapy. Indeed, dyskinesias (DA-2 stimulation) can clearly develop in the absence of a significant anti-akinetic effect. Such an observation may be expected if the experimentally derived concept of different dopamine mechanisms mediating the two effects can be extrapolated to man. It would follow that it may also be possible to design agents to selectively stimulate the DA-1 mechanisms and thus derive a selective anti-akinetic drug, but at the present time this remains a theoretical possibility. Conversely, it may also be possible to design other agents with a partial agonist–antagonist activity to block the DA-2 system and have a potential antidyskinetic activity. It may be that such a potential partly accounts for the clinical observations that some patients with dyskinesias experience some relief from their abnormal involuntary movements when treated with apomorphine, *N*-propylnorapomorphine or lergotrile (Duby *et al.*, 1972; Cotzias *et al.*, 1976; Lieberman *et al.*, 1976; Smith *et al.*, 1976). Alternatively, this may reflect dopamine agonist action on presynaptic DA-2 mechanisms in the absence of an effective response on the postsynaptic DA-2 receptors.

An associated and challenging problem is to dissociate the ability of dopamine agonists to modify motor dysfunction from an ability to stimulate excessively those dopamine systems, probably present within mesolimbic and/or mesocortical circuits, and which modulate thought and emotional behaviour. An improper stimulation of these mechanisms, as evidenced by the indiscriminate cerebral action of L-DOPA, results in psychomotor agitation or schizophreniform symptomology. Bromocriptine and piribedil may have a similar action (Chase *et al.*, 1974; Parkes *et al.*, 1976) but it is interesting that the classical dopamine agonist, apomorphine, fails to induce psychotic disturbance or psychomotor agitation in man (Angrist *et al.*, 1975); indeed, two recent reports suggest that apomorphine may actually *reduce* psychotic disturbance in man (Rostrosen *et al.*, 1976; Smith *et al.*, 1977). (This may be correlated well with the animal observations: apomorphine will actually *reduce* the hyperactivity induced by dopamine from the nucleus accumbens of rat (Costall and Naylor, 1976c).) It is suggested, therefore, that a dissociation between a stimulation of striatal and higher dopaminergic centres may be a realistic concept, and worthy of further study.

The general concepts so far discussed for the activation of different dopamine mechanisms also applies to their antagonism. The dopamine antagonists include, of course, the neuroleptic agents. The classical laboratory tests for these agents have, similarly to those for the agonist drugs, involved gross effects on all dopamine receptors. In inducing catalepsy, for example, neuroleptic agents may be antagonizing different types of dopamine receptors in both the extrapyramidal and mesolimbic areas, and in the stereotypy test then to determine which of the many possible "blocking sites" of the neuroleptics is most important for the antistereotypic action would depend on which stereotypic agent is being employed; for example, there is good evidence to suggest that the changes in striatal activity induced by amphetamine are only partially responsible for its full stereotypic action, and for apomorphine there is some evidence to suggest that the tuberculum olfactorium may be a major site for initiation of stereotypy (see Costall and Naylor, 1977). Further, not only are these tests lacking in specificity as regards the dopamine receptors involved, but even fail to detect reliably all antipsychotic agents, for example, clozapine, sulpiride and thioridazine (Costall and Naylor, 1975c), and indicate drugs which lack proven clinical activity, such as metoclopramide and morphine. The development of the accumbens hyperactivity model would, therefore, appear important. The spectrum of drug action in this model correlates well with the clinic and it would therefore appear possible that those dopamine receptors in the nucleus accumbens (head of the caudate nucleus in man) may malfunction to cause psychomotor agitation related to the disorders of schizophrenia.

Whilst it is impossible to place a value on the antipsychotic action of the neuroleptics, the majority of neuroleptic drugs at present in clinical use do have considerable disadvantages. Firstly, they not only reduce psychomotor agitation, but will also reduce normal movements and with incautious dosage to an unacceptable degree. Secondly, it has become apparent that the long-term therapy with neuroleptics leads to the development of tardive dyskinesias (Marsden *et al.*, 1975). These effects may be clearly expected in so much that the classical neuroleptics, the phenothiazines, butyrophenones and other series, exert an apparently non-selective inhibition of all cerebral dopamine systems. Thus, the inhibition of the DA-1 hyperactivity mechanisms presumably results in the neuroleptic akinesia and, with increasing dosage, a full pseudo-parkinson syndrome may develop as a result of additional effects, direct or indirect, on other neurotransmitter systems. The theoretical basis for dyskinesia induction is not so clearly explicable: one hypothesis which we have considered is that, with time, the sensitivity of the DA-2 mechanism may alter and become, essentially, supersensitive to dopamine. As indicated by the animal models, these DA-2

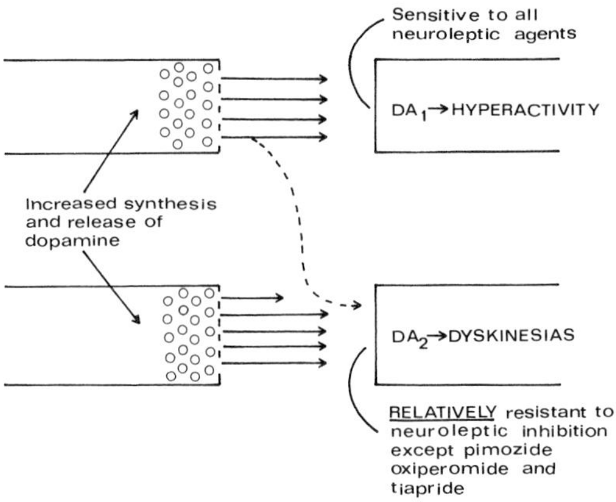

Fig. 2 Diagrammatic representation of the hypothesised DA-1 (hyperactivity) and DA-2 (dyskinesia) mechanisms.

mechanisms are relatively resistant to neuroleptic blockade and, if supersensitivity were to develop, then there would be little opposition to an unusually high reactivity with endogenous dopamine (Fig. 2).

There may even be an "overflow" into the DA-2 system of dopamine synthesized via the DA-1 mechanisms. These responses to dopamine may lead to the development of tardive dyskinesias. Nevertheless, whatever the explanation of the underlying events of dyskinesias, it is clear that they do involve DA-2 systems and it would appear more ideal to use a selective DA-1 antagonist for reducing excessive psychomotor agitation or mania. Presumably, those receptors involved with other psychological disturbances may be located in other areas of the mesolimbic and mesocortical systems, and further analyses are required here before one can predict the possible design of a more specific anti-schizophrenic drug. At the present time it would only appear realistic to consider the development of antipsychotic agents which have a relatively low efficacy to induce pseudo-parkinsonism, agents such as sulpiride and clozapine, and, in the long term, to develop agents which may lack the ability to induce gross tardive dyskinesias—possibly a relative of oxiperomide?

Paradoxically, whilst the neuroleptic agents induce the unwanted and unpleasant tardive dyskinesias, they are also the drugs which are used to relieve these symptoms. The antidyskinetic doses (DA-2 antagonistic doses) of typical neuroleptics are large and invariably cause a poverty of movement (marked DA-1 antagonism). Clearly, it would be advantageous to develop neuroleptic agents with a selectivity of action on the DA-2 mechanisms, since it could be predicted that these agents may have specific antidyskinetic potential.

Our studies in the guinea-pig have indicated three potential DA-2 antagonists: pimozide, oxiperomide and tiapride. However, a complete specificity for the DA-2 mechanisms cannot be claimed for all these agents. Pimozide and oxiperomide can undoubtedly affect DA-1 mechanisms and reduce locomotor activity, but the dosage differential for DA-1 and DA-2 antagonism is narrow for these agents, and markedly contrasts with that of the classical neuroleptics whose affinity appears to be most preferentially DA-1. In contrast, even at high pharmacological doses, we have not as yet obtained any data to indicate an action of tiapride on DA-1 mechanisms. On the basis of this animal experimentation we would suggest that these three agents should exert more selective antidyskinetic activity in the clinic than at present shown by the classical neuroleptics, and evidence is now available to confirm our hypotheses. Pimozide has been shown to abolish drug induced dyskinesias (L-DOPA and neuroleptic induced) without *necessarily* inducing an unacceptable degree of motor impairment (Claveria *et al.*, 1975; Tarsy *et al.*, 1975), although it cannot be ignored that some patients do appear to respond more adversely. Preliminary clinical studies have also established the antidyskinetic potential of oxiperomide (L-DOPA-induced dyskinesias), although careful attention would appear necessary to avoid interference with normal motor behaviour, as one would predict from the animal studies (Marsden, personal communication). Tiapride has also been shown to possess antidyskinetic properties in the clinic with a spectrum of action which could be predicted on the basis of animal observations (L'Hermitte, personal communication).

It is suggested that there is now sufficient evidence available from the use of dopamine agonists and antagonists in animal models to forward differential roles for the major cerebral dopamine systems, and even different dopamine receptor mechanisms within those systems, for the modulation of the various aspects of motor and psychomotor behaviour. Preliminary clinical observations encourage an extrapolation of these findings to man, and the hypothesis has been forwarded that the cerebral dopamine mechanisms may differentially malfunction in movement disorders and during psychotic disturbances. Initial studies have offered some confirmation of the possibility that the different dopamine mechanisms may be stimulated or antagonized by drugs with relative specificity to alleviate the symptoms of motor dysfunction, and it is considered that the future should bring the development of new drugs having specific actions on the different dopamine mechanisms and, therefore, a specificity of effect for the treatment of movement disorders and psychotic diseases for which, so far, drug treatment has not been available or has not been wholly satisfactory.

REFERENCES

Angrist, B., Thompson, H., Shopsin, B. and Gershon, S. (1975). *Psychopharmacologia (Berlin)* **44,** 273–280.
Cannon, J. G., Hsu, F.-L., Long, J. P., Costall, B. and Naylor, R. J. (1978). *J. Med. Chem.* **21,** 248–253.
Cannon, J. G., Lee, T., Goldman, H. D., Costall, B. and Naylor, R. J. (1977). *J. Med. Chem.* **20,** 1111–1116.
Chase, T. N., Woods, A. C. and Glaubiger, G. A. (1974). *Arch. Neurol.* **30,** 383–386.
Chiara, G. Di., Porceddu, M. L., Vargui, L., Argiolas, A. and Gessa, G. L. (1977). *Nature (London)* **264,** 564–566.
Claveria, L. E., Teychenne, P. F., Calne, D. B., Haskayne, L., Petrie, A., Pallis, C. A. and Lodge-Patch, I. C. (1975). *J. Neurol. Sci.* **24,** 393–401.
Cools, A. R. (1973). *In* "The Caudate Nucleus and Neurochemical Control of Behaviour". Drukkerij Brakkenstein, Nijmegen, The Netherlands.
Cools, A. R. (1975). *In* "Animal Models in Neuropsychopathology—Behavioural, Biological and Neurochemical Correlates". Raven Press, New York.
Cools, A. R. (1977). *In* "Cocaine and Other Stimulants" (Eds. E. H. Ellinwood and M. M. Kilbey), 97–141. Plenum Press, New York and London.
Cools, A. R., Hendriks, G. and Korten, J. (1975). *J. Neur. Trans.* **36,** 91–105.
Cools, A. R., Struyker Boudier, A. J. and Van Rossum, J. M. (1976). *Eur. J. Pharmacol.* **37,** 283–293.
Cools, A. R. and Van Rossum, J. M. (1976). *Psychopharmacologia (Berl.)* **45,** 243–254.
Costall, B., Fortune, D. H. and Naylor, R. J. (1977a). *Br. J. Pharmacol.* **60,** 266P–267P.
Costall, B. and Naylor, R. J. (1975a). *Eur. J. Pharmacol.* **33,** 301–312.
Costall, B. and Naylor, R. J. (1975b). *Eur. J. Pharmacol.* **32,** 87–92.
Costall, B. and Naylor, R. J. (1975c). *Psychopharmacologia (Berlin)* **43,** 69–74.
Costall, B. and Naylor, R. J. (1976a). *Eur. J. Pharmacol.* **36,** 423–429.
Costall, B. and Naylor, R. J. (1976b). *Eur. J. Pharmacol.* **40,** 9–19.
Costall, B. and Naylor, R. J. (1976c). *J. Pharm. Pharmacol.* **28,** 592–595.
Costall, B. and Naylor, R. J. (1977a). *In* "Cocaine and Other Stimulants" (Eds. E. H. Ellinwood and M. M. Kilbey), 47–76, Plenum Press, New York and London.
Costall, B. and Naylor, R. J. (1977b). Sem. Hôp. Paris, in press.
Costall, B., Naylor, R. J., Cannon, J. G. and Lee, T. (1977b). *J. Pharm. Pharmacol.* **29,** 337–342.
Costall, B., Naylor, R. J., Cannon, J. G. and Lee, T. (1977c). *Eur. J. Pharmacol.* **41,** 307–319.
Costall, B., Naylor, R. J., Marsden, C. D. and Pycock, C. J. (1976a). *J. Pharm. Pharmacol.* **28,** 523–526.
Costall, B., Naylor, R. J. and Neumeyer, J. L. (1975a). *J. Pharm. Pharmacol.* **27,** 875–877.
Costall, B., Naylor, R. J. and Olley, J. E. (1972). *Eur. J. Pharmacol.* **18,** 83–94.
Costall, B., Naylor, R. J. and Owen, R. T. (1977a). *Eur. J. Pharmacol.* **45,** 357–367.
Costall, B., Naylor, R. J. and Pinder, R. M. (1974). *J. Pharm. Pharmacol.* **26,** 753–762.
Costall, B., Naylor, R. J. and Pinder, R. M. (1975b). *Eur. J. Pharmacol.* **31,** 94–109.
Costall, B., Naylor, R. J. and Pinder, R. M. (1976b). *Psychopharmacologia (Berlin)* **48,** 225–231.

Cotzias, G. C., Papavasiliou, P. S., Tolosa, E. S., Mendez, J. S. and Bell-Midura, M. (1976). *New Eng. J. Med.* **294,** 567–572.
Duby, S. E., Cotzias, G. C., Papavasiliou, P. S. and W. H. Lawrence (1972). *Arch. Neurol. (Chicago)* **27,** 474–480.
Fog, R. (1972). *Acta. Neurol. Scand. Suppl. 48* **50.**
Galey, D., Simon, H. and Le Moal, M. (1977). *Brain Res.* **124,** 83–97.
Gessa, R., Tagliamonte, A. and Gessa, G. L. (1972). *Riv. Farm. Ther. 3* **423–427**.
Grol, C. J. and Rollema, H. (1977). *J. Pharm. Pharmacol.* **29,** 153–156.
Hornykiewicz, O. (1977). *Ann. Rev. Pharmacol. Tox.* **17,** 545–559.
Jackson, D. M., Andén, N.-E. and Dahlström, A. (1975). *Psychopharmacologia (Berlin)* **45,** 139–149.
Kehr, W., Carlsson, A. and Lindqvist, M. (1975). *In* "Advances in Neurology", Vol. 9 (Eds. D. B. Calne, T. N. Chase and A. Barbeau), 185–195. Raven Press, New York.
Lichtensteiger, W. and Lienhart, R. (1977), *Nature (London)* **266,** 635–637.
Lieberman, A. N., Kupersmith, M., Estey, E. and Goldstein, M. (1976). *Lancet* **i,** 515–516.
Marsden, C. D., Tarsy, D. and Baldessarini, R. J. (1975). *In* "Psychiatric Aspects of Neurologic Disease" (Eds. D. F. Benson and D. Blumer), 219–266. Grune and Stratton, New York.
Munday, K. A., Poat, J. A., Watling, K. J. and G. N. Woodruff (1977). *Br. J. Pharmacol.* **57,** 452–453P.
Parkes, J. D., Debono, A. G. and Marsden, C. D. (1976). *J. Neurol. Neurosurg. Psychiat.* **39,** 1101–1108.
Pijnenburg, A. J. J. (1976). *In* "The Nucleus Accumbens—Psychopharmacological Investigations on the Mesolimbic Dopamine System". Druk-Stichting Studentenpers, Nijmegen, The Netherlands.
Pijnenburg, A. J. J., Honig, W. M. M. and Van Rossum, J. M. (1975a). *Psychopharmacologia (Berlin)* **45,** 65–71.
Pijnenburg, A. J. J., Honig, W. M. M. and Van Rossum, J. M. (1975b). *Psychopharmacologia (Berlin)* **41,** 175–180.
Pijnenburg, A. J. J., and Van Rossum, J. M. (1973). *J. Pharm. Pharmacol.* **25,** 1003–1005.
Rotrosen, J., Angrist, M. B., Gershon, S., Sachar, E. J. and Halpern, F. S. (1976). *Psychopharmacologia (Berlin)* **51,** 1–7.
Smith, R. C., Tamminga, C. and Davis, J. M. (1977). *J. Neur. Trans.* **40,** 171–176.
Smith, R. C., Tamminga, C., Haraszti, J., Garver, D. and Davis, J. M. (1976). Paper presented at the 129th annual meeting of the Americaine Psychiatric Association, Miami, Florida.
Strombom, U. (1977). *J. Neur. Trans.* **40,** 191–194.
Tarsy, D., Parkes, J. D. and Marsden, C. D. (1975). *J. Neurol. Neurosurg. Psychiat.* **38,** 331–335.
Tolosa, E. S. and Sparber, S. B. (1974). *Life Sci.* **15,** 1371–1380.
Tuomisto, J. (1977). *Eur. J. Pharmacol.* **42,** 101–106.
Versteeg, D. H., G. Van der Gugten, J. De Jong, W. and Palkovits, M. (1976). *Brain Res.* **113,** 563–574.
Walters, J. R., Bunney, S. B. and Roth, R. H. (1975). *In* "Advances in Neurology", Vol. 9 (Eds. D. B. Calne, T. N. Chase, and A. Barbeau), 273–284. Raven Press, New York.
Woodruff, G. N., Watling, K. J., Andrews, C. D., Poat, J. A. and McDermed, J. D. (1977). *J. Pharm. Pharmacol.* **29,** 422–427.

CHAPTER 32

Central Dopamine Mechanisms and Unconditioned Behaviour

Urban Ungerstedt

Karolinska Institutet, Stockholm, Sweden.

The last decade has seen an explosive development of our knowledge about central dopamine (DA) neurones and their role in behaviour. The great interest in DA mechanisms and their relation to behaviour stems not only from clinical findings indicating an important role for DA in diseases such as schizophrenia and Parkinsonism but also from the availability of a number of excellent experimental tools enabling experimental control over DA neurotransmission.

The widespread termination of DA neurones within the hypothalamus, striatum, limbic and cortical areas (see Lindvall, Chapter 20) forms the basis for an equally extensive involvement in various brain functions. Any global change in DA neurotransmission is thus liable to affect a multitude of different functions subserved by the areas where DA neurones terminate. This is often overlooked when interpreting experimental data and changes in a particular behaviour after alterations in DA neurotransmission may, in fact, be due to manipulation of mechanisms e.g. controlling such functions as sensory-motor integration or hormonal release. It is therefore difficult to delineate the role of DA in the many behavioural functions where it has been implied, for example motility, exploration, sensory and motor control, feeding, sex, aggression, drive, sleep, arousal, stress, learning and attention. In this article I will therefore take a simplistic and pragmatic view on DA and behaviour and confine myself to the effects of experimentally induced increases or decreases in DA neurotransmission on unconditioned behaviour. Such changes have profound influences on animal behaviour which are hard to understand within a general framework of DA function within the brain. However, many contradictions stem from shortcomings in the experimental techniques which are gradually being revealed and today it seems possible to bring some order into the experimental findings and even to start building some explanatory concepts of the functional role of DA neurotransmission in relation to the function of DA-innervated areas.

EXPERIMENTAL TECHNIQUES

Systematic Drug Treatments

Much of our earliest knowledge about the functional significance of DA in the brain arrives from studies where reserpine was used to deplete DA stores (Carlsson *et al.*, 1958; Bertler, 1961). Reserpine was found to induce a state of sedation with loss of motor activity which could be reversed by L-DOPA. Similar effects were obtained with α-methyltyrosine, a selective inhibitor of tyrosine hydroxylase. However, studies with depleting drugs have left many question-marks as they affect DA as well as noradrenaline and serotonin neurones. The same problem of specificity applies to many of the DA receptor-blocking drugs.

Drugs stimulating DA transmission have possibly been of even greater value than the depleting and blocking drugs. Amphetamine has been shown in biochemical and histochemical experiments to release central catecholamines (Carlsson *et al.*, 1966; Glowinski and Axelrod, 1965; Fuxe and Ungerstedt, 1968). The resulting hyperactivity and stereotyped behaviour was prevented by pretreatment with α-methyltyrosine (Randrup and Munkvad, 1966). As in the case with depletors the releasing drugs present problems of interpretation because DA as well as noradrenaline is being released.

The direct DA agonists (e.g. apomorphine) which stimulate DA receptors (Ernst, 1967), have been of great interest. However, many uncertainties stem from problems in interpreting these studies. Firstly, it is clear that direct receptor stimulation is quite unphysiological as it is tonic and not contigent with nervous activity (see Ahlenius *et al.*, 1971). Secondly, the receptor mechanisms seem highly complex and we have to take into account several different DA receptors causing increases as well as decreases in the activity of the postsynaptic neurone depending upon their nature and localization (see below and Cools and van Rossum, 1976).

Intracerebral Drug Administration

The method of administering drugs and transmitters by local injections into specific brain areas increases the specificity of stimulation. However, the doses needed to induce behavioural effects are often within a range that seem unphysiological. The spread within brain tissue is often difficult to control, with concentration gradients extending from the site of injection. Drugs passing the blood–brain barrier may even induce their effects by entering the circulation and by this route stimulate other areas of the brain. Even the injection of the endogenous transmitter DA has several drawbacks. There are powerful inactivation mechanisms due to intraneuronal uptake as well as enzymatic breakdown which limits both the spread and the local effectiveness of the transmitter.

Lesion Studies

The widely used technique of lesioning cell groups, pathways and terminal areas to induce hypofunction within the DA systems has many inherent problems. It is well known that an electrocoagulation has no specificity to a particular DA pathway and the introduction of intracerebral injections of 6-hydroxydopamine (Ungerstedt, 1968, 1971; Hökfelt and Ungerstedt, 1973; Ungerstedt, 1973) has greatly increased the use of lesioning methods. In spite of criticism as regards the specificity of 6-hydroxydopamine (Poirier *et al.*, 1972; Butcher *et al.*, 1974) a wealth of evidence supports the contention that the correct use of this neurotoxin leads to lesions of greater specificity than any other method (Hökfelt and Ungerstedt, 1973; Sotelo *et al.*, 1973; Agid *et al.*, 1973). However, lack of morphological controls, or the use of conventional morphological methods instead of transmitter histochemistry, are major reasons for many interpretational difficulties in the literature. In many studies the entire interpretation rests on the claim that a particular pathway or cell group is lesioned in spite of the fact that this is only inferred from comparisons with other studies where the localization of the DA pathways is described. This is clearly not sufficient.

Behavioural Methods

Changes in DA neurotransmission bring about profound changes in behaviour ranging from akinesia to hyperactive stereotyped behaviour. Many behavioural studies have therefore been concerned with registering spontaneous or drug-induced behaviour. The major problem in such studies concerns the method of recording behaviour which varies greatly from laboratory to laboratory. Automatic instruments are limited by the method for detecting changes in activity. A particular drug-induced behaviour may in fact be recorded as an increase or a decrease from base-line activity just depending upon the type of recording apparatus being used (Ljungberg, 1978a).

One way to by-pass these problems has been the use of rating scales where particular behavioural patterns are rated and the same principle of rating can then be used by many laboratories. However, this technique suffers from the fact that a rating scale is based upon an interpretation of behaviour and it has recently been shown that the commonly assumed intensity relationship between different behavioural patterns in the apomorphine behaviour does not seem correct (Ljungberg and Ungerstedt, 1977a).

BEHAVIOURAL CONSEQUENCES OF INCREASED DA ACTIVITY

Amphetamine and Apomorphine Induced Stereotyped Behaviour

Amphetamine causes a complicated dose-dependent change in animal behaviour (Fig. 1). In lower doses the behaviour is reminiscent of the increased spontaneous activity displayed by a rat in a new environment—that is sniffing, locomotion and rearing. In higher doses the animal develops the well known stereotyped behaviour consisting of continuous sniffing, licking or biting of the cage floor or the animal's own forelegs. The rat sits in a crouched posture and usually presses its body against the cage wall. Normal activities, such as grooming, eating, rearing and forward locomotion are absent (Randrup and Munkvad, 1970). This behaviour has been attributed to the DA releasing action of amphetamine as it is blocked after inhibition of DA and noradrenaline synthesis but restored by DOPA treatment (Randrup and Munkvad, 1966). Furthermore, inhibition of noradrenaline synthesis does not inhibit the amphetamine-induced stereotypies (Scheel-Krüger and Randrup, 1967). A similar behavioural pattern is induced by apomorphine when rats display locomotion as well as oral stereotyped behaviours such as sniffing, licking and gnawing (see Ljungberg and Ungerstedt, 1977a).

In spite of the similarity between the amphetamine- and apomorphine-induced hyperactivity there are obvious differences apparent to an observer. These differences become evident when animals are tested in a hole board, i.e. a test box with a number of holes in the floor. Amphetamine-treated animals react strongly to the holes and show a pronounced increase in the number of head dips into the holes while apomorphine-treated animals

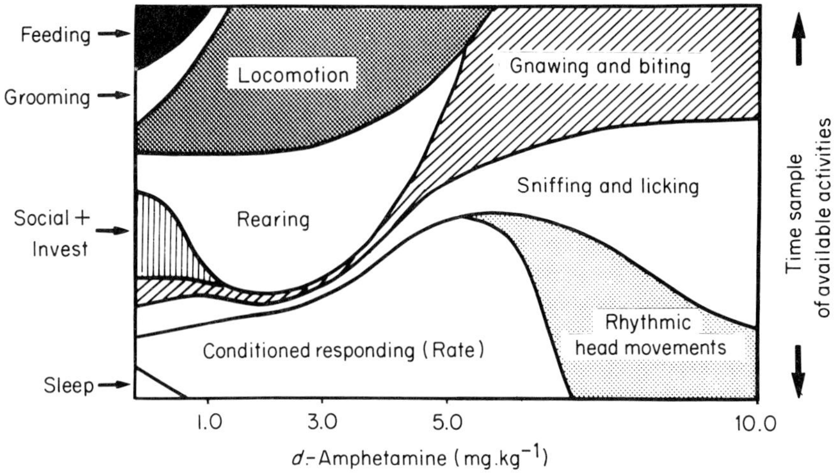

Fig. 1 Illustration from Lyon and Robbins (1975) showing the relative distribution within a given time sample of varying activities in the rat as determined by the increasing dose–response effect of *d*-amphetamine. The diagram serves to illustrate their theory of amphetamine action (see text) as well as the complex effect of increased DA activation on behavioural patterns.

show a strong decrease in head dipping compared to normal animals but increase their locomotion in the box (Ljungberg and Ungerstedt, 1976a).

Even though the behavioural effects elicited by amphetamine may be largely due to DA release (Roberts *et al.*, 1975) it is obvious that the concomitant release of noradrenaline may influence the behaviour accounting for some of the differences between behaviour induced by amphetamine and apomorphine. In a series of papers Svensson and Waldeck (1969, 1970) and Strömberg and Svensson (1971) concluded that both noradrenaline and DA are of importance in controlling motor activity as reduction in brain noradrenaline levels by an inhibitor of DA-β-hydroxylase was accompanied by a reduction in motor activity. Svensson (1970) found that inhibition of noradrenaline synthesis partly inhibited the effect of amphetamine in mice but the amphetamine effect was restored by restoring noradrenaline transmission. The noradrenaline and DA related components in amphetamine-induced behavioural activation were separated by Ljungberg and Ungerstedt (1976a) by the use of a special test situation i.e. an automatic hole board. By comparing the behavioural effects of amphetamine, apomorphine and noradrenaline synthesis inhibition they showed that the increase in the open field activity was related to an increased DA transmission, while the head dipping into holes seemed related to an increased noradrenaline transmission. However, findings like these are purely descriptive and difficult to interpret in functional terms. It is clear, however, that taken together these studies indicate an intricate relationship between DA and noradrenaline neurotransmission in the control of certain aspects of behaviour. This relationship has lately been reviewed extensively by Antelman and Caggiula (1977) where they attempt to bring together the data under an hypothesis assuming the dynamic interrelationship between noradrenaline- and DA-containing systems in the brain.

Separating the Role of Different DA Systems in Drug Induced Behaviour

The combined knowledge of various behavioural patterns induced by DA activation and the localization of different DA-innervated areas raises interesting questions concerning the possibility of allocating behavioural components to anatomical structures. The problem has attracted several groups and resulted in a huge number of studies. The starting point has been the different behavioural components that can be identified after drugs like apomorphine and amphetamine: sniffing, repetitive head and limb movements, biting, gnawing, licking and hyperactivity (usually synonymous with locomotion or running). These behaviours have been related to the most clearly defined DA terminal areas such as the striatum (the caudate putamen in the rat), the nucleus accumbens, the olfactory tubercle and the nucleus amygdaloideus centralis. For a detailed description of the anatomy see Dahlström and Fuxe (1965), Ungerstedt (1971b), Lindvall and Björklund (1974), Berger *et al.* (1976), Hökfelt *et al.* (1976, 1977) and Lindvall (this volume, p. 319). The experiments have generally followed two experimental designs: the DA-related behavioural patterns have been induced by various doses of amphetamine or apomorphine in normal animals and these animals have been compared to experimental animals where various parts of the DA systems have been lesioned by electro-coagulations or 6-hydroxydopamine injections. The other approach has been to stimulate selectively the different DA-innervated areas by local injections of, for example, amphetamine, apomorphine or DA in awake animals.

As was outlined in the beginning of this article much of the confusion in the literature stems from various methodological pitfalls. It is clear that during many years of studies the techniques have been improved and data can be reinterpreted in the light of better knowledge. It may therefore be appropriate to point out some problems that have arisen during the attempts to localize behavioural components to morphological structures.

The major problem is undoubtedly the unspecificity of the commonly used electrocoagulations. It is clear today that many lesions have damaged the DA pathways but also other afferents and efferents to the DA-innervated area. This is in all probability the major reason why results from electrocoagulations are in conflict with those obtained by 6-hydroxydopamine lesions that are considerably more specific (see above). The second problem concerns the completeness of the lesion. There are powerful compensatory mechanisms within DA-innervated areas making necessary depletions amounting to 80% or more in order to see functional changes (see review by Stricker and Zigmond, 1976). The common lack of histochemical or biochemical verification of the extent of the lesion is a common course of misunderstandings. Finally, it is often difficult to compare the behavioural observations between studies. Different authors recognize different behavioural components or use different automatic devices to measure behaviour. Another hardly recognized problem is whether the animals are habituated or not to the test cage when they receive the drug injection. From our own studies (Ljungberg and Ungerstedt, 1978b) it is evident that drugs like amphetamine and apomorphine act differently when they *induce* behaviour in an habituated animal than when the drug actions *interact* with the animal's exploratory behaviour.

The first attempts to abolish amphetamine-induced behaviours by intraventricular injections of 6-hydroxydopamine were surprisingly unsuccessful (Evetts *et al.*, 1970). Electrocoagulations aimed for the DA cell bodies in the substantia nigra were equally unsuccessful (Simpson and Iversen, 1971; Costall *et al.*, 1972). Local injections of 6-hydroxydopamine into the substantia nigra lowered tyrosine-hydroxylase in the striatum to 10–15% causing a decrease of stereotyped activity after amphetamine while the locomotor activity remained (Creese and Iversen, 1972; Fibiger *et al.*, 1973) similar results were obtained after 6-hydroxydopamine injections into the caudate nucleus (Creese and Iversen, 1974). However, the amphetamine effect was found to be completely blocked provided that essentially all DA neurones were destroyed by 6-hydroxydopamine injections in adult animals (Hollister *et al.*, 1974; Ungerstedt, 1974; Price and Fibiger, 1974; Creese and Iversen, 1975a) or by injecting 6-hydroxydopamine in neonatal animals (Creese and Iversen, 1973).

The effect of apomorphine in DA-lesioned animals has been a long-standing controversy in the literature. Ungerstedt (1971), Creese and Iversen (1973), Schoenfeld and Uretsky (1972), Price and Fibiger (1974), Nahorski (1975) and Costall *et al.* (1977a) have reported that the response to apomorphine is strongly potentiated after lesions of the DA pathways. However, the opposite finding, i.e. a reduction in apomorphine stimulation was reported by Costall *et al.* (1972), Costall and Naylor (1973) and Costall *et al.* (1975). The results are directly related to the method of lesioning DA pathways. The effect of apomorphine is potentiated after 6-hydroxydopamine lesions and diminished after electrocoagulations. The obvious explanation is the unspecificity of electrolytic lesions, and the need for biochemical or histochemical verification before interpreting data has, thus, been strongly emphasized by Price and Fibiger (1974). Recent studies have also shown that unspecific lesions along the DA tracts in all probability also damage striatal efferent pathways, probably the descending GABA or substance P neurones (Marshall and Ungerstedt, 1977; Garcia-Munoz *et al.*, 1977).

The lesion studies referred to so far have yielded valuable evidence that the behavioural patterns elicited by amphetamine and apomorphine are due to activation of DA mechanisms. In addition, local lesions or stimulations of DA-innervated nuclei have revealed a relationship between anatomically distinct nuclei and defined behavioural components. The most easily interpreted findings have come from manipulation of the nucleus accumbens septi and its DA innervation. Local 6-hydroxydopamine lesions in the nucleus accumbens were found to block the amphetamine-induced locomotion component and enhanced apomorphine-induced locomotion and sniffing (Kelly *et al.*, 1975). The removal of the DA nerve terminals is thus thought to abolish the possibility for amphetamine-induced DA release while the development of denervation supersensitivity is thought to potentiate the apomorphine effect (Ungerstedt, 1971c). A sufficiently low dose of apomorphine will thus only stimulate the denervated receptors (Ungerstedt, 1971c, 1973) and probably induce a specific component of behaviour related to the function of the nucleus accumbens. In support of the lesion studies several authors have reported locomotion hyperactivity and sniffing behaviour after local injection of DA or DA agonist drugs into the nucleus accumbens (Pijnenburg and van Rossum, 1973; Costall and Naylor, 1975; Costall *et al.*, 1975; Jackson *et al.*, 1975; Pijnenburg *et al.*, 1976; Ungerstedt and Ljungberg, 1977; Costall *et al.*, 1977b).

The olfactory tubercle has been more difficult than the nucleus accumbens to link to a specific DA-related behavioural component. This is partly due to the anatomy of the nucleus. The DA-innervated area extends as a layer along the ventral surface of the brain making it difficult to lesion or stimulate *in extenso*. Costall and Naylor (1974) found that lesions of the olfactory tubercles abolished amphetamine sniffing and the injection of DA into the olfactory tubercle induced sniffing, repetitive head and limb movements as well as hyperactivity (Costall and Naylor, 1975). However, this hyperactivity (i.e. locomotion) was less than that induced by injecting DA into the accumbens. These findings agree with our own recent studies

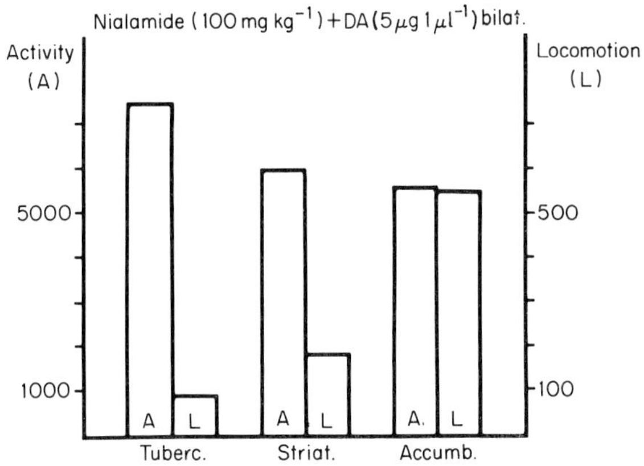

Fig. 2 Different behavioural effects of local injection of DA into the dorsal part of the olfactory tubercle (Tuberc.), the centre of the striatum (Striat.) and the rostral nucleus accumbens (Accumb.). The recordings are made according to Ljungberg and Ungerstedt (1977b) and show how the small stereotyped movements dominate after injection into the dorsal olfactory tubercle while locomotion behaviour dominates after injection into the nucleus accumbens. The variable "activity" is the sum total of movements i.e. small stereotyped movements as well as locomotion, while the variable "locomotion" expresses the actual distance walked by the animal.

using a newly developed test system where different components of stereotypy can be quantified separately. We recorded distinctly different behavioural patterns after inhibition of monoamine oxidase with nialamide and injection of DA into the nucleus accumbens and the olfactory tubercle. The behaviour elicited in the accumbens consisted of strong locomotion accompanied with relatively mild sniffing and repetitive movements of the head. When injected into the olfactory tubercle the animals essentially lacked locomotion but displayed intense sniffing and repetitive head and limb movements (Fig. 2, Ungerstedt and Ljungberg, to be published).

While it seems possible to attribute rather well-defined behavioural patterns to various limbic areas there have been considerable problems to determine a specific behaviour elicited from the striatum in rats. While earlier studies indicated that amphetamine or apomorphine stereotypies could be prevented by large lesions in cortex plus striatum or in striatum alone (Amsler, 1923; Fog *et al.*, 1967) several other studies have reported that extensive electrolesions of the striatum fail to modify any component of amphetamine or apomorphine stereotypy (Divac, 1972; McKenzie, 1972; Costall and Naylor, 1973; Wolfarth *et al.*, 1973; Costall and Naylor, 1974; Neill *et al.*, 1974; Wolfarth, 1974; Costall *et al.*, 1975; Yehuda and Wurtman, 1975). However, in most of these studies the damage to the striatum has not been complete and the behavioural observations often lack details. In studies using local injections of 6-hydroxydopamine into the striatum Creese and Iversen (1974) and Kelly *et al.* (1975) found that the stereotypy after higher doses of amphetamine was abolished while the locomotor response was unchanged. Conversely the apomorphine-induced stereotypies were potentiated in all probability due to the development of denervation supersensitivity of DA receptors. Further evidence for the involvement of the striatum in the intense biting/gnawing stereotypies comes from local injections of apomorphine (Ernst and Smelik, 1966; Costall *et al.*, 1975) and DA (Ungerstedt and Ljungberg, 1977; Costall *et al.*, 1977b). A more detailed analysis of behavioural components elicited by striatal injections of DA has been possible in cats (Cools and van Rossum, 1970).

In spite of the complexity and contradictions in these studies there is an emerging pattern where DA neurotransmission in particular areas may be related to particular behavioural components. The nucleus accumbens seems related to locomotion while the olfactory tubercle may relate to the repetitive movements of the head and limbs and the continuous sniffing stereotypy while the striatum may relate to the gnawing/biting/licking stereotypy. It should, however, be emphasized that the experimental findings do not allow the conclusion that any one area is specifically concerned with a specific behavioural component. Several studies indicate an overlap in this respect where the actual drug used to induce stimulation may work differently in different areas (see e.g. Costall *et al.*, 1975).

In a recent series of studies we have developed and applied a new technique for the automatic recording of different components of drug-induced behaviour (Ljungberg and Ungerstedt, 1978b). The test box consists of a modified open field area. The centre of the box is occupied by a cube thus dividing the box into four alleys, one along each side. Thirty-two holes are symmetrically distributed over the floor. Ten horizontal photobeams cover the entire open field area, four verticle beams cover the corners and other photobeams run horizontally, just underneath the floor, thus revealing every time an animal lowers his nose into a hole. Finally, a sound/vibration sensor detects when the animal gnaws at any part of the cage. The signals from the photobeams and vibration sensor are fed into a Wang 2200S computer via a digital logic. The system simultaneously measures eight components of behaviour.

When studying apomorphine-induced activation of habituated animals we found that the animals first showed a short period of locomotion (defined as the actual distance that the animal walks) which was replaced by a period of gnawing (usually performed on the edges of the holes and registered by the vibration sensor system). These two components of

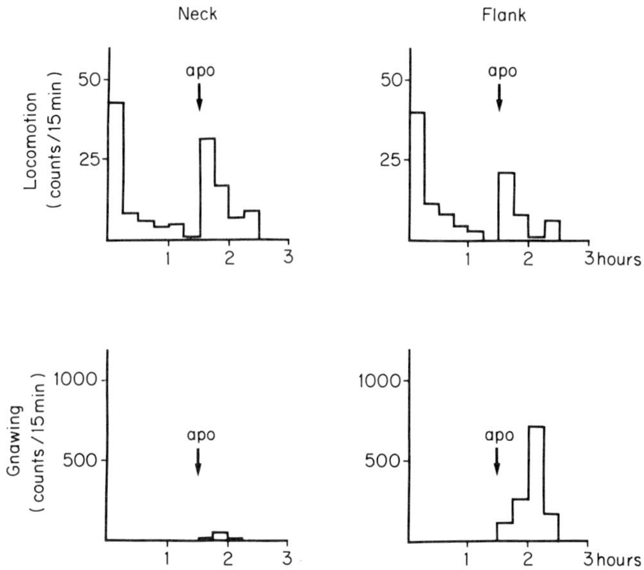

Fig. 3 Behavioural effect of injecting apomorphine 1 mg kg^{-1} s.c. into the neck or the flank of an habituated rat. The neck injection induces strong locomotion without any gnawing while the flank injection induces strong compulsive gnawing after an initial short period of diminished locomotion. Recordings are made in the test system developed by Ljungberg and Ungerstedt (1977b).

apomorphine-induced behaviour could be preferentially induced depending upon how apomorphine was injected. After subcutaneous injection into the neck the animals showed a locomotion behaviour without gnawing while a subcutaneous injection into the flank of the animal induced strong compulsive gnawing (Fig. 3, Ljungberg and Ungerstedt, 1977a). Two important findings emerged from our continued studies. The first finding was that the locomotion and the gnawing patterns of behaviour were differentially sensitive to inhibition by various neuroleptic drugs. While locomotion was strongly inhibited by drugs like sulpiride and clozapine, gnawing was preferentially inhibited by drugs like metoclopramide and haloperidol (Ljungberg and Ungerstedt, 1978c). The other finding concerns the actual mechanism of inducing one or the other of the two behavioural components: by changing the speed of resorption of apomorphine from the injection site it was possible to show that the pattern of behaviour that developed was related to the speed by which apomorphine reached the brain (Ljungberg, 1979; Ungerstedt *et al.*, 1978). When apomorphine was injected in a propylene glycol suspension, causing slow resorption, strong compulsive gnawing was induced while an intravenous injection induced locomotion. These surprising findings have obvious implications for the interpretation of behavioural patterns elicited by DA-stimulating drugs. The common assumption that different behavioural patterns are elicited by different doses of apomorphine does not seem valid because locomotion, which is thought of as a low-intensity behaviour, can be induced by the same dose of apomorphine as gnawing, which is considered a high-intensity stimulation, simply depending upon the site of injection. Furthermore, we found that increasing the dose of the apomorphine treatment that induced locomotion did not result in gnawing. In fact, intravenous administration, which ought to yield a high-intensity stimulation, resulted in locomotion, not gnawing.

Our data seem difficult to explain without assuming the existence of different types of DA receptors. However, at this point it seems premature to favour any particular explanation such as preferential involvement of pre- and postsynaptic DA receptors (see Carlsson, 1975) or different DA receptors in the sense suggested by Cools and van Rossum (1976).

Rotational Behaviour

One of the most widely-used behavioural models in studies of DA mechanisms is the rotational behaviour in rats after a unilateral DA lesion. The model originated from the studies of Andén *et al.* (1966) where rats with unilateral lesions of the striatum were found to rotate in the ipsilateral direction when DA neurotransmission was facilitated on the non-lesioned side of the brain by various drug treatments. After the introduction of the 6-hydroxydopamine lesioning technique (Ungerstedt, 1968, 1971a, 1973) highly reproducible and specific unilateral lesions of the ascending DA pathways could be performed. This prompted the development of a quantitative recording technique for rotation using so-called rotometers (Ungerstedt and Arbuthnott, 1970). After a unilateral lesion of the ascending DA pathways the rat developed a typical motor asymmetry deviating in posture and movements toward the DA denervated side. When such animals were treated with a DA-releasing drug such as amphetamine, they developed a strong rotational behaviour towards the lesioned side (Ungerstedt, 1971d). In contrast, direct DA receptor-stimulating drugs such as apomorphine caused strong rotational behaviour in the opposite direction i.e. towards the intact side (Ungerstedt, 1971c). This was interpreted as a sign that DA receptor supersensitivity had developed in the DA-denervated areas due to the terminal degeneration.

The fact that rotational behaviour after unilateral lesions of the DA pathways is an expression of DA neurotransmission seems well established (Ungerstedt and Arbuthnott, 1970; Arbuthnott and Crow, 1971; Crow, 1971; Voigtländer and Moore, 1973). However, the actual DA terminal area responsible for rotation is still unclear. The fact that unilateral intrastriatal injection of DA and apomorphine in nialamide pretreated rats induce an asymmetric posture and turning contralateral to the side of drug injection indicates that the nigrostriatal DA system is mainly responsible for the rotational behaviour (Ungerstedt *et al.*, 1969). This is further supported by the findings of Kelly (1975) that apomorphine induced rotational behaviour in animals lesioned with 6-hydroxydopamine in the striatum, but not in animals lesioned in the nucleus accumbens. However, the fact that mesolimbic DA nerve terminals in the nucleus accumbens are involved in locomotor behaviour (see above) indicates that this nucleus may at least contribute to the *intensity* of rotational behaviour. This possibility has recently been tested by Kelly and Moore (1976) and Pycock and Marsden (1978). In these studies the striatal DA terminals were lesioned with or without an additional lesion of the nucleus accumbens. The data seem to indicate that striatal DA inbalance causes the postural asymmetry and stimulation of mesolimbic DA systems provide the locomotor component. These findings are also supported by the fact that local injections of DA or its agonists into the normally innervated striatum do not induce the strong rotational behaviour seen after systematic administration in unilaterally lesioned animals (Ungerstedt *et al.*, 1969; Costall *et al.*, 1974). However, the evidence still seems far from conclusive, as strong rotational behaviour can be elicited by local injections of apomorphine into the striatum of DA-denervated animals (Ungerstedt, unpublished). Apomorphine will elicit rotation even in doses as low as 0.1 picogram in 1 μl saline and it seems improbable that this small amount will exert its effect due to widespread diffusion in the brain reaching limbic areas from the site of the striatal injection.

BEHAVIOURAL CONSEQUENCES OF DECREASED DA ACTIVITY

Probably the most important findings relating behavioural depression and sedation to decreased DA transmission stems from the studies showing that reserpine depletes DA stores in the brain (Carlsson *et al.*, 1958; Bertler, 1961). This state of sedation with pronounced akinesia could be reversed by L-DOPA but not 5-HTP, further linking the behavioural syndrome to DA depletion.

The role of DA in this syndrome was further established when it was shown that neuroleptic drugs that induce a behavioural syndrome resembling the reserpine syndrome exert their action by inhibiting DA receptors (Carlsson and Lindqvist, 1963; Andén *et al.*, 1970, 1971).

The stereotaxic mapping of the major DA, noradrenaline and serotonin pathways (Ungerstedt, 1971b) made it possible to apply lesion techniques to study the effect of removing different DA systems on behaviour. The development of a neurochemical lesioning technique, using intracerebral injections of 6-hydroxydopamine, increased the specificity of the denervations and facilitated the interpretations (Ungerstedt, 1968, 1971a, 1973).

Bilateral 6-hydroxydopamine-induced degeneration of the ascending DA neurones induce a severe syndrome of akinesia, adipsia and aphagia. The rats loose all spontaneous behaviour, fail to maintain their body weight and die if not supported by tube feeding (Ungerstedt, 1971e). This severe syndrome is preceded by a short period, limited to day two and three after the lesion, when the animals show slight hyperactivity and irritation. This period coincides with the disappearance of DA from the striatal nerve endings and is probably the combined effect of degeneration release and developing denervation supersensitivity (Ungerstedt, 1971c, 1973; Hökfelt and Ungerstedt, 1973). The severity of the DA degeneration syndrome and the period of adipsia and aphagia are directly related to the degree of DA denervation (Zigmond and Stricker, 1972). It seems necessary to produce more than 85% depletion of DA in order for the syndrome to develop. Although animals may assume spontaneous eating after long periods of tube feeding, we have maintained animals for periods longer than four months without signs of recovery, probably due to essentially zero levels of DA (Ungerstedt *et al.*, 1973). The DA denervation syndrome is in many respects similar to the well-known lateral hypothalamic syndrome after electrocoagulation of the lateral hypothalamus (Ungerstedt, 1971e; Oltmans and Harvey, 1972; Fibiger *et al.*, 1973; Stricker and Zigmond, 1974). The DA axons pass through the lateral hypothalamic area and the classical lateral hypothalamic lesion will, in fact, cause an effective degeneration of the ascending DA axons (Ungerstedt, 1971b) in all probability accounting for the syndrome.

A bilateral injection of 6-hydroxydopamine into the region of the substantia nigra produces a degeneration of nigrostriatal, mesolimbic and mesocortical DA pathways. In addition there is damage to ascending noradrenaline axons due to the specific uptake of 6-hydroxydopamine. The injection may also produce unspecific damage to other axon bundles passing the injection site. By control injections of 6-hydroxydopamine into the ascending noradrenaline pathways as well as into the DA bundles to limbic, septal and cortical areas we have shown that these DA systems contribute little to the akinetic, adipsic and aphagic syndrome. The syndrome develops only when sufficient damage is made to the nigrostriatal neurones (Ungerstedt, 1971e, 1973). However, in view of the complex anatomy of the substantia nigra and the medial forebrain bundle it seems essential to prove that the 6-hydroxydopamine induced damage of DA axons is the true cause of the syndrome. We have therefore attempted to substitute the lost DA by DA receptor-stimulating drugs like apomorphine or by the precursor L-DOPA. These studies show that it is possible to induce locomotion and eating behaviour after apomorphine injections to bilaterally DA-denervated animals (Ljungberg and Ungerstedt, 1976b). The animals consumed food in a coordinated manner and returned to their akinetic state when the effect of the drug subsided. Higher doses of apomorphine-induced stereotyped behaviour and the animals gnawed indiscriminately at other objects in the cage apart from the food pellets. Bilateral DA denervation also causes failure to drink in response to, for example, hypertonic saline (Marshall and Ungerstedt, 1976). Apomorphine also induced drinking in these animals in response to the hypertonic challenge.

The fact that movements, drinking and eating may be induced by substituting the lost DA with DA agonists provides strong support for the conclusion that the behavioural syndrome following a 6-hydroxydopamine denervation is due to the interruption of DA neurotransmission. However, it must be kept in mind that it does not prove that the DA denervation syndrome is the same as the lateral hypothalamic syndrome. In fact, in the substitution studies (Ljungberg and Ungerstedt, 1976; Marshall and Ungerstedt, 1976) it was not possible to substitute animals that were lesioned by lateral hypothalamic electrocoagulations. This finding is analogous to the finding by Costall and Naylor (1973) that the effect of apomorphine is reduced or abolished in lateral hypothalamic lesioned animals. The ineffectiveness of DA agonist drugs is in all probability due to damage of a pallidal efferent system in addition to the damage of the DA afferent pathway (Marshall and Ungerstedt, 1977).

The fact that a lateral hypothalamic manipulation in all probability will involve manipulation of DA pathways implicates DA neurotransmission in a wealth of different behaviours that have been studied with lesions or electrical stimulations of the lateral hypothalamus, e.g. electrical self-stimulation, stimulus-bound eating and drinking, active and passive avoidance as well as a number of other behaviours involving operant conditioning. Theories of reward and punishment as well as motivation have often referred to studies on lateral hypothalamus and it is tempting to relate several of these studies to neurotransmission in the ascending DA systems.

DOPAMINE NEUROTRANSMISSION AND CNS FUNCTION

There are considerable difficulties in reconciling the many diverse effects of changes in dopamine neurotransmission even when our analysis has such a limited focus as changes in unconditioned behaviour, that is spontaneous behavioural activities. However, there is little basis for the assumption that we will ever be able to formulate a global theory for the role of dopamine neurotransmission in the control of behaviour. If anything, our data emphasize the multiplicity of dopamine functions, and as long as we are prepared to ascribe specific functions to specific areas of the brain we have to accept that the particular dopamine systems have to be studied and understood in conjunction with the function of the areas they innervate.

The neostriatal dopamine innervation seems of particular interest as several findings after altering dopamine neurotransmission clearly relate to recent concepts of striatal function (see review by Krauthamer, 1975). It is well known that the striatum is a structure with sensory convergence properties. Electrical stimulation of the striatum also affects polysensory thalamic responses. Studies of Thompson and Bettinger (1970) have linked the polysensory convergence system to the orientating response to novel stimuli and, in fact, recent studies in unilateral dopamine-denervated animals show that they lose their ability to orientate to sensory stimuli presented to the side of the body contralateral to dopamine denervation (Ungerstedt, 1974; Marshall *et al.*, 1974; Ljungberg and Ungerstedt, 1976a). This "sensory neglect" caused by dopamine neurone degeneration is undoubtedly related to the interesting findings of Zimmerberg *et al.* (1974) and Zimmerberg and Glick (1975) that spatial side preference is related to striatal activity and can be correlated to striatal dopamine levels. The involvement of such mechanisms will affect many aspects of behaviour and may form a basis for explaining seemingly unrelated phenomena induced by a dopamine degeneration. Thus, the adipsic and aphagic syndrome after bilateral dopamine denervation may be due to an inability to orient towards any sensory stimuli while the rotational behaviour after a unilateral dopamine lesion may be a sign of a unilateral sensory neglect and a profound shift of cerebral dominance.

The stereotyped behavioural patterns induced by increased dopamine receptor stimulation are more difficult to explain in terms of the same general mechanisms. However, two recent theories on the mechanism by which increased DA stimulation gives rise to stereotyped behaviour relate to the role of DA neurotransmission in sensory–motor integration. Cools (1973) suggests that "the caudate nucleus forms part of a system in which incoming signals are compared with previously established target values, i.e.

links between two or more responses forming an integrated sequence of elements; normal activation of this system produces a breakdown of the established chain of responses, and enables the organism to establish completely new combinations according to available sensory information". The DA innervation of the caudate nucleus may thus influence the ability to break down previously established target values and establish new combinations according to the incoming sensory information. Hyperfunction in DA neurotransmission, as occurs after amphetamine treatment, may evoke the typical repetitive stereotyped behaviour due to an abnormally fast breakdown of target values while hypofunction may prevent new combinations from forming, leading to the situation we see in the Parkinson patient who has difficulties in inducing as well as terminating a behavioural sequence.

Cools' concept of DA function within the caudate nucleus is in many ways related to a theory presented in a recent review of the behavioural actions of amphetamine by Lyon and Robbins (1975) (see Fig. 1). They propose the following: "The action of amphetamine is such that as the dose response within the central nervous system increases the repetition rate of all motor activities will increase with the result that the organism will tend to exhibit increasing response rates within a decreasing number of response categories." This theory thus states that stimulation by amphetamine has two effects: (1) increasing the response rate and (2) causing a repetition within specific response categories. There is an obvious relationship between this theory and Cools' concept where transmitter mechanisms are brought in. The theory of Lyon and Robbins (1975) provides us with an interesting theoretical framework and Cools' (1973) concept of caudate function may develop into a link between the experimental findings reported in this review and DA function within the central nervous system.

ACKNOWLEDGEMENT

This study was supported by grants from the Swedish Medical Research Council (03574, 4575), Karolinska Institutes Fonder, Bergvalls Stiftelse and Ferrosans Jubileums Fond.

REFERENCES

Agid, Y., Javoy, F., Glowinski, J., Bouvet, D. and Sotelo, C. (1973). *Brain Res.* **58,** 291–301.

Ahlenius, S., Andén, N.-E. and Engel, J. (1971). *Physiol. Behav.* **7,** 931–934.

Amsler, C. (1923). Beiträge zur Pharmakologie des Gehirns. *Naun. Schmied. Arch. Exp. Pathol. Pharmacol.* **97,** 1–14.
Andén, N. E., Butcher, S. G., Corrodi, H., Fuxe, K. and Ungerstedt, U. (1970). *Eur. J. Pharmacol.* **11,** 303–314.
Andén, N.-E., Corrodi, H., Fuxe, K. and Ungerstedt, U. (1971). *Eur. J. Pharm.* **15,** 193–199.
Andén, N.-E., Dahlström, A., Fuxe, K. and Larsson, K. (1966). *Acta Pharmacol. Toxicol.* **24,** 263–274.
Antelman, S. M. and Caggiula, A. R. (1977). *Science* **195,** 646–653.
Arbuthnott, G. W. and Crow, T. J. (1971). *Exp. Neurol.* **30,** 484–491.
Berger, B., Thierry, A. M., Tassin, J. P. and Moyne, M. A. (1976). *Brain Res.* **106,** 133–145.
Bertler, Å. (1961). *Acta Physiol.* **51,** 75–83.
Butcher, L. L., Eastgate, S. M. and Hodge, G. K. (1974). *Naun. Schmied. Arch. Pharmacol.* **285,** 31–70.
Carlsson, A. (1975). *In* "Pre- and Postsynaptic Receptors" (Eds. E. Modin and W. E. Bunney), 49–66. Marcel Dekker, New York.
Carlsson, A., Fuxe, K., Hamberger, B. and Lindqvist, M. (1966a). *Acta Physiol. Scand.* **67,** 481–497.
Carlsson, A. and Lindqvist, M. (1963). *Acta Pharmacol. (Kph.)* **20,** 140–144.
Carlsson, A., Lindqvist, M., Magnusson, T. and Waldeck, B. (1958). *Science* **127,** 471.
Cools, A. R. (1973). Thesis, 134–143. University of Nijmegen, The Netherlands.
Cools, A. R. and Van Rossum, J. M. (1970). *Arch. Int. Pharmacodyn. Thér.* **187,** 163–173.
Cools, A. R. and Van Rossum, J. M. (1976). *Psychopharmacologia (Berlin)* **45,** 243–254.
Costall, B., Marsden, C. D., Naylor, R. J. and Pycock, Ch. J. (1977a). *Brain Res.* **123,** 89–111.
Costall, B. and Naylor, R. J. (1973). *Eur. J. Pharmacol.* **24,** 8–24.
Costall, B. and Naylor, R. J. (1974). *Eur. J. Pharmacol.* **25,** 121–129.
Costall, B. and Naylor, R. J. (1975). *Eur. J. Pharmacol.* **32,** 87–92.
Costall, B., Naylor, R. J., Cannon, J. G. and Lee, T. (1977b). *J. Pharm. Pharmacol.* **29,** 337–342.
Costall, B., Naylor, R. J. and Neumeyer, J. L. (1975). *Eur. J. Pharmacol.* **31,** 1–16.
Costall, B., Naylor, R. J. and Olley, J. E. (1972). *Eur. J. Pharmacol.* **18,** 95–106.
Costall, B., Naylor, R. J. and Pinder, R. M. (1974). *J. Pharm. Pharmacol.* **26,** 753.
Creese, I. and Iversen, S. D. (1972). *Nature (London)* **238,** 247–248.
Creese, I. and Iversen, S. D. (1973). *Brain Res.* **55,** 369–382.
Creese, I. and Iversen, S. D. (1974). *Psychopharmacologia (Berlin)* **39,** 345–357.
Creese, I. and Iversen, S. D. (1975). *Brain Res.* **83,** 419–436.
Crow, T. J. (1971). *Exp. Neurol.* **32,** 247–255.
Dahlström, A. and Fuxe, K. (1964). *Acta Physiol. Scand.* **62,** *Suppl. 232,* 1–55.
Divac, I. (1972). *Psychopharmacologia (Berlin)* **27,** 171–178.
Ernst, A. M. (1967). *Psychopharmacologia (Berlin)* **10,** 316–323.
Ernst, A. M. and Smelik, P. G. (1966). *Experientia* **66,** 837–838.
Evetts, K. D., Uretsky, N. J., Iversen, L. L. and Iversen, S. D. (1970). *Nature (London)* **225,** 961–962.
Fibiger, H. C., Fibiger, H. P. and Zis, A. P. (1973). *Br. J. Pharmacol.* **47,** 683–692.
Fibiger, H. C. Phillips, A. G. and Clouston, R. A. (1973). *Am. J. Physiol.* **225,** 1282–1287. 1282–1287.

Fog, R. L., Randrup, A. and Pakkenberg, H. (1967) *Psychopharmacologia* **11,** 179–183.
Fuxe, K. and Ungerstedt, U. (1968). *Eur. J. Pharmacol.* **4,** 135–144.
Garcia-Munoz, M., Nicolaou, N. M. and Tulloch, I. F. (1977). **265,** 363–365.
Glowinski, J. and Axelrod, J. (1965). *J. Pharmacol. Expl. Ther.* **149,** 43–49.
Hökfelt, T., Johansson, O., Fuxe, K., Goldstein, M. and Park, D. (1976). *Med. Biol.* **54,** 427–453.
Hökfelt, T., Johansson, O., Fuxe, K., Goldstein, M. and Park, D. (1977). *Med. Biol.* **55,** 21–40.
Hökfelt, T. and Ungerstedt, U. (1973). *Brain Res.* **60,** 269–297.
Hollister, A. S., Breese, G. R. and Cooper, B. R. (1974). *Psychopharmacologia* 1–16.
Jackson, D. M., Andén, N.-E. and Dahlström, A. (1975). *Psychopharmacologia (Berlin)* **45,** 139–149.
Kelly, P. H. (1975). *Brain Res.* **100,** 163–169.
Kelly, P. H. and Moore, K. E. (1976). *Nature (London)* **263,** 695–696.
Kelly, P. H., Seviour, P. W. and Iversen, S. D. (1975). *Brain Res.* **94,** 507–522.
Krauthamer, G. M. (1975). *In* "Catecholamines and Behaviour", Vol. 1 (Ed. A. J. Friedhoff), 59–87. Plenum Press, New York.
Lindvall, O. and Björklund, A. (1974). *Acta Physiol. Scand. Suppl.* **412,** 1–48.
Ljungberg, T. (1978a). *Pharmacol. Biochem. Behav.* **8,** 191–195.
Ljungberg, T. (1979). *Neuropharmacology*, in press.
Ljungberg, T. and Ungerstedt, U. (1976a). *Eur. J. Pharmacol.* **36,** 181–188.
Ljungberg, T. and Ungerstedt, U. (1976b). *Exp. Neurol.* **53,** 585–600.
Ljungberg, T. and Ungersedt, U. (1976c). *Physiol. Behav.* **16,** 277–283.
Ljungberg, T. and Ungerstedt, U. (1977a). *Eur. J. Pharmacol.* **46,** 41–50.
Ljungberg, T. and Ungerstedt, U. (1977b). *Pharmacol. Biochem. Behav.* **8,** 483–489.
Ljungberg, T. and Ungerstedt, U. (1978c). *Psychopharmacology* **56,** 239–247.
Lyon, M. and Robbins, T. (1975). *Current Devel. Psychopharmacol.* **2,** 81–163.
McKenzie, G. M. (1972). *Psychopharmacologia (Berlin)* **23,** 212–219.
Marshall, J. F., Richardson, J. S. and Teitelbaum, P. (1974). *J. Comp. Physiol. Psychol.* **87,** 808–830.
Marshall, J. F. and Ungerstedt, U. (1976). *Physiol. Behav.* **17,** 817–822.
Marshall, J. F. and Ungerstedt, U. (1977). *Science* **198,** 62–64.
Nahorski, S. R. (1975). *Psychopharmacologia (Berlin)* **42,** 159–162.
Neill, D. B., Boggan, W. O. and Grossman, S. P. (1974). *J. Comp. Physiol. Psychol.* **86,** 1019–1030.
Oltmans, G. A. and Harvey, J. A. (1972). *Physiol. Behav.* **8,** 69–78.
Pijnenburg, A. J. J., Honig, W. M. M., van der Heyden, J. A. M. and van Rossum, J. M. (1976). *Eur. J. Pharmacol.* **35,** 45–58.
Pijnenburg, A. J. J. and van Rossum, J. M. (1973). *J. Pharm. Pharmacol.* **25,** 1003–1004.
Poirier, L., Langelier, P., Roberge, A., Boucher, R. and Kitskis, A. (1972). *J. Neurol. Sci.* **16,** 401–416.
Price, M. T. and Fibiger, H. C. (1974). *Eur. J. Pharmacol.* **29,** 249–252.
Pycock, Ch. J. and Marsden, C. D. (1978). *Eur. J. Pharmacol.* **47,** 167–175.
Randrup, A. and Munkvad, I. (1966). *Nature (London)* **211,** 540.
Randrup, A. and Munkvad, I. (1970). *In* "Amphetamine and Related Compounds". Proc. Mario Negri Inst. for Pharmacological Research, Milan, Italy. Raven Press, New York.
Roberts, D. C. S., Zis, A. P. and Fibiger, H. C. (1975). *Brain Res.* **93,** 441–454.

Schoenfeld, R. and Uretsky, N. (1972). *Eur. J. Pharmacol.* **19,** 115–118.
Scheel-Krüger, J. and Randrup, A. (1967). *Life Sci.* **6,** 1389–1398.
Simpson, B. A. and Iversen, S. D. (1971). *Nature (London)* **230,** 30–32.
Sotelo, C., Javoy, F., Agid, Y. and Glowinski, I. (1973). *Brain Res.* **58,** 264–290.
Stricker, E. M. and Zigmond, M. J. (1976). *Progress in Psychobiology and Physiological Psychology* **6,** 121–188.
Stricker, E. M. and Zigmond, M. J. (1974). *J. Comp. Physiol. Psychol.* **86,** 973–994.
Strömberg, U. and Svensson, T. H. (1971). L-DOPA *Psychopharmacology (Berlin)* **19,** 53–60.
Svensson, T. H. (1970). *Eur. J. Pharmacol.* **12,** 161–166.
Svensson, T. H. and Waldeck, B. (1969). *Eur. J. Pharmacol.* **7,** 278–282.
Svensson, T. H. and Waldeck, B. (1970). *Psychopharmacologia (Berlin)* **18,** 357–365.
Thompson, R. F. and Bettinger, L. A. (1970). *In* "Attention: Contemporary Theory and Analysis" (Ed. D. Mostofsky), 367–401. Appleton Century, Crofts, New York.
Ungerstedt, U. (1968). *Eur. J. Pharmacol.* **5,** 107–110.
Ungerstedt, U. (1971a). *In* "6-Hydroxydopamine and Catecholamine Neurons" (Eds. T. Malmfors and H. Thoenen), 101–127, North-Holland Publishing Company, Amsterdam, London.
Ungerstedt, U. (1971). *Acta Physiol. Scand. Suppl. 367* **82,** 1–48.
Ungerstedt, U. (1971c). *Acta Physiol. Scand. Suppl. 367* **82,** 89–93.
Ungerstedt, U. (1971d). *Acta Physiol. Scand. Suppl. 367* **82,** 49–68.
Ungerstedt, U. (1971e). *Acta Physiol. Scand. Suppl. 367* **82,** 95–122.
Ungerstedt, U. (1973). *Neurosci. Res.* **5,** 73–96.
Ungerstedt, U. (1974). *In* "The Neurosciences" (Eds. F. O. Schmitt and F. G. Worden), 695–711, The MIT Press, Cambridge, Massachusetts and London, England.
Ungerstedt, U. and Arbuthnott, G. (1970). *Brain Res.* **24,** 485–493.
Ungerstedt, U., Avemo, A., Avemo, E., Ljungberg, T. and Ranje, Ch. (1973). *In* "Advances in Neurology" Vol. 3 (Ed. D. B. Calne), 257–271. Raven Press, New York.
Ungerstedt, U., Butcher, L. L., Butcher, S. G., Andén, N.-E. and Fuxe, K. (1969). *Brain Res.* **14,** 461–471.
Ungerstedt, U. and Ljunberg, T. (1977). *In* "Advances in Biochemical Psychopharmacology", Vol. 16 (Eds. E. Costa and G. L. Gessa), 193–199. Raven Press, New York.
Ungerstedt, U., Ljungberg, T. and Schultz, W. (1978). *In* "Advances in Biochemical Psychopharmacology", Vol. 19 (Eds. P. J. Roberts *et al.*), 311–321. Raven Press, New York.
Von Voigtländer, P. F. and Moore, K. E. (1973). *Neuropharmacology* **12,** 451–462.
Wolfarth, S., Grabowska, M., Lacki, M., Dulska, E. and Antkiewicz, L. (1973). *Activitas Nervosa Superior (Praha)* **15,** 132.
Wolfarth, S. (1974). *Pharmacol. Biochem. Behav.* **2,** 181–186.
Yehuda, S. and Wurtman, R. J. (1975). *Eur. J. Pharmacol.* **30,** 154–158.
Zigmond, M. J. and Stricker, E. M. (1972). *Science* **177,** 1211–1214.
Zimmerberg, B. and Glick, S. D. (1975). *Brain Res.* **86,** 335–338.
Zimmerberg, B., Glick, S. D. and Jerussi, T. P. (1974). *Science* **185,** 623–625.

CHAPTER 33

Dopamine and the Neural Mechanisms of Reinforcement

H. C. Fibiger and A. G. Phillips

University of British Columbia, Vancouver, Canada.

The hypothesis that central catecholaminergic (CA) systems mediate reinforcement processes has received considerable experimental attention during the past decade. The paradigm which has been used most often to examine the validity of this hypothesis is the intracranial self-stimulation (ICS) technique first described by Olds and Milner (1954), whereby an animal will perform an operant response in order to obtain electrical stimulation of the brain via a chronically implanted electrode. An underlying assumption of this hypothesis is that the neuronal systems which mediate the reinforcing properties of ICS are those which are activated normally by positively reinforcing environmental stimuli. As originally proposed by Stein (1964, 1968) the CA hypothesis posited that central noradrenergic (NA) neurones mediate the reinforcing properties of ICS. Although early anatomical and pharmacological studies provided a degree of support for this suggestion, more recent investigations have questioned seriously its validity (Clavier *et al.*, 1976; Phillips *et al.*, 1977); furthermore, in a recent review it was concluded that the hypothesis, at least as originally proposed, is no longer tenable (Fibiger, 1978).

As a result of a careful mapping study of ICS sites in the brain, Crow (1972) elaborated upon the CA hypothesis and proposed that dopaminergic (DA) as well as NA systems may be involved in the reinforcing properties of

ICS. Specifically, Crow (1972) reported that high rates of self-stimulation could be obtained from the cell groups in the ventral tegmentum which give rise to the two major ascending DA systems in the brain, the nigrostriatal and the mesolimbic–mesocortical projections. The data upon which the "DA hypothesis" of brain stimulation reward is based, and some of the attempts to tests its predictions, are discussed below.

PHARMACOLOGICAL EVIDENCE

Neuroleptics

There is a considerable pharmacological literature concerning the effects on self-stimulation of compounds which selectively block DA receptors. In an elegant series of experiments Wauquier and Niemegeers (1972) showed that haloperidol, pimozide and pipamperone decreased ICS obtained from lateral hypothalamic electrodes in a dose-related manner. Similar observations have since been made in other laboratories (Rolls *et al.*, 1974; Phillips *et al.*, 1975; Mora *et al.*, 1975), and there is therefore widespread agreement that pharmacological interference with central DA mechanisms can impair self-stimulation behaviour. However, the immediate question that arises from these studies is, to what extent does the reduced rate of responding for ICS reflect a reduction in the reinforcing value of the brain stimulation as opposed to an impaired ability to perform the operant response? This question is relevant particularly in the case of DA systems because these systems are known to serve important motor functions. Wauquier and Niemegeers (1972) favoured the motor deficit interpretation because they pointed out that various learned responses are disrupted by about the same dose levels of neuroleptics. They concluded that these drugs may interfere "with a general system involved in the output of fixed learned behaviour, triggered by various positive and negative reinforcements". The reinforcement versus motor deficit question has since been investigated by several other laboratories. Rolls *et al.* (1974) examined the effects of spiroperidol on operant responding for ICS, food or water. This drug was found to produce a dose-dependent decrease in responding from electrodes in the VTA, hippocampus, anterior hypothalamus, septal area and nucleus accumbens. Interestingly, spiroperidol was just as effective in decreasing the self-stimulation rate from electrodes in an area that did not contain DA neurones (hippocampus) as it was in the other areas where DA neurones were probably directly stimulated. Similarly, Phillips *et al.* (1975) have shown that pimozide and haloperidol decrease ICS from the dorsal tegmentum, a non-dopaminergic site, and from the dopaminergically innervated nucleus accumbens, equally well. Rolls *et al.* (1974) then showed that at the doses that impaired ICS, spiroperidol also produced severe disruptions of bar-pressing for food and water. These same doses had only minor effects on eating and drinking in the home cage after 24 hours of food and water deprivation indicating that the drug did not decrease hunger or thirst. The authors concluded that an impairment of motor function could account for the effects of dopamine-receptor blockade on self-stimulation and that the impairment appeared to increase as a function of response complexity.

In a subsequent experiment from the same laboratory, Mora *et al.* (1975) trained rats either to bar-press or lick a tube for ICS in the lateral hypothalamus. Spiroperidol produced an identical dose-related decrease in either bar-pressing or licking for ICS. At a dose of $0.062\,mg\,kg^{-1}$ the drug reduced both kinds of responding by about 90%. Previously, however, Rolls *et al.* (1974) had shown that a higher dose of the drug ($0.1\,mg\,kg^{-1}$) had negligible effects on licking for water in water-deprived animals. The authors argued therefore that the decrease in licking for brain stimulation was not due to a motor impairment and that DA receptors may be involved in brain stimulation reward. However, neither Mora *et al.* (1975) nor Rolls *et al.* (1974) reported the baseline rates of licking for water or ICS, and as will be discussed below, if the baselines were not the same then this could account for the differential effects of spiroperidol on responding for these two reinforcers. Even if the baseline rates were similar, other interpretative problems exist, and one of these relates to the level of motivation of the animal for two different reinforcers. Thus, although an animal may respond for ICS and water at similar rates, it is conceivable that the organism may be motivated more for one reinforcer than the other. If a neuroleptic did produce a subtle motor deficit, then it is possible that the impairment would be detectable when responding for ICS but might be masked in water-reinforced responding if the organism was relatively more highly motivated for water than for ICS. Perhaps preference tests for ICS and water would serve to resolve this problem.

Fibiger *et al.* (1976) reported that pimozide and haloperidol decreased operant responding both for ICS and for food, but in these experiments it was observed that the relative inhibitory effects of the neuroleptics were significantly greater on ICS than on food-reinforced behaviour. At first glance, this might be taken to indicate a somewhat selective effect of these drugs on brain-stimulation reward. However, as was noted by the authors, the rate of responding for ICS was about six times higher than it was for food. Since baseline response rates are known to influence markedly the effects of drugs (Wauquier and Niemegeers, 1972; Kelleher and Morse, 1968) it is possible that this variable rather than a selective effect on brain stimulation reward was responsible for these observations. To control for this possibility, Fibiger *et al.* (1976) trained animals to respond for brain stimulation or food on a variable interval 60 s schedule. Rates of responding for the two reinforcers did not differ significantly on this schedule and under these circumstances, haloperidol affected bar-pressing for ICS and for food to a similar extent. The doses of pimozide and haloperidol that reduced responding for food did not affect 22 hour deprivation-induced feeding in the home cage, suggesting that the impaired responding was not due to decreased hunger or to decreased reinforcing properties of the food. The authors suggested that neuroleptics may decrease responding for ICS primarily by impairing the function of certain DA systems which are involved critically in the initiation or maintenance of operant behaviour rather than by interfering with central reward processes.

While the above experiments are consistent with the view that central DA systems serve important functions in motor behaviour, *they do not exclude the possibility that these projections also play a role in brain stimulation reward*. However, as was pointed out by Fibiger *et al.* (1976) demonstration of such a role will require techniques which can clearly differentiate between the effects of drugs on motor function and central reinforcement processes. Liebman and Butcher (1973, 1974) attempted to resolve this problem using a number of different techniques. In the first experiments they showed that pimozide decreased bar-pressing for ICS obtained from electrodes in the lateral hypothalamus or mesencephalic central grey (Liebman and Butcher, 1973). These effects were observed at below maximum rates of responding produced by low levels of electrical stimulation. They then showed that the decrease in the rate of responding produced by the lower dose of pimozide could be reversed by increasing the electrical current, and therefore concluded that animals were capable motorically of responding for ICS and that the effect of pimozide was due to a decrease in reward value rather than a motor deficit. However, increasing the current also increased the baseline (non-drug) rate of responding. Inasmuch as it is known that the effects of neuroleptics on ICS can vary as a function of the baseline response rate (Fibiger *et al.*, 1976; Wauquier and Niemegeers, 1972), Liebman and Butcher's observations cannot be regarded as unequivocal evidence for a motivational deficit. In the second study of these authors (Liebman and Butcher, 1974), it was found that using a rate-free measure of ICS, pimozide decreased the amount of time spent on the "ON" side of the apparatus significantly more in animals with electrodes in the substantia nigra than in rats with lateral hypothalamic electrodes. On the basis of these and other data obtained with apomorphine and amphetamine, it was suggested that ICS in the substantia nigra had DA substrates, whereas in the lateral hypothalamus both NA and DA substrates were involved. It should be noted, however, that pimozide also significantly decreased the number of crossings and locomotor activity in the ICS apparatus, and as has been discussed previously, deficits in response initiation or maintenance also could produce changes in performance in the rate-free paradigm (Fibiger, 1978).

In a novel and ingenious approach to the problem of discriminating between motor and reinforcement variables in interpreting the effects of neuroleptics on ICS, Fouriezos and Wise (1976) reasoned that if these drugs blocked the reinforcing value of the brain stimulation then they should produce an extinction curve, similar to that seen when the electrical current is turned off in a well-trained animal. Under this condition animals initially respond at normal rates, and only after several responses fail to be reinforced does the response rate decline. Thus, it was argued that a drug which attenuates the reward value of stimulation would cause a reduced rate of responding only after an initial period of normal responding, whereas a drug that produced motor deficits would result in a reduced but steady rate of responding throughout the session. The authors found that in five out of the seven rats, pimozide treatment resulted in an initially normal rate of responding which was followed by a rapid decline in response rate. These data led the authors to conclude that pimozide decreases ICS by blocking the reinforcing value of the stimulation rather than by producing motor impairment. An alternate explanation might be that pimozide increased motor fatigue in these animals and this caused the animals to cease responding after a short time. In this regard, it is of considerable interest that Parkinsonian patients, in whom the DA nigrostriatal projection has degenerated, often report increased motor fatiguability (Selby, 1968). Fouriezos and Wise attempted to control for this possibility by blocking access to the ICS lever for 10 minutes in one rat that had ceased responding after pimozide. On removing the barrier, this animal briefly resumed responding at a high initial rate. The authors interpreted this to indicate that the animals were capable of responding in the latter part of the session and hence motor deficits could not explain the lack of responding during that period. However, it might be possible that the auditory and visual stimuli associated with the removal of the barrier may have aroused the animal sufficiently to resume responding for a short period of time until motor fatigue again disrupted responding. Although these results are of considerable interest, further work appears necessary before the effects of neuroleptics on the pattern of responding for ICS can be ascribed confidently to a block in the reward value of the brain stimulation.

Further evidence of the complex effects of neuroleptics on ICS is provided by recent data from our laboratory (Phillips and Fibiger, unpublished observations). As in a previous study (Fibiger *et al.*, 1976), animals were trained to respond for electrical stimulation of the LH on a V1-60 schedule, prior to injections of haloperidol (0.1 mg kg^{-1}). The important comparisons in this experiment were between reinforcement and drug; no reinforcement—no drug (i.e. extinction); and no reinforcement—drug. Data from four animals showed very similar patterns of responding during both reinforcement–drug, and extinction conditions, thus apparently confirming Fouriezos and Wise (1976), who argued that neuroleptics block reinforcement thereby generating an extinction curve similar to that seen when the brain stimulation is turned off. If these two manipulations are simply terminating reinforcement, then treatment with neuroleptics during extinction should have no observable effect on the animal's behaviour. However, such is not the case (see Fig. 1). Under the extinction–drug condition, the cumulative record is very different from that seen during extinction–no drug. The subjects respond at near normal rates for the first few minutes of the trial, then when no reinforcement is forthcoming, they virtually cease responding for the remainder of the one-hour trial, only sampling the bar occasionally. These data indicate that neuroleptic-induced reductions in

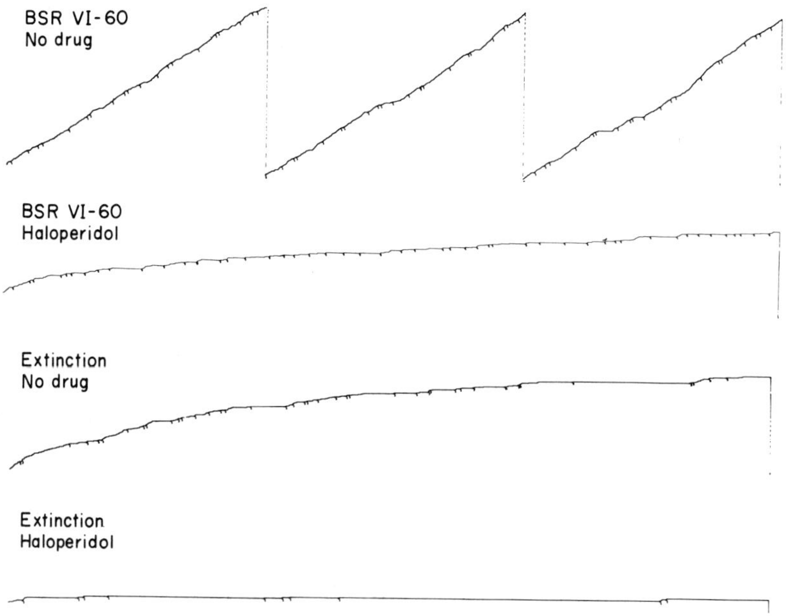

Fig. 1 Cumulative records of an animal bar-pressing for brain-stimulation reward (BSR) on a V1-60 schedule. Compare effect of haloperidol (0.1 mg kg^{-1}) during reinforcement with (a) extinction—no drug and (b) extinction—haloperidol. Session length was 1 h.

rates of responding for ICS cannot be explained solely in terms of a blockade of the reinforcing properties of brain stimulation.

The experiments reviewed so far have all employed acute administration of neuroleptics but there is also evidence that chronic administration may yield further information on the role of DA in brain-stimulation reward. Prolonged blockade of DA receptors by chronic treatment with haloperidol or pimozide has been shown to induce receptor supersensitivity (Gianutsos *et al.*, 1974; Gianutsos and Moore, 1977). Assuming that DA is implicated in ICS, Ettenberg and Wise (1976) reasoned that the development of DA receptor supersensitivity might be accompanied by an enhancement of brain-stimulation reward. Indeed, increased ICS rates were observed in both locus coeruleus and SNC after termination of chronic pimozide treatment. In a subsequent experiment, this treatment was shown to reduce significantly ICS thresholds at sites in the LH (Ettenberg, 1977). Such drug-induced changes in ICS threshold provide evidence for dopaminergic involvement in brain-stimulation reward, but the exact nature of this role remains unanswered.

In a long-term ICS study, Eichler *et al.* (1976) have demonstrated site-specific tolerance to chronic DA receptor blockade. Spiroperidol initially abolished ICS at sites in (1) SNC, (2) far lateral hypothalamus (FLH), and (3) near lateral hypothalamus (NLH) but tolerance was displayed at the first two sites after approximately one week of injections. No tolerance was observed with NLH electrodes. Withdrawal of the drug was accompanied by a significant increase in ICS at all three sites. Again site-specific effects were observed in ICS rates as sites in SNC were elevated until testing was terminated 30 days later, whereas ICS rates in NLH returned to and stabilized at baseline levels. These authors argue that chronic treatment with neuroleptics permits comparison and differentiation between ICS sites and suggest that DA pathways serve different roles in the two regions of the LH.

Further pharmacological evidence which shows that selective blockade of central DA receptors can influence the reinforcing properties of ICS is the recent demonstration by Zarevics and Setler (1978) that pimozide produces a dose-dependent increase in the reward threshold for ICS. In this experiment, these authors utilized a two-lever paradigm first developed by Stein and Ray (1960) in which each response on the first lever resulted in a step-wise decrease in the brain stimulation current, while a response at the second lever reset the current to the maximum value. In this situation, the reset responses presumably occur at current intensities which are no longer reinforcing to the animal. Stein and Ray (1960) originally showed that chlorpromazine increased the reset values and therefore suggested that this drug attenuated the reinforcing effects of ICS. The precise interprctation of these results was hampered because of the relatively non-specific effects of chlorpromazine on central DA receptors. However, Zarevics and Setler (1978) obtained similar dose-related increases in the ICS reset values after injections of pimozide, a drug which at low doses is thought to be very specific for DA receptors. Therefore, to the extent that pimozide produced these effects via DA receptor blockade, these results clearly suggest that DA mechanisms participate in the reinforcing properties of ICS obtained from the posterior hypothalamus.

A new approach in examining the effects of neuroleptics on ICS has been to administer these compounds directly into discrete regions of brain via intracerebral cannulae. In one such study, Broekkamp and Van Rossum (1975) implanted rats with ICS electrodes in the VTA. Injections of haloperidol into the neostriatum ipsilateral or contralateral to the ICS electrode reduced ICS by 30% and 39% respectively. Although the authors ascribed the effects of the contralateral injections to a "general motivational depression", these findings would appear to be more compatible with a motor deficit hypothesis. That is, if the neuroleptic specifically blocked the reinforcing properties of brain stimulation then injections ipsilateral to the electrode would have been expected to have greater effects on ICS than contralateral injections. If, however, these treatments simply impaired operant responding, then contralateral and ipsilateral injections would be expected to have similar effects. In another study, Mora *et al.* (1975) found that intracerebral injections of spiroperidol had greater effects on lateral hypothalamic ICS after injection into the ipsilateral nucleus accumbens when compared with globus pallidus injections. The pallidal injections were found to produce greater motor deficits, as assessed by a battery of simple neurological tests, and therefore the authors suggested that DA receptors in the nucleus accumbens are involved in ICS in a way which is not due just to their role in motor behaviour. It remains possible of course that injections into these different regions produced qualitatively different motor impairments than those measured in the neurological test battery, and the authors conceded that a different and subtle type of motor deficit might account for the effects of spiroperidol on ICS. Unfortunately, the ipsilateral–controlateral paradigm was not utilized in this study. In a subsequent paper, Mora *et al.* (1976b) examined the effects of intracerebral injections of spiroperidol on ICS obtained from a number of brain regions in monkeys. The effects of spiroperidol varied with injection and ICS sites. Generally, amygdala ICS seemed most sensitive and the locus coeruleus ICS least sensitive to the effects of spiroperidol regardless of the injection site. Responding for glucose was not affected by any of the intracerebral spiroperidol injections. Again the authors concluded that decreased self-stimulation found at some sites could not be explained by a single effect such as motor impairment or sedation. While this conclusion may be justified, it is difficult to comment further on this work because all of the drug effects were reported as a percentage of the control rate of responding, and the baseline rates were not stated. If, however, the rates of responding varied at the different ICS sites, then this would complicate the interpretation of these findings.

Stimulants

Recent lesion studies have demonstrated that the well-known stimulant effects of *d*-amphetamine on ICS are mediated by central DA neurones (Cooper *et al.*, 1974; Clavier and Fibiger, 1977). If future research confirms these observations, then a considerable body of evidence on the effects of amphetamines on ICS can be viewed as supporting a role for DA in self-stimulation. Whether the amphetamine-induced facilitation in responding for ICS is due to a potentiation of the reinforcing properties of the brain stimulation, or whether it is due to a more general activation of motor output cannot be answered by most studies. However, some early experiments on the effects of amphetamine on ICS suggest that this drug does in fact have direct effects on the brain-stimulation reward. Stein and Ray (1960) tested rats in a situation in which each response produced a step-wise decrease in ICS current. By pressing a second lever the animals could reset the current to the highest level as described above. The authors found that *d,l*-amphetamine decreased the current at which animals would press the reset lever. These data suggest that the drug decreased the brain-stimulation reward thresholds. In a subsequent study Stein (1964) obtained very low rates of responding for ICS by using current intensities well below those producing maximal rates of responding. In this situation *d*-methamphetamine greatly facilitated the rate of responding. If, however, the ICS current was turned off at the time of the methamphetamine injection no increase in rate occurred. These results clearly show that the facilitation in responding produced by methamphetamine cannot be explained in terms of a non-specific increase in operant responding. Rather, the data are consistent with the hypothesis that amphetamine can decrease ICS reward thresholds. Using a rate-free paradigm, Poschel and Ninteman (1966) found that the amount of time spent in the part of the apparatus in which brain stimulation was delivered increased as a function of ICS current intensity. They then demonstrated that methamphetamine significantly increased the amount of time spent in that part of the apparatus during low current stimulation. Liebman and Butcher (1974) have subsequently reported similar effects with *d*-amphetamine. It is difficult to imagine how an increase in motor activity could account for these observations and the authors' conclusions that methamphetamine and amphetamine increase the rewarding value of brain stimulation appear well justified.

The differential effects of *d*- and *l*-amphetamine on ICS have been taken to implicate DA and NA neurones in self-stimulation. Phillips and Fibiger (1973) observed that *d*- and *l*-amphetamine were approximately equipotent at increasing ICS response rates when electrodes were aimed at a "dopaminergic" site such as the SNC, whereas the *d*-isomer was much more effective than *l*-amphetamine at increasing ICS from a "noradrenergic" site in the lateral hypothalamus. These observations have since been confirmed and extended by other laboratories (Goodall and Carey, 1975; Stephens and Herberg, 1975). The behavioural results appeared to correlate remarkedly well with initial neurochemical findings which indicated that the two isomers were approximately equally effective in inhibiting DA uptake by DA neurones, while *d*- was substantially more potent than *l*-amphetamine in inhibiting NA uptake into NA neurones (Coyle and Snyder, 1968). However, three other laboratories failed to confirm these neurochemical observations (Phillips *et al.*, 1975), and in fact found exactly opposite effects, i.e. equipotentiality of the isomers on NA uptake, and more potent effects of *d*- than of *l*-amphetamine on DA uptake. These latter findings, together with the fact that there is presently no unequivocal evidence for any involvement of NA systems in ICS (Clavier *et al.*, 1976; Phillips *et al.*, 1977) leaves the significance of the differential effects of *d*- and *l*-amphetamine unknown (see Horn, Chapter 16).

Several studies have examined the effect on ICS of the DA-receptor agonist apomorphine with somewhat ambiguous results. Wauquier and Niemegeers (1973) usually observed an inhibition of ICS in the LH, although there was also some evidence to suggest a facilitatory effect at low response rates. Similarly, Broekkamp and van Rossum (1974) reported both facilitation and attenuation which appeared unrelated to the site of brain stimulation. In contrast, several studies have shown a consistent suppressant effect of apomorphine at sites near the A9-A10 DA cell bodies of the ventral tegmentum (Liebman and Butcher, 1973; St. Laurent, *et al.*, 1973; Herberg *et al.*, 1976), and the DA terminal areas of the medial and sulcal prefrontal cortex (Mora *et al.*, 1976a). In the latter experiment, apomorphine produced a facilitatory effect on ICS in the caudate nucleus at one or more doses in the majority of animals, although suppressant effects were also observed.

Self-stimulation can be elicited from the orbitofrontal cortex in primates. This area has a high concentration of DA (MacBrown and Goldman, 1977) and it is therefore important to note that in the primate, ICS at this site is significantly attentuated in a dose-related manner by apomorphine (Phillips *et al.*, 1976c). As in the earlier study with rats (Mora *et al.*, 1976b), apomorphine tended to facilitate ICS in the caudate, thereby discounting motor impairment as an explanation of the suppressant effects of apomorphine on ICS in the orbitofrontal cortex. The attenuation of ICS by apomorphine has been taken as evidence that the release of DA, contingent upon response-related electrical stimulation, mediates ICS in this (Phillips *et al.*, 1976c) and other regions of the brain (Herberg *et al.*, 1976). Such data clearly implicate DA in ICS of the orbitofrontal cortex, but it is also important to note that apomorphine produces a suppression and not an abolition of ICS. Therefore it may be more appropriate to think of DA as a modulator of ICS at this site rather than an exclusive mediator of this phenomenon.

LESION STUDIES

Inasmuch as the pharmacological literature has provided a degree of support for a role of DA systems in brain stimulation reward, the question arises as to whether stimulation of these systems exclusively mediates reward when electrodes are implanted in their vicinity. Breese *et al.* (1971) first showed that intraventricular injections of 6-OHDA, which produced extensive lesions of central DA neurones, resulted in long term reductions in ICS. These observations were confirmed subsequently and extended by others (Cooper *et al.*, 1974; Phillips and Fibiger, 1976). Although these studies implicate DA systems in some aspects of ICS behaviour, they do not permit a distinction to be made between lesion-induced motor impairments and possible changes in the reinforcing value of the brain stimulation. Phillips *et al.* (1976b) attempted to address this question by implanting ICS electrodes in the striatum of rats. After ICS rates had stabilized, 6-OHDA lesions were placed in the SNC either ipsilateral or contralateral to the striatal electrodes. The contralateral lesions were utilized to control for possible motor impairments. The lesions reduced striatal DA by approximately 95%. Intracranial self-stimulation remained at less than 10% of prelesion rates in the ipsilateral group, whereas responding in the contralateral group gradually recovered to reach 72% of control values 28 days postoperatively. The differential effects of ipsilateral and contralateral SNC lesions where interpreted to indicate that in addition to producing motor deficits in operant responding, reduction of striatal DA was accompanied by a loss of brain stimulation reward obtained from this structure.

In a recent experiment, Koob *et al.* (1977) have employed the ipsilateral–contralateral 6-OHDA lesion paradigm to study the role of DA projections in ICS at sites in the LH and locus coeruleus. As in the earlier study by Phillips *et al.* (1976b) the destruction of DA projections at the level of the ventral tegmentum produced a persistent and pronounced deficit in

ICS from ipsilateral electrodes but only transient effects on ICS from contralateral electrodes. The LH electrodes were placed in the trajectory of the ascending DA pathways and the fact that ICS at ipsilateral placements remained suppressed at less than 30%, suggests an important role for DA at this site. Similar effects have been reported by Phillips *et al.* (1976a), with electrodes in the FLH, but ICS at sites in the NLH was only temporarily disrupted. Such a dissociation lends support to the hypothesis that there are multiple substrates for ICS. The finding that ICS from the region of the LC is disrupted by lesions to the ipsilateral DA systems is consistent with the hypothesis that NA-containing neurones do not mediate ICS in the dorsal pons (Clavier *et al.*, 1976). Furthermore, this finding suggests that intact DA systems can be important for brain-stimulation reward sites which do not directly stimulate DA neurones.

Ornstein and Huston (1975) have reported results which appear to be discrepant with those of Koob *et al.* (1977). These authors found that unilateral injections of 6-OHDA into the substantia nigra disrupted ICS equally in the ipsilateral and contralateral lateral hypothalamus. Furthermore, bilateral 6-OHDA injections abolished ICS when lever-pressing was the criterion but not when more simple motor responses were used as operants. The authors concluded that the NSP is not critically involved in the reward process, and that attentuation of ICS by the 6-OHDA lesions was most likely due to disruption of operant performance. However, although no histology was provided by Ornstein and Huston (1975), the coordinates utilized by these workers would have resulted in placements which were medial to those utilized by Koob *et al.* (1977). Inasmuch as previous experiments have demonstrated that 6-OHDA and neuroleptics have effects on hypothalamic ICS which are site-specific (Phillips *et al.*, 1976a; Eichler *et al.*, 1976), this may account for these discrepant results.

In another lesion study, Phillips and Fibiger have studied the role of dopamine in ICS obtained from the A10 dopamine neurones in the ventral tegmentum. Animals were prepared with two ipsilateral electrodes. All had the posterior electrode implanted in A10, while the anterior electrodes were implanted in either the medial prefrontal cortex, or nucleus accumbens, both of which are thought to receive projections from the mesocorticolimbic DA system. 6-OHDA lesions were placed in the ipsilateral FLH as described previously (Clavier and Fibiger, 1977). As may be seen in Figs. 2 and 3, in all animals with extensive depletions of DA in both striatal and mesocorticolimbic regions there was a pronounced disruption of ICS at the A10 placement. A different picture emerged however when the effect of such lesions on the anterior electrode was examined. Those animals with electrodes in the medial prefrontal cortex recovered their ICS rate to 40–50% of prelesion baseline during the three weeks of postlesion testing.

Self-stimulation in the nucleus accumbens was only temporarily disrupted by the lesion, with rates returning to 60–100% of baseline within 5–10 days. These results demonstrate that ICS obtained from electrodes placed in the A10 region has a critical dependence upon ascending DA systems. However, the results indicate equally clearly that dopamine is neither an exclusive nor critical mediator of ICS in the nucleus accumbens and prefrontal cortex.

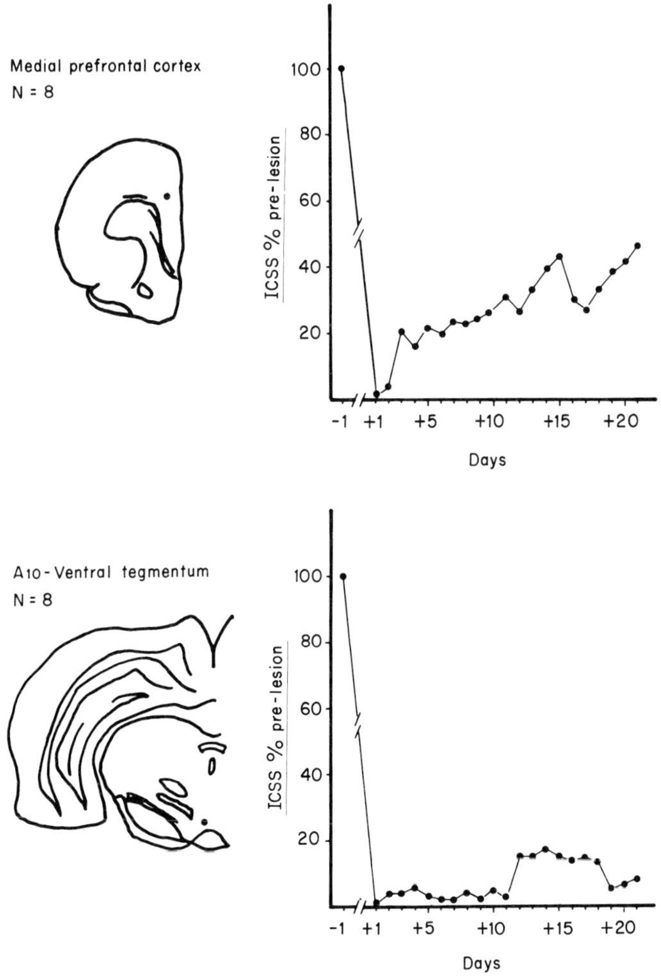

Fig. 2 Effect of unilateral lesion of ascending DA pathways on self-administration at sites in A10 DA region of ventral tegmentum and medial prefrontal cortex. Each subject had electrodes in both sites. Data represent group means ($n = 8$). Location of electrode sites is depicted schematically on the left side of the figure.

It is evident from the above experiments that the involvement of dopaminergic neurones in the ICS varies with electrode placement. Further evidence of this conclusion is found in a recent study by Clavier and Fibiger (1977). In these experiments it was found that ipsilateral 6-OHDA lesions of the NSP resulted in large but temporary reductions in ICS obtained from the SNC which returned to pre-operative rates 8–10 days after the lesions. In this

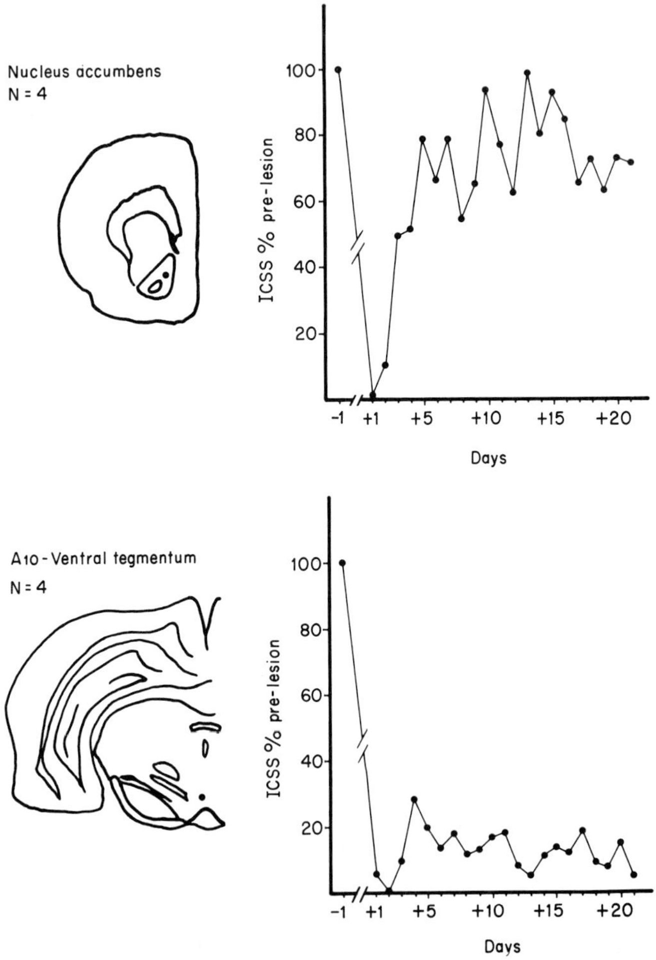

Fig. 3 Effect of unilateral lesion of ascending DA pathways on self-stimulation at sites in A10 DA region of ventral tegmentum and nucleus accumbens. Each subject had electrodes in both sites. Data represent group means ($n = 4$). Location of electrode sites is depicted schematically on the left side of the figure.

instance, however, NSP lesions contralateral to the electrodes had effects identical to the ipsilateral lesions. It was concluded therefore that the deficits in responding for ICS produced by unilateral NSP lesions could be accounted for in terms of transient motor deficits produced by these lesions, and that this dopaminergic projection was not the exclusive or essential neuronal substrate mediating the reinforcing properties of nigral ICS. The observation that nigral ICS could still be obtained despite near-complete destruction of the ascending DA projections implicates other presently unspecified neuronal systems in the vicinity of the SNC. However, Clavier and Fibiger (1977) also noted that it would be incorrect to conclude that the dopaminergic projections are irrelevant to nigral ICS. This conclusion was based on the observation that facilitation of nigral ICS by *d*-amphetamine was abolished by the ipsilateral but not the contralateral NSP lesions. It was therefore suggested that this facilitation of responding was due to the additive effect of electrical stimulation and *d*-amphetamine on the release of DA in the terminal regions of the ascending DA systems. In a similar vein, Cooper *et al.* (1974) have shown that selective depletion of brain DA by intraventricular 6-OHDA injections significantly reduces the rate-increasing effects of *d*-amphetamine on lateral hypothalamic ICS.

INTRAVENOUS SELF-ADMINISTRATION

On reviewing the above evidence, it is possible to make the argument that the problem with the CA hypothesis is not so much with the hypothesis itself but rather with the experimental technique which has most often been used to examine it, namely ICS. Stimulating electrodes will influence many of the neural elements in their vicinity and if, as appears to be the case, there exist multiple and perhaps redundant pathways which can mediate ICS, then it is not surprising that researchers have experienced considerable difficulty in defining the role of specific CA systems. An alternative to this approach has recently been developed which appears to provide strong support for a role of DA in at least some central reinforcement processes. It has been known for some time that animals will self-administer certain psychoactive drugs, but only recently have attempts been made to define the pharmacological substrates of this behaviour. While space does not permit a detailed review of this emerging field, a few observations require comment. Yokel and Wise (1976) demonstrated that low doses of specific DA receptor-blocking agents such as pimozide and (+)-butaclamol increased the rate of responding for intravenous *d*-ampnetamine in rats. The results were interpreted to indicate that low doses of the neuroleptics reduced the reinforcing properties of amphetamine by partial blockade of

DA receptors, so that higher doses of amphetamine were required to maintain the same reinforcing effects. Injections of phentolamine, phenoxybenzamine or propranolol had variable effects on responding and these effects did not appear to be due to blockade in the reinforcing properties of amphetamine. Rather, when effective, these drugs produced performance deficits. These results suggest that DA but not NA mechanisms are involved in the reinforcing effects of amphetamine. Roberts *et al.* (1977) subsequently reported similar findings for intravenous self-administration of cocaine. Furthermore, these authors reported that 6-OHDA lesions of the nucleus accumbens produced long-term decreases in cocaine self-administration while lesions of the ascending NA projections were without effect. These data again point to an involvement of central DA but not of NA neurones in the reinforcing properties of these stimulant drugs.

Further preliminary, but no less intriguing evidence can be found in human psychopharmacological studies. It is well known that drugs such as amphetamine and cocaine are capable of producing global euphoric sensations in man. On this basis it can be hypothesized that these agents may have effects on neuronal systems which may normally be involved in primary positive reinforcement mechanisms. It is of considerable interest therefore that α-MT pretreatment has been reported to block the euphoric effects of amphetamine in man (Jonsson *et al.*, 1971). While the α-MT study could not discriminate between a possible DA or NA mediation of this effect, in a subsequent study by Gunne *et al.* (1972) it was found that pimozide pretreatment reduced amphetamine euphoria whereas neither phenoxybenzamine nor propranolol had this effect. The human studies, like ICS and intravenous self-administration experiments in animals, therefore point to an involvement of central DA but not of NA systems in reward. Given some of the limitations of the ICS technique, it is to be hoped that future research using these other paradigms may further delineate the role of central DA projections in positive reinforcement.

CONCLUSION

Anatomical considerations preclude the possibility that direct stimulation of DA systems exclusively mediates ICS obtained from all regions of the brain. However, this does not rule out an involvement of these systems under some circumstances, and there exist pharmacological data which are consistent with such a role. The interpretation of many of these data is difficult because of the well-established involvement of DA projections in motor output. In attempting to evaluate drug-induced changes in brain-stimulation reward, it therefore becomes necessary to provide rigorous controls for possible modifications in motor performance. To date, there are a few experiments utilizing specific DA receptor blocking agents which appear to provide strong support for participation of DA systems in ICS reward. Certain amphetamine-induced changes in ICS are most likely due to direct effects on brain stimulation reward, and it is probable that these actions of amphetamine have DA substrates. Lesion studies have shown that DA may mediate the reinforcing properties of ICS in some structures (e.g. A10, FLH, striatum). In other regions such as the SNC, NLH, nucleus accumbens and prefrontal cortex DA projections appear to serve a non-exclusive, perhaps modulatory function in ICS, and other non-catecholaminergic systems are almost certainly involved.

All of the evidence relating DA pathways to reinforcement has been gathered from experiments utilizing the useful, but nevertheless artificial paradigms of ICS and intravenous self-administration. If we are to gain a more complete understanding of the function of DA systems, it is essential that their role in natural reinforcement be examined. Given that reinforcement is an important determinant of learning it will also be important to study the function of DA systems from this perspective. Some initial experiments have begun to examine the possible involvement of DA systems in learning and memory but at present their role, if any, is far from clear (Fibiger *et al.*, 1974; Price and Fibiger, 1975; Fibiger and Phillips, 1976; Ranje and Ungerstedt, 1977).

ACKNOWLEDGEMENTS

Original studies in the authors' laboratories were supported by the Medical Research Council, the National Research Council and the Non-Medical Use of Drugs Directorate.

REFERENCES

Breese, G. R., Howard, J. L. and Leahy, J. P. (1971). *Br. J. Pharmacol.* **43,** 255–257.
Broekkamp, C. L. E. and van Rossum, J. M. (1974). *Psychopharmacology* **34,** 71–80.
Broekkamp, C. L. E., van Rossum, J. M. (1975). *Arch. Int. Pharmacodyn.* **217,** 110–117.
Clavier, R. M. and Fibiger, H. C. (1977). *Brain Res.* **131,** 271–286.
Clavier, R. M., Fibiger, H. C. and Phillips, A. G. (1976). *Brain Res.* **113,** 71–81.
Cooper, B. R., Cott, J. M. and Breese, G. R. (1974). *Psychopharmacologia (Berlin)* **37,** 235–248.
Coyle, J. T. and Snyder, S. H. (1968). *J. Pharmacol. Expl. Ther.* **170,** 221–231.
Crow, T. J. (1972). *Psychol. Med.* **2,** 414–421.
Eichler, A. J., Antelman, S. M. and Fisher, A. E. (1976). *Neurosci. Abst.* No. 1225.

Ettenberg, A. (1977). *Proc. Can. Psychol. Assoc.* Vancouver.

Ettenberg, A. and Wise, R. A. (1976). *Psychopharmacol. Commun.* **2,** 117–124.

Fibiger, H. C. (1978). *Ann. Rev. Pharmacol. Toxicol.* **18,** 37–56.

Fibiger, H. C., Carter, D. A. and Phillips, A. G. (1976). *Psychopharmacology* **47,** 21–27.

Fibiger, H. C. and Phillips, A. G. (1976). *Brain Res.* **116,** 23–33.

Fibiger, H. C., Phillips, A. G. and Zis, A. P. (1974). *Pharmacol. Biochem. Behav.* **2,** 87–96.

Fouriezos, G. and Wise, R. A. (1976). *Brain Res.* **103,** 377–380.

Gianutsos, G., Drawbaugh, R. B., Hynes, M. D. and Lal, H. (1974). *Life Sci.* **14,** 887–898.

Gianutsos, G. and Moore, K. E. (1977). *Life Sci.* **20,** 1585–1592.

Goodall, E. B. and Carey, R. J. (1975). *J. Comp. Physiol. Psychol.* **89,** 1029–1045.

Gunne, L. M., Anggard, E. and Jonsson, L. E. (1972). *Psychiat. Neurol. Neurochir. (Amsterdam)* **75,** 225–226.

Herberg, L. J., Stephens, D. N. and Franklin, K. B. J. (1976). *Pharmacol. Biochem. Behav.* **4,** 575–582.

Jonsson, L. E., Anggard, E. and Gunne, L. M. (1971). *Clin. Pharmacol. Ther.* **12,** 889–896.

Kelleher, R. T. and Morse, W. H. (1968). *Ergbn. Physiol.* **60,** 1–56.

Koob, G. F., Fray, P. J. and Iversen, S. D. (1978). *Brain Res.* **146,** 123–140.

Liebman, J. M. and Butcher, L. L. (1973). *Naun. Schmied. Arch. Pharmacol.* **277,** 305–318.

Liebman, J. M. and Butcher, L. L. (1974). *Naun. Schmied. Arch. Pharmacol.* **284,** 167–194.

MacBrown, R. and Goldman, P. S. (1977). *Brain Res.* **124,** 576–580.

Mora, F., Phillips, A. G., Koolhaas, J. M. and Rolls, E. T. (1976a). *Brain Res. Bull.* **1,** 421–424.

Mora, F., Rolls, E. T., Burton, M. J. and Shaw, S. G. (1976b). *Pharmacol. Biochem. Behav.* **4,** 211–216.

Mora, F., Sanguinetti, A. M., Rolls, E. T. and Shaw, S. G. (1975). *Neurosci. Lett.* **1,** 179–184.

Olds, J. and Milner, P. (1954). *J. Comp. Physiol. Psychol.* **47,** 419–427.

Ornstein, K. and Huston, J. P. (1975). *Neurosci. Lett.* **1,** 339–342.

Phillips, A. G., Brooke, S. M. and Fibiger, H. C. (1975). *Brain Res.* **85,** 13–22.

Phillips, A. G., Carter, D. A. and Fibiger, H. C. (1976a). *In* "Brain Stimulation Reward" (Eds. A. Wauquier and E. T. Rolls), 272–280. Elsevier, Amsterdam.

Phillips, A. G., Carter, D. A. and Fibiger, H. C. (1976b). *Brain Res.* **104,** 221–232.

Phillips, A. G. and Fibiger, H. C. (1973). *Science* **179,** 575–577.

Phillips, A. G. and Fibiger, H. C. (1976). *Behav. Biol.* **16,** 127–143.

Phillips, A. G., Mora, F., Koolhaas, J. M. and Rolls, E. T. (1976c). *Neuroscience Abst.* No. 721.

Phillips, A. G., van der Kooy, D. and Fibiger, H. C. (1977). *Neurosci. Lett.* **4,** 77–84.

Poschel, B. P. M. and Ninteman, F. W. (1966). *Psychol. Rep.* **19,** 79–82.

Price, M. T. C. and Fibiger, H. C. (1975). *Pharmacol. Biochem. Behav.* **3,** 285–290.

Ranje, C. and Ungerstedt, U. (1977). *Brain Res.* **134,** 95–111.

Roberts, D. C. S., Corcoran, M. E. and Fibiger, H. C. (1977). *Pharmacol. Biochem. Behav.* **6,** 615–620.

Rolls, E. T., Rolls, B. J., Kelly, P. H., Shaw, S. G., Wood, R. J. and Dale, R. (1974). *Psychopharmacologia (Berlin)* **38,** 219–230.

Selby, G. (1968). *In* "Handbook of Clinical Neurology", Vol. 6 (Eds. P. J. Vinken and G. W. Bruyn), 173–211. North Holland Publishing Co., Amsterdam.
Stein, L. (1964). *Fed. Proc.* **23,** 836–850.
Stein, L. (1968). *In* "Psychopharmacology. A Review of Progress" (Ed. D. H. Efron), 105–123. U. S. Government Printing Office, Washington.
Stein, L. and Ray, O. S. (1960). *Psychopharmacologia (Berlin)* **1,** 251–256.
Stephens, D. N. and Herberg, L. J. (1975). *Brain Res.* **90,** 348–351.
St. Laurent, J., Le Clerc, R. R., Mitchell, M. L. and Miliarresis, T. E. (1973). *Pharmacol. Biochem. Behav.* **1,** 581–585.
Wauquier, A. and Niemegeers, C. J. E. (1972). *Psychopharmacologia (Berlin)* **27,** 191–202.
Wauquier, A. and Niemegeers, C. J. E. (1973). *Psychopharmacologia* **30,** 163–172.
Yokel, R. A. and Wise, R. A. (1976). *Psychopharmacology* **48,** 311–318.
Zarevics, P. and Setler, P. E. (1978). *Brain Res.*, in press.

Clinical Aspects of Dopaminergic Systems

CHAPTER 34

Clinical Chemical Aspects of Dopaminergic Processes

Jakob Korf

University Psychiatric Clinic, Groningen, The Netherlands.

INTRODUCTION

In this chapter a summary is presented of possible ways of measuring biochemical processes in which dopamine plays a role in the body of man. The emphasis is placed more on methodological aspects, and less on the results of the clinical applications. For aspects more concerned with clinical relevance the reader is referred to the Chapters by Hornykiewicz and Van Praag in this volume (i.e. Chapters 35 and 36).

URINE

Most of the dopamine in the body is converted to the acidic metabolites homovanillic acid (HVA), 3,4-dihydroxyphenylacetic acid (DOPAC) and conjugates of the amine and these acids (Armstrong *et al.*, 1956). In man the daily urinary excretion of free HVA is about 3–5 mg (15–30 µmol) (Stott *et al.*, 1975; Roginsky *et al.*, 1974; Muskiet *et al.*, 1978a; Sato, 1965; Yoshida *et al.*, 1976; Wadman *et al.*, 1976; Dziedzic *et al.*, 1972; Hoehn *et al.*, 1976; Goodwin *et al.*, 1974; Crowley *et al.*, 1976). In a group of Parkinson patients about one-third of all HVA excreted was conjugated (Hoehn *et al.*, 1976). The amount of free DOPAC excreted per 24 hours is about 500 µg, and about four times more DOPAC is excreted in a conjugate form (Hoehn *et al.*, 1976; Rutledge and Hoehn, 1973; Swahn and Wiesel, 1976; Muskiet *et al.*, 1978a). The 24-hour excretion of the amine in the adult is about 250 ng, and 4–6 times more conjugates of DA are excreted in the urine (Voorhess, 1967; Mattock *et al.*, 1966; Hoehn *et al.*, 1976; Crowley *et al.*, 1976; Von Studnitz *et al.*, 1963).

The daily urinary excretion of two alcoholic metabolites of DA, namely 3-methoxy-4-hydroxyphenylethanol and of 3,4-dihydroxyphenylethanol is about 100 times lower than that of HVA (Muskiet *et al.*, 1978a). The excretion of DA metabolites is higher in individuals younger than 10 years old, compared to adults (Muskiet *et al.*, 1977, 1978a).

Quantitation of urinary acidic DA-metabolites is a valuable tool in the diagnosis and management of children with neuroblastoma and ganglioneuroblastoma; in both diseases 3-methoxy-4-hydroxy-mandelic acid (a metabolite of noradrenaline) and HVA are commonly elevated (Voorhess, 1966; Von Studnitz *et al.*, 1963; Imashuku *et al.*, 1975; Gitlow *et al.*, 1970, 1973; Wadman *et al.*, 1976; Gjessing, 1968). Successful treatment of these tumours produces, as would be expected, a decrease of the excessive excretion of catecholamine metabolites. Failure of the catabolite excretion to retun to normal following therapy is associated with universal mortality (Gitlow *et al.*, 1973). Some authors have also described abnormal excretion of HVA in adults with malign phaeochromocytoma, but this appears not to be a universal observation (Sankoff and Sourkes, 1963; Robinson *et al.*, 1964; Gitlow *et al.*, 1971; Wadman *et al.*, 1976; Muskiet *et al.*, 1977).

Only part of the HVA in the urine of normal rats originates from dietary sources (Hoeldtke *et al.*, 1972, 1974). In children, however, Weetman *et al.* (1976) found no evidence that the urinary excretion of HVA was much dependent on the various diets. These authors conclude that for the screening of children for neuroblastoma no restricted diet was necessary. Their method for the determination of HVA may, however, not be sufficiently accurate to determine the effects of relatively small differences as the result of various diets.

In non-diseased subjects the origin of the major part of DA and its metabolites in the urine are not well established. In the rat efforts have been made to estimate the contribution of the central nervous system to these levels. After destruction of the cerebral DA-containing neurones with 6-hydroxydopamine a decrease of the excretion of free urinary HVA of about 30 % was observed by one group of investigators (Hoeldtke *et al.*, 1974), while others found no decreased excretion of this metabolite (Peyrin *et al.*, 1978). After unilateral destruction of the nigrostriatal pathway the latter authors observed decreased excretion of the conjugates of both DOPAC (about 25%) and of HVA (about 40%). In none of the above mentioned studies was the excretion of DA or 3-methoxytyramine and their conjugates found to be dependent upon the cerebral dopaminergic structures. The study of Peyrin *et al.* (1978) suggests that measurement of DOPAC and HVA conjugates in the urine of the rat is of interest for assessment of cerebral metabolism of DA. Whether this applies to man is questionable but it is of interest to note that in particular the conjugate formation of DOPAC in the human brain is of significance (Gordon *et al.*, 1976; Swahn and Wiesel, 1976). In the dog the contribution of the HVA in the cisternal CSF to the urine excretion of this metabolite including both free and conjugated has been estimated with a tracer technique. No more than 6% of the urinary HVA was calculated to be produced in the CSF (Raese *et al.*, 1974). However, one has to take into account that not all HVA produced in the brain is eliminated via the CSF (see section on CSF) and that there may be large species differences.

BLOOD

In human blood several parameters related to peripheral or central dopaminergic processes will be considered. We have not included data on the occurrence of the enzymes monoamineoxidase, catechol-*O*-methyltransferase or dopamine-β-hydroxylase, as these enzymes do not represent a specific dopaminergic parameter.

Only a few reports on the levels of HVA in human blood have appeared. Markianos and Beckman (1976) and Markianos *et al.* (1976) found about 90 ng ml^{-1} serum. Diurnal variations of serum HVA were reported: highest levels occurring at 8 p.m. These authors found rather high levels of HVA in CSF, so it is not unlikely that their observed serum levels are also somewhat on the high side. In fact levels as low as 10 ng ml^{-1} plasma were reported by Gordon *et al.* (1976).

In blood two other dopaminergic parameters have been investigated, namely the uptake of DA in platelets and the levels of pituitary hormones. The uptake of exogenous DA by platelets was found to be maximal 90 min after incubation at 37°C. A ratio of platelet to medium content of DA obtained was 130/1 (Boullin and O'Brien, 1970). The uptake was temperature dependent, sensitive to ouabain, and was not inhibited by desipramine, a potent inhibitor of synaptosal uptake of noradrenaline. The effects of more specific uptake inhibitors for DA, such as benztropine or nomifensine, have not yet been reported. With the present knowledge it is not possible to judge whether or not platelet uptake of DA may serve as a model for neuronal uptake mechanisms (Barbeau *et al.*, 1975). For a further discussion of this point see Horn, Chapter 16.

The plasma levels of prolactin and growth hormone (both originating in the pituitary gland) are influenced by some agonists and antagonists of DA. Administration of L-DOPA (with or without a peripheral decarboxylase inhibition) lowers the serum levels of prolactin, and enhances the levels of growth hormone (Kleinberg *et al.*, 1971; Friesen *et al.*, 1972; Frantz *et al.*, 1973). Apomorphine, another DA agonist, also suppresses release of prolactin (Lal *et al.*, 1973), and enhances the secretion of growth hormone in man (Rotrosen *et al.*, 1977).

Serum prolactin levels are increased during treatment with neuroleptics, potent DA-antagonists (Sachar *et al.*, 1976; Clemens *et al.*, 1974; Meltzer and Fang, 1976; Meltzer *et al.*, 1977; Langer *et al.*, 1977; Wilson *et al.*, 1975). The ability of drugs to block central DA receptors in man can therefore be studied by determining their effects on serum prolactin levels. The relative potency of various neuroleptic drugs to increase the release of prolactin correlated well their potency as antipsychotic agents (Langer *et al.*, 1977). During chronic drug treatment differences are observed in the effects of neuroleptics on the prolactin response compared to those on DA systems in the basal ganglia and the mesolimbic areas. The responsiveness of the ascending DA-neurones diminish during repeated administration of the drugs (see, for example, Bowers and Rozitis, 1976; Post and Goodwin, 1975; Scatton *et al.*, 1977) as reflected by a decrease in CSF-HVA content. The effect of neuroleptics on prolactin levels in the blood was not changed during repeated drug administration (see, for example, Van Praag, 1977). Such observations emphasize that one has to be careful when attempting to extrapolate drug effects in one cerebral dopaminergic system to another with an entirely different neuronal organization (see also Westerink and Korf, 1976a; Scatton *et al.*, 1977).

CEREBROSPINAL FLUID

Occurrence of DA Metabolites

In many clinical studies the levels of HVA in CSF have been measured to estimate the metabolism of DA in the brain of psychiatric or neurological patients. In most of these studies lumbar CSF was used. Homovanillic acid in CSF is predominantly of central origin, as systematically applied HVA does not produce significant increases in the CSF levels of the metabolite (Bartholini *et al.*, 1966; Guldberg and Yates, 1968; Van der Poel *et al.*, 1977; Aizenstein and Korf, 1978). In at least two species, the cat and the rat, the spinal cord does not produce significant amounts of HVA (Van der Poel *et al.*, 1977; Westerink and Korf, 1975). In the monkey Kessler *et al.* (1976b) reported that HVA is released into spinal cord perfusates. These authors did not report the spinal cord tissue levels of HVA. Whether or not the human spinal cord releases HVA into spinal fluid is not known, but the spinal contribution if any is likely to be small as patients with a block in the spinal subarachnoid space have much lower levels of HVA in lumbar CSF than those in which HVA diffuses freely along the spinal cord (Curzon *et al.*, 1971; Garelis and Sourkes, 1973; Post *et al.*, 1973; Young *et al.*, 1973).

A possible peripheral source of HVA in CSF is L-DOPA (Bartholini *et al.*, 1971). In subjects not treated with L-DOPA such contribution is likely to be negligible. Besides HVA the occurrence in CSF of two other metabolites of DA, namely DOPAC and its sulphate conjugate has been reported (Gordon *et al.*, 1976). Recently an alcohol metabolite of DA (3-methoxy-4-hydroxyphenylethanol) has been identified in human CSF (Muskiet *et al.*, 1978b).

The main sources of HVA in CSF are the basal ganglia and to a minor extent the mesolimbic areas. Thus destruction of the nigrostriatal pathway results in a decrease of HVA in CSF (Papeschi *et al.*, 1971). The levels of HVA in the human caudate nucleus and putamen are about 3–5 μg (Lloyd *et al.*, 1975). The levels of HVA in the adjacent ventricular CSF are about $0.3\,\mu g\,ml^{-1}$ (Curzon *et al.*, 1976; Chase *et al.*, 1973; Gordon *et al.*, 1975). In cisternal CSF the levels of HVA are about $0.08\,\mu g\,ml^{-1}$, while the lumbar CSF contains about $0.03\,\mu g\,ml^{-1}$ (e.g. Chase *et al.*, 1973; Gordon *et al.*, 1976; Korf *et al.*, 1971; Sourkes, 1973). The concentration gradient between brain tissue and ventricular CSF is presumably due to the fact that only a small proportion of HVA formed in tissue diffuses into CSF. Aizenstein and Korf (1978) calculated that in the rat only 3% of the HVA formed in the brain is released into CSF. Such a small figure does not necessarily apply to the human brain (Sourkes, 1973). The steep gradient between the CSF-compartments is due to the active elimination of HVA, which in particular is localized in the ventricular choroid plexus (Ashcroft *et al.*, 1968) and the subarachnoid spaces (Van der Poel *et al.*, 1977). From the above considerations it is concluded that in human beings no simple quantitative relation exists between the metabolism of DA in the brain and the baseline levels of HVA in the lumbar CSF.

On the Action of Probenecid

In many clinical studies probenecid is used for the study of the metabolism of cerebral DA. Part of the HVA in the central nervous system is transported to blood by a probenecid-sensitive mechanism. In the rat brain Westerink and Korf (1976b) found that the probenecid induced accumulation of HVA corresponds to about 10% of the total turnover of DA. In other species the fraction may be several times higher (e.g. guinea-pig,

Spano and Neff, 1972). In the rabbit HVA does not accumulate in the brain following the administration of probenecid (Extein *et al.*, 1973). It is not known if probenecid causes accumulation of HVA in the human brain. Thus far in all species investigated administration of probenecid induces an elevation of HVA in CSF. The site of action of probenecid has been investigated by a number of studies.

In perfusion studies in the dog Guldberg *et al.* (1966) found evidence that the ventricular choroid plexuses are the site of a probenecid-sensitive transport. The elimination of HVA from the spinal subarachnoid space of the rat and the cat was found to be inhibited by probenecid (Van der Poel *et al.*, 1977; Aizenstein and Korf, 1978). Kessler *et al.* (1976a, b) did not, however, observe a probenecid sensitive transport mechanism in the spinal subarachnoid space of the monkey. Some of the results of Kessler *et al.* were obtained with experiments using a rather high rate of spinal cord perfusion. In experiments with rats Aizenstein and Korf (1978) found that the action of probenecid was more significant in experiments where low rates of perfusion were applied. The rate of perfusion should be rather low for the study of probenecid-sensitive mechanisms, although discrepancies may also be attributed to species differences. High doses of probenecid applied to man cause a rapid elevation of HVA in the lumbar CSF, suggesting the existence of a probenecid sensitive transport mechanism in the human spinal subarachnoid space (Kartzinel *et al.*, 1976).

The variation of the levels of HVA found in the human CSF is rather high, both of the baseline levels as well as of those after probenecid administration. An important factor which influences the CSF-levels of HVA after probenecid treatment is the level of the drug in the central nervous system. Probenecid penetrates poorly into the brain and the CSF (Van Wijk *et al.*, 1979; Korf and Van Praag, 1971; Sjöström, 1972). It is uncertain whether or not in man maximal inhibition of the HVA-egress from CSF can be obtained with probenecid. In studies using relatively low doses of probenecid (3–6 g) a significant correlation between the lumbar CSF-levels of HVA and probenecid was observed (Korf and Van Praag, 1971; Sjöström, 1972). Using higher doses of probenecid (40–70 $mg\,kg^{-1}$ body weight) such a correlation was not observed (Kartzinel *et al.*, 1976; Perel *et al.*, 1974; Goodwin *et al.*, 1977) and it has been suggested that under these conditions CSF-levels of probenecid are sufficiently high to block HVA-egress entirely. But it should be emphasized that even with the highest doses of probenecid the lumbar CSF-levels of HVA (in $ng\,ml^{-1}$) are still about 20 times lower than the caudate levels (in $ng\,g^{-1}$) and about three times lower than the ventricular CSF-levels.

Bowers (1969) reported a case in which the CSF HVA was not transported by a probenecid sensitive mechanism. The levels in lumbar CSF were about 120 mg ml^{-1}. This case suggests, that in man about two-thirds of the HVA in the CSF-system is transported by a probenecid sensitive mechanism, while the remaining one-third of the HVA disappears from the CSF by other transport mechanisms (bulk flow).

Because of the steep gradients of the monoamine metabolites in the various CSF-compartments any perturbation of the CSF-flow influences the levels of HVA in lumbar CSF. In this case changes of the CSF-HVA levels are not related to an altered cerebral metabolism of DA. After physical exercise higher amounts of HVA are found in the CSF (Post *et al.*, 1973). Therefore it is important to collect the lumbar CSF of patients under standard conditions. In spite of the methodological drawbacks mentioned above, several studies indicate that CSF-HVA reflects relatively large changes in cerebral DA metabolism. So the extensive degeneration of the cerebral DA-neurones in Parkinson's disease is reflected in the low HVA levels in the spinal fluid (see for example Olsson and Roos, 1968; Bernheimer *et al.*, 1966; Lakke *et al.*, 1972). Neuroleptics, which are known to increase cerebral metabolism of DA in animals have also been found to increase the human lumbar CSF-levels of HVA during clinical treatment (see for example Post and Goodwin, 1975; Bowers, 1973; see also Van Praag, this volume). After the inhibition of the synthesis of HVA by inhibitors of monoamineoxidase, low levels of the metabolite are found in the CSF of both humans (Bowers and Kupfer, 1971) and animals (Forn *et al.*, 1970). In many other studies the effects of drugs on both cerebral metabolism of DA and on the formation of HVA in CSF have been determined (e.g. Bowers 1970; Bowers and Rozitis, 1976; Guldberg, 1969; Goodwin *et al.*, 1973; Van Praag and Korf, 1975).

POST-MORTEM STUDIES

Various enzymes involved in the synthesis and degradation of DA have been measured in post-mortem brain tissue. The regional distribution of the enzymes synthesizing DA, but not those involved in the degradation of the amine, parallels the distribution of DA. Thus the highest levels of tyrosine hydroxylase and dopadecarboxylase are found in the caudate nucleus, the substantia nigra and the putamen (Lloyd and Hornykiewicz, 1972; McGeer *et al.*, 1971; Fahn, 1976; Fahn *et al.*, 1971). The levels of the DA-degrading enzymes, monoamineoxidase and catechol-*O*-methyltransferase, are mainly localized outside catecholaminergic neurones and are more evenly distributed in the brain. The activity of the enzymes decreases post-mortally, but both the DA-degrading enzymes are still active several days after death (Ganrot *et al.*, 1962; Vogel *et al.*, 1969). In addition to the above enzymes,

DA, HVA and 3-methoxytyramine levels in the human brain have been assayed. An age-dependent decrease of the amines and of tyrosine hydroxylase has been reported (Carlsson, 1976). In animal experiments a marked increase in 3-methoxytyramine levels occurs postmortally, with a concomitant decrease of DA (Carlsson, 1976; Wiesel and Sedvall, 1974). Marked increases in the levels of HVA in the striatum of the rat were obtained in post-mortem dissected tissue. Wiesel and Sedvall (1974) found a seven-fold increase in HVA content when the dissected brain tissue was kept at 4°C for 48 h, while no such increase was noticed when the brain was stored *in situ* under similar conditions. The levels of striatal DA were decreased when the brain tissue was stored for 48 h at 4°C. Again the most pronounced decrease was found in dissected tissue (Wiesel and Sedvall, 1974; McGeer *et al.*, 1962; Bertler and Rosengren, 1959). Moleman *et al.* (1977) showed that shortly after death the rat striatal levels of DOPAC are increased when the brain is dissected, but tend to decrease when the brain was stored *in situ* for 10–20 min. As the time interval between death and collection of the brain samples is rather long in the case of the human brain any post-mortem changes in the levels of DA and its metabolites cannot be excluded.

Direct visualization of nerve terminals containing biogenic amines has been performed in the human brain with a histochemical technique (Chase *et al.*, 1974).

In post-mortem brain tissue Carenzi *et al.* (1975) observed that a DA-sensitive adenylate cyclase was still active. In animal studies no direct relation could be observed between the effect of DA agonists or antagonists and cyclic AMP in cerebrospinal fluid (Sebens and Korf, 1975). Therefore direct evidence for a possible involvement of DA receptors in the action of drugs or in the pathogenesis of neurologic or psychiatric diseases can probably only be obtained by examination of brain tissue obtained *post mortem*. Results thus far are, however, negative (Carenzi *et al.*, 1975).

POSITRON-EMITTING DRUGS

The distribution of positron-emitting compounds in the body can be determined by external (coincidence) detection of the pairs of photons created by annihilation of the positron. Positron-emitting nuclides can be produced by bombardment of stable nuclides by particles with high energy. The relatively simple radioactive molecules thus obtained have subsequently to be incorporated in pharmaceuticals, suitable for clinical use. In the present context two derivatives of DOPA are of interest, namely β-(3,4-dihydroxy,5-fluorophenyl)-D,L-α-alanine-5-^{18}F (fluordopa, Firnau *et al.*, 1973, 1975, 1976) and β-(3,4-dihydroxyphenyl)-D,L-α-alanine-1-^{11}C. (^{11}C-

DOPA, Reiffers *et al.*, 1977a, b, Korf *et al.*, 1978, 1980). Fluorine-18 and ^{11}C isotopes have physical halflives of 108 and 20 min, respectively. To obtain sufficient amounts of radioactive material for clinical studies both synthesis and purification procedures of the mentioned DOPA derivatives should be rapid. Practically, these compounds can only be applied in medical settings which have easy access to facilities, where the radioactive nuclides can be prepared (a cyclotron or a nuclear reactor). Moreover, for clinical studies on dopaminergic processes the radioactive derivatives are useful only when the distribution in the body (or the brain) is highly dependent upon the occurrence or function of cerebral DA-neurones and it is rapid compared to the physical halflife of the nuclides.

After intravenous administration to baboons the cerebral distribution of ^{18}F-DOPA was found to be influenced by reserpine. In the rat relatively more labelled material was found in the caudate nucleus or two hours after systemic application of ^{18}F-DOPA (Firnau *et al.*, 1976). Fluordopa is likely to be converted to ^{18}F-DA in brain areas such as the caudate nucleus (Firnau *et al.*, 1975). In the studies of Firnau *et al.*, however, no peripherally acting inhibitor of DOPA-decarboxylase (such as carbidopa) was used, so the uneven cerebral distribution of radioactivity following ^{18}F-DOPA may be more selective and rapid than reported. Moreover both the biological precursor ^{18}F-DOPA as well as ^{18}F-DA- and ^{18}F-labelled deaminated products are radioactive, and with external detection one cannot monitor the metabolic conversions taking place in the brain.

Experiments with ^{11}C-DOPA indicate that cerebral distribution is rapid and selective only when the drug is intravenously applied in rats pretreated with caridopa (Korf *et al.*, 1978, 1980). In brain areas with high activity of the enzyme DOPA-decarboxylase such as the striatum and the brainstem, relatively low levels of radioactivity were found, presumably because after decarboxylation of the ^{11}C-DOPA the $^{11}CO_2$ produced is rapidly eliminated. The uneven distribution of radioactivity in the brain was not influenced by haloperidol or L-DOPA pretreatment. After destruction of the cerebral dopaminergic systems with 6-hydroxydopamine a more uniform distribution of radioactivity was observed (Korf *et al.*, 1978).

^{11}C-DOPA may be a useful drug for the determination of the regional activity of DOPA-decarboxylase in the brain. Until now no clinical applications of positron-emitting DOPA-derivatives have been described.

REFERENCES

Aizenstein, M. L. and Korf, J. (1978). *Brain Res.* **149**, 129–140.

Armstrong, M. D., Shaw, K. N. F. and Wall, P. E. (1956). *J. Biol. Chem.* **218**, 293–303.

Ashcroft, G. W., Dow, R. C. and Moir, A. T. B. (1968). *J. Physiol.* (*London*) **199,** 397–425.

Barbeau, A., Campanello, G., Butterworth, R. F. and Yamada, K. (1975). *Neurology* **25,** 1–9.

Bartholini, G., Pletscher, A. and Tissot, R. (1966). *Experientia* **22,** 609–610.

Bartholini, G., Tissot, R. and Pletscher, A. (1971). *Brain Res.* **27,** 163–168.

Bernheimer, H., Birkmayer, W. and Hornykiewicz, O. (1966). *Wien Klin. Wschr.* **78,** 417–419.

Bertler, Å. and Rosengren, E. (1959). *Acta Physiol. Scand.* **47,** 350–361.

Boullin, D. J. and O'Brien, R. A. (1970). *Br. J. Pharmacol.* **39,** 779–788.

Bowers, M. B. Jr. (1969). *Brain Res.* **15,** 522–524.

Bowers, M. B. Jr. (1970). *Life Sci.* **9,** 691–696.

Bowers, M. B. Jr. (1973). *Psychopharmacologia* (*Berlin*) **28,** 309–318.

Bowers, M. B. Jr. and Kupfer, D. J. (1971). *Brain Res.* **35,** 561–564.

Bowers, M. B. Jr. and Rozitis, A. (1976). *Eur. J. Pharmacol.* **39,** 109–115.

Carenzi, A., Gillin, J. C., Guidotti, A., Schwartz, M. A., Trabucchi, M. and Wyatt, R. J. (1975). *Arch. Gen. Psychiat.* **32,** 1056–1063.

Carlsson, A. (1976). *In* "The Basal Ganglia" (Ed. M. D. Yahr), 181–189. Raven Press, New York.

Chase, T. N., Balch, T. S., Jacobowitz, D. M. (1974). *J. Psychiat. Res.* **11,** 281–282.

Chase, T. N., Gordon, E. K. and Ng, L. K. Y. (1973). *J. Neurochem.* **21,** 581–587.

Clemens, J. A., Smalstig, E. B. and Sawyer, B. D. (1974). *Psychopharmacologia* (*Berlin*) **40,** 123–127.

Crowley, T. J., Rutledge, C. O., Hoehn, M. M., Stallings, M. A. and Sundell, S. (1976). *Am. J. Psychiat.* **133,** 703–706.

Curzon, G., Gumpert, E. J. W. and Sharpe, D. M. (1971). *Nature* (*London*) **231,** 189–191.

Curzon, G., Kantamaneni, B. D., Barlett, J. R. and Bridges, P. K. (1976). *J. Neurochem.* **26,** 613–615.

Dziedzic, S. W., Bertani, L. M., Clarke, D. D. and Gitlow, S. E. (1972). *Anal. Biochem.* **47,** 592–600.

Extein, I., Korf, J., Roth, R. H. and Bowers, M. B. Jr. (1973). *Brain Res.* **54,** 403–407.

Fahn, S. (1976). *In* "Advances in Neurology", Vol. 14 (Eds. R. Eldridge and S. Fahn), 59–88. Raven Press, New York.

Fahn, S., Libsch, L. R. and Cutler, R. W. (1971). *J. Neurol. Sci.* **14,** 427–455.

Firnau, G., Garnett, E. S., Chan, P. K. H. and Belbeck, L. W. (1976). *J. Pharm. Pharmacol.* **28,** 584–585.

Firnau, G., Garnett, E. S., Sourkes, T. L., and Missala, K. (1975). *Experientia* **31,** 1254–1255.

Firnau, G., Nahmias, C. and Garnett, S. (1973). *J. Appl. Radiat. Isot.* **24,** 182–184.

Forn, J., Maling, H. M. and Gessa, G. L. (1970). *Proc. Soc. Expl. Biol. Med.* **133,** 1310–1313.

Frantz, A. G., Suh, H. K. and Noel, G. L. (1973). *In* "Frontiers in Catecholamine Research" (Eds. E. Usdin and S. H. Snyders), 843–847. Pergamon, Oxford.

Friesen, H., Guyda, H., Hwang, P., Tyson, J. E. and Barbeau, A. (1972). *J. Clin. Invest.* **51,** 706–712.

Ganrot, P. O., Rosengren, E. and Gottfries, C. G. (1962). *Experientia* **18,** 260–261.

Garelis, E. and Sourkes, T. L. (1973). *J. Neurol. Neurosurg. Psychiat.* **4,** 625–629.

Gitlow, S. E., Bertani, I. M., Wilk, E., Li, B. L. and Dziedzic, S. (1970). *Pediatrics* **46,** 513–522.

Gitlow, S. E., Dziedzic, L. B., Strauss, L., Greenwood, S. M. and Dziedzic, S. W. (1973). *Cancer* **32,** 898–905.
Gitlow, S. E., Pertsemlidis, D. and Bertani, L. M. (1971). *Am. Heart J.* **82,** 557–561.
Gjessing, L. P. (1968). *In* "Advances in Clinical Chemistry", Vol. 11 (Eds. O. Bogdanski and C. P. Stewart), 81–131. Academic Press, New York and London.
Goodwin, B. L., Ruthven, C. R. and Sandler, M. (1974). *Clin. Chim. Acta* **55,** 111–112.
Goodwin, F. K., Cowdry, R., Gold, P. W. and Wehr, T. (1977). *In* "Neurotransmission and Disturbed Behaviour" (Eds. H. M. van Praag and J. Bruinvels), 34–59. Bohn, Scheltema and Holkema, Utrecht.
Goodwin, F. K., Post, R. M., Dunner, D. L. and Gordon, E. K. (1973). *Am. J. Psychiat.* **130,** 73–79.
Gordon, E. K., Oliver, J. and Kopin, I. J. (1976). *Life Sci.* **16,** 1527–1532.
Gordon, E. K., Perlow, M., Oliver, J., Ebert, M. and Kopin, I. J. (1975). *J. Neurochem.* **25,** 347–349.
Guldberg, H. C., Ashcroft, G. W. and Crawford, T. B. B. (1966). *Life Sci.* **5,** 1571–1575.
Guldberg, H. C. and Yates, C. M. (1968). *Br. J. Pharmacol.* **33,** 457–471.
Guldberg, H. C. (1969). *In* "Metabolism of Amines in the Brain" (Ed. G. Hooper), 55–64. MacMillan, London.
Hoehn, M. M., Crowley, T. J. and Rudledge, C. O. (1976). *J. Neurol. Neurosurg. Psychiat.* **39,** 941–951.
Hoeldtke, R., Baliga, B. S., Issenberg, P. and Wurtman, R. J. (1972). *Science* **175,** 761–762.
Hoeldtke, R., Rogawski, M. and Wurtman, R. J. (1974). *Br. J. Pharmacol.* **50,** 265–270.
Imashuku, S., Takada, H., Sawada, T., Nakamura, T., Labrosse, E. H. (1975). *Cancer* **36,** 450–457.
Kartzinel, R., Ebert, M. H. and Chase, T. N. (1976). *Neurology*, **26,** 992–996.
Kessler, J. A., Fenstermacher, J. D. and Patlak, C. S. (1976a). *Neurology* **5,** 434–440.
Kessler, J. A., Gordon, E. K., Reid, J. L. and Kopin, L. J. (1976b). *J. Neurochem.* **26,** 1057–1061.
Kleinberg, D. L., Noel, G. L. and Frantz, A. G. (1971). *J. Clin. Endocrinol. Metab.* **33,** 873–876.
Korf, J., Van Praag, H. M. and Sebens, J. B. (1971). *Biochem. Pharmacol.* **20,** 659–668.
Korf, J. and Van Praag, H. M. (1971). *Brain Res.* **35,** 221–230.
Korf, J., Reiffers, S., Beerling-Van der Molen, H. D., Lakke, J. P. W. F., Paans, A. M. J., Vaalburg, W. and Woldring, M. G. (1978). *Brain Res.* **145,** 59–67.
Korf, J., Reiffers, S., Beerling-Van der Molen, H. D., Lakke, J. P. W. F., Paans, A. M. J., Vaalburg, W. and Woldring, M. G. (1980). *In* "Proceedings of the 15th International Annual Meeting of the Society of Nuclear Medicine" (Eds. M. G. Woldring and W. Vaalburg) in press.
Lakke, J. P. W. F., Korf, J., Van Praag, H. M. and Schut, T. (1972). *Nature (London)* **236,** 208–209.
Langer, G., Sacchar, E. J., Gruen, P. H., and Halpern, F. S. (1977). *Nature (London)* **266,** 639–640.
Lloyd, K. G. (1975). *Pharmacol. Ther. B.* **1,** 49–61.

Lloyd, H. G. and Hornykiewicz, O. (1972). *J. Neurochem.* **19,** 1549–1559.
Markianos, E. and Beckman, H. (1976). *J. Neur. Trans.* **39,** 79–93.
Markianos, E. S., Rüther, E. and Gluba, H. (1976). *Neurosci. Lett.* **3,** 37–40.
Mattock, G. L., Wilson, D. L. and Heacock, R. A. (1966). *Clin. Chim. Acta* **14,** 99–107.
McGeer, E. G. and McGeer, P. L. (1962). *Canad. J. Biochem.* **40,** 1141–1151.
McGeer, E. G., McGeer, P. L. and Wada, J. A. (1971). *J. Neurochem.* **18,** 1647–1658.
Meltzer, H. Y. and Fang, V. S. (1976). *Arch. Gen. Psychiat.* **33,** 279–286.
Meltzer, H. Y., Fang, V. S., Simonovich, M. and Paul, S. M. (1977). *Eur. J. Pharmacol.* **41,** 431–436.
Moleman, P., Bruinvels, J. and Westerink, B. H. C. (1977). *J. Neurochem.* **29,** 175–177.
Muskiet, F. A. J., Fremouw-Ottevangers, D. C., Wolthers, B. G., De Vries, J. A. (1977). *Clin. Chem.* **23,** 863–867.
Muskiet, F. A. J., Fremouw-Ottevangers, D. C., Van der Meulen, J. Wolthers, B. G. and De Vries, J. A. (1978a). *Clin. Chem.* **24,** 122–127.
Muskiet, F. A. J., Jeuring, H. J., Adèr, J.-P. and Wolthers, B. G. (1978b). *J. Neurochem.* **30,** 1495–1499.
Olsson, R. and Roos, B.-E. (1968). *Nature (London)* **219,** 502–503.
Papeschi, R., Sourkes, T. L., Poirier, C. J. and Boucher, R. (1971). *Brain Res.* **28,** 527–533.
Perel, J. M., Levitt, M. and Dunner, D. L. (1974). *Psychopharmacology* **35,** 83–87.
Peyrin, L., Cottet-Emard, J. M., Javoy, F., Agid, Y., Herbet, A. and Glowinski, J. (1978). *Brain Res.* **143,** 567–572.
Post, R. M. and Goodwin, F. K. (1975). *Science* **190,** 188–190.
Post, R. M., Goodwin, F. K., Gordon, E. K. and Watkin, D. M. (1973). *Science* **179,** 897–899.
Raese, J., Schmiedek, P. and Ehrlich, E. (1974). *Arzneim. Forsch. (Drug Res.)* **24,** 1088–1093.
Reiffers, S., Beerling-Van der Molen, H. D., Vaalburg, W., Ten Hoeve, W., Woldring, M. G. and Wijnberg, H. (1977a). *J. Labelled Comp. Radiopharm.* **13,** 198–199.
Reiffers, S., Beerling, Van der Molen, H. D., Vaalburg, W., Ten Hoeve, W., Paans, A. M. J., Korf, J., Woldring, M. G. and Wijnberg, H. (1977b). *J. Appl. Rad. Isot.* **28,** 955–958.
Robinson, R., Smith, A. and Whittacker, S. R. S. (1964). *Br. Med. J.* **i,** 1422–1424.
Roginsky, M. S., Gordon, R. D. and Bennett, M. J. (1974). *Clin. Chim. Acta* **56,** 261–264.
Rotrosen, J., Angrist, B. M., Gershon, S., Sachar, E. J. and Halpern, F. S. (1977). *Adv. Biochem. Psychopharmacol.* **16,** 649–653.
Rutledge, C. O., and Hoehn, M. M. (1973). *Nature (London)* **244,** 447–450.
Sachar, E. J., Gruen, P. H., Karasu, T. B., Altman, N. and Frantz, A. G. (1975). *Arch. Gen. Psychiat.* **32,** 885–886.
Sankoff, J. and Sourkes, T. L. (1963). *Can. J. Biochem. Physiol.* **41,** 1381–1388.
Sato, T. L. (1965). *J. Lab. Clin. Med.* **66,** 517–525.
Scatton, B., Dedek, J. and Korf, J. (1977). *Brain Res.* **135,** 374–377.
Sebens, J. B. and Korf, J. (1975). *Expl. Neurol.* **46,** 333–344.
Sjöström, R. (1972). *Psychopharmacologia (Berlin)* **25,** 96–98.
Sourkes, T. L. (1973). *J. Neur. Trans.* **34,** 153–157.
Spano, P. F. and Neff, N. H. (1972). *Brain Res.* **42,** 139–145.

Stott, A. W., Lindsay Smith, J. R., Hanson, P. and Robinson, R. (1975). *Chim. Acta* **63,** 7–12.

Swahn, C.-G., and Wiesel, F. A. (1976). *J. Neur. Trans.* **39,** 281–290.

Van der Poel, F. W., Van Praag, H. M. and Korf, J. (1977). *Psychopharmacology* **52,** 35–40.

Van Praag, H. M. (1977). *In* "Psychobiology of the Striatum" (Eds. A. R. Cools, A. H. M. Lohman and J. H. L. van der Bercken), 155–171. Elsevier/North Holland Biomedical Press, Amsterdam.

Van Praag, H. M. and Korf, J. (1975). *Am. J. Psychiat.* **132,** 593–597.

Van Wijk, M., Sebens, J. B. and Korf, J. (1979). *Psychopharmacology* **60,** 229–235.

Vogel, W. H., Orfei, V. and Century, B. (1969). *J. Pharmacol. Expl. Ther.* **165,** 196–203.

Von Studmitz, W., Käser, H. and Sjoerdsma, A. (1963). *New Engl. J. Med.* **269,** 232–235.

Voorhess, M. L. (1966). *Pediatr. Clin. North Am.* **13,** 3–9.

Voorhess, M. L. (1967). *Pediatrics* **39,** 252–257.

Wadman, S. K., Ketting, D. and Voûte, P. A. (1976). *Clin. Chim. Acta* **72,** 49–68.

Weetman, R. M., Rider, P. S., Oei, T. O., Hempel, J. S., Baekner, R. L. (1976). **88,** 46–50.

Westerink, B. H. C. and Korf, J. (1975). *Biochem. Med.* **12,** 106–112.

Westerink, B. H. C. and Korf, J. (1976a). *Eur. J. Pharmacol.* **40,** 175–178.

Westerink, B. H. C. and Korf, J. (1976b). *Eur. J. Pharmacol.* **37,** 249–255.

Wiesel, F.-A. and Sedvall, G. (1974). *Brain Res.* **65,** 547–550.

Wilson, R. G., Hamilton, J. R., Boyd, W. D., Forrest, A. P. M., Cole, E. N., Boyns, A. R. and Griffith, S. (1975). *Br. J. Psychiat.* **127,** 71–74.

Yoshida, A., Yoshioka, M., Tanimura, T. and Tamura, Z. (1976). *J. Chromatogr.* **116,** 240–243.

Young, S. N., Lal, S. Martin, J. B., Ford, R. M. and Sourkes, T. L. (1973). *Psychiat. Neurol. Neurochir.* (*Amsterdam*) **76,** 439–444.

CHAPTER 35

Brain Dopamine in Parkinson's Disease and Other Neurological Disturbances

Oleh Hornykiewicz

University of Vienna, Vienna, Austria.

PARKINSON'S DISEASE

The crucial role of brain dopamine in the pathophysiology of Parkinson's disease can be regarded as established. A large body of neurochemical, pharmacological and clinical evidence can be quoted in support of a direct relationship between the striatal DA deficiency and the main extrapyramidal symptomatology of the Parkinsonian condition (Hornykiewicz, 1966, 1973; Bernheimer *et al.*, 1973). Besides this direct involvement in the production of the main Parkinsonian symptoms, particularly akinesia and rigidity, the dysfunction of the dopaminergic systems of the brain has two functionally consequential results. First, it leads, within the dopaminergic system itself, to changes in the state of activity of the still functioning DA neurones (presynaptic changes) as well as the state of reactivity of the DA sensitive receptor sites (postsynaptic changes). These alterations can be regarded as representing compensatory measures directed at correcting the dopaminergic dysfunction; they reflect the wide range of the brain's capacity for functional compensation. Secondly, probably as a consequence of the severe dysfunction of the brain's DA systems, there is in Parkinson's disease a derangement of neuronal systems utilizing putative neurotransmitters other than DA, notably noradrenaline (NA), serotonin (5-HT), acetylcholine (ACh) and γ-aminobutyric acid (GABA). Recent experimental evidence has indicated the existence, within the basal ganglia complex, of intricate neurotransmitter interactions; dysfunction of any of these inter-related systems has been observed to elicit a "chain reaction", altering the functional state of the other related neuronal systems. Parkinson's disease represents an ideal human model for the study of these neurotransmitter interactions in the basal ganglia (Hornykiewicz, 1976).

The relevant data concerning Parkinson's disease and brain DA have been summarized in Tables I–IV, and are discussed in the following subsections.

Dopamine

The concentration of DA in those extrapyramidal centres of the basal ganglia which normally contain appreciable amounts of this catecholamine, notably the caudate nucleus, putamen (both these constitute the striatum) and substantia nigra, are severely reduced in patients with Parkinson's disease (Ehringer and Hornykiewicz, 1960; Bernheimer *et al.*, 1963, 1973; Hornykiewicz, 1963; Fahn *et al.*, 1971; Rinne *et al.*, 1974; Lloyd *et al.*, 1975a) (Table I). The reduction of DA in these brain regions is independent of the aetiology of the given Parkinsonian condition: subnormal DA levels in the striatal nuclei have been measured in postencephalitic, idiopathic, senile-arteriosclerotic cases as well as a case of chronic manganese poisoning with Parkinsonian symptomatology (Bernheimer *et al.*, 1973) (Table II). In a case with Hemiparkinsonism, the DA decrease in the striatum contralateral to the side of the symptoms was more pronounced than ipsilaterally (Barolin *et al.*, 1964) (Table II). Whereas in post-encephalitic Parkinsonism both the caudate nucleus and putamen are severely depleted of DA, in idiopathic cases the DA loss in the caudate nucleus is usually distinctly less severe than in the putamen (cf. Table II).

In addition to the changes in the striatal nuclei, DA levels are also low in several limbic forebrain regions which normally contain higher amounts of the amine, notably the nucleus accumbens (septi), the lateral hypothalamus and the parolfactory gyrus (Farley *et al.*, 1977; Price *et al.*, 1978). However, other limbic regions, especially the (medial and lateral) olfactory areas are not affected in patients with Parkinson's disease. Significantly, the DA decrease in the nucleus accumbens was found to be similar in magnitude to the DA decrease measured in the caudate nucleus whereas the DA decrease in the parolfactory gyrus was of the same magnitude as that measured in the putamen (see Table I).

TABLE I Dopamine (DA), homovanillic acid (HVA), L-tyrosine hydroxylase (TH) and L-DOPA decarboxylase (DDC) in discrete brain regions of control and Parkinsonian patients[a]

Brain region	DA	HVA	TH[b]	DDC[b]
Caudate nucleus[b]				
Controls	4.06 ± 0.47 (18)	2.92 ± 0.37 (19)	18.7 ± 2.0 (3)	364.0 ± 95.4 (19)
Parkinson	0.20 ± 0.19 (3)	1.19 ± 0.10 (3)	3.2 ± 0.5 (3)	54.3 ± 14.0 (13)
Putamen[b]				
Controls	5.06 ± 0.39 (17)	4.92 ± 0.32 (16)	17.4 ± 2.4 (3)	431.7 ± 109.1 (18)
Parkinson	0.14 ± 0.13 (3)	0.54 ± 0.13 (3)	3.1 ± 1.2 (3)	32.3 ± 7.1 (13)
Substantia nigra				
Controls	0.46 (13)[c]	2.32 (7)[d]	17.4 (1)	549.3 ± 293.7 (15)
Parkinson	0.07 (10)[c]	0.41 (9)[d]	6.1 ± 1.5 (3)	21.4 ± 6.0 (10)
Globus pallidus				
Controls	0.5 (6)[e]	2.25 (8)[d]	3.5 (1)	22.1 ± 2.7 (9)
Parkinson	0.2 (4)[e]	0.72 (9)[d]	1.7 ± 0.2 (3)	18.0 ± 2.6 (12)
Nucleus accumbens[f]				
Controls	3.79 ± 0.82 (8)	4.38 ± 0.64 (8)	—	—
Parkinson	1.61 ± 0.28 (4)	3.13 ± 0.13 (3)	—	—
Parolfactory gyrus[f]				
Controls	0.35 ± 0.09 (4)	0.98 (2)	—	—
Parkinson	< 0.03 (2)	—	—	—

[a] All results are expressed as means ± S.E.M. (where applicable) (no. of cases in parentheses); DA and HVA in $\mu g\, g^{-1}$ wet tissue; TH in nmol CO_2 $0.5\, h^{-1}$ per 100 mg protein; DDC in nmol CO_2 $2\, h^{-1}$ per 100 mg protein.

[b] From Lloyd *et al.*, 1975a.

[c] From Hornykiewicz, 1963.

[d] From Bernheimer and Hornykiewicz, 1965.

[e] From Ehringer and Hornykiewicz, 1960.

[f] From Price *et al.*, 1978.

TABLE II Relationship between striatal dopamine (DA) and homovanillic acid (HVA) in Parkinsonism and the aetiology, severity of symptoms and response to L-DOPA.

Diagnostic grouping[a]	Caudate DA	Caudate HVA	Putamen DA	Putamen HVA
	% of Control (=100)			
Aetiology[b]				
Senile-arteriosclerotic (5)	32.0	60	15.0	44
Idiopathic (13)	16.0	32	2.0	21
Chronic manganese poisoning (1)	3.5	—	1.2	—
Postencephalitic (6)	1.5	9	0.4	9
Severity of symptoms				
Hemiparkinsonism (right side symptoms) (1)[c]				
right striatum	38.0	—	20.0	—
left striatum	17.0	—	3.5	—
Mild akinesia (13/12)[b]	22.0	52	13.0	37
Marked akinesia (9/7)[b]	8.0	18	1.5	19
Response to L-DOPA (single i.v. test dose)[b]				
Poor response (10)	17.0	29	0.6	19
Good response (10)	1.0	4	0.3	5

[a] No. of cases in parentheses.
[b] From Bernheimer *et al.*, 1973.
[c] From Barolin *et al.*, 1964.

Homovanillic Acid

The concentration of homovanillic acid (HVA), DA's main metabolite in the human brain, is also reduced in all areas showing a DA loss, including the limbic structures (Bernheimer and Hornykiewicz, 1965; Bernheimer *et al.*, 1973; Rinne *et al.*, 1974; Lloyd *et al.*, 1975a; Price *et al.*, 1978) (Table I). In addition, there is also a corresponding reduction of HVA concentrations within the internal capsule, marking, in all probability, the course of the degenerated nigrostriatal DA pathway (Hornykiewicz *et al.*, 1968) (see below, p. 637). Significantly, the average reduction of HVA in the affected areas of the basal ganglia and limbic forebrain is distinctly less severe than that of DA (see Table I). Therefore, the molar ratio of DA to HVA is shifted in favour of the metabolite. This change is probably of considerable functional importance, considering the fact that the DA and HVA found in the Parkinsonian brain must be contained in the still remaining neurones. The shift, in Parkinson's disease, of the DA : HVA ratio in favour of HVA is comparable to a similar shift in animals treated with DA receptor blockers (phenothiazines, butyrophenones) (Andén *et al.*, 1964; Laverty and Sharman, 1965) or animals with experimental lesions of the substantia nigra (Sharman *et al.*, 1967).

Dopamine Synthesizing Enzymes

In analogy to the changes in DA and HVA levels, the activity of the two DA synthesizing enzymes L-tyrosine hydroxylase and L-DOPA decarboxylase, is reduced in the basal ganglia of patients with Parkinson's disease (Lloyd and Hornykiewicz, 1970; Lloyd *et al.*, 1975a; Nagatsu *et al.*, 1977). (No data seem to be available on the limbic structures which are affected in Parkinson's disease.) In one study (Lloyd *et al.*, 1975a), the reduction of L-tyrosine hydroxylase activity in the Parkinsonian striatum was found to be less severe than that of L-DOPA decarboxylase. This has been interpreted as suggesting an induction of the rate-limiting enzyme activity in the still functioning DA neurones. This possibility is consistent with the finding of a shifting of the ratio DA : HVA in favour of the latter, possibly indicating an increased synthesis and release (turnover) of DA in the remaining DA neurones (see below, p. 642).

Morphological-Biochemical Correlations

The most consistent morphological change in the brain of patients with Parkinson's disease is the degeneration of the melanin-containing neurones of the brainstem. In relation to forebrain DA, neuronal cell loss in the compact (melanin-containing) layer of the substantia nigra and the medially located cell groups (paranigral nucleus) is most important (Hassler, 1938). It is now a well-established fact that the neurones of the compact zone of the substantia nigra give rise to a DA pathway mainly innervating the caudate nucleus and the putamen (see Fuxe *et al.*, 1970). There is evidence that at least in the cat the nucleus accumbens also receives DA fibres from the substantia nigra (Moore *et al.*, 1971). In the rat, the main supply of DA fibres to the limbic forebrain stems, however, from an area in the mesencephalic ventromedial tegmentum just rostral and dorsal to the interpeduncular nucleus (area A10 in the Dahlström–Fuxe classification) (see Fuxe *et al.*, 1970); in the human brain, this area most probably is identical with the nucleus parangralis (+parabrachialis pigmentosus) (Olszewski–Baxter classification) (see Farley and Hornykiewicz, 1976).

In the substantia nigra, the neurones projecting to the striatum seem to be arranged topographically. This has been directly demonstrated in laboratory animals and it seems to hold for the human brain (see Bernheimer *et al.*, 1973). Thus, in the idiopathic variety of Parkinson's disease, the distinctly milder reduction of DA in the caudate nucleus as compared with the putamen (see Table II) is paralleled by the usually milder cell loss in the rostromedial portions of the substantia nigra (Hassler, 1938; Bernheimer *et al.*, 1973). From these observations the conclusion has to be drawn that in the human brain the rostromedial portions of the substantia nigra project to the caudate nucleus and the caudolateral portions to the putamen (Bernheimer *et al.*, 1973). This explains why in postencephalitic Parkinsonism the decrease in DA level is equally severe both in the caudate nucleus and putamen (see Table II); in this variety of Parkinsonism the substantia nigra suffers a diffuse cell loss, affecting both the rostral and caudal portions of the nucleus (Hassler, 1938; Bernheimer *et al.*, 1973).

There is no information available regarding the exact site of origin of the DA fibres innervating the human limbic brain. Judged from information obtained in the rat (see above) most of the limbic DA neurones would originate in the paranigral area; however, it has been mentioned, that in the cat the nucleus accumbens seems to receive a DA supply from the substantia nigra. The latter possibility may also apply to the human brain. This has been proposed from observations showing that in Parkinson's disease the limbic regions of nucleus accumbens and parolfactory gyrus suffer a reduction of their DA level similar in degree to the DA loss measured in the caudate nucleus and putamen respectively (Farley *et al.*, 1977; Price *et al.*, 1978).

In conclusion, the consistent cell loss, in Parkinson's disease, in the compact zone of the substantia nigra and, at least in the postencephalitic condition, the paranigral cell groups (Hassler, 1938), satisfactorily explains the reduction of DA, HVA and the synthesizing enzymes in the striatal nuclei as well as the limbic regions of the nucleus accumbens, lateral hypothalamus and parolfactory gyrus. That the degree of nigral cell loss directly determines the degree of striatal DA deficiency has been demonstrated in a combined morphological-biochemical study (Bernheimer *et al.*, 1973). Such a relationship has also been shown to exist in laboratory animals with experimental lesions of the nigrostriatal DA pathway (Poirier and Sourkes, 1965; Goldstein *et al.*, 1969). No data are available on a possible correlation between the morphological state of the substantia nigra and the paranigral area respectively, and the DA deficiency in the limbic forebrain, especially the nucleus accumbens.

Striatal and Limbic (Accumbens) Dopamine Deficiency and Parkinsonian Symptomatology

From a pharmacological point of view the striatal and accumbens DA systems seem to subserve a "kinetic" function in the brain. A large body of pharmacological evidence obtained in laboratory animals strongly suggests that striatal DA is critically involved in a variety of locomotor and stereotyped behaviours (see Cools, 1973). The available evidence for the role of striatal DA assumes special significance in view of observations suggesting a direct cause–effect relationship between the degree of the striatal DA deficiency and the main symptoms of Parkinson's disease, especially akinesia (Bernheimer *et al.*, 1973). Firstly, as mentioned above (see subsection A, and Table II), in Hemiparkinsonism, the neurochemical changes were found to be more pronounced in the striatum contralateral to the affected side of the body; secondly, the degree of striatal DA (and HVA) deficiency has been observed to be positively correlated with the severity of akinesia (Table II); and finally, L-DOPA, DA's immediate precursor substance, is highly efficacious in improving the main clinical features of Parkinson's disease, especially akinesia (Birkmayer and Hornykiewicz, 1961; Barbeau *et al.*, 1962; Cotzias *et al.*, 1967). Considered together with the wealth of pharmacological evidence bearing on DA's role in extrapyramidal motor function (see Hornykiewicz, 1972; Cools, 1973), it seems justified to conclude that the lack of DA in the striatal nuclei is of pathognomonic significance for Parkinson's disease.

Recently, it has been demonstrated that in analogy to the striatal DA, pharmacological manipulations of the accumbens DA (in the rat) also produce marked locomotor effects (Pijnenburg *et al.*, 1976; see also Costa and Gessa, 1977). In view of this evidence the possibility suggests itself that in addition to the role of striatal DA deficiency, the changes in limbic (especially accumbens) DA found in patients with Parkinson's disease may well be causally related to the main motor deficits, especially akinesia, typically observed in this disorder. In this respect, it is interesting to note that from a morphological point of view the nucleus accumbens can be classified as being part of the striatal complex (Ariëns Kappers and Theunissen, 1907).

In addition to these recently discovered motor functions, disturbance of the limbic forebrain mechanisms has been implicated in affective disorders as well as behavioural abnormalities characteristic of the schizophrenic illness (MacLean, 1970). It is possible that the disturbance of DA metabolism in some of the limbic forebrain areas, especially the parolfactory gyrus and the lateral hypothalamus, has something to do with the affective changes commonly seen in patients with Parkinson's disease.

L-DOPA's Metabolism in the Brain of Parkinsonian Patients—Relation to Therapeutic Efficacy

There is evidence to show that in patients with Parkinson's disease, L-DOPA's brain metabolism basically follows the pattern established in laboratory animals. Post-mortem brain analyses in chronically L-DOPA treated patients have shown that the DA and HVA levels in the striatum were several times those of the non-DOPA treated patients with Parkinson's disease (Davidson *et al.*, 1971; Rinne *et al.*, 1974; Lloyd *et al.*, 1975a) (Table III). Particularly noteworthy are the findings that (i) the DA levels were significantly correlated with the time the last dose of L-DOPA was given; from this the conclusion can be drawn that a major portion of this DA must have stemmed from the administered L-DOPA; and (ii) good responders to L-DOPA seemed to accumulate more DA in the striatal nuclei than patients with a poor response to L-DOPA (Lloyd *et al.*, 1975a) (Table III). In addition to DA, low levels of L-DOPA and higher levels of 3-*O*-methyldopa and HVA were measured in the striatum as well as in many extrastriatal brain regions (Davidson *et al.*, 1971; Rinne *et al.*, 1974; Lloyd *et al.*, 1975a). The basis for these metabolic transformations of L-DOPA in the brain of patients with Parkinson's disease is provided by the following biochemical findings: (i) enough L-DOPA decarboxylating activity remains in the Parkinsonian striatum to account for the formation of DA in this brain area (Lloyd and Hornykiewicz, 1970; Lloyd *et al.*, 1975a; Nagatsu *et al.*, 1977) (see Table I); (ii) in extrastriatal brain regions the L-DOPA decarboxylase activity does not change drastically in Parkinson's disease (Lloyd *et al.*, 1975a); and (iii) the DA catabolizing enzymes monoamine oxidase and catechol-*O*-methyltransferase have a widespread distribution in the (human) brain, with little alterations in Parkinson's disease (Lloyd *et al.*, 1975a; Nagatsu *et al.*, 1977). It is interesting that in spite of the elevated HVA levels in many extrastriatal brain regions, the DA levels in these areas were low (Lloyd *et al.*, 1975a). This suggests that the DA storing capacity of extrastriatal brain areas is very low, with the newly formed DA being quickly metabolized to HVA. This conclusion is supported by the observation that in a patient who died very shortly after the last dose of L-DOPA, DA was elevated in many extrastriatal brain regions (Rinne *et al.*, 1971).

The occurrence of only low levels of L-DOPA in the brain of L-DOPA-treated patients indicates that little storage of L-DOPA occurs in the human brain. In addition, the halflife of L-DOPA in the human (Parkinsonian) brain seems to be as short as in the brain of (normal) laboratory animals; thus, detectable L-DOPA levels were measured only in those patients who had received the last dose of L-DOPA within nine hours of death (Lloyd *et*

TABLE III L-DOPA metabolism in the brain of patients with Parkinson's disease—dependence of striatal dopamine (DA) and homovanillic acid (HVA) levels on time of L-DOPA administration—relation to therapeutic response.[a]

Brain region		Untreated	L-Dopa 2–9 hrs before death	L-Dopa 10–72 hrs before death	Response to L-dopa good	poor
Caudate nucleus	DA	0.20±0.19 (3)	1.57±0.78 (5)	0.94±0.32 (6)	1.85 (3)	0.22 (3)
	HVA	1.19±0.10 (3)	7.67±2.48 (5)	1.58±0.18 (6)	—	—
	Dopa	<0.02 (3)	0.53±0.09 (5)	0.04±0.04 (6)	0.50 (3)	0.25 (3)
	3-*O*-Methyldopa	<0.02 (3)	2.76±1.14 (5)	0.87±0.87 (6)	4.65 (3)	1.28 (3)
Putamen	DA	0.14±0.13 (3)	1.68±0.68 (5)	0.15±0.13 (6)	1.16 (3)	0.05 (3)
	HVA	0.54±0.13 (3)	10.19±3.17 (5)	2.17±0.96 (5)	—	—
	Dopa	<0.02 (3)	0.58±0.14 (5)	0.09±0.06 (6)	0.62 (3)	0.23 (3)
	3-*O*-Methyldopa	<0.02 (3)	3.85±1.78 (5)	1.06±1.06 (6)	6.19 (3)	1.70 (3)
Temporal cortex	DA	0.05 (3)	0.05 (3)	0.05 (5)	—	—
	HVA	0.05 (3)	3.60±0.52 (3)	0.48±0.18 (6)	—	—
	Dopa	<0.02 (3)	0.45±0.14 (4)	0.06±0.04 (5)	—	—
	3-*O*-Methyldopa	<0.02 (3)	3.08±1.83 (4)	0.87±0.87 (5)	—	—

[a] Results are expressed (in μg/g wet tissue) as mean ± S.E.M. (no. of cases in parentheses). Data taken from : Lloyd *et al.*, 1975a.

al., 1975a). The presence, in the brain of L-DOPA treated patients, of 3-*O*-methyldopa in concentrations several times those of its parent compound is consistent with observations made in laboratory animals (Bartholini *et al.*, 1970). The suggestion has been made that the 3-*O*-methyldopa pool formed in the brain from L-DOPA may serve as a possible depot of precursor for further DA formation; this suggestion is based on the intriguing observation that, in the rat brain, 3-*O*-methyldopa is slowly demethylated to L-DOPA (Bartholini *et al.*, 1974).

Besides being decarboxylated to DA, L-DOPA may, to some degree, be transformed to complex tetrahydroisoquinoline and tetrahydropapaveroline derivatives. This has been shown to take place in vivo in the rat (Turner *et al.*, 1974; Coscia *et al.*, 1977), and can be expected also to occur in the body of L-DOPA-treated patients (Sandler *et al.*, 1973). The pharmacological spectrum of activity of these complex L-DOPA condensation products ranges from weak indirect sympathomimetic actions to weak sympatholytic properties (Hjort *et al.*, 1938; Fasset and Hjort, 1938; Holtz *et al.*, 1964). Thus, accumulation of high enough amounts of these compounds in the organs (including the brain) of Parkinsonian patients could potentially influence the spectrum of activity of L-DOPA as a DA-replenishing drug. However, there is no evidence to support speculations that the main antiparkinson activity of L-DOPA may be due to the formation of these derivatives (see Hornykiewicz, 1974).

Functional Consequences of the Striatal Dopamine Deficiency

Overactivity of the remaining DA neurones

The above-mentioned observation (see p. 636) of a shifting of the ratio DA:HVA in the striatum in favour of the metabolite has been interpreted (Hornykiewicz, 1966; Bernheimer *et al.*, 1973) as indicating an increased DA turnover, i.e. synthesis and release, as well as a corresponding decrease in the actual storage capacity in the functionally intact nigrostriatal DA neurones. It is possible that this observation is directly related to the finding (in one of two studies) that the L-tyrosine hydroxylase activity is less severely reduced in the Parkinsonian striatum than the activity of L-DOPA decarboxylase (see above, p. 635). These observations seem to support the possibility that in Parkinson's disease the loss of a large portion of the dopaminergic innervation of the striatum triggers a compensatory activation of the remaining neurons (Hornykiewicz, 1966; Bernheimer *et al.*, 1973). An analogous conclusion has been reached from experiments in rats with partial (6-OH-DA induced) lesions of the nigrostriatal pathway (Agid *et al.*, 1973). From a pathophysiological point of view, the formation, in L-DOPA-treated patients, of increased amounts of DA in, and increased release from, the remaining overactive neurones may play an important role for L-DOPA's therapeutic efficacy (see below, p. 645).

Supersensitivity of the dopaminergic receptors

Indirect clinicopharmacological evidence has suggested that in Parkinson's disease the striatum develops a supersensitivity to DA analogous to the well-known phenomenon of denervation supersensitivity. First, patients with more severe akinesia, and correspondingly higher degrees of striatal DA decrease, have been observed to respond more sensitively to an i.v. test dose of L-DOPA than patients with mild akinesia (Bernheimer *et al.*, 1973) (see Table II); secondly, it seems that only in Parkinsonian subjects do therapeutic doses of (chronically administered) L-DOPA produce dyskinetic (hyperkinetic) side effects; these L-DOPA induced dyskinesias in all probability are striatal in origin and mediated by DA (see below, p. 648).

Recently, direct evidence has been obtained which supports the notion of a dopaminergic supersensitivity in the caudate nucleus and putamen in Parkinson's disease. Using the specific binding of ^{3}H-haloperidol and ^{3}H-apomorphine to tag the DA receptors in the brain, it has been shown that in the Parkinsonian caudate nucleus and putamen (i) ^{3}H-apomorphine binding was significantly reduced, and (ii) ^{3}H-haloperidol binding was significantly increased (Lee *et al.*, 1978). Since ^{3}H-apomorphine seems to

TABLE IV Noradrenaline (NA), serotonin (5-HT), glutamic acid decarboxylase (GAD) and choline acetyltransferase (CAT) in discrete brain regions of control and Parkinsonism patients[a].

Brain region	NA	5-HT[e]	GAD[f]	CAT[g]
Caudate nucleus				
Controls	0.09 ± 0.01 (11)[b]	0.33 (6)	659 ± 93 (13)	1460 (3)
Parkinson	0.04 (6)[c]	0.12 (5)	321 (2)	694 (3)
Putamen				
Controls	0.10 ± 0.01 (12)[b]	0.32 (6)	622 ± 110 (13)	1656 (3)
Parkinson	0.05 (6)[c]	0.14 (5)	292 (2)	780 (3)
Substantia nigra				
Controls	0.23 ± 0.05 (9)[b]	0.55 (6)	637 ± 149 (13)	63 (6)
Parkinson	0.11 ± 0.03 (3)[b]	0.26 (5)	263 (2)	12 (3)
Globus pallidus				
Controls	0.06 ± 0.02 (13)[d]	0.23 (6)	553 ± 15 (13)	231 (2)
Parkinson	0.05 ± 0.03 (5)[d]	0.13 (5)	388 (2)	39 (3)
Nucleus accumbens				
Controls	1.29 ± 0.14 (7)[b]	—	—	—
Parkinson	0.52 ± 0.11 (4)[b]	—	—	—
Hypothalamus				
Controls	1.83 ± 0.25 (4)[b]	0.29 (6)	—	—
Parkinson	0.33 ± 0.09 (4)[b]	0.12 (5)	—	—

[a] All results are expressed as means $\pm$S.E.M. (where applicable) (no. of cases in parentheses); NA and 5-HT in $\mu g\ g^{-1}$ wet tissue; GAD in nmol $CO_2\ h^{-1}$ per 100 mg protein; CAT in pmol min^{-1} per mg protein.

[b] From Farley and Hornykiewicz, 1976.

[c] From Ehringer and Hornykiewicz, 1960.

[d] From Rinne *et al.*, 1974.

[e] From Bernheimer *et al.*, 1961.

[f] From Lloyd and Hornykiewicz, 1973.

[g] From Lloyd *et al.*, 1975b.

bind preferentially to presynaptic DA receptors (Nagy *et al.*, 1978) and ^{3}H-haloperidol to postsynaptic DA receptors (Creese *et al.*, 1977), the above results indicate a decrease in presynaptic DA receptors (corresponding to the severe loss of the "presynaptic" DA neurones) and an increased number (or increased affinity for DA) of the postsynaptic DA receptors.

In conclusion, it would appear that in addition to the compensatory overactivity of the remaining DA neurones, the striatum develops another compensatory mechanism, namely the supersensitivity of the specific postsynaptic receptor sites to DA, thus providing for a maximal degree of utilization of the endogenous DA as well as the DA produced from exogenous L-DOPA.

Changes in other putative neurotransmitter systems

In Parkinson's disease, the morphologically determined striatal DA deficiency is accompanied by a decreased level of 5-HT (Bernheimer *et al.*, 1961; Rinne *et al.*, 1974) and the GABA-synthesizing enzyme glutamic acid decarboxylase (GAD) in the nigrostriatopallidal complex (Bernheimer and Hornykiewicz, 1962; McGeer *et al.*, 1971; Lloyd and Hornykiewicz, 1973; Rinne *et al.*, 1974) as well as NA in the nucleus accumbens (and in several other limbic-hypothalamic regions) (Farley and Hornykiewicz, 1976). The number (or affinity) of specific GABA binding sites (GABA receptors) in the substantia nigra (and hippocampus) also seems to be reduced in Parkinson's disease (Lloyd *et al.*, 1977). In addition, there is probably present a functional overactivity of the striatal cholinergic mechanisms (although the activity of the ACh-synthesizing choline acetyltransferase has been found to be either unchanged (McGeer *et al.*, 1973) or even subnormal (Lloyd *et al.*, 1975b) in the Parkinsonian striatum); this can be inferred from animal studies which show that normally the nigrostriatal DA exerts an inhibitory influence on the striatal ACh neurones (see Bartholini *et al.*, 1975). The functional significance of the decrease in 5-HT and GAD may be that normally both these systems exert an inhibitory influence on motor behaviour, probably by inhibiting the dopaminergic system at the level of the substantia nigra (Grabowska and Michaluk, 1974; Maj *et al.*, 1974; Hollister *et al.*, 1975; Andén, 1974; Lahty and Losey, 1974; Tarsy *et al.*, 1975). Accordingly, in Parkinson's disease, both 5-HT and GABA systems, when activated, may aggravate the condition by further diminishing the already reduced dopaminergic influence on the striatum. Thus, a functional deactivation of these chemical influences on the affected striatum may be an important compensatory mechanism in this respect. In contrast, the functional consequence of the decrease in brain NA and the probable striatal cholinergic overactivity may be an aggravation of the extrapyramidal symptomatology, especially akinesia and rigidity. This is suggested by animal experiments demonstrating that (i) brain NA is synergistic with brain DA in respect to locomotor activation (e.g. by dopaminergic drugs) (for references and discussion, see Lloyd and Hornykiewicz, 1975), and (ii) striatal cholinergic hyperactivity definitely results in an akinetic-rigid syndrome and worsens the corresponding symptoms in patients with Parkinson's disease (Duvoisin, 1967).

It is noteworthy in this context that apart from increasing the striatal DA levels, chronic L-DOPA therapy seems to normalize the subnormal GAD activity in the basal ganglia (Lloyd and Hornykiewicz, 1973). Since L-DOPA is converted (via DA) to NA and also has been found to release brain 5-HT from its stores (Bartholini *et al.*, 1968), an activation of brain noradrenergic and serotonergic mechanisms can be expected to take place in L-DOPA treated patients. Further, L-DOPA probably reduces the cholinergic overactivity in the striatum by reinstating the inhibitory dopaminergic influence on the striatal cholinergic neurones.

L-DOPA Therapy—Mechanisms and Limitations

At present, L-DOPA is the most efficacious drug in the chemotherapy of Parkinson's disease. There is little doubt that this is so because L-DOPA functionally replaces the missing DA in the Parkinsonian brain. All the evidence discussed in the preceding sections is consistent with this thesis.

Three factors seem to determine L-DOPA's efficacy as an antiparkinson drug. These are: (i) preservation of a minimum number of DA (and possibly 5-HT) terminals in the striatum containing the enzymatic machinery for the transformation of L-DOPA to DA; (ii) the compensatory overactivity of these remaining nigrostriatal DA neurones; and (iii) the development of supersensitivity to DA of the striatal postsynaptic DA receptors. With respect to point (i), the high degree of divergence of the dopaminergic innervation of the striatum (Andén, 1966) safeguards, up to a certain critical point, the basal ganglia against the consequences of loss of nigral DA neurones. It seems that these three factors are responsible for the ability of the Parkinsonian striatum to compensate, over longer periods of time, functionally lesser degrees of DA deficiency. This probably explains the observation apparent from Table II that even mild, clinically just detectable degrees of akinesia were associated with disproportionately high degrees of striatal DA deficiency (80% DA decrease) (Bernheimer *et al.*, 1973). Thus, from the neurochemical point of view, clinically manifested Parkinsonism would appear to represent the second, decompensated stage of a progressive deterioration of striatal DA mechanisms. From this, it seems reasonable to postulate that the main pharmacological feature of L-DOPA as a specific, though predominantly symptomatically acting antiparkinson drug is its ability to revert the decompensated stage of Parkinson's disease to that of functional recompensation (Hornykiewicz, 1974).

The crucial role that these compensatory mechanisms probably play in determining the efficacy of L-DOPA in Parkinson's disease clearly shows the limitations of successful chronic L-DOPA medication. It is obvious that in untreated patients the degree of the overactivity of the remaining DA neurones as well as supersensitivity of the DA receptors will directly be determined by the magnitude of loss of nigral DA neurones. The obvious consequence of increased striatal DA levels during a successful chronic L-DOPA administration will be a gradual reduction, and eventual loss, of these compensatory mechanisms. In this respect, a loss of the postsynaptic supersensitivity of the striatal DA receptor sites in L-DOPA treated patients has recently been directly shown by means of the specific assay of ^{3}H-haloperidol binding to striatal membrane preparations. In contrast to the increased ^{3}H-haloperidol binding in striatal homogenates of untreated patients (see above, p. 642), this measure of postsynaptic receptor sensitivity was found to be within normal limits in L-DOPA-treated patients (Lee *et al.*, 1978). It is evident that with the nigral DA neurones progressively being reduced in number in the course of the disease process, the L-DOPA-induced abolition of the compensatory mechanisms in the diseased striatum will eventually result in a substantial reduction of the drug's effectiveness. It has been suggested that at this stage a temporary discontinuation of L-DOPA may restore the compensatory overactivity of the DA neurones as well as the supersensitivity of the DA receptors, thus rendering the patients again sensitive to L-DOPA's antiparkinson effects (Lee *et al.*, 1978).

NEUROLOGICAL CONDITIONS OTHER THAN PARKINSON'S DISEASE

Apart from Parkinson's disease, there is at present no other well-defined brain disease known to involve clear-cut alterations of brain DA metabolism. Based on mostly indirect, clinicopharmacological evidence, brain DA has been suspected to participate in several neurological conditions, such as Huntington's chorea and drug-induced dyskinesias.

Huntingdon's Chorea

Clinicopharmacologically, Huntington's chorea displays several features suggesting a dopaminergic hyperactivity in the basal ganglia. Thus, drugs which either decrease the brain DA levels (e.g. reserpine, tetrabenazine, α-methyl-*p*-tyrosine) or block (postsynaptic) DA receptors (e.g. phenothiazines and butyrophenones) ameliorate the hyperkinetic movements; on the other hand, L-DOPA worsens the choreatic condition and may provoke it in patients at risk (for reference see Barbeau, 1973).

In view of our present knowledge regarding neurohumoral interactions in the basal ganglia, a hyperactivity of the nigrostriatal DA neurones would not be unexpected to occur in Huntington's chorea. Recent neurophysiological studies have demonstrated that the activity of the nigrostriatal DA neurones is under the control of an inhibitory GABAergic striatonigral (feedback) pathway (Precht and Yoshida, 1971). Since in Huntington's chorea the prominent neurochemical alteration is (as far as we know) a marked loss of GABA (and ACh) neurones in the basal ganglia (Perry *et al.*, 1973; Bird and Iversen, 1974), a defect in the striatonigral GABA pathway may be expected to release the nigrostriatal DA neurones from the GABAergic inhibition. However, studies of DA metabolism did not reveal any significantly increased striatal DA turnover (as judged from the ratio DA:HVA) in patients with Huntington's chorea (Bernheimer and Hornykiewicz, 1973; Bird, 1976). Although this negative result may appear disappointing, it does not preclude the participation of striatal DA mechanisms in the choreatic condition. Since in this disorder there is a severe loss of striatal neurones, the remaining neurones may find themselves under a relatively increased dopaminergic influence. This would be equivalent to relative predominance of striatal DA, a factor which may contribute to the choreatic (hyperkinetic) condition. This possibility is consistent with the exacerbation of symptoms by L-DOPA and the (limited) beneficial effect of antidopaminergic agents. (The fact that loss of the striatonigral GABAergic inhibition probably does not result in increased actity of the nigrostriatal DA neurones is not necessarily in contradiction to the above-mentioned neurophysiological observation. Since Huntington's chorea is a slowly progressing, chronic disease, an initial increase of dopaminergic activity would be expected to activate negative feedback control mechanisms reducing the overactivity to normal levels. However, admittedly nothing is known about the anatomical and/or functional intactness of such regulatory mechanisms in the brain of patients with Huntington's chorea.)

Drug-Induced Dyskinesias

Drug-induced Parkinsonism

Drug-induced Parkinsonism may be caused by two types of drugs: compounds that deplete the brain of monoamines (e.g. reserpine, α-methyl-*p*-tyrosine) and compounds that block monoamine receptors (e.g. neuroleptics of the phenothiazine and butyrophenone type) (see Hornykiewicz, 1975). There apparently are no published reports on chemical changes in the brain accompanying Parkinsonism produced by these drugs. In contrast, their actions on brain monoamines and/or their specific receptors have been amply documented in laboratory animals (for references and discussion, see Lloyd and Hornykiewicz, 1975). From the animal studies, and in conjunction with what is known regarding the neurochemical pathology of Parkinson's disease (see above, p. 634), the conclusion can be drawn that the common denominator for the Parkinsonian condition produced by these drugs (a condition which sometimes is indistinguishable clinically from the genuine syndrome) is either their ability to deplete the brain of DA or block the brain DA receptors; the result being either an actual striatal DA deficiency (reserpine, α-methyl-*p*-tyrosine) or a functional lack of DA at the specific receptor sites (phenothiazines, butyrophenones). Thus, it seems that the phenomenon of drug-induced Parkinsonism finds its full explanation within the framework of the "Striatal DA Deficiency" hypothesis of Parkinson's disease.

Drug-induced hyperkinetic conditions

L-DOPA-induced hyper(dys-)kinesias

In addition to its beneficial effects, L-DOPA may induce, usually after several months of therapy, in many patients with Parkinson's disease (40–60%) an array of hyperkinetic symptoms; these most frequently consist of oral-buccal-facial dyskinesias (grimacing, tongue protrusion, lip smacking, chewing, head rolling) as well as choreoathetotic movements of the extremities (see Barbeau, 1969). The following evidence seems to suggest that these symptoms may be due to overstimulation of the DA receptors (which may be supersensitive to DA) in the striatum of the treated patients: (i) The dyskinesias disappear when the dose of L-DOPA is decreased (see Calne, 1970; Barbeau, 1969); (ii) usually they may be reversed by DA-receptor blocking agents (see Barbeau, 1969; Postma, 1972); (iii) their reversal is frequently followed by worsening of the Parkinsonian condition (see Barbeau, 1969); (iv) in Hemiparkinsonian patients the dyskinesias are more severe on the affected side of the body (Duvoisin *et al.*, 1969; Klawans, 1973); (v) there appears to be a certain parallelism between the beneficial responses to L-DOPA therapy and the severity of the resultant dyskinesias (Markham, 1971; Mones *et al.*, 1971); and (vi) L-DOPA-induced dyskinesias in many respects resemble the stereotyped behaviour produced by high doses of amphetamine (Rubovits *et al.*, 1973).

This clinicopharmacological evidence obtained in chronically L-DOPA treated Parkinsonian patients is supplemented by observations made in laboratory animals showing that both naive animals (monkey, rats) (Sassin *et al.*, 1972; Bieger and Hornykiewicz, unpublished observations) and animals with specific lesions of the nigrostriatal tract (Goldstein *et al.*, 1973) will develop dyskinesias upon administration of single (acute) doses of L-DOPA. Thus, in monkeys with tegmental lesions, high doses of L-DOPA or DA receptor-stimulating agents (piribedil) resulted in a dose-dependent oral-buccal dyskinetic syndrome and chorea while simultaneously producing relief of the experimental Parkinsonism (mainly tremor). These dyskinetic effects were blocked by haloperidol (Goldstein *et al.*, 1973). Also, in cats, unilateral injection of L-DOPA, DA, methoxytyramine, or *d*-amphetamine into the rostromedial caudate nucleus resulted in a choreoathetosis-like syndrome (Cools, 1973).

Thus, it would appear that on the whole a good case can be made for the possibility that the L-DOPA-induced dyskinesias are indeed due to an overstimulation of striatal (and/or limbic) DA receptors. However, since in the body, L-DOPA may be converted to NA and some "minor" complex metabolites (tetrahydroisoquinolines, tetrahydropapaverolines), participation of these or similar metabolites should not completely be overlooked. This is especially so in view of the fact that in general the L-DOPA dyskinesias manifest themselves rather late during chronic L-DOPA therapy.

Neuroleptic-induced tardive hyper(dys-)kinesias

Chronic use of neuroleptic drugs with antipsychotic activity (e.g. phenothiazines, butyrophenones) frequently results in practically irreversible chorea-like hyper(dys-)kinetic symptoms, commonly termed tardive dyskinesias (see Crane, 1972). The pathophysiological explanation for the occurrence of tardive dyskinesias is that they are part of the brain's homeostatic response to the Parkinsonism usually induced by the neuroleptic drugs in the early stages of chronic treatment. It has already been mentioned that all Parkinsonism-inducing neuroleptic drugs possess a high degree of specificity in blocking central DA receptors.

Two responses elicited by the neuroleptic-induced blockade of DA receptors appear particularly important in relation to the concept of tardive dyskinesias as a homeostatic response to the Parkinsonism-inducing activity of these drugs: (i) blockade of striatal DA receptors (both pre- and postsynaptic) results in an increase in the synthesis and release of the amine (Nybäck and Sedvall, 1970; Carlsson, 1975); and (ii) as a consequence of prolonged blockade, the (postsynaptic) striatal DA receptors may become supersensitive to the physiological actions of the amine ("disuse" type of supersensitivity) (Ungerstedt, 1971); this has recently been directly demonstrated by increased ^{3}H-haloperidol binding in striatal tissue from rats treated chronically with neuroleptic drugs (Burt *et al.*, 1977). The combination of these two effects on the striatal DA system could indeed account for the hyperkinetic side effects of chronic neuroleptic treatment; experiments in laboratory animals have demonstrated that increased brain (striatal) DA levels are consistently accompanied by locomotor and stereotyped hyperkinesias (see Hornykiewicz, 1973). The observation that the dyskinesias usually become worse on discontinuation of the neuroleptic drug may be due to the possibility that the neuroleptic-induced activation of the presynaptic DA turnover (presynaptic DA receptor blockade?) may outlast the postsynaptic receptor blockade. Thus, increasing the dose of the neuroleptic usually suppresses the dyskinetic condition (Crane, 1972).

The hypothesis linking the phenomenon of tardive dyskinesias to a neuroleptic-induced hyperactivity and hyper-reactivity of the striatal DA system is indirectly supported by the following clinicopharmacological observations: (i) the incidence of dyskinesias produced by a given neuroleptic drug is paralleled by its Parkinsonism-inducing potency in man and DA receptor blocking potency in animals (Klawans, 1973); (ii) patients who initially more readily develop Parkinsonian side effects are more likely to become dyskinetic later in the course of the treatment (Crane, 1972; Klawans, 1973); (iii) there is a striking resemblance between the L-DOPA-induced dyskinesias in Parkinsonian patients and the neuroleptic-induced tardive dyskinesias; in addition, L-DOPA acutely worsens the dyskinetic condition produced by chronic neuroleptic treatment (Calne and Reid, 1972).

In addition to their receptor-blocking activity, the neuroleptics (of the phenothiazine type) have the ability to block catecholamine uptake into the presynaptic terminals (Axelrod *et al.*, 1961), thus blocking one of the most important inactivating mechanisms for synaptically released transmitter. This action of neuroleptics requires higher doses of the drug than that required for blockade of receptors. Thus, in the patients, the blockade of DA re-uptake would be expected to occur after a prolonged period of high-dose treatment, permitting a cumulative effect. In conjunction with the neuroleptic-induced increase in DA turnover, this mechanism could significantly contribute to reversal of the blockade of the postsynaptic DA receptor sites. It should be noted in this respect that neuroleptic drugs (especially those of the phenothiazine series) are effective blockers of the DA-induced activation of striatal adenylate cyclase (Clement-Cormier *et al.*, 1974). Most significantly, this blockade of the enzyme can be reversed by increasing the concentration of DA. (DA-sensitive adenylate cyclase in the striatum has been postulated as representing the immediate receptor for the synaptically released DA (Kebabian *et al.*, 1972).) Recently a supersensitivity to DA of this DA-specific striatal enzyme has been demonstrated in rats treated subacutely with haloperidol (Gnegy *et al.*, 1977).

The above evidence seems to justify the conclusion that striatal dopaminergic mechanisms may play an important role in the development of dyskinetic-hyperkinetic reactions during the course of chronic neuroleptic treatment. That this may not be the full explanation of this phenomenon is suggested by the observations that (a) in chronically treated laboratory animals the striatal dopaminergic mechanisms usually become insensitive over time (tolerant) to the turnover-stimulating activity of neuroleptics (Asper *et al.*, 1973; Scatton, 1977); and (b) accepting the existence of truly irreversible tardive dyskinesias, the underlying pathophysiological mechanisms in such cases can hardly be purely functional–biochemical in nature.

REFERENCES

Agid, Y., Javoy, F. and Glowinski, J. (1973). *Nature (London)* **245,** 150–151.
Andén, N.-E. (1974). *Arch. Pharmacol.* **283,** 419–424.
Andén, N.-E., Roos, B.-E. and Werdinius, B. (1964). *Life Sci.* **3,** 149–158.
Andén, N.-E., Dahlström, A., Fuxe, K., Larsson, K., Olson, L. and Ungerstedt, U. (1966). *Acta Physiol. Scand.* **67,** 313–326.
Ariëns Kappers, C. U. and Theunissen, W. F. (1907). *Folia Neurobiol. (Leipzig)* **1,** 173–288.
Asper, H., Baggiolini, M., Bürki, H. R., Lauener, H., Ruch, W. and Stille, G. (1973). *Eur. J. Pharmacol.* **22,** 287–294.
Axelrod, J., Whitby, C. G. and Hertting, G. (1961). *Science* **133,** 383–384.
Barbeau, A. (1969). *Can. Med. Assoc. J.* **101,** 791–800.
Barbeau, A. (1973). *In* "Huntington's Chorea" (Eds. A. Barbeau, T. N. Chase and G. . Paulson), 473–516. Advances in Neurology, Vol. 1. Raven Press, New York.
Barbeau, A., Sourkes, T. L. and Murphy, G. (1962). *In* "Monoamines et Système Nerveaux Central" (Ed. J. de Ajuriaguerra), 247–262. Georg, Génève and Masson, Paris.
Barolin, G. S., Bernheimer, H. and Hornykiewicz, O. (1964). *Schweiz. Arch. Neurol. Psychiat.* **94,** 241–248.
Bartholini, G., DaPrada, M. and Pletscher, A. (1968). *J. Pharm. Pharmacol.* **20,** 228–229.
Bartholini, G., Kuruma, I. and Pletscher, A. (1970). *Br. J. Pharmacol.* **40,** 461–467.
Bartholini, G., Lloyd, K. G. and Pletscher, A. (1974). *Life Sci.* **14,** 323–328.
Bartholini, G., Stadler, H. and Lloyd, K. G. (1975). *In* "Cholinergic Mechanisms" (Ed. P. G. Waser), 411–418. Raven Press, New York.
Bernheimer, H. and Hornykiewicz, O. (1962). *Arch. Expl. Pathol. Pharmak.* **243,** 295–296.
Bernheimer, H. and Hornykiewicz, O. (1965). *Kiln. Wschr.* **43,** 711–715.
Bernheimer, H. and Hornykiewicz, O. (1973). *In* "Huntington's Chorea" (Eds. A. Barbeau, T. N. Chase and G. W. Paulson), 525–531. Advances in Neurology, Vol. 1. Raven Press, New York.
Bernheimer, H., Birkmayer, W. and Hornykiewicz, O. (1961). *Klin. Wschr.* **39,** 1056–1059.
Bernheimer, H., Birkmayer, W. and Hornykiewicz, O. (1963). *Klin. Wschr.* **41,** 465–469.

Bernheimer, H., Birkmayer, W., Hornykiewicz, O., Jellinger, K. and Seitelberger, F. (1973). *J. Neurol. Sci.* **20,** 415–455.

Bird, E. D. (1976). *In* "Biochemistry and Neurology" (Eds. H. F. Bradford and C. D. Marsden), 83–92. Academic Press, London and New York.

Bird, E. D. and Iversen, L. L. (1974). *Brain* **97,** 457–472.

Birkmayer, W. and Hornykiewicz, O. (1961). *Wien. Klin. Wschr.* **73,** 787–788.

Burt, D. R., Creese, I. and Snyder, S. H. (1977). *Science* **196,** 326–328.

Calne, D. B. (1970). "Parkinsonism". Arnold, London.

Calne, D. B. and Reid, J. L. (1972). *Drugs* **4,** 49–74.

Carlsson, A. (1975). *In* "Pre- and Postsynaptic Receptors" (Eds. E. Usdin and W. E. Bunney), 49–63. M. Dekker, New York.

Clement-Cormier, Y. C., Kebabian, J. W., Petzold, G. L. and Greengard, P. (1974). *Proc. Natl. Acad. Sci., U.S.A.* **71,** 1113–1117.

Cools, A. R. (1973). "The Caudate Nucleus and Neurochemical Control of Behavior". Brakkenstein, Nijmegen.

Coscia, C. J., Burke, W., Jamroz, G., Lasala, J. M., McFarlane, J., Mitchell, J., O'Toole, M. M. and Wilson, M. L. (1977). *Nature (London)* **269,** 617–619.

Costa, E. and Gessa, G. L. (Eds.) (1977). "Nonstriatal Dopaminergic Neurons" Advances in Biochemical Psychopharmacology, Vol. 16. Raven Press, New York.

Cotzias, G. C., Van Woert, M. H. and Schiffer, L. M. (1967). *New Eng. J. Med.* **276,** 374–379.

Crane, G. E. (1972). *Arch. Neurol.* **27,** 426–430.

Creese, I., Burt, D. R. and Snyder, S. H. (1977). *Science* **197,** 596–598.

Davidson, L., Lloyd, K. G., Dankova, J. and Hornykiewicz, O. (1971). *Experientia* **27,** 1048–1049.

Duvoisin, R. C. (1967). *Arch. Neurol.* **17,** 124–136.

Duvoisin, R., Barrett, R., Schear, M., Hoehn, M. Yahr, M. (1969). *In* "Third Symposium on Parkinson's Disease" (Eds. F. J. Gillingham and I. M. L. Donaldson), 185–192. Livingstone, Edinburgh.

Ehringer, H. and Hornykiewicz, O. (1960). *Klin. Wschr.* **38,** 1236–1239.

Fahn, S., Libsch, L. R. and Cutler, R. W. (1971). *J. Neurol. Sci.* **14,** 427–455.

Farley, I. J. and Hornykiewicz, O. (1976). *In* "Advances in Parkinsonism" (Eds. W. Birkmayer and O. Hornykiewicz), 178–185. Editiones Roche, Basel.

Farley, I. J., Price, K. S. and Hornykiewicz, O. (1977). *In* "Nonstriatal Dopaminergic Neurons" (Eds. E. Costa and G. L. Gessa), 57–64. Advances in Biochemical Psychopharmacology, Vol. 16. Raven Press, New York.

Fassett, D. W. and Hjort, A. M. (1938). *J. Pharmacol. Exp. Ther.* **63,** 253–271.

Fuxe, K., Hökfelt, T. and Ungerstedt, U. (1970). *In* "International Review of Neurobiology" (Eds. C. C. Pfeiffer and J. R. Smythies), Vol. 13, 93–126. Academic Press, New York and London.

Gnegy, M., Uzunov, P. and Costa, E. (1977). *J. Pharmacol. Exp. Ther.* **202,** 558–564.

Goldstein, M., Anagnoste, B., Battista, A. F., Owen, W. S. and Nakatani, S. (1969). *J. Neurochem.* **16,** 645–653.

Goldstein, M., Battista, A. F., Ohmoto, T., Anagnoste, B. and Fuxe, K. (1973). *Science* **179,** 816–817.

Grabowska, M. and Michaluk, J. (1974). *Pharmacol. Biochem. Behav.* **2,** 263–266.

Hassler, R. (1938). *J. Psychol. Neurol. (Leipzig)* **48,** 387–476.

Hjort, A. M., deBeer, E. J. and Fassett, D. W. (1938). *J. Pharmacol. Exp. Ther.* **62,** 165–173.

Hjort, A. M., deBeer, E. J., Buck, J. S. and Randall, L. O. (1942). *J. Pharmacol.* **76,** 263–269.
Hollister, A. S., Breese, G. R., Kuhn, C. M., Cooper, B. R. and Shanberg, S. M. (1975). *Neurosci. Abstr.* **1,** 297.
Holtz, P., Stock, K. and Westermann, E. (1964). *Arch. Expl. Path. Pharmak.* **248,** 387–405.
Hornykiewicz, O. (1963). *Wien. klin. Wschr.* **75,** 309–312.
Hornykiewicz, O. (1966). *Pharmacol. Rev.* **18,** 925–962.
Hornykiewicz, O. (1972). *In* "The Structure and Function of Nervous Tissue" (Ed. B. H. Bourne) Vol. VI, 367–415. Academic Press, New York and London.
Hornykiewicz, O. (1973). *Fed. Proc.* **32,** 183–190.
Hornykiewicz, O. (1974). *Life Sci.* **15,** 1249–1259.
Hornykiewicz, O. (1975). *In* "Dopaminergic Mechanisms" (Eds. D. B. Calne, T. N. Chase and A. Barbeau), 155–164. Advances in Neurology, Vol. 9. Raven Press, New York.
Hornykiewicz, O. (1976). *In* "The Basal Ganglia" (Ed. M. D. Yahr), 269–278. Association for Research into Nervous and Mental Diseases, Research Publications, Vol. 55. Raven Press, New York.
Hornykiewicz, O., Lisch, H.-J. and Springer, A. (1968). *Brain Res.* **11,** 662–671.
Kebabian, J. W., Petzold, G. L. and Greengard, P. (1972). *Proc. Natl. Acad. Sci., U.S.A.* **69,** 2145–2149.
Klawans, H. L. (1973). *Am. J. Psychiat.* **130,** 82–86.
Lahti, R. A. and Losey, E. G. (1974). *Res. Commun. Chem. Pathol. Pharmacol.* **7,** 31–40.
Laverty, R. and Sharman, D. F. (1965). *Br. J. Pharmacol.* **24,** 759–772.
Lee, T., Seeman, P., Rajput, A., Farley, I. J. and Hornykiewicz, O. (1978). *Nature (London)* **273,** 59–61.
Lloyd, K. G. and Hornykiewicz, O. (1970). *Science* **170,** 1212–1213.
Lloyd, K. G. and Hornykiewicz, O. (1973). *Nature (London)* **243,** 521–523.
Lloyd, K. G. and Hornykiewicz, O. (1975). *In* "Catecholamines and Behavior" (Ed. A. J. Friedhoff), Vol. 1, 41–57. Plenum Press, New York.
Lloyd, K. G., Davidson, L. and Hornykiewicz, O. (1975a). *J. Pharmacol. Exp. Ther.* **195,** 453–464.
Lloyd, K. G., Möhler, H., Hertz, Ph. and Bartholini, G. (1975b). *J. Neurochem.* **25,** 789–795.
Lloyd, K. G., Shemen, L. and Hornykiewicz, O. (1977). *Brain Res.* **127,** 269–278.
MacLean, P. (1970). *In* "Physiological Correlates of Emotion" (Ed. P. Black), 129–146. Academic Press, New York and London.
Maj, J., Pawlowski, L. and Sarnek, J. (1974). *In* "Serotonin—New Vistas" (Eds. E. Costa, G. L. Gessa and M. Sandler), 253–256. Advances in Biochemical Psychopharmacology, Vol. 10. Raven Press, New York.
Markham, C. H. (1971). *In* "Monoamines Noyaux Gris Centraux et Syndrome de Parkinson" (Eds. J. de Ajuriaguerra and G. Gauthier), 485–490. Masson, Paris.
McGeer, P. L., McGeer, E. G., Wada, J. A. and Jung, E. (1971). *Brain Res.* **32,** 425–431.
McGeer, P. L., McGeer, E. G. and Fibiger, H. C. (1973). *Lancet* **ii,** 623–624.
Mones, R. J., Elizan, T. S. and Siegel, G. J. (1971). *J. Neurol. Neurosurg.* **34,** 668.
Moore, R. Y., Bhatnagar, R. K. and Heller, A. (1971). *Brain Res.* **30,** 119–135.
Nagatsu, T., Kato, T., Numata (Sudo), Y., Ikuta, K., Sano, M., Nagatsu, I., Kondo,

Y., Inagaki, S., Iizuka, R., Hori, A. and Naragayashi, H. (1977). *Clinica Chemica Acta* **75,** 221–232.

Nagy, J., Lee, T., Seeman, P. and Fibiger, C. H. (1978). *Nature (London)* **274,** 278–281.

Nybäck, H. and Sedvall, G. (1970). *Eur. J. Pharmacol.* **10,** 193–205.

Perry, T. L., Hansen, S. and Kloster, M. (1973). *New Eng. J. Med.* **288,** 337–342.

Pijnenburg, A. J. J., Honig, W. M. M., Van der Heyden, J. A. M. and Van Rossum, J. M. (1976). *Eur. J. Pharmacol.* **35,** 45–58.

Poirier, L. J. and Sourkes, T. L. (1965). *Brain* **88,** 181–192.

Postma, J. U. (1972). *Psychiat. Neurol. Neurochir.* **75,** 69–71.

Precht, W. and Yoshida, M. (1971). *Brain Res.* **32,** 229–233.

Price, K. S., Farley, I. J. and Hornykiewicz, O. (1978). *In* (Eds. P. J. Roberts and G. N. Woodruff), "Dopamine", Advances in Biochemical Psychopharmacology, Vol. 19. 293–300. Raven Press, New York (in press).

Rinne, U. K., Sonninen, V. and Hyyppa, M. (1971). *Life Sci.* **10,** 549–557.

Rinne, U. K., Sonninen, V., Rickkinen, P. and Laaksonen, H. (1974). *Med. Biol.* **52,** 208–217.

Rubovits, R., Patel, B. C. and Klawans, H. L. (1973). *In* "Huntington's Chorea" (Eds. A. Barbeau, T. N. Chase and G. W. Paulson), 671–679 Advances in Neurology, Vol. 1. Raven Press, New York.

Sandler, M., Carter, S. B., Hunter, K. R. and Stern, G. M. (1973). *Nature (London)* **241,** 439–443.

Sassin, J. F., Taub, S. and Weitzman, E. D. (1972). *Neurology* **22,** 1122–1125.

Scatton, B. (1977). *Eur. J. Pharmacol.* **46,** 363–369.

Sharman, D. F., Poirier, L. J., Murphy, G. F. and Sourkes, T. L. (1967). *Can. J. Physiol. Pharmacol.* **45,** 57–62.

Tarsy, D., Pycock, C., Meldrum, B. and Marsden, C. D. (1975). *Brain Res.* **89,** 160–165.

Turner, A. J., Baker, K. M., Algeri, S., Frigerio, A. and Garattini, S. (1974). *Life Sci.* **14,** 2247–2257.

Ungerstedt, U. (1971). *In* "6-Hydroxydopamine and Catecholamine Neurons" (Eds. T. Malmfors and H. Thoenen), 315–332. North-Holland Publishing Co., Amsterdam.

CHAPTER 36

Dopamine and the Development of Disorders of Human Behaviour

H. M. van Praag

University Psychiatric Clinic, University of Utrecht, The Netherlands.

Do central dopaminergic systems play a role in the pathogenesis of certain disorders of human behaviour? I shall first of all discuss the data which justify raising this question (Van Praag, 1967, 1975), then the question whether it is at all possible to obtain information on the central dopamine (DA) metabolism in human individuals, and finally the data obtained in this context in psychiatric patients.

INCENTIVES TO STUDY THE CENTRAL DA METABOLISM IN PSYCHIATRIC PATIENTS

DA-Ergic Hyperactivity and Psychoses

Central stimulants

Amphetamine derivatives and related compounds such as methyl phenidate (Ritalin) cause an intracerebral release of DA from DA-ergic nerve endings to the synaptic cleft; moreover, they block the reuptake of DA into the nerve endings and inhibit DA degradation via monoamine oxidase (MAO). Possibly, they also activate postsynaptic DA receptors directly. All these actions enhance the central DA-ergic activity (Randrup and Munkvad, 1970).

Central stimulants exert a more or less characteristic influence on human behaviour. In small doses, they have an activating, fatigue-reducing effect, particularly when performance has diminished as a result of an unusual exertion. In large doses, they have a disintegrating effect. The so-called amphetamine psychosis was first described in individuals addicted to stimulants (Young and Scoville, 1938; Connell, 1958). Two types of psychosis are possible: "common" toxic psychosis (usually delirious) and "true" amphetamine psychosis: a paranoid syndrome with ideas of reference, largely auditory hallucinations, ideas of persecution and unclouded, sometimes even unusually lucid consciousness (as can also be observed in LSD psychosis). The toxic psychosis chiefly develops after one or two very large doses, whereas the "true" syndrome is more likely to develop after repeated parenteral administration. All authors agree that amphetamine psychosis closely resembles paranoid types of schizophrenia and could be regarded as a model for this type of schizophrenia (Randrup and Munkvad 1972; Snyder, 1973).

In normal volunteers with a "clean" history and without schizoid personality features, too, a "true" amphetamine psychosis can be induced within 1–4 days by giving increasing doses of amphetamines at short intervals. The amount required is 300–500 mg (Griffiths *et al.*, 1968; Angrist *et al.*, 1974). Amphetamine psychosis, therefore, is not simply a latent schizophrenia provoked by drugs. The fact that some test subjects become psychotic after only 24 hours rules out lack of sleep as a decisive factor.

Finally, in schizophrenic patients, even small doses of amphetamine derivatives and methylphenidate cause aggravation of the schizophrenic symptoms (Janowsky *et al.*, 1972, 1973). Lysergic acid diethylamide (LSD) also causes exacerbation of these patients' condition, but the LSD syndrome is superimposed on the schizophrenic syndrome, so to speak. The patient himself recognizes a difference between LSD phychosis and the original syndrome (Hollister, 1962). This effect of amphetamine is not a result of a non-specific stimulation of the CNS: caffeine in large doses has no psychotogenic effect (Angrist *et al.*, 1974).

If the psychosis-inducing effect of amphetamines and related compounds is indeed related to enhanced DA-ergic activity in the brain, then other DA agonists, e.g. L-DOPA, bromocriptine and apomorphine, can be expected to have a similar disintegrating effect.

L-DOPA

Exogenous L-DOPA is centrally converted largely to DA, and only to a small extent to noradrenaline (NA). As a therapeutic agent in Parkinson's disease, this compound induces psychotic reactions in about 3.5% of cases

(Goodwin, 1971, 1972). These are of a diverse nature, lack specific characteristics, and show no similarity to "true" amphetamine psychosis. The reasons could be: (a) that therapeutic doses of L-DOPA cause a DA increase which is probably much smaller than the increased dopaminergic activity after amphetamines in doses which induce psychosis (Snyder *et al.*, 1974), and (b) that the DA-producing apparatus is partly destroyed in Parkinson patients.

Furthermore, L-DOPA seems capable of actualizing an existing psychotic predisposition. For example, (hypo)manic reactions after L-DOPA are seen in particular in patients with manic periods in the history, and in patients with a history of psychosis the risk of relapse is increased during DOPA medication (Goodwin, 1972).

L-DOPA has also been given to schizophrenic patients, in daily doses up to 5 g (without peripheral decarboxylase inhibitor). This led to clinical deterioration in all cases. Most patients showed exacerbation of existing pathology, but some showed only an increased degree of agitation. A few patients developed new symptoms, particularly auditory hallucinations (Angrist *et al.*, 1973; Yaryura-Tobias *et al.*, 1970). Sathanathan *et al.* (1973) found that tolerance to the psychosis-inducing effect of L-DOPA is much higher in non-psychotic patients than in schizophrenics. On this basis they postulated a continuum of threshold values for the amounts of DA-ergic stimulation which are tolerated without psychotic symptoms—an attractive concept because it can serve as a kind of biological explanation of the clinical notion that susceptibility to schizophrenic reactions shows marked interindividual variations.

So far as I know, there has been no systematic research into the behavioural effects of large doses of L-DOPA in normal test subjects.

Bromocriptine

This compound enhances central DA-ergic activity by direct stimulation of postsynaptic DA receptors. It has been successfully used in Parkinson patients, but the clinical literature so far is limited. There has so far been only one report on "paranoid schizophreniform psychosis" occurring in temporal association with bromocriptine treatment (in combination with L-DOPA plus carbidopa) (Lipper, 1976). The patient was a woman aged 47 with a five-year history of ideopathic parkinsonism and no previous psychiatric impairment. The psychosis was characterized by a pervasive, systematized paranoid delusional system, and initially by auditory and possibly visual hallucinations, somatic delusions, inappropriate affect and Capgras phenomenon (delusional belief that familiar persons have "doubles").

Neuroleptics could not be given in view of aggravation of the Parkinson symptoms. Psychotherapy was unsuccessful. One year after its onset the psychosis continued without remission. The last-mentioned fact raises doubts about the DA-ergic nature of the syndrome. Typical amphetamine and DOPA psychoses are self-limiting and disappear soon after discontinuation of medication.

Apomorphine

Like bromocriptine, apomorphine is a direct stimulator of postsynaptic DA receptors. Studies of this compound are much impeded by its emetic effect. Given in small doses to normal test subjects, it has no psychosis-inducing effect (Lal and De La Vega, 1975). No aggravation of psychotic symptoms has been observed in schizophrenic patients (Angrist *et al.*, 1975).

Conclusion

The so-called amphetamine psychosis seems to be a strong argument in support of the hypothesis that DA-ergic hyperactivity in the brain can play a role in the pathogenesis of certain schizophrenic symptoms. The strength of this argument is reduced, however, by the fact that there is no convincing evidence that other DA agonists induce similar symptoms.

DA and the Mechanism of Action of Neuroleptics

In animals, all neuroleptics, i.e. antipsychotic agents, increase central DA turnover (Carlsson and Lindqvist, 1963) regardless of their chemical structure. There are sound reasons for the postulate that this effect is secondary, not to increased but to impaired transmission in DA-ergic synapses. Perhaps it can be regarded as a compensatory mechanism aimed at breaking the blockade (Bartholini and Pletscher, 1972). The inhibition of transmission is believed to result from diverse effects: (a) blockade of postsynaptic DA receptors (Van Rossum, 1967); (b) inhibition of DA release in the synaptic cleft after stimulation of the axon (Seeman and Lee, 1975).

The increase in DA turnover probably also involves several mechanisms. To begin with, the fact that neuroleptics inhibit presynaptic as well as postsynaptic DA receptors plays a role (Kehr *et al.*, 1972). It is assumed that their activation normally leads to reduced DA synthesis, and that the amount of DA in the synaptic cleft is regulated in this way. Their inactivation abolishes a brake on the DA production. Secondly, the possibility of a negative feedback mechanism from the blocked postsynaptic

DA receptors is taken into account. The manner of mediation of this information, however, is controversial.

The mechanism of action of neuroleptics need not be further discussed here. Suffice it to say that blockade of transmission in human DA-ergic neurones is considered plausible as well, in view of two observations. (1) Neuroleptics of the phenothiazine, thioxanthene and butyrophenone type increase the baseline concentration of homovanillic acid (HVA) in cerebrospinal fluid (CSF), and induce an increased HVA response after inhibition of HVA transport by probenecid (see p. 664) (Persson and Roos, 1969; Bowers, 1973; Van Praag and Korf, 1975). Both phenomena indicate that an abnormally large amount of DA is degraded in the CNS. (2) The release of prolactin from the anterior pituitary lobe is subject to the influence of DA-ergic neurones of the tuberoinfundibular system. Suppression of their activity leads to increased release of prolactin and an increased prolactin concentration in blood and CSF. The reverse effect is seen when these neurones are activated. Neuroleptics cause an increase in the prolactin concentration in blood and CSF (Wilson *et al.*, 1975; Post and Goodwin, 1975; Van Praag, 1977a). This indicates that the increased DA turnover they produce, in fact results from transmission *blockade*; not from *increased* transmission activity in DA-ergic synapses.

Numerous animal experiments indicate the plausibility of a relation between the therapeutic activity of neuroleptics in psychoses and their DA-antagonistic activity (for review see Crow *et al.*, 1976). The plausibility of a similar relation in the human organism has also been demonstrated (Van Praag and Korf, 1976; Sedvall *et al.*, 1977): the degree of improvement of psychiatric patients in response to neuroleptics proved to correlate with the degree of increase of the HVA response in the CSF after probenecid; and this response is considered to be an indicator of DA turnover in the CNS. Additional evidence is provided by the fact that the therapeutic effect of neuroleptics is potentiated by α-methyl-*p*-tyrosine—a compound which blocks the synthesis of DA (as well as that of NA) via inhibition of the enzyme tyrosine hydroxylase (Carlsson *et al.*, 1972).

If it is justifiable to assume that the therapeutic effect of neuroleptics in psychoses is related to suppression of neuronal activity in central DA-ergic systems, then it is justifiable to raise the question whether central DA-ergic activity is perhaps increased in (certain) psychoses.

DA and Motor Pathology

Drugs which enhance central DA activity—e.g. L-DOPA (combined with an MAO inhibitor), amphetamines and related stimulants, and apomorphine—induce a hyperactivity syndrome in test animals. This has been most amply studied in rats (Randrup and Munkvad, 1974). After moderate doses, the "*DA-ergic hyperactivity syndrome*" is characterized by apparently aimless increased locomotor activity. After larger doses of the DA agonists, one observes stereotyped behaviour patterns such as sniffing, rearing, licking, gnawing and spurts of backward locomotion. The stereotyped behaviour can become so predominant that hyperactivity in the sense of increased ambulatory locomotor activity disappears, so that in fact the behavioural repertory is drastically reduced. There are indications that the mesolimbic DA system is required for enhancement of locomotor activity, and the nigrostriatal DA system for the production of stereotyped behaviour (Iversen, 1975; Moore, 1977).

Increased activity of central serotonergic systems likewise leads to motor hyperactivity (Grahame-Smith, 1971; Green and Grahame-Smith, 1976; Jacobs, 1977). The "*serotonergic hyperactivity syndrome*" can be induced (a) by combining a 5-hydroxytryptamine (5-HT; serotonin) precursor—tryptophan, or 5-hydroxytryptophan—with an MAO inhibitor; and (b) with the aid of a direct 5-HT receptor stimulator such as 5-methoxy-*N*,*N*-dimethyltryptamine.

Neuroleptics suppress not only DA-ergic but also serotonergic hyperactivity syndromes. This led Green and Grahame-Smith (1974) to the conclusion that a dopaminergic system is involved in the behavioural expression of the 5-HT-induced hyperactivity.

If, inversely, the activity in the nigrostriatal and/or the mesolimbic/mesocortical DA system is eliminated—either by chemical or electrical destruction or with the aid of receptor blockers such as neuroleptics, or store depletors such as reserpine—then there is generally a marked reduction of motor activity, to the point of total inactivity, without induction of sleep. One recent report would seem to contradict this rule. According to Le Moal *et al.* (1977), rats show *hyper*activity when a particular group of DA-ergic cell bodies is damaged, namely the A10 group in the ventral mesencephalic tegmentum. However, this area is also rich in serotonergic and noradrenergic fibres, and it is therefore not certain that the hyperactivity syndrome in fact results from the destruction of the DA-ergic cells.

Be this as it may, motor regulation can be seriously disturbed in some psychiatric syndromes. Motor agitation can be observed in schizophrenic and other types of psychosis; and also in certain types of vital (endogenous) depression. The reverse can also be observed in these syndromes i.e. motor inhibition which may go as far as a state of almost total immobility (stupor). All these observations justify the question whether dysregulation of central DA systems is involved in the pathogenesis of these phenomena.

Dopamine, Depression and Parkinson's Disease

In Parkinson's disease, the DA and 5-HT concentrations in the basal ganglia are decreased. The DA deficiency plays a role in the pathogenesis of the hypokinesia, and possibly also in that of the rigidity. The significance of the 5-HT deficiency for the pathogenesis of the motor symptoms is still obscure.

Depressive syndromes are significantly more common in Parkinson patients than in a comparable control group (Hoehn *et al.*, 1976; Asnis, 1977). The increased incidence of depression can result from psychosocial factors: from the disability which the disease causes, and from the social isolation which usually results from this. A biological factor may equally well be involved: the central dysfunctions which occur in Parkinson's disaese could produce neurological as well as behavioural disorders. So far, a correlation between central 5-HT deficiency and depressive symptoms has not been demonstrated in Parkinson patients (Puite *et al.*, 1973). It therefore makes sense to raise the question whether the DA deficiency could be significant in this respect. In Parkinson patients it is impossible to distinguish the contribution of the DA deficiency to the motor pathology from its possible contribution to the mood pathology. The only possibility is to study the DA metabolism in patients with "pure" depressions.

The hypothesis of a relation between DA deficiency and depressive symptoms receives little support from psychopharmacology. Tricyclic antidepressants inhibit the uptake of 5-HT and NA into the corresponding neurones. The resulting serotonergic and noradrenergic hyperactivity is believed to be related to their antidepressant potency. However, these drugs have been found to be only weak inhibitors of DA uptake in rat brain synaptosomes (Halaris *et al.*, 1975), and there is only minimal evidence that they do so *in vivo* (Fuxe and Ungerstedt, 1968). The fact that they exert no influence on human serum prolactin points in the same direction (Meltzer *et al.*, 1977). Nomifensine, a new non-tricyclic antidepressant, was originally considered to be a specific DA-reuptake inhibitor. However, it appeared to have an inhibiting effect on NA uptake as well. Finally, a mood-elevating effect of specific DA agonists, such as bromocriptine, has so far not been demonstrated.

Conclusions

Research into DA metabolism in psychiatric patients is of value. Various scattered data, derived chiefly from psychopharmacological research and studies of Parkinson's disease, warrant the hypothesis that disorders of the central DA-ergic systems can play a role in the pathogenesis of psychotic and depressive symptoms.

RESEARCH INTO HUMAN CENTRAL DA METABOLISM

Peripheral Indicators

The periphery affords three possibilities of gaining an impression of human central DA metabolism. To begin with, measurement of the excretion of DA and its metabolites in the urine. Studies of this kind have been criticized as reflecting peripheral rather than cerebral events, and as being subject to interference from incidental stress, physical activity, the rate of diuresis, and dietary factors. A second possibility is measurement in the blood of the activity of enzymes involved in DA metabolism. This strategy is more valid than the former, because an enzyme defect (if any) is more likely to be generalized than confined to the brain. A third possibility is measurement of plasma prolactin concentrations. This yields a fairly accurate indication of the activity of DA-ergic cells in the tubero-infundibular DA system (see p. 659).

Post-Mortem Examination of the Brain

The most direct approach is of course to measure DA, its metabolites and the relevant enzymes post mortem in the brain of psychiatric patients, and to compare the results with those in a comparable control group of non-psychiatric patients. Such a comparison, however, is beset by numerous uncontrollable variables liable to influence the results: age, sex, nutritional condition at the time of death and during the preceding period, the duration of the death agony, premortal medication, the duration of the interval between death and examination, the time at which death ensued (the concentration of monoamines in the brain is subject to diurnal variation), etc.

CSF Studies Without Probenecid

Practically the most important source of information on human central DA metabolism is the concentration of homovanillic acid (HVA) in the CSF. The second DA metabolite (3,4-dihydroxyphenylacetic acid) is present in only insignificant amounts in human CSF. This strategy is based on the hypothesis that DA degradation in the CNS is related to HVA concentration in the CSF. This is a plausible postulate, for after pharmacological manipulation of central DA metabolism the HVA concentration in the CSF changes in the expected direction. For example, the DA precursor L-DOPA stimulates central DA synthesis and causes an increase in the HVA concentration in the CSF (Goodwin *et al.*, 1970). The reverse of this effect is produced by α-methyl-*p*-tyrosine, an inhibitor of tyrosine hydroxylase and therefore of DA synthesis (Brodie *et al.*, 1971). In Parkinson patients, a decrease in the DA concentration in the basal ganglia is associated with a decrease of the HVA concentration in the CSF (Olsson and Roos, 1968).

Homovanillic acid in lumbar CSF is largely of cerebral origin, which is not surprising because, so far as we know, few or no DA-ergic neurones or nerve endings are contained in the spinal cord (Carlsson *et al.*, 1964). Accordingly, the gradient between the ventricular and the lumbar HVA concentration is high (10:1) (Sourkes, 1973). There are other strong arguments in support of the view that lumbar HVA is representative of cerebral DA metabolism. Partial block (Curzon *et al.*, 1971; Garelis and Sourkes, 1973; Post *et al.*, 1973) and total block of the spinal subarachnoid space (Young *et al.*, 1973) leads to a decrease or total disappearance, respectively, of HVA in the CSF caudal to the block. After occlusion of both interventricular foramina, only very little HVA is demonstrable in the lumbar CSF (Garelis *et al.*, 1974). This demonstrates that HVA originates largely from structures in the vicinity of the lateral ventricles. These findings substantiate the view that HVA in the human lumbar CSF originates from DA-rich areas close to the lateral ventricles, and mainly from the caudate nucleus.

As a method, however, measurement of the baseline HVA concentration is not very reliable. To begin with, HVA is only partly transported to the bloodstream via the CSF. A (probably large) proportion disappears directly in the blood stream. A substantial transport of HVA directly to the bloodstream does not as such devalue its determination in CSF, provided the ratio between directly and indirectly transported material remains constant; it is precisely on this point, however, that we have no information whatever. Another factor detrimental to the reliability of this method is that the flow rate of the CSF is subject to the influence of CSF pressure (Davson, 1967). Because of the gradient, the amount of HVA encountered in the lumbar CSF at a given moment is in part determined by the CSF flow. This means that the concentration is affected by a variety of factors which influence CSF pressure, e.g. postural changes, straining, sneezing, coughing — all factors which it is difficult either to eliminate or to normalize.

The above-mentioned factors probably explain the substantial day-to-day differences in the concentration of HVA in CSF in a given individual.

CSF Studies after Probenecid Administration

A number of the flaws of the method of measuring the baseline HVA concentration in CSF can be overcome by premedication with probenecid. Probenecid blocks the transport of HVA from the CNS to the blood stream, and thus induces accumulation of HVA in the brain and the CSF. Of course there is a relation between DA degradation and HVA accumulation. If HVA accumulation after probenecid is small, then this indicates a low HVA production and therefore a low DA degradation. A large HVA accumulation on the other hand indicates a high degradation of the patent amine DA. Determination of the HVA concentration in CSF before and after probenecid administration, therefore, gives an indication of the amount of DA degraded during a given period in the CNS as a whole. This technique has been described in detail in another publication (Van Praag, 1977b).

Probenecid blocks both the direct HVA transport to the blood stream and the indirect transport via the CSF. In this way, therefore, we eliminate the uncertainty factor which lies in our lack of knowledge of the amounts of HVA transported directly and indirectly to the blood stream. Moreover, HVA is so to speak trapped in the CNS as a result of the transport block; the ventriculolumbar concentration gradient is therefore reduced, and the influence of local changes in CSF pressure on the HVA concentration in the lumbar CSF diminishes accordingly. Consequently the post-probenecid HVA concentrations show less day-to-day variation than the baseline concentrations.

This method, too, has its flaws (e.g. it is uncertain whether HVA transport is really completely blocked by the amounts of probenecid which the human organism tolerates), but nevertheless it is a substantial improvement compared with baseline CSF determination.

DA AND AFFECTIVE DISORDERS

Studies in the Periphery

Consistent changes in the *pattern of excretion* of DA and its metabolites in the urine have not been found.

The results of measuring the blood *prolactin level* (indicator of central DA activity) are inconclusive. Sachar *et al.* (1973) found increased baseline values both in unipolar and in bipolar depressions, which might indicate reduced DA-ergic activity in the brain. Others, however, found normal values (Maeda *et al.*, 1975; Ehrensing *et al.*, 1974; Legros *et al.*, 1975). In depressive patients the prolactin response to TRH can be either increased

(Maeda *et al.*, 1975) or decreased (Legros and Breulet, 1975). Inadequate classification of the depressive syndromes can perhaps explain this discrepancy. Another possibility is infrequent blood sampling: plasma prolactin shows diurnal fluctuations.*

The findings concerning the *enzymes* involved in DA metabolism are not unequivocal either. Murphy and Weiss (1972) found decreased MAO activity in the blood platelets in patients with bipolar depressions, and normal activity in patients with unipolar depressions. In the bipolar group the urinary tryptamine excretion was increased. Since tryptamine degradation is totally dependent on MAO, this phenomenon indicates an overall deficiency of MAO activity. The age distribution in the two groups was comparable—an important point because MAO activity increases with increasing age (Robinson *et al.*, 1971). Unlike Murphy and Weiss (1972), Klaiber *et al.* (1972) found higher plasma MAO values in depressive patients than in a comparable control group, and Belmaker *et al.* (1976a) reported the same for MAO values in blood platelets.

Catechol-*O*-methyltransferase (COMT) is the enzyme mainly involved in extraneuronal DA degradation. Two studies (Cohn *et al.*, 1970; Dunner *et al.*, 1971) disclosed about 50% less COMT activity in the erythrocytes in women with unipolar depressions than in a control group. In women with bipolar depressions the COMT activity was likewise decreased, but less markedly so. Women under other diagnostic headings, including schizophrenia, showed no abnormalities of COMT activity. Surprisingly, decreased COMT activity was found to be definitely sex-linked. It has not been observed in men with bipolar or unipolar depressions.

The significance of this phenomenon is obscure. It has so far been observed exclusively in depressions but is not syndrome-linked: COMT activity was found to be decreased also when the depressive symptoms had subsided.

The serum activity of DA-β-hydroxylase (the enzyme which converts DA to NA) was not related to the mood level in a study by Shopsin *et al.* (1972). In another study in which enzyme activity was measured indirectly—by the conversion of infused ^{3}H-DA to radioactive NA metabolites—such a relation was indeed found: in bipolar patients NA synthesis proved to be increased in the manic and normal in the depressive phase (Rosenblatt *et al.*, 1973).

Postmortem Studies

In a study of suicide victims the DA concentration in the caudate nucleus was found to be normal (Pare *et al.*, 1969). Gottfries *et al.* (1974) found that

* At asterisks throughout this chapter, refer to notes added in proof, p. 677.

MAO activity in various areas of the brain was 20–40% lower in suicide victims ($n = 15$) than in a comparable group ($n = 20$) with a different cause of death. The decrease was demonstrable regardless of whether tryptamine or β-phenylethylamine was used as substrate for MAO. It seems questionable, however, whether these observations really corroborate those reported by Murphy and Weiss (1972; see p. 665). It is unknown whether the patients in the suicide group had all been suffering from depressions, let alone whether these depressions were of the bipolar type. The number of post-mortem studies is too small to warrant any conclusion.

CSF Studies

The data on HVA concentrations in CSF are fairly consistent, at least so far as certain categories of depressive patients are concerned. A majority of the investigators found decreased values in these patients, both baseline and post-probenecid values, although the decrease was not always statistically significant (for references see Van Praag, 1977b). The phenomenon is most pronounced when the depressive *syndrome* has vital (endogenous) features and shows a unipolar course (which is to say that several depressive but no manic phases occur in the course of life) (Ashcroft *et al.*, 1973). According to Van Praag and Korf (1971), the decreased HVA concentration is related not so much to the disturbance of mood as such as to the factor of motor inhibition. Their observations were corroborated by Banki (1977). This raises the question: is the biochemical disorder of importance in the pathogenesis of the motor inhibition or is it a consequence of this inhibition, e.g. a consequence of insufficient "mixing" of CSF from higher levels (with a high HVA concentration) with CSF from lower levels (with a lower HVA concentration)? The former explanation is more plausible than the latter, for three reasons:

1. Not only the baseline but also the post-probenecid HVA concentration is decreased in retarded depressions (for references see Van Praag, 1977b). After probenecid the concentration gradient of HVA diminishes, and consequently the influence of the "mixing effect" decreases.
2. L-DOPA—a DA precursor—normalizes both the motor status and the HVA concentration in the CSF in these patients (Van Praag, 1974; Bunney *et al.*, 1972). The mood level, however, is hardly influenced.
3. Nomifensin—a compound with a marked inhibitory effect on DA uptake into DA-ergic neurones—is therapeutically more effective in depressive patients with decreased HVA accumulation after probenecid than in patients with normal values (Van Scheyen *et al.*, 1977).

There is no agreement on the question whether the CSF changes are syndrome dependent or syndrome independent. According to Ashcroft *et al.* (1973) these changes persist after abatement of the depressive symptoms; according to Van Praag (1977c) they disappear with the depression.

The situation in the manic syndrome is less clear. Several authors maintain that the HVA concentration is decreased in this category of patients as well (references: Van Praag, 1977b). According to Post *et al.* (1973), normal and increased values are just as likely to be found, depending on the degree of motor unrest. Perhaps the "mixing effect" manifests itself in the agitated patient.

Conclusions

The results of enzyme studies are too heterogeneous to warrant conclusions. Post-mortem studies are under-represented, and do not allow conclusions either. The CSF findings are fairly consistent and indicate the probability that the DA turnover is decreased in certain categories of depressive patients, and possibly also in manic patients. This phenomenon is more likely to be a factor in the pathogenesis of the depressive syndrome than a consequence of it. To which symptoms the phenomenon is related remains obscure. On the one hand there are indications that it is related to motor retardation, but on the other hand its occurrence in manic as well as in depressive syndromes is more suggestive of a relation with mood dysregulation.

DA AND SCHIZOPHRENIC PSYCHOSES

Studies in the Periphery

Like depressive patients, schizophrenic patients have shown no consistent changes in the *renal excretion pattern* of DA and its metabolites. Both in acute and in chronic forms of schizophrenia, the baseline plasma *prolactin* level is normal (Meltzer *et al.*, 1974; Brambilla *et al.*, 1976). In chronic schizophrenic patients, however, the prolactin response to intravenous TRH was found to be increased (Brambilla *et al.*, 1976). This might indicate an insufficient brake on prolactin release, and therefore on a hypoactive DA system, at least in the tuberoinfundibular system.*

Enzyme studies

Enzyme studies in peripheral blood have not yielded less unequivocal results than in depressive patients. Murphy and Wyatt (1972) and Murphy *et al.* (1976) reported that MAO activity in blood platelets from chronic schizophrenic patients was reduced by about 50%. This phenomenon was not caused by psychotropic drugs, and the MAO activity per individual was fairly stable in time, independent of therapeutic results. In monozygotic twins of whom one was schizophrenic and the other not, moreover, the MAO activity in the platelets was found to be closely correlated (Wyatt *et al.*, 1973). Decreased MAO activity, therefore, is not associated with schizophrenia as such but might be a genetic marker which determines the "vulnerability" to schizophrenic psychoses.

Several authors have confirmed this observation (Carpenter *et al.*, 1975; Meltzer and Stahl, 1974; Nies *et al.*, 1974; Schildkraut *et al.*, 1976). Zeller *et al.* (1975) maintained that there was no decrease in the concentration of the enzyme but an abnormality in its structure. According to Garelis *et al.* (1975), the serum 5-HT concentration is increased in chronic schizophrenic patients, as could be expected in view of a decreased MAO activity. On the other hand, Shaskan and Becker (1975) found MAO values described as decreased by Murphy and Wyatt (1972b) also in chronic alcoholics and normal test subjects. The results reported by Owen *et al.* (1976) and Belmaker *et al.* (1976a) are likewise negative. Belmaker *et al.* (1976b) found no genetic mutant forms of MAO in chronic schizophrenics. In schizophrenia, the serum activity of the second enzyme involved in DA degradation—COMT—does not deviate from the normal (Dunner *et al.*, 1971).

The plasma DA-β-hydroxylase activity shows no abnormalities in schizophrenic patients (Dunner *et al.*, 1973). Conversion of infused ^{3}H-DA to radioactive NA metabolites, however, was found to be increased in acute schizophrenia (Rosenblatt *et al.*, 1973); this might indicate that the activity of this enzyme is indeed increased.

Post-Mortem Studies

From a study of the brains of post-mortem schizophrenic and non-schizophrenic patients it was found that the DA content in the former group was 48% higher in the nucleus accumbens but normal in the putamen (Bird *et al.*, 1977). The authors stressed, however, that it was unclear whether or not these findings were related to the illness or were a consequence of prolonged treatment with neuroleptic drugs.* In another study MAO activity in the brain was found to be normal (Schwartz *et al.*, 1974); the same applies to the activity of DA-sensitive adenylate cyclase in the caudate nucleus (Carenzi *et al.*, 1975). Wise and Stein reported in 1973 that the activity of DA-β-hydroxylase in the brain, particularly in diencephalon and hippocampus, is reduced by about 50% in chronic schizophrenics. This enzyme is present exclusively in NA-ergic neurones. Noradrenaline deficiency can be expected as a result of the reduced activity: (a) because NA production decreases, and (b) because DA accumulates in the NA-ergic neurone and may come to function as a false transmitter. Another possible consequence of the reduced activity is that excess DA "leaks" from the NA-ergic neurones, is taken up in DA-ergic cells and enhances their activity. Noradrenaline is an important transmitter in the so-called pleasure or reward systems. According to Stein and Wise (1971), hypoactivity of these systems—as a result of the supposed NA deficiency—plays an important role in the pathogenesis of schizophrenia. The deficiency in DA-β-hydroxylase is believed to be related to lesions of NA-ergic neurones due to the production of a toxic DA metabolite: 6-hydroxy-DA. This substance is in fact capable of damaging the NA-ergic apparatus, but its production *in vivo* has not so far been demonstrated.

The only reduplication study so far made (Wyatt *et al.*, 1975), however, failed to corroborate the findings of Wise and Stein. The DA-β-hydroxylase values were indeed decreased in schizophrenics (average 77–89% of control values), but the differences were not significant. The heterogeneousness of the test group may have played a role: it seems as if the enzyme activity showed a significantly more marked decrease in the (small) subgroup of patients with a chronic undifferentiated form of schizophrenia than in the other categories of schizophrenia. This demonstrates once again the importance of careful psychopathological differentiation of test subjects for the interpretation of biological data. Repetition of this investigation is certainly advisable.*

CSF Studies

These studies have so far yielded no sensational results. The majority of authors have reported a normal baseline HVA concentration (for references see Van Praag, 1977b). In psychotic patients without first-rank Schneider symptoms, post-probenecid HVA accumulation was found to be slightly more marked than that in patients with these classical symptoms, and that in controls (Bowers, 1973). Post *et al.* (1975) found normal HVA accumulation in patients with acute schizophrenia. Yet the DA metabolism had not remained entirely unaffected, for after recovery the accumulation was significantly lower than before. Van Praag and Korf (1975) studied a group of patients with psychoses of mixed aetiology but always with manifest delusions and/or hallucinations. Homovanillic-acid accumulation was slightly but not significantly increased. After classification of the patients according to the degree of motor hyperactivity, HVA accumulation in agitated patients proved to be significantly higher than that in patients without these symptoms. It was concluded that an increase in DA turnover can occur in psychoses but is dependent on the motor activity factor rather than on "true" psychotic symptoms such as delusions and hallucinations. This observation was considered to support the hypothesis that disorders of central DA metabolism are related not so much to a given nosological entity (schizophrenia, Parkinson's disease) or a given syndrome (vital depression) as to a given functional condition, i.e. motor hyperactivity or motor hypoactivity (Van Praag *et al.*, 1975a).

Bowers (1974) found decreased post-probenecid values in lumbar CSF in one recent study of schizophrenic patients, and increased values in another (1976). This variability of Bowers' results also suggests that the disturbances in DA metabolism correlate less with the syndrome or disease entity schizophrenia than with a particular component (or with a particular subgroup, still to be identified).

Conclusions

The view that reduction of central DA-ergic activity can result in a therapeutic effect on psychosis is a fairly sound one. But there is no convincing evidence to support its complement: the hypothesis that DA-ergic hyperactivity plays a role in the pathogenesis of schizophrenia or, more widely, psychotic symptoms. Even though it is quite possible that in fact there is no relation between DA-ergic systems and schizophrenic symptoms, other explanations of the negative findings are conceivable. (1) The HVA concentration in CSF reflects mainly the DA metabolism in the nigrostriatal DA system, whereas serum prolactin is a function of the DA activity in the tuberoinfundibular system. There are no methods which supply information on the human limbic DA system, and it may be precisely this system that is involved in the pathogenesis of psychotic symptoms. (2) There is no hyperproduction of DA, but the postsynaptic DA receptors are hypersensitive. In this case one might even expect decreased DA synthesis via a feedback mechanism from the receptors. Meltzer *et al.* (1976) regard this as unlikely because in schizophrenic patients they were unable to demonstrate hypersensitivity to various effects of the direct DA agonist apomorphine. According to Rotrosen *et al.* (1976), however, this generalization is not justifiable, for they found that the plasma growth hormone response to apomorphine was markedly increased in schizophrenic patients who were later found to be resistant to the therapeutic effect of neuroleptics. The growth hormone response to L-DOPA was normal. The authors conclude that, in a *particular* subgroup of schizophrenics, *particular* DA receptors are hypersensitive. (3) The primary defect is not an excess of activity in a DA-ergic system but a deficit in a closely related but antagonistic system. In this case DA-ergic activity, although within normal limits or even somewhat reduced, would be in relative excess; and this excess would be reduced by the effects of neuroleptic drugs (Crow *et al.*, 1976). There are no arguments either in favour of this theory or against it. (4) As a last but by no means the least plausible explanation I should mention the primitive state of the classification of psychoses. A label such as "schizophrenia" covers several quite different syndromes—different in symptomatology as well as aetiology and clinical course. Labels such as organic psychosis, psychogenic psychosis and schizoaffective psychosis are not much better: they encompass a multiplicity of syndromes and clinical courses. The diagnosis of psychoses should be consistently three-dimensional and based on three criteria: symptomatology, aetiology and course (Van Praag, 1976, 1977d). The current system is totally unsuitable as a starting-point for biological research. This contention implies that it is still too early for any conclusion about the question whether DA metabolism can be disturbed in "schizophrenia".

DISCUSSION

The principal conclusions so far to be drawn from DA research in psychiatry are the following.

1. There are strong indications that the therapeutic effect of neuroleptics is related to their ability to suppress the transmission in central DA-ergic systems. However, neuroleptics inhibit not only DA-ergic systems but, in varying degrees, NA-ergic systems as well. Only recently has this action component been given proper attention. There are sound reasons to assume that they are of importance for the therapeutic profile of neuroleptic drugs, and that the ratio DA-ergic block/NA-ergic block is a better predictor of their clinical efficacy than is their chemical structure (Van Praag *et al.*, 1975b, 1976).

The complementary hypothesis, which postulates that DA-ergic hyperactivity plays a role in the pathogenesis of schizophrenia, has so far been based exclusively on indirect arguments (amphetamine psychosis, DOPA psychosis). So far, no consistent changes in DA metabolism have been demonstrated in schizophrenic patients.

2. In certain types of depression central DA metabolism can be disturbed. Characteristics of these "certain types": retarded vital depressive symptomatology, unipolar course, and demonstrable hereditary features in the aetiology. It is uncertain whether the metabolic disorder is related more to mood disturbances or more to motor changes.

This correlation between biochemical and psychopathological variables would never have been found if the group of depressions had been regarded as a homogeneous entity or had been at best divided into acute and chronic forms. Only on the basis of a three-dimensional diagnosis (criteria: symptomatology, aetiology and course) was it possible to discern the contours of a fourth principle of classification: the cerebral substrate underlying the disturbed behaviour.

The principle of three-dimensional diagnosis is not consistently applied in biological schizophrenia research. This omission might well be an important factor in the disappointing results which this research has so far produced.

As in medicine in general, the nosological entity or syndrome is the focus of biological-psychiatric research. Investigators hope to find the cause of *the* schizophrenia or of *the* vital depressive syndrome. A psychiatric syndrome, however, is a composite of many disturbances of psychological function. There are indications that cerebral functional disorders (regardless of whether they are of a physiological or of a biochemical nature) can be correlated more readily with psychological dysfunctions than with the syndrome as such which, as I said, is a conglomerate of psychological dysfunctions (Van Praag *et al.*, 1975a).

This statement implies that the three-dimensional diagnosis which I urge, is only a first phase in the further elaboration of psychiatric classification. A subsequent phase is: decomposition of the psychiatric syndrome into its component psychological dysfunctions, and the development of methods to measure these dysfunctions. In the elaboration of such a *functional psychopathology*, intensive cooperation between experimental psychiatrist and experimental psychologist is a primary requirement. I consider this approach to be an indispensable prerequisite for enhancing the scientific standard of psychiatry (and certainly not only of biological psychiatry)—an evolution which this discipline urgently needs.

SUMMARY

A survey is given of the findings which have stimulated studies of central DA metabolism in psychiatric patients. Methods available for evaluation of the human central DA metabolism are discussed. The principal results of DA research in psychiatry are the indications that (1) the therapeutic effect of neuroleptics is related to suppression of transmission activity in central DA-ergic systems, and (2) a decreased DA turnover plays a role in the pathogenesis of certain symptoms in certain types of depression. The inadequacy of current schizophrenia diagnosis is emphasized, and is held responsible for the low yield of biological schizophrenia research so far. An urgent plea is made for (1) consistent three-dimensional diagnosis of behaviour disorders, and (2) the decomposition of psychiatric syndromes into the component psychological dysfunctions, and therewith the elaboration of a really functional psychopathology.

ACKNOWLEDGEMENT

This study was written while the author was working as Lady Davis Visiting Professor in the Department of Psychiatry, Hadassah University Hospital, Jerusalem, Israel.

REFERENCES

Angrist, B. M., Sathanathan, G. and Gershon, S. (1973). *Psychopharmacologia* **31,** 1–12.

Angrist, B. M., Sathanathan, G., Wilk, S. and Gershon, S. (1974). *J. Psychiat. Res.* **11,** 13–23.

Angrist, B. M., Thampson, H., Shopsen, B. and Gershon, S. (1975). *Psychopharmacologia* **44,** 273–280.

Ashcroft, G. W., Blackburn, I. M., Eccleston, D., Glen, A. I. M., Hartley, W., Kinloch, N. E., Lonergan, M., Murray, L. G. and Pullar, I. A. (1973). *Psychol. Med.* **3,** 319–325.

Asnis, G. (1977). *Am. J. Psychiat.* **134,** 191–195.

Banki, C. M. (1977). *J. Neurochem.* **28,** 255–257.

Bartholini, G. and Pletscher, A. (1972). *In* "Studies of Neurotransmitters at the Synaptic Level" (Eds. E. Costa, L. L. Iversen and R. Paoletti), 135–148. Raven Press, New York.

Belmaker, R. H., Ebbesen, K., Ebstein, R. and Rimon, R. (1976a). *Br. J. Psychiat.* **129,** 227–232.

Belmaker, R. H., Ebstein, R., Rimon, R., Wyatt, R. J. and Murphy, D. L. (1976b). *Acta Psychiat. Scand.* **54,** 67–72.

Bird, E. D., Barnes, J., Iversen, L. L., Spokes, E. G., Mackay, A. V. P. and Shephard, M. (1977). *Lancet* 1157–1159.

Bowers, M. B., Jr. (1973). *Psychopharmacologia* **28,** 309–318.

Bowers, M. B., Jr. (1974). *Arch. Gen. Psychiat.* **31,** 50–54.

Bowers, M. B., Jr. (1976). *Biol. Psychiat.* **11,** 245–249.

Brambilla, F., Guastalla, A., Guerrini, A., Rovere, C., Legnani, G., Sarno, M. and Riggi, F. (1976). *Acta Psychiat. Scand.* **54,** 275–286.

Brodie, H. K. H., Murphy, D. L., Goodwin, F. K. and Bunney, W. E., Jr. (1971). *Clin. Pharm. Ther.* **12,** 219–224.

Bunney, W. E., Jr., Gershon, E. S., Murphy, D. L. and Goodwin, F. K. (1972). *J. Psychiat. Res.* **9,** 207–226.

Carenzi, A., Gillin, J. C., Guidotti, A., Schwarts, M. A., Trabucchi, M. and Wyatt, R. J. (1975). *Arch. Gen. Psychiat.* **32,** 1056–1059.

Carlsson, A. and Lindqvist, M. (1963). *Acta Pharmacologia (Kbh)* **20,** 140–144.
Carlsson, A., Falck, B., Fuxe, K. and Hillarp, N. A. (1964). *Acta Physiol. Scand.* **60,** 112–119.
Carlsson, A., Persson, T., Roos, B.-E. and Wålinder, J. (1972). *J. Neurol. Trans.* **33,** 83–90.
Carpenter, W. T., Murphy, D. L. and Wyatt, R. J. (1975). *Am. J. Psychiat.* **132,** 438–441.
Cohn, C. K., Dunner, D. L. and Axelrod, J. (1970). *Science* **170,** 1323–1324.
Connell, P. H. (1958). "Amphetamine Psychosis". Maudsley Monographs, No. 5. Oxford University Press, Oxford, London.
Crow, T. J., Deakin, J. F. W., Johnstone, E. C. and Longden, A. (1976). *Lancet* **ii,** 563–566.
Curzon, G., Gumpert, E. J. W. and Sharpe, D. M. (1971). *Nature (London)* **231,** 189–191.
Davson, H. (1967). "Physiology of the Cerebrospinal Fluid", Churchill, London.
Dunner, D. L., Cohn, C. K., Gershon, E. S. and Goodwin, F. K. (1971). *Arch. Gen. Psychiat.* **25,** 348–353.
Dunner, D. L., Cohn, C. K., Weinshilboum, R. M. and Wyatt, R. J. (1973). *Biol. Psychiat.* **6,** 215–220.
Ehrensing, R., Kastin, A., Schalch, D., Friesen, H., Vargas, R. and Schally, A. (1974). *Am. J. Psychiat.* **131,** 714–718.
Fuxe, K. and Ungerstedt, A. (1968). *Eur. J. Pharmacol.* **4,** 135–144.
Garelis, E. and Sourkes, T. L. (1973). *J. Neurol. Neurosurg. Psychiat.* **4,** 625–629.
Garelis, E., Young, S. N., Lal, S. and Sourkes, T. L. (1974). *Brain Res.* **79,** 1–8.
Garelis, E., Gillin, J. C., Wyatt, R. J. and Neff, N. (1975). *Am. J. Psychiat.* **132,** 184–186.
Goodwin, F. K., Brodie, H. K. H., Murphy, D. L. and Bunney, W. E. Jr. (1970). *Biol. Psychiat.* **2,** 341–366.
Goodwin, F. K. (1971). *J. Am. Med. Assoc.* **218,** 1915–1920.
Goodwin, F. K. (1972). *In* "Psychiatric Complications of Medical Drugs" (Ed. R. I. Shader), 149–174. Raven Press, New York.
Gottfries, C. G., Oreland, L. and Wiberg, A. (1974). *Lancet* **ii,** 360.
Grahame-Smith, D. G. (1971). *J. Neurochem.* **18,** 1053–1066.
Green, A. R. and Grahame-Smith, D. G. (1974). *Neuropharmacology* **13,** 949–959.
Green, A. R. and Grahame-Smith, D. G. (1976). *Nature (London)* **260,** 487–491.
Griffith, J. J., Oates, J. and Cavanaugh, J. (1968). *J. Am. Med. Assoc.* **205,** 39–44.
Halaris, A. E., Belendink, K. T. and Freedman, D. X. (1975). *Biochem. Pharmacol.* **24,** 1896–1898.
Hoehn, M. M., Crowley, T. J. and Rutledge, C. O. (1976). *J. Neurol. Neurosurg. Psychiat.* **39,** 941–951.
Hollister, L. E. (1962). *Ann. N. Acad. Sci.* **96,** 80–92.
Iversen, L. L. (1975). *Science* **188,** 1084–1089.
Jacobs, B. L. (1976). *Life Sci.* **19,** 777–786.
Janowski, D. S., El-Yousef, M. K. and Davis, J. M. (1972). *Compr. Psychiat.* **13,** 83.
Janowski, D. S. and Sekerke, H. J. (1973). *Arch. Gen. Psychiat.* **28,** 542–552.
Kehr, W., Carlsson, A., Lindqvist, M., Magnusson, T. and Atack, C. V. (1972). *J. Pharm. Pharmacol.* **24,** 744–747.
Klaiber, E. L., Broverman, D. M., Vogel, W., Kobayashi, Y. and Moriarty, D. (1972). *Am. J. Psychiat.* **128,** 42–48.
Lal, S. and De La Vega, C. E. (1975). *J. Neurol. Neurosurg. Psychiat.* **38,** 722–726.

Legros, J. J. and Breulet, M. (1975). *Lancet* **ii,** 663.
Le Moal, M., Stinus, L., Simon, H., Tassin, J. P., Thierry, A. M., Blanc, G., Glowinski, J. and Cardo, B. (1977). *In* "Advances in Biochemical Psychopharmacology", Vol. 16 (Eds. E. Costa and G. L. Gessa). Raven Press, New York.
Lipper, S. (1976). *Lancet* **ii,** 571–572.
Maeda, K., Kato, Y., Ohgo, S., Chilara, K., Yoshimoto, Y., Yamaguki, N., Kuromaru, S. and Imara, H. (1975). *J. Clin. Endocr.* **40,** 501–505.
Meltzer, H. Y. and Stahl, S. M. (1974). *Rex. Commun. Chem. Pathol. Pharmacol.* **7,** 419–431.
Meltzer, H. Y., Goode, D. J., Fang, V. S., Schijve, P. and Young, M. (1976). *Lancet* **ii,** 1142.
Meltzer, H. Y. and Stahl, S. M. (1974). *Res. Commun. Chem. Pathol. Pharmacol.* **7,** *cology* **51,** 185–187.
Moore, K. E. (1977). *Biol. Psychiat.* **12,** 451–462.
Murphy, D. L. and Weiss, R. (1972). *Am. J. Psychiat.* **128,** 35–41.
Murphy, D. L. and Wyatt, R. J. (1972). *Nature (London)* **238,** 225–226.
Murphy, D. L., Donnelly, C. H. Miller, L. and Wyatt, R. J. (1976). *Arch. Gen. Psychiat.* **33,** 1377–1381.
Nies, A., Robinson, B. S., Harris, L. S. and Lambourn, K. L. (1974). *In* "Advances in Biochemical Psychopharmacology" (Ed. E. Usdin), 59–71. Raven Press, New York.
Olsson, R. and Roos, B.-E. (1968). *Nature (London)* **219,** 502–503.
Owen, F., Bourne, R., Crow, T. J., Johnstone, E. C., Bailey, A. R. and Hershon, H. I. (1976). *Am. J. Psychiat.* **33,** 1370–1373.
Pare, C. M. B., Yeung, D. P. H., Price, K. and Stacey, R. S. (1969). *Lancet* **ii,** 133–135.
Persson, T. and Roos, B.-E. (1969). *Br. J. Psychiat.* **115,** 95–98.Post, R. M., Goodwin, F. K., Gordon, E. and Watkin, D. M. (1973a). *Science* **179,** 897–899.
Post, R. M., Kotin, J., Goodwin, F. K. and Gordon, E. K. (1973b). *Am. J. Psychiat.* **130,** 67–72.
Post, R. M., Fink, E., Carpenter, W. T. and Goodwin, F. K. (1975). *Arch. Gen. Psychiat.* **32,** 1063–1069.
Post, R. M. and Goodwin, F. K. (1975). *Science* **190,** 488–489.
Praag, H. M. van (1967). *Psychiat. Neurol. Neurosurg.* **70,** 361–379.
Praag, H. M. van and Korf, J. (1971). *Psychopharmacologia* **19,** 199–203.
Praag, H. M. van (1974). *Pharmacopsychiatry* **7,** 281–292.
Praag, H. M. van (1975). *Compr. Psychiat.* **16,** 7–22.
Praag, H. M. van, Korf, J., Lakke, J. P. W. F. and Schut, T. (1975a). *Psychol. Med.* **5,** 138–146.
Praag, H. M. van, Dols, L. C. W. and Schut, T. (1975b). *Compr. Psychiat.* **16,** 255–263.
Praag, H. M. van and Korf, J. (1975). *Am. J. Psychiat.* **132,** 593–597.
Praag, H. M. van and Korf, J. (1976). *Am. J. Psychiat.* **133,** 1171–1177.
Praag, H. M. van (1977a). *Br. J. Psychiat.* **130,** 463–474.
Praag, H. M. van (1977b). *Biol. Psychiat.* **12,** 101–131.
Praag, H. M. van, Korf, J. and Dols, L. C. W. (1977). *Br. J. Psychiat.* **129,** 547–555.
Praag, H. M. van (1976). *Compr. Psychiat.* **17,** 481–497.
Praag, H. M. van (1977c). "Depression and Schizophrenia. A Contribution on their Chemical Pathologies". Spectrum, New York.

Praag, H. M. van (1977d). "Psychotropic Drugs. A Guideline for the Practicing Physician. Brunner Mazel, New York.
Puite, J. K., Schut, T., Praag, H. M. van, and Lakke, J. P. W. F. (1973). *Psychiat. Neurol. Neurosurg.* **76,** 61–70.
Randrup, A. and Munkvad, I. (1970). *In* "Amphetamines and Related Compounds" (Eds. E. Costa and S. Garattini), 695–713, Raven Press, New York.
Randrup, A. and Munkvad, I. (1972). *Orthomol. Psychiat.* **1,** 2–27.
Randrup, A. and Munkvad, I. (1974). *J. Psychiat. Res.* **11,** 1–10.
Robinson, D. S., Davis, J. M., Nies, A., Ravaris, C. G. and Sylwester, D. (1971). *Arch. Gen. Psychiat.* **24,** 536–539.
Rosenblatt, S., Leighton, W. P. and Chanley, J. D. (1973). *Science* **182,** 923–924.
Rossum, J. M. van (1967). *In* "Neuropsychopharmacology" (Ed. H. Brill), 321–329. Excerpta Medica Foundation, Den Haag.
Rotrosen, J., Angrist, B. M., Gershon, S., Sachar, E. J. and Halpern, F. S. (1976). *Psychopharmacology* **51,** 1–7.
Sachar, E., Frantz, A., Altman, N. and Sassin, J. (1973). *Am. J. Psychiat.* **130,** 1362–1366.
Sathanathan, G., Angrist, B. M. and Gershon, S. (1973). *Biol. Psychiat.* **7,** 139–146.
Scheyen, J. D. van, Praag, H. M. van, and Korf, J. (1977). *Br. J. Clin. Pharmacol.* **4,** 179S–184S.
Schildkraut, J. J., Herzog, J. M., Orsulak, P. J., Edelman, S. E., Shein, H. M. and Frazier, S. H. (1976). *Am. J. Psychiat.* **133,** 438–440.
Schwartz, M. A., Wyatt, R. J., Yang, H. Y. and Neff, N. H. (1974). *Arch. Gen. Psychiat.* **31,** 557–560.
Seeman, P. and Lee, T. (1975). *Science* **188,** 1217–1219.
Shaskan, G. E. and Becker, R. E. (1975). *Nature (London)* **253,** 659–660.
Sedvall, G., Alfredson, G., Bjerkenstedt, L., Eneroth, P., Fyrö, B., Härnryd, C. and Wode-Helgodt, B. (1977). *In* "The Impact of Biology on Modern Psychiatry" (Eds. E. S. Gershon, R. H. Belmaker, S. S. Kety and M. Rosenbaum), Plenum Press, New York and London.
Shopsin, B., Freedman, L. S. and Goldstein, M. (1972). *Psychopharmacologia* **27,** 11–16.
Snyder, S. H. (1973). *Am. J. Psychiat.* **130,** 61–66.
Snyder, S. H., Banerjee, S. P., Yamamura, H. I. and Greenberg, D. (1974). *Science* **184,** 1243–1253.
Stein, L. and Wise, C. D. (1971). *Science* **171,** 1032–1036.
Sourkes, T. L. (1973). *J. Neur. Trans.* **34,** 153–157.
Wilson, R. G., Hamilton, J. R., Bord, W. D., Forrest, A. F. M., Cole, E. N., Boyns, A. R. and Griffiths, K. (1975). *Br. J. Psychiat.* **127,** 71–74.
Wise, C. D. and Stein, L. (1973). *Science* **181,** 344–347.
Wyatt, R. J., Murphy, D. L., Belmaker, R., Cohen, S., Donelley, C. H. and Pollin, W. (1973). *Science* **179,** 916–918.
Wyatt, R. J., Schwartz, M. A., Erdelyi, J. D. and Barchas, J. D. (1975). *Science* **187,** 368–370.
Yaryuara-Tobias, J. A., Diamond, B. and Merlis, S. (1970). *Curr. Ther. Res.* **12,** 528–531.
Young, D. and Scoville, W. B. (1938). *Med. Clin. N. Am.* **22,** 637–646.
Young, S. N., Lal, S., Martin, J. B., Ford, R. M. and Sourkes, T. L. (1973). *Psychiat. Neurol. Neurosurg.* **76,** 439–444.
Zeller, E. A., Boshes, B., Davis, J. M. and Thorner, M. (1975). *Lancet* **i,** 1385.

Notes added in proof

p. 665. In 24 H plasma-profile studies of prolactin it was recently found that prolactin can be elevated in depression (Halbreich *et al. Arch. Gen. Psychiat.* (1979). in press). This finding might indicate dopaminergic hypoactivity.

p. 667. The present situation is confusing. Neuroendocrinological data in favour of normal, hypo- and hypersensitivity of DA receptors have been published. Perhaps the sub-grouping of schizophrenic patients has been deficient. (For a review, see Van Praag, *Int. J. Neurol.*, in press.)

p. 668. The inherent difficulties of the interpretation of such studies has been stressed by Clow *et al.* (1979). (*Lancet* **i**, 934).

p. 669. Recently, Owen *et al.* (1978), *Lancet* **ii**, 223–225, demonstrated an increased DA receptor density in the brains of some schizophrenic patients, drug-free for more than a year. This is a most important observation in favour of the DA hypothesis.

Index

F

I

K

L

Q

R

S

T